Plastics Flammability Handbook

Principles, Regulations, Testing, and Approval
3rd Edition

Edited by Jürgen Troitzsch

HANSER
Hanser Publishers, Munich

Hanser Gardener Publications, Inc., Cincinnati

The Editor:
Dr. Jürgen Troitzsch, Adolfsallee 30, 65185 Wiesbaden, Germany

Distributed in the USA and in Canada by
Hanser Gardner Publications, Inc.
6915 Valley Avenue, Cincinnati, Ohio 45244-3029, USA
Fax: (513) 527-8801
Phone: (513) 527-8977 or 1-800-950-8977
Internet: http://www.hansergardner.com

Distributed in all other countries by
Carl Hanser Verlag
Postfach 86 04 20, 81631 München, Germany
Fax: +49 (89) 98 12 64
Internet: http://www.hanser.de

The use of general descriptive names, trademarks, etc., in this publication, even if the former are not especially identified, is not to be taken as a sign that such names, as understood by the Trade Marks and Merchandise Marks Act, may accordingly be used freely by anyone.

While the advice and information in this book are believed to be true and accurate at the date of going to press, neither the authors nor the editors nor the publisher can accept any legal responsibility for any errors or omissions that may be made. The publisher makes no warranty, express or implied, with respect to the material contained herein.

Library of Congress Cataloging-in-Publication Data

Brandverhalten von Kunststoffen. English
  Plastics flammability handbook : principles, regulations, testing, and
approval / edited by Jürgen Troitzsch.-- 3rd ed.
      p. cm.
Rev. ed. of: International plastics flammability handbook / Jürgen
Troitzsch. c1990.
Includes bibliographical references and indexes.
  ISBN 1-56990-356-5 (hardcover)
  1. Plastics--Flammability--Handbooks, manuals, etc. I. Troitzsch,
Jürgen. Brandverhalten von Kunststoffen. English. II. Title.
  TH9446.5.P45B7313 2004
  628.9'222--dc22

                        2003021691

Bibliografische Information Der Deutschen Bibliothek
Die Deutsche Bibliothek verzeichnet diese Publikation in der Deutschen Nationalbibliografie;
detaillierte bibliographische Daten sind im Internet über <http://dnb.ddb.de> abrufbar.
ISBN 3-446-21308-2

© Carl Hanser Verlag, Munich 2004
Production Management: Oswald Immel
Coverconcept: Marc Müller-Bremer, Rebranding, München, Germany
Coverdesign: MCP • Susanne Kraus GbR, Holzkirchen, Germany
Typeset, printed and bound by Kösel, Kempten, Germany

Contributors List

Diplom-Physikerin Edith Antonatus
BASF AG
Aweta Brandschutztechnik
KTE/SB – A 521
D-67056 Ludwigshafen
Germany

Dr. Tamás Bánky
NPC for Quality Control and Innovation
in Building
Diószegi út 37
H-1113 Budapest
Hungary

Thierry Bonnaire
Formerly: CSTB
84, Avenue Jean Jaurès
Champs-sur-Marne
F-77421 Marne-La-Vallée
France

Prof. Dr. Serge Bourbigot
ENSCL
Cité Scientifique, Bâtiment C 7
BP 108, F-59652 Villeneuve D'Ascq, CEDEX
France

Prof. Dr. Jean-Claude Brosse
Université du Maine
Laboratoire de Chimie et Physicochimie
Macromoléculaire
B.P. 35
F-72017 Le Mans
France

Prof. Dr. Joseph Davidovits
Institut Géopolymère
16 rue Galilée
F-02100 Saint-Quentin
France

Dr. Daniel Derouet
Université du Maine
Laboratoire de Chimie Organique Macro-
moléculaire
Avenue Olivier Messiaen
F-72085 Le Mans Cedex 9
France

Prof. Dr. Colomba Di Blasi
Università degli Studi di Napoli
"Federico II"
Dipartimento di Ingegneria Chimica
P. le V. Tecchio
I-80125 Napoli
Italy

Vincent P. Dowling
CSIRO Manufacturing and Infrastructure
Technology
Fire Science & Technology Laboratory
PO Box 56
Highett, Victoria 3190
Australia

Dr. Axel Ebenau
BASF Future Business GmbH
Rathausplatz 10
D-67059 Ludwigshafen
Germany

Dr. Thomas Eckel
Bayer Polymers
Innovation - Polymer Alloys
D-41538 Dormagen
Germany

Dr. Jadviga Fangrat
Building Research Institute
Fire Reseach department
Filtrowa 1
00-611 Warsaw
Poland

Dr. Uwe Fink
SRI Consulting
Katharinenweg 7
CH-8002 Zürich
Switzerland

Santiago Garcia Alba (deceased)
LICOF/AFITI
Ctra. Valencia Km 23,400
E-28500 Arganda del Rey
Madrid
Spain

Dr. Ondrej Grexa
State Forest Products Research Institute
Dubravská Cesta 1
83330 Bratislava
Slovakia

Hartmut Grupp
Allianz Zentrum für Technik
Krausstraße 22
85737 Ismaning
Germany

Dr. Patrick van Hees
SP – Swedish National Testing & Research
Institute
Dept. of Fire Technology
P.O. Box 857
S-50115 Borås
Sweden

Rüdiger Hoffmann
Swiss Institute of Safety and Security
Nüschelerstrasse 45
CH-8001 Zurich
Switzerland

Prof. Dr. A. Richard Horrocks
Bolton Institute
Deane Road
Bolton, BL3 5AB
UK

Dr. Baljinder K. Kandola
Bolton Institute
Deane Road
Bolton, BL3 5AB
UK

Dr. Björn Karlsson
Icelandic Fire Authority
Skulagata 21
101 Rekjavik
Island

Ir. Mathijs F.M. Koppers
GE Plastics bv
1 Plasticslaan PO Box 117
NL-4600 AC Bergen op Zoom
The Netherlands

Dr. Michel Le Bras
ENSCL
Cité Scientifique, Bâtiment C7
BP 108, F-59652 Villeneuve D'Ascq,
CEDEX
France

Dr. Sergei V. Levchik
Akzo Nobel Chemicals Inc.
1 Livingstone Ave.
Dobbs Ferry, NY 10522
USA

Dr. Maryline Lewandowski
GEMTEX
ENSAIT
9, rue de l'Ermitage BP 30329
F-59056 Roubaix
France

Dr. Christine Lukas
Dow Construction Products
2 Heathrow Boulevard
284 Bath Road
West drayton
Middlesex UB7 0DQ
England

Dr. Michael A. McKinney
Marquette University
Department of Chemistry
P. O. Box 1881
Milwaukee, WI 53201
USA

Dr. Jim Mehaffey
Forintek Canada Corp.
Suite 4100 CTTC
1125 Colonel By Drive
Ottawa, Ontario K1S 5R1
Canada

Silvio Messa
LSF Srl – Laboratorio di Studi e Ricerche
sul Fuoco
Via Garibaldi 28a
I-22070 Montano Lucino (Como)
Italy

Dr. Michael Mitzlaff
Siemens Axiva GmbH
Brandhaus Hoechst
Industriepark Hoechst - C 369
D-65926 Frankfurt
Germany

Dipl.-Ing. Berthold Müller
Bayer AG
Bayer Industry Services
SUA-SPA-Brandtechnologie
Geb. 411
D-51368 Leverkusen
Germany

Janet Murrell
Warrington Fire Research Centre Ltd.
Holmesfield Road
Warrington
Cheshire WA1 2DS
UK

Dr. Keith Paul
9, Birch Drive
Shawbury, Shrewsbury
Shropshire SY4 4HZ
UK

Prof. Dr. Jürgen Pauluhn
Bayer HealthCare / Toxicology Build. 514
D-42096 Wuppertal
Germany

Dipl.-Ing. Dr. Christian Pöhn
MA 39 Versuchs- und Forschungsanstalt der
Stadt Wien
Rinnböckstrasse 15
A-1110 Wien
Austria

Prof. Dr. Dennis Price
Salford University
Cockroft Building
Fire Chemistry Research Group
Salford M5 5WT
UK

Jian-Min Qian
Sichuan Fire Research Institute
266 Waibei Street
Dujiangyan City
Sichuan Province, 611830
China

Dr. Kurt A. Reimann
BASF Corporation
1419 Biddle Avenue
Wyandotte, MI, 48192
USA

Dr. Maryline Rochery
GEMTEX
ENSAIT
F-59056 Roubaix
France

Dr. Herman Stone
115 Cimarand Dr
Williamsville, NY 14221
USA

Björn Sundström
SP – Swedish National Testing & Research
Institute
Dept. of Fire Technology
P.O. Box 857
S-50115 Borås
Sweden

Guy Touchais
Formerly: CSTB
84, Avenue Jean Jaurès
Champs-sur-Marne
F-77421 Marne-La-Vallée
France

Dr. Jürgen Troitzsch
Fire Protection Service
Adolfsallee 30
D-65185 Wiesbaden
Germany

Dr. Heinz Ulrich Werther
Dr. Hans Hoffmann Strasse 12
D-67157 Wachenheim
Germany

Prof. Dr. Charles A. Wilkie
Marquette University
Department of Chemistry
P. O. Box 1881
Milwaukee, WI 53201
USA

Prof. Dr.-Ing. Friedrich-Wilhelm Wittbecker
Bergische Universität Wuppertal
Fachbereich 14 Sicherheitstechnik
Brand- und Explosionsschutz
Gaußstraße 20
D-42119 Wuppertal
Germany

Koichi Yoshida
National Maritime Research Institute, Japan
(NMRI)
6-38-1 Shinkawa
Mitaka City
Tokyo 181-0004
Japan

Dr. Roman Zoufal
Bulharská 38/1401
10100 Prague 10
Czech Republic

Preface to the Third Edition

None of the many publications on the reaction of plastics to fire provides a comprehensive review of the fundamentals as well as of the relevant regulations and test methods. The "International Plastics Flammability Handbook" was first published in 1983 to fill this gap. In the 1980s, major changes occurred in the field of plastics fire behavior ratings on national and increasingly on international levels. These changes made it necessary to prepare a completely revised 2nd edition of the handbook, which was published in 1990. The last thirteen years saw a breakthrough in the internationalization of fire testing and classification, particularly in building, electrical engineering, and transportation. At the same time, the perception of how to assess the main parameters governing a fire and the role of combustible materials like plastics were redefined and led to new approaches particularly in the fields of smoke development and toxicity of fire effluents. All these developments required a comprehensive revision of the handbook.

To fulfill these demands, the various chapters were revised by experts in the relevant fields. I should like to express my gratitude to my responsible editors *M. Le Bras* and *S. Bourbigot* (I Fundamentals), *M. Mitzlaff* (II Fire protection regulations and test procedures), *H. U. Werther* (III Fire effluents) and to the more than 40 co-authors, whose expertise and commitment made the revision of this handbook feasible.

The handbook consists of four parts: After a historical review and a detailed synopsis of the market situation, Part I describes the basic principles of the burning process, covers the thermal properties and burning behavior of thermoplastics, foams, thermosets, and elastomers in depth. Chapters on flame retardants, their mode of action and flame retardant plastics, the burning behavior of textiles and flame retardant textiles, smoke development and suppression follow. It is hoped that this will facilitate the reader's introduction to this complex subject and also provide the background to a better understanding of the fire test procedures, regulations, and approval criteria covered in the second part.

Part II starts with an introduction to the methodology of fire testing and describes fire protection regulations, the fire test methods introduced to satisfy these regulations, and product approval procedures for combustible products and plastics components in various applications.

The most extensive section in the handbook is devoted to the building sector for which numerous regulations and test methods were developed in all industrialized countries and where a tremendous harmonization effort is under way or was already completed.

Further sections cover transportation and electrical engineering where international harmonization of regulations and test methods has made greater progress. Their number and variety is thus less extensive. The chapter on furnishings focuses on developments in the European Union and the US.

Part III deals with the smoke development, toxicity, and corrosivity of fire effluents. These topics are of increasing public interest and thus covered in some depth.

Part IV, the Appendix, contains listings intended to assist the reader in his daily work.

It is hoped that this book will be of interest to all those concerned with plastics, flame retardancy, fire testing as well as fire protection and will help to better understand this complex matter.

Jürgen Troitzsch

How to use this book

Structure

The book commences with an historical synopsis on the topic of fires and an account of the present situation in fire protection. The market situation for plastics and flame retardants is covered in some depth. Subsequently the basic principles of the burning process and the combustion of individual plastics are considered. The thermal properties and burning behavior of plastics and textiles are examined in the light of the physical and chemical processes taking place. Methods of rendering plastics and textiles flame retardant and reducing smoke emissions are covered and the modes of action of flame retardants and smoke suppressants are explained.

Part II begins with an introduction to the methodology of fire testing followed by discussion of national and international fire protection regulations and test methods for combustible materials and plastics. Building fire protection occupies the largest part of the book and is arranged according to a uniform scheme for all 20 countries, the European Union and ISO dealt with in Chapter 10. Each section for the respective country commences with an account of the statutory regulations and continues with a summary of the relevant test methods in the form of diagrams and tables of test specifications. The reader is thus able to grasp the essentials at a glance. Further details should be taken from the original standards listed in the bibliography at the end of each section.

Officially recognized test institutions and procedures for obtaining official product approval are listed under the heading "Official approval". Each section ends with a look at future developments in the relevant country.

Chapters 11 to 13 are arranged in a similar fashion except that the divisions are according to subject rather than country.

Fire effluents are dealt with together for the sake of clarity in Part III in Chapters 14 to 17. Chapter 15 "Smoke development of fire effluents" thus contains all the test methods relevant to building, transportation and electrical engineering. Chapter 16 covers the basics of the toxicity of fire effluents and testing principles; Chapter 17 deals with corrosivity assessment methods and the practice of clean-up procedures of fire effluents.

Part IV "Appendix" contains various lists including a suppliers' index for flame retardants and smoke suppressants, addresses of standards and electrical engineering organizations. The glossary of technical fire protection terminology and the English, German, and French equivalents of the most important terms will be extremely useful to workers in the field. The Appendix closes with a list of all the standards and guidelines mentioned in the book and a comprehensive name and subject index.

References

A bibliography can be found at the end of each chapter and the principal journals and books on the subject are listed in the Appendix.

Illustrations and tables

Illustrations and tables are numbered consecutively in each chapter. Diagrams of test equipment include only those features necessary for understanding the method.

The tables of test specifications summarize details of test specimens, specimen position, ignition source, test duration and conclusions.

Contents

I

Fundamentals

1 Introduction

S. BOURBIGOT, M. LE BRAS, AND J. TROITZSCH

1.1 Historical Review

Few discoveries have had as significant an influence on the development of mankind as the skill of generating and using fire. This capability has been a basic requirement for all civilizations, as it enabled Man to reduce his dependence on a hostile environment. With the aid of fire he learned to prepare meals, fire pottery, and extract metals from their ores. Gradually people ceased to live nomadically and built settlements that developed into large towns with the advent of the first great cultures. There is, however, another side to fire, namely a deadly uncontrolled natural force against which Man is helpless.

Historically, fires, wars, and epidemics were considered to be the principal scourges of the human race. Owing largely to urban design, towns and cities were frequently close to each other in a veritable maze of streets, and, particularly during the Middle Ages, were constructed almost entirely of wood.

Some of the catastrophes that have achieved particular notoriety are listed here as examples of the vast number of fires that have occurred throughout history:

- In *A.D.* 64, during Nero's reign, Rome burnt down in 8 days, and 10 of the 14 districts were completely destroyed.
- In 1666, the Great Fire of London destroyed 13,200 houses, 94 churches, and countless public buildings in the space of 3 days. Unfortunately, flame retardant materials were unheard of at this time.
- In 1812, the Russians set fire to Moscow to repel Napoleon's army. The conflagration raged for 5 days and destroyed 90 % of the city.
- In 1842, 4200 buildings were destroyed in Hamburg by a fire, which cost 100 lives and rendered 20 % of the population homeless.
- In 1906, following an earthquake, fire almost totally destroyed San Francisco and killed 1000 persons. The fire extended over an area of almost 10 km^2.
- In 1923, Tokyo and Yokohama were almost completely devastated by fires following earthquakes.
- World War II added a new dimension to fire, the firestorm: In 1945 the Allied attack on Dresden resulted in the most devastating fire catastrophe on record. Approximately 35,000 people were killed in the firestorm unleashed by bombardment and the city was completely devastated.

Throughout history attempts have been made to limit this uncontrolled aspect of fire in two ways.

First, fire fighting has been organized using fire fighters or brigades, an example of the latter existed in ancient Egypt, and the "Cohortes Vigilum" in the Rome of the Caesars can be considered as forerunners of the modern fire service. In addition, fire prevention measures have also been practiced. These include passage of fire regulations, and attempts to protect materials more effectively against attack by fire, for example, with flame retardants. Different types of the latter have been known since ancient times and a few are mentioned in the paragraph that follows.

In 360 *B.C.*, a treatise on fortifications recommended that timbers should be protected against fire by painting them with vinegar. In the battle for the town of Piraeus in 83 *B.C.*, wooden siege towers were impregnated with alum to prevent their being set alight by the Romans and remained successfully in service without catching fire. In *A.D.* 77, Pliny the Elder described asbestos tablecloths that could be cleaned by heating to red heat.

In 1638, Nicolas Sabbatini suggested that clay and gypsum could be used as flame retardants for theatre scenery made of painted canvas. The first patent (no. 551), in which textiles and papers treated with flame retardants are described, was published in England in 1735 by Obadiah Wyld. The mixture used consisted of alum, borax, and vitriol (probably impure iron sulfate). In 1783, the Montgolfier brothers used a flame retardant coating of alum on the envelope of their hot air balloon.

Gay-Lussac was commissioned in 1786 to reduce the flammability of textiles used in French theaters following several fires. In 1820, he suggested the use of mixtures of ammonium phosphate, ammonium chloride, and borax. These proved highly effective and are still used on occasion.

Toward the middle of the 19th century, numerous flame retardants for textiles were tested. Ammonium phosphate, ammonium sulfate, and a mixture of ammonium phosphate and chloride were found to be the most effective.

At the beginning of the 20th century, William Henry Perkin carried out fundamental research into flame retarding textiles. He was the first person to study systematically the mechanisms of flame retardancy. Modern flame retardants for wood, cotton, paper, and especially plastics are based on his studies.

1.2 The Present Situation

Today, the probability of a catastrophic fire razing an entire town in peacetime is remote. Technical advances such as powerful water jets and fire engines have considerably improved the effectiveness of fire fighting. Owing to the plethora of mainly governmental legislation, fire protection today plays an extremely important role in reducing fire risk. Housing areas, for example, are separated from each other by open spaces. Individual buildings have outer walls built of fire-resistant materials (so-called fire walls); only building materials and components that have passed certain fire tests and therefore meet fire performance requirements are used. Many buildings are protected by fire alarms and sprinkler systems.

One would think, therefore, that Man has mastered the uncontrolled aspect of fire. Nothing could be further from the truth. Major fires occur daily and, although they are not on the same scale as earlier catastrophes, they are nevertheless increasing in size and frequency as the following examples show.

In Europe

- On May 22, 1967, the A l'Innovation department store in Brussels burnt down, leaving 270 dead.
- In a discotheque fire in Saint-Laurent-du-Pont in France on November 1, 1970, 146 young people met their death.
- The huge fire that destroyed the Summerland entertainment center on the Isle of Man on August 2, 1973 caused 49 fatalities.
- The fire at the Ford Spare Parts Center in Cologne in the Federal Republic of Germany on October 20, 1977 fortunately did not cause any fatalities. The fire damage, however, amounted to some DM 240 million (total damage was estimated at DM 425 million). Damage of this order was a new experience in Germany in peacetime.

- The fire in the Stardust Disco on February 14, 1981 in Dublin caused the deaths of 48 young people.
- A fire on November 18, 1987 in the London Underground at King's Cross Station killed 31 people. Investigators think it started when a cigarette dropped onto the old wooden escalator fell into a pile of rubbish.
- More than 400 people died as a result of a subway fire on October 28, 1995 in Baku, Azerbaijan. The Azeri deputy prime minister in charge of the investigation stated that the fire was caused by "outdated Soviet equipment (the rail cars were built in the 1960s) and the absence of any flame retardant material of construction."
- In a freight carry, a truck burnt on November 18, 1996 in the Channel tunnel. Fortunately, no one was killed but all 33 people aboard suffered from smoke inhalation. The Channel tunnel exposed fundamental problems with the operators' safety procedures. According to a report: "The incident was more serious than it should have been because the emergency procedures were too complex and demanding and the staff on duty had not been adequately trained to carry them out."
- A Belgian truck carrying flour and margarine burnt in the Mont-Blanc Alpine Tunnel connecting France and Italy on March 24, 1999. Forty-two people died in this disaster. The truck exploded and most of the victims were found in or around their vehicles. Only a few had been able to reach the 17 heat-resistant bunkers lining the tunnel.

In the Americas

- February 1974: The Cresiful Building in Sao Paulo, Brazil, was consumed by fire and 200 people died.
- April 1987, Minnesota: Electrical malfunction of a computer component caused overheating and ignited internal wiring. The damage to the building and contents was in excess of $ 450,000.
- December 1989, Ohio: An electrical short circuit in a copier triggered over $ 2 million in damage.
- October 1990, Arkansas: An electrical malfunction in a computer caused it to ignite. Fire and smoke spread and caused property damage of $ 2 million.
- January 1992, Michigan: A fire in a microfilming room, suspected to have been initiated by a malfunctioning microfilm printer, caused $ 1.5 million in property loss.
- March 1994, Wisconsin: A fire started in a computer's uninterrupted power supply, but was contained by a water sprinkler system. The total loss was $ 125,000.
- May 1995, Florida: A malfunction in a computer monitor caused fire damage of $ 125,000 to the building and $ 275,000 to the contents.
- October 1995, Texas: A fire from an undetermined problem in electronic audio equipment ignited the combustible housing of the unit in a home, resulting in five deaths.

In Asia

- December 11, 1996, Hong Kong: In Hong Kong's worst fire since 1957, a fire at the 16-story Garley Building burned for 21h and killed 39 people. A spark from construction welders is thought to have started the blaze in the building, situated in the crowded Kowloon district.
- February 23, 1997, Thailand: Helicopters plucked more than 93 survivors from the roof of the fire-ravaged Meridien President Tower in central Bangkok, but three people lost their lives there. Fire broke out on the seventh floor of the 36-story building. Workers told reporters that they heard two blasts on a floor where air-conditioning work was going on before the fire broke out. Work was in progress on some new tenant offices due to open the following day. Owing to a loophole in the building control act, fire prevention systems were not required to be in operation before buildings were officially opened.

- December 8, 1997, Indonesia: Fifteen people died in Jakarta in a fire that engulfed the top floors of a new 26-story tower for Indonesia's central bank, Bank Indonesia. Firemen said that a short circuit might have been responsible for the blaze in the mostly unoccupied building, which was just being completed. Helicopters rescued several people from a rooftop heli-pad.

As a result of the turbulent development of science and technology, we are confronted with unfamiliar circumstances. One of these is scale. Houses and storage sheds have been replaced by high-rise buildings or skyscrapers and huge warehouses. There has been an unprecedented development in the size and number of methods of transport. A second factor is the vast increase in gross national product of the industrialized nations, which has resulted in increased wealth and living standards.

Both of these aspects and a range of related factors have resulted in an increase in the risk of fire:

- Many houses and apartments are furnished luxuriously. Carpeting, furnishings, equipment, and oil and gas for heating all increase the fire load in a building.
- In industry and commerce, ever-increasing amounts of goods are manufactured and stored in larger and larger buildings.
- The concentration of people in towns owing to the availability of jobs in itself attracts new industries, increases consumption, and results in a further concentration of population and materials.
- New processes and applications of technology introduce many new fire hazards (e.g., new ignition sources such as welding sparks and short circuits).
- New technologies increasingly involve new types of material, as traditional materials do not meet certain technical requirements.

The exponentially increasing cost of fire damage is a further factor. In the United States, in 1989–1993, fires that started in TVs, radios, video-players, and phonographs accounted for approx. 2400 home fires every year. Thanks to improved fire safety, the number of deaths declined to 35 per year, injuries to 166 per year, and direct property damage to less than $35 million per year.

In 1990–1994, fires originating from electronic equipment rooms or areas caused an average of 1179 fires every year, resulting in one death per year, 36 injuries per year, and nearly $29 million in direct damages per year.

In Germany, direct fire losses are estimated at approx. 3.3 billion € each year. In the United Kingdom, statistics from the UK Home Office show that the daily rate of domestic fires can be up to 54 % higher than average in December, no doubt due in great part to accidents involving Christmas trees and decorations hung too near to fires and lights, wrapping paper and presents lying near fire and other heat sources, and Christmas cards on mantelpieces and TVs that come into contact with heat. In Sweden, what happens when fire safety standards are relaxed may be illustrated by data reported by Swedish authorities (Table 1.1). In just 5 years, the number of fires attributed to TV and audio equipment increased by more than 30 %.

Prosperity could be said to be characterized on the one hand by the availability of sophisticated and therefore expensive products such as machines and, on the other, by increasing use of combustible materials. The combination of scale, high-value products, and combustible materials is encountered with increasing frequency in fires with ensuing high financial losses. The often expressed opinion that the type of combustible material involved is the sole factor in determining the extent of losses is unfounded.

Table 1.1 TV/Audio equipment fires according to the Swedish Insurance Federation

Year	Number of TV/Audio equipment fires	Percent increase
1990	1176	
1991	1263	7.4%
1992	1545	22.3%
1993	1898	22.8%
1994	2471	30.2%

2 The Market Situation

U. FINK AND J. TROITZSCH

2.1 Plastics

J. TROITZSCH

The introduction of plastics has made a decisive contribution to the present high living standard of the industrialized countries. Historically, plastics have been developed in several stages. Those based on natural products appeared first, with vulcanized fiber in 1859 followed by celluloid and artificial horn in 1870 and 1897, respectively. The development of thermosetting plastics commenced in the early part of the 20th century with the introduction of Bakelite in 1910. The real breakthrough occurred with the event of polymerization plastics in 1930.

After World War II development was vigorous and production of plastics increased exponentially, interrupted only briefly by slumps such as the oil crisis in 1973. Figures 2.1–2.9 show the main plastics application sectors, as well as plastics production and consumption in general and broken down by regions, countries, and plastics types as well as a consumption forecast for 2010 followed by the increasing trend to concentration in the plastics producing industry [1].

The main plastics application sectors for Germany in 2000 are building and packaging, as shown in Fig. 2.1. Worldwide plastics production reached a new record of 180 million tons in 2000 as seen in Fig. 2.2. In the 2000 worldwide plastics production broken down by producer countries and regions, Western Europe has a share of 26.5%. The 2010 forecast for world plastics production is 300 million tons, corresponding to an annual average growth rate of 5.3% (Fig. 2.3–2.4). The main plastics produced worldwide in 2000 are broken down by type in Fig. 2.5.

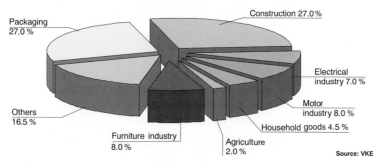

Figure 2.1   Fields of use for plastics in Germany in 2000 in percent

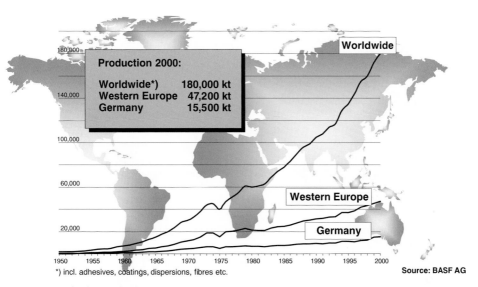

Figure 2.2 Plastics production 1950–2000

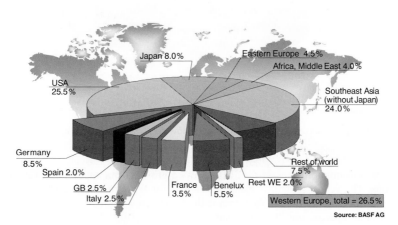

Figure 2.3 World plastics production broken down by producer countries in 2000

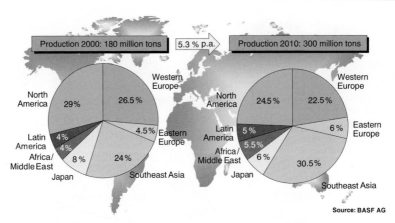

Figure 2.4   Plastics production forecast by regions in percent and growth p.a. (including coatings, adhesives, dispersions, etc.)

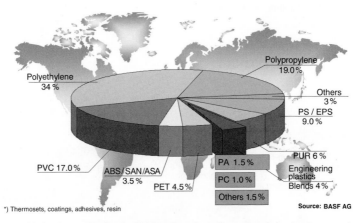

Figure 2.5   World plastics production in 2000 broken down by types of plastics (tonnage in percent)

The plastics consumption forecast for 2010 shows a growth from approx. 150 (2000) to 258 million tons. The yearly growth is estimated at 4% for Western Europe, 3% for Japan and 4.5% for the United States. Southeast Asia is the fastest growing region, with 7.5%; in 2010, it will consume approx. one third of total world plastics (Fig. 2.6). The world per capita consumption of plastics materials will rise from 24.5 kg (54 lb) in 2000 to 37 kg (81.5 lb) per capita in 2010 which is an increase of 51% (Fig. 2.7). The worldwide annual growth of the main types of plastics is shown in Fig. 2.8.

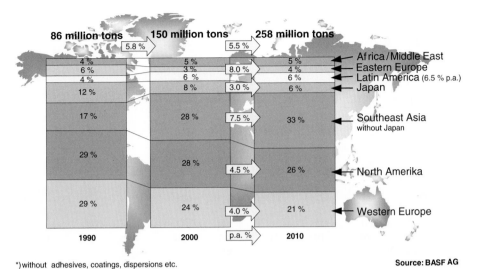

*)without adhesives, coatings, dispersions etc.

Source: BASF AG

Figure 2.6   Consumption of plastics materials forecast for 2010 by consumer regions in percent and growth p. a.

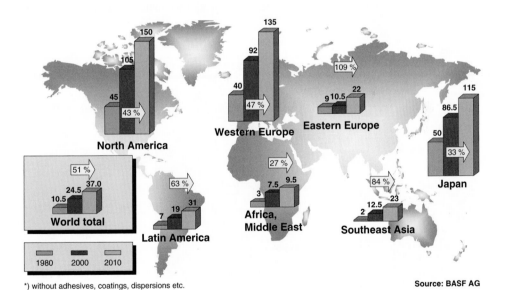

*) without adhesives, coatings, dispersions etc.

Source: BASF AG

Figure 2.7   Per capita consumption of plastic materials in kg per capita 1980–2010 (without adhesives, coatings, dispersions, etc.)

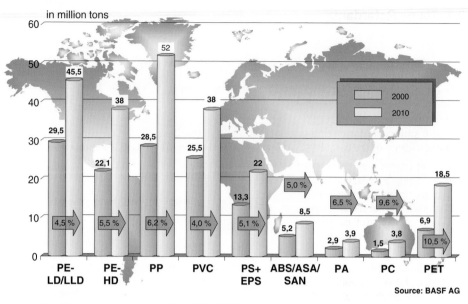

Figure 2.8   World plastics consumption 2000–2010 (growth p.a.)

In Fig. 2.9, the ongoing concentration process in the plastics manufacturing industry and a forecast for 2005 are shown for the capacities and shares of the major West European manufacturers.

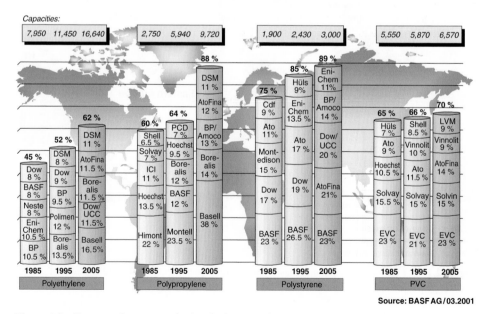

Figure 2.9   Concentration process in the plastics manufacturing industry. Share of the respective five major manufacturers in Western Europe 1985–2005

2.2 Flame Retardants

U. Fink

Markets for flame retardants were developed after governmental regulatory action; loss of life and property because of fire resulted in public pressure to provide safer materials [2]. Insurance companies also exerted pressure by increasing payment rates in unprotected environments.

Pressure still exists for environmentally safe products with lower levels of smoke and with reduced toxicity of the combustion products. Important areas for flame retarded products are:

- Construction materials (thermal insulating materials [foams], mattresses, furniture cushioning, chip boards, laminates).
- Transportation components (polymer parts of airplanes, trains, subways, busses) must meet flammability standards. Flame retardants are especially important in seating materials.
- Fabrics and apparel; carpets and draperies as well as children's sleepwear are furnished with flame retardant chemicals.
- Other applications are in adhesives, paper and pulp products, and upholstery.

Adding flame retardant chemicals in sufficient amounts can nearly always adequately retard the flammability of a combustible material. The challenge, however, is to optimize the use of flame retardant chemicals to achieve a cost-effective flame retardant end-use product without seriously compromising that product's desired physical properties. A successful manufacturer offers the customer a solution to this manufacturing problem and not just a chemical product.

The total market for flame retardants in the United States, Western Europe, and Asia in 1998 amounted to more than 1.14 million metric tons and was valued at almost $2.1 billion. This market is expected to grow at an average annual rate of about 3.5-4.0 % per year on both a value and a quantity basis over the 1998–2003 period, approaching 1.4 million metric tons valued at $2.6–2.7 billion. Table 2.1 shows the breakdown of this market by major geographical region for 1995 and 1998, and projected growth rates through 2003.

Table 2.1   Consumption of flame retardants by major region (millions of dollars)

	1995	1998	Average annual growth rate, 1998–2003 (%)
United States	585	630	2.8–3.6
Western Europe	631[a]	685[a]	3–4
Japan	348[b]	373[b]	3.8
Other Asian countries	>244	>390	5.1
Total	**1,808**	**2078**	**3.5–4.0 %**

[a] Based on exchange rates of DM 1.54 and DM 1.75 per dollar, the average exchange rates in effect during 1995 and 1998, respectively
[b] Based on exchange rates of ¥ 108 and ¥ 131 per dollar for 1995 and 1998, respectively.
Source: SRI Consulting.

The flame retardant business has emerged as a result of requirements that manufacturers of plastics, textiles, and other materials meet various safety standards and government regulations by improving the flame retardant characteristics of their products. Because most flame retardants contribute no other useful properties to a product (and often compromise other performance characteristics), their use is almost entirely driven by legislation. Indeed, growth in this business can often be impacted far more dramatically by new regulations than by growth in the end-use markets. Because many flame retardants (e.g., halogenated compounds) are subject to scrutiny either for their own suspected toxicity or for that of their combustion products, current or potential health and environmental regulations are also important determinants of the specific types of flame retardant used. An understanding of current regulations and an awareness of potential new ones is an important requirement for success as a participant.

The flame retardant business is highly internationalized. Not only do many companies participate on a worldwide basis, but the impact of regulations in one geographical area also often has reverberations throughout the world. Manufacturers of end-use products (e.g., consumer electronics and automobiles), wherever located, must comply with regulations in destination countries for any products they export. Thus, manufacturers will insist that their raw material suppliers (e.g., resin manufacturers, custom compounders, or flame retardant producers) help them meet these requirements. Manufacturers with significant exports follow regulatory developments throughout their market areas closely. Because of economies in production and distribution, they may not wish to vary their flame retardant formulations within a specific product line (e.g., computer housings), no matter where its intended destination. Therefore, they will utilize flame retardant formulations that meet the most stringent regulations of any region where their product is to be sold.

Flame retardants historically entered the business from a product-oriented view (i.e., manufacturers generally produced similar products for other applications, frequently – at least historically – of much larger volume). Manufacturing companies have begun to take a broader, market-oriented view of the plastics additive business as a whole, but can still be constrained by technology, market access, and manufacturing cost considerations when competing with companies that are basic in key raw materials.

During the last several years major global producers of halogenated compounds have been adding antimony and organophosphorus compounds to their product line, largely through acquisition, but also by adding new manufacturing capacity. The plastics industry is the largest consumer of flame retardants, most of which are sold to basic resin manufacturers, custom compounders, or plastics fabricators. However, smaller volumes of flame retardants are also sold to the textile and paper industries. The ability to identify and anticipate customer problems and provide solutions is also an essential requirement for a strong competitive position in the flame retardant business. This requires well-directed applied research, highly effective technical service capabilities, and a willingness to invest in the facilities and people required to provide them.

All major classes of flame retardants – brominated compounds, organophosphorus products, chlorinated chemicals, aluminum trihydrate, antimony oxides, and various other flame retardants – are covered in the following sections for the United States, Western Europe, and Japan.

2.2.1 Flame Retardants Market in the United States of America

In the United States the market for flame retardant resins, and hence for flame retardant chemicals, has grown at an average annual rate of approx. 4.5 % in the past decade. However, some product segments may have experienced differences in their volume and value growth rates because of interproduct substitution (e.g., increased use of newer, higher priced brominated or phosphorus flame retardants, or lower priced aluminum trihydrate) or different growth rates of the specific applications that a particular class of flame retardant is most frequently used for. It is estimated that during the last several years, tonnage growth has been slightly higher than value growth because of severe price pressure, especially for halogenated products and antimony. Based largely on the anticipated growth rates for materials that use flame retardants, mainly plastic resins, growth is forecast to slow slightly and exhibit a rate of 2.8–3.6% annually through the year 2003. Table 2.2 shows US consumption of flame retardants by product category for 1995 and 1998, with forecasts of average annual growth for 1998–2003 for all applications.

Table 2.2   U.S. consumption of flame retardants[a]

	Value (millions of dollars) 1998	Quantity (thousands of metric tons)		Average annual growth rate quantity basis (%) 1998–2003
		1995	1998	
Brominated compounds				
Additive	135	34.6	41.3	2.5–3.0
Reactive	85	25.4	27.0	4.0
Organophosphorus compounds				
Nonhalogenated	90	26.4	31.1	3–4
Halogenated	70	19.8[b]	26.0[b]	3–4
Chlorinated compounds	50	17.5	18.5	1.5
Aluminum trihydrate	85	232[c]	259[c]	3–4
Antimony oxides	75	22.5	28	2–2.5
Boron compounds	17	13.5	29.7	1.5–2.0
Other	23	11.5	13	4.0
Total	**630**	**403.2**	**473.6**	**2.8–3.6**

[a] Includes all flame retardant markets (textiles, paper, paint, plastics, other).
[b] Includes pentabromodiphenyl ether and nonhalogenated phosphates used in bromine/phosphate mixtures.
[c] Includes approx.35,000 metric tons of white hydrate consumption where flame retardancy is of secondary importance.
Source: SRI Consulting.

Table 2.3 provides a breakdown of the 1998 US consumption of flame retardants by major application. A number of changes are occurring in the US flame retardant markets, as follows:

Table 2.3   U.S. consumption of flame retardants by application – 1998 (thousands of metric tons)

Application	Brominated compounds	Organophosphorus compounds		Chlorinated compounds	Aluminum trihydrate	Antimony oxides	Boron compounds
		Non-halogenated	Halogenated				
ABS	11.1	–	–	–	–	2.9	–
Engineering resins	7.4	11.4[a]	0.2	1.1	5	2.0	0.2
Epoxy	14.0	0.2	–	0.5	12	1.0	0.2
Poly-acrylates	–	–	–	–	50[b]	–	–
Poly-olefins	3.2	–	–	3.5	28	2.4	0.3
Poly-styrene	20.7	–	–	0.1	–	4.8	–
Poly-urethane	3.0	–	24.5	4.2	11	–	–
Poly(vinyl chloride)	2.0	12.5	0	4.5	23	8.8	4.0
Unsaturated polyester	2.9	0.2	0.3	1.5	46	1.8	small
Other[c]	4.0	6.8	1.0	3.1	84[d]	4.3	25[e]
Total[f]	**68.3**	**31.1**	**26.0**	**18.5**	**259**	**28.0**	**29.7**

[a] Mostly polyphenylene oxide and polycarbonate/ABS.
[b] Includes some material used more for filler/pigment applications than for flame retardancy.
[c] Includes rubber and elastomers, phenolics, paper and textiles, paint, adhesives, cellulose acetate, vinyl/nitrile sponge and intumescent paint.
[d] Used primarily in styrene-butadiene latex carpet backings, rubber, phenolics and textiles.
[e] Used primarily in cellulose insulation.
[f] Does not include other flame retardants (e. g., Mg[OH]$_2$) included in Table 2.2 so the sum of the totals in 13 thousand metric tons less than Table 2.2.
Source: SRI Consulting.

- The US Consumer Products Safety Commission (CPSC) is evaluating tests to measure the resistance of upholstered furniture and mattresses to small open flames. If either voluntary or mandatory standards for flame retarding residential furniture result, there will be increased opportunities for use of flame retardants.
- Industry globalization and competition will increase as flame retardant producers realize opportunities both to supply indigenous producers in rapidly growing offshore markets and

to support their existing customers as they invest in offshore resin production/compounding facilities.

- Asia, excluding Japan, represents the most rapidly growing market for flame retardants. Accordingly, some US flame retardant resin production destined for this market may shift to overseas production sites.
- Product support (i.e., technical service, including assistance with performance testing and regulatory compliance) is increasingly important. Flame retardant producers will continue to expand their businesses, through acquisition or joint ventures, into complementary plastic additives lines, spreading costs over a broader product base.
- Although, with the exception of penta- and octabromodiphenyl ether, no government restriction on the use of polybrominated diphenyl ethers is expected, alternative brominated flame retardants will probably continue to be increasingly utilized for products serving global markets. This practice permits a manufacturer to use a single formulation for a specific item sold worldwide, independent of possible pressures in any specific region. Major US automotive concerns have begun to restrict use and are likely to eliminate the use of brominated bisphenols.
- Engineering plastics and resins markets (including acrylonitrile-butadiene-styrene [ABS]-polycarbonate blends) will achieve much higher growth rates than those projected for more traditional flame retardant resins. This market will continue to offer good opportunities for new flame retardant types that can meet the required high performance standards.
- As the industry matures, companies are consolidating. As a result, more suppliers are basic in a broader range of chemical compounds. This strategy is both defensive and responsive to customer demands for a broader range of products. This trend is likely to continue.
- In certain environments (e.g., electronic manufacturing clean rooms), preference will be for low-smoke, low-conversion flame-retarded products to protect expensive manufacturing equipment.
- Consumers of plastic compounds increasingly demand a full line of products from individual suppliers. Flame retardant suppliers must increasingly supply lines of compounds or concentrates incorporating a full line of additives.

2.2.2 Flame Retardants Market in Western Europe

The 1998 Western European market for flame retardants is estimated at 355,000–365,000 metric tons valued at $670–700 million. Growth from 1995 to 1998 averaged 3 % per year, with aluminum trihydrate representing the largest growth in this period in the major flame retardant categories. Table 2.4 shows the estimated Western European consumption of flame retardants in 1995 and 1998 by volume and value as well as their estimated average annual growth rates for the next 5 years.

The major market areas for flame retardants in Western Europe are plastics, rubber and elastomers, paints and coatings, textiles and nonwovens, and paper and paperboard, in declining order of importance.

Plastics represent by far the most important market for flame retardants in Western Europe, accounting for nearly 90 % of overall consumption. The most important plastics for flame retardants are poly(vinyl chloride) (PVC), polyurethanes, unsaturated polyesters, polyolefins, and polystyrene. Together, PVC and polyurethane foams account for almost 50 % of total consumption. Apart from these two main uses, a large number of other resins exist that also consume flame retardants, although in much smaller quantities. The textile and paper segments together accounted for approx. 20,000 metric tons of flame retardants consumed in 1998.

Some flame retardant chemicals are used almost exclusively in a few types of plastics, and others can be used in many different polymers. Some chemicals are used mainly in foams,

Table 2.4   Western European consumption of flame retardants – 1998

	Value[a] (millions of dollars)		Volume (thousands of metric tons)		Average annual growth rate, quantity basis (%)
	1995	1998	1995	1998	1998–2003
Brominated compounds	160–170	180–200	43–45	48–55	3–4
Organophosphorus compounds	160	170	66	71	2–4
Chlorinated compounds	44.3	28	28.8	24.7	(–3)–(–4)
Aluminum trihydrate	117.5	145	120	160	4–5
Antimony trioxide	90.9	70–75	20.0	22–24	0
Zinc/boron compounds	7.3	7–8	2.1	3.5	3
Magnesium hydroxide	10	22	3.5	7.3	15
Melamine	14.3	23	8.5	14	3–4
Ammonium polyphosphate	12	15	3	3–4	1–2
Red phosphorus	10	11	1.1	1.5	1–2
Total	**626–636**	**671–697**	**296–298**	**355–365**	**3–4**

[a] Exchange rates used were DM 1.54 per dollar in 1995 and DM 1.75 in 1998. Values shown are only approximations, as prices within any one group of products vary significantly.
Source: SRI Consulting

others in film and sheet or in molded products. The choice of product for use in foams is often critical because the flame retardant can affect the cell structure of the foam (e.g., chlorinated organophosphorus compounds have been used successfully in foams). In all cases the compatibility of the flame retardant chemicals with other additives must be tested. With only a few exceptions, it is quite impossible to tie a single flame retardant to a special plastic type.

Halogenated flame retardants are very effective additives and they can easily be made compatible with various polymer systems. Their main application is in engineering thermoplastics. Because they are effective in the gas phase they can generate corrosive gases and higher smoke densities and carbon monoxide contents than nonhalogenated flame retardants. Phosphorus-based flame retardants are very effective in the condensed phase. Therefore, they have to be tailor made to the individual polymer compositions to melt and decompose at a given temperature range. Preferred applications are in polyamides and polyurethanes. Inorganic flame retardants such as aluminum trihydrate are used because of their low price and smoke density levels. High-fill grades of up to 50 % or more are required. They are used mainly in elastomers. Table 2.5 shows Western European consumption of flame retardants by polymer.

The largest textile use of flame retardants in Western Europe is for cellulosic fibers and fabrics, mainly cotton and cotton/polyester blends with up to 20 % polyester fiber. These are usually flame retarded by means of organic phosphorus-nitrogen compounds. About 10,000

Table 2.5   Western European consumption of flame retardants by polymer type – 1998[a]

	PVC	PU	UP	PS	PE	PP	ABS	PET/PBT	PA	PC	Epoxy	Rubber	Other[b]
Brominated compounds	○	●	○	●	○	●	●	●	●	○	●		○
Organophosphorus compounds													
Nonhalogenated	●	●	○	○		○	○					○	○
Halogenated		●	●									○	○
Chlorinated compounds	●	○	●	○	○	○	○				○	●	○
Aluminum trihydrate	●	○	●	○	●	○	○		○		○	○	○
Antimony trioxide	●		○	●	○	●	●	○	○	○	○	●	○
Boron compounds	●											●	○
Magnesium hydroxide	○			○	●	●			●				○
Melamine		●		○		○			○			○	○
Ammonium polyphosphate			○			●							○
Red phosphorus			○			○			●		●	○	○

[a] ● indicates >10 % of consumption   ○ indicates <10 % of consumption
PVC = poly(vinyl chloride); ABS = acrylonitrile-butadiene-styrene; PU = polyurethane; PET = polyethylene terephthalate; UP = unsaturated polyester; PBT = polybutylene terephthalate; PS = polystyrene; PA = polyamide; PE = polyethylene; PC = polycarbonate; PP = polypropylene
[b] Includes specialty polymers, textiles and paper.
Source: SRI Consulting.

metric tons of permanent flame retardant formulations, including those for synthetic fibers and for wool, and about 7000 metric tons of temporary flame retardants were used for flame retarding textile fabrics in Western Europe in 1998. These formulations include, in addition to the flame retardant chemical itself, additives such as pigments and binders. The main use of permanent flame retardant fabrics is for bedding in hospitals, for public buildings, and for flame-resistant work clothing. Temporary flame retardant fabrics are used mainly for curtains and textile furnishings in public buildings. Future demand will depend to a large extent on regulations mandating flame retardant textile products for specific applications.

Inherently flame retardant synthetic fibers, such as polyester, polyacrylamide, or modacrylic fibers, are produced in Western Europe in relatively small quantities by incorporating various reactive flame retardant comonomers, usually halogenated hydrocarbons and organic phosphorus derivatives.

This market is not expected to gain particular importance in Western Europe in the near future unless new regulations mandate the use of these fibers on a larger scale. At present, many synthetic fibers meet the existing flammability standards for most uses without a flame retardant.

Low profitability in flame retardants has led to mergers and joint ventures and to vertical integration. The major suppliers of brominated flame retardants in Western Europe are Great Lakes, followed by Eurobrom and Albemarle. Great Lakes, which acquired FMC's activities in 1999, is particularly strong in phosphate esters, whereas Akzo Nobel and Bayer are major suppliers of tris(chloropropyl) phosphate. Martinswerk (now Albemarle) is the largest Western European supplier of aluminum trihydrate, while DSM and Agrolinz are the leaders in melamine flame retardants.

The future of the flame retardant industry in Western Europe will be determined mainly by regulatory factors, as follows:

- New regulations in individual countries regarding the fire prevention and flame retardant requirements of specific products, as part of the harmonization of laws and regulations within the European Union,
- Voluntary restrictions or legal prohibitions on the use of certain types of flame retardants for environmental protection or labor hygiene/safety.

Whereas these factors are mainly external and difficult to predict or even influence, a number of other key trends are expected to be crucial in the Western European flame retardants industry over the next few years, as follows:

- Acquisitions: Several acquisitions have occurred in the flame retardant industry in Western Europe in the last few years. The buyers included Albemarle, Akzo Nobel, Clariant, Great Lakes, DSM, and Rhodia. In many cases these companies could complement their product range, establish a manufacturing base in Western Europe, or gain market share through these acquisitions. In a very competitive and research-intensive environment, size and financial power are certainly important factors, and further acquisitions may be seen in the future.
- Consolidation in chlorinated compounds: Several companies have exited the chlorinated compounds business in Western Europe, including the former companies Hüls, Hoechst, and CECA. Ongoing concerns about environmental problems of short-chain chlorinated paraffins (e.g., bioaccumulation) have reduced demand in Europe. Future consolidation of the few remaining players is not unlikely.
- Growth of nonhalogenated flame retardant markets: Inorganic flame retardants, particularly aluminum trihydrate (ATH) and magnesium hydroxide, belong to the fastest-growing group of flame retardants. Several producers reported expansions of their ATH plants, and even companies strongly rooted in brominated flame retardants such as Eurobrom are offering magnesium compounds.
- Shift away from brominated flame retardants: In spite of their effective mechanism, there are several drawbacks in using brominated products. First, there are some technical disadvantages such as corrosivity and sensitivity to light, and thus yellowing and surface embrittlement of the plastic part. At the high processing temperatures of engineering thermoplastics, brominated products may cause problems because of decomposition. Because of the gas phase mechanism of their flame retardancy, brominated flame retardants cause a higher smoke density when burning. Also, the corrosive gases in a fire may cause additional damage to buildings and electronic installations. In addition, environmental pressure and the uncertainty of future government regulations have caused manufacturers Albemarle and Great Lakes Chemicals to diversify. In mid-1999 Albemarle announced plans to produce phosphorus-based, nonhalogen flame retardants. Great Lakes Chemical is adding phosphate ester flame retardants through its acquisition of FMC's additives division.

- A continued shift toward specialty business: Several of the traditional flame retardants (e.g., certain brominated compounds) have come under regulatory pressure, which has forced users to find alternative products. Whereas in the past universally applicable products were often available, today many of the new alternative flame retardants are developed specifically for certain niche applications. Apart from the increased R&D effort required, this shift requires an even more flexible marketing, production, and sales approach than in the past.
- Continued concern regarding environmental issues: Although the main focus has been on brominated compounds, other flame retardants have also come under scrutiny, including chlorinated paraffins, antimony trioxide, and phosphorus-based products. Sweden is reportedly still pushing for a total ban on tetrabromobisphenol A (TBBA) and polybrominated diphenyl ethers. Although, with the exception of penta- and octabromodiphenyl ether and short-chain chloroparaffins, the imminent danger of an EU-wide ban on certain flame retardants has been alleviated at least temporarily, flame retardant products are likely to be on the environmental agenda of many government decision makers in the future.
- Globalization of the industry: Driven by the trend of OEMs to demand a global presence of their suppliers, many of the players in the flame retardant industry have increased their international activities. In Europe this has been mainly through acquisitions; in Asia it has been through sales offices and local partnerships.
- Growth of the independent compounder and masterbatch producer: Profit optimization has forced many of the large, captive compounders to focus their sales activities on large key accounts. Independent compounders that offer customer-specific solutions for small clients have picked up much of the detail business. These independents have become a customer segment of growing importance for European flame retardant producers.

More stringent fire regulations will lead to systems that offer slower heat release rates in fires, along with slow smoke generation, and low flame retardant additive toxicity and smoke toxicity.

The need for miniaturization in the electric and electronics industries has obliged resin producers to replace traditional flame retardant systems with melt-blendable and high-flow brominated flame retardant systems. The political threat against brominated flame retardant systems still exists in the European market. It did not materialize in terms of overall consumption, but resulted in a shift from polybrominated diphenyl ethers to other brominated compounds such as TBBA in styrene copolymers high-impact polystyrene ([HIPS], ABS). Although brominated flame retardants are still considered an optimal combination of effective flame retardancy and thermomechanical properties, several producers are replacing halogenated ABS with polycarbonate alloys that have been flame retarded by phosphate-based systems.

Because of price and health reasons related to antimony trioxide, alternative synergists for use with halogenated flame retardants are increasingly being used. These include iron oxides and zinc compounds such as borates, oxides, stannates and phosphates.

2.2.3 Flame Retardants Market in Japan

The 1995 Japanese market for flame retardant chemicals amounted to approx. 122,000 metric tons valued at about $348 million (based on a stabilized exchange rate of ¥ 108 per dollar for mid-1996). The market is projected to grow at an average annual rate of 2.4 % over the 1998–2003 period. Japanese consumption of flame retardant chemicals is summarized in Table 2.6. The table shows that consumption of flame retardants is generally expected to grow over 1998–2003. Nonhalogenated organophosphorus compounds and magnesium hydroxide are expected to experience the best growth.

Table 2.6 Japanese consumption of flame retardants

	Value[a) (millions of dollars) 1998	Quantity (thousands of metric tons) 1998[a]	Quantity (thousands of metric tons) 2003	Average annual growth rate, quantity basis, (%) 1998–2003
Brominated compounds	147.2	47.8	52.2	1.8
Organophosphorus compounds				
Nonhalogenated	67.0	22.0	27.7	4.7
Halogenated	9.2	4.0	4.3	1.5
Chlorinated compounds	7.4	2.1	2.1	0
Aluminum trihydrate	25.2	42.0	43.0	0.5
Antimony oxides	85.8	15.5	17.2	2.1
Other				
Magnesium hydroxide	6.1	4.0	8.5	16.3
Guanidines	13.4	5.0	5.5	1.9
Miscellaneous	11.5	1.5	1.6	1.3
Total	**372.8**	**143.9**	**162.1**	**2.4**

[a) The exchange rate for 1998 was ¥ 131 per dollar.
Source: SRI Consulting.

The major end uses for flame retardants in Japan are as follows:

- Epoxy resins, especially for printed circuit boards,
- Styrenic resins, especially ABS resins and high-impact polystyrene for housings of home electronics and business machines,
- PVC, especially for equipment housing applications,
- Polyurethane foam, especially for automotive seats,
- Paper backing, PVC wall coverings,
- Polyolefins, especially polypropylene,
- Phenolic resins, especially for printed circuit boards and other electronic applications,
- Unsaturated polyester and poly(methyl methacrylate), especially artificial marble in homes,
- Photographic film, especially acetate film,
- Textiles, especially for automotive seats and carpet, and for carpets and draperies in public buildings,
- Engineering plastics, especially for polycarbonate and polycarbonate-ABS blends, nylon, and poly(butylene terephthalate),
- Rubber.

The Japanese flame retardants supply can be summarized as follows:

- Most brominated flame retardants are imported from Great Lakes Chemical (United States), Dead Sea Bromine (Israel), and Albemarle Corporation (United States). Imports were estimated to be 80–90 % of the total supply in 1998.

- About 15 % of the supply of chlorinated flame retardants is imported from the United States.
- About 45–50 % of antimony oxide was imported from the People's Republic of China in recent years.
- All other flame retardants are produced and marketed in Japan.

The Asia/Pacific region, excluding Japan, represents the most rapidly growing market for flame retardants. Currently, the greatest use is in printed circuit boards and housings for consumer electronics and business machines for export, so these products must meet the flame retardant regulations of the destination countries. However, as the standards of living in the Asia/Pacific countries continue to rise and the area recovers from recent economic downturns, both rapidly increasing regional demand for the products themselves and domestic regulations requiring flame retardancy are expected to follow. The major flame retardant resins in this region are epoxy and phenolic resin (for printed circuit boards) plus ABS and polystyrene (for consumer electronics and business machine housings). The growth in Japanese flame retardant consumption is expected to be less than in other Asian countries because of a shift in production to those areas.

Japan is currently a major exporter of consumer electronics (television sets, videocassette recorders, etc.), business machines (plain paper copiers, facsimile machines, personal computers, etc.), and automobiles to Western Europe and the United States, where flame retardant regulations are more stringent than in Japan. These exported products comply with all regulations of the countries to which they are exported. Thus, the consumption of flame retardants in Japan has been heavily dependent on regulations and growth in the export markets.

There is a strong movement among the Japanese manufacturers to supply the domestic market with products with flame retardant characteristics similar to those of exported products. This trend is expected to support the growth of the Japanese flame retardant market even if export markets for electronic goods and automotive products do not grow significantly through 2003.

The following flame retardant end-use applications are expected to grow significantly in the near future:

- Engineering thermoplastics, for example, poly(butylene terephthalate), nylon, polyphenylene oxide (PPO), polycarbonate,
- Polymer alloys (although the technology of flame retardants for polymer alloys is more complex than that for homopolymers, and a substantial effort will be necessary to develop this technology,
- Communication and power cables,
- Wall coverings, carpets, draperies, and upholstery for public buildings – although more consideration is being paid to safety in the construction of public and office buildings, actual consumption of flame retardants for these applications is increasing rather slowly, and the growth rate will probably be at 1–2 % per year through 2003.

The market for flame retardants for household products will not grow substantially because there is no strong movement to strengthen current regulations.

There is worldwide concern about dioxin production from the combustion of halogen-containing plastics. In addition, there is a general concern about potential carcinogenicity and mutagenicity of all chemical compounds. Japan is no exception. Thus, the controversy may continue in Japan about the future use of specific flame retardants.

Low-smoke flame retardants are increasingly preferred for telephone cable, power cable, and similar products. This trend will lead to increased consumption of aluminum trihydrate and magnesium hydroxide in Japan.

The overall Japanese flame retardant market will continue to grow during the 1998–2003 period, in parallel with Japan's gross national product or its total consumption of plastics.

2.3 Overview of the Flame Retardants Industry

U. FINK

2.3.1 Flame Retardants Industry in the United States of America

The flame retardant market has emerged largely as a result of regulatory pressure on manufacturers of plastics, textiles, and other materials to improve the safety of these products by increasing their flame retardant properties [2]. Most flame retardant market participants either already were manufacturing products with flame retardant characteristics (although they were not produced for that application) or could adapt their core chemistries to produce related chemicals with flame retardant properties. Thus, the entrance of most flame retardant manufacturers into the business was a logical extension of existing product lines to meet an emerging market need. For example, most of the producers of phosphate ester flame retardants [e.g., Akzo Nobel, Albright & Wilson Americas (now Rhodia), Great Lakes (formerly FMC), and Solutia (now Ferro)] or their corporate predecessors had long produced various phosphate derivatives for use in agriculture, detergents, and other markets. Nearly all of these companies were once basic in elemental phosphorus, although some have lost their basic raw material position as a result of acquisitions and corporate restructuring. These latter companies have attempted to secure an equivalent raw material position by long-term arrangements with surviving basic producers.

Similarly, Great Lakes Chemical, Albemarle Corporation, and Dead Sea Bromine, which dominate the brominated flame retardant business, are all back-integrated into bromine, producing numerous brominated chemicals for a variety of applications. The basic aluminum companies (e.g., Kaiser Aluminum Co., Alcoa Inc.) are the dominant producers of ATH for flame retardants (although they do not necessarily perform finishing operations and supply the end-use market directly), and US Borax Inc. dominates the market for borate flame retardants.

The largest producers of the major classes of flame retardant chemicals are shown in Table 2.7 in the order of their market share.

The majority of flame retardants, especially the additive types, are sold to companies that compound various plastics additives, including flame retardants, into the base resin to obtain the desired characteristics for end users' requirements. The compounding can be carried out by the base resin manufacturers or by independent specialty compounders (both those that compound and market their own formulations and those that are custom compounders). Resin manufacturers that compound in-house include many of the producers of thermoplastic resins. Independent compounders are numerous and include large companies such as A. Schulman, Polyone, and many smaller companies.

Systems suppliers are another category of customer, primarily for flame retardants used by the polyurethane industry. Systems suppliers provide a two-component flame-retarded polyol and isocyanate system. These systems are purchased by manufacturers of polyurethane foam or by contractors. Examples of systems suppliers include Freeman and Stepan Chemical.

Another special category of customer is the processors of ATH. These companies either produce their own ATH (e.g., Alcoa Inc.) or purchase crude ATH from basic manufacturers and further process it for sale to compounders, resin manufacturers, and end users. An example of the latter type is the Engineered Minerals Division of J.M. Huber Corporation.

Some flame retardant producers also produce compounds/concentrates to facilitate use of the product; an example is Anzon's antimony oxide concentrate.

Finally, some flame retardants are sold to end-product manufacturers or plastics fabricators.

Table 2.7   Major US producers of flame retardants by product category – 1998 (millions of dollars)

	Market size	Major producers	Remarks
Brominated compounds	220	Great Lakes Chemical Corporation	Has a little less than half of market
		Albemarle Corporation	Second-largest producer of brominated products; about 40 % market share
		Ameribrom Inc.[a]	Has a little over 10 % US market share. Imported material
		Ferro Corporation (now Albemarle)	Offers brominated polystyrenes
Organophosphorus compounds			
Nonhalogenated	90	Akzo Nobel Chemicals Inc. Great Lakes Solutia (now Ferro) Albright & Wilson Americas (now Rhodia)	First three are largest suppliers, with a total of 90 % of market
Halogenated	70	Akzo Nobel Chemicals Inc. Albright & Wilson Americas (now Rhodia) Great Lakes	Akzo has highest share of the halogenated market; Albright & Wilson has leading position in textiles
Chlorinated compounds	50	Occidental Chemical Corporation Dover Chemical Corporation Velsicol Chemical Corporation	Top two dominate the market

Table 2.7   (Continuation)

	Market size	Major producers	Remarks
Aluminum trihydrate	85	Kaiser LaRoche Hydrate Partners Alcoa Inc. Alcan Aluminum Corporation[b] Reynolds Metals Company	Consists of basic manufacturers of ATH.
		J. M. Huber Corporation, Engineered Materials Division Alcoa Inc. Alcan Aluminum Corporation[b]	Offer processed product.
Antimony oxides	75	Anzon Incorporated[c] Laurel Industries Inc.[d]	Imports from the People's Republic of China account for 25 % of the market.
Boron compounds	17	U.S. Borax Inc. Anzon Incorporated[c]	U.S. Borax dominates cellulosic insulation segment; Anzon supplies plastics segment.
Other	23		
Total	**630**		

[a] Subsidiary of Dead Sea Bromine, with production in Israel.
[b] Subsidiary of Alcan Aluminum Ltd. with production at Jonquière, Québec, Canada
[c] Owned by Great Lakes
[d] Subsidiary of Occidental Chemical.
Source: SRI Consulting.

The relative sales volumes for many of the major flame retardant product types in various customer categories are shown in Table 2.8.

These channels of distribution have not changed dramatically during the last several years. The percentages reflect to a large extent where compounding in general takes place for the resins using the various flame retardant categories. For instance, brominated flame retardants are used in many commodity resins, and a significant portion is compounded by the polymer manufacturer, whereas a good portion of nonhalogenated organophosphorus compounds are used in engineering resins where the independent compounder plays a larger role. If any trend is occurring, it is a slight shift out of compounding by polymer producers.

Table 2.8 US end users of selected flame retardants – 1999 (percent of total volume)

	Bromi- nated com- pounds	Organophosphorus compounds		Chlorin- ated com- pounds	Aluminum trihydrate	Antimony oxides
		Nonhalo- genated	Halogen- ated			
Resin manu- facturers	75	30	–	25	–	65[a]
Specialty com- pounders[b] (systems suppliers)	15–20	35	20–30	40	30	30
End- product manu- facturers (including PVC fabri- cators)	5–10	35	70–80	35	70	5
Total	**100**	**100**	**100**	**100**	**100**	**100**

[a] Antimony oxide manufacturers do a majority of compounding themselves.
[b] Also includes compounders that toll for resin manufacturers.
Source: SRI Consulting.

2.3.2 Flame Retardants Industry in Western Europe

All major manufacturers of flame retardant chemicals in Western Europe are part of a larger product portfolio within their parent companies. For some companies, such as Akzo Nobel, Bayer AG, Clariant, Imperial Chemical Industries PLC, or Solvay SA, flame retardants account for only a small share of their total business. For other companies, such as Albright & Wilson (now Rhodia), Eurobrom BV, and Great Lakes Chemical Corporation, these products are fairly important. No Western European company is exclusively active in flame retardants. In addition to flame retardant producers, there are numerous retailers and compounders on the distribution side.

With regard to their product range, most producers focus on one or two families of flame retardants such as brominated or organophosphorus compounds or inorganic flame retardants. There is a rather clear distinction between producers of organic and inorganic flame retardants, although some producers of organic compounds have recently started to offer inorganic flame retardants as well.

In line with the general trend of large polymer producers to focus on their key clients, inde- pendent compounders and master batch producers have recently gained in importance and are a significant and growing customer segment for flame retardant manufacturers in Europe.

The manufacturers of flame retardants sometimes sell the same products or compounds for other uses (e.g., as fillers and plasticizers), either to improve the utilization of their production facilities or to broaden the product range they can offer to customers.

Several plastics compounders are involved in the manufacture of flame retardant compounds and masterbatches in Western Europe. The main compounders include Albis Plastics, Schulman, Ferro and Cabot Plastics. These independent compounders tend to focus on special polymer grades, whereas the large polymer producers such as Bayer, BASF and GE Plastics act as compounders for standard grades.

Table 2.9 lists the major European suppliers of flame retardants in each main product category.

The dominant chemical classes of flame retardants (on a volume basis) are ATH followed by the phosphorus esters and the brominated compounds. In value, the brominated compounds rank first, followed by the organophosphorus compounds and ATH. Boron compounds and antimony trioxide are produced mainly by companies engaged in metallurgy and mining but also by a few chemical companies. The major manufacturers of ATH are the aluminum producers and some specialized chemical companies with a limited line of more sophisticated products.

The most important customers of the flame retardant industry in Western Europe are resin producers that make flame-retarded plastic grades; the second largest group are plastic compounders and processors that are involved in the manufacture of flame-retarded compounds, foams, films, cables, engineering plastics for the electrical/electronics industries, and reinforced plastic parts; producers of wall and floor coverings; and producers in the textile and related industries. Master batch producers are a growing customer segment for the Western European flame retardant manufacturers.

The major raw material suppliers to the producers of organic flame retardants are

- Oil and petrochemical producers that supply the basic aromatic and aliphatic compounds,
- Halogen producers that supply the bromine and chlorine for the manufacture of halogenated flame retardants,
- Producers of phosphorus and derivatives (i. e., phosphorus trichloride and oxychloride) for organophosphorus flame retardants.

A number of major flame retardant manufacturers are back-integrated into specific raw materials (Bayer and Solvay in chlorine; Eurobrom BV and Great Lakes in bromine; and Clariant in elemental phosphorus). Companies that are integrated in bromine and elemental phosphorus have a cost advantage. The producers of inorganic flame retardants and synergists draw primarily on captive raw materials for their manufacturing processes; these companies are involved in mining, metal refining or related activities.

Table 2.9   Major Western European suppliers of flame retardants by product category – 1998 (millions of dollars)

	Market size	**Major suppliers**
Brominated compounds	180–200	Great Lakes Chemical Eurobrom BV Albemarle
Organophosphorus compounds		
Nonhalogenated	100	Akzo Nobel NV Clariant Bayer AG Albright & Wilson (now Rhodia) Solutia (now Ferro)

	Market size	Major suppliers
Halogenated	70	Great Lakes Chemical Akzo Nobel NV
Chlorinated compounds	28	Imperial Chemical Industries PLC Caffaro Occidental Chemical Europe
Aluminum trihydrate	145	Martinswerk GmbH (now Albemarle) Alcan Chemicals Europe[a] Nabaltech GmbH
Antimony oxides	70–75	Anzon Ltd. (GLCC) Campine
Zinc and boron compounds	7–8	Borax Joseph Storey & Company Ltd.
Magnesium hydroxide	22	Martinswerk GmbH (now Albemarle) Incemin (now Ankerpoort)
Melamine	23	DSM BASF Aktiengesellschaft Agrolinz Melamin GmbH
Ammonium polyphosphate	15	Clariant Albright & Wilson (now Rhodia)
Red phosphorus	11	Clariant
Total	**670–700**	

[a] Stopped production in 2002
Source: SRI Consulting.

2.3.3 Flame Retardants Industry in Japan

There are no producers that cover all categories of the variety of flame retardants, as in the other regions.

The majority of brominated and chlorinated hydrocarbons, organophosphorus compounds, and antimony oxide flame retardants are sold to resin manufacturers. In contrast, magnesium hydroxide, ATH, and guanidines are often supplied directly to plastics processors from producers. Magnesium hydroxide and ATH flame retardants are also sold to compounders for wire and cables. The wire and cable manufacturers are large companies that have been in the business for a long time; thus, they have the technical capabilities needed to establish appropriate formulations for flame-retarded finished products. The wire and cable manufacturers purchase resins from resin producers through compounders. Guanidines (usually in the form of guanidine salts) are also supplied directly to the finished product manufacturers, such as PVC wall covering manufacturers.

Table 2.10 lists the major Japanese suppliers of flame retardants by product category and their approximate market shares.

Table 2.10   Major Japanese producers/suppliers of flame retardants by product category – 1998 (millions of dollars)

	Market size[a]	suppliers	Market share (%)
Brominated compounds	147.2	Great Lakes Chemical Corporation	30–35
		Albemarle Corporation	25–30
		Dead Sea Bromine Ltd.	
		Bromokem (Far East) Ltd.	25–30
		Tosoh Corporation	10
Organophosphorus compounds			
Nonhalo-genated	67.0	Daihachi Chemical Industry Co., Ltd.	40–45
		Akzo Kashima Ltd.	30–40
		Ajinomoto Co., Inc.	15–25
		Other	5–10
Halogenated	9.2	Daihachi Chemical Industry Co., Ltd.	45–50
		Akzo Kashima Ltd.	40–45
		Other	
Chlorinated compounds	7.4	Ajinomoto Co., Inc.	30–35
		Asahi Denka Kogyo K.K.	20–25
		Tosoh Corporation	5–10
		Other	40–50
Aluminum trihydrate	25.2	Showa Denko K.K.	45
		Sumitomo Chemical Co., Ltd.	35
		Nippon Light Metal Company, Ltd.	20

Table 2.10 (Continuation)

	Market Size[a)]	Suppliers	Market Share (percent)
Antimony oxides	85.8	Nihon Seiko Co., Ltd.	20–22
		Sumitomo Metal Mining Co., Ltd.	14–16
		Mikuni Smelting & Refining Co., Ltd.	9–11
		Other	5–8
		Imports	47–50
Others			
Guanidines	13.4	Nippon Carbide Industries Co., Inc.	
		Sanwa Chemical Company, Ltd., subsidiary	90–95
		Other	5–10
Magnesium hydroxide	6.1	Kyowa Chemical Industry Co., Ltd.	80–90
		Asahi Glass Company, Ltd.	5–10
		Other	1–5
Miscellaneous	11.5		
Total	**373.0**		

[a)] The average 1998 exchange rate was ¥ 131 per dollar.
Source: SRI Consulting.

The flame retardant resins are sold first to compounders, which are usually pigment manufacturers or subsidiaries of wire and cable manufacturers. Occasionally, compounders have detailed technical expertise about flame-retarded resins, but this is relatively rare. In many cases, compounders simply color the resins and sell them to resin processors or wire and cable manufacturers.

It is common practice for flame retardant manufacturers to use distributors or trading houses to sell their products. The manufacturers often have technically trained people on their staffs; these people should be knowledgeable about the manufacture of resins and plastics, as well as about flame retardants, polymer reactions, and the properties of resins.

Table 2.11 shows the major flame retardant distributors in Japan and the producers they represent.

Table 2.11   Major Japanese distributors of flame retardants – 1998

Distributor	Producer represented	Country of origin	Product/ market share
Albemarle Asano Corp.	Albemarle Corporation	United States	Brominated/>25 %
Miki & Co., Ltd.	Great Lakes Chemical Corporation	United States	Brominated/>30 %
Mori	Dead Sea Bromine Ltd.		
	Bromokem (Far East) Ltd., subsidiary	Israel	Brominated/>25 %
Somar Corp.	Occidental Petroleum Corporation	United States	Chlorinated/<10 %

Source: SRI Consulting.

One of the most successful non-Japanese flame retardant manufacturers, Great Lakes Chemical, sells its products through Miki & Company Limited, a trading house. Although Great Lakes Chemical provides Miki with technical information, Miki has its own technical service people who are familiar not only with technical matters but also with the market. The relationship between Great Lakes Chemical and Miki has been quite good. Albemarle Corporation and Dead Sea Bromine also have relationships with their distributors that are beneficial to all participants.

The principal manufacturers of flame retardants are listed in Section 1 of the Appendix.

References

[1] *Anon.:* Economic Data and Graphics Regarding Plastics. VKE Statistics and Market Research Board. Verband Kunststofferzeugende Industrie e.V., Frankfurt, Germany, April 2001
[2] R. E. Davenport, U. Fink, Y. Ishikawa: 1999 Flame Retardants Report. Specialty Chemicals Update Program, SRI Consulting, Zurich, Switzerland, 1999

3 The Burning Process and Enclosure Fires

B. KARLSSON

3.1 Introduction

Fire safety of plastic products has been a matter of great concern to legislators, authorities, and manufacturers since the advent of building fire regulations. Most such products are found in buildings, transport vehicles, or enclosures of some kind, and considerable effort is dedicated to minimizing the material reaction-to-fire and the resulting hazard to humans.

Fire is a physical and chemical phenomenon that is strongly interactive by nature. The interactions between the flame, its fuel, and the surroundings can be strongly nonlinear, and quantitative estimation of the processes involved is often complex. The burning process and the processes of interest in an enclosure fire mainly involve mass fluxes and heat fluxes to and from the fuel and the surroundings. Figure 3.1 shows a schematic of these interactions, indicating the complexity of the mass and heat transfer processes discussed here.

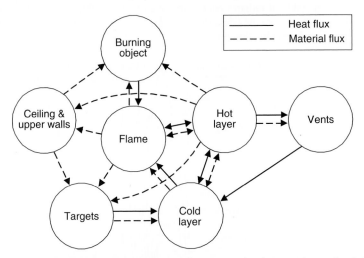

Figure 3.1 Schematic of the heat fluxes and mass fluxes occurring in an enclosure fire (adapted from Friedman [1])

It is important to introduce the reader to the physical and chemical processes that occur during an enclosure fire and the terminology used to describe it, as the environmental conditions in the enclosure are strongly interactive with the burning process. This chapter starts by generally describing the burning process, using the burning of a candle as an example. A qualitative description of enclosure fire development is then given, introducing the physical and chemical processes involved and the terminology used to describe the enclosure fire. Subsequently, the factors influencing the fire development are discussed. Finally, a brief description of enclosure fire simulation models is given.

3.2 General Description of the Process of Combustion

The study of combustion is a complex subject; it includes a number of disciplines such as fluid mechanics, heat and mass transport, and chemical kinetics. Here, we give only an introductory discussion on combustion and use a candle as our fuel; the discussion is therefore limited to a laminar, steady flame on a solid substrate. Later chapters describe the burning process in far greater detail.

The study of a burning candle, however, is very illustrative with regard to the natural processes we are interested in. Michael Faraday, the 19th century scientist, gave lectures on "The Chemical History of a Candle" [2] at the so-called Christmas Lectures at the Royal Institution in London. He claimed that there was no better way by which one could introduce the study of natural philosophy than by considering the physical phenomena of a candle.

3.2.1 A Burning Candle

Consider Fig. 3.2, which shows an illustration of a burning candle and the temperature distribution through the flame. An ignition source, a match for example, heats up the wick and starts melting the solid wax. The wax in the wick vaporizes and the gases move, by the process of diffusion, out into a region where oxygen is found. The gases are oxidized in a complex series of chemical reactions, in regions where the oxygen-fuel mixture is flammable. The candle flame is then stable; it radiates energy to the solid wax, which melts. Because the wax vaporizes and is removed from the wick, the melted wax moves up the wick, vaporizes, burns, and the result is a steady combustion processes.

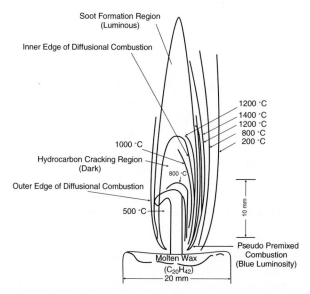

Figure 3.2   A burning candle and the temperature distribution in the flame (from [3])

Because the processes occurring in the flame involve mainly the flow of energy and the flow of mass, we shall briefly discuss these.

The flow of energy occurs by the processes of radiation, convection, and conduction. The dominant process is that of radiation; it is mainly the soot particles produced by combustion that glow and radiate heat in all directions. The radiation down toward the solid is the main heat transfer mode that melts the solid, but convection also plays a role. The convective heat flux is mainly upwards, transferring heat up and away from the combustion zone. The larger and more luminous the flame, the quicker the melting process is.

The radiative energy reaching the solid is, however, not sufficient to vaporize the wax, only to melt it. The wick is therefore introduced as a way to transport the melted wax up into the hot gases, where the combined processes of radiation, convection, and conduction supply sufficient energy to vaporize the melted wax.

The mass transfer and the phase transformations are also exemplified by the burning candle. The fuel transforms from solid to liquid state. The mass balance requires that the mass, which disappears from the wick by vaporization, be replaced and thus the liquid is drawn up into the wick by capillary action. Once there, the heat transfer from the flame causes it to vaporize and the gases move away from the wick by the process of diffusion. The inner portion of the flame contains insufficient oxygen for full combustion, but some incomplete chemical reactions occur, producing soot and other products of incomplete combustion. These products move upwards in the flame due to the convective flow and react there with oxygen. At the top of the flame nearly all the fuel has combusted to produce water and carbon dioxide, the efficiency of the combustion can be seen by observing the absence of smoke emanating from the top of the candle flame.

This self-sustained combustion process can be changed most easily by changing the dimensions and properties of the wick, and thereby the shape and size of the flame. A longer and thicker wick will allow more molten wax to vaporize, resulting in a larger flame and increased heat transfer to the solid. The mass and heat flows will quickly enter a balanced state, with steady burning as a result.

To a chemist, fuel consists of the whole of the gaseous products evolving from combustible (solid or molten) materials. These gases arise from pyrolysis or incomplete oxidation of the initial polymeric materials. The general public, however, would use the word fuel for various combustible materials that can also be in solid or liquid state (gasoline, coal, etc). We shall here use the latter to describe enclosure burning.

3.2.2 Other Solid Fuels

Without the wick the candle will not sustain a flame, as is true for many other solid fuels. Factors such as the ignition source, the type of fuel, and the amount and surface area of the fuel package determine whether the fuel can sustain a flame or not. A pile of wooden sticks may sustain a flame while a thick log of wood may not do so. Once these factors are given, the above-discussed processes of mass and energy transport will determine whether the combustion process will decelerate, be steady, or accelerate.

Also, the phase transformations of other solid fuels may be much more complicated than the melting and vaporizing of the candle wax. The solid fuel may have to go through the process of decomposition before melting or vaporizing, and this process may require considerable energy. The chemical structure of the fuel may therefore determine whether the burning is sustained or not.

For fuels more complex than the candle it is difficult to predict fire growth. The difficulty is due not only to the complexity of the physical and chemical processes involved, but also due to the dependence of these processes on the geometric and other fuel factors mentioned earlier and the great variability in these.

When the fuel package is burnt in an enclosure, the fire-generated environment and the enclosure boundaries will interact with the fuel, as seen in Fig. 3.1, making predictions of fire growth even more complex.

It is currently beyond the state of the art of fire technology to predict the fire growth in an enclosure fire with any generality, but reasonable engineering estimates of fire growths in buildings are frequently obtained using experimental data and approximate methods. Such test methods are described in later sections of this book.

3.3 General Description of Fire Growth in an Enclosure

A fire in an enclosure can develop in a multitude of different ways, mostly depending on the enclosure geometry and ventilation and the fuel type, amount, and surface area. The following is a general description of the various phenomena that may arise during the development of a typical fire in an enclosure. Methods for quantitatively estimating the variables mentioned are generally well known and are discussed in several handbooks, for example, [4], [5], and [6].

Ignition: After ignition, the fire grows and produces increasing amounts of energy, mostly due to flame spread. In the early stages the enclosure has no effect on the fire, which then is fuel controlled. Besides releasing energy, a variety of toxic and nontoxic gases and solids are produced. The generation of energy and combustion products is a very complex issue, as mentioned previously, and the engineer must rely on measurements and approximate methods in order to estimate energy release rates and the yield of combustion products. The issue of energy release rate and combustion product yields measurements is dealt with in later chapters.

Plume: The hot gases in the flame are surrounded by cold gases and the hotter, less dense mass will rise upwards due to the density difference, or rather, due to buoyancy. The buoyant flow, including any flames, is referred to as a fire plume.

As the hot gases rise, cold air will be entrained into the plume. This mixture of combustion products and air will impinge on the ceiling of the fire compartment and cause a layer of hot gases to be formed. Only a small portion of the mass impinging on the ceiling originates from the fuel; the greatest portion of this mass originates from the cool air entrained laterally into the plume as it continues to move the gases toward the ceiling. As a result of this entrainment the total mass flow in the plume increases and the average temperature and concentration of combustion products decrease with height.

Ceiling Jet: When the plume flow impinges on the ceiling the gases spread across it as a momentum driven circular jet. The velocity and temperature of this jet is important because quantitative knowledge of these variables will allow estimates to be made on the response of any smoke and heat detectors and sprinkler links in the vicinity of the ceiling.

The ceiling jet eventually reaches the walls of the enclosure and is forced to move downwards along the wall as seen in Fig. 3.3. However, the gases in the jet are still warmer than the surrounding ambient air and the flow will turn upwards due to buoyancy. A layer of hot gases will thus be formed under the ceiling.

Gas Temperatures: Experiments have shown for a wide range of compartment fires, that it is reasonable to assume that the room becomes divided into two distinct layers: a hot upper layer consisting of a mixture of combustion products and entrained air and a cold lower layer consisting of air. Further, the properties of the gases in each layer change with time but are assumed to be uniform throughout each layer. For example, it is commonly assumed when using engineering methods that the temperature is same throughout the hot layer at any given time.

The Hot Layer: The plume continues to entrain air from the lower layer and transports it toward the ceiling. The hot upper layer therefore grows in volume and the layer interface descends toward the floor. The smoke filling process and methods for calculating smoke filling times are discussed in handbooks, for example [4], [5] and [6].

Figure 3.3   Schematic of the development and descent of a hot smoke layer (from [4])

Heat Transfer: As the hot layer descends and increases in temperature, the heat transfer processes are augmented. Heat is transferred by radiation and convection from the hot gas layer to the ceiling and walls, which are in contact with the hot gases. Heat from the hot layer is also radiated toward the floor and the lower walls, and some of the heat will be absorbed by the air in the lower layer. In addition, heat is transferred to the fuel bed, not only by the flame, but also to an increasing extent by radiation from the hot layer and the hot enclosure boundaries. This leads to an increase in the burning rate of the fuel and the heating up of other fuel packages in the enclosure.

Vent Flows: If there is an opening to the adjacent room or out to the atmosphere, the smoke will flow out through it as soon as the hot layer reaches the top of the opening. Often, the increasing heat in the enclosure will cause the breakage of windows and thereby create an opening. Methods for calculating the mass flow rates through vents are discussed in handbooks (e.g., [4], [5], and [6]).

Flashover: The fire may continue to grow, either by increased burning rate, by flame spread over the first ignited item, or by ignition of secondary fuel packages. The upper layer increases in temperature and may become very hot. As a result of radiation from the hot layer toward other combustible material in the enclosure, there may occur a stage at which all the combustible material in the enclosure is ignited, with a very rapid increase in energy release rates. This very rapid and sudden transition from a growing fire to a fully-developed fire is called flashover. The fire can thus suddenly jump from a relatively benign state to a state of awesome power and destruction.

The solid line in Fig. 3.4 shows the initiation of the transition period at point A, resulting in a fully-developed fire at point B. Once point B has been reached (the fully-developed fire), flashover is said to have taken place.

The Fully-Developed Fire: At the fully-developed stage flames extend out through the opening and all the combustible material in the enclosure is involved in the fire. The fully-developed fire can go on for a number of hours, as long as sufficient fuel and oxygen are available for combustion.

Oxygen Starvation, the Underventilated Fire: For the case where there are no openings in the enclosure or only small leakage areas, the hot layer will soon descend toward the flame region and eventually cover the flame. The air entrained into the combustion zone now contains little oxygen and the fire may die out due to oxygen starvation. The dotted line in Fig. 3.4 shows that a fire may reach point A and start the transition period toward flashover, but due to oxygen depletion the energy release rate decreases, as well as the gas temperature.

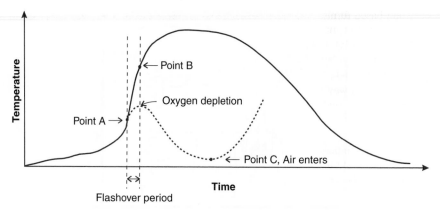

Figure 3.4   Enclosure fire development in terms of gas temperatures: some of the many possible paths a room fire may follow (from Karlsson and Quintiere [6])

Even though the energy release rate decreases, the pyrolysis may continue at a relatively high rate, causing the accumulation of unburnt gases in the enclosure. If a window breaks at this point, or if the fire service creates an opening, the hot gases will flow out through the top of the opening and cold and fresh air will flow in through its lower part. This may diminish the thermal load in the enclosure but the fresh air may cause an increase in the energy release rate. The fire may then grow toward flashover as shown by the dotted line in Fig. 3.4.

3.4 Factors Influencing the Fire Development in an Enclosure

The factors that influence the development of a fire in an enclosure can be divided into two main categories: those that have to do with the enclosure itself and those that have to do with the fuel. We can specifically mention

- The size and location of the ignition source,
- The type, amount, position, spacing, orientation and surface area of the fuel packages,
- The geometry of the enclosure,
- The size and location of the compartment openings,
- The material properties of the enclosure boundaries.

Ignition Source: An ignition source can consist of a spark with very low energy content, a heated surface, or a large pilot flame, to name a few examples. The source of energy is chemical, electrical, or mechanical. The greater the energy of the source, the quicker the subsequent fire growth on the fuel source is. A spark or a glowing cigarette may initiate smoldering combustion, which may continue for a long time before flaming occurs, often producing low heat but considerable amounts of toxic gases. A pilot flame usually produces flaming combustion directly, resulting in flame spread and fire growth.

The location of the ignition source is also of great importance. A pilot flame positioned at the lower end of, say, a window curtain may cause rapid upward flame spread and fire growth. The same pilot flame would cause much slower fire growth were it placed at the top of the curtain, resulting in slow, creeping, downward flame spread.

Fuel: The type and amount of combustible material is of course one of the main factors determining the fire development in an enclosure. In building fires the fuel usually consists of solid materials; in some industrial applications the fuel source may also be in liquid state.

Heavy, wood-based furniture usually causes a slow fire growth but can burn for a long time. Some modern interior materials include porous lightweight plastics, which cause more rapid fire growth but burn for a shorter time. A high fire load therefore does not necessarily constitute a greater hazard; a rapid fire growth is more hazardous in terms of human lives.

The position of the fuel package can have a marked effect on the fire development. If the fuel package is burning away from walls, the cool air is entrained into the plume from all directions. When placed close to a wall, the entrainment of cold air is limited. This causes not only higher temperatures but also higher flames, as combustion must take place over a greater distance.

Figure 3.5 shows temperatures measured above fires in 1.22 m (4 ft) high stacks of wood pallets [7]. Curve A shows temperatures as a function of height above pallets burning without the presence of walls. Curve B is for a similar stack near a wall and curve C is for a stack in a corner.

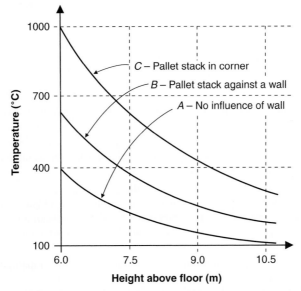

Figure 3.5   The temperature in the fire plume as a function of height above a burning stack of wood pallets. A, B, and C indicate burning away from walls, by a wall, and in a corner, respectively (adopted from Alpert et al. [7])

The spacing and orientation of the fuel packages are also important. The spacing in the compartment determines to a considerable extent how quickly the fire spreads between these. Upward flame spread on a vertically oriented fuel surface will happen much more rapidly than lateral spread along a horizontally oriented fuel surface.

Combustible lining materials, mounted on the compartment walls and/or ceiling, can cause very rapid fire growth. Figure 3.6 shows results from a small-scale room test where the lining material is mounted on the walls only (with noncombustible ceiling) and when the material is mounted on both walls and ceiling [8]. In both cases an initial flame was established on the lining material along one corner of the room. With material mounted on the ceiling, the flame spreads with the flow of gases (concurrent-flow flame spread), causing rapid growth. With noncombustible ceiling the flame spreads horizontally (opposed-flow flame spread) across the material, a process that is much slower and requires the lining material to be heated consider-

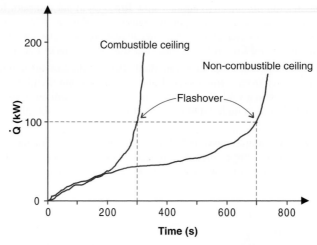

Figure 3.6   Energy release rate versus time in a small room with a combustible material mounted on walls only and with combustible material mounted on walls and ceiling. Flashover occurs when the energy release rate is 100 kW in this compartment.

ably before the flame can spread rapidly over it. As a result, the time to flashover is 4 min in the former case and 12 min in the latter.

A fuel package of a large surface area will burn more rapidly than an otherwise equivalent fuel package with a small surface area. A pile of wooden sticks, for example, will burn more rapidly than a single log of wood of the same mass.

Enclosure Geometry: The hot smoke layer and the upper bounding surfaces of the enclosure will radiate toward the burning fuel and increase its burning rate. Other combustible items in the room will also be heated up. The temperature and thickness of the hot layer and the temperature of the upper bounding surfaces thus have a considerable impact on the fire growth.

A fuel package burning in a small room will cause relatively high temperatures and rapid fire growth. In a large compartment, the same burning fuel will cause lower gas temperatures, longer smoke filling times, less feedback to the fuel, and slower fire growth.

The fire plume entrains cold air as the mixture of combustion products and air moves upwards toward the ceiling. The amount of cold air entrained depends on the distance between the fuel source and the hot layer interface. In an enclosure with a high ceiling this causes relatively low gas temperatures, but owing to the large amount of air entrained, the smoke filling process occurs relatively rapidly. The smaller the floor area, the faster the smoke filling process.

With a low ceiling the heat transfer to the fuel will be greater. In addition, the flames may reach the ceiling and spread horizontally under it. This results in a considerable increase in the feedback to the fuel and to other combustibles, and a very rapid fire growth is imminent.

For enclosures with a high ceiling and a large floor area the flames may not reach the ceiling and the feedback to the fuel is modest. The fire growth rather occurs through direct radiation from the flame to nearby objects, where the spacing of the combustibles becomes important.

In buildings with a large floor area but a low ceiling height the feedback from the hot layer and ceiling flames can be very intensive near the fire source. Further away, the hot layer has entrained cold air and has lost heat to the extensive ceiling surfaces, and the heat flux to the combustible materials, in the early stage of the fire, is therefore lower than the heat flux closer to the fire source.

We can conclude that the proximity of ceilings and walls can greatly enhance the fire growth. Even in large spaces, the hot gases trapped under the ceiling can heat up the combustibles beneath and result in extremely rapid fire spread over a large area. The fire that so tragically engulfed the Bradford City Football Stadium in England, 1985 is a clear reminder of this [9].

Compartment Openings: Once flaming combustion is established the fire must have access to oxygen for continued development. In compartments of moderate volume that are closed or have very small leakage areas, the fire soon becomes oxygen starved and may self-extinguish or continue to burn at a very slow rate depending on the availability of oxygen.

For compartments with ventilation openings, the size, shape, and position of these become important for the fire development under certain circumstances. During the growth phase of the fire, before it becomes ventilation controlled, the opening may act as an exhaust for the hot gases, if its height or position is such that the hot gases are effectively removed from the enclosure. This will diminish the thermal feedback to the fuel and cause slower fire growth. For other circumstances the geometry of the opening does not have a very significant effect on fire growth during the fuel-controlled regime.

It is first when the fire becomes controlled by the availability of oxygen that the opening size and shape becomes all-important. Kawagoe [10] found, mainly through experimental work, that the rate of burning depended very strongly on the "ventilation factor," defined as $A_o\sqrt{H_o}$, where A_o is the area of the opening and H_o is its height. The importance of this factor can also be shown by theoretically analyzing the flow of gases in and out of a burning compartment. It can be shown that the rate of burning is controlled by the rate at which air can flow into the compartment. An increase in the factor $A_o\sqrt{H_o}$ will lead to an equal increase in the burning rate. This is valid up to a certain limit when the burning rate becomes independent of the ventilation factor and the burning becomes fuel controlled.

Properties of Bounding Surfaces: The material in the bounding surfaces of the enclosure can affect the hot gas temperature considerably and thereby the heat flux to the burning fuel and other combustible objects. Certain bounding materials designed to conserve energy, such as mineral wool, will limit the amount of heat flow to the surfaces so that the hot gases will retain most of their energy.

The material properties controlling the heat flow through the construction are the conductivity (k), the density (ρ), and heat capacity (c). These are commonly collected in a property called thermal inertia and given as the product $k\rho c$. Insulating materials have a low thermal inertia; materials with relatively high thermal inertia, such as brick and concrete, allow more heat to be conducted into the construction, thereby lowering the hot gas temperatures.

3.5 Engineering Models for Enclosure Fires

The rapid progress in the understanding of fire processes and their interaction with buildings has resulted in the development of a wide variety of models that are used to simulate fires in compartments. The models can be classified as either deterministic or probabilistic. Probabilistic models do not make direct use of the physical and chemical principles involved in fires, but make statistical predictions about the transition from one stage of fire growth to another. Such models are not discussed further here. The deterministic models can roughly be divided into three categories: CFD models; zone models; and hand-calculation models.

CFD Models: The most sophisticated of these are termed "field models" or "CFD models" (Computational Fluid Dynamics models). The CFD modeling technique is used in a wide range of engineering disciplines. Generally, the volume under consideration is divided into a very large number of subvolumes and the basic laws of mass, momentum, and energy conservation are applied to each of these. Figure 3.7 shows a schematic of how this may be done for a fire in an enclosure. The governing equations contain as further unknowns the viscous stress

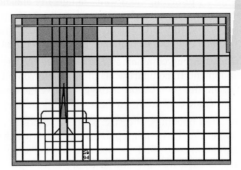

Figure 3.7   CFD models divide the enclosure into a large number of subvolumes.

components in the fluid flow. Substitution of these into the momentum equation yields the so-called Navier-Stokes equations, and the solution of these is central to any CFD code.

The myriad of engineering problems, which can be addressed by CFD models, is such that no single CFD code can incorporate all of the physical and chemical processes that are important. There exist, therefore, only a handful of CFD codes that can be used for problems involving fire. These, in turn, use a number of different approaches to the subprocesses that need to be modeled. Some of the most important of these subprocesses can be considered to be turbulence modeling, radiation and soot modeling, pyrolysis and flame spread modeling, and combustion modeling. The subprocesses are usually modeled at a relatively fundamental level and an understanding of these requires expert knowledge in a number of specialized fields of physics and chemistry. Cox [11] provides an excellent summary of the main issues.

With regard to the burning process, CFD models generally need as input the fuel supply rate to calculate the energy release rate. Further, various assumptions are made to estimate species produced. Recently, however, some work has been published in which the burning process and the environmental conditions are linked by incorporating flame spread, pyrolysis, and fire growth models into CFD codes. An example of such work is given by Yan [12] and Holmstedt, who developed and described a flame spread model and by Tuovinen et al. [13], who incorporated the model into the CFD code SOFIE [14]. Such models are still under development and they currently offer the only practical way to link the burning process with the fire-induced environment in the enclosure.

Using CFD models requires considerable computational capacity as well as expert knowledge, not only in physics and chemistry, but also in numerical methods and computer science. In addition, to set up the problem, run it on the computer, extract the relevant output, and present the results is very time consuming and costly, so practical use for fire safety engineering design is relatively rare. However, such a modeling methodology can be very useful when dealing with complex geometries and may be the only way to proceed with certain design problems.

Two-Zone Models: A second type of deterministic fire models are those that divide the room into a limited number of control volumes or zones. The most common type is termed "two-zone models" in which the room is divided into an upper, hot zone and a lower, cold zone. The equations for mass and energy conservation are solved numerically for both zones for every time step. The momentum equation is not explicitly applied; instead, information needed to calculate velocities and pressures across openings come from analytically derived expressions where a number of limiting assumptions have been made. Several other subprocesses are modeled in a similar way, such as plume flows and heat transfer. The section on hand calculations below lists a number of these processes.

In general, zone models describe the production of species in terms of yields per mass loss rate of the pyrolyzing material. The energy release rate is required as input but is usually expressed as a mass loss rate times the actual heat of combustion. Both terms are very much dependent on the ventilation conditions. For well-ventilated fires the energy release rate and the species production can be reasonably constant for a given fuel. In general, these can vary with time and can significantly vary as ventilation-limited conditions are approached and achieved. In particular, they vary with the equivalence ratio, defined as

$$\phi = \frac{(\dot{m}_f / \dot{m}_{air})_{act}}{(\dot{m}_f / \dot{m}_{air})_{stoich}} \tag{1}$$

where ϕ is the equivalence ratio, $(\dot{m}_f / \dot{m}_{air})_{act}$ is the actual relationship between the mass flow rate of fuel and the mass flow rate of air, and $(\dot{m}_f / \dot{m}_{air})_{stoich}$ is the same relationship at stoichiometry. Stoichiometry gives the exact proportions of the reactants (fuel and air) for complete conversion to products, where no reactants are remaining. The equivalence ratio may be computed in a zone (or upper layer) in which combustion has occurred by computing the mass concentration of the available fuel and oxygen in the zone. This method of predicting concentrations is still under study, but it currently offers the only practical zone model approach for estimating species, such as CO, under ventilation-limited conditions in enclosure fires.

Many two-zone models (see Fig. 3.8) have been described in the literature. Some of these only simulate a fire in a single compartment; others simulate fires in several compartments, linked by doors, shafts or mechanical ventilation. In addition, the degree of verification, documentation, and user friendliness vary greatly between these models.

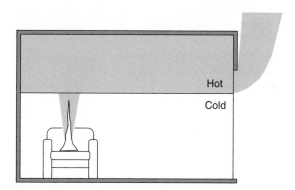

Figure 3.8   Two-zone modeling of a fire in an enclosure

In recent years there has been an upsurge in the use of two-zone models in fire safety engineering design. This is partly due to the increasing availability and user-friendliness of the computer programs. However, any serious use of such models requires that the user is well acquainted with the assumptions made and the limitations of the models, that is, that the user has had some training in the subject of enclosure fire dynamics.

Hand-Calculation Models: A third way of analytically describing some basic fire processes is to use simple hand-calculation methods. These are basically a collection of simplified solutions and empirical methods to calculate flame heights, mass flow rates, temperature and velocities in fire plumes, time to sprinkler activation, room overpressure, and many more variables.

The remainder of this section describes summarily the hand-calculation models. The methods discussed below can, for convenience, be divided into three categories: Those that deal with combustion, those that estimate the resulting environmental conditions, and those that have to do with heat transfer. These methods have highly varying limits of applications, and the user must have some knowledge of classical physics to apply them correctly.

3.5.1 Energy Evolved and Species Generated

Calculating fire growth and the amount to energy evolved from the primary fire source requires knowledge of the heat flux to the fuel bed, the flame heat flux to the fuel bed, and a large number of material parameters. Later chapters describe fundamental models for the burning process when the external conditions are given as well as methods for determining material parameters.

However, the heat fluxes are mainly a function of the enclosure fire environment, the ventilation conditions, and enclosure geometry. Models that link the burning process with the fire-induced environment are scarce.

In Fire Safety Engineering, a rough estimate of the energy release rate from a fire source requires that the type and amount of fuel involved be known. Typical burning rates and the heat of combustion for a range of liquid fuels, burning in the open, have been experimentally determined and are given in the literature. This allows the energy evolved to be roughly estimated if the area of the liquid spill is known. If the amount of spilt liquid is known then the time to burnout can also be calculated. Fire growth information for solids and other burning objects is available from several sources. Energy release rates for many items of furniture, curtains, and different types of materials are available. Such values are also available for species production rates, allowing species concentrations to be calculated.

The rate of energy evolved in a compartment is also dependent on the rate of supply of oxygen. Knowledge of the ventilation conditions can therefore be used to evaluate the maximum rate of energy release inside a compartment. Any excess, unburnt fuel will then be burnt outside the fire compartment, where oxygen is available.

Computer programs with material databases are also available where the user is assisted in choosing an appropriate energy release rate curve.

3.5.2 Fire-Induced Environment

The basic principles used for calculating the environmental conditions due to a fire in a compartment are the conservation of mass, energy, and momentum. The application of the conservation laws will lead to a series of differential equations. By making certain assumptions on the energy and mass transfer in and out of the compartment boundaries, the laws of mass and energy conservation can result in a relatively complete set of equations. Owing to the complexity and the large number of equations involved, a complete analytical solution is not possible and one must resort to numerical analysis through computer programs.

However, analytical solutions can be derived by making use of results from experiments and a number of limiting approximations and assumptions. Such solutions have generated innumerable expressions that may be used to predict a variety of environmental factors in a fire room. Examples of a number of these are given below.

The buoyant gas stream rising above a burning fuel bed is often referred to as the fire plume. The properties of fire plumes are important in dealing with problems related to fire detection, fire venting, heating of building structures, smoke filling rates, and so forth. By using dimensional analysis, the conservation equations, and data from experiments, expressions for various plume properties have been developed. These include expressions for plume temperature, mass flow and gas velocities at a certain height above the fire as well as flame height. Similar expressions have been derived for the jet that results when the plume gases impinge on a ceiling.

Mass flow in and out of compartment openings can be calculated, as the pressure differences across the opening can be estimated. The use of classical hydraulics and experimentally determined flow coefficients has resulted in hand-calculation expressions for such mass flows.

The gas temperature in a naturally or mechanically ventilated compartment can be calculated by hand, using regression formulas. These are based on experimentally measured gas temperatures, in a range of fire scenarios, and a simplified energy and mass balance. Such expressions are available for both pre- and post-flashover fires. By using similar expressions, the onset of flashover can be estimated.

By combining the expressions for gas temperature, plume flows, and vent flows, the descent of the smoke layer as a function of time can be calculated. The resulting expressions are usually in the form of differential equations but certain limiting cases can be solved by hand. Such solutions usually require an iteration process or the use of precalculated curves or tables.

Several other types of hand-calculation expressions have been developed. These include expressions for mass flow through roof openings, buoyant pressure of hot gases, species concentration, fire-induced room pressures, flame sizes from openings, and so forth. Some of such expressions have been collected in relatively user-friendly computer programs.

3.5.3 Heat Transfer

There are three mechanisms by which heat is transferred from one object to another: radiation, convection, and conduction. Classical textbooks on heat transfer provide innumerable hand-calculation expressions for calculating heat fluxes to and from solids, liquids, and gases as well as expressions for estimating the resulting temperature profiles in a target. These analytical expressions are usually arrived at by setting up the energy balance, by assuming constant properties and homogeneity in the media involved and by ignoring the heat transfer mechanisms that seem to be of least importance in each case.

The radiative heat flux from flames, hot gases, and heated surfaces impinging on a solid surface can be estimated using classical heat transfer and view factors. The same applies for convective heat transfer to solids and conductive heat transfer through solids. The surface temperature of a solid subjected to a radiative, convective, or conductive heat flux can be calculated by hand assuming the solid to be either semiinfinite or to behave as a thermally thin material. Numerous types of heat transfer problems can be solved in this way; a few examples are given below.

Assuming that a secondary fuel package is subjected to a known heat flux, and that it has a certain ignition temperature and constant thermal properties, then the time to ignition can be calculated. Similarly, if the activation temperature of a sprinkler bulb is known, the activation time can be estimated. Several other problems can be addressed in this way such as temperature profiles in building elements, flame spread over flat solids, heat detector activation, spread of fire from one building to another, and so forth. Analytical solutions to such problems can be found in common textbooks on heat transfer.

3.6 Summary

Most plastic products are found in buildings, transport vehicles, or enclosures of some kind, and the enclosure fire environment has a considerable and decisive effect on the burning process of such products. This chapter has given a description of enclosure fire development and the models used to predict its environmental consequences.

In an enclosure fire, two environmental factors have a decisive influence on the burning behavior of a combustible material: the flame heat flux (from the flames above the burning combustible) and the external heat flux (from hot gases, hot surfaces, and flames on other burning combustibles). The former is very much controlled by ventilation conditions and possible oxygen starvation; the latter has much to do with enclosure geometry, ventilation conditions, and surface materials. Before any reasonable prediction of real burning behavior can be made, these factors must be quantified.

A number of different models exist for predicting the environmental consequences of an enclosure fire, but most of these require the fire properties of the combustible (energy release rate and species production) be given as input. Only a few models exist that allow the prediction of both, in which the burning process is predicted and the enclosure variables calculated; the variables are then coupled to the burning process calculations. These models are under development and are still not used in engineering.

Most fuels are complex in nature, and the conditions they burn in are not often well defined. Engineers therefore often rely on measurements and tabulated data from small- or large-scale experiments to make estimates of energy release rate and species production as a function of time. Later chapters describe such experiments and the resulting data.

References

[1] R. Friedman: Status of Mathematical Modeling of Fires, FMRC Technical Report RC81-BT-5, Factory Mutual Research Corporation, Boston, 1981
[2] M. Faraday: The Chemical History of a Candle, Thomas Y. Crowell, New York, 1957 (First published in London, 1861)
[3] J.W. Lyons: Fire Research on Cellular Plastics: The Final Report of the Products Research Committee (PRC), Library of Congress Catalog Card Number 80–83306, Products Research Committee, 1980
[4] P.J. DiNenno (Ed.): SFPE Handbook of Fire Protection Engineering, 2nd edit., National Fire Protection Association, Quincy, MA, 1995
[5] D. Drysdale: An Introduction to Fire Dynamics, Wiley-Interscience, London, 1992
[6] B. Karlsson and J.Q. Quintiere: Enclosure Fire Dynamics, CRC Press , Boca Raton, FL, 1999
[7] R.L. Alpert and E.J. Ward: Fire Safety J. 7 (1984) pp 127–143
[8] B. Andersson: Model Scale Compartment Fire Tests with Wall Lining Materials, Report LUTVDG/ (TVBB-3 041), Department of Fire Safety Engineering, Lund University, Lund, 1988
[9] Fire Prevention No. 181, July/August, Fire Protection Association, London, 1985
[10] K. Kawagoe: Fire Behaviour in Rooms, Report No. 27, Building Research Institute, Tokyo, 1958
[11] G. Cox: in G. Cox (Ed.) Combustion Fundamentals of Fire, Academic Press, 1995
[12] Z. Yan: Numerical Modeling of Turbulent Combustion and Flame Spread, Department of Fire Safety Engineering, Lund University, TVBB-1018, Lund, 1999
[13] H. Tuovinen, J. Axelsson, P. Van Hees, B. Karlsson: Implementation of a physical flame spread model in the SOFIE CFD model, SP Report 1999: 32, SP, Borås, Sweden, 1999
[14] P. Rubini: SOFIE – Simulation of Fires in Enclosures, Proceedings of 5th International Symposium on Fire Safety Science, Melbourne, Australia, International Association for Fire Safety Science, 1999

4   The Burning of Plastics

4.1 The Combustion Process

C. DI BLASI

Three classes of polymers can be identified [1] on the basis of their thermal response:

- Those that degrade completely with breaking of the main chain (melting or thermoplastic polymers)
- Those that undergo rupture of side fragments with the formation of both volatiles (aromatics) and char
- Cross-linked polymers, whose main degradation product is char. A more schematic representation [2] can be made simply in terms of thermoplastic and char-forming polymers.

An understanding of polymer degradation kinetics is the first step in the development of rational polymer processing technologies, but there are other fields where this process is of paramount importance. Solid thermal degradation is the first step in fire initiation and growth; thus systematic studies on high-temperature degradation are important for both predicting the reaction to fire of construction products [2] and for synthesizing less combustible polymers [3]. After use, plastics are returned as waste, causing serious concern over environmental issues because suitable landfill sites are difficult to procure in industrial countries. Alternatives are available by means of recycling, through direct reuse of plastic by molding companies, incineration for heat recovery, pyrolysis, and gasification and liquefaction for oil and gas production. In all these technologies degradation plays a central role. Other applications [4] are related to hybrid rocket motors, solid propellant burning, and ablation of reentry vehicles.

The flammability behavior of polymers is defined on the basis of several processes and/or parameters, such as burning rates (solid degradation rate and heat release rate), spread rates (flame, pyrolysis, burnout, smolder), ignition characteristics (delay time, ignition temperature, critical heat flux for ignition), product distribution (in particular, toxic species emissions), smoke production, and so forth. In this chapter, the processes associated with fire initiation and growth are revised in terms of controlling mechanisms and mathematical modeling.

4.1.1 Heating

Combustion processes of charring polymers, exposed in an oxidizing flow environment, could proceed by two alternative pathways involving flaming combustion and smoldering (or glowing) combustion. Conditions of flaming combustion are achieved when the heat released by gas phase combustion of volatile products provides the heat flux needed for solid fuel degradation and flame spread. When the temperature or the intensity of the heat flux is below certain levels, oxidation of the char could result in smoldering combustion. Indeed, for porous materials, air may diffuse inside the solid matrix and cause slow oxidation: the low heat release rate, in the absence of significant losses, provides the heat flux needed for further charring and propagation of the smoldering combustion. For thermoplastics, solid phase processes are somewhat simplified because vaporization and pyrolysis are confined to a thin layer of the fuel sample, at the condensed-phase/gas-phase interface, which continuously regresses. As polymer molecules are too large for direct vaporization, thermal degradation to relatively low molecular weight species occurs first, as a result of heat transfer to the unreacted polymer. Moreover, the volatile release rate is also greatly affected by mass transport of volatiles through the reacting layer (bubble formation and dynamics).

Flaming combustion processes are the result of complex interactions of transport phenomena in the gas phase (momentum and mass and heat transfer) and in the solid phase, thermal degradation of the solid and fuel vapor oxidation chemistry. Processes that can lead to gas phase ignition include the vaporization of the solid, the formation of a flammable mixture adjacent to the solid surface, and initiation and the sustenance of oxidation reactions. The characteristics of these processes determine whether ignition will occur and the ignition delay time. In general, for flaming ignition to take place, three conditions must be met [5]:

1. Fuel and oxidizer must be available at a proper concentration to give a mixture within the flammability limits.
2. The gas phase temperature must become sufficiently high to initiate and accelerate the combustion reaction.
3. The size of the heated zone must be sufficiently large to overcome heat losses.

The temperature of the mixture above the solid surface plays a key role. An increase of this temperature above certain levels can occur by heat transfer from the hot degrading surface and/or by devices capable of creating a region of very high temperature in the gas phase, such as pilot flames, sparks, and hot wires (piloted ignition). Ignition can also be caused by a hot air stream.

After ignition, proper conditions may allow flame spread and solid burning which, among others, determine the heat release rate. In general, the flame spread rate is determined by the energy feedback (radiation, convection, and conduction) from the burning region to the unburned solid ahead of the flame while the burning rate is determined by the rate of energy transfer from the flame to the degrading solid beneath the flame. Combustion kinetics may also become important for near limit flame spread processes.

4.1.2 Decomposition

Different regimes of polymer degradation can be established [6] on the basis of the dependence of the heating exposure (temperature and global heat transfer coefficient and/or heat transfer mechanism) and polymer properties:

- Kinetic regime,
- Ablative regime (surface degradation or linear pyrolysis),
- Heat and/or mass transport controlled regime.

A kinetic regime can be achieved with both thermoplastic and char forming polymers if working with very small samples and moderate heating conditions (low temperatures and/or heating rates), to avoid significant intraparticle temperature and concentration gradients. For thick thermoplastics, heat transfer conditions at the condensed-phase/gas-phase interface do not change, thereby allowing degradation processes to be studied under quasi-stationary conditions (constant solid regression rate). As the external heating conditions are made successively more severe, the thickness of the reaction zone in the solid phase decreases and an ablative regime is established, in which internal heat transfer (or volatile evolution from the thin molten layer) is the controlling mechanism. Quasi-stationary conditions, of course, cannot be established for char forming polymers because heat and mass transfer processes are strongly affected by the formation of the char layer. In this case, the propagation rate of the reaction front through the solid decreases as time increases and both internal and external heat transfer are important.

Degradation reactions can be classified roughly as primary, which are related to the decomposition of the virgin polymer and the formation of wax-like intermediates, condensable and noncondensable volatiles, and/or char, and secondary and tertiary reactions (cracking and/or polymerization) of primary volatile products. Primary solid degradation is strictly coupled with heat transfer, whereas secondary and tertiary degradation is affected by both heat and mass transfer. Therefore, conversion characteristics, product distribution, and flammability behavior are, in general, the result of a strong interaction between chemical and physical processes. Detailed reaction mechanisms have been investigated for relatively simple polymers, such as poly(methyl methacrylate) (PMMA), polyethylene (PE), and polystyrene (PS), through the numerical solutions of large systems of ordinary differential equations [3]. Given the high number of unknown variables, however, such a treatment has never been coupled with the energy conservation equation or the description of other physical processes for engineering applications.

Numerous analyses of the decomposition of synthetic polymers are based on a one-step reaction mechanism. Early investigations tried to simulate fast heating rates through the application of the hot-plate technique, first introduced by Schultz and Dekker [7] and successively applied in numerous other studies. These and others, which tried to simulate the fast heating rates typical of combustion processes through heat flow from a massive conducting block, thermal radiation, convective heating by a hot gas jet, and heat flow from a diffusion flame, have been reviewed by Khalturinskii and Berlin [1] and Beck [4]. Analysis of several experimental data indicates the desorption of monomer units formed in the subsurface thermal layer of the polymer as the rate-limiting step, although noticeable efforts have been devoted to ascertain the relationship between the kinetics of ablative and isothermal bulk degradation. In general, it is not possible to assume that linear pyrolysis (apparent) kinetics are the same as those of bulk degradation [8], given the important role played by in-depth degradation, especially for thick samples.

Polymer degradation kinetics have also been investigated through measurements of weight loss curves, by thermogravimetric analysis (TGA) or specifically designed reactors, and/or volatile species evolution, with the formulation of semiglobal mechanisms, based on a few reactions and lumped product classes. Screen heater experiments [10] for nylon, mylar and PE, with heating rates of 100 K/s in the temperature range 300–375 °C and sample thickness of 25 μm (0.001 in.) have been interpreted by kinetic mechanisms consisting of several steps.

These represent a sequence of processes involving solid depolymerization, to give a molten polymer, and further reactions of depolymerization and devolatilization. More recent analyses are based on conventional TGA or TG-mass specroscopy systems, such as those by Conesa et al. [11] for PE, Knumann and Bockorn [12] for poly(vinyl chloride) (PVC), polyamide 6 (PA6), PE, polypropylene (PP), polystyrene (PS), and by Anthony [13] for PVC. The slow heating rates of TG systems avoid significant heat and mass transfer limitations, while the products of the different degradation steps are determined. However, as it is often pointed out that the corresponding heating rates are much slower than those typical of fire conditions, fast heating systems have also been developed, in which possible heat and mass transfer limitations and activity of secondary reactions are kept to a minimum. An example is the flash pyrolysis of thin films of PMMA (effective sample heating rates of 600–1000 K/s) in which the evolution rate of each devolatilization compound is used to obtain kinetic information through T-jump/FTIR [14]. It is noted that, for temperatures below 500 °C, the kinetic constants correspond to those previously determined for random C-C scission kinetics (MMA evolution, activation energies in the range 43–66 kcal/mol) and for decarboxylation (CO_2 evolution with activation energies in the range 39–46 kcal/mol). However, for higher temperatures, desorption/diffusion of devolatilization products becomes the limiting step, with low values of the apparent activation energy.

Product evolution and composition from polymer degradation have been examined mainly in relation to recycling and energy recovery. The most widely recycled plastics are poly(ethylene terephthalate) (PET) and high-density polyethylene (HDPE). Through solvolysis, condensation polymers, such as PET, can be chemically decomposed into raw materials for reuse [15]. This technology is applicable only to polymers, which in their functional groups contain weak chemical bonds that can be dissociated by some chemical agents. Addition polymers, such as polyolefins, PS, and PVC, decompose only at high temperatures into a mixture of monomers, liquid and gaseous fuels, and chars. Also, in general, plastic wastes are mixed and contaminated and thus expensive to separate and recycle, so that other treatments are preferred. In contrast to burning, pyrolysis [16] offers the facility of preserving the available polymer hydrocarbons, producing valuable petrochemicals, reducing the volume of the product gas by factors of 5–20 (lower costs in gas conditioning), and concentrating pollutants in the solid char. This process has been examined in melting vessels, blast furnaces, autoclaves, tube reactors, rotatory kilns, coking chambers, and fluidized-bed reactors. Rotatory kiln processes are particularly numerous as they can treat a variety of materials requiring only coarse grinding. The main characteristic is the long waste residence time (20 min) in contrast to fluidized-bed reactors with residence times from few seconds to 1.5 min. Owing to the large temperature gradient inside the rotatory kiln, the product spectrum is very wide and the corresponding heating value is usually exploited by direct combustion.

Fluid-bed units [7, 16, 17] can be operated to maximize the yields of a given product. Summary of the pyrolysis product yields and gas composition from some polymers and for a temperature of 550 °C are reported in Tables 4.1 and 4.2, reproduced from Williams and Williams [9]. The main gases consist of hydrogen, methane, ethane, ethene, propane, propene, butane, butene, and for PET carbon monoxide and carbon dioxide. Hydrogen chloride is the main product of PVC. As expected, because of the increased activity of secondary reactions, the gas yields increase at the expense of oil and wax, as the temperature is increased [17]. Analysis of oils and waxes [9] shows that PE and PP give mainly aliphatic composition consisting of alkanes, alkenes, and alkadienes, while PVC, PS, and PET give a mainly aromatic oil. On a rough basis, yields of ethene and propene from olefins do not exceed a maximum of 60 %. About 64 % of the monomeric styrene can be recovered from PS, but purification to obtain polymerization-grade monomer is expensive. An exception is PMMA, which gives up to 97 % monomers.

Modeling of the chemical reactions and transport processes taking place during polymer degradation is very difficult; thus several simplifications are made [2], some of which are still retained in the transport models recently proposed [6–18]. Depolymerization and melting are followed by complete devolatilization. Surface regression, property variation, heat convection, and conduction through the virgin solid and the molten layer are properly taken into account. The dependence of process dynamics, regression rate, surface temperature, and thickness of the reaction zone on the heating conditions is simulated and the results are applied to clarify the role of internal and external heat transfer mechanisms and interaction between transport phenomena and chemical reactions. Examples of the temperature profiles simulated for PE particles convectively heated, for two external temperatures, are shown in Fig. 4.1.

Table 4.1   Product yields from fluid-bed pyrolysis of some plastics at 550 °C in wt% [9]

Product	HDPE	LDPE	PP	PS	PVC	PET
Gas	11.4	21.4	6.5	0.7	3.9	49.1
Oil	36.8	17.8	31.5	59.0	22.1	23.5
Wax	29.9	35.4	38.3	12.4	0.0	15.9
Char	0.0	0.0	0.0	0.0	13.5	12.8
HCl	0.0	0.0	0.0	0.0	31.7	0.0

Table 4.2 Yields of gas species from fluid-bed pyrolysis of some plastics at 550 °C in wt% [9]

Gas	HDPE	LDPE	PP	PS	PVC	PET
Hydrogen	0.31	0.23	0.24	0.01	0.20	0.06
Methane	0.86	1.52	0.44	0.08	0.79	0.41
Ethane	0.90	1.71	0.45	<0.01	0.55	0.02
Ethene	3.01	5.33	1.48	0.09	0.51	1.27
Propane	0.79	0.84	0.67	<0.01	0.28	0.00
Propene	2.26	4.80	1.08	0.02	0.92	0.00
Butane	0.35	0.55	0.26	0.00	0.11	0.00
Butene	2.34	6.40	1.95	0.02	0.92	0.00
Carbon dioxide	0.00	0.00	0.00	0.00	0.00	24.28
Carbon monoxide	0.00	0.00	0.00	0.00	0.00	21.49
Hydrogen chloride	0.00	0.00	0.00	0.00	31.70	0.00

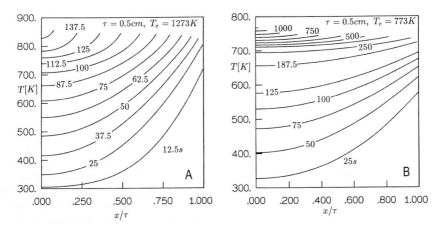

Figure 4.1 Spatial profiles of temperature along a PE particle (a slab with half thickness τ of 0.5 cm [0.20 in.]), exposed to convective heating for temperatures of 1273 K (A) and 773 K (B) [6]

The effects of bubbles inside the molten layer on the steady-state transport of volatiles, during degradation of thermoplastic polymers, have been predicted by means of a one-dimensional model by I.S. Wichman [19]. The model includes the description of individual bubble characteristics in terms of velocity and growth rate. Expressions for the regression rate, the nondimensional melt velocity, and the liquid volume fraction are obtained by means of the method of matched asymptotic expansions. The regression rate is the result of a balance between the surface heating rate, the rate of heat removal into the condensed phase, and the rate at which the monomer is lost from the polymer with bubble formation.

Despite the large number of kinetic investigations, reliable mechanisms and kinetic constants for polymer degradation are still needed for applications in both fire safety science and reactor design and development, when, after use, plastics are returned as waste to be eliminated. Also, models with a comprehensive treatment of the dynamics of both heat and mass transfer should be developed, which together with experimental measurements can improve the understanding of the degradation behavior of different polymer classes.

4.1.3 Ignition and Smoldering

A clear-cut division between autoignition and piloted ignition does not exist because, for fire safety science, the system of interest is open. In general, piloted ignition is assumed to be caused by a device capable of creating a region of very high temperature in the gas phase (pilot flames, sparks, and hot wires). Autoignition is caused by thermal radiation, hot air streams, or hot surfaces [20]. These different modes of ignition have been investigated experimentally as each of them corresponds to a physical situation of interest. For instance, cases of piloted ignition are related to fires caused by accidental localized heat sources. Thermal radiation is the primary mode of heat transfer from large fires. A hot-air stream is also important in fire conditions because it simulates ignition caused by hot combustion products. A hot surface is related to room fires when solid materials can become combustible as a consequence of temperature increase caused by near flames. The simplest criterion for ignition is based on the surface temperature reaching a critical value. In this way all the processes of interest are localized in the solid phase and the mathematical description of the problem reduces to the energy conservation equation. Closed-form solutions for the ignition delay time [20–22] and empirical correlation for piloted ignition [22] are available.

The ignition delay time increases by lowering the radiation heat flux and decreasing the heating area [20], with minimum radiative heat flux intensities of 160–250 kW/m^2s for ignition. Lower heat fluxes result only in polymer degradation. Radiative ignition has also been modeled [23, 24] through one-dimensional equations both for the solid phase (energy balance) and gas phase (chemical species mass and energy) for noncharring materials (absence of exothermic surface reaction). The only mechanism that can lead to ignition is heat conduction from the solid surface, which is radiatively heated, to the adjacent gas layer. Consequently such models predict ignition only if very high surface temperatures are reached or if high (unrealistic) combustion reaction rates are simulated. Successive experimental investigations [25] showed evidence of the absorption of some of the incident radiation by gas phase (pyrolysis products or particulate), thus reducing the radiant heat flux reaching the solid surface and increasing the gas phase temperature. This mechanism is also included in a one-dimensional transport model for PMMA ignition [26]. The ignition process occurs in the gas phase in a premixed fashion, rapidly followed by the transition to a diffusion flame. As the radiative heat flux is increased, higher surface temperatures and pyrolysis mass fluxes are reached, ignition occurs closer and closer to the fuel surface, and ignition delay times decrease. Gas phase absorption of radiation plays a fundamental role in the predicted ignition phenomenon and ignition delay times.

Hot-air ignition processes are qualitatively similar to radiative ignition and again simple solid phase thermal models have been proposed [20, 21] for the ignition delay time as a function of the ignition temperature, although validity is limited to relatively low temperatures. Some models, based on a boundary layer formulation, indicate that the gas phase reaction is the controlling mechanism with ignition delay times increasing as the flow velocity is decreased (e.g., Kashiwagi et al. [27]). Hot-air ignition of several polymers has also been investigated by Nioka et al. [28] for different velocities and temperatures. Analysis indicates that ignition delay is controlled by the rate of solid degradation for low flow velocity and by gas phase combustion at large flow velocities.

Smoldering combustion is one of the most hazardous modes of fire initiation, because of the possible transition to flaming combustion, and an efficient generator of carbon monoxide with consequent health damage. Products of the smoldering combustion are char and combustible volatiles. The global kinetics of the smoldering combustion, according to Ohlemiller [29], can be broadly described as endothermic solid pyrolysis and exothermic oxidation of solid and char. Moreover, another contribution to positive heat release may come from oxygen reactions with the devolatilization products, within the pores of the solid matrix. These reactions are considered of secondary importance in smoldering combustion; however, they may acquire a controlling role during the transition from smoldering to flaming combustion.

Most of the experimental analyses have been focused on the measurement of the spread rate of the thermal wave through one-dimensional systems, under forced flow conditions. Oxygen is forced to flow through porous particle beds in the same direction as or in the direction opposite to the smolder spread rate. Smolder spread in the direction opposite to the oxygen flow is indicated as reverse smolder. According to Ohlemiller [29], the structure of the smoldering wave closely resembles that of a laminar diffusion flame: oxidative and pyrolytic degradations coexist and become active as the temperature becomes high enough, as a consequence of heat transfer from the adjacent hot region. Smolder spread in the same direction as the oxygen flow is indicated as forward smolder. The structure of the wave is, in some way, more complex because the solid is degraded essentially through pyrolysis. Oxygen is completely depleted by char oxidation, which drives the smolder process, because, through exothermic heat release, it also determines the rate of heat transfer to the virgin solid and the rate of solid pyrolysis. In this case, ahead of the pyrolysis front, there is a region where water driven out from the smoldering region or formed through it, undergoes evaporation and/or condensation (on dependence of the temperature profile). Therefore, in the forward smolder there are three distinct fronts (water movement, solid pyrolysis, and char oxidation) whereas in the reverse smolder there is not a clear distinction among the regions where these processes are underway.

The difference in the configuration between reverse and forward smolder also leads to different controlling mechanisms. Forward smolder is supported by char oxidation, whereas reverse smolder is controlled by oxidative solid degradation [29]. Furthermore, forward smoldering combustion is more critical for the transition to self-sustained flaming combustion [30]. Smolders that are purely forward or reverse have been studied through a number of experiments because they can be used as a first approximation of parts of real fires and can be modeled easily. However, multidimensional structures of the smoldering front are more representative of the real fire-related hazards [31]. Oxygen from the top surface diffuses toward both the virgin solid and the charred residual, making both exothermic solid and char oxidation active, which are the driving mechanisms for reverse and forward smolder, respectively.

Some conclusions can be drawn in relation to the current state of the art and future developments in relation to polymer ignition and smoldering. Almost all the ignition and smoldering models, apart from a few exceptions [32, 33], are formulated for one-dimensional systems. Further developments should consider a wider range of materials from both the experimental and the mathematical points of view. Other aspects that deserve further investigation, are related to exothermic reaction processes of the solid phase, transition from smoldering to flaming ignition, piloted ignition, and multidimensional flow characteristics.

4.1.4 Flame Spread

After ignition, the exothermicity of solid and gas phase reactions causes a heat flux that, through solid and gas phase mechanisms, heats the unburned fuel. For flame spread to occur, the burning region must supply enough heat to the unburned solid to cause degradation. At the same time, proper conditions in the gas phase should be met. In fact, flame spread characteristics are affected not only by the mechanisms of solid degradation but also by other factors, such as the flow configuration, oxygen level, and orientation of the solid fuel. Flame spread over solid fuels can be classified into two main categories according to flow conditions [22, 34]. One mode occurs when the flame spread is in the same direction as the oxidizing flow (flow-assisted flame spread). The second mode occurs when the flame spreads against the oxidizing gas flow (opposed-flow flame spread). In the flow-assisted mode of flame spread, the concurrent flow pushes the flame ahead of the degrading fuel surface. The heat transfer from the hot mixture of reacting gases and the combustion products above the degraded region to the unburned fuel surface favors the propagation of the flame. The resulting flame spread process is very rapid and consequently of great importance to fire safety science. For a very large range of flow velocities and oxygen concentrations, this phenomenon appears to be controlled simply by heat transfer mechanisms [35]. However, the flow remains laminar only in the initial stage, at which heat transfer from the flame to the fuel is due mainly to convection. When the dimensions of the flame increase, the flow becomes turbulent and flame radiation appears to be the dominant mode of heat transfer. In the case of opposed-flow flame spread, the heat transfer to the unburned fuel is more difficult, as the flame and pyrolysis fronts are in the same location. The opposed flow pushes the flame into the burning region, and heat transfer to the unburned fuel, which occurs through solid or gas phase conduction, is very slow. The flame front is well defined and generally the size of the fire is easily controlled. The phenomenon shows an interesting dependence on environmental conditions [34]: It is dominated by heat transfer mechanisms at relatively low opposed-flow velocities and high oxygen concentrations and by chemical kinetics at high flow velocities or low oxygen concentrations.

Ignition and flame spread processes strongly depend on the presence of gravity and, in a normal gravity environment, on the orientation of the solid [34]. Upward flame spread, that is, in the direction opposed to the gravity vector, is faster than downward flame spread, that is, in the same direction as the gravity vector. In the case of upward flame spread the heat transfer from the hot combustible gases to the fuel is enhanced by natural convection whereas, in the opposite case, natural convection slows the spread process, taking away the hot combustion products from the unburned fuel. Upward and downward modes of flame spread show the same characteristics of flow-assisted and opposed-flow flame spread.

Another important factor for flame spread processes is fuel thickness. In general the definitions "thermally thin" and "thermally thick" fuel are used [34]. The term thermally thin indicates that the fuel thickness is small compared to the characteristic thickness of the thermal diffusion layer in the solid, along which no significant variation of solid properties is observed. Depending on the thickness of the solid fuel, the role played by the solid fuel thickness in the path for heat transfer to the unburned fuel changes from being of no importance for thin fuels to becoming of primary importance for thick fuels. If the combustible is thin, flame spread is characterized further by the consumption of the solid fuel in the burning region. This produces a propagating burnout front, which affects the flame and pyrolysis front propagation. Material properties influence the behavior of the sample under fire conditions. Most experiments on flame spread over thick solid fuels have been carried out with PMMA, while experiments related to thin fuels have used mainly paper.

An extensive literature on simplified models, reviewed in [22, 35–38] is available. Sometimes explicit expressions for spread rates have been obtained. Generally, these formulae are able to describe qualitatively the dependence on environmental conditions and property values of the solid fuel, when phenomena are controlled by heat transfer mechanisms. More comprehensive mathematical models include balance equations for both gas and solid phases and need a numerical treatment. These do not give explicit expressions for global parameters, which can be derived from the predicted time and space evolution of the phenomenon, often by means of the same approaches used in the experiments. A review on the numerical modeling of flame spread process has been presented by Di Blasi [2].

More recent and advanced models are two-dimensional and quasi-steady or unsteady balance equations for the gas phase coupled through the boundary conditions at the interface to solid phase balance equations. As for the gas phase, they include momentum, energy, and chemical species mass balance equations (Navier-Stokes equations for reactive flows). All the analyses are for laminar flow, and finite rate combustion kinetics are described through an overall, second-order reaction. Viscous dissipation and compressive work are neglected. Furthermore, the coupling between the momentum equations and the state equation due to pressure terms, when momentum balance equations are included in the mathematical formulation of the problem, is neglected. Pressure variations in space are very small and, because in general the system is open, the mean pressure reduces to the specified ambient pressure. Therefore, pressure excess with respect to the ambient value is neglected in the state equation, while it is retained in the momentum equations. The decoupling of momentum equations from the state equations cuts off the acoustic waves and the determination of the pressure field becomes an elliptic problem. Because the models refer to small-scale configurations and are aimed at the simulation of flammability characteristics, radiative heat transfer is generally not taken into account. Few exceptions are represented by models applied to study microgravity flame spread [39], where radiation is controlling (e.g., Di Blasi [40]) even for small-scale configurations. From a mathematical point of view, models of thermally thin fuels can take advantage of the uniformity of variable distribution along the fuel thickness, observed in the experiments, and use one-dimensional balance equations. However, fuel consumption (burnout) cannot be neglected. Two-dimensional heat transfer must be taken into account for thermally thick fuels but, in this case, fuel consumption is generally neglected. Models for fuels of intermediate thickness should account for both two-dimensional and solid consumption effects. Most recent models are applicable for fuels of widely variable thicknesses and also for composite materials [41–43]. However, simplifications are still retained in relation to regression rate, chemical kinetics (a one-step Arrhenius reaction for thermal degradation and absence of char combustion/oxidative degradation), melting behavior (absence or no transport phenomena), and properties (assumed to be a linear function of conversion).

Numerical simulations of the flame spread process have contributed to the understanding of chemical and physical mechanisms controlling flame spread, and thus to verify the phenomenological arguments used to explain the experimental results, sometimes with quantitative predictions of the spread rates. An example of the simulated flame structure is given in Fig. 4.2 for downward flame spread over composite materials [42]. The solid, made of a thermoplastic polymer and inert additives, undergoes in-depth endothermic pyrolysis, for the active part, with volatile monomer formation. The effective thermal conductivity of the composite material depends on the content of inert and the variable content of the active part, whose thermal conductivity varies between that of the polymer and that of the melted phase monomer. Even though surface regression is not taken into account, polymer consumption is modeled through a mass balance. The spread process is strongly affected by the solid perpendicular (to the spread direction) thermal conductivity which, when decreased, causes a continuous increase in the spread rate. On the contrary, both numerical and analytical solutions give no dependence of the spread rate on the solid parallel (to the spread direction) thermal conductivity for a wide range of variation. In agreement with previous experimental results, at very large values of the latter, the finite-rate reaction model predicts a decrease in the spread rate.

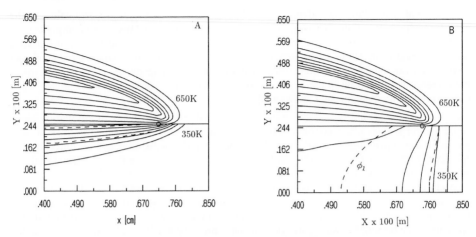

Figure 4.2   Gas and solid phase isotherms (solid lines, values from 650 K and with step 250 K, and from 300 K with step 50 K, respectively) and solid phase constant contour levels of polymer mass fraction, Φ_1, (dashed lines, values of 0.01 and 0.89) for two different values of the solid phase thermal conductivity in the direction perpendicular to the flame spread rate (0.0126 W/mK (A) and 0.230 W/mK (B). The position of the flame is indicated by a circle [42].

Both thermodiffusive models and more complete models, based on the solution of the Navier-Stokes equations, have been used to simulate flow-assisted flame spread for different fuel thicknesses [2, 44]. For thick solids, the flame and pyrolysis spread rates increase with oxygen concentration and flow velocity and vary over a very large range of values. Both the increase of oxygen concentration and concurrent flow velocity increase the heat flux from the flame to the fuel and the spread process is enhanced. This indicates that, unlike the opposed-flow case, flow-assisted flame spread is controlled essentially by heat transfer. The effects due to finite-rate kinetics are of increasing importance as extinction is approached, as a consequence of low oxygen concentrations or very large flow velocities. These effects appear mainly at the flame leading edge. Here the extinction length, that is, the distance of the flame leading edge from the edge of the fuel slab, increases as the flow velocity is increased or the oxygen concentration is lowered. Again, the extinction process can be explained in terms of a decrease, below a critical value, of the Damkohler number at the flame leading edge. The flame counteracts the decrease by moving in a zone of high fuel concentration. The extinction process begins in this way, but the downstream spread process goes on increasing flame and pyrolysis lengths until the pyrolysis spread rate is greater than the upstream extinction rate. Complete extinction occurs when the extinction distance extends to the position of the pyrolysis front.

Numerical simulation could be a powerful tool for the prediction of reaction to fire of materials, provided all the main chemical and physical processes are taken into account even with some simplifications. Further improvements in the flame spread models should deal with a better description of chemical processes, radiative heat transfer, and solid phase processes. A wider variety of material classes should also be examined.

4.1.5 Heat Release Rate

Following ignition and flame spread, solid burning takes place, which provides heat for solid pyrolysis and the continuation of the process. Burning processes of a combustible material are classified [22] as wall and pool fires, dependent on the material (solid or liquid) and orientation (vertical or horizontal surfaces). From the mathematical point of view, the processes are described by the same equations, despite differences in physical and chemical properties.

The rate of heat generation is indicated as heat release rate and is defined as the product of the heat of combustion and the burning rate. This can be expressed [45] as the ratio between the net flux reaching the condensed phase and the heat of vaporization. It is a simplification of the energy conservation equation, written at the solid-gas interface, and requires the different contributions (external heat flux, flame radiative heat flux transferred to the surface, flame convective heat flux transferred to the surface, surface re-radiation loss) to be specified through experimental analysis or the solution of the solid and gas phase conservation equations for the variables of interest. It should be noted that, apart from the difficulties for a correct evaluation of the different contributions, incomplete combustion is also an important point. Indeed, the emission rate of soot and chemical species (essentially carbon monoxide and hydrocarbons) is affected by flame temperature, oxygen availability, entrainment of air by the flame, and pyrolysis rate.

Both empirical correlations, based on experimental data, and analytical or numerical solutions of model equations have been proposed (reviews on the subject are reported, among others, by Drysdale [46], Sibulkin [47], Fernandez-Pello [22], and Joulain [48]. Laminar, nonradiative wall fires are usually described by boundary layer theories coupled with infinite-rate kinetics, so allowing closed-form solutions to be obtained for the burning rate. This shows a power law dependence on the Grashof number (1/4) for natural convection and the Reynolds number (1/2) for forced convection [22], results that are in agreement with experimental observation. Large-scale processes are turbulent and are dominated by radiative heat transfer (mainly from soot emissions), characteristics that make the analysis very difficult. Burning rate measurements are again correlated with the 1/4 power of the Grashof number. The complex flow field associated with the development of a buoyant plume over pool fires requires the solution of the reactive Navier-Stokes equations, which for large-scale (diameter of the pool) processes are again dominated by turbulence and radiation. Therefore, simplified analytical solutions are not applicable and nondimensional analysis and experimental measurements are applied mainly to provide correlations for the burning rate.

References for Section 4.1

[1] N.A. Khalturinskii and A.A. Berlin: in H.H.G. Jellinek (Ed.), Degradation and Stabilization of Polymers, Vol. 1, Elsevier, Amsterdam, 1983
[2] C. Di Blasi: *Prog. Energy Combust. Sci.* 19 (1993) 71
[3] T. Kashiwagi: Twenty-Fifth Symposium (Int.) on Combustion, The Combustion Institute, Pittsburgh, p. 1423–1437, 1994
[4] W.H. Beck: *Combust. Flame* 70 (1987) 171
[5] T. Kashiwagi: *Fire Safety J.* 3 (1981) 185
[6] C. Di Blasi: Polym. Degrad. Stab. 64 (1999) 359
[7] R.D. Schultz and A.O. Dekker: 5th (Int.) Symposium on Combustion, Reinhold, New York, pp. 260–267, 1955
[8] R.F. Chaiken, W.H. Andersen, M.K. Barsh, E. Mishuck, G. Moe and R.D. Schultz: *J. Chem. Phys.* 32 (1960) 141
[9] P. T. Williams, E. A. Williams: *J. Inst. Energy* 71 (1998) 81
[10] A.D. Baer: *J. of Fire Flamm.* 12 (1981) 214
[11] J.A. Conesa, A. Marcilla, R. Font, J.A. Caballero: *J. of Appl. Analyt. Pyrol.* 36 (1996) 1
[12] R. Knumann, H. Bockhorn: *Combust. Sci. Technol.* 101 (1994) 285
[13] G.M. Anthony: *Polym. Degrad. Stab.* 64 (1999) 353
[14] H. Arisawa and T. B. Brill: *Combust. Flame* 109 (1997) 415
[15] S.W. Ng: *Energy Fuels* 9 (1995) 216
[16] W. Kaminsky: *J. Analyt. Appl. Pyrol.* 8 (1985) 439
[17] D.S. Scott, S. R. Czernik, J. Piskorz, D. Radlein: *Energy Fuels* 4 (1990) 407
[18] C. Di Blasi: *J. Analyt. Appl. Pyrol.* 40–41 (1997) 463
[19] I.S. Wichman: *Combust. Flame* 63 (1986) 217

[20] K. Akita: in H.H.G. Jellinek (Ed.): Aspects of Degradation and Stabilization of Polymers, Elsevier, p. 500–525, 1978
[21] A.M. Kanury: *Combust. Sci. Technol.* 97 (1994) 469
[22] A.C. Fernandez-Pello, in G. Cox (Ed.): Combustion Fundamentals of Fire, Academic Press, London, p. 31–100, 1995
[23] T. Kashiwagi: *Combust. Sci. Technol.* 8 (1974) 225
[24] M. Kindelan, F.A. Williams: *Combust. Sci. Technol.* 16 (1977) 47
[25] T. Kashiwagi: *Combust. Flame* 44 (1982) 223
[26] C. Di Blasi, S. Crescitelli, G. Russo, G. Cinque: *Combust. Flame*, 83 (1991) 333
[27] T. Kashiwagi, G.G. Kotia, M. Summerfield: *Combust. Flame* 24 (1975) 357
[28] T. Nioka, M. Takahashi, M. Izumikawa: Eighteenth Symposium (Int.) on Combustion, The Combustion Institute, Pittsburgh, pp. 741–747, 1981
[29] T.J. Ohlemiller: *Prog. Energy Combust. Sci.* 11 (1985) 277
[30] R.A. Anthenien, A.C. Fernandez-Pello: 27th (Int.) Symposium on Combustion, The Combustion Institute, p. 2683–2690, 1998
[31] T.J. Ohlemiller: *Combust. Flame* 81 (1990) 354
[32] C. Di Blasi: *Combust. Sci. Technol.* 106 (1995) 103
[33] W.E. Mell, T. Kashiwagi: 27th (Int.) Symposium on Combustion, The Combustion Institute, Pittsburgh, p. 2635–2641, 1998
[34] A.C. Fernandez-Pello, T. Hirano: *Combust. Sci. Technol.* 32 (1983) 1
[35] A.C. Fernandez-Pello: *Combust. Sci. Technol.* 39 (1984) 119
[36] F.A. Williams: Sixteenth Symposium (Int.) on Combustion, The Combustion Institute, p. 1281–1294, 1976
[37] W.A. Sirignano: *Combust. Sci. Technol.* 6 (1972) 95
[38] I.S. Wichman: *Prog. Energy Combust. Sci.* 18 (1992) 553
[39] P.D. Ronney: 27th (Int.) Symposium on Combustion, the Combustion Institute, p. 2485–2506, 1998
[40] C. Di Blasi: *Combust. Flame* 100 (1995) 332
[41] C. Di Blasi: *Combust. Flame* 97 (1994) 225
[42] C. Di Blasi: *Polym. Degrad. Stab.* 54 (1996) 241
[43] C. Di Blasi, I.S. Wichman: *Combust. Flame* 102 (1995) 229
[44] C. Di Blasi: *Fire Safety J.* 25 (1995) 287
[45] A. Tewarson, R.F. Pion: *Combust. Flame* 26 (1976) 85
[46] D. Drysdale: An introduction to fire dynamics, John Wiley, New York, 1985
[47] M. Sibulkin: Progr. *Energy Combust. Sci.* 14 (1988) 195
[48] J.P. Joulain: 27th (Int.) Symposium on Combustion, the Combustion Institute, pp. 2691–2706, 1998

4.2 Thermal Properties and Burning Behavior of the Most Important Plastics

4.2.1 Thermal Properties of Thermoplastics

C.A. WILKIE AND M.A. MCKINNEY

The combustion of any organic polymer involves several steps. The solid polymer must first be degraded to form small fragments, which can escape to the vapor phase. This involves an input of heat, either from an external source or by the feedback of energy from the already combusting material. These small molecules must now diffuse to the surface where they can escape and undergo further reactions to yield the actual species, usually hydrogen and hydroxy radicals, which make up the flame. The energy generated in this step is now fed back to continue the process.

This review is devoted largely to the chemistry that is involved in the degradation step of the combustion process. It is generally believed that oxygen plays an important role in the combustion step, but not in the degradation step; the presence or absence of oxygen in the degradation step is unimportant. It is certainly true that degradation studies are quite dependent on the atmosphere in which the degradation is performed, and attention is paid to degradation in an inert atmosphere as well as in air.

4.2.1.1 Scope

This section covers the thermal degradation of various classes of thermoplastics as well as the methods used to study these degradations. The coverage is in no case encyclopedic, rather the section tries to provide enough information on the degradation of particular polymers or classes of polymers so that the reader will be able to decide if further literature work is warranted.

Thermoplastics include materials that soften when they are heated. Polymers that would be included are polyolefins, polybutadiene and other elastomers; PS and styrene-containing co- and terpolymers; poly(meth)acrylates; halogen-containing polymers such as PVC; and the fluorine-containing polymers such as polytetrafluoroethylene, polyamides, polyesters, and polycarbonate. The thermal characteristics of selected thermoplastics are summarized in Section 3 of the Appendix.

In general, the degradation of thermoplastics occurs by four mechanisms: random-chain scission, end-chain scission, chain-stripping, and other processes such as cross-linking. Scission involves the cleavage of a carbon-carbon bond in the backbone of the polymer to generate two radicals. This may be initiated at random positions throughout the polymer and give rise to a monomer and oligomers or it may be initiated strictly at the ends of the chain, end-chain scission. End-chain scission results in the exclusive formation of a monomer and is initiated at unsaturated chain ends.

In a random-scission reaction two radicals are formed from cleavage of the backbone of the polymer. One of these radical centers typically will be a methylene group, a primary radical, while the other will likely be on a carbon that whatever substituents are present in the polymeric unit will result in a secondary or tertiary radical. The reactive primary radical ordinarily will abstract hydrogen from a neighboring position to give a more stable secondary or tertiary radical, or from a more removed position, even on another polymer strand. This new secondary or tertiary radical can now undergo further degradation, typically by the formation of a new primary radical and an unsaturated species, as shown in Fig. 4.3.

End-chain scission occurs when hydrogen transfer is inhibited, for instance, if the substituents adjacent to the primary radical site are groups other than hydrogen atoms. Polymers such as

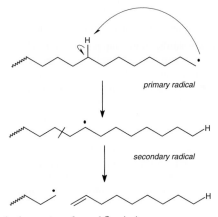

Figure 4.3   Intramolecular hydrogen transfer and β-scission process

PMMA, with methyl and carboxyl substituents, polymethacrylonitrile, with nitrile and methyl groups, poly(-methyl styrene), phenyl and methyl, are examples of polymers that undergo end-chain scission. If at least one of the substituents is a hydrogen, then random scission can occur with hydrogen transfer, and oligomers as well as monomers are produced. One method to differentiate between end-chain and random scission is based on the identification of the products. If only monomer is produced, it is quite likely that only end-chain scission has occurred.

The chain stripping process occurs when a small molecule, such as HCl or acetic acid, can be eliminated from the polymer. Thus polymers such as PVC, poly(vinyl acetate), and poly(vinyl alcohol) can eliminate HCl, $HOOCCH_3$, and H_2O, respectively, with the formation of a double bond in the main chain. One should recognize that these new polymeric structures containing double bonds can undergo subsequent reactions that may, or may not, be advantageous in the overall degradation scheme.

The other mechanism of degradation is a cross-linking process. The best known example of this process is in the degradation of polyacrylonitrile, which is discussed later in this chapter. Related to this cross-linking degradation pathway is the formation of char. One of the areas of active interest in the fire retardant community is the promotion of char formation because char offers the opportunity to insulate the polymer from the source of the heat and thus to inhibit the degradation processes.

4.2.1.2 Techniques to Study the Thermal Degradation of Thermoplastics

The techniques that are used in the study of thermal degradation pathways may be separated into two distinct categories based on the information desired: Scientific techniques are used to assess thermal stability and to provide data which can be used to determine the process by which a polymer degrades and commercial tests, which are performed in order to provide data, which can be used to enhance the salability of a particular product. The scientific techniques include TGA and related proceduresand differential scanning calorimetry (DSC) and related procedures. One can couple any of these as a means of identifying products that evolve during the degradation to derive information on the course of the degradation.

In TGA, a sample of milligram size is heated, usually under an inert atmosphere, either isothermally or at a constant rate, and the mass of sample is followed as a function of either time or temperature. The degradation depends somewhat on the rate at which the temperature is increased and this must be taken into account in any investigation. In some cases the evolved gases are then analyzed, typically by mass spectroscopy or infrared spectroscopy. If the derivative of the mass loss versus temperature is recorded, this is called differential thermogravimetry (DTG). The main advantage of DTG is that the derivative enables one to determine more exactly the temperature at which some degradation is occurring.

Related to TGA is thermal volatilization analysis (TVA). The major practitioners of this technique are Norman Grassie and Ian McNeill. Here the sample is heated in a vacuum system, which has a liquid nitrogen trap between the sample and the vacuum pump. Any volatiles produced will increase the pressure in the system. Thus pressure is related to the rate of volatilization, and a sample transducer, not a balance, is used to measure the rate of degradation. The use of suitable traps enables one to trap all of the products, which are evolved for later analysis.

Another related technique is pyrolysis gas chromatography. The pyrolysis is carried out in the injector of the gas chromatograph and the volatile products that are formed flow through the chromatography column and can give a fingerprint of the degradation products. This technique is used for the identification of polymers and is used less often for mechanistic studies.

The aforementioned techniques give information on the rate of degradation but it is also of interest to obtain information on the heat involved in various thermal processes, that is, its endo- or exothermicity. Two techniques, differential thermal analysis (DTA) and DSC, are used for this purpose. The measurement is slightly different but both techniques provide information on the heat effects, which occur when a polymeric sample is heated.

Two more tests that more or less fall into the realm of scientific tests are the oxygen index (OI), also known as the limiting oxygen index (LOI), and cone calorimetry. Oxygen index measures the ease of extinction of a fire. A sample is ignited in an atmosphere that contains a known composition of nitrogen and oxygen and one measures the minimum oxygen concentration that is required for flame extinction, called the oxygen index. This can be very useful because by substituting another oxidizer, for instance nitrous oxide (N_2O), for oxygen, one can determine if an additive functions in the condensed or the vapor phase. A condensed mechanism will not show any variation with the identity of the oxidizer while a vapor phase reaction will be dependent on the oxidizer. Cone calorimetry measures the depletion of oxygen as a sample is combusted. One is able to measure the rate at which heat is released and mass is lost, as well as the evolution of smoke. Both of these tests are used primarily for commercial evaluation but there is some scientific component, as one can develop information that relates to the degradation pathway.

The commercial tests include measurement of ease of ignition, flame spread, rate of heat release, rate of extinction, smoke evolution, and evolution of toxic gases. Only one of these, the UL 94 test, which measures the ease of ignition of a polymeric sample, is referred to in this chapter. In this test, a series of samples are ignited and one observes if the sample continues to burn when the ignition source is removed; in addition dripping of the sample is important. The results of this test lead to a classification.

4.2.1.3 Thermal Degradation Pathways

Polyolefins. We consider only two systems in this classification, polyethylene (PE) and polypropylene (PP). Because both of these have hydrogen atoms on all carbons, one can expect that random scission will be the dominant pathway of the degradation and therefore one can expect to find monomer and oligomers as the products of the degradation. A TGA curve of low-density PE [1] shows that in a nitrogen atmosphere, degradation does not commence until the temperature is above 400°C while complete volatilization has occurred by 500°C. By contrast, in air the degradation begins at 235°C and rapid weight loss occurs above this temperature. The TGA of high-density PE [2] appears to show similar thermal stability characteristics. The onset of PP degradation begins at a somewhat lower temperature [2]. The principal products observed during the degradation of PE include ethylene and higher oligomers, which arise from hydrogen transfer to different positions along the polymer chain, thus one observes propylene, butene, pentene, hexene, and so forth, as well as a series of alkanes [3]. A degradation pathway for PE is shown in Fig. 4.4. The degradation of PP is more complex and a wider variety of products is produced. The major volatile products found in the degradation of PP, in order of importance are 2,4-dimethyl-1-heptene, 2-pentene, propylene, 2-methyl-1-pentene, and in much smaller amount, isobutene [4].

Recently Hedrick and Chuang [5] reported on a temperature-programmed reaction technique, coupled to *in situ* infrared and mass spectral analysis, to study the thermal degradation of PP. The degradation reaction is performed within the infrared cell and the evolved gases are analyzed by mass spectroscopy, infrared analysis, and gas chromatography. At the same time, infrared spectra of the solid material can be obtained. They observed similar products to those previously found but suggested a slightly different mechanistic pathway for the degradation.

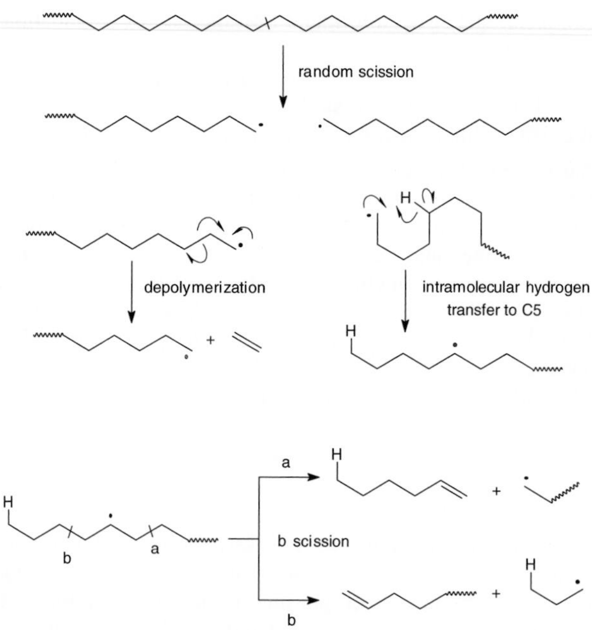

Figure 4.4 Degradation of PE showing initial random scission, depolymerization, intramolecular hydrogen transfer, and β-scission of the macroradical

Camino, Costa, and co-workers [6–13] investigated the effects of a chlorinated paraffin, either with or without a metal salt, on the thermal degradation of PE and PP. Several conclusions may be drawn from this work. The presence of the chlorinated paraffin alone increases the fraction of low molecular weight (C-1 to C-4) saturated hydrocarbons and decreases the fraction of olefins formed from both PP and PE. It has been speculated that this may occur by the reaction of hydrogen, which is eliminated from the degrading chloroparaffin, interacting with olefins to give saturated species. In the presence of the chlorinated paraffin, PE undergoes a cross-linking reaction while PP becomes subject to degradation. The observation that the N_2O index shows the same slope as the concentration of the chlorinated paraffin is changed for PE as does the oxygen index indicating that fire retardant effects occur in the condensed phase. For PP, the slope is similar but not exactly the same and this implies that both condensed phase as well as vapor phase effects are occurring. The degradation of PP in the presence of both the chlorinated paraffin and a metal salt, either antimony oxide or bismuth carbonate, shows that synergism occurs between these components. Surprisingly the combination with antimony oxide functions in the vapor phase while that with bismuth carbonate shows condensed phase activity.

The Italian Turin group has also extensively investigated an intumescence system with PP [14–19]. The French group from Lille has also played a major role in developing the understanding of intumescent systems [20, 21]. Members of both of these two groups have recently authored an authoritative review on intumescence [22]. The normal intumescent system consists of an inorganic acid or its precursor, a polyhydric compound, which acts as the source of carbon for char formation, an organic amine or amide, and a halogenated organic compound. Intumescence involves more than a single physical effect and it is believed that one of these effects is to produce char by dehydration of the polyhydric compound catalyzed by the acid, and this char is then blown by volatile components evolving as the amine or halogenated compound are heated. In this work they used the combination of ammonium polyphosphate and pentaerythritol, sometimes also using melamine or its salts. If one compares the TGA curve of PP with that of a mixture of PP with ammonium polyphosphate and pentaerythritol, one sees that the calculated and experimental mass loss curves are in agreement up to 460°C. At higher temperatures a lower mass loss is found than calculated and a larger amount of char is formed at 500°C. The fraction of PP that degrades is reduced in the presence of the additive. This is believed to be strictly a physical effect in which the foamed char prevents heat from causing further degradation and a chemical effect in which the char traps the polymer degradation products and as a consequence prevents the fuel transfer to the flame.

There have been two reports on PP nanocomposites [23, 24]. These materials are formed using a montmorillonite clay together with PP-graft-maleic anhydride (PP-g-MA). The maleic anhydride is necessary to make the polymer and the clay compatible. The peak rate of heat release is lower by more than 50 % in the nanocomposite compared to virgin PP-g-MA.

Diene Polymers: Polybutadiene (PBD), Polyisoprene (PIP), and Polychloroprene (PCP). Once again, these polymers have at least one hydrogen on all carbons and degradation will occur by the random scission process so one expects to observe monomer and oligomers. In the case of PCP, the presence of the chlorine atom means that the possibility to eliminate a small molecule, HCl, is present, so one may expect that chain-stripping will occur for that polymer. The three polymers show quite different thermal behavior. PBD is the most stable, with degradation commencing at about 450°C [25]. The TGA curve for PIP shows that degradation begins below 350°C [26] while in PCP there is some degradation as early as 200°C but the main degradation occurs between 300 and 450°C [27]. The lower thermal stability of PCP is likely related to the possibility of chain-stripping as a degradation pathway for this polymer but not for the other polymers.

The products from the thermal degradation of PBD include monomeric butadiene and a small amount of aromatics arising by coupling of the oligomeric fractions that are produced [28–30]. Similar products are to be expected from PIP. The DSC curve of PBD shows that *cis-trans* isomerization occurs at a temperature as low as 200°C, well below the temperature at which mass loss is observed. On the other hand, PIP shows negligible isomerization below 350°C. It has clearly been shown that cross-linking of PBD occurs during the course of a TGA [29, 31]. The degradation of PCP has been studied by thermal volatilization analysis and it was found that the major product that evolved was hydrogen chloride [27]. Ultimately, about 90 % of the available chlorine is lost as HCl and this will lead to the formation of double bonds along the polymer chain, which can cross-link [32]. The scheme shown in Fig. 4.5 has been proposed to account for the depolymerization and cross-linking of PBD [29].

Work has been carried out on the cross-linking of PBD and related polymers using chemical initiation [31] and high-energy irradiation [33]. When PBD is cross-linked, regardless of the process used, it undergoes an earlier thermal degradation than does the virgin polymer. It has been shown in this work that PBD will cross-link under TGA conditions and the resulting material has a higher cross-link density than is obtained by either irradiation or chemical cross-linking.

Polystyrene (PS) and Its Terpolymer Acrylonitrile-Butadiene-Styrene (ABS). Coverage here is limited to PS and the terpolymer acrylonitrile-butadiene-styrene (ABS). The degradation of PS has been extensively studied; again the degradation proceeds by a random scission process. The degradation begins at about 360°C and is complete by 450°C; the observed products are styrene and its oligomers along with benzene and toluene [30, 34–38]. The degradation of ABS begins at 340°C with the formation of butadiene monomer and aromatics are first noted at 350°C. As the temperature is raised, styrene and its oligomers become more prominent. At the highest temperatures, aromatics decrease in intensity but butadiene remains strong throughout the course of the degradation. The evolution of acrylonitrile is also noted [30]. As monomeric acrylonitrile is not seen in the degradation of polyacrylonitrile, its observation here may be a surprise. As discussed later, the degradation of polyacrylonitrile occurs by a cross-linking pathway in which one nitrile nucleophilically attacks the adjacent unit to give a cyclized structure. In styrene-acrylonitrile, which is the precursor of ABS, acrylonitrile units are not adjacent so such attack cannot occur and the monomer is observed instead.

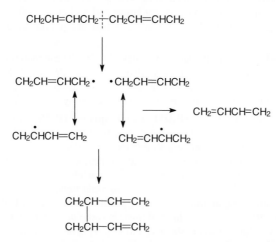

Figure 4.5   Formation of monomer and cross-linking from degradation of polybutadiene

A recent article covers the thermal degradation of cross-linked PS, prepared using varying amounts of divinylbenzene together with styrene [39]. The onset temperature of the degradation is increased as the cross-link density increases and up to 10 % of these cross-linked polymers are non volatile at 600°C.

Halogenated compounds are the normal additive for PS and the work discussed in the preceding paragraphs using chlorinated paraffin for PE and PP has also been applied to PS. The behavior of PS is identical with that of PE in this regard.

Recently a new procedure has been introduced to enhance the fire retardancy of PS, based on Friedel-Crafts chemistry. In early work, it had been shown that one can cross-link PS using Friedel-Crafts chemistry but at low temperatures where processing will be impaired [40–42]. Recent work has shown that one can cause the cross-linking reactions to occur only when the PS is heated to 250 or 300°C [43–45]. Cone calorimetry and radiative gasification studies on this system have shown that the rate of heat release is decreased by 60 % and that this effect occurs because char is formed at the surface, which protects the underlying polymer from further degradation.

Weil et al. [46] have shown that potassium carbonate is a good flame retardant, as measured by oxygen index, for both ABS and high-impact polystyrene (HIPS). The effect is due to oxidative crosslinking of the butadiene component of both polymeric systems. Gilman et al. [23, 24] have prepared intercalated nanocomposites of PS and again observed that the peak heat release rate is significantly lower for the nanocomposite than for the virgin polymer.

Poly(meth)acrylates. The degradation of PMMA gives essentially only monomer by a random-chain scission. At least two and sometimes three stages of degradation are evident from a TG study of the degradation. From the earliest work in this area, these stages have been attributed to reactions at the chain ends and a random scission process producing only monomer [47–50]. All of the available information leads one to believe that degradation of PMMA proceeds to form only monomer, but little information is available to explain how the reaction occurs. Hodder et al. [51] examined the thermal degradation of a copolymer of MMA and MMA-d$_8$ to see if scrambled products were produced. If the two ends of the monomer that were initially joined become reattached in the degradation, no scrambled product results; however, if the tail from one unit is attached to the head from a different unit, scrambled product would result. The monomer products that they obtained were completely deuterated or completely undeuterated; therefore no scrambling had occurred.

Kashiwagi et al. [52] have shown by TGA that a radically polymerized sample degrades in three stages, at approx. 165, 270, and 360 °C, whereas an anionically polymerized sample degrades in only one stage, at 360 °C. Because the anionically polymerized sample should contain no weak links, they propose that the single degradation pathway observed for this sample is the result of the random scission process. They suggest that the first step is due to the presence of head-to-head linkages present in the sample and attribute the second step to unsaturated chain ends. The instability of the unsaturated chain ends and of the head-to-head linkages has been confirmed by Meisters et al. [53]; however, they report that the head-to-head linkage is more stable than the unsaturated end groups. Another problem regarding the instability of the head-to-head linkage is the question of diffusion. For degradation to occur from a head-to-head linkage, the two radicals must diffuse away rather than recombine. There is some evidence to suggest that the diffusion of these radicals is very slow and that recombination is much more likely [54].

Any scheme that invokes random scission as the initial step must produce both a primary radical as well as a tertiary radical and both must degrade to give identical products. The tertiary radical will likely unzip, but unzipping of the primary radical is considered unlikely. Both Kashiwagi [55] and [56] have considered this problem and both have proposed Manring that side-chain scission is a viable initial step in the degradation process. Kashiwagi suggests that the primary radical undergoes β-scission with the formation of a methoxycarbonyl radical and unsaturated PMMA oligomer. This last oligomer then will degrade by the route known for PMMA molecules with terminal unsaturation. On the other hand, Manring suggests that the original step in the degradation process is side-chain cleavage with the direct formation of a methoxycarbonyl radical and the unsaturated polymer.

These mechanisms are virtually identical. Kashiwagi's process begins with a main-chain scission followed by a side-chain scission whereas Manring's mechanism requires an initial side-chain scission followed by the main-chain scission. The degradation pathway, following the Kashiwagi scheme, is shown in Fig. 4.6.

Degradation of cross-linked MMAs has also been studied [39]. These cross-linked materials were prepared using four different dimethacrylates, three of which are aliphatic and one, bisphenol A dimethacrylate, aromatic. The onset temperature of the degradation is lower for the cross-linked polymers than for PMMA. Insignificant char formation is observed for the polymers cross-linked using the aliphatic dimethacrylates but the amount of char does appear to increase for the aromatic system. As carbonaceous char is graphitic in character, the aromatic material is closer to what is required and produces char more easily.

When esters other than methyl are studied, ester decomposition with the formation of olefin and the acid becomes an important reaction. For poly(n-butyl methacrylate), approx. 30–50 % of monomer is formed and poly(ethyl methacrylate) behaves similarly [57]. Ester decomposition is the principal degradation pathway for poly(tert-butyl methacrylate) and the major volatile products are isobutene and water (from the degradation of poly(methacrylic acid) (PMAA) [58]. The degradation pathway for all esters is a tradeoff of these two processes.

The degradation of PMAA occurs in two stages. In the first stage, at 150–250°C, water is lost and poly(methacrylic anhydride) is formed. The second stage, which involves degradation of this polymeric anhydride, begins at about 300°C and is complete by 450°C. Ho et al. [59] have studied both stages kinetically and have also proposed a mechanistic scheme for the degradation of the anhydride.

There have been several publications on the degradation of salts of PMAA [60–63]. In general, the salts are stable up to approx. 350°C. The principal degradation products include monomer and the corresponding isobutyrate, carbonate, oxide, carbon dioxide, and a collection of volatiles, which include several aldehydes and ketones. The stability of the metal carbonate controls the course of the reaction to a certain extent.

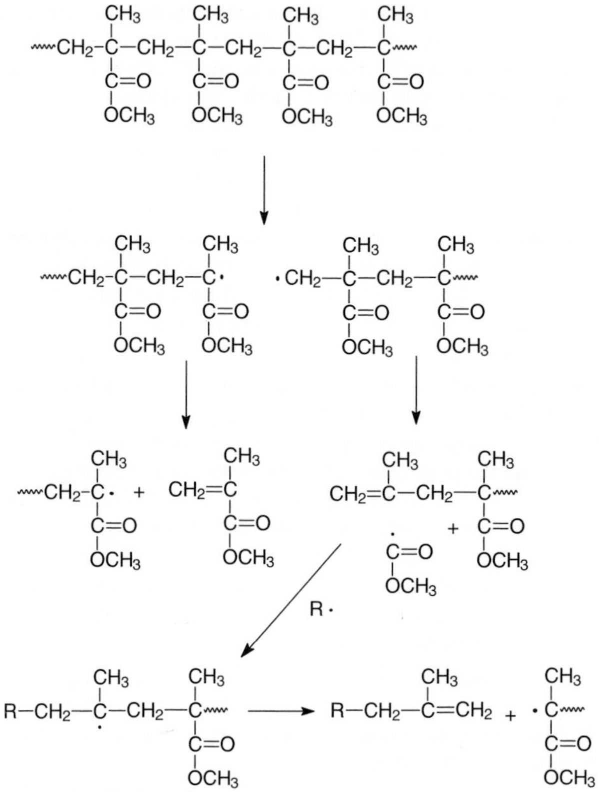

Figure 4.6   Degradation pathway of PMMA

The degradation of polyacrylates is quite different from that of the corresponding methacrylates because of the difference between a methine carbon in the acrylate and the quaternary carbon in the methacrylate. The degradation of PMMA produces 100 % monomer whereas acrylate gives perhaps 1 % monomer, and the main products are those due to random chain scission, namely oligomeric fragments. Just as for methacrylates, higher acrylate esters also undergo the ester cleavage reaction to give olefin and poly(acrylic acid) (PAA).

The degradation of PAA is qualitatively similar to that of the methacrylic acid in that water loss with the formation of anhydrides occurs. Various oligomeric acid and anhydride structures are also produced [64]. The degradation of salts appears to follow a similar pathway [65].

A significant amount of work has been carried out on the interaction of PMMA with additives; additives that have been examined include silver acetate [66], red phosphorus [67, 68], zinc bromide [69, 70], Wilkinson's salt, $(PPh_3)_3RhCl$ [71, 72], Ph_xSnCl_{4-x} (x = 0–4) [73, 74], Ph_2S_2 [75], Nafions [76], various transition metal halides [77–79], metal acetylacetonates [80,

81], and copolymers of MMA with 2-sulfoethylmethacrylate [82]. The focus of this work has been to develop a mechanistic understanding of the effects of the additive on the degradation of the polymer. To enhance the thermal stability of the polymer, one must either prevent the initial degradation or capture the products of this degradation so they cannot undergo further reaction.

Polyacrylonitrile (PAN) and Polymethacrylonitrile (PMAN). An examination of the formulas of these two compounds explains the great differences in their degradation pathways. PMAN has two substituents on a carbon atom and thus will undergo end-chain scission with the formation of a large fraction of monomer, while PAN degrades via a cross-linking pathway to give only very little monomer [83]. The degradation of PAN has been of interest owing to the possibility of generating carbon fibers from its degradation. We have recently reviewed the literature on this degradation and proposed a pathway by which the volatile species, ammonia and hydrogen cyanide, are evolved [84]. The essentials of the scheme to account for the evolution of ammonia involve cyclization by nucleophilic attack of an end group on an adjacent nitrile to give a ring structure. This is followed by tautomerization, isomerization, and elimination of ammonia and hydrogen to give the graphite-like char known to be produced. Only small amounts of HCN are produced and this may arise from either the presence of head-to-head linkages or a radical process in which cyanide radical is eliminated. It should be noted that the degradation of styrene-acrylonitrile copolymer is quite different as the acrylonitrile units are randomly arranged and the nucleophilic attack that leads to ring formation cannot occur and monomer is instead eliminated.

The thermal degradation of PMAN has been studied by Metcalfe et al. [85] and they report that, in addition to methacrylonitrile, hydrogen cyanide, propyne, propane, butene, acetonitrile, acrylonitrile, propionitrile, benzene, crotonitrile, pyridine, toluene, cyanobutene, styrene, benzonitrile, indene, methyl benzonitrile, and ethyl styrene are produced. It is interesting that the fraction of methacrylonitrile formed decreases as the temperature at which the combustion is performed is increased while hydrogen cyanide evolution increases. At 700°C 64% acrylonitrile and 2% HCN are produced, while at 900°C 64% HCN and 0.1% methacrylonitrile are obtained. This suggests that HCN is a secondary product that arises from the degradation of the primary products, and several mechanisms were proposed to account for its production.

Poly(vinyl chloride) (PVC) and Related Materials. It is quite simplistic to state that PVC degrades by chain-stripping with the loss of hydrogen chloride, as this reaction yields additional products that can participate in further reaction, mainly cross-linking. Degradation and stabilization of PVC has been reviewed by Tran [86]. The chain-stripping commences at temperatures as low as 100°C and is autocatalytic. The double bonds that are formed are conjugated and can give rise to aromatic structures that will burn with the formation of significant quantities of smoke. A scheme has been proposed to account for the autocatalytic character of the degradation, as shown in Fig. 4.7 [87].

The cross-linking reaction can be prevented by heating in the presence of maleic anhydride, which suggests that a Diels-Alder type of cross-linking occurs, as shown in Fig. 4.8.

Further reactions of the polyenes produced by loss of HCl determine the course of the PVC degradation. If the cyclization reactions noted in the preceding occur, the aromatics, largely benzene, that are produced, burn in the vapor phase with the formation of large amounts of smoke and the heat generated in the combustion facilitates further degradation. On the other hand, if the polyene sequences continue to undergo cross-linking, then char formation occurs and the production of aromatics, with their inherent char, is reduced and the material is stabilized [88–93].

Anthony [94] has recently constructed a kinetic model that fits the TGA curve reasonably well. The model involves the initial loss of hydrogen chloride to give a polyolefin. The poly-

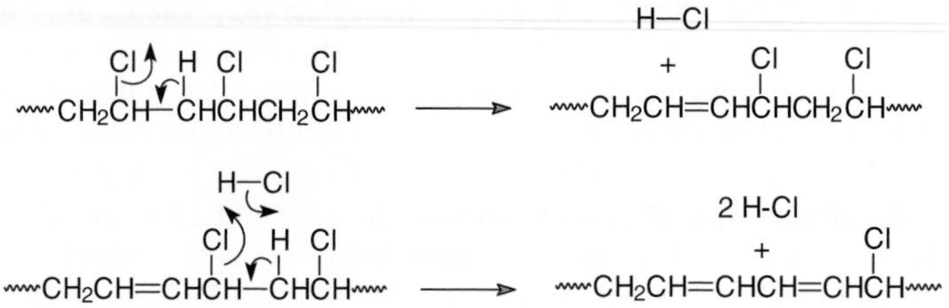

Figure 4.7   Autocatalytic degradation of PVC

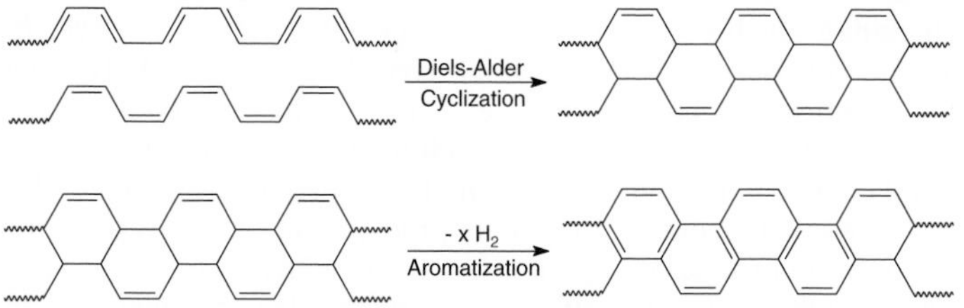

Figure 4.8   Cyclization and aromatization of PVC

olefin can either cyclize and fragment to give benzene or it can cross-link. The cross-linked material, in the same way, can cyclize and fragment to give substituted benzenes, or it can form a char.

It is believed that the mixed *cis,trans*-polyene gives rise to aromatics and smoke, whereas if isomerization occurs to an all -*trans* form, cross-linking and hence char formation will occur. This isomerization is apparently promoted by Lewis acids; the most common materials in use for PVC smoke suppression are molybdenum salts which promote the isomerization [95–97]. A recent publication focuses on a new mechanism for the cross-linking of PVC, reductive coupling [98]. Low valent metal salts, usually Cu(I), promote gel formation and reduce the evolution of hydrogen chloride. PVC is inherently flame retardant owing to its very high chlorine content; the oxygen index has been measured at between 45 and 49 [99].

The degradation of poly(vinylidene chloride) has recently been reviewed by Šimon [100] and Howell [101]. The degradation occurs in two stages, with the first stage completed by 220°C. The major product is HCl but there is also a small amount of monomer produced. The presence of chlorine permits the loss of HCl by chain-stripping while the presence of disubstituted carbons means that end-chain scission, and hence monomer formation, can occur. This first step leads to the formation of conjugated double bonds, just as in PVC, but there is no apparent autocatalysis for this polymer. The polychloroacetylene units that are produced give rise to color, and even at 1 % degradation the polymer is insoluble. The second step of the degradation involves some cross-linking, probably by a Diels-Alder pathway, to give a large surface area, highly absorbent form of carbon.

Stabilization of poly(vinylidene dichloride) can occur by several means [102]. Howell et al. have studied the effects of metal formiates on the degradation of vinyldiene chloride copolymers [103]. Some salts have little effect while others can markedly decrease the rate of both initiation and of propagation of the degradation. This seems to depend on the ability of the metal ion to interact with chlorine. The presence of some comonomer with the vinyldiene chloride can enhance thermal stability. This interrupts the conjugation that is formed by HCl loss and enhances the utility of the polymer. The presence of groups that can consume evolved HCl may be beneficial. Uhl [104] has examined the effect of magnesium hydroxide on the degradation and concluded that it does not enhance the thermal stability of the polymer.

Polytetrafluoroethylene (PTFE). The thermal degradation of PTFE has recently been examined by Simon and Kaminsky [105], who have reviewed the extensive literature on this topic. Because of the high strength of the carbon-fluorine bond, it is not easily cleaved and reaction occurs solely by breaking of carbon-carbon bonds. In this regard it may be considered to fall into the classic pattern noted earlier, because the carbon is disubstituted, one would expect that end-chain scission would occur with the formation of monomer. Indeed up to 97 % monomer is observed when the degradation is performed in a vacuum. The yield drops significantly at higher pressures and products such as hexafluoropropene, perfluorocyclobutane, and other fluorocarbons are formed. These are considered to arise from secondary reactions from the principal products of the degradation, tetrafluoroethylene and difluorocarbene. Both of these primary products arise from simple carbon-carbon bond cleavage. Some of these secondary reactions are shown in Fig. 4.9.

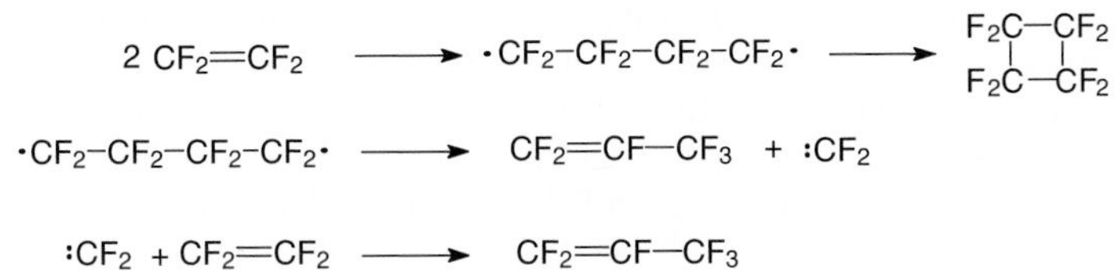

$$2\ CF_2{=}CF_2 \longrightarrow \cdot CF_2{-}CF_2{-}CF_2{-}CF_2 \cdot \longrightarrow \begin{matrix} F_2C{-}CF_2 \\ |\ \ \ \ \ | \\ F_2C{-}CF_2 \end{matrix}$$

$$\cdot CF_2{-}CF_2{-}CF_2{-}CF_2 \cdot \longrightarrow CF_2{=}CF{-}CF_3 \ + \ :CF_2$$

$$:CF_2 \ + \ CF_2{=}CF_2 \longrightarrow CF_2{=}CF{-}CF_3$$

Figure 4.9   Secondary reactions of tetrafluoroethylene to form observed degradation products

Polyamides. The thermal degradation of polyamides has been studied using pyrolysis–gas chromatography [106], infrared spectroscopy of condensates [107, 108], and mass spectroscopy [109, 110]. The products of the degradation of aliphatic polyamides consist of several components. The types of products that are formed and their relative amounts depend on the particular type of polyamide that is studied. For polyamide 6, the major product of the degradation is caprolactam. This is formed by cleavage of the relatively weak C-N bond followed by cyclization as shown in Fig. 4.10.

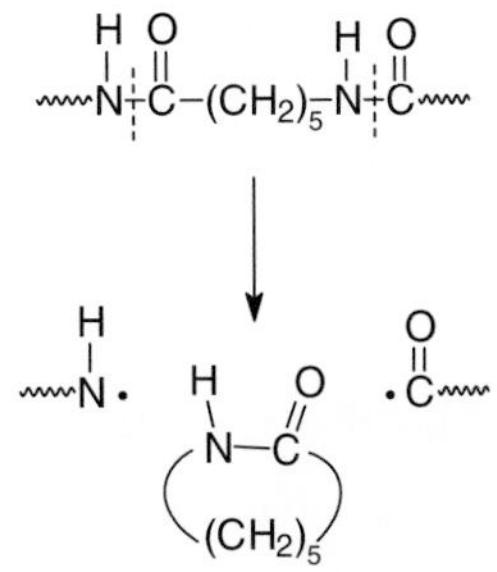

Figure 4.10   Formation of caprolactam from the degradation of polyamide 6

For larger polyamides, the lactam product is much less important, probably because of the reduced chance of ring formation with the longer chain. The major products are now mononitriles and -olefins have been observed in some cases. A scheme to account for these products has been suggested and is shown in Fig. 4.11. In addition to the products mentioned earlier, hydrocarbons are also noted among the products. These include alkanes, α-olefins, and α, ω-diolefins, with the olefins being the major component. These are also observed for those polyamides formed from diacids and diamines.

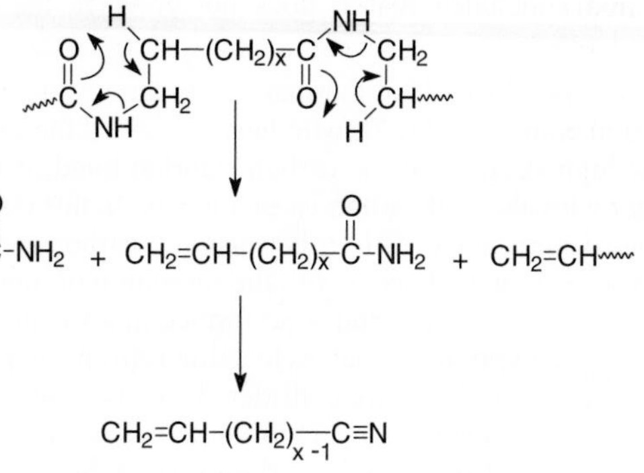

Figure 4.11 Formation of mononitriles from the degradation of polyamides

The major product observed in the degradation of polyamide 6,6 is cyclopentanone. This is formed by cleavage of the two C-N bonds to give a diradical at the two carbonyl centers. This can then rearrange to give cyclopentanone and carbon monoxide. A small dinitrile peak is also observed and this is more important in larger polyamides. This is believed to be due to a pathway similar to that shown previously for the mononitrile. When the polyamide is prepared from a diacid and a diamine, two amide end groups can be formed, in contrast to only one when a half-acid, half-amine compound is used. Figure 4.12 leads to the formation of the diolefin as well as the dinitrile via a -hydrogen shift.

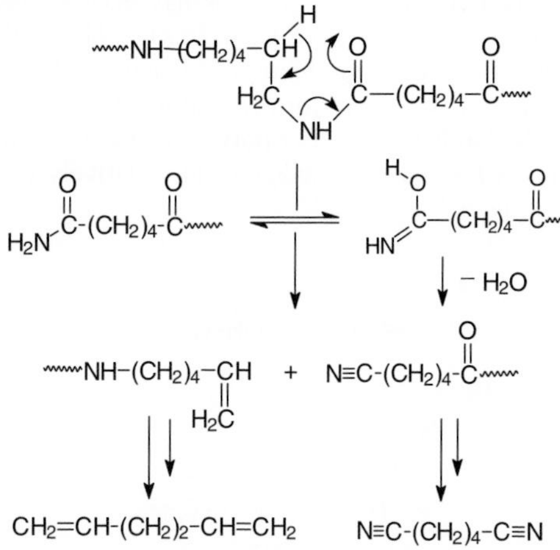

Figure 4.12 Formation of 1,5-hexadiene and 1,4-dicyanobutane from polyamide 6,6

Levchik et al. [111–116] studied the thermal degradation of various polyamides in the presence of ammonium polyphosphate and other inorganic phosphorus compounds. The presence of ammonium polyphosphate changes the degradation process by lowering the temperature of decomposition and changing the composition of the volatile products. This is an intumescent process in which the volatile products that are produced from the decomposition of the polyamide can blow the layer of polyphosphoric acid that is mixed with the residue of the degradation. Wheeler, Zhang, and Tebby [117] have described the utilization of organic phosphates to enhance the thermal stability of polyamides. Mateva [118] has reported that a flammability and thermal behavior of polyamide 6 is changed when the polymerization of caprolactam is carried out in the presence of some N-substituted phosphorus-containing lactams.

Polyamide 6 has been irradiated both in the presence and absence of flame retardants and radiation promoters [119]. The gel content and cross-link density are increased in the presence of a radiation promoter, such as melamine cyanurate. Invariably the onset temperature of the degradation is decreased after irradiation. Polyamide nanocomposites show a reduction of 30 % or more in the peak heat release rate [23, 24].

Polyesters. The degradation of poly(ethylene terephthalate) (PET) has been studied most extensively and only that material is covered herein. It is believed that the initial step in the degradation of PET involves a typical ester degradation as shown in Fig. 4.13 [120–126].

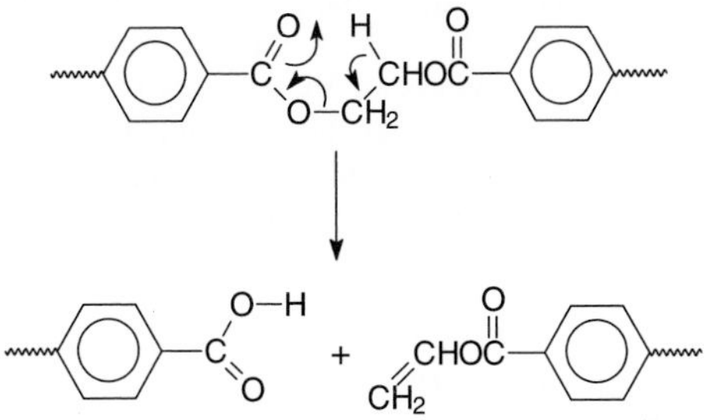

Figure 4.13   Degradation of PET to a carboxylic acid and a vinyl ester

These initial products can then undergo secondary reactions to produce the final, volatile products of the degradation; a pathway to these products is shown in Fig. 4.14.

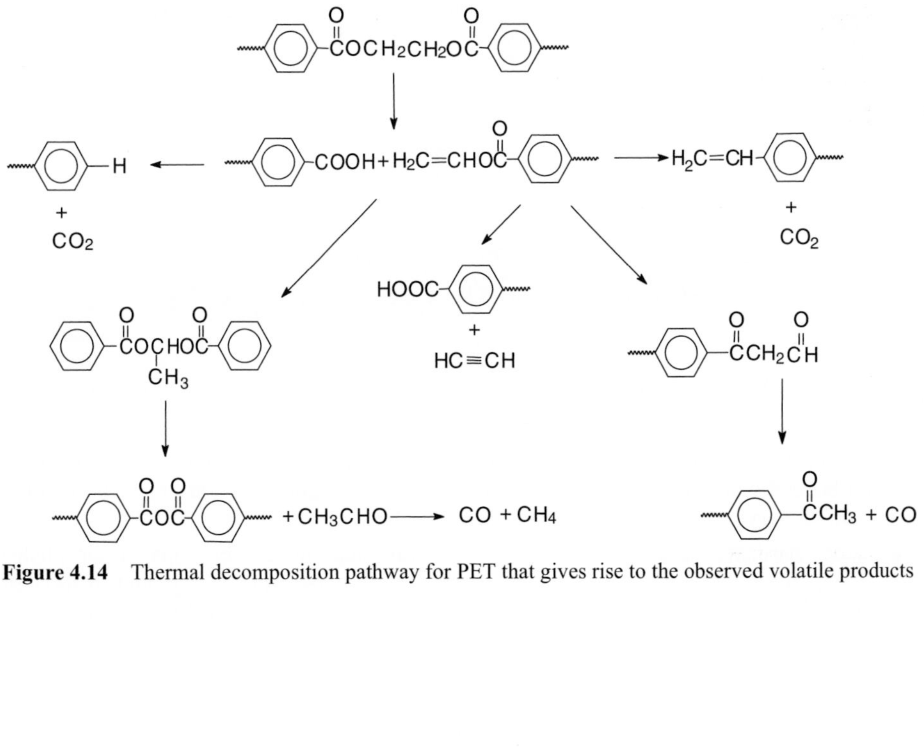

Figure 4.14   Thermal decomposition pathway for PET that gives rise to the observed volatile products

Chang [126] has suggested a four-step pathway by which the thermal stability of PET may be enhanced. The initial step makes use of the known initial step of the degradation reaction, that is, the ester decomposition reaction to give a carboxylic acid and a vinyl ester. Homopolymerization of the vinyl ester, followed by chain-stripping with the loss of a substituted carboxylic acid, will produce unsaturation along the main chain of the polymer. These unsaturated linkages can then cyclize to give a highly cross-linked and aromatic polymer.

Flame retardancy of PET is usually achieved by vapor phase processes using halogen or phosphorus compounds.

Polycarbonates. The first article on the degradation of polycarbonates appeared in the early 1960s and it has continued to be an active area of investigation up to this time [127–144]. In this review attention is devoted only to bisphenol A polycarbonate. The major gaseous product evolved in the thermal degradation is carbon dioxide, which is taken to indicate that the carbonate linkages are easily cleaved. Additional gaseous products are carbon monoxide and methane, and these increase in abundance as the temperature is raised. Phenols are also evolved. It is believed that cyclic oligomers are the primary products of polycarbonate degradation, but there is controversy about whether this proceeds by an ionic or a radical process. Figure 4.15 shows the radical pathway, after McNeill [138], for the formation of the cyclic oligomers as well as phenol. The radical scheme is used here for convenience; this should not be taken as a preference for this pathway.

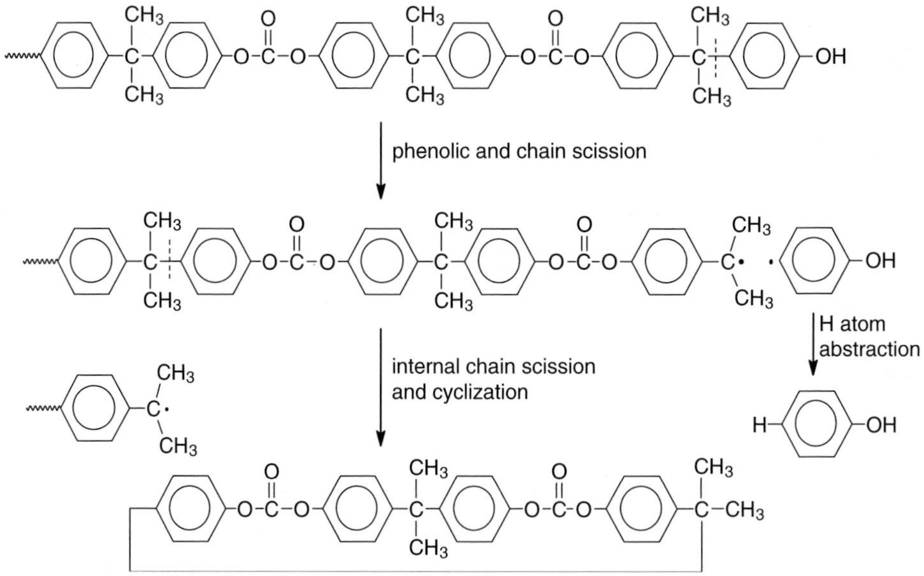

Figure 4.15   Radical pathway for the formation of cyclic oligomers from bisphenol A polycarbonate

Polycarbonates undergo extensive gel formation when heated in an open vessel or under vacuum but not when heated in a sealed tube. The cross-linking reaction proceeds with the formation of diaryl ester, ether, and carbonaceous bridges.

The flame retardants that have been used for polycarbonate include brominated phosphates [145–147], zinc borate [148], aromatic sulfonates (intumescent) [149], and resorcinol bis(diphenyl)phosphate [144, 150]. All, except the intumescent system, function in the gas phase as radical traps.

4.2.1.4 Concluding Comments

In this section we have attempted to delineate the various pathways by which thermoplastics may undergo thermal degradation as well as to give some indication of the types of investigations one may use to develop this type of information. In addition, information is provided on processes, which can be used to change the degradation pathway and render the polymer less flammable. We hope that the reader will be able to use this as a guide to the literature. We have not attempted to cover every article on any of the polymers, rather we have tried to give a sampling of the work which has been already accomplished and the new worker in this area is invited to build on this work so that a better understanding of the thermal degradation of polymers, and the means to prevent this degradation, can be developed.

References for Section 4.2.1

[1] H.E. Blair in E. Turi (Ed.): Thermal Characterization of Polymeric Materials, 2nd edit., Academic Press, p. 2289, 1997
[2] L. Costa, G. Camino: *Polym. Degrad. Stab.* 12 (1985) 297
[3] S.L. Madorsky, S. Strauss: *J. Res. Natl. Bur. Stand.* (US) 53 (1954) 161
[4] J.K.Y. Kiang, P.C. Uden, J.C.W. Chien: *Polym. Degrad. Stab.* 2 (1980) 113
[5] S. A. Hedrick, S. S. C. Chuang: *Thermochim. Acta* 315 (1998) 159
[6] G. Camino, L. Costa: *Polym. Degrad. Stab.* 3 (1980–81) 423
[7] L. Costa, G. Camino: *Polym. Degrad. Stab.* 12 (1985) 105
[8] L. Costa, G. Camino: *Polym. Degrad. Stab.* 12 (1985) 125
[9] L. Costa, G. Camino: *Polym. Degrad. Stab.* 12 (1985) 297
[10] L. Costa, G. Camino, M.P. Luda di Cortemiglia: *Polym. Degrad. Stab.* 14 (1986) 113
[11] L. Costa, G. Camino, M.P. Luda di Cortemiglia: *Polym. Degrad. Stab.* 14 (1986) 159
[12] L. Costa, G. Camino, M.P. Luda di Cortemiglia: *Polym. Degrad. Stab.* 14 (1986) 165
[13] G. Camino: *Devel. Polym. Degrad.* 7 (1987) 221
[14] G. Camino, L. Costa, L. Trossarelli: *Polym. Degrad. Stab.* 7 (1984) 25
[15] G. Camino, L. Costa: *Rev. Inorg. Chem.* 8 (1986) 69
[16] G. Camino, R. Arnaud, L. Costa, J. Lemaire: *Angew. Makromol. Chem.* 160 (1988) 203
[17] *G. Camino, L. Costa, G. Martinasso:* Polym. Degrad. Stab., 23 (1989) 359
[18] G. Bertelli, G. Camino, E. Marchetti, L. Costa, E. Casorati, R. Locatelli: *Polym. Degrad. Stab.* 25 (1989) 277
[19] G. Camino, M.P. Luda in M. Le Bras, G. Camino, S. Bourbigot, R. Delobel (Eds.): Fire Retardancy of Polymers: The Use of Intumescence, Roy. Soc. Chem., p. 48, 1998
[20] M. Le Bras, S. Bourbigot in M. Le Bras, G. Camino, S. Bourbigot, R. Delobel (Eds.): Fire Retardancy of Polymers: The Use of Intumescence., Roy. Soc. Chem., p. 64, 1998
[21] S. Bourbigot, M. Le Bras in M. Le Bras, G. Camino, S. Bourbigot, R. Delobel (Eds.): Fire Retardancy of Polymers: The Use of Intumescence, Roy. Soc. Chem., p. 222, 1998
[22] *G. Camino, R. Delobel in A.F. Grand, C. A. Wilkie Eds: .Fire Retardancy of Polymeric Materials,* Marcel Dekker, 2000, pp. 217
[23] J.W. Gilman, T. Kashiwagi, E.P. Giannelis, E. Manias, S. Lomakin, J.D. Lichtenbaum, P. Jones in M. Le Bras, G. Camino, S. Bourbigot, R. Delobel (Eds.): Fire Retardancy of Polymers: The Use of Intumescence, Roy. Soc. Chem., p. 203, 1998
[24] J.W. Gilman, T. Kashiwagi, M. Nyden, J.E.T. Brown, C.L. Jackson, S. Lomakin, E.P. Giannelis, E. Manias in S. Al-Malaika, A. Golovoy, C.A. Wilkie (Eds.): Chemistry and Technology of Polymer Additives, p. 249, 1999
[25] K. McCreedy, H. Keskkula: *J. Appl. Polym. Sci.* 22 (1978) 999
[26] I.C. McNeill in Encyclopedia of Polymer Science and Technology, Pergamon Press, 1989, p. 451
[27] D.L. Gardner, I. C. McNeill: *Eur. Polym. J.* 7 (1971) 569
[28] C.F. Cullis, H. S. Laver: *Eur. Polym. J.* 14 (1978) 571
[29] M.A. Golub, R.J. Gargiulo: *J. Polym. Sci.* B10 (1972) 41
[30] M. Suzuki, C.A. Wilkie: *Polym. Degrad. Stab.* 47 (1995) 217
[31] D.D. Jiang, G.F. Levchik, S.V. Levchik, C.A. Wilkie: *Polym. Degrad. Stab.* 66 (1999) 387

[32] D.D. Jiang, G.F. Levchik, S.V. Levchik, D. Dick, J.J. Liggat, C.E. Snape, C.A. Wilkie: Polym. Degrad. Stab., 68 (2000) 75

[33] W. Schnabel, G.F. Levchik, C.A. Wilkie, D.D. Jiang, S.V. Levchik: *Polym. Degrad. Stab.* 63 (1999) 365

[34] O. Chiantiore, G. Camino, L. Costa, N. Grassie: *Polym. Degrad. Stab.* 3 (1981) 209

[35] L. Costa, G. Camino, A. Guyot, M. Bert, A. Chiotis: *Polym. Degrad. Stab.* 4 (1982) 245

[36] A. Guyot*: Polym. Degrad. Stab.* 15 (1986) 219

[37] G. Audiso, F. Bertini: *J. Analyt. Appl. Pyrolysis*, 24 (1992) 61

[38] I.C. McNeill, M. Zulfiqar, T. Kousar: *Polym. Degrad. Stab.* 28 (1990) 131

[39] G.F. Levchik, K.Si, S.V. Levchik, G. Camino, C.A. Wilkie: *Polym. Degrad. Stab.* 65 (1999) 395

[40] N. Grassie J. Gilks: *J. Polym. Sci.: Polym. Chem. Ed.* 11 (1973) 1985

[41] S.K. Brauman: *J. Polym. Sci.: Polym. Chem. Ed.* 17 (1979) 1129

[42] J.F. Rabek: *J. Polym. Sci.: Polym. Chem. Ed.* 26 (1988) 2537

[43] J. Li, C.A. Wilkie: *Polym. Degrad. Stab.* 57 (1997) 293

[44] Z. Wang, D.D. Jiang, M.A. McKinney, C.A. Wilkie: *Polym. Degrad. Stab.* 64 (1999) 387

[45] Z. Wang, D.D. Jiang, C.A. Wilkie, J.W. Gilman: *Polym. Degrad. Stab.* 66 (1999) 373

[46] E.D. Weil, W. Zhu, H. Kim, N. Patel, L. Rossi di Montelera in M. Le Bras, G. Camino, S. Bourbigot, R. Delobel (Eds.): Fire Retardancy of Polymers: The Use of Intumescence, Roy. Soc. Chem., p. 35, 1998

[47] N. Grassie, H.W. Melville: *Proc. Roy. Soc.* (Lond.), 199 (1949) 1

[48] A. Brockhaus, E. Jenckel: *Makromol. Chem.* 18 (1956) 262

[49] N. Grassie, E. Vance: *Trans. Faraday Soc.* 49 (1953) 184

[50] G.G. Cameron, D.R. Kane: *J. Polym. Sci., Polym. Lett.* 2 (1964) 693

[51] A.N. Hodder, K.A. Holland, I.D. Rae: *J. Polym. Sci., Polym. Lett. Ed.* 21 (1983) 403

[52] T. Kashawagi, A. Inaba, J.E. Brown, K. Hatada, T. Kitayama, E. Masuda: *Macromolecules.* 19 (1986) 2160

[53] A. Meisters, G. Moad, E. Rizzardo, D.H. Solomon: *Polym. Bull.* 20 (1988) 499

[54] L.E. Manring, D.Y. Sogah, G.M. Cohen: *Macromolecules* 22 (1989) 4654

[55] T. Kashiwagi, A. Inabi, A. Hamins: *Polym. Deg. Stab.* 26 (1989) 161

[56] L. Manring: *Macromolecules* 24 (1991)

[57] N. Grassie, J. R. MacCallum: *J. Polym. Sci.* 2 (1964) 983

[58] D. Grant, N. Grassie: *Polymer* 1 (1960) 445

[59] B.-C. Ho, Y.-D. Lee, W.-K. Chin: *J. Polym. Sci.: Part A: Polym. Chem.* 30 (1992) 2389

[60] I.C. McNeill, M. Zulfiqar: *J. Polym. Sci.: Polym. Chem. Ed.* 16 (1978) 3201

[61] I.C. McNeill, M. Zulfiqar: *Polym. Degrad. Stab.* 1 (1979) 89

[62] G. Cardenas T., C. Retamal, L.H. Tagle: *Thermochim. Acta* 176 (1991) 233

[63] M. Zulfiqar, R. Hussain, S. Zulfiqar, D. Mohammad, I.C. McNeill: *Polym. Degrad. Stab.* 41 (1993) 45

[64] I.C. McNeill, S.M.T. Sadeghi: *Polym. Degrad. Stab.* 29 (1990) 233

[65] I.C. McNeill, S.M.T. Sadeghi: *Polym. Degrad. Stab.* 30 (1990) 213

[66] A. Jamieson, I.C. McNeill: *J. Polym. Sci.: Polym. Chem. Ed.* 16, (1978) 2225

[67] C.A. Wilkie, J.W. Pettegrew, C.E. Brown: *J. Polym. Sci.: Polym. Lett. Ed.* 19 (1981) 409

[68] E. Brown, C.A. Wilkie, J. Smukalla, R.B. Cody, Jr., J.A. Kissinger: *J. Polym. Sci. Polym. Chem. Ed.* 24 (1986) 1297

[69] I.C. McNeill, R.C. McGuiness: *Polym. Degrad. Stab.* 9 (1984) 167

[70] I.C. McNeill, R.C. McGuiness: *Polym. Degrad. Stab.* 9 (1984) 209

[71] S.J. Sirdesai, C.A. Wilkie: *J. Appl. Polym. Sci.* 37 (1989) 863

[72] S.J. Sirdesai, C.A. Wilkie: *J. Appl. Polym. Sci.* 37 (1989) 1595

[73] J.A. Chandrisiri, C.A. Wilkie: *Polym. Degrad. Stab.* 45 (1994) 83

[74] J.A. Chandrisiri, C.A. Wilkie*: Polym. Degrad. Stab.* 45 (1994) 91

[75] J.A. Chandrisiri, C.A. Wilkie*: Polym. Degrad. Stab.* 46 (1994) 275

[76] C.A. Wilkie, J.R. Thomsen, M.L. Mittleman: *J. Appl. Polym. Sci.* 42 (1991) 901

[77] J.A. Chandrasiri, D.E. Roberts, C.A. Wilkie: *Polym. Degrad. Stab.* 45 (1994) 97

[78] C.A. Wilkie, J.T. Leone, M.L. Mittleman: *J. Appl. Polym. Sci.* 42 (1991)

[79] R.S. Beer, C.A. Wilkie, M.L. Mittleman: *J. Appl. Polym. Sci.* 46 (1992) 1095

[80] I.C. McNeill, J.J. Liggat: *Polym. Degrad. Stab.* 29 (1990) 93

[81] I.C. McNeill, J.J. Liggat: *Polym. Degrad. Stab.* 37 (1992) 25

[82] S.M. Hurley, M.L. Mittleman, C.A. Wilkie: *Polym. Degrad. Stab.* 39, (1992) 353
[83] C.F. Cullis, M.M. Hirschler in: The combustion of organic polymers, Clarendon Press, Oxford, p. 119, 1981
[84] T.J. Xue, M.A. McKinney, C.A. Wilkie: *Polym. Degrad. Stab.* 58 (1997) 193
[85] E. Metcalfe, A.R. Harman, P.J. Fardell: *Fire Mater.* 11 (1987) 45
[86] V.H. Tran: *J. Macromol. Sci. – Rev. Macromol. Chem. Phys.* C38 (1998) 1
[87] A. Jamieson, I.C. McNeill: *J. Polym. Sci. Part A-1,* 12 (1974) 387
[88] W.H. Starnes Jr., J.A. Wallach, H. Yao: *Macromolecules* 29 (1996 7671)
[89] T. Hjertberg, E.M. Srvik in E.D. Owen (Ed): Degradation and Stabilization of PVC, Chapter 2, Elsevier, New York, 1984
[90] H. Starnes Jr., D. Edelson: *Macromolecules* 12 (1980) 797
[91] R.P. Lattimer, W.J. Kronke: *J. Appl. Polym. Sci.* 25 (1980) 101
[92] G. Camino, L. Costa, M.P. Luda di Cortemiglia: *Polym. Degrad. Stab.* 33 (1991) 131
[93] R.P. Lattimer, W.J. Kronke: *J. Appl. Polym. Sci.* 26 (1981) 1191
[94] G.M. Anthony: *Polym. Degrad. Stab.* 64 (1999) 353
[95] W.H. Starnes Jr., L.D. Wescott Jr., W.D. Reents, R.E. Cais, G.M. Villacorta, I.M. Piltz, L.J. Anthony in J.E. Kresta (Ed.): Polymer Additives, Plenum, New York, p. 237, 1984
[96] W.H. Starnes, Jr., L.D. Wescott, Jr., W.D. Reents, R.E. Cais, G.M. Villacorta, I.M. Piltz, L.J. Anthony: *Org. Coat. Appl. Polym. Sci. Proc.* 46 (1982) 556
[97] L.D. Wescott Jr., W.H. Starnes Jr., A.M. Mujsce, P.A. Linxwiller: *J. Analyt. Appl. Pyrol.* 8 (1985) 163
[98] R.D. Pike, W.H. Starnes, Jr., J.P. Jeng, W.S. Bryant, P. Kourtesis, C.W. Adams, S.D. Bunge, Y.M. Kang, A.S. Kim, J.A. Macko, C.P. O'Brien: *Macromolecules* 30 (1997) 6957
[99] C.J. Hilado in Flammability Handbook for Plastics, Technomic, Stamford, CT, 1969
[100] P. Šimon: *Polym. Degrad. Stab.* 43 (1994) 125
[101] B. A. Howell: *J. Polym. Sci.: Part A: Polym. Chem. Ed.* 25 (1987) 1681
[102] B.A. Howell, B.S. Warner, C.V. Rajaram, S.I. Ahmed, Z. Ahmed: *Polym. Adv. Tech.* 5 (1994) 485
[103] B.A. Howell, C.V. Rajaram: *J. Vinyl Tech.* 15 (1993) 202
[104] F. Uhl: M. S. Thesis, Central Michigan University, 1998
[105] C. M. Simon, W. Kaminsky: *Polym. Degrad. Stab.* 62 (1998) 1
[106] H. Ohtani, T. Nagaya, Y. Sugimura, S. Tsuge: *J. Analyt. Appl. Pyrol.* 4 (1982 117)
[107] M. Svoboda, B. Schneider, J. Štokr: *Collect. Czech. Chem. Commun.* 56 (1991) 1461
[108] C.H. Do, E.M. Pearce, B.J. Bulkin, H.K. Reimschussel: *J. Polym. Sci.: Part A: Polym. Chem.* 25 (1987) 2409
[109] U. Bahr, I. Luederwald, R. Mueller, H.-R. Schulten: *Angew. Makromol. Chem.* 120 (1984) 163
[110] A. Ballistreri, D. Garozzo, M. Giuffrida, G. Montaudo: *Macromolecules* 20, (1987) 2991
[111] S.V. Levchik, L. Costa, G. Camino: *Polym. Degrad. Stab.* 36 (1992) 31
[112] S.V. Levchik, L. Costa, G. Camino: *Polym. Degrad. Stab.* 36 (1992) 229
[113] S.V. Levchik, L. Costa, G. Camino: *Polym. Degrad. Stab.* 43 (1994) 43
[114] S.V. Levchik, L. Costa, G. Camino, G. F. Levchik: *Fire Mater.* 19 (1995) 1
[115] G.F. Levchik, S.V. Levchik, A.F. Selevich, A.I. Lesnikovich, A.V. Lutsko, L. Costa in M. Le Bras, G. Camino, S. Bourbigot, R. Delobel (Eds.): Fire Retardancy of Polymers The Use of Intumescence, Roy. Soc. Chem., p. 280, 1998
[116] G.F. Levchik, S.V. Levchik, G. Camino, E.D. Weil in M. Le Bras, G. Camino, S. Bourbigot, R. Delobel (Eds.): Fire Retardancy of Polymers: The Use of Intumescence, Roy. Soc. Chem., p. 304, 1998
[117] J.W. Wheeler, Y. Zhang, J.C. Tebby in M. Le Bras, G. Camino, S. Bourbigot, R. Delobel (Eds.): Fire Retardancy of Polymers: The Use of Intumescence, Roy. Soc. Chem., p. 252, 1998
[118] R.P. Mateva, N.V. Dencheva: *J. Appl. Polym. Sci.* 47 (1993) 1186
[119] A.I. Balabanovich, W. Schnabel, G.F. Levchik, S.V. Levchik, C.A. Wilkie in M. Le Bras, G. Camino, S. Bourbigot, R. Delobel (Eds.): Fire Retardancy of Polymers: The Use of Intumescence, Roy. Soc. Chem., p. 236, 1998
[120] E.P. Goodings: *Soc. Chem. Ind.* (Lond.): Sci. Monograph No. 13 (1961) 211
[121] Y. Sugimura: *J. Chrom. Sci.* 17 (1971) 269
[122] M. Day, K. Ho, D.M. Wiles: *J. Appl. Polym. Sci.* 25 (1980) 2943
[123] D.J. Carlsson, M. Day, T. Suprunchuk, D.M. Wiles: *J. Appl. Polym. Sci.* 28 (1983) 715

[124] T. Suebsaeng, C.A. Wilkie, V.T. Burger, J. Carter, C.E. Brown: J. *Polym. Sci., Polym. Chem. Ed.* 22 (1984) 945

[125] T. Suebsaeng, C.A. Wilkie, V.T. Burger, J. Carter, C.E. Brown: J. *Polym. Sci., Polym. Lett. Ed.* 22 (1984) 625

[126] P.-H. Chang, C.A. Wilkie: *J. Appl. Polym. Sci.* 38 (1989) 2245

[127] L.-H. Lee: *J. Polym. Sci.* 2 (1964) 2859

[128] A. Davis, J.H. Golden: *Makromol. Chem.* 78 (1964) 16

[129] A. Davis, J.H. Golden: *Nature* 206 (1965) 397

[130] A. Davis, J.H. Golden: *J. Macromol. Sci. – Revs. Macromol. Chem.* c3 (1969) 49

[131] R. H. Wiley: *Macromolecules* 4 (1971) 254

[132] K. B. Abbas: *Polymer* 21 (1980) 936

[133] S. Foti, M. Giuffrida, P. Maravigna, G. Montaudo: *J. Polym. Sci.: Polym. Chem. Ed.* 21 (1983) 1581

[134] G. Montaudo, C. Publisi, F. Samperi: *J. Polym. Sci.: Part A: Polym. Chem.* 31 (1983) 13

[135] A. Rincon, I.C. McNeill: *Polym. Degrad. Stab.* 18 (1987) 99

[136] A.S. Politou, C. Morterra, M.J.D. Low: *Carbon* 28 (1990) 529

[137] A.S. Politou, C. Morterra, M.J.D. Low: *Carbon* 28 (1990) 855

[138] I.C. McNeill, A. Rincon: *Polym. Degrad. Stab.* 31 (1991) 163

[139] G. Montaudo, C. Puglisi, F. Samperi: *Polym. Degrad. Stab.* 31 (1991) 291

[140] G. Montaudo, C. Puglisi: *Polym. Degrad. Stab.* 37 (1992) 91

[141] I.C. McNeill, A. Rincon: *Polym. Degrad. Stab.* 39 (1993) 13

[142] I.C. McNeill, S. Basan: *Polym. Degrad. Stab.* 39 (1993) 145

[143] H.G. Schild, M.G. Horner: *J. Macromol. Sci. – Pure Appl. Chem.* A31, (1994) 1955

[144] E. A. Murashko, G. F. Levchik, S. V. Levchik, D. A. Bright, S. Dashevsky: *J. Appl. Polym. Sci*, 71 (1998) 1863

[145] J. Green: *J. Fire Sci.* 12 (1994) 257

[146] J. Green: *J. Fire Sci.* 12 (1994) 551

[147] J. Green: *J. Fire Sci.* 14 (1996) 426

[148] R. Benrashid, G.L. Nelson, D.J. Ferm, L.W. Chew: *J. Fire Sci.* 13 (1995) 224

[149] A. Ballistreri, G. Montaudo, E. Scamporrino, C. Puglisi, D. Vitalini, S. Cucinella: *J. Polym. Sci.: Part A: Polym. Chem.* 26 (1988) 2113

[150] E.A. Murashko, G.F. Levchik, S.V. Levchik, D.A. Bright, S. Dashevsky: *J. Fire Sci.* 16 (1998) 278

4.2.2 Thermal Properties of Foam Plastics

H. STONE

Virtually any plastic can be produced with a foam structure, but only a few of these have achieved commercial significance. Depending on the polymer used they can be classified in the same manner as elastomeric, thermoplastic, or thermoset materials. Commercially the most important thermoplastic foams are those derived from polystyrene, PVC, and poly-alkenes (PE and PP). Phenolic [1], urea, and rigid polyurethane foams are the most important thermoset materials. Some specialty thermoset polymers such as polyimides, polybenzimida-zoles, polymeric silicones, and other highly aromatic polymers are finding uses in applications requiring high heat stability and extreme resistance to ignition, such as for the passenger compartments of passenger airplanes. The family of elastomeric foams includes mechanically frothed and chemically blown natural, synthetic (chloroprene and butyl) rubbers and styrene-butadiene rubbers. Filled chloroprene types find specialty uses in products requiring high resistance to ignition, such as in mattresses in prisons and other institutions. Depending on their composition, polyurethane foams can fit into any of these classifications.

A distinction is made between rigid and flexible foams, particularly in the case of poly-urethanes. The definition is based on the ability or lack of ability to regain its initial shape after mechanical deformation. Rigid foams can be produced with an open or closed cell structure, relying on the rigidity of the polymer to permit the foam to retain its shape. The closed-cell types are generally used for applications requiring good thermal insulation characteristics. Flexible polyurethane foams have open cells to varying degrees (depending on the desired end-use application ranging from gasketting to air filters). The wide range of resilience and flexibility available together with good strength at low densities has made them the preferred material in upholstery, mattresses, automotive products, and carpet padding as well as numerous other household and industrial uses.

The burning behavior of foams differs from that of the solid polymer from which they are derived [2, 3]. Because combustion of a solid requires degradation of the surface, the total surface area has a major influence on the ignition and burning performance. With their low bulk density (ranging down to about 8 kg/m^3 for some rigid foams and to 11 kg/m^3 for some flexible foams), foams have a very high ratio of surface to mass. This results in rapid and almost total pyrolysis of combustible material on exposure to radiant heat or flame, while the material is in contact with atmospheric oxygen, particularly for open-cell structures in which the inside layers of cells are also exposed to air. In the case of closed-cell systems heat buildup also occurs due to the low thermal conductivity of the polymer. As a consequence of these factors, foams generally are more easily ignited and will burn more rapidly than the solid polymers.

Because of the cellular structure and consequent low density of foams the mass consists of less than 5 % polymer and more than 95 % air or inert gases (in the case of closed-cell systems) and therefore they contribute a much lower fire load per unit volume than the corresponding solid polymers. Foams therefore contribute little to heat radiation. The combination of high surface area and access to oxygen, however, often will result in more rapid evolution of the available heat of combustion. Much research and development has led to a variety of additives and polymer modifications to increase the resistance to ignition and reduce the rate of pyrolysis without major deterioration of the desirable physical properties of the foams. Some of these are discussed in more detail in the following paragraphs.

Burning behavior of thermoplastic and thermosetting polymers including foams varies according to their chemical structure:

- Pure thermoplastic foams such as PS foam withdraw rapidly from the ignition source due to melting and shrinking so that ignition may not occur at all. Some of these materials are treated with additives to increase this tendency to melt and shrink.
- Thermoset foams do not withdraw from the ignition source and often do not ignite from a small flame. Depending on chemical structure and/or additives, they may form a solid char layer protecting the underlying material from direct exposure to the ignition source.
- Flexible polyurethane foams fall into an intermediate category. They may melt with decomposition (and withdraw from a small ignition source), or they may char and behave somewhat like the thermoset type mentioned previously.

The following section deals in some detail with the structures and burning behavior of PS and PVC thermoplastic foams, polyurethane foams, and thermoset foams (phenolic and urea).

4.2.2.1 PS Foam

PS foam is a tough, closed-cell thermoplastic foam. It is produced as molded bead foam (bulk density 15–30 kg/m^3) and extruded foam, mainly in sheets (15–60 kg/m^3). It is used mainly for packaging, eating supplies (such as disposable cups and plates), and for thermal insulation in buildings.

On exposure to a flame, PS fuses and burns like solid PS with a luminous yellow, sooty flame and a sweetish odor of styrene monomer. It continues to burn after removal of the ignition source. On short exposure to a small flame it frequently shrinks from the flame so rapidly that it does not ignite. Additives to increase the speed of melting are sometimes used. It is resistant to short exposures to about $100\,°C$ and decomposes at about $300\,°C$. When used as insulating panels in buildings these are generally covered with a film or treated paper and are usually enclosed by the building walls. Under these conditions PS panels are not expected to be either the initial item ignited in a fire or to be significant contributors to fire growth until possibly late in the development of a fire.

4.2.2.2 PVC Foams

PVC foams are available in rigid (unplasticized) and as flexible (plasticized) forms. Rigid PVC foam is a tough, closed-cell product manufactured with bulk densities between 40 and $130\,kg/m^3$. It is used mainly in structural sandwich panels and molded components.

Rigid PVC foam is resistant to temperatures of about $70\,°C$. It decomposes and shrinks at about $220\,°C$ but ignites only on exposure to an external flame. The flame is usually green and the product chars and liberates HCl. Flash- or self-ignition occurs only at elevated temperatures.

Elasticized (flexible) PVC foam is available in closed-cell (bulk density of 40–$130\,kg/m^3$) or open-cell (bulk density of 60–$130\,kg/m^3$) types. Plasticized PVC foams usually have a high plasticizer content. The most commonly used plasticizers are aliphatic and aromatic esters, and such foams continue to burn with a sooty flame once ignited. Some PVC foams are plasticized with phosphate ester plasticizers, and these are much more resistant to ignition. Because of the wide variety of products with different types and amounts of plasticizer, burning performance cannot be generalized.

4.2.2.3 Polyurethane Foams

The chemistry of the polyurethanes is extraordinarily versatile owing to the very large choices of starting materials and combinations. With suitable choices of these, thermoplastic, thermosetting, and elastomeric foams can be produced. Foams that are rigid, semirigid, and flexible can be manufactured in a wide range of density and both in open-cell and closed-cell variations [4]. Foams represent only one type of polyurethane polymer, which also include solid plastics and elastomers as well as coatings and fibers and the family of reaction injection molded (RIM) polymers. A short summary of polyurethane chemistry follows to illustrate the potential variations for foams.

Polyurethanes are produced by the reaction of di- or polyisocyanates with polyols ranging in functionality from two to eight and in molecular weight from several hundred to about $6\,000$. The isocyanates are mainly aromatic, although there are some specialty products based on aliphatic ones. The isocyanates most commonly used for foam manufacture are mixtures of toluene-2,4-diisocyanate

$$(4.0)$$

toluene-2,4-diisocyanate   toluene-2,6-diisocyanate

and toluene-2,6-diisocyanate (TDI) (see 4.0), most often in an 80:20 ratio but also in a 65:35 ratio and sometimes in blends between those ranges.

For some rigid foam products the undistilled blend with various condensed products is also used. The other major isocyanate used in foams is based on diphenylmethane-4,4'-diisocyanate (MDI) (4.1), which may also contain varying amounts of the 2,4' isomer. The variation used in foams is a blend of these monomers with higher condensed, higher functionality MDI, referred to as polymeric or sometimes as crude MDI (4.2). A large variety of these is available in a range of functionality and average molecular weight.

OCN—⟨benzene⟩—CH₂—⟨benzene⟩—NCO

$$OCN-\bigcirc-CH_2-\bigcirc-NCO \qquad (4.1)$$

diphenylmethane-4,4'-diisocyanate (MDI)

$$ \qquad (4.2)$$

polymethylene polyphenylisocyanate (pMDI)

The polyols used as the second major reaction component are numerous and include the following:

Polyether alcohols based on polymers of propylene oxide and ethylene oxide and blends thereof. A variety of low molecular weight initiators are used ranging from glycerol to sucrose to determine the functionality of the end product. The degree of polymerization determines the molecular weight and the processing conditions for the alkylene oxides can be used to influence the relative number of primary and secondary end groups.

The trifunctional or higher polyols of relatively low molecular weight are used in rigid foams and the higher molecular weight lower functionality ones are used for flexible foams.

Modified polyethers containing grafted styrene and some acrylonitrile and dispersions of PS are used extensively in flexible foams, particularly to control firmness (load bearing). These are used alone or in blends with the standard polyethers.

Polyester alcohols manufactured by polycondensation of di- or polycarboxylic acids with di- or higher functional alcohols are used for elastomeric foams. The most widely used polyesters are based on adipic acid and diethylene glycol in a range of molecular weights and functionality by blending in small amounts of glycerol. These are the basis for a large family of polyester-based polyurethane foams with a wide variety of end uses except for furniture or bedding.

Various polyols such as glycerol, pentaerythritol, sorbitol, and modified natural products such as castor oil and tall oil and nitrogen-containing ones such as diethanolamine are added for various purposes such as chain lengthening, additional cross-linking and so forth.

The stoichiometry between polyols (and water for flexible foams) and the isocyanate can also be varied to control physical properties of the foams. Most commonly an excess of isocyanate (ranging from 1% to 20%) is used but it is also possible to produce special foams with an excess of up to 30% hydroxyl groups. The main reaction in polyurethane chemistry is that of the isocyanate group with the hydroxyl group of a polyol to produce urethane groups which give the polymer its name. For rigid foams this is the dominant reaction.

This reaction is thermally reversible starting at about 135 °C and increasing at higher temperatures. Gases used to generate the foam structure are low-boiling organic materials, which evaporate during the foaming reaction from the heat generated by the reaction. The other blowing agent is carbon dioxide, which is generated by the reaction of isocyanate with water.

In addition it can also be introduced as a liquid in solution to the polyol. The reaction of isocyanate with water also generates aromatic urea groups, which increase the firmness and rigidity of foams, as illustrated in (4.3):

$$-N=C=O \ + \ HO- \ \rightleftarrows \ -N-C-O-$$
$$\qquad\qquad\qquad\qquad\qquad\ \ | \quad \|$$
$$\qquad\qquad\qquad\qquad\qquad\ H \quad O$$

isocyanate   alcohol   hydroxyl urethane

$$2-N=C=O \ + \ HOH \ \longrightarrow \ H_2N-C-NH_2 + CO_2$$
$$\qquad\qquad\qquad\qquad\qquad\qquad\qquad \|$$
$$\qquad\qquad\qquad\qquad\qquad\qquad\qquad O$$

isocyanate   water   urea   carbon dioxide

(4.3)

Rigid polyurethane foams can be produced without the use of water. The preferred blowing agent at one time had been a chlorofluorocarbon R-12, the use of which has been discontinued. Other blowing agents are some fluorocarbons containing hydrogen, some chlorinated hydrocarbons such as methylene chloride and some hydrocarbons such as isobutane pentanes or cyclopentane. Water can also be used, producing the usual carbon dioxide, but these foams tend to be more brittle, more open-cell, and poorer insulating materials.

Flexible polyurethane production requires the use of water in the formulation to achieve desirable load-bearing ability and the needed open-cell structure. The amount of water used also determines the density of the foam produced. Because of the large difference in equivalent weight of water (9.0) and that of the polyols used (averaging about 1000+) the chemical structure of the polymer actually contains a much larger number of urea groups than urethane linkages. To produce lower densities without the effect of producing firmer and more brittle foams inherent in the use of water, auxiliary blowing agents are added. The preferred one used to be a chlorofluorocarbon R-11, now discontinued. Blowing agents used now are mainly methylene chloride and liquid carbon dioxide dissolved under pressure in the polyol. Another technique for producing lower densities is a process under which the polymerization reaction is carried out under reduced atmospheric pressure.

Other reactions used to increase firmness and thermal stability, particularly for rigid foams, is reaction of isocyanate with some added carboxylic acids to produce aromatic amides and carbon dioxide and catalytic trimerization of isocyanate to produce highly stable isocyanurate rings (shown in (4.4)).

$$3 \ -N{=}C{=}O \qquad\qquad \longrightarrow$$

(4.4)

Rigid polyurethane foams can be produced as closed-cell (> 90%) foams, usually in the absence of water or as open-cell foams, in a wide range of densities depending on end use. The closed-cell varieties are preferred for thermal insulation applications. Densities range as low as about 8 kg/m^3 for packaging materials. Applications for thermal insulation may range from about 16 kg/m^3 to about 50 kg/m^3. Applications for sprayed systems such as roof insulation vary from about 20 kg/m^3 to about 48 kg/m^3.

The mechanism of the thermal degradation of foams is not different from that of the solid polymers [5]. Because of the generally good insulating properties of foams thermal degradation is usually slower than that of the solid polymers. Closed-cell foams are good insulators and even the open-cell foams will transmit heat more slowly.

Thermoplastic foams will usually degrade to volatile monomers, which will burn in the vapor phase. The case for polyurethanes is slightly different. Thermal degradation starting at temperatures of approx. 130 °C will emit TDI and subsequently degrade the polyol portion of the polymer. TDI is most likely emitted from the polyurethane portion of the polymer. The aromatic urea portion is thermally quite stable. More rapid heating rates can change this behavior. Depending on the structure of the foam and particularly on the presence of flame retardant additives, the decomposition may proceed to volatile degradation products including the polyol portion. This will cause the normal free radical combustion process, which in turn can be influenced by the usual halogen-containing additives. Some of the phosphorus-containing additives, some melamine derivatives, some aldehyde additives, and some fillers such as aluminum trihydrate can change the degradation process to produce a solid char layer interfering with fire propagation and inhibiting release of volatile degradation products.

These foams burn with a yellow luminous flame and tend to continue to burn after removal of the ignition source, producing a char. The presence of isocyanurate rings (see earlier) increases thermal stability and reduces smoke formation during burning. The more brittle nature of such foams can be modified by blending the two technologies of producing polyurethanes and polyisocyanurates, or by addition of highly branched polyols. The use of polymeric MDI is generally preferred because it produces higher heat stability and less brittleness than TDI. Although rigid foams are flammable and ignited fairly readily, this can also be mitigated by addition of combustion modifying additives. In any case a large percentage of rigid foams is used in applications where they are not likely to become directly exposed to an ignition source (as insulation in home appliances such as refrigerators and freezers, in building panels inside of walls, or as roof insulation covered by water proofing roof coverings). In these applications the foam would not be expected to become involved in combustion unless a significantly large fire has developed from some other source.

Flexible polyurethane foams are produced in such a large variety of formulations that it is difficult to generalize about their properties or flammability [6]. It is always open-cell at least to some degree, varying from very low air permeability for such uses as gaskets to totally open, having only the skeletal cell structure for such uses as air filters. The large end-use applications such as in furniture and bedding will range between these two extremes. Bulk densities vary from about 10 kg/m^3 to as high as 100 kg/m^3. It is stable to about 130 °C at which point it begins to dissociate to starting materials. Flexible polyurethane foam is not completely elastomeric and melts with some decomposition. Foams based on polyester polyols melt more cleanly than those based on polyether polyols. Because of these wide variations, generalizations about flammability behavior have to be made carefully.

Unmodified foams will ignite from exposure to small flame ignition sources applied either vertically or horizontally. They are generally resistant to ignition by smoldering ignition sources such as cigarettes. Addition of flame retardants (usually halogenated phosphate esters or halogenated aromatics blended with aromatic phosphates) enables such foams to pass small-scale laboratory ignition tests such as those prescribed by the US Federal Motor Vehicle Safety Standard (FMVSS 302), by the State of California furniture standard (T.B. 117) or the Underwriters Laboratory Standard for use in electrical appliances (UL 94 HF-1) and similar tests. Foams of this type will melt and drip on exposure to the ignition source and, depending on foam type and size and duration of the ignition exposure, will not ignite or will extinguish once the ignition source is removed. Great caution must be observed in interpreting performance under laboratory conditions as applying to performance under ignition conditions likely to prevail in real fire conditions. Once ignited, foam will continue to burn and may burn rapidly. It is the rapid evolution of the available heat of combustion that presents the real life hazard of such fires.

In the case of closed-cell rigid polyurethane foams the blowing agent is retained in the closed cells. It is present as a vapor at significantly less than atmospheric pressure. The foam is hot at the end of the polymer production from the heat of reaction. The gases are trapped at the elevated temperature and remain once the foam has cooled. The trapped gases do not affect the thermal degradation of the foam. The halogen-containing ones (hydrofluorocarbons, chlorinated hydrocarbons, etc.) do not contribute to combustion. The hydrocarbon ones, now used to some extent (isobutane, cyclopentane), are combustible. However, the weight involved compared to the weight of the polymer means that the potential heat contribution of the blowing agent to a fire is significantly less than 10 % of the potential total heat of combustion.

For flexible or rigid open-cell foams, the blowing agents do not contribute to potential heat of combustion. Because of the open-cell structure the blowing agents are emitted at the time of foam manufacture and during the cooling-curing cycle. In the case of flexible foams, where the use of blowing agents replaces some of the water used in the polymer formation, there is an effect on the polymer structure producing somewhat more urethane linkages compared to urea structures. This theoretically can influence the thermal degradation route as discussed previously, but the effect at most is small. The other effect is a small lowering of the melting (degradation) point.

Because of these factors, a number of approaches has been developed to modify flexible polyurethane foams for much greater resistance to ignition by larger ignition sources. These characteristics are measured by more intense laboratory ignition scenarios and by measurements of the rate of heat release. Approaches include the use of high levels of conventional flame retardants together with high levels of melamine dispersed in the foam. This produces a rapidly melting material, with the melt including the melamine being highly resistant to ignition. Other approaches are designed to produce an insulating char layer at the flame front, with the char layer tightly bound to the foam substrate. Such modifications include use of expandable graphite and use of aluminum trihydrate (producing both char and evolving water vapor on exposure to flame). Depending on the type of ignition used, the melting or the char-forming approach may perform better.

One major problem in judging potential hazard is the fact that in most applications of flexible foams the product is not used alone but is covered by a fabric or a film. It has long been recognized that testing of such composites is likely to produce different and unpredictable results from testing the components separately. The current trend in developing test protocols and performance criteria is to use composite tests simulating the actual uses. Particularly in the application of foams in furniture it has become obvious that the fabric used to cover the foam plays a dominant role in determining performance regardless of the nature (as is or flame retarded) of the foam used.

4.2.2.4 Phenolic Foams

Phenolic foams are hard, brittle, mixed-cell (partly open and partly closed) materials supplied in bulk densities of 10–100 kg/m^3. The major application is as thermal insulation in buildings. Very low density foams are used in plant and flower merchandising. Temperature resistance extends to 130 °C, although the foams will withstand short exposures to 250 °C. Decomposition starts at 270 °C. On exposure to flame, the foams burn for a few seconds. Above 400 °C they incandesce and glow. Flaming or self-ignition does not occur. Phenolic foams tend to continue smoldering and glowing. There is little smoke and the foams tend to char completely.

The lower flammability compared to solid phenolic resins is probably due to liberation of volatiles at temperatures and concentrations below the flash point. The carbonaceous residue is very stable but continues to incandesce until completely consumed.

4.2.2.5 Urea Foams

Urea foams are hard and very brittle with low bulk densities of 5–15 kg/m³. They have low heat resistance (50–90 °C). Depending on the process used, they can be open or closed cell. Brittleness and tendency to abrasion can be reduced by addition of additives such as glycols, but at a sacrifice in ignition resistance. Decomposition temperatures are in the range of 250–300 °C. On exposure to an ignition source, progressive charring occurs. The foamed char layer shields the rest of the foam from direct exposure to the flame.

One of the major applications for these foams has been direct foaming by introducing froth between walls for insulation. This was made possible by the fact that such a process generates little pressure and no special need existed to keep the walls from distortion. Within the last few years much of this market has disappeared because of concerns about continuing release of traces of formaldehyde into the living areas of buildings treated in this manner.

References for Section 4.2.2

[1] R.W. Martin (Ed.): Chemistry of Phenolic Resins, John Wiley, New York, 1976
[2] K.C. Frisch (Ed.): Plastic Foams, Marcel Dekker, New York, 1973
[3] R. Vieweg, A. Hchtlen (Eds.): Kunststoff Handbuch, Vol. VII, Carl Hanser, Munich, 1993
[4] R. Herrington: Flexible Polyurethane Foams, Dow Chemical, London, 1991
[5] Overview of the Combustibility and Testing of Filling Materials and Fabrics for Upholstered Furniture (prepared for Consumer Products Safety Commission), Polyurethane Foam Association Proceedings, London, 1998
[6] J.D. Saunders, K.C. Frisch (Eds.): Polyurethanes Chemistry and Technology, Wiley Interscience, New York, 1962 and 1964

4.2.3 Thermal Degradation of Thermosetting Polymers

S.V. LEVCHIK

A polymer is defined as a thermoset if it can be transformed by heat or a chemical curing process into an infusible and insoluble (cross-linked) product. In many respects, combustion and thermal decomposition of thermoset polymers is similar to that of the thermoplastic resins; however, there are some unique features resulting from the cross-linked structure of thermosets. Thermoset resins are usually more resistant to heating than thermoplastics with similar structure because volatilization of chain fragments from thermosets requires breaking of more chemical bonds. Thermoset resins do not melt before thermal decomposition; therefore in the burning resin heat transfer conditions from the flame to the polymer bulk are different in thermosets and thermoplastics because of the lack of a melt layer on the surface of the polymer. Furthermore, thermosets normally do not drip during combustion.

The application of thermosets is very broad and ranges from everyday consumable goods to high technological products made by injection molding, coating, pultrusion, and other processes. In the building industry, transportation, electronic and electrical applications, thermosets often require flame retardant properties. In this section, we will discuss the combustion performance and thermal decomposition behavior of phenolic resins, unsaturated polyesters, epoxy resins, polyimide resins, and amino resins.

4.2.3.1 Phenolic Resins

Phenolic resin is a generic name for a variety of thermosetting polymers prepared by curing of phenol-formaldehyde precursors. Basically there are two types of phenolic resins: resols, able to self-cure upon heating, and novolacs, which require a hardener to become cross-linked. Some novolacs are prepared from phenols with aliphatic substituents. This helps in compatibility of phenolic resins with low polar or nonpolar matrices. Phenolic resins are one of the oldest commercial polymeric products, first commercialized almost simultaneously in the United States and Germany in 1909–1910. They are mostly used as protective coatings, bonding resins for composite materials, adhesives, molding resins, foams, and high performance fibers (Kynol) [1].

Phenolic resins are considered to be materials of lower flammability. They show a high oxygen index, typically above 35 [2]. Depending on their structure, phenolic resins either self-extinguish quickly or burn very slowly when they are ignited in air in a vertical position. The self-ignition temperature of phenol-formaldehyde resin is about 480 °C [3]. In spite of the high content of aromatic structures, phenolic resins generate very little smoke. This explains the frequent utilization by airspace producers of phenolic resins as binders for composite materials in aircraft interiors.

Both cured and uncured phenol-formaldehyde resins start to decompose above 250 °C [4–8]. Insignificant volatilization may occur at lower temperature because of the presence of low molecular weight byproducts or unreacted phenol. Resol type phenolic resins are typically less stable than novolacs, because the resols contain relatively weak bismethylene ether linkages and methylol pendant groups. Regardless of whether decomposition proceeds in an inert or an oxidizing atmosphere, phenolic resins evolve significant amounts of water at the first step of thermal decomposition [9]. The simple explanation for this phenomenon is phenol-phenol condensation, which yields diphenyl ether cross-links [10, 11], see (4.5).

$$(4.5)$$

Chiantore et al. [12] showed infrared evidence of the diphenyl ether formations; however, Jackson and Conley [6] and Morterra and Low [13] did not find any significant amount of diphenyl ether functionalities in solid residues or chars produced by phenolics. An alternative explanation for the evolution of water was suggested by Morterra and Low [13], see (4.6).

$$(4.6)$$

Apart from water, the principal gaseous degradation products of phenolic resins are carbon monoxide, carbon dioxide, acetone, and low molecular weight hydrocarbons. The aromatic fragments, which volatilize at relatively high temperature, mostly consist of phenol, cresol, methylenebisphenol, and dimethylphenol [14]. Carbon monoxide and acetone can derive from carbonyl groups that were suggested [6, 10, 13] to appear due to the internal oxidation of methylene groups by evolved water, see (4.7).

$$ (4.7) $$

On the other hand, carbon dioxide is assumed to be a product of the thermal decomposition of benzoate groups derived from the internal oxidation of methylene groups by hydroxy radicals ($\cdot$OH) [15], see (4.8)

$$ (4.8) $$

Hydrocarbons appear from free-radical scission of polymer chains and recombination of methyl radicals [12].

Phenolic resins produce a significant amount of char on thermal decomposition in an inert atmosphere or in air [3]. The structure-charring relationships of various novolac type and resol type phenol-formaldehyde resins was studied by Costa et al. [9] and it was shown that the amount of char depends on the structure of phenol, initial cross-links, and tendency to cross-link during decomposition. Comparative study of thermal degradation of o-, m- and p-resol types showed [16] that o-cresol novolac tends to produce larger volume of H_2O and CO_2 than p- or m-cresol novolacs. Structural changes during heating of phenolic resins depend on the geometry of the sample, as restricted diffusion of the evolved gases in the bulky samples provokes secondary reactions and retardation of carbonization [17]. High char yields allow the use of phenolic resins as a component of ablative fire retardant coatings or as a starting material for preparation of special quality carbons, for example, porous carbons for chromatography [12, 15] or glassy carbon coating [17].

Phenolics are also known as a char source in multicomponent flame retardant additives for low charring polymers. Often synergistic combinations consist of a phosphorus-containing additive and a novolac type phenolic resin. It was reported [18–20] that a combination of red phosphorus and phenolic resins can provide flame retardancy in a wide range of polymers. In some cases, red phosphorus can be substituted by white colored phospham [21] or phosphorus oxynitride. It is known that novolac type phenolic resins can serve as a char source also in the combination with inorganic phosphates [22]. Most often phenolic resins are suggested as synergistic coadditives for aromatic phosphates in styrenic plastics [22–26] or polyolefins [27]. Combination with other high charring additives [26] or char improving catalysts [24] enhances the flame retardant action of novolacs with aromatic phosphates. Esters of phosphinic, phosphonic, or phosphoric acids mixed with the novolac type phenolic resin provide nondripping self-extinguishing thermoplastic polyesters [28]. There are also known applications of phenolic resins together with classical halogenated flame retardants, for example, tetrabromobisphenol A and antimony trioxide [29]. Flame retardancy of PVC can be significantly improved by the addition of small amounts of novolac salts with polyvalent metals [30].

4.2.3.2 Unsaturated Polyesters

Unsaturated polyester resins are made from stoichiometric mixtures of unsaturated and saturated dibasic acids or anhydrides and diols or oxides. This low molecular weight polymer is typically dissolved in a liquid monomer such as styrene to decrease viscosity and increase reactivity. Addition of a free-radical catalyst and heating initiate curing, which can be done at room temperature in the presence of an activator. Unsaturated polyesters are the principal products on the thermoset market [31].

Cured unsaturated polyesters are highly flammable showing a LOI typically about 20 [32]; however, there are many applications in which cross-linked polyesters should be flame retarded. The major applications for flame retardant resins are in the marine industry for small and medium size boat bodies, in transportation for automotive parts; and in the construction industry for artificial stone panels, curtain walls, and ductwork [33]. Most unsaturated polyesters are used with fillers or glass fibers to form composite structures. Recently Weil and Kim [34] reviewed flame retardant unsaturated polyesters.

Products of thermal decomposition of unsaturated polyesters made from different anhydrides and different glycols with various ratios of the components were studied by Luce et al. [35]. Braun and Levin [36] reviewed the literature on products of thermal decomposition and combustion of polyesters in terms of toxicity. Because the most abundant product evolved from unsaturated polyesters is styrene, polyesters produce significant amount of smoke on combustion [37–39]. Unexpectedly, it was found [39] that propylene glycol contributes significantly to smoke formation. A brominated glycol used to provide flame retardancy in polyesters enhances smoke production, whereas Al(OH)$_3$ [40] or Mg(OH)$_2$ [41] decrease smoke, see (4.9).

$$(4.9)$$

Because the composition of unsaturated polyesters is rather complex, it has been suggested [42–47] to consider that gases evolved during thermal decomposition originate from two distinct parts: PS (cross-links) and linear polyester. Most polyesters start to decompose at above 250°C, whereas the main step of weight loss usually occurs between 300 and 400°C. Because mostly styrene is volatilized at the initial steps of thermal decomposition, it is generally accepted that PS cross-links start to decompose first.

It is believed that the polystyrene cross-links decompose similar to linear polystyrene, the mechanism of which is well documented [48].

The linear polyester portion undergoes chain scission similar to the commodity thermoplastic polyesters. General purpose unsaturated polyesters are made from maleic anhydride, phthalic anhydride, and propylene glycol. The most likely position of the first chain scission in this type of polyester is the alkyl ester bond. It was hypothesized that chain scission proceeds through a *cis*-elimination mechanism, see (4.10).

$$\text{(4.10)}$$

On further heating the vinyl ester chain ends undergo decarbonylation, decarboxylation, or splitting off of methylacetylene.

In an oxidative atmosphere of air, unsaturated polyesters start to decompose at lower temperatures (approx. 220°C), however, the main degradation step remains in the same temperature region as in the inert atmosphere. In earlier work, it was suggested [42] that oxygen attacks the α-carbon in the polystyrene cross-link and this facilitates its decomposition, see (4.11).

$$\text{(4.11)}$$

Later this mechanism was doubted [49] because of the low concentration of benzaldehyde found in the products of thermal oxidation of the polyester resin; however, no alternative mechanism was suggested.

Kinetic studies showed [50, 51] two distinctive steps of the thermal oxidative degradation of the polyesters, which might be the oxidation of cross-links and polyester chains, respectively. A two-step process was also found in an inert atmosphere for the thermal decomposition of unsaturated polyesters end capped with dicyclopentadiene. This was also attributed to the decomposition of cross-links and the polyester chains [52].

Flame retardant unsaturated polyesters usually contain chlorinated or brominated segments in the polymer chain. It is interesting to note that the original idea of making such resins was primarily related to improving corrosion resistance, whereas flame retardancy came as a secondary advantageous property. Chlorendic acid (known as HET acid) made by addition of maleic anhydride to hexachlorocyclopentadiene is the most popular halogenated component of unsaturated polyesters [32, 34]. The thermal decomposition study of the HET-containing resin showed a mechanism similar to regular resins [53, 54]. Chlorinated species mostly evolved in the form of HET anhydride or hexachlorocyclopentadiene, which is likely to be responsible for the flame retardant effect [55]. Agrawal and Kulkarni [56] studied the flame retardancy and thermal decomposition behavior of the resin prepared with tetrachlorophthalic anhydride. In the resin containing dibromoneopentyl glycol, benzaldehyde was detected as the major product of thermal decomposition at the early stage along with some styrene [45]. This led the authors to the conclusion that the thermal decomposition starts in the main chain, but not in the cross-links as mentioned earlier. The resin containing bromostyrene as a partial substitution of the regular styrene shows a much lower exotherm of oxidation in air measured by DTA experiments [57]. There are reports in the literature of the effect of halogenated [58] or phosphorus-containing [59, 60] flame retardant additives or molybdenum trioxide [61] on the thermal decomposition of unsaturated polyesters.

4.2.3.3 Epoxy Resins

Epoxy resins are characterized by the presence of epoxide groups, and they might contain aliphatic, aromatic, or heterocyclic structures in the backbone. Epoxy resins are relatively expensive; however, their long service time and good physical properties often provide a favorable cost-performance ratio. The main fields where flame retardancy of epoxy resins is required are in electronics, for printed wiring boards and encapsulation of semiconductors, and in transportation (automotive, high-speed trains, and military and commercial aircraft) for composite structural and furnishing elements [62].

Epoxy resins are very reactive and allow the use of versatile curing agents, either catalytic or reactive. The catalytic curing agents do not build themselves into the thermoset structure and therefore do not much affect the flammability of the resin. For example, it has been shown [63] that epoxy resins catalyzed by various boroxines have essentially the same oxygen index (OI) because the cross-linking densities are comparable. On the other hand, some enhancement of the OI is observed upon increasing the amount of the catalyst and a very significant increase is observed upon increasing of curing time, which is attributed to the increase of the cross-linking density [64].

In contrast, reactive curing agents mostly represented by amines, anhydrides, or phenolic resins strongly modify flammability. The combustion behavior of similar epoxy resins depends on the ratio of oxygen to carbon atoms in the polymer structure. Epoxy resins cured with amines tend to produce more char, and they apparently are less flammable than acid or anhydride cured resins at comparable cross-linking density [65]. However, nitrogen can also be supplied from the epoxy monomer. For example, self-cured tetraglycidyl diaminodiphenyl methane (TGDDM) type resin containing nitrogen in its structure is less combustible [66] than TGDDM cured by diaminodiphenyl sulfone [67].

Pearce and co-workers [68, 69] showed that the OI of epoxy resins strongly correlates with the charring performance of the resins. Epoxy resins cured by phenol-formaldehyde resins are inherently flame retardant because of the significant charring tendency of the phenolic component. The char yield can be increased by using special epoxy monomers containing highly aromatic bisphenols (e.g., phenolphthalein, bisphenol-fluorenone [63, 68, 70], bisphenol-anthrone, tetraphenol-anthracene [69]) or containing double bonds able to undergo Diels-Alder reaction (e.g., mono-, di-, or trihydroxystyrylpyridine) [64]. These monomers, cured either alone or in combination with the regular grade epoxy monomers (e.g., commercial diglycidyl ether of bisphenol A [DGEBA]), allowing to obtain higher cross-linking density, show an increase of OI from 20 for DGEBA to almost 40 for highly aromatic monomers.

Thermal stability of the epoxy resins, as well as flammability depends on the structure of the monomer, the structure of the curing agent, and the cross-linking density. In general, the thermal stability increases with increasing cross-linking density for the analogous resins [71]. Meta-substituted aromatic amines as curing agents were found to impart greater thermal stability to epoxies than para-substituted aromatic amines, see (4.12).

$$\mathasymp\bigcirc\!-\!O\!-\!CH_2\!-\!\underset{\underset{OH}{|}}{CH}\!-\!CH_2\!-\!O\!-\!\bigcirc\!\mathasymp \xrightarrow{-H_2O} \mathasymp\bigcirc\!-\!O\!-\!CH_2\!-\!CH\!=\!CH\!-\!O\!-\!\bigcirc\!\mathasymp$$

(4.12)

Independently of the epoxy resin structure, its thermal decomposition starts from the glycidyl group. Dehydration of the secondary alcohols leading to the formation of double bonds is generally accepted in the literature [66, 67, 72–82] as the low temperature step of the decomposition of epoxy resins.

The resulting allylic C-O bond is thermally less stable than the original C-O; therefore chain scission mostly occurs at the allylic position. In epoxies copolymerized with amines, the allylic C-N is less stable than the C-O bond; therefore in general, amine cured epoxy resins are less stable than anhydride cured epoxies [83]. Both homolytic and heterolytic chain scission of allylic C-N (C-O) bonds were suggested in the literature [77, 80, 81, 84], which led essentially to the same result, see (4.13).

$$\text{[structure]} -O-CH=CH-CH_2-N-\text{[structure]} \xrightarrow{+H} \text{[structure]} -O-CH=CH-CH_3 \;+\; HN-\text{[structure]}$$

$$(4.13)$$

If decomposition proceeds in the same way with the secondary amide group (C-NH), evolution of the curing agent is expected. However, this seems not to be the case, as the curing agent is usually present as a minor product of thermal decomposition [85]. It mostly volatilizes as a part of the chain fragments or remains in the solid residue. On the other hand, phthalic anhydride was regenerated in large quantities on thermal decomposition of the anhydride cured epoxy resins [83], see (4.14).

$$\text{[structure]} -O-CH=CH-CH_3 \xrightarrow{+H} \text{[structure]} \;+\; CH_2=CH-CH_2-OH$$

$$(4.14)$$

On further decomposition, aliphatic chain ends produce light combustible gases. The presence of allylic alcohol apart from hydrocarbons indicates splitting of aromatic C-O bond weakened by the double bond.

However, significant concentration of acetone in the light volatile products from the thermal decomposition of epoxies also proves splitting of amine (C-N) or ether bonds (C-O) prior the dehydration of the secondary alcohol [66, 67, 79, 80, 81, 83], see (4.15).

$$\text{[structure]} -O-CH_2-\underset{\underset{OH}{|}}{CH}-CH_2-N-\text{[structure]} \xrightarrow{+H} \text{[structure]} -O-CH_2-\underset{\underset{O}{\|}}{C}-CH_3 \;+\; HN-\text{[structure]}$$

$$\downarrow +H$$

$$\text{[structure]} -OH \;+\; CH_3-\underset{\underset{O}{\|}}{C}-CH_3$$

$$(4.15)$$

In addition to the main processes of chain scission discussed earlier, many secondary processes, which lead to minor products of thermal decomposition, were reported in the literature [76–81]. For example, cyclization of aliphatic chain ends formed instead of their splitting off positively contributes to charring and fire retardancy [66, 67, 72–74, 77–79, 81], see (4.16).

$$\text{[structure]} -O-CH_2-\underset{\underset{OH}{|}}{CH^{\bullet}} \xrightarrow{-H_2O} \text{[structure]}$$

$$(4.16)$$

$$\text{[structure]} -N-CH=CH-CH_2^{\bullet} \xrightarrow{-H} \text{[structure]}$$

Thermal oxidative decomposition of the epoxy resins is extensively published in the literature as well [75, 86–90]. Basically, three mechanisms for oxidation of epoxies are suggested:

1. Attack of oxygen to the allylic methylene group [75],

2. Oxidation of the tertiary carbons in the aliphatic portion of the chain, which is usually an ester type cross-link in the anhydride cured resins [88],
3. Oxygen attack on the nitrogen in the amine cured epoxies [86, 90].

Any of these mechanisms leads to the formation of carbonyl groups (isomerization in the case of the third mechanism), which further decompose and result in chain splitting.

Kinetics of the thermal decomposition of epoxy resins in relation to their flammability and to model combustion have been reported in the literature [91–93]. On the basis of statistical selection of the kinetic function, better prediction of the shape of the thermogravimetry curve was shown by assuming that the rate of thermal oxidative decomposition of epoxies is controlled by diffusion. Similarly, Le Huy et al. [89] showed that thermal oxidative decomposition of epoxies at relatively low temperatures is diffusion controlled because of the need of oxygen penetration to the deep layers of the resin.

Paterson-Jones et al. [79] showed that commonly used inorganic fillers, alumina or silica, accelerate thermal decomposition of epoxies, probably because of catalysis. Kinetic evidence of the catalytic action of Al, Cu, or Zn used as fillers in epoxy resins were provided by G. Sanchez et al. [94].

Few studies reported in the literature concern epoxy resins cured with bis(aminophenyl)methylphosphine oxide, as well as thermal and combustion performance of these types of resins [95–100]. Mechanistic studies suggested [97, 99] that bis(aminophenyl)methylphosphine oxide at low concentration provides most of its flame retardant action in the condensed phase, increasing char yield and aromatization of the char, whereas at relatively high concentrations, phosphorus species tend to volatilize and become active in the gas phase. In line with general observations in epoxy resins, it was found that completely aromatic bis(aminophenyl)phenylphosphine oxide tends to produce more char than bis(aminophenyl)methylphosphine oxide [100] or bis(aminophenyl)phenylphosphinate [101, 102]. The use of tris(aminophenyl)phosphine oxide for curing of various epoxy resins, which led to substantial increase of char yield compared to the epoxies cured by regular hardeners, was reported in the literature [103]. A succinic anhydride with a pendant cyclic arylphosphinate group was used to cure either DGEBA or novolac type epoxy resins [104]. Increase of char yield from 8–10 % for the non-phosphorus-containing resin to 25–30 % for the flame retardant resins led to significant improvement of the flame retardancy without loss of mechanical properties, as phosphorus was in the pendant group. Thermal decomposition of the brominated epoxy resin was studied by Creasy [105]. Evolution of Br was detected mostly in the form of HBr or CH_3Br.

4.2.3.4 Amino Resins

Amino resins are made from products containing two or more amino groups and from an aldehyde. Urea-formaldehyde and melamine-formaldehyde are the largest amino resins; all other resins are specialty products made in relatively small volumes. The resins are usually sold in the form of low molecular weight methylol condensates, which are cured by heating usually with an acid catalyst. Adhesives to make plywood, chipboard, and sawdust board represent the largest market of amino resins. Those are also the main applications in which the flame retardancy of amino resins is required. Finishing of textile fabrics is another application of amino resins where flame retardancy is important [106].

Both urea-formaldehyde and melamine-formaldehyde resins are difficult to ignite in air and they self-extinguish quickly on removal of the ignition source [3]. The flash ignition temperature of melamine-formaldehyde resin is 602°C [107]. Amino resins produce little smoke and exhibit an intumescence behavior on combustion, which made these resins useful flame retardant additives themselves [108]. Flame retardancy of polyurethanes was improved by urea-formaldehyde resin [109]. Similarly, polycarbonate was flame retarded by melamine-formaldehyde resin [110]. Melamine-formaldehyde resin was also found particularly useful for coating of red phosphorus, providing protection and synergistic flame retardant action [111, 112]. In combination with various phosphoru-containing products, amino resins form long-term flame retardant finishes on the surface of cellulosic or synthetic fibers [113–122], see (4.17).

$$\text{(4.17)}$$

In spite of the great importance of amino resins, information on their thermal decomposition is rather scarce. Basically, two mechanisms of chain scission are suggested in the literature. Caruso et al. [123] studied the fragmentation of the series of N-substituted polyureas, which are considered as model compounds for amino resins, by direct pyrolysis in the ion source of the massspectrometer. N-H hydrogen transfer yielding an isocyanate chain end was suggested as the primary mechanism of thermal decomposition of NH containing polyureas, see (4.18)

$$\text{(4.18)}$$

If N-H hydrogen is not available, it was suggested [124] that C-H hydrogen transfer occurs from the α-methyl or methylene group. Two competitive reactions of primary chain scission were considered for N-substituted polyureas. These reactions model the highly cross-linked amino resin.

On the other hand, Camino et al. [125, 126] suggested a free radical mechanism of primary chain scission in urea-formaldehyde resin, see (4.19).

$$\text{(4.19)}$$

These reactions led to simple products detected in the evolved gases. Macroradicals formed in the polymer chain yield cyclic structures, for example, ethylene urea-formaldehyde, see (4.20)

$$(4.20)$$

or tetra-*sym*-hydrotriazin structures (4.21).

$$(4.21)$$

These ring structures are more stable than the original resin; therefore the thermal decomposition of urea-formaldehyde resin is deferred to higher temperature. This is important for the char-forming mechanism.

On thermal decomposition, ethylene urea-formaldehyde gives a high char yield which in combination with ammonium polyphosphate provides an intumescent structure [127]. A mechanistic study of the thermal decomposition of ethylene urea-formaldehyde resin showed that double bonds are produced on opening of the ethylene urea ring, and those mostly contribute to the char formation because they undergo polymerization at high temperature [128].

It was found [129] that the methylene groups are attacked by oxygen in the thermal oxidative decomposition of amino resins. Decomposition of peroxides formed led to the chain scission of the amino resin.

4.2.3.5 Curable Polyimides

Polyimides are condensation polymers made by the reaction of bifunctional carboxylic acid anhydrides and primary diamines. Some polyimides are thermoplastic, although they are usually processed from solutions. However, there is also a significant portion of polyimides, which are made by polymerization of unsaturated reactants and are considered to be thermosets. They use monomers (bismaleimides or bisnadimides) or low molecular weight oligomers end capped with unsaturated groups (acetylenic) able to further polymerization and cross-linking. Because of outstanding high performance and engineering properties, polyimides are mostly used in aerospace applications where high thermal stability and low flammability are required [130].

Polyimides are inherently flame retardant because of a high concentration of aromatic and heterocyclic structures tending to carbonization upon heating. Typically, polyimides produce more than 40 % char as measured by thermogravimetry. Thermal decomposition [131–133] and carbonization [134–138] processes of polyimides are comprehensively reported in the literature. Comparison of the flammability of composite panels prepared from commercial bismaleimides and epoxies showed a significant advantage of bismaleimides both in OI and cone calorimeter tests [139].

Thermal stability of the curable imides depends on the structure of resin and curing agent (if any is used). In general bisnadimides are more thermally stable than bismaleimides [140]. Bismaleimides [141, 142] or bisnadimides [142] with a broad variety of aromatic bridges between the imide groups were thermally cured and then decomposed in an inert atmosphere. It was shown that independently of the aromatic bridge most of the resins started to decompose at 400–450°C, which indicates that either imide group or aliphatic cross-links are the weakest sites in the resin network. However, the amount of char produced by these resins strongly depends on the structure of the aromatic bridge and the char yield could reach as much as 65 % for the resins that form the most char. Composite materials based on these resins and graphite laminates showed LOI ≥ 70 [141].

To overcome the adverse affect of the aliphatic cross-links on the thermal stability and flammability of the imide resin, ethynyl-terminated resins undergoing trimerization in curing were designed [143] (4.22).

$$(4.22)$$

These resins showed superior thermal stability and high char yield. Polyimides with aryl-alicyclic bridges between the imide groups were studied by Jakab et al. [144]. The first group of the volatile products detected on thermal decomposition of these resins clearly indicated the primary scission of the alicyclic fragments. Bismaleimide resins containing a sulfone bridge in the backbone showed improved physical properties but they were less stable compared to the thermoplastic polysulfone polymer [145]. The imide resins cured by amines are less stable than self-cured resins, and they decompose with partial recovery of the amine or its fragments [146]. Nevertheless, the char yield strongly depends on the structure of the amine [133, 147].

Thermal decomposition of the imide group in the imide polymers and in the model compounds was extensively documented in the literature [140, 144, 146, 148–153]. CO and CO_2 are the most abundant gases evolved during the thermal decomposition of imides. It is generally accepted that CO is produced by the cleavage of the imide ring, probably via the radical mechanism (4.23).

$$(4.23)$$

On the other hand, the mechanism of CO_2 evolution is not clear and the literature suggests few alternative pathways. Imide-isoimide summarization followed by the cleavage of the ring was considered in refs. [140, 148, 152, 153].

This route of CO_2 evolution was disproved by Fedotova et al. [154] and Zubakovska-Orszagh et al. [155] on the basis of infrared and ^{13}C-nuclear magnetic resonance studies. With model compounds it was shown that isoimide tends to isomerize to imide at high temperature, instead of imide isomerization to isoimide. Other mechanisms considered were an intramolecular rearrangement involving ionic [148] or radical structures [144]; however, details of these mechanisms are not clear.

Alternatively, it was suggested that CO_2 evolves in the course of the secondary reactions. These might be dimerization of isocyanates formed by the thermal decomposition of the imide ring [140] or interaction of isocyanates with the imide [149] (4.24).

(4.24)

Another mechanism suggested [140] was a hydrolytic attack on the imide group leading to the formation of an acid that decarboxylates with evolution of CO_2 (4.25).

(4.25)

Hydrolytic attack to $R-\overset{.}{C}=O$ radical resulting in evolution of CO_2 was also suggested [144].

The thermal oxidative decomposition of linear polyimides showed fast cross-linking on heating to 400°C in air [156]. Saponification treatment of the cross-linked residue indicated that most cross-links are likely to be different from the amide linkages. It is likely that the thermoset imide resins undergo similar transformations; however they are masked by the existing cross-links. Phosphorus-containing additives were shown [157] to accelerate the cross-linking on thermal oxidative decomposition of polyimides.

Various authors [158–160] studied kinetics of thermal or thermal oxidative decomposition of polyimides. It was found [159] in the aging experiments that the mechanism of thermal decomposition changes on advancement of thermal decomposition, therefore, various kinetic models have been suggested for use at different degrees of decomposition. Nevertheless, the kinetic approach was found [160] to be satisfactory for the prediction of the lifetime of the composites based on the polyimide resins.

For further improvement of the flame retardant properties, phosphorus-containing bisimides were prepared from bis(aminophenylmethyl)phosphine oxide and maleic anhydride [100, 150]. The imide resins can be further flame retarded by simultaneous incorporation in the molecule of phosphine oxide and bisphenylhexafluoropropane structures [161]. Triphenylenephosphine oxide or diphenylenemethylphosphine oxide structures were incorporated into the ethynyl-terminated imide resins by Alam and Varma [143]. It was also suggested [162] to use direct phosphorylation of the soluble imide resins by dichlorophenylphosphine. In all of the aforementioned cases, phosphorus-containing polyimides showed significant increase of char yield and as a result significant improvement of the flame retardancy. For example, it was shown [150] that graphite cloth laminates based on the bismaleimide resin with the diphenylenemethylphosphine oxide structures provide a LOI of 100.

References for Section 4.2.3

[1] P.W. Kopf, A.D. Little: Encyclopedia of Polymer Science and Engineering, Vol.11, John Wiley, New York, p. 45, 1988
[2] D. W. van Krevelen: *Polymer* 16 (1975) 615
[3] N.B. Sunshine in W.C. Kuryla, A.J. Papa (Eds.): Flame Retardancy of Polymeric Materials, Vol.2, Marcel Dekker, New York, p. 201, 1973
[4] R.T. Conley, J.F. Bieron: *J. Appl. Polym. Sci.* 7 (1963) 103
[5] R.T. Conley, J.F. Bieron: *J. Appl. Polym. Sci.* 7 (1963) 171
[6] W.M. Jackson, R.T. Conley: *J. Appl. Polym. Sci.* 8 (1964) 2163
[7] R.T. Conley: *J. Appl. Polym. Sci.* 9 (1965) 1117
[8] H.W. Lochte, E.L. Straus, R.T. Conley: *J. Appl. Polym. Sci.* 9 (1965) 2799
[9] L. Costa, L. Rossi di Montelera, G. Camino, E.D. Weil, E.M. Pearce: *Polym. Degrad. Stab.* 56 (1997) 23
[10] V. Jha, A.K. Banthia, A. Paul: *J. Thermal Anal.* 35 (1989) 1229
[11] M.A. Serio, S. Charpenay, R. Bassilakis and P.R. Solomon: *ACS Prepr. Div. Fuel. Chem.* 36 (1991) 66
[12] O. Chiantore, I. Novak, D. Berek: *Anal. Chem.* 60 (1988) 638
[13] C. Morterra, M.J.D. Low: *Carbon* 23(1985)525
[14] E.A. Sullivan: *J. Appl. Polym. Sci.* 42 (1991) 1815
[15] G. Camino, M.P. Luda de Cortemiglia, L. Costa, L. Trossarelli in B. Miller (Ed.): Thermal Analysis. Proc. 7th ICTA, Vol. 2, John Wiley, Chichester, p.1137, 1982
[16] I.V. Lampe, R. Kunze, D. Neubert, W. Gunther: *Polym. Degrad. Stab.* 50 (1995) 337
[17] Z. Lausevic, S. Marinkovic: *Carbon* 24 (1986) 575
[18] E.N. Peters, A.B. Furtek, D.I. Steinbert, D.T. Kwiatkowski: *J. Fire Retard. Chem.* 7 (1980) 69
[19] M.T. Huggard: *Polym. Eng.* (1993)(11)29
[20] M. Suzuki, N. Saiki: WO Patent Appl., (to Teijin Ltd.), WO 98 30,632-A1, 01.06.1998
[21] E.D. Weil, N.G. Patel: *Fire Mater.* 18 (1994) 1
[22] V. Muench, J. Hambrecht, A. Echte, K.H. Illers, J. Swoboda: US Pat. (to BASF), 4,632,946, 30.12.1986
[23] L. Costa, L. Rossi di Montelera, G. Camino, E.D. Weil, E.M. Pearce: *J. Appl. Polym. Sci.* 68 (1998) 1067
[24] M. Suzuki, H. Itoh, K. Sumi, Y. Kamoshida, S. Abe, S. Atomori, T. Furuama: US Pat. 5,605,962 (to Japan Synthetic Rubber), 25.02.1997
[25] J.H. Truen: US Pat. 5,290,836 (to DSM), 01.03.1994
[26] E.D. Weil, W. Zhu, N. Patel, S.M. Mukhopadhyay: *Polym. Degrad. Stab.* 54 (1996) 125
[27] S. Gandi, L. Delfosse: Canadian Pat. Appl. 2,039,013 (to Atochem), 27.09.1991
[28] K. Fuhr, F. Mueller, K.H. Ott: EP Pat. Appl. 458,137 (to Bayer), 27.11.1991
[29] K. Seki, T. Nakagawa: UK Pat. Appl. 2,054,610 (to Idemitsu Kosan), 30.06.1980
[30] N. Fishman, D.B. Parkinson: US Pat. 3,943,188 (to Stanford Research Institute), 09.03.1976
[31] J. Selley in: Encyclopedia of Polymer Science and Engineering, Vol. 12, Polyesters, John Wiley, New York, p.256, 1988
[32] A. Ram, A. Calahorra: *J. Appl. Polym. Sci.* 23 (1979) 797
[33] H.E. Stepniczka: *J. Fire Retard. Chem.* 3 (1976) 5
[34] E.D. Weil, H.K. Kim: in. Prog. VIII BCC Conf. On Recent. Adv. Flame Retard. Polym. Mater., Stamford, CT, 1997

[35] C.C. Luce, E.F. Humphrey, L.V. Guild, H.H. Norrish, J. Coul, W.W. Castor: *Anal. Chem.* 36 (1964) 483
[36] E. Braun, B.C. Levin: *Fire Mater.* 10 (1986) 107
[37] A.M. Calcraft, R.J.S. Green, T.S. McRoberts: *Plast. Polym.* (Oct. 1974) 200
[38] R.J.S. Green, J. Hume, S. Kumar: *Fire Mater.* 1 (1976) 36
[39] D.P. Miller, R.V. Petrella, A. Manca: *Modern. Plast.* (Sept. 1976) 95
[40] R.C. Stauffer, E.R. Larsen, R.V. Petrella, A. Manca, D.P. Miller: *J. Fire Ret. Chem.*, 3 (1976) 34
[41] E. Kicko-Walczak: *Fire Mater.* 22 (1998) 253
[42] D. Anderson, E.S. Freeman: *J. Appl. Polym. Sci.* 1 (1959) 192
[43] G.S. Learmonth, A. Nesbit: *Br. Polym. J.* 4 (1972) 317
[44] A.N. Das, S.K. Baijal: *J. Appl. Polym. Sci.* 27 (1982) 211
[45] G.A. Skinner in G. Pritchard (Ed.): Developments in Reinforced Plastics, Vol. 3, Elsevier, London, p.123, 1984
[46] R.K. Bansal, J. Mittal, P. Singh: *J. Appl. Polym. Sci.* 37 (1989) 1901
[47] N. Regnier and B. Mortaigne: *Polym. Degrad. Stab.* 49 (1995) 419
[48] N. Grassie, G. Scott: Polymer Degradation and Stabilisation, Cambridge University Press, Cambridge, p.25, 1985
[49] G.A. Skinner, P.J. Haines, T.J. Lever: *J. Appl. Polym. Sci.* 29 (1984) 763
[50] P. Budrugeac, E. Segal: *J. Thermal Anal.* 49 (1997) 183
[51] J.P. Agrawal, D.B. Sarwade, P.S. Makashir, R.R. Mahajan, P.S. Dendage: *Polym. Degrad. Stab.* 62 (1998) 9
[52] A. Baudry, J. Dufay, N. Regnier, B. Mortaigne: *Polym. Degrad. Stab.* 61 (1998) 441
[53] G.H. Irzl, C.T. Vijayakumar, J.K. Fink, K. Lederer: *Polym. Degrad. Stab.* 16 (1986) 53
[54] G.H. Irzl, C.T. Vijayakumar, J.K. Fink, K. Lederer: *Polym. Degrad. Stab.* 16 (1986) 73
[55] C.T. Vijayakumar, J.K. Fink: *J. Appl. Polym. Sci.* 27 (1982) 1629
[56] J.P. Agrawal, K.S. Kulkarni: *J. Appl. Polym. Sci.* 32 (1986) 5203
[57] M. Prins, G. Marom: M. Levy: *J. Appl. Polym. Sci.* 20 (1976) 2971
[58] G.S. Learmonth, A. Nesbit, D.G. Thwaite: *Br. Polym. J.* 1 (1969) 149
[59] S.K. Brauman: *J. Fire Retard. Chem.* 4 (1977) 18
[60] S.K. Brauman: *J. Fire Retard. Chem.* 4 (1977) 38
[61] A.N. Das, P.J. Hainnes, T.J. Lever, G.A. Skinner: *Fire Mater.* 7 (1983) 41
[62] L.V. McAdams, J.A. Gannon in: Encyclopedia of Polymer Science and Engineering, Vol. 6, John Wiley, New York, p.322, 1988
[63] C.S. Chen, B.J. Bulkin, E.M. Pearce: *J. Appl. Polym. Sci.* 27 (1982) 1177
[64] H.J. Yan, E.M. Pearce: *J. Appl. Polym. Sci.* 22 (1984) 3319
[65] F.J. Martin, K.R. Price: *J. Appl. Polym. Sci.* 12 (1968) 143
[66] S.V. Levchik, G. Camino, M.P. Luda, L. Costa, B. Costes, Y. Henry, E. Morel, G. Muller: *Polym. Adv. Technol.* 6 (1995) 53
[67] S.V. Levchik, G. Camino, M.P. Luda, L. Costa, B. Costes, Y. Henry, G. Muller, E. Morel: *Polym. Degrad. Stab.* 48 (1995) 359
[68] S.C. Lin, E.M. Pearce: *J. Polym. Sci., Polym. Chem.* 17 (1979) 3095
[69] C.S. Chen, B.J. Bulkin, E.M. Pearce: *J. Appl. Polym. Sci.* 27 (1982) 3289
[70] B. Arada, S.C. Lin, E.M. Pearce: *Int. J. Polym. Mat.* 7 (1979) 167
[71] T. Dyakonov, P.J. Mann, Y. Chen, W.T.K. Stevenson: *Polym. Degrad. Stab.* 54 (1996) 67
[72] L.-H. Lee: *J. Appl. Polym. Sci.* 9 (1965) 1981
[73] L.-H. Lee: *J. Polym. Sci., Part A* 3 (1965) 859
[74] M.A. Keenan, D.A. Smith: *J. Appl. Polym. Sci.* 11 (1967) 1009
[75] D.P. Bishop, D.A. Smith: *Ind. Eng. Chem.* (August 1967) 32
[76] J.C. Paterson-Jones, D.A. Smith: *J. Appl. Polym. Sci.* 12 (1968) 1601
[77] E.C. Leisegang, A.M. Stephen, J.C. Paterson-Jones: *J. Appl. Polym. Sci.* 14 (1970) 1961
[78] J.C. Paterson-Jones, V.A. Percy, R.G.F. Giles, A.M. Stephen: *J. Appl. Polym. Sci.* 17 (1973) 1867
[79] J.C. Paterson-Jones, V.A. Percy, R.G.F. Giles, A.M. Stephen: *J. Appl. Polym. Sci.* 17 (1973) 1877
[80] J.C. Paterson-Jones: *J. Appl. Polym. Sci.* 19 (1975) 1539
[81] I.D. Maxwell, R.A. Pethrick: *Polym. Degrad. Stab.* 5 (1983) 275
[82] N. Grassie, M.I. Guy, N.H. Tennent: *Poly. Degrad. Stab.* 14 (1986) 125
[83] D.P. Bishop, D.A. Smith: *J. Appl. Polym. Sci.* 14 (1970) 205

[84] M.B. Neiman, B.M. Kovarskaya, L.I. Golubenkova, A.S. Strizhkova, I.I. Levantovskaya and M.S. Akutin: *J. Polym. Sci.* 56 (1962) 383

[85] J.M. Stuart, D.A. Smith: *J. Appl. Polym. Sci.* 9 (1965) 3195

[86] M.F. Dante: *ACS Preprints, Div. Org. Coatings Plastics Chem.* 24 (1964) 135

[87] E. Cerceo: *Ind. Eng. Chem. Prod. Res. Develop.* 9 (1970) 96

[88] H.M. Le Huy, V. Bellenger, J. Verdu, M. Paris: *Polym. Degrad. Stab.* 35 (1992) 77

[89] H.M. Le Huy, V. Bellenger, M. Paris, J. Verdu: *Polym. Degrad. Stab.* 35 (1992) 171

[90] B.L. Burton: *J. Appl. Polym. Sci.* 47 (1993) 1821

[91] N. Rose, M. Le Bras, R. Delobel, B. Costes, Y. Henry: *Polym. Degrad. Stab.* 42 (1993) 307

[92] N. Rose, M. Le Bras, S. Bourbigot, R. Delobel, B. Costes: *Polym. Degrad. Stab.* 54 (1996) 355

[93] M. Le Bras, N. Rose, S. Bourbigot, Y. Henry, R. Delobel: *J. Fire Sci.* 14 (1996) 199

[94] G. Sanchez, Z. Brito, V. Mujica, G. Perdomo: *Polym. Degrad. Stab.* 40 (1993) 109

[95] I.K. Varma and U. Gupta: *J. Macromol. Sci., Chem.* A23 (1986) 19

[96] W.-K. Chin, M.-D. Shau, W.-C. Tsai: *J. Polym. Sci., Polym. Chem.* 33 (1995) 373

[97] S.V. Levchik, G. Camino, M.P. Luda, L. Costa, G. Muller, B. Costes, Y. Henry: *Polym. Adv. Technol.* 7 (1996) 823

[98] S.V. Levchik, G. Camino, L. Costa, M.P. Luda: *Polym. Degrad. Stab.* 54 (1996) 317

[99] S.V. Levchik, G. Camino, M.P. Luda, L. Costa, G. Muller, B. Costes: *Polym. Degrad. Stab.* 60 (1998) 169

[100] H. Zhuang, B. Tan, C. Tchatchoua, Q. Ji, H. Ghassemi, J.E. McGrath: *ACS Polym. Repr. Div. Polym. Chem.* 38 (1997) 304

[101] Y.-L. Liu, G.-H. Hsiue, R.-H. Lee, Y.-S. Chiu: *J. Polym. Sci., Polym. Chem.*, 35 (1997) 895

[102] Y.N. Liu, Q. Ji, E. McGrath: *Polymer. Prepr.* 38 (1997(1)) 223

[103] T.-S. Wang, J.-F. Yeh, M.-D. Shau: *J. Appl. Polym. Sci.* 59 (1996) 215

[104] C.S. Cho, S.-C. Fu, L.-W. Chen, T.-R. Wu: *Polym. Intern.* 47 (1998) 203

[105] W.R. Creasy: *Polymer* 33 (1992) 4486

[106] I.H. Updegraff in: Encyclopedia of Polymer Science and Engineering, Vol. 1, John Wiley, New York, p.752, 1988

[107] J.E. Hauck: *Mater. Design. Eng.* 60 (1964) 83

[108] C.F. Cullis, M.M. Hirschler in: The Combustion of Organic Polymers, Clarendon Press, Oxford, 1981

[109] G. Inverarty, G.P. Twiss: US Pat. 5,100,936 (to BIP Chemical Ltd.), 31.03.1992

[110] J.B. Williams: US Pat. 4,209,427 (to General Electric), 24.06.1980

[111] G. Albanesi, G. Rinaldi: US Pat. 4,440,880 (to Saffa S.p.A.), 03.04.1984

[112] I. Sakon, M. Sekiguchi, A. Kanayama: US Pat. 5,041,490 (to Rinkagaku Kogyo), 20.08.1991

[113] S.J. O'Brien: US Pat. 3,816,212 (to American Cyanamid), 11.06.1974

[114] J.M. Thomson: US Pat. 3,859,124 (to Proctor Chemical Co.), 07.01.1975

[115] M.W. Duke, R.S. Gregorian: US Pat. 3,936,562 (to United Merchants and Manufacturers), 03.02.1976

[116] M.W. Duke: US Pat. 4,028,052 (to United Merchants and Manufacturers), 07.06.1977

[117] D.A. LeBlanc, R.B. LeBlanc: US Pat. 4,020,262 (to LeBlanc Research Co.), 26.04.1977

[118] D.A. LeBlanc, R.B. LeBlanc: US Pat. 4,148,602 (to LeBlanc Research Co.), 10.04.1979

[119] R.G. Weyker, W.F. Baitinger Jr.: US Pat. 4,026,711 (to American Cyanamid Co.), 31.05.1977

[120] R.J. Berni, M.M. Smith, R.R. Benerito: US Pat. 4,084,027 (to Secretary of Agriculture of USA), 11.04.1978.

[121] K. Umetani, M. Date: US Pat. 4,095,945 (to Toyo Boseki Kabushiki Kaisha), 20.06.1978

[122] E.N. Walsh, T.A. Hardy: US Pat. 4,292,036 (to Stauffer), 29.09.1981

[123] S. Caruso, S. Foti, P. Maravigna, G. Montaudo: *J. Polym. Sci., Polym. Chem.* 20 (1982) 1685

[124] G. Montaudo, E. Scamporrino, D. Vitalini: *J. Polym. Sci., Polym. Chem.* 21 (1983) 3321

[125] G. Camino, L. Operti, L. Costa, L. Trossarelli in B. Miller (Ed.): Thermal Analysis. Proc. 7th ICTA, Vol. 2, John Wiley, Chichester, p.1144, 1982

[126] G. Camino, L. Operti, L. Trossarelli: *Polym. Degrad. Stab.* 5 (1983) 161

[127] G. Camino in: Proc. VIIth Conf. Recent Adv. Flame Retardancy Polym. Mater. Stamford, CT, p. 1, 1996

[128] G. Camino, M.P. Luda, L. Costa, M. Guaita: *Macromol. Chem. Phys.* 197 (1996) 41

[129] R.T. Conley (Ed.) in: Thermal Stability of Polymers, Vol. 1, Marcel Dekker, New York, p.457, 1970

[130] J.W. Verbicky in: Encyclopedia of Polymer Science and Engineering, Vol. 12, John Wiley, New York, p.364, 1988
[131] Yu.N. Sazanov: *J. Therm. Anal.* 34 (1988) 1117
[132] Yu.N. Sazanov: L.A. Shibaev: *Acta Polym.* 31 (1988) 1
[133] I.K. Varma, S. Sharma: *Indian J. Technol.*, 25 (1987) 136
[134] A.V. Gribanov, L.A. Shibaev, A.I. Koltsov, R.E. Teeiaer, Yu.N. Sazanov, N.G. Stepanov, E.P. Lippmaa, T. Szekely: *J. Appl. Polym. Sci.* 32 (1987) 815
[135] Yu.N. Sazanov, N.G. Stepanov, L.A. Shibaev, M.I. Tsapovetsky, T.A. Antonova, A.V. Gribanov, T. Szekely, I. Bertoti, A. Toth, M. Blazso, E. Jakab: *Acta Polym.* 39 (1988) 422
[136] Yu.N. Sazanov, N.G. Stepanov, L.A. Shibaev, T.A. Antonova, A.V. Gribanov, I. Bertoti, A. Toth, M. Blazso, E. Jakab, T. Szekely: *Acta Polym.* 39 (1988) 422
[137] Yu.N. Sazanov, T. Szekely, A.V. Gribanov, I. Bertoti, T.A. Antonova, A. Toth: *Acta Polym.* 39 (1988) 516
[138] T. Takeichi, H. Takenoshita, S. Ogura, M. Inagaki: *J. Appl. Polym. Sci.* 54 (1994) 361
[139] D.A. Kourtides: *J. Thermoplast. Compos. Mater.* 1 (1988) 12
[140] W.W. Wright in N. Grassie (Ed.): Developments in Polymer Degradation, Vol. 3, Appl. Sci. Publishers, London, p.1, 1981
[141] I.K. Varma, G.M. Fohlen: *J. Polym. Sci., Polym. Chem.*, 20 (1982) 283
[142] D. Kumar, J. Kaur, D. Gupta: *J. Appl. Polym. Sci.*, 48 (1993) 453
[143] S. Alam, I.K. Varma: *Angew. Makromol. Chem.*, 236 (1996) 55
[144] E. Jakab, F. Till, T. Szekely, S.S. Kozhabekov, B.A. Zhubanov: *J. Anal. Appl. Pyrol.* 23 (1992)229.
[145] G.T. Kwiatkowski, L.M. Roberson, G.L. Brode, A.W. Bedwin: *J. Polym. Sci. Polym. Chem.* 13 (1975) 961
[146] B. Crossland, G.J. Knight, W.W. Wright: *Brit. Polym. J.* 19 (1987) 291
[147] I.K. Varma, S.P. Gupta, D.S. Varma: *Thermochim. Acta* 93 (1985) 217
[148] J.L. Cotter, G.J. Knight, W.W. Wright in: Therm. Anal. Proc. 4th ICTA, Budapest, Vol. 2, p. 163, 1974
[149] L.A. Oksant'evich, M.M. Badaeva, G.I. Tuleninova, A.I. Pravednikova: *Vysokomol. Soedin.* 19 A (1977) 553
[150] I.K. Varma, G.M. Fohlen, M.-T. Hsu, J.A. Parker: *Contemp. Topics Polym. Sci.* 4 (1984) 115
[151] I.E. Simanovich, T.I. Zhukova, G.N. Fedorova, A.Yu. Elkin, I.V. Kalinina, N.V. Mikhailova, Yu.N. Sazanov, A.V. Gribanov, M.M. Koton: *Vysokomol. Soedin.* 31 A (1989) 966
[152] H. Hatori, Y. Yamada, M. Shiraishi, M. Yoshihara, T. Kimura: *Carbon* 34 (1996) 201
[153] R. Torrecillas, N. Regnier, B. Mortaigne: *Polym. Degrad. Stab.* 51 (1996) 307
[154] O.Ya. Fedotova, V.I. Gorokhov, O.I. Paresishvili, G.S. Karetnikov, G.S. Kplesnikov: *Vysokomol. Soedin.* 14 A (1972) 1256
[155] J. Zubakowska-Orszagh, T. Chreptowicz, J. Kaniski: *Europ. Polym. J.* 15 (1979) 409
[156] R.A. Dine-Hart, D.B.V. Parker, W.W. Wright: *Br. Polym. J.* 3 (1971) 222
[157] V.V. Guryanov, N.G. Annenkova, T.N. Novotortseva, A.B. Blyumerfield, V.A. Baranova, F.Sh. Malyukova, L.A. Shesternina, B.M. Kovarskaya: *Vysokomol. Soedin.* 20 A (1978) 207
[158] J.M. Mazon-Arechederra, J.M. Barrales-Rienda: *Polym. Degrad. Stab.* 15 (1986) 357
[159] B.L. Stump, W.J. Snyder: *High Perform. Polym.* 3 (1989) 247
[160] I. Salin, J.C. Seferis, C.L. Loechelt, R. Rothschilds in: Proc. 37th Intern. SAMPE Symposium, Vol. 37, p. 1365, March 1992
[161] I.K. Varma, G.M. Fohlen, J.A. Parker: *J. Polym. Sci., Polym. Chem.* 21 (1983) 2017
[162] Y.-L. Liu, G.-H. Hsiue, C.W. Lan, J.-K. Kuo, R.-J. Jeng, Y.-S. Chiu: *J. Appl. Polym. Sci.*, 63 (1997) 875

4.2.4 High Temperature Resistant Plastics

M. ROCHERY AND M. LEWANDOWSKI

High temperature resistant plastics designate polymers that are suitable for use above a temperature of 150 °C. Their development started at the end of the 1950s with the event of space travel, and continued with military research. For example, several polymers among the so-called rigid-rod polymers – a relatively new class of heat resistant polymeric material – have been developed by the US Air Force Ordered Polymer Program (polyoxazole or polythiadiazole).

These polymers are relatively difficult to manufacture and process, leading to high production costs. Their use is hence limited to specialized applications only – for example, in high-performance protective apparel such as firemen's turnout coats and astronaut space suits (aramid and polybenzimidazole fibers), or in the aerospace and electronic industries (such as polyimides, polyaryl ethers).

Thermal properties of these polymers can be discussed in terms of the following: What is their maximum use temperature? Do they melt or drip? Are the degradation products toxic? What is their reaction to flame? Depending on the final usage, some of these characteristics will be more or less relevant. It is, for instance, well known that during a fire incident, smoke obscuration can lead to a large panic effect, and it will hence be preferable to use low-smoking materials in this type of application. Another example is the generation of toxic byproducts during combustion, which represents a serious impediment for the use of some of these polymers in building.

The temperature resistance of linear polymers increases with rigidity, and the incorporation of aromatic or heterocyclic rings is one way to achieve this stiffening. Aromatic condensation polymers also generate less smoke, while aliphatic-aromatic polymers are less flammable and produce even less smoke. The limiting oxygen index (LOI) values of aromatic polymers confirm this property: they are higher than those obtained with aliphatic polymers such as polyamide-66, polyethylene, or poly(ethylene terephthalate).

In the following sections, the high temperature resistant organic and inorganic polymers are presented. Among the organic polymers, the best heat properties are obtained with aromatic and heterocyclic polymers and are discussed below.

4.2.4.1 Organic Polymers

Heat Resistant Organic Plastics

Table 4.3 presents the thermal properties of the main heat-resistant organic plastics. The tendency to char increases with the degree of linking in these polymers: the formation of gaseous decomposition products and the tendency to burn with an open flame diminish in consequence. Polyimides (PI), for example, cannot be ignited and incandesce in the presence of a flame. They represent an important class of heterocyclic polymers and are used in the most diverse applications. They are, for example, widely used in components in the electrical and electronic industries.

Among these selected polymers, polyphenylene ether (PPE), usually modified with high-impact polystyrene (HIPS), is commercially the most important of all high temperature resistant plastics.

Polyaramids are widely used in the form of fibers. Well-known commercial names are Kevlar® or Technora®, which are p-aramids (PPTA), or Nomex®, which is an *m*-aramid. Examples of heat-resistant applications are fire protective clothing and textiles in the aviation industry.

Further details and developments of these plastics and of the whole field of aromatic and heterocyclic polymers are given in a series of comprehensive monographs and articles [1–8], while Yang [9] reports on the properties of aramids, Kevlar fibers in particular.

A special mention can be made of polymeric carbon, which does not appear in the table. Flame-resistant carbon fibers are partially carbonized fibers, which give the following transformation when submitted to very high temperatures (4.26).

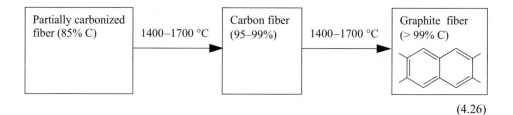

$$(4.26)$$

Carbon fibers are extremely resistant to high temperatures: their melting temperature is 4000 °C. They can also be considered as flame resistant, as they will burn only at very high temperatures. They hence represent a choice material for applications at extremely high temperatures, for example, in the filtration of molten iron (see [10], p. 365). They have a bulk density of 1.8 g/cm^3.

Polyketones represent another interesting group of polymers that possess useful high-temperature properties. Polyketones are found in the form of cables and composites in specialist applications in the chemical, electronics, and nuclear industries.

Poly(ether etherketone) (PEEK, bulk density of 1.4 g/cm^3) was the first polyketone synthesized in 1979, and other related materials have also been developed, such as poly(ether ketone) (PEK), poly(ether ether ketoneketone) (PEEKK), or poly(ether diphenyletherketone) (PEDEK), and that are based upon ether (E), ketone (K), or diphenyl (D) links between the aromatic rings (4.27).

$$(4.27)$$

PEEK

The melting temperature ranges from 330 °C (PEEK) to 400 °C (for PEDEK). The presence of aromatic rings (diphenyl groups) leads to an increase in melting temperature, but may, however, represent a setback in processability.

Other organic polymers can be found in high temperature resistant applications, such as phenol-formaldehyde and melamine-formaldehyde compounds.

Phenol-formaldehyde can be obtained by two reactions: the first one involves the presence of a basic catalyst and an excess of formaldehyde, giving rise to a product called resol, while the second one utilizes an acid catalyst in the presence of excess phenol, giving an initial product called novolac. This second category of phenolic resins presents good thermal properties: flame retarded fibers based on cross-linked novolacs (trade name Kynol®, for example) are used for thermal insulation and protective clothing [11].

Table 4.3 Thermal and flame resistance of the main aromatic polymers used in high temperature applications

Polymer	Thermal resistance	Reaction to flame	Thermal decomposition
Polyphenylene ether (PPE) Bulk density = 1.06 g/cm^3	Good resistance up to 110 °C, and even 190 °C in the absence of air.	Ignites with difficulty. Burns with a bright, sooty flame. Melts and chars. Continues to burn upon removal of flame.	Occurs at around 200 °C in air. Slight phenol smell. Decomposition products consist of acetone, phenol and higher cross linked structures.
Polyphenylene sulfide (PPS) Bulk density = 1.36 g/cm^3	Resistant up to 260 °C. Melts in the range 270–285 °C.	Burns with a luminous orange-yellow flame and gray smoke, forming a black carbonized residue. Extinguishes immediately when flame is removed.	Up to 500 °C, no volatile products formed. Above 700 °C, degradation to hydrogen, methane, and carbon oxides.
Polyether sulfones (PES) or polysulphones (PSO) Bulk density = 1.24–1.37 g/cm^3 PES PSO	Good resistance to temperature. Can be exposed for long periods to 160–250 °C.	Difficult to ignite. Burns with a yellow, sooty flame, even after removal of ignition source. Leaves a black residue after melting.	Slight smell of hydrogen sulfide. Above 380 °C some polysulfones give rise to sulfur dioxide, methane, carbon oxides, hydrogen, and phenol or its derivatives.
Polyimides (PI) Bulk density = 1.42 g/cm^3	Resistant up to 260–320 °C, and to 500 °C for short periods.	Cannot be ignited by an external flame, does not melt but incandesces and chars, almost without smoke.	Slight phenol smell. Water initially eliminated during thermal decomposition. At higher temperatures, carbon oxides,; ammonia; small amounts of hydrogen; aromatics such as aniline, phenol, benzene, and higher condensed products formed.
Polyaramids (PPTA) Bulk density = 1.38–147 g/cm^3	Excellent thermal stability. Resistant up to 250 °C. Melting temperature around 550 °C.	Flame-resistant. Does not melt, but glows during ignition. No after burning after removal of flame. Chars above 450 °C.	Produces few toxic gases such as carbon monoxide. Between 300 and 360 °C, produces carbon dioxide and water. Above 500 °C, pyrolysis produces benzonitrile, aniline, and benzanilide and other byproducts.

Melamine is an aromatic, heterocyclic compound that undergoes condensation reactions with formaldehyde to give rise to melamine-formaldehyde resins. These melamine resins find their largest use in the manufacture of decorative dinnerware, or as laminated counter and table tops (trade name Formica®). A modified melamine-formaldehyde resin can be found under the commercial name of Basofil® (BASF) and is used as a flame-retardant reinforcing fiber.

Rigid-Rod Polymers

Rigid-rod polymers represent the upcoming class of heat resistant materials and are thus reviewed in detail in this section. Some of these ladder polymers were first developed by the US Air Force researchers who were looking for super heat-resistant polymers that would surpass the traditional aramid fibers.

The most important polymers are polybenzimidazoles (PBI), polyoxazoles (PBO), and poly-benzothiazoles (PBZT), which are distinguished by high temperature resistance and low tendency to burn.

These polymers – also known as polybenzazoles – have been extensively investigated because they present outstanding mechanical properties as well as excellent heat and flame resistance. The recent literature [12–15] reports these properties, sometimes in comparison between them or with other high performance polymers such as polyaramids.

The development of these rigid-rod polymers has recently led to a novel polymer called poly(2,6-diimidazol [4,5-b:4′,5′-e]pyridinylene-1,4(2,5-dihydroxy)phenylene) or "PIPD," also known as "M5" during the development stage. The design of this polymer combines the high stiffness and tenacity of the rigid-rod polymer family with extensive possibilities to form hydrogen bonds (see structure (4.28)):

(4.28)

PIPD

Recent work on the synthesis and properties of this polymer can be found in the literature [16–18]. Klop and Lammers [17, 18] have hightlighted the exceptionally good compression performance of heat-treated PIPD fibre: this is due the monoclinic crystal structure with its bi-directional hydrogen bonding network. PIPD flame resistance properties are currently under investigation. However, good performance in this field can be expected, with regard to the excellent heat properties of the previous polymers of the same class. The LOI value measured on PIPD exceeds 50 % [16] (see Table 4.4 for comparison with other polymers).

Table 4.4   LOI values of some heat resistant polymers

Polymer	PPTA	PBI	PBO	PIPD
LOI (vol%)	29	41	68	> 50

The combustion characteristics of the three main rigid-rod polymers are described in more detail below.

Polybenzimidazoles (PBI)

$$(4.29)$$

PBI

The bulk density of PBI is $1.46 g/cm^3$. PBIs are particularly stable thermally: their continuous temperature resistance in air lies in the range 300–350 °C, and in nitrogen, increases to 480–500 °C.

PBI cannot be ignited by a flame but chars and incandesces. Pyrolysis products consist of hydrogen, hydrogen cyanide, ammonia, water, methane, nitrogen, and traces of aromatics such as benzene, aniline, and phenol, while 70–80 % by weight of the polymer remains as a carbonaceous residue.

Polybenzothiazoles (PBZT)

$$(4.30)$$

PBZT

Polybenzothiazoles represent the first class of heat resistant polymers developed by the US Air Force. They have a bulk density of $1.57 g/cm^3$. Their maximum use temperature is 350 °C.

Polyoxazoles (PBO)

$$(4.31)$$

PBO

Poly(p-phenylene-2,6-benzobisoxazole) or PBO is a rigid-rod isotropic crystal polymer, which was developed after PBZT: PBO was advantageous in performance and cost, and became the main center of investigation as it showed the most attractive properties and the greatest economic potential among the polybenzazoles. After much research in production technology and fiber spinning, PBO is now commercialized by Toyobo (Japan) under the trademark name of Zylon®.

The bulk density of PBO is $1.57 g/cm^3$. It contains an aromatic heterocyclic ring as shown above.

PBO presents good flame resistance and thermal stability, particularly in comparison with p-aramid fibers [13,14]. Degradation occurs at about 600 °C in air and at 700 °C in nitrogen. Combustion gases consist of carbon oxides and water, with smaller amounts of toxic gases such as hydrogen cyanide, sulfur or nitrogen oxides compared to p-aramid fibers [14]. PBO also has a much higher LOI value compared to other heat resistant polymers (Table 4.4).

With regard to these exceptional thermal properties, PBO is, for example, expected to replace asbestos, which is still used as a heat resistant cushion material for glass and metal manufacturing processes.

Some interesting comparison work on the thermal properties of PBO and traditional *p*-aramid (PPTA) fibers are reported in the rest of this section. They represent two important high performance polymers because they have outstanding mechanical properties as well as being high temperature resistant. They hence cover a wide range of applications, especially where both thermal stability and good mechanical performance are required

Kuroki et al. [14] have studied the high temperature properties of these two fibers, in particular the temperature dependence of tensile strength and modulus. They found that even at 400 °C, PBO fibers retained 70 % of their modulus at room temperature. In fact, the modulus started to decrease gradually from around 300 °C. In the case of the aramid fiber, the modulus decreased to 70 % of that of room temperature at 300 °C, that is, at a much lower temperature than PBO.

The thermal properties of PBO and PPTA fibers have also been recently compared in cone-calorimetry experiments, while their thermal stability under pyrolytic or thermooxidative conditions was evaluated through TG/DSC tests [13]. These results are reported below.

The cone calorimeter is used as a fire model [19] and measures the rate of heat release (RHR). These tests were performed on PPTA (Kevlar and Technora) and PBO-based knitted structures according to the standard test method ASTM 1354–90a (for measuring heat and visible smoke release for materials and products using an oxygen depletion calorimeter).

Figure 4.16 shows the RHR curves under two different external heat fluxes (50–100 kW/m²): an external heat flux of 50 kW/m² corresponds to a mild fire scenario, whereas a more important heat flux (100 kW/m²) corresponds to a post-flashover [20]. These results confirm that the fire behavior of PBO fibers is better than that of PPTA fibers. The RHR peaks under 50 kW/m² are 60 kW/m² for PBO and 400 kW/m² for PPTA, illustrating the higher fire resistance of PBO. Similar results are obtained under an external heat flux of 100 kW/m².

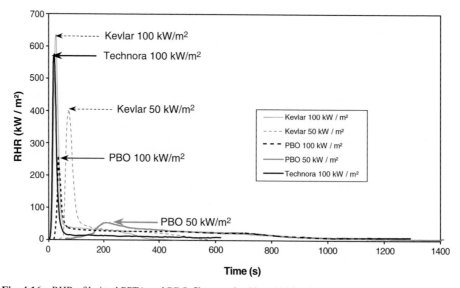

Fig. 4.16   RHR of knitted PPTA and PBO fibers under 50 or 100 kW/m² external heat fluxes

The TG curves obtained with a heating rate of 10 °C/min are presented in Fig. 4.17: they show that PBO fibers have a better heat resistance than PPTA fibers. Whatever the atmosphere, the degradation occurs in one step for both polymers, but at a higher temperature for PBO. For PBO, thermal degradation starts at around 600 °C in air, while for PPTA, degradation starts at around 450 °C. A 3 wt% residue is obtained in the end with both fibers at 1200 °C. A similar behavior is obtained in a nitrogen atmosphere (degradation temperature of approximately 700 °C for PBO and 550 °C for PPTA) but the residues left are 65 wt% for PBO and 38 wt% for PPTA. These results emphasize the importance of the presence of oxygen during thermal degradation.

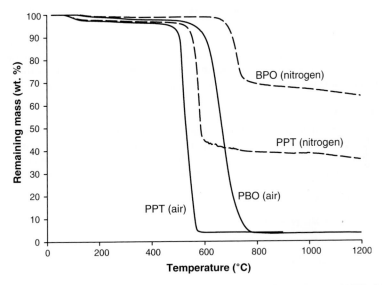

Fig. 4.17   TG curves of PPTA and PBO fibers under nitrogen and air (heating rate = 10 °C/min)

4.2.4.2 Inorganic Polymers

Although the major studies in the field of heat-resistant materials have focused on organic polymers, a considerable volume of work dealing with inorganic and partially inorganic polymers can nevertheless be found [21–23]. Silicones and polyphosphazenes are the inorganic polymers with elastomeric and thermoplastic properties close to organic ones. A detailed description of silicone and polyphosphazene rubbers can be found in Section 4.2.6.

Silicones

Silicones cover the more extended period: they were discovered in the 1930s and now offer an alternative to "organics" in a wide range of novel materials in industry sectors such as cosmetics, electronics, aerospace, and automotive. The dominant polymer in the silicone industry is polydimethylsiloxane (PDMS), which consists of a distribution of trimethylsiloxy end-blocked species:

$$H_3C-\underset{\underset{CH_3}{|}}{\overset{\overset{CH_3}{|}}{Si}}-O-\left[\underset{\underset{CH_3}{|}}{\overset{\overset{CH_3}{|}}{Si}}-O\right]_n\underset{\underset{CH_3}{|}}{\overset{\overset{CH_3}{|}}{Si}}-CH_3 \qquad (4.32)$$

PDMS

They are recognized for their unique fire properties: the combustion of long-chain PDMS exhibits a low heat release rate with the unique property that this heat release rate does not increase significantly when the external applied thermal radiant flux increases [24]. This is in contrast to most hydrocarbon materials and is intrinsic to dimethyl-substituted siloxanes. The lower burning rate is partially explained by the accumulation of a silica ash layer at the silicone fuel surface [25]. The gasification of silicone fluids under external thermal radiation in a nitrogen atmosphere results in identical combustion products – SiO_2, CO_2, and H_2O. It occurs via two modes: (1) the volatilization of molecular species native to the polymer, and (2) the volatilization of molecules resulting from thermally induced degradation of the polymer via siloxane rearrangement. The former process is dominant for short-chain oligomers ($\eta < 10$ cS) and has to be limited whereas the latter one is dominant in all higher molecular weight polymers ($\eta > 100$ cS) [26].

Modification of polysiloxanes is a fruitful area of research. The development of graftable and polymerizable polyhedral oligomeric silesquioxane (POSS) are examples of new polymer building blocks which promise the preparation of copolymers with excellent fire properties: loss of the silesquioxane "cage" structure occurs on heating from 450 to 650 °C [27].

Polyphosphazenes

Polyphosphazenes or phosphonitrilic polymers are among the most interesting inorganic-type polymers developed in recent years [28,29]. They are notably used as oil-resistant hoses or gaskets. The best known are the polyaryloxyphosphazenes, which form amorphous, clear, colorless films with elastomeric properties despite the alternation of single and double bonds.

$$*\left[\begin{array}{c} OR \\ | \\ P=N \\ | \\ OR \end{array}\right]_n *$$ (4.33)

Polyphosphazene

Polyphosphazenes are thermally stable up to approx. 200 °C and even up to 300 °C for fluoro-alkoxy-substituted polymers. They are useful as flame-retardant materials, not only because they are either difficult to ignite and self-extinguishing but because they burn with only moderate evolution of smoke. They have limited flammability due to the phosphorus and synergistic nitrogen components. With polyaryloxyphosphazenes, the aromatic side chain groups cause softening, dense smoke evolution in some cases, and charring. Aluminum trihydrate filled formulations have a more favorable burning behavior.

Miscellaneous

There are several kinds of inorganic fibers, which are sometimes called ceramic fibers on account of their high temperature resistance. Ceramic fiber is defined as a polycrystalline refractory fiber composed of metal oxide, metal carbide, metal nitride or other mixtures. In this context, silicon and boron are regarded as metals. Most of them are heat resistant up to 1200 °C. They are expensive and are generally used to reinforce composite structures, or to save weight for aircraft, aerospace, and automotive parts. Discussions on the formation, classification, and novel applications of the various types of ceramic fibers are available in [30–32].

References for Section 4.2.4

[1] H. Lee, D. Stoffey, K. Neville: New Linear Polymers. McGraw-Hill, New-York, 1967
[2] A.H. Frazer: High Temperature Resistant Polymers. Wiley-Interscience, New York, 1968
[3] V.V. Korshak : Heat-Resistant Polymers. Israel Program for Scientific Translation, Jerusalem, 1971
[4] R.D. Deanin (Ed.): New Industrial Polymers. ACS Symposium Series 4, Washington, D.C., 1972
[5] D.A. Kourtides: *Polym.-Plast. Technol. Eng.* 11 (1978) p. 159
[6] H.G. Elias: *Kunststoffe* 70 (1980), p. 699
[7] M. Itoh, M. Mitsuzuka, K. Iwata, K. Inoue: *Macromolecules* 27 (1994) 7917–7919
[8] G.W. Meyer, S.J. Pak, Y.J. Lee, J.E. Mc Grath: *Polymer* 36(11) (1995) 2303–2309
[9] H.H. Yang (Ed.): Kevlar Aramid fiber. John Wiley, Chichester, 1993
[10] H. Saechtling: International Plastics Handbook. 2nd edit., Carl Hanser, Munich, Vienna, New York, 1987
[11] J. Economy, L.C. Wohrer, F.J. Frechette, G.Y. Lei: *Appl. Polym. Symp.* 21 (1973) 81
[12] T. Kitagawa, H. Murase, K. Yabuk: *J. Polym. Sci. Part B : Polym. Phys.* 36 (1998) 39–48
[13] X. Flambard, S. Bourbigot, S. Duquesne, F. Poutch: *Polym. Int.* 50 (2001) 157–164
[14] T. Kuroki, Y. Tanaka, T. Hokudoh, K. Yabuki: *J. Appl. Polym. Sci.* 65 (1997) 1031–1036
[15] S. Yukoyame, T. Ueda, M. Watanabe, K. Sanui, N. Ogata: *New Polym. Mater.* 2(1) (1990) 67–74
[16] D.J. Sikkema: *Polymer* 39(24) (1998) 5981–5996
[17] E.A. Klop, M. Lammers: *Polymer* 39(24) (1998) 5987–5998
[18] M. Lammers, E.A. Klop, M.G. Northolt: *Polymer* 39(24) (1998) 5999–6005
[19] V. Babrauskas: Development of Cone Calorimeter – A bench scale rate of heat release based on oxygen consumption, NBS-IR 82–2611, US Nat. Bur. Stand., Gaithersburg, 1982
[20] V. Babrauskas: *Fire Mater.* 19 (1995) 243
[21] J. Mark, H. R. Allcock, R. West: Inorganic Polymers. Prentice-Hall, Englewood Cliffs, NJ, 1992
[22] N.H. Ray: Inorganic Polymers. Academic Press, New York, 1978
[23] I. Manners: *Angew. Chem., Int. Ed. English* 35 (1996) 1602
[24] R.R Buch: *Fire Safety J.* 71 (1991), p. 1
[25] M. Kanakia: Characterization of transformer fluid pool fires by heat release rate calorimetry. Southwest Research Institute, Project No. 03-5344-001, Presented at the 4th Int. Conf. Fire safety, Univ. of San Francisco (1979)
[26] P.J. Austin, R.R. Buch, T. Kashiwagi: *Fire Mater.* 22 (1998) 221–252
[27] R.A. Mantz, P.F. Jones, K.P. Chaffee, J.D. Lichtenhan, J.W. Gilman, I.M.K. Ismail, M.J. Burmeister: *Chem. Mater.* 8 (1996) 1250–1259
[28] H.R. Allcock: Phosphorus-Nitrogen Compounds, Academic Press, New York, 1972
[29] H.R. Allcock: *Angew. Chem., Int. Ed. English* 16 (1977) 147
[30] N. Anderson: America's Textiles Inst. 16 (1987) 110
[31] Advanced Composites Bull., Dow Corning. (1991), p. 2
[32] T. Yogo, H. Iwahara: *J. Mater. Sci.* 27(6) (1992) 1499

4.2.5 Geopolymers

J. DAVIDOVITS

Geopolymers result from geosynthesis and technologies derived from geochemistry [1]. For the chemical designation of geopolymers based on silico-aluminates, the term poly(sialate) was suggested. Sialate is an abbreviation for silicon-oxo-aluminate. The sialate network consists of SiO_4 and AlO_4 tetrahedra linked alternately by sharing all the oxygens. Positive ions (Na^+, K^+, Ca^{++}, H_3O^+) must be present in the framework cavities to balance the negative charge of Al^{3+} in IV-fold coordination. The sialate link or sialate bridge designates the bridge Si-O-Al between two polysialate or polysilicate chains. It is the cross-linking or networking element [2–4] (4.34).

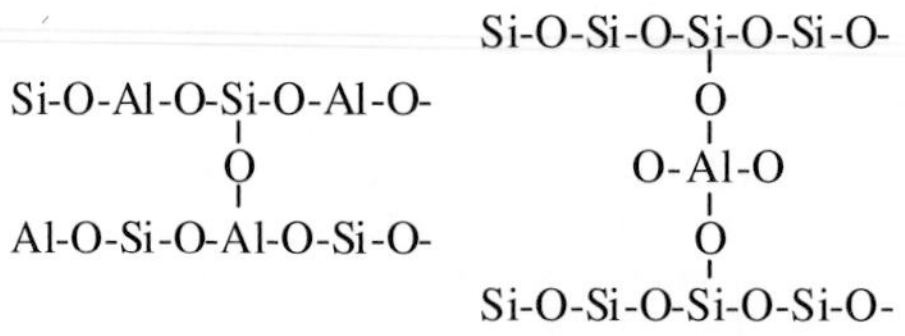

$$
\begin{array}{c}
\text{Si-O-Si-O-Si-O-Si-O-}\\
\text{Si-O-Al-O-Si-O-Al-O-} \qquad\qquad \text{O}\\
\text{O} \qquad\qquad\qquad\qquad \text{O-Al-O} \\
\text{Al-O-Si-O-Al-O-Si-O-} \qquad\qquad \text{O}\\
\text{Si-O-Si-O-Si-O-Si-O-}
\end{array}
\qquad (4.34)
$$

the sialate link

In polysialate empirical formula $M_n\{-(SiO_2)_z-AlO_2\}_n$, $_wH_2O$, M is a cation such as potassium, sodium or calcium, and n is a degree of polycondensation; z is 1, 2, 3 or higher, up to 32.

Several poly(sialate)-nanopolymers are being evaluated for fireproof aircraft cabin interior panels, marine structural composites, and infrastructure applications. One geopolymer of this type is a potassium poly(sialate)-nanopolymer, with the polymeric structure shown in 4.35

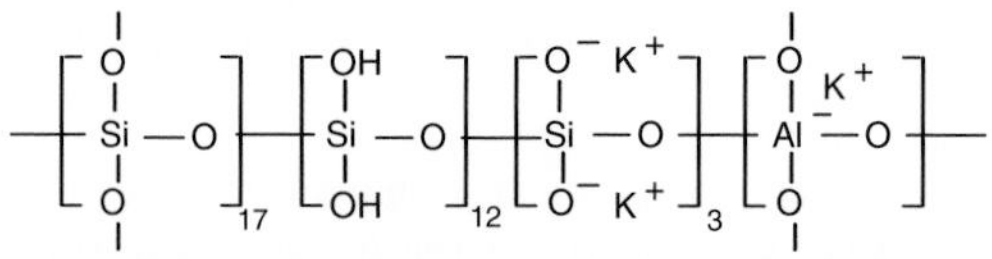

$$
\qquad (4.35)
$$

poly(sialate)-nanopolymer structure

This particular resin hardens to an amorphous or glassy material at moderate temperatures with a density of $2.14\,g/cm^3$ and is one of a family of geopolymeric poly(sialate)-nanopolymers with the ratio Si:Al ranging from 10 to 35. It is used in the making of poly(sialate)/carbon composite cured in a vacuum bag at 80 °C.

Table 4.5 lists the inplane shear, interlaminar shear, warp tensile, and flexural properties of the poly(sialate)-carbon fiber/fabric cross-ply laminates. The room temperature strengths of the poly(sialate)-carbon fiber/fabric laminates are 343, 245, and 14 MPa for warp tensile, flexure, and interlaminar shear, respectively. The corresponding values for a phenolic resin-T-300 carbon fabric 0–90 ° cross-ply laminate are 436 and 290 MPa for warp tensile and flexural strength, respectively, and 24 MPa for interlaminar (short beam) shear strength. Moduli for the poly(sialate) resin cross-ply fabric laminate in the warp tensile and flexure tests are 79 GPa and 45 GPa, respectively, compared to 49 GPa and 29 GPa for the corresponding moduli of a phenolic resin composite.

Flashover is a phenomenon unique to compartment fires in which incomplete combustion products accumulate at the ceiling and ignite causing total involvement of the compartment materials and signaling the end to human survivability. Consequently, in a compartment fire the time to flashover is the time available for escape and this is the single most important factor in determining the fire hazard of a material or set of materials in a compartment fire. The US Federal Aviation Administration has used the time-to-flashover of materials in aircraft cabin tests as the basis for heat release and heat release rate acceptance criteria for cabin materials for commercial aircraft. Figure 4.18 shows the calculated time to flashover of various 6 mm (0.24-in.) thick composite materials if they were used as wall linings in an 2.4m x 3.6m (8 ft x 12 ft) room that is 2.4m (8 ft) high.

Table 4.5 Mechanical properties of poly(sialate) carbon fiber composites, T-300 carbon fabric, 0–90 °C cross-ply [1]

Property	Test Temperature (°C)	n	Modulus (GPa)	Strength (MPa)
Inplane shear	22	3	4.0 ± 0.1	30.5 ± 1.2
Interlaminar shear	22	5		14.1 ± 0.6
	200	5		12.5 ± 0.3
	400	5		6.8 ± 0.4
	600	5		4.6 ± 0.1
	800	5		4.6 ± 0.2
	1000	5		5.6 ± 0.5
Warp tensile	22	5	79 ± 2	343 ± 31
Flexure	22	5	45.3 ± 0.9	245 ± 8
	200	5	36.5 ± 4.0	234 ± 10
	400	5	27.5 ± 2.5	163 ± 6
	600	5	18.3 ± 1.4	154 ± 24
	800	5	12.3 ± 0.5	154 ± 9

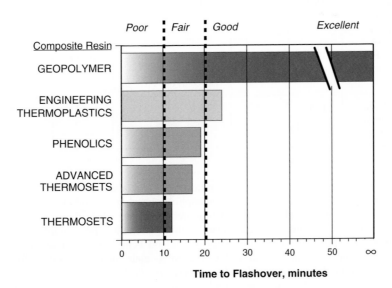

Figure 4.18   Predicted time to flashover in ISO 9705 corner/room fire test with various structural composites as wall materials [1]

Table 4.6 compares some thermomechanical properties of fiber-reinforced concrete, structural steel, a 7000-series aluminum used in aircraft structures, a phenolic-E glass fabric cross-ply laminate, a phenolic-carbon fabric cross-ply laminate, and the poly(sialate)-carbon fabric cross-ply laminate. Maximum temperature capability is defined as the temperature in air at which the nominal tensile or flexural strength falls to one half of its room temperature value. Specific flexural strength is the flexural strength of the material divided by the bulk density and is the figure of merit for weight-sensitive applications such as aerospace and surface transportation vehicles. Similarly, specific modulus is defined here as the tensile (Young's) modulus of the material divided by its bulk density. In the case of the anisotropic cross-ply laminates, the warp tensile modulus is used for the calculation. The poly(sialate) composite is superior to all of the materials listed with regard to specific modulus and is second only to the phenolic-carbon cross-ply laminate in specific strength. However, the poly(sialate)-carbon fabric laminate is unique in its high temperature structural capability and fire resistance.

Table 4.6   Typical properties of structural materials [1]

Material	Density kg/m^3	Tensile modulus (GPa)	Specific modulus (MPa-m^3/kg)	Flexural strength (MPa)	Specific Flexural strength (Mpa-m^3/kg)	Maximum temperature capability (°C)
Fiber-Rein-forced Concrete	2300	30	13.0	14	0.006	400
Structural Steel	7860	200	25.4	400	0.053	500
7000 Series Aluminum	2700	70	25.9	275	0.102	300
Phenolic-Carbon Fabric Laminate	1550	49	31.6	290	0.187	200
Phenolic-E Glass Fabric Laminate	1900	21	11.0	150	0.074	200
Poly(sialate)-carbon Fabric Laminate	1850	76	41.0	245	0.132	= 800

Poly(sialate)-carbon composites are noncombustible materials that are ideally suited for construction, transportation, and infrastructure applications where a combination of fire endurance, non-combustibility, and specific flexural strength is needed. Carbon fabric reinforced poly(sialate) cross-ply laminates have comparable initial strength to fabric reinforced phenolic resin composites but have higher use temperatures and better strength retention after fire exposure. In comparison to structural steel the poly(sialate)-nanocomposite falls short in flexural strength, modulus, and cost but the temperature capability is superior. Consequently in applications requiring fire endurance, replacement cost or the added cost of a fire barrier must be figured into the material cost for metallic structures.

References for Section 4.2.5

[1] R. Lyon, A. Foden, P. N. Balaguru, M. Davidovics, J. Davidovits: *J. Fire Mater.* 21 (1997) 67–73
[2] J. Davidovits: *J. Thermal. Anal.* 37(8) (1991) 1633–56
[3] J. Davidovits: *J. Mater. Educ.* 16(2–3) (1994) 91–137
[4] J. Davidovits, R. Davidovits, C. James (Eds.): Geopolymere'99, Proceedings of the Second International Conference on Geopolymers, Université de Picardie, France, 1999

4.2.6 Thermal Properties of Elastomers

D. DEROUET AND J.-C. BROSSE

4.2.6.1 Introduction

Elastomers, as plastomers, are composed of long, flexible, and more or less mixed-up macromolecular chains that present only relatively low physical mutual interactions. This low level of interactions between the chains governs the elasticity of the material. Moreover, the mobility of the macromolecules depends on the temperature. As any polymer, an elastomer is characterized by its glass transition temperature T_g which is necessarily always widely lower than room temperature.

Crude rubbers, whether natural or synthetic, are plastic-like materials that can be deformed at high temperatures. They are generally not suitable for use in the form in which they are supplied. The elastic properties have to be developed by further compounding (i.e., incorporation of various additives in the crude rubber). To avoid the irreversible fluage phenomenon and to reach high elasticity properties, the flexible chains must be interconnected by chemical bonds, that is the case of *thermoset elastomers,* or only by physical process due to the formation of microdomains, that is the case of *thermoplastic elastomers.*

Unlike crude rubbers, elastomers are elastic materials, that is, they have the ability to deform substantially under the application of a force and then snap back to almost their original shape when the force is removed. Compared to thermoplastics, elastomers present various specific characteristics:

- At ambient temperature, their rigidity (or module) is low (a few megapascals).
- They are deformable, which means they can sustain large reversible deformation (up to 100 %) without breaking.
- They are resilient, which means they are able to recover their initial geometry after repeated deformations, by releasing quantitatively, on the outside, the energy that was previously supplied to deform it. This last characteristic highly depends on the elastomer nature, the temperature, and the number of repeated deformations.

4.2.6.2 Vulcanizable Elastomers

All thermoset elastomers are manufactured from a combination of ingredients, the most important being the crude polymer because it provides the basic properties of the vulcanizates such as ozone and oil resistance, low-temperature flexibility, flammability, and so on. However, the other ingredients, such as fillers, plasticizers, cure system components, aging protectors, and other miscellaneous additives, also have a great influence on the final product properties.

In most cases, the flame retardancy of elastomers is very low. It depends on their chemical structure. When exposed to an ignition source, they tend to soften and melt, with emission of important black smoke and formation of a char residue. However, the use of flame retardants or specific plasticizers may strongly increase their flame retardant behavior.

The purpose of the following part is to give a brief, up-to-date summary of the principal commercially available types of raw (crude) elastomers with their principal characteristics, including special purposes on their thermal and flame retardant properties (Table 4.7), as well as the means used to improve their flame retardancy. The topics that are considered for each rubber type include their chemistry [1–3] and their flame retardant properties [4–7]. For this purpose the elastomers can be classified in two categories [2–4]:

- Non flame-retardant elastomers, which necessitate incorporation of flame retardants to improve their flammability properties,
- Inherently flame-retardant elastomers that are based on halogen (or phosphorus)-containing polymers.

Thermal decomposition, combustion, and flame retardancy of elastomers have been discussed in detail in various books and reviews [8–15]. Some of the material in this chapter has been drawn from the following sources: technical documentation from the suppliers, reviews [6, 7], conferences, and books [1–5, 13].

Non Flame-Retardant Elastomers

According to their flame behavior, the non flame-retardant elastomers or rubbers can be classified into three categories in relation to the nature of their atoms.

Hydrocarbon Rubbers that Contain Only Carbon and Hydrogen

Chemistry

Natural rubbers (NR). Natural rubber, which is issued from the latex (emulsion in water) produced by laticiferous cells inside the rubber tree *Hevea brasiliensis*, consists primarily of high molecular weight *cis*-1,4-polyisoprene molecules. Its virtually entirely *cis*-1,4 configuration explains its rubbery properties. Because of its natural origin, it contains also small amounts of nonrubber substances, notably fatty acids, proteins, salts, and lipids.

The main characteristics of crude rubbers are summarized in Table 4.7.

Natural rubber
(cis-1,4 polyisoprene)

(4.36)

Like hydrocarbons, natural rubber burns quite readily with a sooty flame. Flame retardants can reduce this propensity, but where fire is a serious potential problem, alternative rubbers are usually preferred.

Synthetic polyisoprene (IR) [trade names: Afprene (Carbochem); Natsyn (Goodyear); Isoprene, Cariflex IR (Shell); Hartex (Firestone); Nipol IR (Nippon Zeon); Europrene IP (Enichem); Caron (Danubiana); Isolene (Hardman)] (4.37).

(cis -1,4)   (trans -1,4)   (3,4)   (1,2)

(4.37)

Table 4.7 Principal characteristics of crude rubbers

Designation	Chemical name	Tg (°C)	Service temperature range (°C)	Heat resistance	Flame resistance (LOI)
ACM	acrylic rubbers	−22 to −40	−10 to 170	G–E	P
AFMU	carboxy-fluoro-nitroso rubbers	−54		E	E
AU/EU	polyurethane	−35/−55	−22/−35 to 75/75	F–G	P–G
BIIR	bromobutyl rubbers	−66	−38 to 150	G	P
BR	1,4-polybutadiene	−112 (cis); −106 (trans)	−72 to 90	F–G	P
CIIR	chlorobutyl rubbers	−66	−38 to 150	G	P
CM	chlorinated polyethylene	−25	−12 to 150	G	P
CO / ECO	epichlorohydrin rubbers	−26/−45	−10/−25 to 150/135	G–E	G–E
CR	neoprene rubbers (polychloroprene)	−45 (trans); −20 (cis)	−25 to 125	F–G	G–E (26)
CSM	chlorosulfonated polyethylene	−25	−10 to 150	G	G–E
EAM	ethylene-acrylate rubbers	−40	−20 to 175	G–E	F
EPM and EPDM	ethylene-propylene rubbers	−56[a]	−35 to 150	G–E	P
EVM (EVA)	ethylene-vinyl acetate rubbers	−30	−18 to 160	G–E	F
FPM pr FKM	fluorocarbon rubbers	−18 to −50	(−10 to −35) to 260	E	G–E (22)
FFKM	perfluoro rubbers	−10	0 to 290	E	E

Table 4.7   (Continuation)

Designation	Chemical name	Tg (°C)	Service temperature range (°C)	Heat resistance	Flame resistance (LOI)
FMQ and FVMQ	fluoromethyl(vinyl)silicone rubbers	–70	–45 to 215	E	G–E
FZ	polyfluoroalkoxyphosphazene rubbers	–66	–42 to 175	E	G–E
GPO	propylene oxide rubbers	–72	–60 to 150	G–E	P (15)
HNBR	hydrogenated-acrylonitrile-butadiene rubbers	–30	–18 to 160	E	P
IIR	butyl rubbers	–66	–38 to 150	G	P
IR	1,4-polyisoprene	–73 (cis); –60 (trans)	–55 to 100	P–F	P (18,5)
MQ (VMQ)	methyl(vinyl)silicone rubbers	–120	–85 to 225	E	E (30)
NBR	acrylonitrile-butadiene rubbers	–20 to –45[b]	(–10 to –28) to 125	G	P–F
NR	natural rubber	–72	–45 to 100	F	P (17)
PNR	polynorbornene	+25 (–60[c])	–30 to 100	F	F
PZ	polyaryloxyphosphazene rubbers	–66	–20 to 125	G	E (28)
SBR	styrene-butadiene rubbers	–59 to –64[d]	–28 to 110	F–G	P (17)
TM(T)	polysulfide rubbers	–59[e]	–45 to 100	F	P
XNBR	carboxylated acrylonitrile-butadiene rubbers	–30	–18 to 120		

Glossary: P = poor; F = fair; G = good; E = excellent; a. lowest value corresponding to a propylene content of about 40 %; b. depending on acrylonitrile content. $T_g = 0 °C$ for a 50/50 copolymer; c. after adding a wide variety of mineral oils or ester plasticizers; d. vary almost linearly with the monomer ratios in the polymer. $T_g = –60 °C$ for a typical 75/25 styrene/butadiene composition; e. T_g of poly(ethyl formal disulfide) which is the basic polymer of current commercial formulations.

When isoprene is polymerized, three types of monomer units can be noticed in the chain (*cis*-1,4; *trans*-1,4; 1,2, and 3,4), and further variations in the chain structure are possible because of the head-to-head/head-to-tail alternative of monomer polymerization. In the synthetic rubbers, the above forms coexist in a variety of combinations and proportions. For synthetic rubbers, interest is confined to polymers with a *cis*-1,4 content above 90 %.

Synthetic polyisoprene rubbers (IR) are prepared by anionic polymerization (polyisoprene with about 92–93 % of *cis*- and 7 % of 3,4-units) or Ziegler-Natta type polymerization (polyisoprene with about 96–98 % of *cis*- and the rest of 3,4 units).

Polybutadiene (BR) [trade names: Afdene (Carbochem); Budene (Goodyear); Duragene (General Tyre); Buna CB (Bayer); Nipol BR (Nippon Zeon); Ameripol CB (BF Goodrich Chemical); Trans 4, Cis 4 (Phillips Chemical); Diene (Firestone); Cariflex BR (Shell); Europrene Cis PB (Enichem)].

In comparison with polyisoprene, the absence of a methyl side group in the butadiene monomer unit decreases the number of possible structures. In this case three repeating monomer units are possible: *cis*-1,4; *trans*-1,4 and -1,2. Commercial polymers are prepared in solution using either alkyl lithium, or Ziegler-Natta catalysts namely, titanium, cobalt, nickel, and neodymium compounds. Their properties are very similar to those of natural rubber.

Random styrene-butadiene rubbers (SBR) [trade names: Buna SL, Baystal (Bayer); Plioflex, Pliolite (Goodyear); Cariflex S (Shell); Solprene (Phillips Chemical); Butaprene (BF Goodrich Chemical); Europrene SBR (Anic/Enichem); Nipol SBR (Nippon Zeon); Krylene (Polysar)].

Commercial random styrene-butadiene rubbers (SBR) are prepared mainly by emulsion polymerization. SBRs of first generation known as "hot rubbers" are prepared at about 50 °C, using a water-soluble initiator such as potassium persulfate, the average molecular weight being controlled by using a chain transfer agent such as *t*-dodecyl mercaptan. SBRs of second generation are known as "cold rubbers" and prepared at about 5 °C, using a redox initiating system (4.38).

butadiene   styrene

copolymerization
(emulsion)

styrene - butadiene rubber
(with y from 16 to 40 %)

$$\mathsf{\sim\!\!\!\left(CH_2\!-\!CH\!=\!CH\!-\!CH_2\right)_x\!\!\!\sim\!\!\!\left(CH_2\!-\!CH\right)_y\!\!\!\sim}$$

(4.38)

Polynorbornene (PNR) [trade names: Norsorex (Elf Atochem)].

Polynorbornene rubber is obtained by ring-opening polymerization of norbornene catalyzed by a [RuCl$_3$] n H$_2$O complex, in butyl alcohol between 50 and 100 °C. Both *cis* and *trans* structures are obtained (4.39).

(4.39)

The basic polymer is a thermoplastic powder with a T_g of about 35 °C. By adding a wide variety of mineral oils or esters plasticizers, the T_g can be lowered to about −60 °C and the polymer acquires rubbery properties.

Butyl rubbers (IIR) [trade names: Exxon Butyl (Exxon Chemical); Bucar Butyl (Columbian Chemicals); Polysar Butyl (Bayer)].

Butyl rubbers (IIR), or isobutene-isoprene rubbers, are copolymers containing mostly isobutene monomer units with just a small percentage of isoprene monomer units to obtain the necessary unsaturation for vulcanization. They are produced at low temperature by suspension cationic polymerization in chloromethane initiated by $AlCl_3$ (4.40).

isobutene + isoprene $\xrightarrow{\text{copolymerization}}$ isobutene - isoprene rubber (IIR)

$$(4.40)$$

The presence of unsaturation in small content is sufficient to allow the vulcanization.

Ethylene-propylene rubbers (EPM and EPDM) [trade names: Vistalon (Exxon Chemical); Royalene, Royaltherm (Uniroyal); Epcar (BF Goodrich Chemical); Nordel (DuPont Dow Elastomers); Keltan, Epsyn (DSM); Buna EP, Polysar EPDM, Polysar EPM (Bayer)].

The ethylene-propylene rubbers are of two types:

• EPM: fully saturated copolymer of ethylene and propylene (4.41)

ethylene + propylene $\xrightarrow{\text{copolymerization}}$ ethylene - propylene rubber (EPM)

$$(4.41)$$

• EPDM: terpolymers of ethylene, propylene and a small percentage of a nonconjugated diene (the three basic monomers used are 1,4-hexadiene; dicyclopentadiene; and 5-ethylidene norbornene), which provides unsaturation in side chains pendent from the fully saturated backbone (4.42).

ethylene + propylene + 5-ethylidene norbornene (ENB) $\xrightarrow{\text{copolymerization}}$ ethylene - propylene rubber (EPDM)   vulcanization site

$$(4.42)$$

EPM and EPDM are produced by a polymerization process using anionic coordinated catalyst systems based on a combination of a derivative of titanium or vanadium as catalyst in conjunction with an alkyl aluminum as cocatalyst [e.g., $Al(C_2H_5)_2Cl$]. Polymerization may be carried out in solution or in suspension.

The completely saturated EPMs require organic peroxides or radiation for vulcanization, while EPDM terpolymers can be vulcanized with peroxides, radiation or sulfur.

Chemically Cross-Linked Polyethylene (PE): PE is a thermoplastic homopolymer (plastomer) produced by polymerization of ethylene gas which, like rubbers, can be converted to an elastomeric material by cross-linking the polymer chain using a cross-linking agent, usually organic peroxides or electron beams (radiation cure).

Flammability and Means to Improve Flame Retardancy

Combustion products formed during hydrocarbon elastomer burning are carbon, carbon dioxide, carbon monoxide, and water.

Natural rubber and synthetic hydrocarbon elastomers (polyisoprene, polybutadiene, styrene-butadiene rubber), which are used in large volumes, all burn readily with a sooty flame (LOI = 17 vol%) with much smoke in the absence of flame retardants. They are characterized as high-volatile loss, low-char polymers, relative to polymers containing acrylonitrile, halogens, or main-chain aromatic groups.

Because vulcanization formulations of these rubbers generally include fillers, the use of aluminum trihydrate [14–16] or magnesium hydroxide [16–18] as fillers offers a significant opportunity to reduce flammability and smoke evolution. The effect of fillers on flammability has been investigated extensively [7, 8]. At high aluminum trihydrate loadings (> 20 wt%), mass dilution and/or endothermic dehydration are thought to slow the buildup of heat, favoring char-forming processes and reducing flammability and smoke generation in the rubber [14, 16].

Other approaches to flame retardance include halogenation and incorporation of chlorinated paraffins or halogenated additives [8], but generally increase the formation of smoke and various toxic products [19], as well as charring. Several metal compounds derived from antimony (antimony oxide) [20–23], molybdenum [24], and iron [Fe(III) oxide] [25, 26], metallocenes [27]; and various metal chelates [28, 29] exhibit synergistic effect with halogens for flame retardancy and smoke suppression.

Phosphorus compounds such as tributyl phosphate and triphenyl phosphate have been suggested, and are reported to be less objectionable than halogen compounds in terms of their effects on the aging properties of the elastomer [8].

Chemical modification of the chains has also been demonstrated to be an efficient method to improve the flame retardancy of hydrocarbon rubbers [6, 30, 31]. The method consists in the reaction of the polymer with reactive flame retardants.

EPDM and other terpolymers are slightly less flammable (for EPDM, LOI = 22 vol% [32]) but still require the addition of flame retardants, for instance $Mg(OH)_2$ [17].

Flame retardancy is needed especially for applications where physical properties must be preserved and resistance to aging must not be impaired.

Elastomers Containing Only Carbon, Hydrogen, and Oxygen

Chemistry

Acrylic rubbers (ACM) [trade names: Cyanacryl (American Cyanamid); Europrene AR (Enichem); Hytemp, Nipolar (Nippon Zeon); Hycar (BF Goodrich Chemical); Paracril OHT (Uniroyal)] (4.43). Nowadays, acrylic rubbers are saturated copolymers of acrylate monomers and reactive cure site monomers. They are prepared by radical emulsion copolymerization.

$$\text{CH}_2{=}\text{CH} + \text{CH}_2{=}\text{CH} \xrightarrow[\text{(emulsion)}]{\text{copolymerization}} \sim\!\!\left(\text{CH}_2\text{-CH}_2\right)_x\!\!\sim\!\!\left(\text{CH}_2\text{-CH}\right)_y\!\!\sim$$

CH₂=CH (C=O, O, R) acrylate monomer (R : ethyl, butyl, methoxyethyl, ethoxyethyl)

CH₂=CH (R*) cure site monomer
R = -O-CH₂-CH₂-Cl
    -C(O)-OH
    -O-C(O)CH₂-Cl
    -CH₂-O-CH₂-CH-CH₂ (epoxide O)

polyacrylate rubber (with C=O, O, R side group; R* side group)

$$(4.43)$$

The most recent types of ACM rubbers contain minor amount (less than 5 % by weight) of cure site monomers with functional groups whose presence increases the cure rate and improves the vulcanizate's properties.

Ethylene-acrylate rubbers (EAM) [trade name: Vamac (DuPont)]. EAM is a terpolymer of ethylene and methyl acrylate with a cure site monomer that contains a carboxylic acid group to permit cross-linking by reaction with some bifunctional reagents such as diamines (4.44).

$$\text{CH}_2{=}\text{CH}_2 + \text{CH}_2{=}\text{CH} + R \xrightarrow{\text{polymerization}} \sim\!\!\left(\text{CH}_2\text{-CH}_2\right)_x\!\!\sim\!\!\left(\text{CH}_2\text{-CH}\right)_y\!\!\sim\!\!\left(R\right)_z$$

ethylene

CH₂=CH (C=O, OCH₃) methyl acrylate

CH₂=CH (C=O, OH) cure site monomer

ethylene - acrylic rubber (side groups: C=O, OCH₃; C=O, OH)

$$(4.44)$$

Ethylene-vinyl acetate rubbers (EVM or EVA) [trade names: Levapren (Bayer); Elvax (DuPont); Vynathene (USI)]. Ethylene-vinyl acetate rubbers are copolymers of ethylene and vinyl acetate with fully saturated backbones, prepared by ethylene-vinyl acetate copolymerization initiated by free radicals, using azo-compounds or peroxides as initiators. The commercially available EVM rubbers have vinyl acetate contents ranging from 40 % to 70 % by weight (4.45).

$$\text{CH}_2{=}\text{CH}_2 + \text{CH}_2{=}\text{CH} \xrightarrow{\text{polymerization}} \sim\!\!\left(\text{CH}_2\text{-CH}_2\right)_x\!\!\sim\!\!\left(\text{CH}_2\text{-CH}\right)_y\!\!\sim$$

ethylene

CH₂=CH (O, C=O, CH₃) vinyl acetate

ethylene - vinyl acetate rubber (side group: O, C=O, CH₃)

$$(4.45)$$

Because EVM rubbers have a saturated backbone, vulcanization is achieved either by exposure to high-energy radiation or by reacting with peroxides, usually in conjunction with a coagent to increase degree of cure.

Propylene oxide rubbers (GPO) [trade name: Parel (Nippon Zeon)].

Polypropylene oxide rubbers (GPO) are prepared by copolymerization of propylene oxide and allylglycidylether, which provides carbon-carbon pendant double bonds in side chains to enable vulcanization with sulfur and conventional accelerators (4.46).

$$(4.46)$$

Flammability and Means to Improve Flame Retardancy

Elastomers containing oxygen in addition to carbon and hydrogen burn more slowly than the simple hydrocarbon ones. This is because the presence of oxygen in the molecules replaces potentially available fuel so that it is not available to support combustion. Consequently, these elastomers show lower heat of combustion than hydrocarbon elastomers. Their oxygen index (LOI according to ASTM D2863) is intrinsically relatively high, and depends on the oxygenated monomer concentration in the copolymer; for instance, EVM grades with 40% vinyl acetate have an oxygen index of 35 vol%, while those with 70% vinyl acetate reach a value of higher than 50 vol% [33].

Combustion products formed during burning are the same as observed in the case of hydrocarbon elastomers.

Acrylic elastomers have a very wide ignition temperature range, depending on the particular elastomer. That range extends from 430 °C to 560 °C, which is the reported temperature range for polymethylmethacrylate. They burn with a slightly crackling, luminous yellow-topped blue flame and may emit a sweet fruit-like odor. Smoke production may be very low. Below the ignition temperature (from 275 °C to 400 °C), the elastomer depolymerizes, producing a gaseous monomer.

In applications with higher fire safety requirements, oxygen-containing rubbers such as EVM (EVA) and EAM have been recently specified. Indeed, in the event of fire, the EVM and EAM vulcanizates create less toxicity and corrosion hazard. Flame retardancy of EAM can be improved by blending with polysiloxanes (LOI values of 27 vol% can be reached, 33.5 vol% if lead stearate is simultaneously incorporated). On the other hand, with the aid of aluminum trihydrate, it is possible to develop EMA compounds with good flame retardancy for halogen-free cable jackets. In electric applications where there is a demand for low-smoke, halogen-free jacketing compounds, attention has turned toward EVA rubbers and the performances of magnesium hydroxide were demonstrated [34]. It was found that magnesium hydroxide gives significantly higher LOI values than the same amount of aluminum trihydrate, which is attributed to the relative tendency of these fillers to influence decomposition of this polymer and promote char formation [35].

Elastomers Containing Carbon, Hydrogen, and Nitrogen (and Occasionally Oxygen, Sulfur)

Chemistry

Acrylonitrile-butadiene rubbers (NBR) [trade names: Chemigum, Cheminic (Goodyear); Paracril, Nitrex (Uniroyal); Hycar (BF Goodrich Chemical); Krynac (Polysar); Nysyn, Nysynblack (DSM); Perbunan N (Bayer); Breon (Nippon Zeon); Elaprim (Montedison)].

Acrylonitrile-butadiene rubbers (NBR) are prepared by emulsion copolymerization of butadiene and acrylonitrile at 5 or 30 °C with per-salts or redox initiators. Five grades of NBR based on the acrylonitrile content are commercialized (4.47).

$$CH-CH \atop CH_2 \quad\quad CH_2 \quad + \quad CH_2=CH \atop \overset{C\equiv N}{|} \quad \xrightarrow{\text{copolymerization}} \quad \sim\!\!\left[CH_2\!-\!CH_2\right]_x\!\!\sim\!\!\sim\!\!\left[CH_2\!-\!CH\right]_y\!\!\sim \atop \overset{C\equiv N}{|}$$

butadiene   acrylonitrile   acrylonitrile - butadiene rubber (NBR)

(4.47)

Carboxylated acrylonitrile-butadiene rubbers (XNBR). Carboxylated acrylonitrile-butadiene rubbers (XNBR) are terpolymers of butadiene, acrylonitrile, and an acidic monomer with a carboxylic group such as methacrylic or acrylic acid.

Hydrogenated acrylonitrile-butadiene rubbers (HNBR) [trade names: Zeoforte, Zeptol (Nippon Zeon); Tornac, Therban (Bayer)].

Hydrogenated acrylonitrile-butadiene rubbers (HNBR) are currently produced by selective and controlled hydrogenation of NBR. The increase in the degree of hydrogenation results in improved heat-resistance.

Polyurethane rubbers (AU/EU) [trade names: Texin (Mobay Chemical); Vulkollan, Urepan, Desmopan (Bayer); Pellethane (Dow Chemical); Adiprene (DuPont); Cyanaprene (American Cyanamid); Roylar (Uniroyal)].

The formation of polyurethane rubbers by a polycondensation process requires bifunctional reagents to enable the building of large chains.

The basic components used to prepare polyurethanes are:

- A diisocyanate (generally a small aromatic molecule, for instance 1,5-naphthalene diisocyanate),
- A long-chain macroglycol of polyester type (formation of polyester urethanes *au*) or polyether (formation of polyether urethanes *eu*),
- A short-chain glycol (HO-R'-OH) or amine (H_2N-R'-NH_2) used as chain extender and cross-linker.

$$HO-R-OH + O=C=N-R-N=C=O + HO-R-OH$$
Polyol   Diisocyanate   Polyol

$$\downarrow$$

$$\sim\!\!\sim\!\!\sim\; O-R-O-\underset{\underset{O}{\parallel}}{C}-NH-R-\underset{\underset{O}{\parallel}}{C}-O-R-O \;\sim\!\!\sim\!\!\sim$$

Polyurethane   (4.48)

In a first step, the macroglycol derivative is transformed into a diisocyanate telechelic prepolymer by reaction with an excess of the diisocyanate reagent, and then, in a next stage, with the chain extender, to realize the chain extension.

They are vulcanized either by organic peroxides, either by sulfur/accelerators systems (in this case, an unsaturated diol such as glycerol monoallyl ether is introduced as a co-monomer to create unsaturation on the chains).

Polysulfide rubbers (TM or T) [trade names: FA and ST Polysulfide Rubbers, LP Liquid Polysulfide Rubbers (Thiokol Chemical)].

Polysulfide rubbers are produced by reacting a suitable aliphatic dichloride with aqueous solution of sodium polysulfide (4.49).

$$Cl - R - Cl + Na_2S_x \rightarrow - RS_x - + 2\,NaCl \text{ (where } x = 2\text{–}4) \tag{4.49}$$

They are available in two forms: solids and liquids. Their heat resistance is relatively low.

Flammability and Means to Improve Flame Retardancy

The heats of combustion of this group fall in about the middle of the range of all the elastomers. They burn at about the same rate as the elastomers containing carbon, hydrogen, and oxygen for the same reason: the nitrogen that occupies space on the polymer chain does not burn. The rate of burning also is about the same for both groups.

In addition to the combustion products observed during burning of the elastomers from the two previous groups, hydrogen cyanide is formed because of the presence and subsequent release of nitrogen from the macromolecules. Like carbon monoxide, and any intermediate combustion products, hydrogen cyanide is flammable, and the majority of hydrogen cyanide produced is burned.

NBR, which represents an extremely large volume for a material that was originally classified as an engineering plastic, is highly flammable. It burns with a luminous yellow flame and supports its own combustion (formation of toxic fumes during burning). NBR flame retardancy can be increased by blending with PVC [26]: NBR and PVC form one of the relatively few pairs of distinct, but miscible, polymers (when acrylonitrile content in NBR is higher than 25 %) and can thus form homogeneous blends. Phosphates such as tritolyl and trixylyl phosphates are also of great interest because of their plasticizing properties and the great interaction with NBR [3].

Elastomers Containing Carbon, Hydrogen, and Halogen: Halobutyl Rubbers (BIIR and CIIR)

$$\text{(4.50)}$$

Halobutyl rubbers include bromobutyl rubbers (bromine content: 1–1.3 %) (BIIR) [trade name: Polysar Bromobutyl (Bayer)] and chlorobutyl rubbers (chlorine content: 1.9–2.1 %) (CIIR) [trade name: Polysar Chlorobutyl (Bayer)]. They are produced by reacting bromine or chlorine, respectively, with IIR dissolved in a light aliphatic hydrocarbon such as hexane (4.50). Carefully controlled conditions are used to ensure that the original double bonds are largely retained.

By comparison with nonmodified IIRs, halogenated IIRs show a greater tendency to thermal degradation and are more flammable [36].

Flame Retardant Elastomers

In this category of elastomers [15], it is possible to distinguish two families by their burning characteristics:

Elastomers that Burn and Extinguish Slowly Outside of the Flame

Chloroprene rubbers (CR) [trade names: Denka (Denka Chemical); Duprene, Neoprene (DuPont); Baypren (Bayer); Butaclor (Distugil/Schulman)] (4.51).

Chloroprene rubbers (CR) are essentially homopolymers of chloroprene (chlorobutadiene). Commercial polymerization is carried out in emulsion at 40 °C using free-radical initiators, such as potassium and ammonium persulfate.

$$\text{(4.51)}$$

Flammability characteristics of polychloroprene have been discussed [37]. The chlorine atoms along the polymer chains impart flame retardancy to the vulcanizates, as well as a better resistance to oxygen and ozone, due to a reduction of the chemical reactivity at and around the carbon-carbon double bond. When CRs burn, they develop great quantities of toxic and corrosive fumes of hydrogen chloride, which are extremely dangerous and cannot be tolerated in areas where there is a high demand for safety. The pyrolysis and combustion tests showed that hydrogen chloride is evolved at 300–350 °C. Moreover, they have a greater tendency to char than do IR, SBR, NBR, or EPDM [38].

CR vulcanizates with a higher degree of flame retardancy than can be obtained by using flame retarded plasticizers, such as phosphate esters. This property can be also further improved by compounding with appropriate fillers including calcium carbonate, clay, aluminum trihydrate, and carbon black [8, 15], as well as a variety of other additives based on antimony oxide, zinc borate, and ammonium sulfate. Another solution to enhance flame retardancy of CRs is to copolymerize chloroprene with dichlorobutadiene. However, for most uses, for example, as molded or extruded articles, no additional flame retardancy is required.

Chloroprene rubbers show better heat resistance than diene hydrocarbon rubbers (but inferior to IIR, EPDM, NBR, and to most of the specialty engineering rubbers).

Chlorosulfonated and chlorinated polyethylene rubbers (CSM and CM) [trade names CM: Tyrin (DuPont); Plaskon (Dow Plastics); Kelrinal (DSM). CSM: Hypalon (DuPont); Noralon (Denki Kagaku)].

Chlorosulfonated polyethylene rubbers (CSM) result from the treatment of polyethylene in tetrachloromethane by chlorine and SO_2 (Reed-Horn reaction that involves co-reaction of sulfur dioxide and chlorine in presence of light irradiation) (4.52).

$$\text{(4.52)}$$

The chlorine content of commercially available CSM grades ranges from 24–43 % by weight (around one chlorine for seven carbon atoms). The sulfur content is approx. 1.0–1.5 % by weight (one SO_2Cl group for around 60–90 carbon atoms), the majority having a sulfur content of about 1 %.

An alkylated CSM version (ACSM) with better ageing resistance properties at 140 °C than that of HNBR, is supplied by DuPont (trade name: Hypalon).

The CSM vulcanizates demonstrate low flammability. The best flame retardancy is obtained with the high-chlorine grades: when more than 35 wt% chlorine is present (LOI = 26), no further flame retardancy needs to be imparted, but otherwise fillers, antimony oxide, or other additives have to be incorporated.

Chlorinated rubbers (CM) are structurally very similar to the chlorosulfonated polyethylene rubbers, the main difference being the absence of the sulfonyl chloride cure sites. The range of chlorine content is very similar to that available for CSM rubbers. Increasing the chlorine content improves flame retardancy but lowers heat resistance.

Epichlorohydrin rubbers (CO, ECO, and GECO) [trade names: Hydrin (BF Goodrich Chemical); Epichloromer (Osaka Soda); Herclor (Hercules)].

Epichlorohydrin rubbers are saturated polymers of aliphatic polyethers with chloromethyl side chains. They are prepared by ring-opening polymerization with alkyl aluminum derivatives such as tributyl aluminum used in conjunction with water and in the presence of chelating agents such as acetylacetone and zinc acetyl acetonate (4.53).

$$
\begin{array}{c}
\underset{\substack{| \\ CH_2\,Cl}}{CH} \!\!-\!\!-\!\! CH_2 \qquad\longrightarrow\qquad \left[\underset{\substack{| \\ CH_2\,Cl}}{CH} - CH_2 - O \right]_n
\end{array} \qquad (4.53)
$$

epichlorohydrin rubber (CO)

Three types are currently available:

- Homopolymers of epichlorohydrin (CO),
- Copolymers of epichlorohydrin and ethylene oxide (ECO),
- Terpolymers of epichlorohydrin, ethylene oxide, and an unsaturated monomer (allyl glycidyl ether) introduced to obtain sulfur-curable grades to avoid the problems encountered in the softening of ECO rubber in contact with hydroperoxide-containing gasoline ("sour gas").

CO rubbers show high temperature resistance and good flame retardancy. The ECO copolymers have 1:1 molar ratios and differ mainly from the homopolymers that they are not self-extinguishing. These rubbers are inherently flame retardant but still slightly inferior to polychloroprenes. Their performance may be improved by incorporation of the usual flame retardants, but this is not generally necessary.

Elastomers Difficult to Ignite, Extinguishing Immediately Outside of the Flame

Optimum fire retardancy in elastomers can be achieved through the synthesis of specialty elastomers, which are inherently flame retarded because part or all of the carbon in the polymer backbone is replaced by inorganic elements: silicone elastomers [39] and phosphonitrile elastomers [40] are examples of such advanced elastomeric materials.

When the highest possible levels of heat resistance in elastomers are required, the choice lies between fluoro rubbers, silicone rubbers, fluorosilicone rubbers and polyphosphazenes.

Fluoro Rubbers (FKM). Fluoro rubbers [41], which correspond to highly fluorinated polymers, are available in various grades, differing mainly in the polymer composition and fluorine content. They are classified in several categories of products:

- Copolymers based on vinylidene fluoride and hexafluoropropylene (FPM or FKM) [trade names: Viton (DuPont Dow Elastomers); Fluorel (3M); Dai-el (Daikin Kogyo); Tecnoflon (Montedison)].

They are prepared by free-radical emulsion polymerization, usually using peroxide initiators at temperatures from 80 to 125 °C and pressures from 2 to 10 MPa. On the market, about 70 grades of FPM are available, depending on the fluorine and hydrogen contents (Table 4.8).

- Copolymers based on vinylidene fluoride and chlorotrifluoroethylene (CFM) [trade names: Voltalef 3700, Voltalef 5500 (Ugine Kuhlmann)].
  CFM are inferior to FKM types in respect of heat resistance and the market for this category of rubbers is small compared with that of FKM.
- Perfluorocarbon rubbers (FFKM) [trade names: Kalrez (DuPont Dow Elastomers); Dai-el Perfluor (Daikin Kogyo)].
  These rubbers, which contain no C-H groups, were developed to improve the thermal stability of fluorine-containing elastomers. They are prepared by high pressure emulsion copolymerization of tetrafluoroethylene with perfluoromethyl vinyl ether [$CF_2 = CF (OCF_3)$], and a small amount of a perfluorinated cure site monomer.
- Tetrafluoroethylene-propylene rubbers (TFE/P) [trade name: Aflas (3M)].
  TFE/P is an alternated copolymer of tetrafluoroethylene and propylene with a fluorine content of only 53.5 %.

Table 4.8   Various copolymers based on vinylidene fluoride and hexafluoropropylene

Elastomer type	Fluorine content (%) (approx.)	Hydrogen content (%) (approx.)
$\left[CF_2 - CH_2 \right]_x \left[CF(CF_3) - CF_2 \right]_y$ copolymer	65	1.9
$\left[CF_2 - CH_2 \right]_x \left[CF(CF_3) - CF_2 \right]_y \left[CF_2 - CF_2 \right]_z$ terpolymer	68	1.4
$\left[CF_2 - CH_2 \right]_x \left[CF(CF_3) - CF_2 \right]_y \left[CF_2 - CF_2 \right]_z CS^* -$ tetrapolymer	70	1.1
$\left[CF_2 - CH_2 \right]_x \left[CF(CF_3) - CF_2 \right]_y \left[CF_2 - CF_2 \right]_z CS^* -$ LT tetrapolymer	67	1.1

CS^* = cure site monomer [e.g., trifluorobromoethylene ($-CF_2$-$CFBr$-)]

The working temperature range of fluorocarbon rubbers is considered to be 29–250 °C, but they can be used at temperatures up to 316 °C for short times, and DuPont Kalrez is normally recommended up to 260 °C. Their stability is mainly due to the high strength of the C-F bond compared with the C-H bond. They melt and decompose at elevated temperatures (> 500 °C). These elastomers are generally nonflammable owing to their high fluorine content and do not need further flame retardant treatment.

Silicone rubbers (Q) [trade names: Silastic (Dow Corning); Silopren (Bayer); Rhodia, Rhodorsil, Silbione (Rhône Poulenc); Eposilrub, Isochemsilrub (Isochem)] (4.54).

Silicone rubbers or polysiloxanes are generally prepared by reacting chlorosilanes with water to give hydroxyl compounds, which then condense to give a polymer.

$$Cl-\underset{\underset{R'}{|}}{\overset{\overset{R}{|}}{Si}}-Cl \ + \ H_2O \longrightarrow HO-\underset{\underset{R'}{|}}{\overset{\overset{R}{|}}{Si}}-OH \longrightarrow \left[R-\underset{\underset{R'}{|}}{\overset{\overset{R}{|}}{Si}}-O-\underset{\underset{R'}{|}}{\overset{\overset{R}{|}}{Si}}-O \right]_n$$

R = methyl
R' = vinyl and/or phenyl or trifluoropropyl (fluorosilicones)

(4.54)

Silicone rubbers are classified as follows:

- Silicone rubbers having only methyl substituent groups on the polymer chain (polydimethylsiloxanes) (MQ),
- Silicone rubbers having both methyl and vinyl substituent groups on the polymer chain (VMQ),
- Silicone rubbers having both methyl and phenyl substituent groups on the polymer chain (PMQ),
- Silicone rubbers having methyl, phenyl, and vinyl substituent groups on the polymer chain (PVMQ),
- Silicone rubbers having both fluoro and methyl substituent groups on the polymer chain (FMQ),
- Silicone rubbers having fluoro, vinyl, and methyl substituent groups on the polymer chain (FVMQ).

The phenyl groups in PMQ and PVMQ improve low-temperature flexibility. Vulcanization is possible only when vinyl substituents are present on the silicone chains.

Silicone rubbers possess excellent resistance to extreme temperature and burn less easily than many organic rubbers. The level of combustible components produced by these elastomers during burning is low. If they are involved in a fire and burn, the silicone materials have the advantage to generate nonconducting silica ash, thus providing additional safety in electric equipment. Their flame resistance may be improved further by use of flame retardant additives, including platinum compounds (sometimes in combination with titanium dioxide), carbon black, aluminum trihydrate, or zinc or ceric [42] compounds.

Polyphosphazenes (FZ and PZ) [trade names: Eypel F, Eypel A (Ethyl Corporation); PNF Rubber, APN Rubber (Firestone); Orgaflex (Atochem)].

- Polyphosphazenes are similar to silicon rubber in that they are semiorganic rubbers (i.e., partly organic and partly inorganic in nature). The polymer backbone is composed of alternate phosphorus and nitrogen atoms with pendant organic groups linked to phosphorus atoms.
- Polyphosphazene rubbers are made from an inorganic polymer, polydichlorophosphazene, by replacing chlorine atoms with organic groups. The replacement of chlorine atoms by organic groups is realized to improve the hydrolytic stability of the material (4.55).

$$n \ PCl_3 + n \ NH_4Cl \ \xrightarrow[120\ °C]{solvent} \ n/3 \ (NPCl_2)_3 + 4n \ HCl$$

(4.55)

The fluoroalkoxy-substituted polyphosphazenes (FZ) show a large range of temperature (-60 to +175 °C) and good flame retardancy [43–45]. On the other hand, the phenol-substituted polyphosphazenes offer excellent flame retardancy without the presence of halogen. An oxygen index of 28 for the raw polymer may be raised to 46 vol% by appropriate compounding [46]. Rather more importantly, the toxicity of the combustion products appears to be considerably less than for many other commonly used flame retardant elastomers [47]. Although there is some smoke evolution, the compounds char on burning rather than dripping or flowing.

Fluoro-Carboxy-Nitroso Rubbers (AFMU)

AFMUs result from copolymerization of trifluoronitrosomethane and tetrafluoroethylene, in the presence of a third monomer, nitrosoperfluorobutyric acid ($HOOC-CF_2-CF_2-CF_2-NO$), to provide cure sites for vulcanization (4.56).

$$\text{\textasciitilde\textasciitilde\textasciitilde} - NO - CF_2 - CF_2 - NO \text{\textasciitilde\textasciitilde} \atop CF_3 (CF_2)_3\text{-COOH}$$

(4.56)

carboxy-nitroso rubbers (AFMU)

AFMUs have good thermal stability. The copolymers do not burn (even in pure oxygen), but their decomposition in toxic gaseous products begins at about 200 °C.

4.2.6.3 Thermoplastic Elastomers (TPEs)

TPEs [48] are polymers that combine the processability of thermoplastics (e.g., polyethylene, PVC) and the functional performance of conventional thermoset elastomers (i.e., the chemically cross-linked rubbers). Unlike conventional rubbers, TPEs need no vulcanization and yet they show elastomeric properties within a certain temperature range. They are composed of structures including two noncompatible phases, the one assembling chains with thermoplastic sequences dispersed in the elastomer phase (this last phase can be partially vulcanized). At the usage temperature (lower to the glass temperature T_g of the thermoplastic phase), the thermoplastic domains constitute reversible physical reticulation bonds that confer the same behavior as classical elastomers after vulcanization. Because of the absence of chemically bonded chains in TPEs, the scrap generated during processing can be recycled in blends with virgin material without any loss in physical properties.

There are two main groups of commercially available TPEs (Table 4.9) [49]:

- *Block copolymers* prepared by copolymerizing two or more monomers, using either block or graft polymerization techniques. One of the monomers develops the hard or crystalline segment that functions as a thermally stable component (which softens and flows under shear, as opposed to the chemical cross-links between polymer chains in a conventional vulcanized rubber); the other monomer develops the soft, or amorphous segment, which contributes to the rubbery characteristics. They can be classified in four groups of materials:
  - *Thermoplastic polyurethane elastomers* [50] (TPU) [trade names: Desmopan (Bayer); Pellethane (Dow Plastics); Estane (BF Goodrich Chemical); Roylar (Uniroyal); Elastollan (Elastogran)],
    which are block copolymers consisting in soft amorphous segments elastomers [long flexible polyester (TPU-AU), polyether (TPU-EU) or caprolactone chains] and hard crystalline segments formed through the reaction between the diisocyanate and the low molecular weight chain extender.

– *Polyether(or ester)-ester thermoplastic elastomers* [51, 52] (CPE) [trade names: Hytrel (DuPont); Riteflex (Celanese); Ecdel (Eastman Chemical); Arnitel (Akzo)],

which are block copolymers constituted of aromatic polyester hard segments   and aliphatic polyether soft segments. They are typically prepared by a melt transesterification process involving a phthalate ester, usually dimethyl terephthalate, and a low molecular weight polyether, usually polytetramethylene ether glycol of molecular weight 6000–30,000. Various flame retardant and high heat resistance grades of copolyester-based thermoplastic elastomers have been developed by General Electric Plastics since 1985 (trade name: Lomod).

– *Polyether-amide block copolymers* (PEBA) [trade names: Pebax (Atochem); Vestamid (Creanova); Grilamid Ely, Grilon ELX (EMS); Jeffamine (Huntsman)],

which are prepared by reacting a polyether diol (or diamine) with a dicarboxylic polyamide [53]. They are composed of alternating linear rigid polyamide segments and flexible polyether segments.

– *Styrene block copolymers* (SBR, SIS, or SEBS) [trade names: *SBR*: Kraton D in the United States and Cariflex TR in Europe (Shell); Solprene T (Phillips Chemical); Styroflex [54] (BASF); SEBS: Kraton G, Elexar (Shell); Thermolast K (Kraiburg)],

which most commonly are composed of long styrene-diene triblock copolymer chains with an elastomeric central block [polybutadiene (SBR), polyisoprene (SIS) or polyethylene-butylene (SEBS)] and polystyrene end blocks. They are prepared using a process of sequential anionic polymerization. Standard grades burn easily, but fire safety can be improved by using flame retardants.

• *Olefin-based elastomeric alloys* (rubber/plastic blends with separate phases) often referred to as "thermoplastic polyolefin elastomers (TPO)," which consist of a mixtures of two or more polymers that have received appropriate treatment to give them properties significantly superior to those of simple blends of the same constituents. Linear elastomeric molecules or cross-linked (vulcanized) very small elastomeric particles, most frequently ethylene-propylene EPR or EPDM, are blended with a compatible thermoplastic (most commonly: polypropylene) in ratios that determine the stiffness of the resulting elastomer. The two types of commercial elastomeric alloys are melt-processable rubbers (MPRs) and thermoplastic vulcanizates (TPVs).

TPVs [48] are essentially a fine dispersion of highly vulcanized rubber in a continuous phase of a polyolefin. Three major commercial lines of TPVs have gained large acceptance:

• *PP/NR-VD* are blends of *in situ* cross-linked NR and PP [trade names: Vyram (Advanced Elastomer System); Thermoplastic natural rubber (MRPRA)]

• *PP/EPDM-VD* are blends of *in situ* cross-linked EPDM and polypropylene (PP) [trade names: Santopren (Monsanto); Levaflex (Bayer); Sarlink (DSM)]. Flame retardant grades are available

• *PP/NBR-VD* are blends of *in situ* cross-linked NBR and PP [trade name: Geolast (Monsanto)]

• *PP/IIR-VD* are blends of *in situ* cross-linked IIR PP [trade name: Trefsin (Advanced Elastomer System)].

MPRs are single-phase materials introduced by DuPont in 1985 (EVA/VC) called Alcryn, which is a heat-resistant plasticized alloy of partially cross-linked ethylene interpolymer (EVA) and a chlorinated polyolefin (vinylidene chloride). Its working temperature range is considered to be -40 to 107 °C. MPRs based on alloys associating high molecular weight PVC with NBR [trade names: Sunprene, Vaycron] or PP and EPDM [trade names: Vistaflex (Exxon Chemical, Advanced Elastomer System)] have also been developed.

Heat resistance and flammability of TPEs are generally comparable to that of the corresponding vulcanized rubbers. However, TPEs have the disadvantage that they soften and melt abruptly at a specific temperature range.

Table 4.9   Typical properties of thermoplastic elastomers (TPEs)

Designation	Phase nature		Hardness (Shore A or Shore D)	Service temperature range[a] (°C)	Heat resistance	Flame resistance
	elastic	thermoplastic				
Styrene block copolymers:						
SBS, SIS	polybutadiene or polyisoprene	polystyrene	35 A / 51 D	−50 to 50 / −60 to 75	P	P
SEBS	polyethylene-butylene	polystyrene	25 A / 95 A	−40 to 80 / −50 to 120	P	P
Thermoplastic polyurethane elastomers (TPU):						
TPU-EU (or AU)	polyether (or polyester)	polyurethane	62 A / 75 D	−30 to 80 / −50 to 120	F	P-G[b]
Polyether (or ester)-ester thermoplastic elastomers:						
CPE	polyether (or polyester)	polyethylene or poly-butylene terephthalate	80 A / 82 D	−40 to 110 / −60 to 130	G	F-G[b]
Polyether-amide block copolymers:						
PEBA	polyether	polyamide	75 A / 70 D	−40 to 100 / −60 to 130	G	P
Thermoplastic polyolefin elastomers (TPO):						
PP/NR-VD	cross-linked NR	polypropylene	60 A / 50 D	−35 to 80 / −50 to 120	P	P

Designation	Phase nature		Hardness (Shore A or Shore D)	Service temperature range[a] (°C)	Heat resistance	Flame resistance
	elastic	thermoplastic				
PP/EPDM-VD	cross-linked EPDM	polypropylene	45 A 50 D	−30 to 100 −60 to 140	F	P[b]
PP/NBR-VD	cross-linked NBR	polypropylene	70 A 50 D	−20 to 100 −40 to 120	F	P
PP/IIR-VD	cross-linked IIR	polypropylene	45 A 65 A	−30 to 100 −60 to 130	F	P
EVA/VC (Alcryn)	EVA	chlorinated polyolefin	60 A 80 A	−40 to 80 −50 to 120	G	F
TPE/PVC	NBR	PVC	51 A 69 A	−20 to 70 −55 to 85	F	G

Glossary: P = poor; F = fair; G = good; E = excellent
a. operating limiting temperatures vary with the grade, and essentially depend on TPE hardness;
b. flame-retardant grades available

Most of TPE standard grades burn easily, except NBR/PVC vulcanizates, which show improved flame retardancy. However, some flame retardant grades are commercially available, for instance in the category of TPU (LOI 28–30) and CPE (LOI 20–21 for some polyester-ester thermoplastic grades).

4.2.6.4 Outlook

In addition to the well-known commercially available rubbers and/or TPEs, it is possible to relate new families of derivatives that were or could be developed in the near future:

- Polyketones [55–59] [strictly alternating copolymers prepared by copolymerization of ethylene (or 1-olefins) with carbon monoxide initiated by novel palladium catalyst systems] that were placed on the market by Shell and RTP Company and recently withdrawn (trade name: Carilon),
- Elastomers and TPEs synthesized by using new generations of metallocene initiators, able to promote stereospecific (co)polymerization of α-olefins or styrene, and also capable to tolerate polar comonomers still known as severe poisons of conventional Ziegler catalysts [56, 60],
- Multiblock copolymers derived from triazinedithiols and the polyesteramide elastomers which are promising for applications in automotive [61],
- TPEs of semiinterpenetrating polymer network type synthesized using a simultaneous interpenetrating network technique [61].

As shown, most of the commercially available rubbers are generally easily flammable. Recent research has demonstrated that it is possible to prepare new rubbers with reduced flammability but the uneconomically small-scale production of such products, the difficulties of synthesis and reprocessing into products, and finally the high cost all limit the applications of these materials. For these reasons, it appears less expensive to combine existing rubbers rather than to develop totally new ones with improved flame retardancy and heat resistance. An example is given by the recent development of dynamically vulcanized blends of PVC with chloroprene rubber (CR), which show superior flame retardancy [62]. Moreover, because of the numerous demands for high temperature resistant elastomers, recent investigations were realized toward new polysiloxane rubbers including silylphenylene groups [63], as well as polyesteramide elastomers based on terephthalamide monomer and poly(tetramethylene ether)glycol [64–66]. The blending of completely miscible rubber-thermoplastic systems can be a convenient solution to increase thermal stability of individual polymers [67].

On the other hand, the modification of conventional elastomers, either by simple copolymerization techniques or by post-polymerization chemical modification, is another alternative to impart flame retardancy or improve heat resistance. For instance, grafting of char-forming monomer or reagent onto an elastomer backbone is a reasonable approach to flame retardancy. Good results were obtained by grafting vinyl monomers, such as methacrylic acid, onto styrene-butadiene and acrylonitrile-butadiene-styrene copolymers [30], as well as by covalent bonding of dialkyl (or aryl) phosphate groups along the chains of unsaturated hydrocarbonated rubbers (IR and BR) [31]. The introduction along a rubber backbone of complexing groups, able to form relatively stable complexes with convenient transition metal compounds, well known for their flame retardant and smoke suppressant properties, could be a promising approach [68].

References for Section 4.2.6

[1] W. Hofmann (Ed.): Rubber Technology Handbook, Carl Hanser, Munich, Vienna, New York, 1989
[2] K. Nagdi in: Rubber as an Engineering Material. Guideline for Users, Carl Hanser, Munich, Vienna, New York, Barcelona, 1993

[3] J.A. Brydson in: Rubbery Materials and their Compounds, Elsevier, London and New York, 1988
[4] F.L. Fire in: Combustibility of Plastics, Van Nostrand Reinhold, New York, 1991
[5] C.F. Cullis, M.M. Hirschler in: The Combustion of Organic Polymers, Clarendon Press, Oxford, UK, 1981
[6] G.C. Tesoro: *J. Polym. Sci., Macromol. Rev.* 13 (1978) 283
[7] D.F. Lawson: *Rubber Chem. Technol.* 59 (1986) 455
[8] M. Lewin, S.M. Atlas, E.M. Pearce (Eds.) in: Flame Retardant Polymeric Materials, Plenum Press, New York, 1975
[9] H.J. Fabris, J.G. Sommer: *Rubber Chem. Technol.* 50 (1977) 523
[10] P.C. Warren in W.L. Hawkins (Ed.): Polymer Stabilization, Wiley-Interscience, New York, 1972
[11] S.L. Madorsky in: Thermal Degradation of Organic Polymers, Wiley-Interscience, New York, 1974
[12] R.M. Aseeva, G.E. Zaikov in: Combustion of Polymer Materials, Carl Hanser, Munich, Vienna, New York, 1985
[13] J.M. Charrier in: Polymeric Materials and Processing, Carl Hanser, Munich, Vienna, New York, 1991
[14] D.F. Lawson in M. Lewin, S.B. Atlas, E.M. Pearce (Eds.): Flame-Retardant Polymeric Materials, Vol. 3, Chap. 2, Plenum Publishing, New York, 1982
[15] H.J. Fabris, J.G. Sommer in W.C. Kuryla, A.J. Papa (Eds.). Flame Retardancy of Polymeric Materials, Vol. 2, Ch.3, Marcel Dekker, New York, 1973
[16] D.F. Lawson, E.L. Kay, D.T. Roberts Jr.: *Rubber Chem. Technol.* 48 (1975) 124
[17] M. Moseman, J.D. Ingham: *Rubber Chem. Technol.* 51 (1978) 970
[18] R.J. Ashley, R.N. Rothon in: *Proc. MOFFIS 91, Le Mans,* p. 87, 1991
[19] D.F. Lawson: *Org. Coat. Plast. Chem.* 43 (1980) 171
[20] D.A. Dalzell, R.J. Nulph in: Proc. Soc. Plast. Eng., Annual Tech. Conf., Tech. Pap., 28th, p. 215, 1970
[21] H.E. Trexler in: *Rubber Chem. Technol.* 46 (1973) 1114
[22] K.C. Hecker, R.E. Fruzzetti, E.A. Sinclair: *Rubber Age* (NY) 105 (1973) 25
[23] R.L. Clough: *J. Polym. Sci., Polym. Chem. Ed.* 21 (1983) 767
[24] F.W. Moore, C.J. Hallada, H.F. Barry: U.S. 3,956,231 (1976) [Chem. Abstr. 85: 64122d]
[25] C.F. Cullis, M.M. Hirschler, T.R. Thevaranjan: *Eur. Polym. J.* 20 (1984) 841
[26] D.M. Florence: U.S. 3,993,607 (1976) [Chem. Abstr. 86: 56265h]
[27] J.J. Kracklauer, C.J. Sparkes, E.V. O'Grady in: Saf. Health Plast., Natl. Tech. Conf., Soc. Plast. Eng. 129, 1977
[28] C.F. Cullis, A.M.M. Gad, M.M. Hirschler*: Eur. Polym. J.* 20 (1984) 707
[29] C.F. Cullis, M.M. Hirschler: *Eur. Polym. J.* 20 (1984) 53
[30] C.A. Wilkie, M. Suzuki, X. Dong, C. Deacon, J.A. Chandrasiri, T.J. Xue: *Polym. Degrad. Stab.* 54 (1996) 117
[31] D. Derouet, F. Morvan, J.C. Brosse: *J. Nat. Rub. Res.* 11 (1996) 9
[32] J.L. Isaacs: *J. Fire Flamm.* 1 (1970) 36
[33] E. Rohde: Isolier- und Mantelwerkstoffe aus Elastomeren In: Kabel- und isolierte Leitungen. VDI-Verlag, Düsseldorf, p. 33, 1984
[34] L.R. Holloway: *Rubber Chem. Technol.* 61 (1988) 186
[35] J. Rychly, K. Vesely, E. Gal, M. Kummer, J. Jancar, L. Rychla: *Polym. Degrad. Stab.* 30 (1990) 57
[36] G. Janowska, L. Slusarski, M. Koch, U. Wincel: *J. Therm. Anal.* 50 (1997) 889
[37] C.E. McCormack: *Rubber Age* (NY) 104 (1972) 27
[38] C.J. Hilado, C.J. Casey: *J. Fire Flamm.* 10 (1979) 227
[39] T.L Laur, L.B. Guy: *Rubber Age* (NY) 102 (1970) 63
[40] G.L. Hagnauer, N.S. Schneider: *J. Polym. Sci., Part A-2,* 10 (1972) 699
[41] *Elastomerics* 120 (1988) 18
[42] P. Peccoux, F. Joachim in: Proc. MOFFIS 91, Le Mans, p. 259, 1991
[43] P.H. Potin, R. de Jaeger: *Eur. Polym. J.* 27 (1991) 341
[44] S.H. Rose: *J. Polym. Sci., Part B,* 6 (1968) 837
[45] S.H. Rose: US Patent 3,515,688 (1970) [Chem. Abstr. 73: 36299a]
[46] P. Hubin-Eschger, F. Claverie in: Proc. MOFFIS 91, Le Mans, p. 261, 1991
[47] P.J. Lieu, J.H. Magill, Y.C. Alarie: *J. Combust. Technol.* 8 (1981) 242
[48] A.Y. Coran in N.R. Legge, G. Holden, H.E. Schroeder (Ed.) in Thermoplastic Elastomers, 2nd edit., Chap. 7, Carl Hanser, Munich, 1996

[49] M. Biron, C. Biron: Thermoplastiques et TPE. Eléments de Technologie pour l'Utilisateur de Pièces Plastiques, LRCCP, p. 229, 1998
[50] R.B. Seymour in: Encyclopedia of Physical Science and Technology, Vol. 11, Academic Press, New York, 1987
[51] W.K. Witsiepe: Ger. Offen. 2,213,128 (1972) [Chem. Abstr. 78: 17337y]
[52] W.K. Witsiepe: *Nuova Chim.* 48 (1972) 79
[53] D. Judas in: Proc. Internat. Rubber Conf. IRC 98, Paris, p. 147, 1998
[54] K. Knoll, N. Niessner: *Macromol. Symp.* 132 (1998) 231
[55] E. Drent: P.H.M. Budzelaar: *Chem. Rev.* 96 (1996) 663
[56] R. Mülhaupt, F.G. Sernetz, J. Suhm, J. Heinemann: *Kautsch. Gummi Kunstst.* 51 (1998) 286
[57] A.S. Abu-Surrah, R. Wursche, B. Rieger, G. Eckert, W. Pechhold: *Macromolecules* 29 (1996) 4806
[58] A. Sommazzi, F. Garbassi: *Prog. Polym. Sci.* 22 (1997) 1649
[59] A.S. Abu-Surrah, B. Rieger in: Proc. Internat. Symp. Polycondens., Polycondensation '96, Paris, p. 362,(1996
[60] G. Xu, S. Lin: *Macromolecules* 30 (1997) 685
[61] V. Duchacek: *J. Macromol. Sci. - Phys.* B37 (1998) 275
[62] K. Ji, X. Meng, A. Du, Y. Lui in: Proc. Internat. Rubber Conf., Kobe, p. 279, 1995
[63] R. Zhang, A.R. Pinhas, J.E. Mark: *Macromolecules* 30 (1997) 2513
[64] H. Yamakawa, I. Kirikihira, Y. Kubo, K. Mori in: Proc. Internat. Rubber Conf., Kobe, p. 523, 1995
[65] H. Yamakawa, K. Isamu, J. Kubo: Jpn. Kokai Tokkyo Koho JP 06,228,310 (1994) [Chem. Abstr. 122: 163227f]
[66] H. Yamakawa, I. Kirikihira, J. Kubo, S. Shimozato: Jpn. Kokai Tokkyo Koho JP 07,313,201 (1995) [Chem. Abstr. 124: 178697z]
[67] P.P. Lizymol, S. Thomas: *Polym. Degrad. Stab.* 41 (1993) 59
[68] J.R. Ebdon, L. Guisti, B.J. Hunt, M.S. Jones: *Polym. Degrad. Stab.* 60 (1998) 401

5 Flame Retardant Plastics

5.1 Flame Retardants

S. BOURBIGOT AND M. LE BRAS

Events (see Section 1.1), such as the polypropylene fire at the BASF plant (March 1995, in Teeside, UK) [1] has shown the need to improve flame retardancy of materials and, in particular, of plastics. This clearly entails a more comprehensive appraisal of all aspects of the combustion of polymers and the ways to prevent it. Reductions in the propensity of organic materials to ignite or emit dense and/or toxic fumes are equally important.

In the last 10 years, rapid progress in the field of flame retardancy of polymers has occurred. On the market appeared new additives, new application systems leading to an ever increasing diversity of products for which flame retardancy is a dominant requirement, as well as new standards and testing methods and instruments. The developments were accompanied by a pronounced effort to gain a better understanding of the underlying principles and mechanisms governing flammability and flame retardancy and to develop new mechanistic approaches for the emerging new flame retardancy systems [2].

5.1.1 Mode of Action

A flame retardant should inhibit or even suppress the combustion process. Depending on their nature, flame retardants can act chemically and/or physically in the solid, liquid, or gas phase. They interfere with combustion during a particular stage of this process, for example, during heating, decomposition, ignition, or flame spread. The various ways in which a flame retardant can act physically or chemically are described in the following subsections. They do not occur singly but should be considered as a complex process in which many individual stages occur simultaneously with one dominating (e.g. in addition to an endothermic reaction, dilution of the ignitable gas mixture, due to the formation of inert gases, may also occur).

5.1.1.1 Physical Action

There are several ways in which the combustion process can be retarded by physical action:

- *By formation of a protective layer.* The additives can form under an external heat flux a shield with a low thermal conductivity that can reduce the heat transfer from the heat source to the material. It then reduces the degradation rate of the polymer and decreases the "fuel flow" (pyrolysis gases issue from the degradation of the material) able to feed the flame. It is the principle of the intumescence phenomenon [3]. Phosphorus additives may act in a similar manner. Their pyrolysis leads to thermally stable pyrophosphoric or polyphosphoric compounds, which form a protective vitreous barrier. The same mechanism can be observed using boric acid based additives, inorganic borates, or low melting glasses.
- *By cooling.* The degradation reactions of the additive can play a part in the energy balance of combustion. The additive can degrade endothermally, which cools the substrate to a temperature below that required for sustaining the combustion process. Aluminum trihydrate (ATH) acts under this principle, and its efficiency depends on the amount incorporated in the polymer.

- *By dilution.* The incorporation of inert substances (e.g., fillers such as talc or chalk) and additives, which evolve inert gases on decomposition, dilutes the fuel in the solid and gaseous phases so that the lower ignition limit of the gas mixture is not exceeded. In a recent work, G. Marosi et al. [4] showed the insulating effect of a high amount of ash formed from certain silica-based fillers in flame retarded systems. Moreover, these authors point out that an opposite effect occurs as well, when the thermal degradation of the polymer in the bulk is increased due to the increased heat conductivity of the filled material.

5.1.1.2 Chemical Action

The most significant chemical reactions interfering with the combustion process take place in the condensed and gas phases.

- *Reaction in the condensed phase.* Here, two types of reaction can take place. Firstly, breakdown of the polymer can be accelerated by the flame retardant, causing pronounced flow of the polymer and, hence, its withdrawal from the sphere of influence of the flame that breaks away. Second, the flame retardant can cause a layer of carbon (charring) to form on the polymer surface. This can occur, for example, through the dehydrating action of the flame retardant generating double bonds in the polymer. These form the carbonaceous layer via cyclizing and cross-linking processes.
- *Reaction in the gas phase.* The radical mechanism of the combustion process, which takes place in the gas phase, is interrupted by the flame retardant or its degradation products. The exothermic processes that occur in the flame are thus stopped, the system cools down, and the supply of flammable gases is reduced and eventually completely suppressed. In particular, metallic oxides can act as flame inhibitors. The active radicals $HO\cdot$ are adsorbed on the surface of oxide particles. A part of the collision energy is transferred to the oxides and $HOO\cdot$ radicals, which are less reactive than the initial $HO\cdot$ radicals are formed. As another example, the "hot" radicals $HO\cdot$ and $H\cdot$ can react in the gas phase with other radicals, such as halogenated radicals $X\cdot$ issued from the degradation of the flame retardant, to create less energetic radicals [5].

Flame retardant additive systems may be used alone or in association with other systems (sometimes using low amounts of the last system) in polymeric materials to obtain a synergistic effect, that is, the protective effect is higher than this is assumed from addition of the separate effects of each system. M. Lewin has recently given a general approach for this effect [6]. Phosphorus-nitrogen, halogen-antimony, bromine-chlorine, bromine-phosphorus and intumescent coadditive synergisms in particular have been discussed elsewhere by this author [2].

5.1.2 The Most Important Flame Retardants and Their Mode of Action

The flame retardant families consist of organic and inorganic compounds based on phosphorus, halogen, nitrogen, and of inorganic compounds such as aluminum trihydrate (ATH), magnesium hydroxide, and antimony trioxide, that may act in the gas and/or condensed phase [6–8].

5.1.2.1 Flame Retardants Acting Through Gas Phase Mechanism

The most effective commercial flame retardants in general use are halogen-containing compounds often used in combination with metal compounds. The widely accepted mechanism of fire retardancy of these systems is based on chemical action, which occurs mainly in gas phase [2], although some studies have shown that condensed phase reactions with the polymer may also be involved [6–10].

Radical Transfer

Due to the release of hydrogen halide during decomposition, halogen-containing compounds interrupt the chain reaction of combustion by replacing the highly reactive $\cdot$OH and H$\cdot$ radicals by the less reactive halogen X$\cdot$ according the following reactions (1–3) where RX and PH represent the flame retardant and the polymer, respectively:

$$R\text{-}X + P\text{-}H \rightarrow H\text{-}X + R\text{-}P \tag{5.1}$$

$$H\text{-}X + H\cdot \rightarrow H_2 + X\cdot \tag{5.2}$$

$$H\text{-}X + \cdot OH \rightarrow H_2O + X\cdot \tag{5.3}$$

By dissipating the energy of the $\cdot$OH radicals by trapping, the thermal balance is modified. This strongly reduces the combustion rate.

Flame inhibition studies have shown that the effectiveness of halogens decreases in the order: HI > HBr > HCl > HF [11, 12]. Brominated and chlorinated organic compounds are generally used because iodides are thermally unstable at processing temperatures and the effectiveness of fluorides is too low. The choice depends on the type of polymer, for example, in relation to the behavior of the halogenated flame retardant under processing conditions (stability, melting, distribution) and/or the effect on properties and long-term stability of the resulting material. In particular, it is advisable to use an additive that supplies the halide to the flame in the same temperature range at which the polymer decomposes to combustible volatile products. Thus, fuel and inhibitor would reach the gas phase together according to the "right place at the right time" principle [13].

In this connection, it is important to note that the assessment of the thermal degradation behavior of the flame retarded material is carried out on the material itself and should not be deduced solely from that of the polymer and flame retardant heated separately. This is because chemical interactions can take place between the degrading polymer and the additive, which are not predictable from the degradation of the pure polymer or the sole additive.

In a thorough study of the mechanism of action of a typical flame retardant, chloroparaffin (CP) (Cl = 70 %), it was shown that the CP modified the thermal degradation of polyethylene (PE), polypropylene (PP), and polystyrene (PS) [14]. It was shown that while CP undergoes dehydrochlorination, thermal volatilization of the polymers takes place, whereas under the same experimental conditions, the virgin polymers do not give way to measurable volatilization. Moreover, it was found, that under the same conditions, CP promotes extensive chain scission in PP and PS, but cross-linking in PE.

Destabilization of these polymers by CP should be beneficial in terms of flame retardancy because it induces the formation of the fuel at the temperature at which HCl is evolved. In the absence of interaction, thermal volatilization of the polymers would occur at a temperature at least 100 °C higher than that of CP dehydrochlorination. Thus, the occurrence of a polymer-additive interaction in the condensed phase may simply optimize the gas phase flame retardant action of HCl. Nevertheless, the evolution on heating of a potential flame inhibitor does not ensure that the flame retardant action depends solely on flame poisoning. Indeed, halogenated flame retardants are thermally reactive species, which can react with the polymeric matrix.

Radical Recombination

Several metal compounds, which when used alone do not impart significant flame retardant properties to polymers, can strongly enhance the effectiveness of halogenated compounds [10, 11, 14–18]. These mixtures evolve on heating metal halides, which are well-known flame inhibitors with greater effectiveness than hydrogen halides. The metal halides give metal oxides in the flame, with elimination of hydrogen halide. The most widely used synergistic agent is antimony trioxide. In the mechanism of action, it is assumed that the flame retardancy is due to SbX_3 according to the following reactions:

$$R\text{-}X + P\text{-}H \rightarrow H\text{-}X + R\text{-}P \tag{5.4}$$

$$Sb_2O_3 + 2\,HX \rightarrow 2\,SbOX + H_2O \tag{5.5}$$

$$5\,SbOX \rightarrow Sb_4O_5X_2 + SbX_3 \tag{5.6}$$

$$4\,Sb_4O_5X_2 \rightarrow 5\,Sb_3O_4X + SbX_3 \tag{5.7}$$

$$3\,Sb_3O_4X \rightarrow 4\,Sb_2O_3 + SbX_3 \tag{5.8}$$

Among all species formed, SbX_3, HX, and H_2O can evolve under the conditions of the combustion process. Therefore, these compounds play the role of an inhibitor in the gas phase.

A similar catalyzed radical combination process has also been suggested to explain the flame inhibition effect of volatile phosphorus-containing flame retardants, which are oxidized to PO in the flame [11, 19, 20].

As an illustration, Table 5.1 shows typical examples of using halogen-containing compounds in PP. The thermal stability of these compounds compared to PP pyrolysis temperature is satisfactory for delivering HBr in high enough concentrations during burning. But when the processing temperatures of PP are over 230 °C, aromatic brominated flame retardants are recommended for PP because of their better thermal stability. Nevertheless, brominated compounds are generally more expensive than chlorinated ones and the latter can constitute an interesting alternative when flame retardancy and rheological (loss due to higher loading) performances are not very strictly required.

5.1.2.2 Flame Retardant Polymer Interaction in the Condensed Phase

As a rule, in the condensed phase, there is a chemical interaction between the flame retardant, which is usually added in substantial amounts, and the polymer. This interaction occurs at temperatures lower than those of the pyrolytic decomposition. Two principal modes of this interaction have been suggested: dehydration and cross-linking. They have been established for a number of polymers including cellulosics and synthetics [21, 22]. Another route to modify the condensed phase of the burning material is the poly-blending approach [2, 23–26].

Dehydration

The varying efficiency of phosphorus compounds in different polymers has been related to their susceptibility to dehydration and char formation. Cellulosics are adequately flame retarded with approx. 2 % of P, whereas 5–15 % are needed for polyolefins [16]. The interaction of P-derivatives with the polymers not containing hydroxyls is slow and has to be preceded by an oxidation. It has been suggested that 50–99 % of the P derivatives are being lost by evolution, possibly of P_2O_5 or other oxides formed from the pyrolysis of the P derivative [27]. This may be one of the reasons for the low yield.

Two alternative mechanisms have been proposed for the condensed phase dehydration of cellulosics with acids and acid-forming agents of phosphorus and sulfur derivatives. Both mechanisms lead to char formation [28]: (a) esterification and subsequent pyrolytic ester decomposition (see Eq. 5.9) and (b) carbonium ion catalysis (Eq. 5.10):

$$R_2CH\text{-}CH_2OH + ROH \text{ (acid)} \rightarrow R_2CH\text{-}CH_2OR + H_2O \rightarrow R_2C\text{=}CH_2 + HOR \text{ (acid)} \tag{5.9}$$

$$R_2CH\text{-}CH_2OH + H^+ \rightarrow R_2CH\text{-}CH_2OH_2^+ \rightarrow R_2CH\text{-}CH_2^+ + H_2O \rightarrow R_2C\text{=}CH_2 + H^+ \tag{5.10}$$

Table 5.1 Performance of flame retarded PP loaded with typical halogen-containing additive systems

Brominated aliphatic flame retardants						
Ingredients	**Percentage (wt%)**					
PP	100	97.15	58			
HBCD		2.15				
EBTPI			22			
Sb_2O_3		0.7	6			
Talc			14			
Properties						
UL 94 V Rating [1.6 mm (0.06 in.)]	NR	V-2	V-0			
Dripping	Yes	Yes	No			
Brominated aromatic flame retardants						
Ingredients	**Percentage (wt%)**					
PP	100	91.1	93.0	90.9	88.0	64.3
TBBA		5.2				
DBDPO			4.8			23.8
INDAN				5.8		
PBB-PA					8.5	
Sb_2O_3		1.7	1.6	2.6	2.8	11.9
Additives		2.1	0.6	0.7	0.6	
Properties						
MFI (g/10 min)	6	7	4	–	5	–
UL 94 V Rating [3.2 mm (0.13 in.)]	NR	V-2	V-2	V-2	V-2	V-0 (1.6 mm)
Dripping	Yes	Yes	Yes	Yes	Yes	No
Notched Izod impact (J/m)	21.5	26.7	21.5	21.5	21.5	21.5
Oxygen index (%)	18	–	–	–	–	22

Table 5.1 (Continuation)

Chlorinated flame retardants					
Ingredients	**Percentage (wt%)**				
PP	100	55	61	60	63
Dechlorane		35	30	27	
Chlorinated paraffin					27
Sb_2O_3		4	6	13	10
Zinc borate		6	4		
Properties					
UL 94 V Rating (1.6 mm)	NR	V-0	V-0	V-0	V-0
UL 94 V Rating (3.2 mm)	NR	V-0	V-0	V-0	V-0
Oygen index (%)	18	–	–	–	27

Glossary: NR = No Rating; MFI = melt flow index; PP = homopolymer polypropylene; HBCD = hexabromocyclododecane; EBTPI = ethylene bis(tetrabromophthalimide); TBBA = tetrabromobisphenol A; DBDPO = decabromodiphenyl oxide; INDAN = octabromotrimethylphenylindan; PBB-PA = poly(pentabromobenzylacrylate); dechlorane = 2 mol. hexachlorocyclopentadiene + 1 mol. cyclopentadiene

P-compounds reduce the flammability of cellulosics. The flame retardancy is achieved primarily by an esterification-ester decomposition mechanism (Eq. 5.9), which, being relatively slow, is affected by the fine structure of the polymer. Less ordered regions pyrolyze at a lower temperature than the crystalline regions and decompose before all of the phosphate ester can decompose, which decreases the flame retarding effectiveness and necessitates a higher amount of P. Sulfated celluloses, obtained by sulfation with ammonium sulfamate, are dehydrated by carbonium ion disproportionation (Eq. 5.10) and show a strong acid activity, which rapidly decrystallizes and hydrolyzes the crystalline regions.

Cross-Linking

It was early recognized that cross-linking might reduce the flammability of polymers. In particular, cross-linking promotes char formation in pyrolysis of cellulose [29] and is assumed to be operative in P-N synergism [30]. Curing increases the limiting oxygen index (LOI) of phenolic resins, but it does not markedly alter the flammability of peroxy resins [31] although studies of Rose et al. [32] showed that charring and flame retardant properties of epoxy resins depend on stoichiometry and curing conditions of the resins. In the case of cross-linked PS (obtained by copolymerizing it with vinylbenzyl chloride) a drastic increase in char formation is observed when comparing to un-cross-linked PS. PS pyrolyzes predominantly to monomer and dimer units almost without char. Cross-linked PS yielded 47% of char [33]. Cross-linking and char formation were obtained by an oxidative addition of organometallics to polyester [34].

The flame retardancy of a polymer is in some way connected to both the number of cross-links and to the strength of the bonds that make up the cross-linked structure [35]. It was observed [36] that cross-linked methyl methacrylates copolymers prepared by the radical polymerization of various dimethacrylates with methyl methacrylate, degrade at essentially the same temperature as does the homopolymer, poly(methyl methacrylate) (PMMA). The most likely explanation is that the cross-linked structure is produced by very weak bonds that may be relatively easily cleaved thermally. On heating the cross-linked structure is lost, and the resulting polymer does not differ from that of methyl methacrylate and its degradation proceeds in the same way [36].

Cross-linking promotes the stabilization of the cellulose structure by providing additional, covalent bonds between the chains, which are stronger than the hydrogen bonds, and which have to be broken before the stepwise degradation of the chain occurs on pyrolysis. However, low degrees of cross-linking can decrease the thermal stability by increasing the distance between the individual chains and consequently weakening and breaking the hydrogen bonds. Thus, although the LOI of cotton increases marginally with increasing formaldehyde cross-linking, that of rayon markedly decreases [37].

Recently, cross-linking of PS by Friedel-Crafts chemistry under fire conditions was investigated by Wilkie et al. [38]. The system is a copolymer of styrene and p-vinylbenzyl alcohol, which can be cross-linked with PS in the presence of a catalyst. The cross-linking temperature can be controlled using a particular catalyst such as 2-ethyldiphenyl phosphate. Cone calorimetry of these systems shows a significant decrease of the heat release rate, that is, of the degradation and/or combustion rate.

A recent trend in flame retardancy is a modification of polymer physical structure (morphology) by means of polymer blends. Some examples are poly(vinyl chloride) (PVC) – acrylonitrile-butadiene-styrene (ABS) blends and the blends of poly(2,6-dimethyl-1,4-phenylene) oxide (PPO) with PS. In the case of PPO-PS blends, it has been shown that the good char forming ability of the PPO greatly helps flame retardancy, requiring only the addition of a triaryl phosphate, which serves to flame retard the polystyrene pyrolysate, which reaches the vapor phase [39]. Another example is the use of polyphosphazene synthesized from hexachlorophosphazene as a flame retardant blended with polymers [40].

5.1.2.3 Intumescence Flame Retardants and Char Formation

Flame retarding polymers by intumescence is essentially a special case of a condensed phase mechanism. Intumescent systems interrupt the self-sustained combustion of the polymer at its earliest stage, that is, the thermal degradation with evolution of the gaseous fuels. The intumescence process results from a combination of charring and foaming of the surface of the burning polymer (observed between 280 °C and 430 °C under air) using the PP-ammonium polyphosphate (APP)-pentaerythritol (PER) model system [41]. The resulting foamed cellular charred layer, the density of which decreases with temperature [42], protects the underlying material from the action of the heat flux or of the flame.

Because thermal protection is the main purpose of intumescent materials, several models have been developed to study the effects of intumescence on heat transfer to the underlying surface. These are primarily one-dimensional in nature, and concentrate on the effects of swelling on the thermal properties of a coating.

The one-dimensional models [43–48] that have been developed to investigate intumescent behavior apply the equations of energy and mass conservation to some variation of the geometry. The thickness of virgin material and char layers and the location of the pyrolysis zone are functions of time, and each layer is assigned its own values of thermodynamic parameters. All of these one-dimensional models rely on empirical information about the amount and rate of expansion. Their scope is limited to heat transfer, and they are not capable of supplying insights into the swelling process itself.

Another approach has been used to investigate the heat transfer for the specialized geometry of an intumescent penetration seal [49]. A three-dimensional model was then developed. Temperature as a function of time is predicted at locations within the caulk and along the concrete and pipe walls. Limitations of the model are due to the control volume approach, the simple representation of intumescent expansion, and to the difficulty of obtaining accurate thermal property data for the intumescent char.

As demonstrated by the one-dimensional heat transfer models, swelling is central to the fire protective capabilities, and a fundamental understanding of the mechanisms that cause expansion is important. Unlike foams and softening coals, temperature gradients and heat transfer play a central role in intumescent behavior. In particular, the effect of the growing bubbles on the temperature field cannot be neglected. The sizes of nearly all bubbles may be quite different due to the large temperature gradients within the intumescent melt. Considering this, a promising three-dimensional model is being developed at NIST [50–52] that incorporates bubble and melt hydrodynamics, heat transfer, and chemical reactions (Fig 5.1).

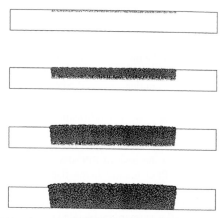

Figure 5.1   Bubble model (development of 10,000 bubbles with time, when a heat flux is applied to the upper surface of a volume) [52]

In this model, the intumescent system is represented as a highly viscous incompressible fluid containing a large number of expanding bubbles. The bubbles obey equations of mass, momentum, and energy on an individual basis according to the values of local parameters, and their collective behavior is responsible for the swelling and flame retardant properties of the material.

Effect of ventilation inside the bubbles has been considered by I.S. Reshetnikov and N.A. Khalturinskii in modeling heat transfer in combustion of polymeric materials leading to foamed cokes [53, 54] or intumescence [55]. These authors have proposed the high porosity in terms of bubble size does lead to the best resistance of the intumescent protection during a fire. This trend has been recently verified by M. Le Bras et al. [56], who showed, studying the dynamic properties of an intumescent system (viscosity and expansion of the coating), that expansion of the coating is a two-step process and that the first expansion (at temperatures lower than 430 °C) is the only step responsible of the protective behavior.

Chemistry of Intumescence

Generally, intumescent formulations contain three active additives: an acid source (precursor for catalytic acidic species), a carbonific (or polyhydric) compound, and a spumific (blowing) agent (Table 5.2). In a first stage ($T < 280\,°C$), the reaction of the acidic species with the carbonization agent takes place with formation of esters mixtures. The carbonization process then takes place at about 280 °C (via both Friedel-Crafts reactions and a free radical process [57]). In a second step, the spumific agent decomposes to yield gaseous products, which cause the char to swell ($280 \leq T \leq 350$ °C). This intumescent material then decomposes at the highest temperatures and loses its foamed character at about 430 °C (temperature ranges are characteristic of the extensively studied PP-APP-PER system, in which APP plays two roles: acid source and blowing agent [41, 58, 59, and references therein]).

Table 5.2   Examples of components of intumescent coatings [60]

a) Inorganic acid source
1. *Acids*
  Phosphoric
  Sulfuric
  Boric
 	
2. *Ammonium salts*
Phosphates, polyphosphates
Sulfates
Halides
 	
3. *Phosphates of amine or amide*
Products of reaction of urea or guanidyl urea with phosphoric acids
Melamine phosphate
Product of reaction of ammonia with P$_2$O$_5$
 	
4. *Organophosphorus compounds*
Tricresyl phosphate
Alkyl phosphates
Haloalkyl phosphates
	b) Polyhydric compounds
	Starch
	Dextrin
	Sorbitol
	Pentaerythritol, monomer, dimer, trimer
	Phenol-formaldehyde resins
	Methylol melamine
	
	c) Amines/amides
	Urea
	Urea-formaldehyde resins
	Dicyandiamide
	Melamine
	Polyamides
	
	d) Others
	Charring polymers (PUR, PA-6)

The carbonaceous material formed from the additives plays two different chemical roles in the flame retardancy process:

- It contains free radical species, which react with the gaseous free radical products formed during the degradation of the polymer. These species may also play a part in termination steps in the free radical reaction schemes of the pyrolysis of the polymer and of the degradation of the protective material in the condensed phase.
- It is a support for acidic catalytic species, which react with the oxidized products formed during the thermooxidative degradation of the material.

The material resulting from the degradation of an intumescent formulation is a heterogeneous substance, composed of "trapped" gaseous products in a phosphocarbonaceous cellular material, that is, the condensed phase.

This condensed phase is a mixture of solid and liquid phases (acidic tars) possessing the dynamic properties of interest, which allows the trapping of the gaseous and liquid products resulting from the degradation of the polymer. The carbonaceous fraction of the condensed phase consists of polyaromatic species that are organized in stacks characteristic of a pregraphitization stage (Fig. 5.2).

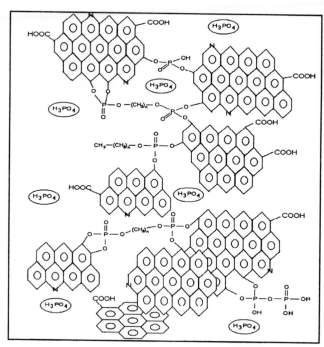

Figure 5.2   Intumescent coating resulting from a polyethylenic formulation (additive: APP/PER) heat treated at 350 °C [61]

The phosphocarbonaceous material is then a multiphase material composed of crystalline macromolecular polyaromatic stacks bridged by polymer links and phosphate (poly-, di-, or orthophosphate) groups, crystalline additive particles, and an amorphous phase that encapsulates the crystalline domains. This amorphous phase is composed of small polyaromatic molecules, easily hydrolyzed phosphate species, alkyl chains formed via the degradation of the additive system, and fragments of the polymer chain. It governs the protective behavior of the coating: this phase has to be voluminous enough to coat the crystalline domains perfectly and has to show an adequate rigidity/viscosity, which yields the dynamic properties of interest (it avoids dripping and accommodates stress induced by solid particles and gas pressure).

Protection from Intumescence

The protection mechanism is based on the charred layer acting as a physical barrier that slows down heat and mass transfer between the gas and the condensed phases. The limiting effect for fuel evolving is proved by the presence of the polymer chains in the intumescent material. In addition, the layer inhibits the evolution of volatile fuels via an "encapsulation" process related to its dynamic properties. Finally, it limits the diffusion of oxygen to the polymer bulk. The limitation of the heat transfer (thermal insulation) has been discussed in recent studies [42, 62, 63].

As an example (Fig. 5.3), it is shown that the temperature is reduced using intumescent systems in PP.

In the case of the PP-AP 750 (AP 750 is ammonium polyphosphate with an aromatic ester of tris(2-hydroxyethyl)-isocyanurate and bound by an epoxy resin [64] supplied by Clariant as Exolit AP 750) system in comparison with P-APP/PER at $t = 1600$ s, for example, the temperature (depending on the distance) lies between 200 and 350 °C in the case of PP-APP/PER,

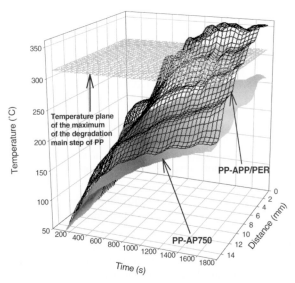

Figure 5.3   Temperature profiles [temperatures measured under the conditions of the cone calorimeter with a heat flux equaling 50 kW/m² and with a sheet 2 cm (0.80 in.) thick] of PP-APP/PER and PP-AP750 [62]

and lies only between 200 and 250 °C in the case of PP-AP 750. One remarks that the temperatures of PP-AP 750 are always below the temperature (approx. 310 °C) where the degradation rate of PP is at maximum. In the case of PP-APP/PER, the temperature reaches 310 °C at about 900 s and the degradation rate can increase fast. This can therefore explain why the PP-AP 750 system is more efficient than the PP-APP/PER one.

Consequently, the stability of the intumescent material limits the formation of fuels and leads to self-extinction in standard conditions. Oxygen consumption calorimetry in a cone calorimeter confirms the low degradation rate related to the presence of a surface intumescent material (typical examples are presented in Figure 5.4 [62]).

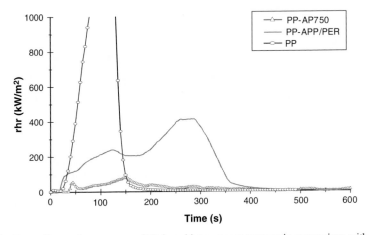

Figure 5.4   Rate of heat release curves of PP-based intumescent systems in comparison with virgin PP (loading = 30 wt%) [62]

Synergy and Char Formation

The intumescent additive systems usually developed are mixtures of at least three ingredients:

- A precursor of a carbonization catalyst such as APP [41], diammonium diphosphate (PY) [41], or diammonium pentaborate (APB) [57],
- A carbonization agent such as PER, xylitol (XOH), mannitol (MOH), *d*-sorbitol (SOH), β-cyclodextrine (BCOH) [65, 66] or polyamide 6 (PA 6) [67] and polyurethanes (PUR) [68, 69],
- A synergistic agent such as zeolites [70], clay materials [71], BCOH [65] and/or zinc borate [4, 72],
- A compatibilizer such as an ethylene-vinyl acetate (8 %) copolymer (EVA-8) or other functionalized polymers [73] and borosiloxane elastomer as an encapsulation agent [74, 75].

The functionalized copolymer generally plays two roles in the formulations: first, it increases the LOI and, second, it is a compatibilizer of the additives in the polymer (as an example, EVA-8 maximizes the interfacial bonding and so prevents rejection of the mineral additive throughout a polymer matrix [70]).

Addition of an active agent in an additive system may lead either to an additional [65], antagonistic [71], or a synergistic effect [65–76]. The part played by zeolites in intumescent formulations has been extensively studied by Bourbigot et al. [76–80]. A typical example of a synergistic effect obtained using a zeolite 13X in a PS-based formulation is presented in Fig. 5.5 (the LOI being used as the rating test). It shows that the addition of a low 13X synergist amount (about 0.9 wt%) leads to a sharp increase and that addition in excess of 1 wt% leads to a decrease of the flame retardant performance [81].

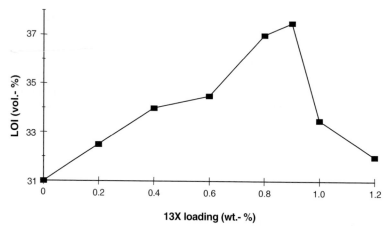

Figure 5.5   LOI values of PS-APP/PER/13X formulations vs. the amount of zeolite 13X [81]

A recent study of Le Bras et al. [65] concerns the use of BCOH as a carbonization agent in intumescent polyol-based formulations. The thermal degradation of this starch derivative leads to high amounts of carbonaceous residue and so may be a typical carbon source for intumescence. As a typical example, the study shows that in low-density polyethylene (LDPE) formulations, BCOH is not a carbon source of interest.

Nevertheless, it may be added either as a synergist or as an antagonistic agent in LDPE-based formulations in association with other polyols produced by the agrochemical industry (flame retardant performances of two typical intumescent systems are presented in Fig. 5.6).

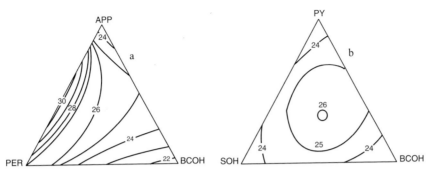

Figure 5.6   LOI values of LDPE-APP/PER/BCOH (a) and LDPE-PY/SOH/BCOH (b) formulations in the Scheffé {3–3} lattice used for experiments with mixtures [65]

These studies show that an additive may play very different roles depending on its concentration in the polymeric material and its association with the other different additives or fillers.

The part played by the synergists is not well known. A direct relationship between the LOI and the carbonization mode and/or the amount of stable carbonaceous materials formed from flame retardant formulations has been previously reported [82, 83]. A recent study of Le Bras et al. [57], which considers several new intumescent formulations or carbonizing additive systems, shows that this relationship does not exist whatever is the considered carbonaceous material (intumescent, unexpanded or "high temperature") and proposes that the flame retardant performance depends on the "quality of the carbon."

5.1.2.4 Flame Retardants Acting Through Physical Effects

The flame retardant effect of additives undergoing an endothermic decomposition in the solid phase has been modeled in the literature [84–87]. The first approach was to consider thermally thin thermoplastics [85–87] and to investigate the effect of additives (inert and heat sink) on the flammability of the material. In a first paper [85], Nelson et al. analyzed the variation of the critical heat flux for ignition with the decomposition kinetics of the fuel and with the properties of inert and heat sink additives. They suggested that experimental results should be presented in terms of a dilution factor rather than by the usual wt% additive. They then proposed a transient theory for the critical heat flux for ignition and developed supercritical flammability properties [88]. They investigated a method to define the flame retardancy effect of inert additives [86]. Finally, they extended their model to include post-ignition behavior and to investigate the effect of adding heat sink additives to the polymer (fuel), identifying the key parameters governing additive performance [88].

Staggs [84] modeled the effect of a simple solid-phase additive on the thermal degradation of a solid under thermally thick and thin conditions. The model simulates the conditions of the cone calorimeter. He assumed that the thermal parameters of the two solid phases were identical and he discussed the role of the components' degradation mechanisms in relation to the overall degradation of the solid.

Flame retardancy due to physical effects usually requires relatively large amounts of additives: 50–65 % in the case of ATH and magnesium hydroxide ($Mg(OH)_2$). The activity of these additives consists in:

- Dilution of the polymer in the condensed phase,
- Decreasing the amount of available fuel,
- Increasing the amount of thermal energy needed to raise the temperature of the composition to the pyrolysis level, due to the high heat capacity of the fillers,
- Increasing the enthalpy of decomposition – emission of water vapor,

- Dilution of gaseous phase by water vapor – decrease of fuel and oxygen amount in the flame,
- Possible endothermic interactions between water and decomposition products in the flame [reactions (5.11) and (5.12)],
- Decrease of feedback energy to the pyrolyzing polymer,
- Insulating effect of the oxides remaining in the char,
- Charring of the materials [89–94].

$$2\,Al(OH)_3 \rightarrow Al_2O_3 + 3\,H_2O \quad \Delta H = 298\ kJ/mol \qquad (5.11)$$

$$Mg(OH)_2 \rightarrow MgO + H_2O \qquad \Delta H = 380\ kJ/mol \qquad (5.12)$$

The decompositions of these hydrates are in the temperature range 180–240 °C for $Al(OH)_3$ and 330–460 °C for $Mg(OH)_2$. This implies that $Al(OH)_3$ cannot be used in thermoplastics other than polyethylenics and PVC because of its relatively low processing temperature. It has been shown that in these kinds of polymers, no interaction and degradation occur during compounding [95, 96].

Considerable attention is given to the particle size-surface area, as it is known to influence greatly the melt flow of the treated polymer, and to the crystal size of the additive. Hydrophobic coatings on the particles of the additives are applied to facilitate dispersion of the highly hydrophilic metallic oxides in the polymer. The possible effect of these coatings on the combustion behavior of the polymers is not clear. Another important property appears to be the "surface free energy" determining the reactivity of the additive with acidic groups in polymers, which might produce cross-linking and reduce melt flow, as well as with acidic groups in air [91–93].

Some evidence has been presented pointing to the possibility of an interaction between the endothermic additives and several polymers during pyrolysis and combustion, in the presence of some metallic catalysts [97, 98], such as nickel(II) oxide, manganese borate, 8-hydroxychinolino-copper, and ferrocene. These interactions are believed to be the cause of a highly significant increase in char and LOI. Other compounds, that is, nickel(III) oxide and metal acetylacetonates, gave negative results.

More recently, it has been shown that zinc borates (FB415: $4ZnO \cdot B_2O_3 \cdot H_2O$; FBZB: $2ZnO \cdot 3B_2O_3 \cdot 3.5H_2O$) can be used as synergists in EVA-ATH and EVA-Mg(OH)$_2$ flame retardant formulations (Figs. 5.7 and 5.8) and as smoke suppressants (Fig. 5.9) [94, 99, 100].

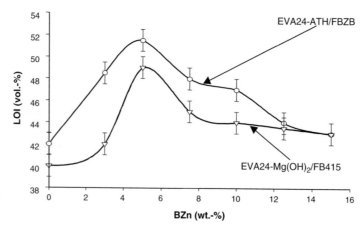

Figure 5.7   LOI values versus the substitution in zinc borates of the formulations EVA-24-ATH/FBZB and EVA-24-Mg(OH)$_2$/FB415 (total loading remains constant) [94]

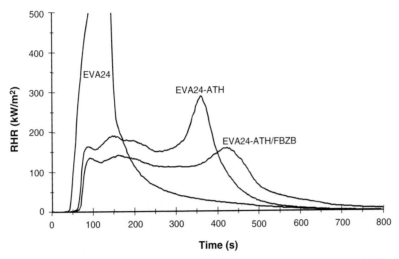

Figure 5.8   Rate of heat release values versus time of the formulations EVA-24-ATH and EVA-24-ATH/FB415 in comparison with the virgin EVA-24 [94]

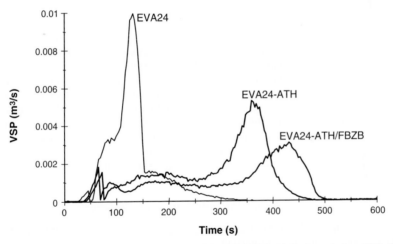

Figure 5.9   Volume of smoke production (VSP) values versus time of the formulations EVA-24-ATH and EVA-24-ATH/FB415 in comparison with the virgin EVA-24 [94]

It has been demonstrated that the decomposition of ATH to Al_2O_3 during the heating of the polymer results in an increase of the ignition time. Moreover, the formation of Al_2O_3 *in situ* from ATH during the combustion of the polymer promotes charring, which reduces the flammability of the material. Concurrently, zinc borate degrades and it is proposed that a vitreous protective coating is created, which yields a more efficient char [94–100].

The effect of zinc hydroxystannate (ZHS, a classical flame retardant and smoke suppressant additive for halogen-containing polymers) also should be mentioned because of the high number of relevant patents and articles [101 and references therein]. In particular, its association with ATH and $Mg(OH)_2$ in PVC leads to improved FR properties (increase of LOI values, Rate of Heat Release reduction) and suppression of smoke generation.

5.1.2.5 Phosphorus-Containing Flame Retardants

Phosphorus-containing additives are active mainly in the condensed phase and influence pyrolysis and char formation. In the condensed phase, they form phosphoric and related acids, which act as "heat sink" as they undergo endothermic reaction. They then form a thin glassy coating, which is a barrier that lowers the evolution of combustible gases in the gas phase, and limits the diffusion of oxygen and the heat transfer. In the gas phase, they stop the free radical oxidation process of the carbon at the carbon monoxide stage, preventing the highly exothermic reaction of carbon dioxide formation. When cellulosics or functionalized polymers such as polyamide 6 (PA 6) [102] or polyurethane (PUR) [103] are subjected to heat, a charred layer is formed at the surface of the material insulating the substrate (similar mechanism as the intumescence phenomenon).

Red phosphorus is recommended in polymers containing oxygen such as polyamides (PA), poly(ethylene terephthalate) (PET), or polycarbonate (PC), but can be used in polyolefins, ABS, high-impact polystyrene (HIPS), epoxy resins, or polybutylene terephthalate (PBT), as well. Under the action of oxygen red phosphorus is transformed, mainly by oxidation, into phosphoric acid or phosphoric anhydride, which can act as a physical barrier when the polymer undergoes thermal degradation. Upon heating, these acids generate polyphosphoric acid, which catalyzes the dehydration of organic compounds and then promote charring. The mechanism of action has been extensively investigated in the literature for various polymers [104–112].

Other phosphorus compounds can be used as flame retardants (as additives or grafted on the polymer chains) such as mineral phosphates, organic phosphates, phosphites, phosphonates, or phosphorinanes [113–115].

Organophosphorus additives work well in oxygen-containing polymers [116], such as polyesters, PA [115], PET [117], or PURs, but are not effective as such in hydrocarbon-based polymers. The use of organophosphorus additives and organophosphorus functionality to modify the thermal and flame retardant behavior of polymers has been recently reviewed by J. W. Wheeler et al. [116]. They propose that these flame retardant additives work in the condensed and the solid phases, promoting and yielding the formation of large amounts of char. The authors then discuss the use of organophosphorus functionality on the polymer chain, particularly phosphine oxide polyamide, aromatic phosphine oxide sulfone polyamide, and aromatic-aliphatic phosphine oxide polyamides. They show that the resulting materials present a wide processability range and increased thermal stability and suggest their use as flame retardant materials.

5.1.2.6 Flame Retardants Acting Through Combined Mechanism

In the pursuit of approaches for flame retarding polymers, the use of inorganic compounds or hybrid compounds seems to be promising.

Low-Melting Glasses

Low-melting phases (glasses) as polymer additives have been proposed in the literature [118–120]. Low-melting glasses can improve the thermal stability and flame retardancy of polymers by:

*   Providing a thermal barrier for both non-decomposed polymer and the char, if any, forming as a combustion product,
*   Providing a barrier to retard oxidation of the thermally degrading polymer and combustion of char residue,
*   Providing a "glue" to hold the combustion char together and giving it structural integrity,
*   Providing a coating to cover or fill in voids in the char, thus providing a more continuous external surface with a lower surface area,
*   Creating potentially useful components of intumescent polymer additive systems.

Kroenke et al. [119] developed a low-melting glass flame retardant system for PVC based on $ZnSO_4$ - K_2SO_4 - Na_2SO_4. This system proves to be an excellent char former and a smoke depressant. Flame retardant low-melting glass systems with the transition metals Al, Ca, Ce, Ni, Mn, Co, and V were also tested.

Gilman et al. [121] have shown that silica gel in combination with potassium carbonate is an effective flame retardant (at mass fraction of only 10 % total additive) for a wide variety of common polymers such as PP, PA, PMMA, poly(vinyl alcohol), cellulose, and to a lesser extent PS and styrene-acrylonitrile. The authors proposed a mechanism of action for these additives through the formation of a potassium silicate glass during combustion.

Levchik et al. [122, 123] studied the action of ammonium pentaborate (APB) in PA 6 as a glass former. They found that the degradation of APB into boric acid and boron oxide provides a low-melting glass at the temperature of interest and that in addition the char formed by the degradation of PA 6 is protected by the glass providing the flame retardant effect.

Zinc phosphate glasses have recently been developed with glass transition temperatures in the range of 280–370 °C [120, 124]. These glasses melt at sufficiently low temperatures to be melt mixed with engineering thermoplastic polymers such as polyaryl ether ketones, aromatic liquid polyesters, polyarylsulfones, perfluoroalkoxy resins, or polyetherimides, via conventional plastics compounding techniques. In addition, there is a high increase in LOI with glass loading when compared to the pure polymers. However, the incorporation of conventional low-temperature glasses as fillers in commodity polymers can produce measurable flame retardancy as well. Nevertheless, the loading must be high enough (> 60 wt% in polycarbonate) for increasing significantly the LOI.

Other work on inorganic glass-forming flame retardants examined analogous borate/carbonate systems. These compounds were found to form an inorganic, glassy foam as a surface barrier, which insulates and lowers the escape of volatile decomposition products [125]. As an example, it was shown that the efficiency of boric acid in ethylene-methylacrylate copolymer (EMA) forms a viscous layer at the surface of the sample when burning and, in addition, acts in the gas phase [126].

Silicone Additives

Silicon-based materials are potential flame retardants as they produce protective surface coatings during a fire, resulting in a low rate of heat release. Low levels of silicone in certain organic polymer systems have been reported to improve their LOI and UL 94 performance [75, 127–135]. Page et al. [132] have compounded silicone [poly(dimethylsiloxane)-type] containing dry powders with a variety of organic plastics. In particular in PS, they have shown that additive levels as low as 1–3 % reduced the rate of heat release by 30–50 %. They reported similar improvements in HIPS, PP, PS blends, PP, and EVA.

Nelson et al. [136] have studied silicone-modified PUR and shown a significant rate of heat release decrease from these materials when compared to unmodified PURs. The proposed mechanism is the formation of a silicon dioxide layer on the surface of the burning material, acting as a thermal insulator and preventing energy feedback to the substrate by reflecting the external heat flux radiation [137].

Recently, Iji et al [138] have proposed new silicone derivatives for polycarbonate (PC) and PC/ABS resins, which offer both good mechanical properties (strength, molding) and high flame retardancy (UL-94, 1/16 in. V-0 at 10 phr). They studied linear and branched chain-type silicone with (hydroxy or methoxy) or without (saturated hydrocarbons) functional reactive groups. They showed that the silicone, which has a branched chain structure and contains aromatic groups in the chain and nonreactive terminal groups, is very effective. In this case, the silicone is finely dispersed in the PC resin and the authors proposed that it could move to the surface during combustion to form a highly flame-retarding barrier on it.

Nanocomposites

Polymer-clay nanocomposites are hybrid materials consisting of organic polymer and inorganic layers materials with unique properties when compared to conventional filled polymers. Their synthesis (Fig. 5.10) [41] and fine structure [139, 140] are well known.

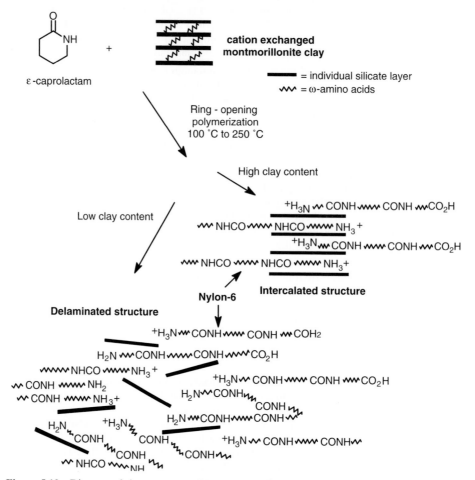

Figure 5.10 Diagram of the process used to prepare polymer layered silicate nanocomposites with either a delaminated structure or an intercalated structure [143]

The mechanical properties for nylon-6-clay nanocomposite, with clay mass fraction of 5 %, show excellent improvement over those for the pure nylon-6 [141, 142]. Gilman et al. [143, 144] and Giannelis [145, 146] reported on the flammability properties of nylon-6-clay nano-composites. The cone calorimeter data (Fig. 5.11) indicate that the peak heat release rate (HRR) is reduced by 63 % in a nylon-6-clay nanocomposite containing a clay mass fraction of only 5 %. These samples were exposed to a heat flux of 35 kW/m^2. The authors proposed that the delaminated hybrid structure collapses as the nylon-6 decomposes. This forms a reinforced char layer which acts as an insulator and a mass transport barrier, slowing the escape of the volatile products (e.g., ε–caprolactam) generated as the nylon-6 decomposes.

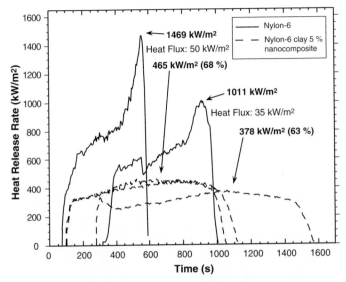

Figure 5.11 Rate of heat release versus time plot for nylon-6-clay nanocomposite (5 wt%) and pure nylon-6

Other polymers as polymer-clay nanocomposites were also investigated by the authors. They showed that the flame retardancy performances of the PS-clay nanocomposite and of the PP-g-MA-clay nanocomposite are very similar to that for the nylon-6-clay nanocomposite. They suggested that the mechanism for the PS-clay nanocomposite and the PP-g-MA-clay nanocomposite is a condensed phase mechanism in which a reinforced char layer forms, which acts as an insulator and a mass transport barrier, slowing the escape of the volatile decomposition products generated as the polymer decomposes.

The data for the 35 kW/m^2 and 50 kW/m^2 flux exposures are shown. Two experiments at the 50 kW/m^2 flux exposure are included to show the typical reproducibility. The nanocomposite has a 63 % lower rate of heat release (HRR) at 35 kW/m^2 and a 68 % lower HRR at 50 kW/m^2 [141].

Preceramic Additives

Lichtenhan et al. [147, 148] mention a new generation of silicon-based plastics such as polysilsesquioxane (PSS), polycarbosilane copolymers (PCS), and polyhedral oligomeric silsesquioxane (POSS) nanocomposites that can be used as blendable resins and can improve the flame retardancy of polymers such as PP, styrene-butadiene-styrene (SBS, Kraton™), or tetramethylene ether-polyamide copolymer (Pebax™). As an example, the latter blended with 10 wt% of PSS exhibits 72 % lower HRR than the pure polymer. In the case of PP blended with 20 wt% PSS, HRR peak decreases from 1466 kW/m^2 to 892 kW/m^2.

Marosi et al. [4, 135, 149] have tried to form the preceramic structure *in situ* during compounding of PP-APP-PER-vinyl triethoxy-silane (VTS). They proposed that the structure forms via a three-step reaction: coupling reaction between PER and VTS, chemical bounding of an interlayer of silicone-based compounds around the APP particles, and finally, radical coupling to the polymer matrix. The resulting PP-based materials have UL 94 V-0 rating [135].

Expandable Graphite

Expanding up to 100 times their original thickness, expandable graphite flakes generate an insulating layer similar to intumescence [149–151].

Crystalline graphite consists of stacks of parallel planes of carbon atoms. Because no covalent bonding exists between the planes, other molecules can be inserted between them. In the commercial process, sulfuric acid is inserted into the graphite, after which the flake is washed and dried. Because the intercalated species is sealed within the graphite lattice, the expandable graphite is a dry, pouring material with only minimal acidity.

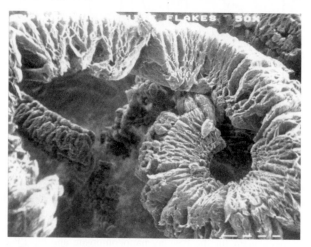

Figure 5.12   View of expanded graphite after thermal treatment, magnified 50x [152]

When the intercalated graphite is exposed to heat or flame, the inserted molecules decompose to generate gas. The graphite layer planes are forced apart by the gas and the graphite expands. The expanded graphite is a low-density, nonburnable thermal insulation that can reflect up to 50 % of radiant heat. It is often referred to as a "worm" because of its long, twisting shape. A typical segment of an expanded graphite worm is shown in Fig. 5.12.

The expandable graphite can be used in thermoplastic resins and thermosetting resins [153–157] and presents good flame retardancy properties at low loading. It is often combined with other flame retardants such as phosphorus-based additives (APP, red phosphorus), boron compounds (zinc borates), antimony trioxide, or magnesium hydroxide [155–157]. A survey of the patent literature [158–163] shows that expandable graphite finds its main application in foams, particularly in PUR foams.

Recently, Nyden et al. [164] suggested the use of graphite nanocomposites in PP. When using molecular dynamics simulations, a thermal stabilization of the material was shown.

5.1.3 Outlook

There are many flame retardants with numerous different chemical structures acting at different levels of combustion and/or pyrolysis. Flame retardants or flame retardant formulations exist for almost all commercial polymers. Nevertheless, the market is very complex because of the existing different national regulations. As part of the European harmonization effort, European standards will replace national standards. The new standards will take into account smoke emission, toxicity, and corrosivity. In addition, industrial hygiene measures with regard to additives will focus on staff protection, particularly during production and processing of plastics. Although halogenated flame retardants are highly effective for reducing the flammability of polymers, the future use of some of these compounds is under scrutiny in the context of health and environmental risk assessments. Public perception of the environmental impact of certain halogenated flame retardants during incineration has become an issue in Europe because of the possible production of dioxins and furans [165–173].

The trend is to flame retardant plastics containing flame retardants produced at reasonable cost and preserving or even improving the polymer properties. Very promising new approaches are made to find adequate solutions. The "nanocomposite approach" is one of them. For designing new materials with the required properties, it allows improvement of both flame retardancy and mechanical performance of the polymer by working at a nanoscale for additive incorporation. The silicone chemistry is another powerful approach to synthesize flame retardants, which already allows to propose solutions for polycarbonate. Finally, the intumescence concept using charring polymers as additives (blend technology) leads to acceptable materials that can pass major fire tests in different industrial sectors.

References for Section 5.1

[1] P. Carty: *Fire Mater.* 20 (1996) 158
[2] M. Lewin in M. Le Bras, G. Camino, S. Bourbigot, R. Delobel (Eds.): Fire Retardancy of Polymers: The Use of Intumescence, Royal Soc. Chem., p. 3–34, 1998
[3] H.R. Vandersall: *J. Fire Flamm.* 2 (1971) 97
[4] G. Marosi, I. Csontos, I. Ravadits, P. Anna, G. Bertalan, A. Toth in M. Lewin (Ed.): Recent Advances in Flame Retardancy of Polymeric Materials, Vol. 10, BCC,, Norwalk, CT, 2000
[5] J. Brossas: Techniques de l'Ingénieur, Paris, AM 3 237 (1998) 1
[6] M. Lewin (Ed.) in: Recent Advances in Flame Retardancy of Polymeric Materials, Vol. 9, BCC, Norwalk CT, 1998
[7] J. Green in M. Lewin (Ed.): Recent Advances in Flame Retardancy of Polymeric Materials, Vol. 7, BCC, Norwalk, CT, 1998
[8] G. Pál, H. Macskásy in: Studies in Polymer Science 6 - Plastics, their Behaviour in Fires, Elsevier, New York, pp. 304–332, 1991
[9] R.R. Hindersinn, G. Witschard in W.C. Kuryla, A.J. Papa (Eds.): Flame Retardancy of Polymeric Materials, Vol. 4, Chapter 1, M. Dekker, New York, 1978
[10] G. Camino, L. Costa: Polym. Degrad. Stab. 20 (1988) 271
[11] C.F. Cullis, M.M. Hirschler in: The Combustion of Organic Polymers, Clarendon Press, Oxford, 1981
[12] R.J. Schwarz, W.C. Kuryla, A.J Papa in W.C. Kuryla, A.J. Papa (Eds.): Flame Retardancy of Polymeric Materials, Vol. 2, Chapter 2, M. Decker, New York, 1973
[13] J.A. Rhys: *Chem. Ind.* (Lond.), (1969) 187
[14] G. Camino, N. Grassie (Eds.) in: Developments in Polymer Degradation-7, Chapter 7, Applied Science, London, 1987
[15] M. Lewin, S.M. Atlas, E.M. Pearce in: Flame Retardant Polymeric Materials, Vol. 3, Plenum Press, New York, 1982
[16] J.W. Lyons in: The Chemistry and Uses of Fire Retardants, Wiley, New York, 1970
[17] R. Smith, P. Georlette, I. Finberg, G. Reznick: *Polym. Degrad. Stab.* 54 (1996) 167
[18] B. Touré, J.M. Lopez-Cuesta, M. Longerey, A. Crespy: *Polym. Degrad. Stab.* 54 (1996) 345
[19] J.W. Hastie: *Combust. Flame* 21 (1973) 49
[20] J.W. Hastie: *J. Res. Nat. Bur. Stand. – A. Phys. Chem.* 77 A (1973) 733
[21] M. Lewin, A. Basch in M. Lewin, S.M. Atlas, E.M. Pearce (Eds.): Flame Retardant Polymeric Materials, Vol. 2, Plenum, New York, p.1, 1978
[22] E. D. Weil in M. Lewin, S.M. Atlas, E.M. Pearce (Eds.): Flame Retardant Polymeric Materials, Vol. 2, Plenum, New York, p.103, 1978
[23] E.D. Weil, R.H. Hansen, N. Patel in G.L. Nelson (Ed.): Fire and Polymers I, Chapter 8, ACS Symp. Series 425, Washington DC, 97, 1990

[24] G.E. Zaikov, S.M Lomakin in M. Le Bras, G. Camino, S. Bourbigot, R. Delobel (Eds.): Fire Retardancy of Polymers: The Use of Intumescence, Roy. Soc. Chem., Cambridge, p. 421, 1998
[25] L. Richard-Campisi, S. Bourbigot, M. Le Bras, R. Delobel: *Thermochim. Acta* 275 (1996) 37
[26] S. Bourbigot, L. Richard-Campisi, M. Le Bras, R. Delobel: *J. Textile Inst.* 88(1) (1997) 64
[27] S.K. Brauman: *J. Fire Retard. Chem.* 4 (1977) 18
[28] M. Lewin: A. Basch: *Textile Res. J,* 43 (1973) 693
[29] E.L. Back: *Pulp Paper Mag. Canada, Tech. Sec.* 68 (1967) 1
[30] J.E. Hendrix, G. L. Drake, R. H. Barker: *J. Appl. Polym. Sci.* 16 (1972) 257
[31] J. Economy in M. Lewin, S.M. Atlas, E.M. Pearce (Eds.): Flame Retardant Polymeric Materials, Vol. 2, Plenum, New York, p.203, 1978
[32] N. Rose, B. Costes, M. Le Bras, R. Delobel in: Proceeding of GFP Conference Polymérisations – Mécanismes – Méthodes - Procédés, GFP, Bordeaux, pp. 73–74, 1991
[33] Y.P. Khanna, E.M. Pearce in M. Lewin, S.M. Atlas, E.M. Pearce (Eds.): Flame Retardant Polymeric Materials, Vol. 2, Plenum Press, New York, p. 43, 1982
[34] S.J. Sirdesai, C.A Wilkie: *Polymer Preprints, ACS Polym. Chem. Ed.* 28 (1987) 149
[35] A.I. Balabanovich, G.F. Levchik, S.V. Levchik, W. Schnabel, CA. Wilkie in M. Le Bras, G. Camino, S. Bourbigot, R. Delobel (Eds.): Fire Retardancy of Polymers: The Use of Intumescence, Roy. Soc. Chem., Cambridge, p. 236, 1998
[36] S.V. Levchik, C.A. Wilkie in A.F. Grand, C.A. Wilkie (Eds.): in Fire Retardancy of Polymeric Materials, Marcel Dekker, New York, 2000
[37] C. Roderig, A. Bash, M. Lewin: *J. Polym. Sci., Polym. Chem.* 15 (1975) 1921
[38] C.A. Wilkie, M.A. McKinney in K.A. Beall (Ed.): Annu. Conf. Fire Res. NISTR 6242, NIST, Gaithersburg, MD, pp. 49–50, 1998
[39] J. Carnahan, W. Haaf, G. Nelson, G. Lee, V. Abolins, P. Shank in 4th Proc. Int. Conf. Flamm. Safety, San Francisco, 1979
[40] H.R. Allcock, R.L. Kugel, K.J., Valan*: Inorg. Chem.* 5 (1966) 1709
[41] R. Delobel, M. Le Bras, N. Ouassou, F. Alistiqsa: *J. Fire Sci.* 8(3–4) (1990) 85
[42] S. Bourbigot, J.M. Leroy in M. Le Bras, G. Camino, S. Bourbigot, R. Delobel (Eds.): Fire Retardancy of Polymers: The Use of Intumescence, Roy. Soc. Chem., Cambridge, pp. 129–139, 1998
[43] D.E. Cagliostro, S.R. Riccitiello, K.J. Clark, A.B. Shimizu: *J. Fire Flamm.* 6 (1975) 205
[44] C.E. Anderson, D.K. Wauters: *Intern. J. Engineer. Sci.* 22 (1984) 881
[45] J. Buckmaster, C.E. Anderson, A. Nachman: *Intern. J. Engineer. Sci.* 24 (1986) 263
[46] C.E. Anderson, J. Dziuk, W.A. Mallow, J. Buckmaster: *J. Fire Sci.* 3 (1985) 1985
[47] V.G. Zverev, G.N. Isakov, V.V. Nesmelov, V.A. Nazarenko: *Int. J. Polym. Mater.* 20 (1993) 91
[48] V. Sh. Mamleev, E.A. Bekturov, K.M. Gibov: *J. Appl. Polym. Sci.* 70 (1998) 1523
[49] R. Pehrson, J.R. Barnett: *J. Fire Protect. Engr.* 8 (1996) 13
[50] K.M. Butler, H.R. Baum, T. Kashiwagi in: Proc. Int. Conf. Fire Res. Eng., Orlando, p.261, 1995
[51] K.M. Butler in F.A.A. Biannual Report on Fire Resistant Materials for Aircraft Cabin Interiors, 1997
[52] K.M. Butler in K. C. Khemani (Ed.): Polymeric Foams: Science and Technology, ACS Symposium Series 669, Chapter 15, ACS, pp. 214–230, 1997
[53] I.S. Reshetnikov, N.A. Khalturinskii: *Chem. Phys. Rep.* 16(10) (1997) 1869–1875
[54] I.S. Reshetnikov, N.A. Khalturinskii: *Chem. Phys. Rep.* 16(3) (1997) 499–506
[55] I. S. Reshetnikov, N. A. Khalturinskii, M.Yu. Yablokova, Al. Al. Berlin in M. Le Bras, G. Camino, S. Bourbigot, R. Delobel (Eds.): Fire Retardancy of Polymers: The Use of Intumescence, Roy. Soc. Chem., Cambridge, pp. 104–112 and pp.152–158, 1998
[56] M. Bugajny, , M. Le Bras, S. Bourbigot: *Fire Mater.,* 23 (1999) 49–51
[57] M. Le Bras, S. Bourbigot, C. Delporte, C. Siat, Y. Le Tallec: *Fire Mater.* 20 (1996)191
[58] G. Camino, L. Costa, L. Trossarelli: *Polym. Deg. Stab.* 12 (1985) 213
[59] M. Le Bras, S. Bourbigot in J. Karger-Kocsis (Ed.): Polypropylene: An A-Z Reference, Chapman & Hall, London, 357–365, 1998
[60] H.R. Vandersall: *J. Fire Flamm.* 2 (1971) 97
[61] S. Bourbigot, M. Le Bras, R Delobel, P. Bréant, J.-M. Trémillon: *Carbon* 33 (1995) 283
[62] L. Morice, S. Bourbigot, J.M. Leroy: *J. Fire Sci.* 15 (1997) 358
[63] M. Le Bras, S. Bourbigot, C. Siat, R. Delobel in M. Le Bras, G. Camino, S. Bourbigot, R. Delobel (Eds.): Fire Retardancy of Polymers: The Use of Intumescence, Roy. Soc. Chem., Cambridge, pp. 269–279, 1998
[64] E. Jennewein, W. D. Pirig, Eur. Pat. EP 0 735 119 A1 (Hoechst AG), 1996

[65] M. Le Bras, S. Bourbigot, Y. Le Tallec, J. Laureyns: *Polym. Degrad. Stab.* 56 (1997) 11
[66] A. Marchal, R. Delobel, M. Le Bras, J.M. Leroy, D. Price: *Polym. Degrad. Stab.* 44 (1994) 263
[67] S. Bourbigot, M. Le Bras, C. Siat in M. Lewin (Ed.): Recent Advances in Flame Retardancy of Polymeric Materials, Vol. 8, BCC, Norwalk,146–160, 1998
[68] S. Bourbigot, M. Le Bras, M. Bugajny, F. Dabrowski in K.A. Beall (Ed.): Proc. Annu. Conf. Fire Res. NISTR 6242, NIST, Gaithersburg, MD, pp. 43–44, 1998
[69] M. Bugajny, M. Le Bras, S. Bourbigot, R. Delobel: *Polym. Degrad. Stab.* 64 (1999) 157–163
[70] S. Bourbigot, M. Le Bras, R. Delobel, P. Bréant, J.-M. Trémillon: *Polym. Degrad. Stab.* 54 (1996) 275
[71] M. Le Bras, S. Bourbigot: *Fire Mater.* 20 (1996) 39
[72] Gy. Marosi, I. Balogh, P. Anna, I. Radavits, A. Tohl, M. A. Maatoug, K. Swentirmai, I. Bertoti, A. Toth: Muänyag ès Gumi 34 (1997) 265
[73] C. Siat, S. Bourbigot, M. Le Bras in M. Lewin (Ed.): Recent Advances in Flame Retardancy of Polymeric Materials, Vol. 7, BCC, Norwalk, CT, p. 318, 1997
[74] Gy. Marosi, P. Anna, I. Balogh, Gy. Bertalan, A. Tohl, M. A. Maatoug: *J. Therm. Anal.* 48 (1997) 717
[75] Gy. Marosi, I. Ravadits, P. Anna, I. Balogh, Gy. Bertalan, A. Tohl, M. Botreau, M.D. Dran in M. Lewin (Ed.): Recent Advances in Flame Retardancy of Polymeric Materials, Vol. 9, BCC, Norwalk, CT, p. 81, 1999
[76] S. Bourbigot, M. Le Bras, P. Bréant, J.-M. Trémillon: R. Delobel: *Fire Mater.* 20 (1996) 145
[77] S. Bourbigot, M. Le Bras, R. Delobel, R. Decressain, J.-P. Amoureux: *J. Chem. Soc. Faraday Trans.* 92 (1996) 149
[78] S. Bourbigot, M. Le Bras, R. Delobel, J.-M. Trémillon: *J. Chem. Soc. Faraday Trans.* 92 (1996) 3435
[79] S. Bourbigot, M. Le Bras, R. Delobel, L. Gengembre: *Appl. Surface Sci.* 120 (1997) 15
[80] S. Bourbigot, M. Le Bras in M. Le Bras, G. Camino, S. Bourbigot, R. Delobel (Eds.): Fire Retardancy of Polymers - The Use of Intumescence, Roy. Soc. Chem., Cambridge, pp. 222–235, 1998
[81] S. Bourbigot, Le Bras, M.: unpublished results
[82] G. Montaudo, E. Scamporino, D. Vitalini: *J. Polym. Sci., Polym. Chem.* 21 (1983) 3361
[83] M. Le Bras, S. Bourbigot in M. Le Bras, G. Camino, S. Bourbigot, R. Delobel (Eds.): Fire Retardancy of Polymers - The Use of Intumescence, Roy. Soc. Chem., Cambridge, pp. 64–75, 1998
[84] J.E.J. Staggs: *Polym. Degrad. Stab.* 64 (1999) 369–378
[85] M.I. Nelson, J. Brindley, A.C. McIntosh: *Fire Safety J.* 24 (1995) 107
[86] M.I. Nelson, J. Brindley, A.C. McIntosh: *Polym. Degrad. Stab.* 54 (1996) 255
[87] M.I. Nelson, J. Brindley, A.C. McIntosh A.C.: *Fire Safety J.* 28 (1997) 67
[88] M.I. Nelson, J. Brindley, A.C. McIntosh: *Combust. Sci. Technol.* 113 (1996) 221
[89] P. R. Hornsby: *Fire Mater.* 18 (1994) 269
[90] P. R. Hornsby, C. L. Watson: *Polym. Degrad. Stab.* 30 (1990) 73
[91] F. Molesky in M. Lewin, G. Kirshenbaum (Eds.): Recent Advances in Flame Retardancy of Polymeric Materials, Vol. 1, BCC, Norwalk, CT, p.92, 1990
[92] J. Levesque in M. Lewin, G. Kirshenbaum (Eds.): Recent Advances in Flame Retardancy of Polymeric Materials, Vol. 1, BCC, Norwalk, CT, p.102, 1990
[93] O. Kalisky, R. J. Mureinik, A. Weismann, E. Reznik in M. Lewin, G. Kirshenbaum (Eds.): Recent Advances in Flame Retardancy of Polymeric Materials, Vol. 4, BCC, Norwalk, CT, p.140, 1993
[94] S. Bourbigot, M. Le Bras, R. Leeuwendal, K.K. Shen, D. Schubert: *Polym. Degrad. Stab.* 64 (1999) 419–425
[95] N. Pécoul, S. Bourbigot, B. Revel: *Macromol. Symp.* 119 (1997) 309
[96] S. Bourbigot, F. Carpentier, M. Le Bras, C. Fernandez, J.P. Amoureux, R. Delobel in J. Stanford, R. Rothon (Eds.): Extended Abstracts of Eurofillers'97, British Plastics Federation, London, p.419, 1997
[97] C.F. Cullis, M. M. Hirschler: *Eur. Polymer J.* 20 (1984) 53
[98] K. Kanemitsuya in M. Lewin, G. Kirshenbaum (Eds.): Recent Advances in Flame Retardancy of Polymeric Materials, Vol. 2, BCC, Norwalk, CT, p.220, 1991
[99] K.K. Shen, D.F. Fern in M. Lewin, G. Kirshenbaum (Eds.): Recent Advances in Flame Retardancy of Polymeric Materials, Vol. 7, BCC, Norwalk, CT, 1996
[100] S. Bourbigot, M. Le Bras, R. Leeuwendal, D. Schubert in M. Lewin (Ed.): Recent Advances in Flame Retardancy of Polymeric Materials, Vol. 10, BCC, Norwalk, CT, 1999
[101] R. G. Baggaley, P. R. Horsby, R. Yahya, P.A. Cusack, A.W. Monk: *Fire Mater.* 21 (1997) 179–185

[102] S.V. Levchik, L. Costa, G. Camino: *Polym. Degrad. Stab.* 36 (1992) 229
[103] N. Grassie, M. Zulfiqar in G. Scott (Ed.): Developments in Polymer Stabilization, Chapter 6, Applied Science, London, pp. 197–217, 1978
[104] E.N. Peters, A.B. Furtec, D.I. Steinbert, D.T. Kwiatkowski: *J. Fire Retard. Chem.* 7 (1980) 69
[105] A. Granzow, R.G. Ferrillo, A. Wilson: *J. Appl. Polym. Sci.* 21 (1977) 1687
[106] E.N. Peters: *J. Appl. Polym. Sci.* 24 (1979) 1457
[107] C.A. Wilkie, J.W. Pettegrew, C.E. Brown: *J. Polym. Sci.: Polym. Lett.* 19 (1981) 409
[108] T. Suebsaeng, C.A. Wilkie: *J. Polym. Sci.: Polym. Chem.* 22 (1984) 945
[109] C.E. Brown, C.A. Wilkie, J. Smukalla, R.B. Cody Jr., J.A. Kinsinger: *J. Polym. Sci.: Polym. Chem.* 24 (1986) 1297
[110] A. Ballistreri, S. Foti, G. Montaudo, E. Scamporrino, A. Arnesano, S. Calgari: *Makromol. Chem.* 182 (1981) 1301
[111] A. Ballistreri, G. Montaudo, C. Puglisi, E. Scamporino, D. Vitalini, S. Calgari: *J. Polym. Sci.: Polym. Chem.* 21 (1983) 679
[112] G.F. Levchik, S.V. Levchik, G. Camino, E.D. Weil in M. Le Bras, G. Camino, S. Bourbigot, and R. Delobel (Eds.): Fire Retardancy of Polymers – The Use of Intumescence, Roy. Soc. Chem., Cambridge, pp. 304–315, 1998
[113] E.D. Weil in: Encyclopedia of Polymer Science and Technology, Vol. 11, Wiley-Interscience, New York, 1986
[114] A.M. Aaronson: Phosphorus Chemistry, ACS Symposium Series 486, Chapter 17, 218, 1992
[115] T. Kashiwagi, J.W. Gilman, M.R. Nyden, S.M. Lomakin in M. Le Bras, G. Camino, S. Bourbigot, R. Delobel (Eds.): Fire Retardancy of Polymers – The Use of Intumescence, Roy. Soc. Chem., Cambridge, pp. 175-202, 1998
[116] J. W. Wheeler, Y. Zhang, J. C. Tebby in M. Le Bras, G. Camino, S. Bourbigot, R. Delobel (Eds.): Fire Retardancy of Polymers – The Use of Intumescence, Roy. Soc. Chem., Cambridge, pp. 252–265, 1998
[117] J. W. Hastie, C. L. McBee in: Mechanistic Studies of Triphenyl Oxide- Poly(ethylene terephthalate) and Related Flame Retardant Systems, NBISR 75–741, 1975
[118] J. Troitzsch (Ed.) in: International Plastics Flammability Hand Book 2nd edition, Hanser, Munich, pp. 50-51, 1990
[119] W. J. Kroenke: *J. Mater. Sci.* 21 (1986) 1123
[120] C.J. Quinn, G.H. Beall in M. Lewin, G. Kirshenbaum (Eds.): Recent Advances in Flame Retardancy of Polymeric Materials, Vol. 4, BCC, Norwalk, CT, p.62, 1994
[121] J. W. Gilman, S. J. Ritchie, T. Kashiwagi, S. M. Lomakin: *Fire Mater.* 21 (1997) 23
[122] G.F. Levchik, S.V. Levchik, A.F. Selevich, A.I. Lesnikovich: *Vesti AN Belarusi., Ser. Khim.* 3 (1995) 34
[123] S.V. Levchik in M. Lewin (Ed.): In *Proc. 8th BCC Rec. Adv. Flame Retard. Polym. Mater.*, BCC, Norwalk, CT, 1997
[124] W.A. Bahn, C.J. Quinn in: Proc. ANTEC, p 2370, 1991
[125] R. E. Myers, E. Licursi: *J. Fire Sci.* 3 (1985) 415
[126] S. Bourbigot: unpublished results
[127] G. Nelson in: Proc. Fire Safety Conf., St. Petersburg Beach, FL, 1991
[128] M. Huber: *Plastics Compound.* Sept./Oct. 1990, pp. 124–128
[129] M.R. McLaury, A.L. Schroll: *J. Appl.Polym. Sci.* 30 (1985) 461
[130] P.G. Page, D.J. Romenesko in M. Lewin (Ed.): Recent Advances in Flame Retardancy of Polymeric Materials, Vol. 7, BCC, Stamford , CT, 1997
[131] T. Kashiwagi, T. Cleary, G. Davis, J. Lupinski in: Proc. Int. Conf. Promot. Adv. Fire Resist. Aircraft Interior Mater., Atlantic City, NJ, 1993
[132] B. Page, B. Buch, D. Romenesko in M. Lewin (Ed.): Recent Advances in Flame Retardancy of Polymeric Materials, Vol. 4, Lewin, M. (Editor), BCC, Norwalk, CT, pp. 322–329, 1994
[133] P. Pawar: *Plast. Techn.* 36(3) (1999) 75–79
[134] *Gy. Marosi, I. Radavits, G. Bertalan, P. Anna, M. A. Maatoug, A. Tohl, M. D. Tran* in: Flame Retardant'96, Interscience Comm. Ltd, London, 1996, p. 115.
[135] Gy. Marosi, G. Bertalan, I. Balogh, A. Tohl, P. Anna, G. Budai, S. Orban in M. Le Bras, G. Camino, S. Bourbigot, R. Delobel (Eds.): Fire Retardancy of Polymers - The Use of Intumescence, Roy. Soc. Chem., Cambridge, pp. 325–341, 1998
[136] G.L. Nelson, C. Jayakody, U. Sorathia in M. Lewin (Ed.): Proceedings of Recent Advances in Flame Retardancy of Polymeric Materials, BCC, Stamford, CT, 1997

[137] F.-Y. Hshieh: *Fire Mater.* 22 (1998) 69–76
[138] M. Iji, S. Serizawa: *Nec Res. Dev.* 39(2) (1998) 82
[139] Y. Kojima, A. Usuki, M. Kawasumi, A. Okada, T. Kurauchi, O. Kamigaito: *J. Polym. Sci., Part A*, 31 (1993) 983
[140] Y. Kojima, A. Usuki, M. Kawasumi, A. Okada, T. Kurauchi, O. Kamigaito: *J. Polym. Sci., Part B*, 32 (1993) 625
[141] Y. Kojima, A. Usuki, M. Kawasumi, A. Okada, Y. Fukushima, T. Kurauchi, O. Kamigaito: *J. of Mater. Res.* 8 (1993) 1185
[142] A. Usuki, A. Koiwai, Y. Kojima, M. Kawasumi, A. Okada, T. Kurauchi, O. Kamigaito: *J. Appl. Polym. Sci.* 55 (1995) 119
[143] J.W. Gilman, T. Kashiwagi, S. Lomakin, J.D. Lichtenhan, P. Jones, E.P. Giannelis, E., Manias in M. Le Bras, G. Camino, S. Bourbigot, R. Delobel (Eds.): Fire Retardancy of Polymers - The Use of Intumescence, Roy. Soc. Chem., Cambridge, pp. 203–221, 1998
[144] J.W. Gilman, T. Kashiwagi, J.D. Lichtenhan: *SAMPE J.* 33(4) (1998) 40
[145] E.P. Giannelis in K.A. Beall (Ed.): Proc. Annual Conf. Fire Res., NIST, Gaithersburg, MD, 1998
[146] E.P. Giannelis: *Adv. Mater.* 8 (1996) 29
[147] J.D. Lichtenhan, T.S. Haddad, J.J. Schwab, M.J. Carr, K.P. Chaffee, P.T. *Mather: Polym. Prep. (ACS - Division of Polymer Chemistry)* 39 (1998) 489
[148] J.D. Lichtenhan, J.W. Gilman: US Pat. 1997
[149] G. Marosi, A. Tohl, G. Bertalan, P. Anna, M.A. Maatoug, G. Radavits, G. Budai, A. Toth: Composites, Pt. A, 29 (1998) 1305
[150] D.W. Krassowski, D.A. Hutchings, S.P. Qureshi in: Proc. Fire Retard. Chem. Assoc. Fall Mtg., Naples FL, pp. 137–146, 1996
[151] F. Okisaki in: Proc. Fire Retard. Chem. Assoc. Spring Meeting, San Francisco, pp. 11–2, 19974
[152] D.W. Krassowski, D.A. Hutchings, Qureshi, S.P.: http://www.ucar.com/ucarcarb/grafoil/ggpaper1.htm
[153] S.P. Qureshi, D.W. Krassowski in: Proc. Int. SAMPE Technical Conference, Society for the Advancement of Material and Process Engineering, pp. 625–634, 1997
[154] B. Schilling: *Kunststoffe*, 87 (1997) 8
[155] F. Okisaki, A. Hamada, M. Obasa: Eur. Pat. EP 794 229 A2, 1997
[156] F. Okisaki, A. Hamada, S. Endo, G. Ochiai: Eur. Pat. EP 730 000 A1 960 904, 1996
[157] M. Goto, Y. Tanaka, K. Koyama: Eur. Pat. EP 0 824 134 A1, 1997
[158] U. Heitmann, H. Rossel: Eur. Pat. EP 0 450 403 A3, 1991
[159] I. Yoshizawa, T. Shibata, K. Kaji, M. Hatada: German Pat. DE 4 121 831 C2, 1991
[160] H.-D. Lutter, H.J. Gabbert, V. Haase, K. Fimmel: Eur. Pat. EP 0 482 507 A3, 1986
[161] R.W.H. Bell: UK Pat. 2 168 706 A, 1985
[162] P. Haas, H. Hettel: Eur. Pat. EP 0 337 228 A1, 1989
[163] H. Horacek: German Pat. DE 3 813 251 A1, 1988
[164] M.R. Nyden, J.W. Gilman: *Comput. Theoret. Polym. Sci.* 7 (1997) 191
[165] Fifth Draft Status Report, OCDE Workshop on the Risk Reduction of Brominated Flame Retardant, Neufchâtel (Switzerland) (May 26, 1992, February 22–25, 1993), OECD - Direction de l'Environnement, April 1993
[166] Preliminary first draft report, International Program on Chemical Safety - Environmental Health Criteria for Brominated Diphenylethers (January 1993), first draft report, International Programme on Chemical Safety - Environmental Health Criteria for Tris(2,3-dibromopropyl) phosphate and Bis(2,3-dibromopropyl) phosphate (January 1993), United Nations Environmental Programme, PCS/EHC/92.45
[167] D. Bieniek, M. Bahadir, F. Korte: *Heterocycles* 28 (1989) 719
[168] M.N. Pinkerton, R.J. Kociba, R.V. Petrella, D.L. McAllister, M.L. Willis, J.C. Fulfs, H. Thoma, O. Hutzinger: *Chemosphere* 18 (1989) 1243
[169] R. Dumler, D. Lenoir, H. Thoma, O. Hutzinger: *Chemosphere* 20 (1990) 1867
[170] E. Clausen, E.S. Lahaniatis, M. Bahadir, D. Bieniek: *Z. Anal. Chem.* 327 (1987) 297
[171] D. Lenoir, K. Kampke-Thiele, A. Kettrup in M. Lewin (Ed.): Recent Advances in Flame Retardancy of Polymeric Materials, Vol. 8, BCC, Norwalk, CT, 1997
[172] G.L. Nelson in: Recycling of Plastics - A New FR Challenge, The Future of Fire Retarded Materials: Applications & Regulations, FRCA, Lancaster, PA, p.135, 1994
[173] H.C.H.A. Van Riel in: Is There a Future in FR Material Recycling; The European Perspective, FRCA, Lancaster, PA, p.167, 1994

5.2 The Most Important Flame Retardant Plastics

T. ECKEL

The main objective of flame retarding polymers is to increase the ignition resistance of the material and to reduce flame spread in case of burning. All flame retardants have effects on the technical properties of plastics. Therefore, the task for product development is to find a compromise between the desired improvement in fire safety and a minimal reduction of other properties. The demands on an ideal flame retardant package are:

- High efficiency to minimize negative effects on properties and costs,
- Compatibility with polymer and other additives,
- Thermal stability at processing temperature,
- Low tendency to bleed out or to evaporate,
- UV-stability,
- Color- and odorless,
- Resistance to ageing and hydrolysis,
- No corrosion of equipment,
- Environmental safety and no toxicity,
- Recyclable.

In practice, all flame retardant polymers are compounded materials. In addition, these materials may contain additives, for example, impact modifiers, stabilizers, plasticizers, fillers, reinforcing additives, mold release agents, or color masterbatches. All these components are incorporated in known manner by melt-compounding the mixtures at temperatures up to 300 °C in conventional units such as kneaders and twin-screw extruders. It is evident that the flame retardant additives have to be adjusted to each polymer system to match its decomposition characteristics. Their effect must start to act below the decomposition temperature of the plastic and continue over the whole range of decomposition. The higher the efficiency of a flame retardant for a particular resin, the less will be needed in the blends and the broader are the options for a blend design. Their amount also depends on application and specifications.

To approach the desired range of properties, numerous optimized formulations for flame retarded plastics have been worked out. These may be based on additive or reactive flame retardants to which synergists may be added frequently. The effectiveness of bromine (especially aromatically bound) flame retardants, for example, can be considerably increased without significantly affecting processability by the addition of synergists such as antimony trioxide.

The most versatile flame retardants belong to the group of *additive flame retardants*. Their application ranges from polymers with a pure carbon chain structure such as polyolefins, PVC, and PS to polymers with a heterogeneous chain such as polyurethanes, polyesters, polycarbonates, polyphenylene ethers, and polyamides. Compared to reactive flame retardants, they can be easily and cheaply incorporated into the polymers and do not change their characteristics significantly. *Reactive flame retardants* are tailor-made products for certain polymers, preferred for polymers with a heterogeneous chain such as epoxy resins or polyurethanes. This class of flame retardants is incorporated into the polymer chain by chemical reactions and becomes a permanent part of the polymer chain which ensures a durable flame retardant effect as they cannot bloom out, plate out, or evaporate. Reactive flame retardants are also used for the preparation of oligomeric or polymeric flame retardants, that is, a brominated polycarbonate oligomer.

Commercially available flame retardants include halogenated (chlorine and bromine), phosphorus-containing, and chlorine- and bromine-containing phosphorus compounds, red phosphorus, nitrogen-containing, silicone, and inorganic metal compounds (such as antimony trioxide used as a synergist); hydrated minerals; and others.

In the past, halogen-containing products have been used as the most versatile flame retardants. The efficiency of halogenated flame retardants is assessed as follows:

$$I \; > \; Br \; > \; Cl \; > \; F$$

Brominated products are about two times more effective than chlorine compounds corresponding to their atomic weights. Aliphatic halogen compounds are more effective than aromatic products, with the alicyclic materials in between.

Commercial *chlorine-containing flame retardants* are chlorinated hydrocarbons and chlorinated cycloaliphatics. The general benefit of this group is a good light stability combined with low raw material costs. A disadvantage is that to provide a sufficient level of flame retardancy they have to be utilized in quantities, which adversely affect the properties of the polymer materials. The most commonly utilized chlorinated hydrocarbons are *chloroparaffins* which are available in liquid or solid form depending on their chlorine content (between 40 % and 70 %). Their poor thermal resistance (dehydrochlorination starts at 180 °C) restricts their utilization to PVC, LDPE and polyurethanes.

Cycloaliphatic chlorine compounds have a considerably better thermal stability (up to 260 °C) and have found widespread applications as flame retardants in polymers. The most important products are *hexachloroendomethylenetetrahydrophthalic acid (HET acid)* and the Diels-Alder diadduct of hexachlorocyclopentadiene and cyclooctadiene, which is known as *Dechlorane Plus^{TM}* and stable up to its melting point of about 350 °C.

Numerous *bromine-containing flame retardants* are known and have been commercialized since years. They show a higher effectiveness compared to chlorine compounds and allow a user to incorporate them in lower concentrations with less effect on technical properties of a plastic material. They can be readily incorporated in nearly all polymers and have low tendency to bleed out. The effectiveness can be enhanced by antimony oxide. Disadvantages are that they exhibit a poorer light stability and are more expensive than chlorine compounds. They can be applied both as reactive and additive flame retardants.

The commercially available bromine compounds can have aliphatic, alicyclic, or aromatic structures depending on processing temperatures and polymer systems encountered. According to their structure the following groups may be distinguished:

- *Aliphatic and cycloaliphatic compounds* with hexabromobutene, dibromoethyl-dibromocyclohexane, and *hexabromocyclododecane* as the most important representatives primarily used in PS foam. Reactive aliphatic flame retardants are *dibromoneopentyl glycol* used in unsaturated polyesters, *tribromoneopentyl alcohol* in polyurethane foam as well as *brominated aliphatic polyol and polyethertriol* in polyurethane rigid foams and spray can foams.
- *Polybrominated diphenyl ethers* consist of *decabromodiphenyl ether,* octabromodiphenyl ether, pentabromodiphenyl ether and a fully brominated diphenoxy benzene. Decabromodiphenyl ether, with the highest bromine content of 83 %, is one of the largest volume organic flame retardants currently used in HIPS, PE, PP, engineering plastics, and other products. Octa- and pentabromodiphenylether are being phased out for environmental reasons.
- As a *replacement for decabromodiphenylether (bis-pentabromophenyl)ethane* has been developed and is used mainly in high-impact polystyrene (HIPS), polyolefins, and engineering plastics. *Brominated trimethylphenylindan* finds its application in polyolefins, styrenics, engineering plastics (nylons, PC, PC/ABS), elastomers, and textiles.
- *Brominated phenol derivatives* comprising *tetrabromobisphenol A (TBBA),* the largest volume organic flame retardant. It is used as an additive (i.e., for ABS) and mainly as a reactive flame retardant for epoxy resins (printed circuit boards). It has numerous derivatives such as different chain-capped TBBA polycarbonate oligomers and reaction products with ethylene oxide and acrylic acid. *TBBA-bis(2,3-dibromopropyl ether)* is mainly used in German Class B1 PP pipes and in UL 94 V-2 PP and HIPS. *TBBA-based epoxy resins* low

molecular weight grades can be used as reactive components in unsaturated polyesters (UP), phenolics (PF), and epoxy resins. The high molecular weight grades are recommended as additives for engineering thermoplastics such as nylons, linear polyesters (PBT), PC, and PC-ABS and for standard thermoplastics such as HIPS and ABS.

Other flame retardants in this group are *bis-(tribromophenoxy)ethane* used mainly in ABS and HIPS, as well as *polydibromophenylene oxide* for applications in engineering plastics.

- *Tetrabromophthalic anhydride* (reactive for flame retarding unsaturated polyesters) and its derivatives such as *1,2-bis(tetrabromophthalimide) ethane* (the reaction product with ethylene diamine) as the most important representatives. The benefits of the bis-imide are UV and thermal stability. It is applied, for example, in styrenics (HIPS, ABS), polyolefins (PP) and engineering plastics (particularly PBT). Other flame retardants are *tetrabromophthalate diols* used as polyol components in rigid polyurethane foam and *tetrabromophthalate esters*, liquid flame retardants with plasticizing properties applied in PVC, elastomers, adhesives, and coatings.

- *Brominated oligomers and polymers* such as *tetrabromobisphenol-A polycarbonate oligomers, polydibromophenylene oxide, brominated polystyrene,* and *poly(pentabromobenzyl) acrylate* are advantageous as nonvolatile flame retardants.

The range of *phosphorus-containing flame retardants* is extraordinarily versatile, as it extends over several oxidation stages of the phosphorus, phosphines, phosphine oxides, phosphonium compounds, phosphonates, elemental red phosphorus, phosphates, and phosphates. All are used as flame retardants. In general, phosphorus compounds are less versatile than brominated flame retardants, which limits their utilization to certain groups of polymers. Phosphorus-containing flame retardants are effective in oxygen-containing polymers and show only little performance in polyolefins and styrenics. The effectiveness can be enhanced if the phosphorus compounds contain halogens, in particular bromine. A synergistic effect with brominated phosphates is reported for several polymer systems [1,2]. Many phosphorus-containing flame retardants are liquid or have low melting points and can act as plasticizers for the polymer matrix. According to their structure the following groups of phosphorus-containing flame retardants may be distinguished:

- Halogen-free aromatic phosphate esters with *triphenyl phosphates*, their alkyl-substituted derivatives (isopropyl and *t*-butyl), tricresyl, cresyl diphenyl phosphates, and the less volatile compounds *resorcinol bis-(diphenyl phosphate) (RDP)* and *bisphenol A bis-(diphenyl phosphate) (BDP)*. All are used commercially predominantly in PVC, polyphenylene oxide-HIPS, and polycarbonate-ABS blends. RDP and BDP have virtually substituted brominated flame retardants in computer housings in Europe.

- *Chlorinated phosphate esters* such as *tris-(2-chloropropyl) phosphate, tris-(1,3-dichloro-2-propyl) phosphate,* and *tris-(chloroethyl) phosphate derivatives* coupled with pentaerythritol or diethylene glycol represent the second largest group of phosphorus compounds mainly used as flame retardants for polyurethane foam.

- *Salts of organic phosphinic acids* have been described as effective flame retardants in engineering plastics such as ABS, PET, PBT, and PA [3].

- *Inorganic phosphorus compounds* such as *ammonium polyphosphate* are commercially used as an acid source in intumescent systems (in combination with a char former and a blowing agent) for plastics. Best results are obtained in PP. Several intumescent systems are known [4]. *Red phosphorus* is used in engineering thermoplastics, mainly in nylons, and further in polyurethanes, unsaturated polyester resins, rubber, and adhesives.

An overview on commercial phosphorus-containing flame retardants is given in [5].

In general, *nitrogen-containing flame retardants* show a lower efficiency than other flame retardant additives. Therefore, their application is limited to certain polymer systems. Commercial nitrogen compounds are melamine, melamine cyanurate, and melamine phosphates. *Melamine* is used in large quantities for the production of combustion modified high resilient foam (CMHR) polyurethane foam in home furnishings to comply with the demanding fire safety requirements in the United Kingdom. In a microencapsulated formulation, it may also be used in thermoplastic polyurethane. *Melamine cyanurate (MC)* is mainly used in unfilled PA 6 and 66 UL 94 V-0 grades. It is also used in thermoplastic polyurethane and HIPS. The use of *melamine-based HALS products* as flame retardants in polyolefins is relatively new and has interesting perspectives.

Metal hydrates are one important group in the segment of *inorganic flame retardants.* Most important commercially available representatives are *aluminum trihydrate (ATH), magnesium hydroxide,* and *magnesium carbonate hydroxide*, which are applied in EVA, PP, and PA. These materials are filler-type flame retardants, which decompose endothermically and release water at pyrolysis temperatures. Their properties are compared in [6].

Synergists for other flame retardants such as *antimony oxides, aluminum oxide* and smoke suppressants such as *zinc borates* and molybdenum and tin compounds represent a second important group of inorganic flame retardants [7–9]. Zinc borate is used in flexible and rigid PVC formulations, partly substituting antimony trioxide. It is increasingly used as a component of halogenated and halogen-free formulations in polyolefins, nylons, thermoplastic polyesters, epoxies, and rubber.

Expandable graphite (EG) is used in special flexible polyurethane formulations for high fire safety seating applications and has been proposed for polyolefins and nylons. Because of its color, its application field is restricted to black formulations.

Silicon flame retardants, for example organofunctional modified silicon resin have proved as effective flame retardants in several engineering thermoplastics (i.e., PPO/HIPS, PC, PP) [10].

As a consequence of the discussion of the risk potential and environmental protection issue of certain chlorinated and brominated flame retardants that took place during the last 15 years in Europe, many polymer producers, compounders and end-product manufacturers have been seeking nonhalogen flame retardants. Nonhalogen solutions had to overcome technical problems such as lower heat distortion temperatures and difficult processing or handling problems (red phosphorus), and were successfully commercialized in a number of polymer systems (Table 5.3).

Table 5.3 Polymers with non-halogenated flame retardants [11]

Polymer	UL 94 Rating	Compromise
Polyolefins	V-0	ATH, Mg hydroxide 60% addition level; severe effect on physical properties Intumescent systems 25% addition level; cost, physical properties
Styrenics HIPS ABS	 V-2 V-0	 organophosphates, effect on HDT*, processing
Alloys PC/ABS; PPO/HIPS	V-0	Organophosphates
Polyamides	V-0	Glass reinforced – red P; color, corrosion, processing
Epoxy	V-0	ATH at high levels, effect on physical properties

* Heat Distortion Temperature

Flame retardant plastics are discussed in detail by Lyons [12, Chapters 7 and 8] and in the first three volumes of Kuryla and Papa [13–15]. Further compilations of flame retardants producers and distributors in the United States are given in [16] and [17]. Other reviews are given by Lewin [18, Vol. 1)], Kirk-Othmer [19], Katz and Milewski [20], and various other workers [21–31].

5.2.1 Plastics with All-Carbon Backbones

5.2.1.1 Polyolefins

For flame retarding *polyethylene* various commercial halogen-containing flame retardants such as chlorinated paraffins, for example, Dechlorane Plus[TM], the Diels-Alder adduct of hexachlorocyclopentadiene and cyclooctadiene, decabromodiphenyl ether, and ethylene bis-tetrabromophthalimide are applied in combination with antimony oxide [23, 26, 32]. The effective amount of flame retardants necessary to match a certain specification depends on the melt index of the polymer. PE grades with higher melt index (better flow) require less flame retardant (heat loss by dripping flaming). On the other hand, cross-linked PEs need significantly more flame retardant because they do not drip.

Chlorinated paraffins are used as very efficient and low-cost flame retardants in LDPE. Disadvantages are poor thermal stability and their tendency to bloom out from the polymer.

Commercial flame retarded cross-linked PEs contain Dechlorane, decabromodiphenyl ether or ethylene bis-tetrabromophthalimide in amounts of 30 % in combination with antimony oxide as additives. The latter, although more expensive than decabromodiphenyl ether, is used for wire and cable applications and shows no blooming.

In recent years, hydrated fillers such as aluminum trihydrate or magnesium hydroxide have been utilized as halogen-free flame retardants for high-density PE. To get a V-1 or V-0 rating to UL-94, levels between 40 % and 60 % are necessary.

Commercially available flame retarded *polypropylene* can meet either the UL 94 V-0 or V-2 rating, which is acceptable for many electrical applications including monitor housings, electrical conductors, or for PP pipes.

For a V-2 rating, smaller amounts of flame retardants, usually between 3 % and 5 % plus 2–3 % antimony oxides are required. With the low loading level, the properties of the virgin resins remain mainly unchanged. Here, brominated phosphates and tetrabromobisphenol A derivatives are used as flame retardants.

For commercial V-0 rated products, higher amounts (40 %) of halogenated flame retardants such as Dechlorane Plus or decabromodiphenyl ether in combination with antimony oxide are required. The impact strength of flame retarded PP can be increased significantly by using a flame retardant package containing a mixture of certain silicon polymers (e.g., SFR-100, silicon resin solved in a high viscosity silicon fluid) [33].

Attempts for flame retarding PP with nonhalogen flame retardant systems have been made. It was found that a UL 94 V-0 rating is achieved for PP with hydrated fillers such as magnesium hydroxide [34]. Loading levels of 50–60 % are necessary. An additional benefit of these formulations is reduced smoke emission when burning.

For PP, several intumescent additives based on a nitrogen-phosphorus system such as ammonium polyphosphate are available. With additive levels of at least 25–30 % UL 94 V-0 is obtained. A characteristic of intumescent systems is their rapidly declining effect at lower concentrations. This is in contrast with halogenated flame retardants, where a gradual reduction in flame retardant effectiveness can be observed with decreasing concentration (Fig. 5.13).

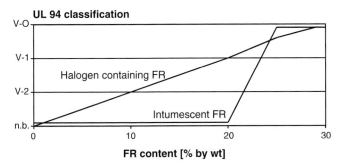

Figure 5.13   UL 94 rating and flame retardant content in PP (schematic) [35]

Typical advantages of PP flame retarded with intumescent systems are the thermal stability (up to 260 °C), UV-stability (compared to halogen systems), and a reduced smoke density when burning [35, 36].

Bromine-containing phosphorus compounds with high heat resistance and good resistance to hydrolysis have been described [37, 38]. Formulations with red phosphorus have also been developed but have not gained acceptance owing to the intrinsic color of this compounds [39]. A detailed description of flame retardant polyolefins is given by Schwarz in [14, Chapter 2] and Green in [40, Chapter 1].

5.2.1.2 Poly(vinyl chloride)

Virgin (rigid) PVC contains nearly 57 % chlorine, is inherently flame retardant, and thus meets the usual fire performance requirements for combustible materials. In case of burning, PVC pyrolyzes to form HCl and volatile aromatic compounds, which burn and evolve considerable amounts of smoke. When PVC is plasticized to make high-impact products, as used in electrical wire and cable insulation or sheathing, its self-extinguishing properties become less favorable depending on the amount and kind of plasticizer and other additives used. The impact modifiers and plasticizers burn more easily than the PVC matrix itself and also dilute the total chlorine content.

Commercial plasticized PVC contains up to 50 % plasticizer so that flame retardants must be incorporated for many applications. This is achieved by partly or totally replacing the plasticizer with a flame retardant phosphate plasticizer or chloroparaffin and adding antimony trioxide as a synergist. For many applications such as cables, large quantities of fillers such as chalk or aluminum trihydrate [41, 42] are used. Finely divided aluminum trihydrate is cheap and even large amounts do not significantly affect the impact strength.

Triaryl, alkyl diaryl, and trialkyl phosphates such as tricresyl phosphate, isopropylphenyl phosphate, trisisopropylphenyl phosphate, isodecyldiphenyl phosphate, and others are used as flame retardant phosphate plasticizers. Their different flame retardant performance is shown in Fig. 5.14.

Metal compounds that function as Lewis acids are effective as smoke suppressants. Zinc borate, copper compounds, and molybdenum oxides frequently used in combination with antimony oxide and aluminum trihydrate are known systems [8].

A review on fire performance of PVC is given in [43]. Other aspects of the fire performance of PVC and related polymers are described by O'Mara in [13, Chapter 3], Burn and Martin in [44] and Troitzsch in [45, p. 736].

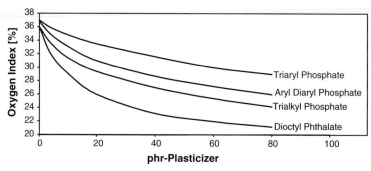

Figure 5.14   Effect of phosphate ester type on the oxygen index of PVC

5.2.1.3 Polystyrene and Styrene Copolymers

The use of general purpose *polystyrene* grades in countless applications has been known for many years. Flame retarded products are requested for foam insulation, in which two types of products, extruded foam and expandable bead boards, can be distinguished. Compared to injection molded PS, which is processed at temperatures up to 260 °C, the manufacture of PS foam with physical blowing agents is carried out at approximately 120 °C for bead foam (expansion by pressurized steam) and about 200 °C for extruded foam. These lower processing temperatures allow the use of the highly effective but thermally less stable alicyclic bromine compounds as flame retardants. The major flame retardant is hexabromocyclododecane, followed by tetrabromobisphenol A bis-allylether. Because of the high effectiveness of these compounds, they can be used in fairly small amounts (1–2 %). The level can be reduced further (< 1 %) in bead foam by the addition of radical initiators such as organic peroxides. Organic peroxides such as dicumyl peroxide or compounds such as oligomeric diisopropyl benzene are used as "synergistic" radical initiators.

Because of its rubber content (butadiene rubber), *high impact polystyrene (HIPS)* is much more difficult to flame retard and needs higher amounts of flame retardants. Commercially available flame retardant HIPS grades achieve UL 94 V-0 with decabromodiphenyl ether or ethylene bis-tetrabromophthalimide in loading of more than 10 % in combination with antimony oxide. The decisive disadvantage of decabromodiphenyl ether is its poor UV stability; even the usual UV stabilizers do not show their known effectiveness. As a consequence, many molded parts in typical applications such as TV housings are produced in dark colors or even painted. Ethylene bis-tetrabromophthalimide, although twice as expensive as decabromodiphenyl ether, represents an alternative with a significantly better UV stability.

The search for nonhalogen flame retardants for HIPS resulted in new developed blends containing polyphenylene oxide (up to 35 %) as char former and triarylphosphates such as triphenylphosphate or resorcinol bis-(diphenyl phosphate). With different amounts of polyphenylene oxide, flowability and heat distortion temperature can be adjusted to meet the requirements.

In many applications, flame retarded *acrylonitrile-butadiene-styrene copolymer (ABS)* acts as a competitive material for flame retardant HIPS. In general, ABS has a better impact strength, a higher heat distortion temperature, and a better surface quality (i.e., gloss); on the other hand, it is more expensive. Typical commercial flame retardants utilized for the past several years have been bis-(tribromophenoxy) ethane, octabromodiphenyl ether, and tetrabromobisphenol A combined with antimony oxide as a synergist. The bromine compounds are applied in levels between 10 % and 15 %. PVC is also frequently used for flame retarding ABS.

The level of antimony oxide is critical because the impact behavior of ABS is sensitive to filler particles. As a compromise and to obtain acceptable mechanical properties and ignition resistance, the antimony oxide level has to be kept at a low level (4 %) and small particle sizes are preferred.

Brominated flame retardants affect the ABS properties in a characteristic way. Bis-(tribromophenoxy) ethane has the tendency to plate out, which may increase the molding cycle times. With octabromodiphenylether, the plate-out problem can be minimized at the expense of poorer UV stability; this FR is to be banned for environmentals reasons. Tetrabromobisphenol A reduces the heat distortion temperature to a level, which can be even below flame retarded HIPS.

To overcome these general disadvantages, several attempts to replace these bromine compounds have been made in recent years. New bromine-containing flame retardants such as brominated epoxy oligomers produced from tetrabromobisphenol A [46] have been successfully incorporated in ABS; bromostyrene has been copolymerized with styrene and acrylonitrile to make ABS inherently flame retardant.

5.2.2 Plastics with a Heterogeneous Backbone

5.2.2.1 Polyurethane Foam

Polyurethane foam is one of the most widely used foam materials and belongs to the highest volume plastics. The main areas of applications range from building, to transportation, refrigeration, and furniture. Usage is split between flexible foam and rigid foam. Technical fire protection requirements have to be met for certain applications so that in many cases a flame retardant treatment is necessary.

Numerous additive and reactive flame retardants are in use to improve the flame retardancy of rigid and flexible foams (Table 5.4).

Table 5.4   Flame retardants in polyurethanes [47]

	Additive flame retardants	**Reactive flame retardants**
Halogenated flame retardants	Tris-(chloropropyl) phosphate (TCPP) Tris-(chloroethyl) phosphate (TCEP) Tris-(dichloropropyl) phosphate (TDCPP) Halogenated diphosphates	Brominated /chlorinated polyol Tetrabromophthalic acid diol Dibromoneopentyl glycol Chlorinated phosphorus polyol
Nonhalogenated flame retardants	Dimethyl methyl phosphonate (DMMP) Melamine Ammonium polyphosphate (APP) Aluminum trihydrate (ATH) Red phosphorus	Phosphorus polyol

In contrast to rigid PUR foam, it is difficult to provide flexible foams with a really effective flame retardant treatment, as factors such as open-cell structure, low degree of cross-linking, and chemical structure impair the flame-retardant effect. The physical properties of the foam are also adversely affected if incompatible additives such as fillers are incorporated. The use of flame retardants introduces a whole series of effects detrimental to the properties of the foam including reduced resistance to heat and hydrolysis and increased smoke evolution. Foams with reduced flammability can be obtained without adding flame retardants by modification with aromatic (highly aromatic polyols) or cross-linking components such as isocyanurates or carbodiimide. Such foams exhibit a strong tendency to char. They are thus better protected against the effects of flame [48].

The most frequently used reactive flame retardants are polyols that have been modified with groups containing phosphorus, halogens (usually bromine) or both. Examples are brominated polyols, chlorinated phosphorus polyols, dibromoneopentyl glycol, tribromoneopentyl alcohol, or phosphate polyols. Of the phosphorus/halogen polyols those based on tetrabromophthalic anhydride-polyol adducts containing phosphorus are usually used in rigid polyurethane foam. These products are less suitable for flexible foams owing to their lower effectiveness.

Halogenated polyols are advantageously used in flexible foams. A highly effective example is dibromoneopentyl glycol, which is relatively stable to hydrolysis and contains aliphatically bound bromine. Flame retardants containing aromatically bound bromine are of limited effectiveness because they do not decompose at the lower temperature range of the burning foam.

Phosphate polyol is a suitable and effective flame retardant for flexible polyurethane foam. Used in amounts of 7–10 %, the material reacts with the PUR matrix and is superior to chlorinated phosphates in terms of migration to the surface, fogging, and aging behavior. In addition, smoke density is reduced in comparison to halogenated products.

Polyesters containing hexachloroendomethylenetetrahydrophthalic acid (HET acid) structures as flame retardant components are frequently used as polyol components. Antimony trioxide is added to increase the flame retardant effect.

In general, additive flame retardants can be more easily incorporated into PUR foams than reactive ones. A major disadvantage is that they frequently cause shrinkage particularly in flexible foams, where they are preferred. The phosphate esters, the most popular group of additive flame retardants, also act as plasticizers. For flame retarding flexible foam, an increase in the flame retardant efficiency is achieved by using halogen-containing phosphorus compounds. Commercially used halogenated phosphates are tris-(chloropropyl) phosphate, tris-(dichloropropyl) phosphate, and halogenated diphosphates. Because of possible health impairment, tris-(chloroethyl) phosphate is no longer used. Current formulations contain more than 10 % of these chlorinated phosphates.

As in other polymer systems the more effective bromine compounds are generally used. Depending on requirements, aliphatic, cycloaliphatic and aromatic bromine compounds such as dibromoneopentyl glycol, hexabromocyclododecane and tetrabromophthalic acid esters have substituted pentabromodiphenyl ether. Aromatic bromine compounds are, however, preferred, as they do not impair the activity of the amine catalysts by elimination of hydrogen halide. Antimony trioxide is used in many cases as a synergist for a further improvement of the flame-retardant effect.

Melamine is used in large quantities for the production of CMHR polyurethane foam in home furnishings to comply with the demanding fire safety requirements in the United Kingdom.

The use of inorganic flame retardants such as aluminum trihydrate causes problems because they are difficult to incorporate and to distribute homogeneously in the foam; furthermore they provide only limited flame retardancy. Some fillers can even act as fire promoters owing to a wicking effect. Typical inorganic compounds in use after aluminum trihydrate are red phosphorus and salts such as calcium phosphate. Ferrocene has also proved to be a good flame retardant and smoke inhibitor. A more recent development is the use of expandable graphite in special flexible polyurethane formulations for high fire safety seating applications in aircraft and its use for flame retarding insulation tubes.

PUR foams are described in detail by Frisch and Reegen in [18] and Papa in [15, Chapter 1]. Frisch discusses mainly flame retardants containing reactive phosphorus, isocyanurate modified foams, and high temperature resistant PUR foams. *Papa*, in contrast, gives a general review of the additive and reactive flame retardants described in the literature and patents.

5.2.2.2 Thermosetting Resins

The most important representatives of the thermosetting resins group are unsaturated polyesters, epoxies, phenolic, urea, and melamine resins. Owing to their different chemical structures, their inherent flame retardancy may vary considerably. For most applications of phenolic, urea, or melamine resins additional flame retardants are not needed. Melamine resins are even added to other plastics to reduce their flammability. In contrast, unsaturated polyesters and epoxy resins are used commercially in applications such as electronics, transportation or the building industry, which require flame retardant treatment to meet fire safety requirements.

Polyester and epoxy resins are flame retarded by incorporating reactive or additive flame retardants and synergists. Reactive flame retardants are halogen compounds to which antimony trioxide must be added as a synergist to satisfy the most stringent requirements. Thus, tetrabromobisphenol A is frequently used in polyester and mainly in epoxy resins. Other flame retardants, which can be utilized in glass fiber reinforced polyesters, are hexachloroendomethylenetetrahydrophthalic acid (HET acid) or its anhydride, dibromoneopentyl glycol, and tetrabromophthalic anhydride. Alkylene oxide addition products of tetrabromobisphenol A are used for corrosion-resistant grades. Halogen-free flame retardant systems based on phosphorus or nitrogen have been developed; however, a broad application is limited because of the higher cost of these systems.

Large quantities of additive flame retardants and fillers are frequently incorporated in *unsaturated polyester resins*. These include chloroparaffins and various bromine compounds such as brominated polystyrene or poly(pentabromobenzyl) acrylate. Antimony trioxide and large quantities of calcium carbonate or aluminum trihydrate are added to these. Tin oxides have been described as flame retardants and smoke suppressants [49]. The most frequently used flame retardant for halogen-free polyesters and epoxy resins is aluminum trihydrate, which can be incorporated in amounts up to 400 % of the resin level. These extremely high loadings have negative effects on processability (high viscosity of resin mixture) and mechanical properties. Combinations of aluminum trihydrate with ammonium polyphosphate or red phosphorus have been developed, which need up to 50 % reduced amounts of flame retardants and show advantages in case of burning (reduced smoke density and smoke toxicity). The more expensive, coated (e.g., with silane) aluminum trihydrate is easier to incorporate in glass fiber reinforced types. An overview of the flame retardant systems used in unsaturated polyester resins is given by Krolikowski et al. [50].

In the few cases in which *phenolic resins* require flame retardant treatment, additive or reactive flame retardants have proved effective. Besides familiar flame retardants such as tetrabromobisphenol A and various organic phosphorus compounds, halogenated phenols and aldehydes (e.g., *p*-bromobenzaldehyde) are specifically utilized as reactive flame retardants. Phosphorus can be introduced by direct reaction of the phenolic resin with phosphorus oxychloride.

Chlorine compounds (e.g., chloroparaffins) and various thermally stable aromatic bromine compounds are utilized as additive flame retardants. Antimony trioxide is added as a synergist. Suitable phosphorus compounds include halogenated phosphoric acid esters such as tris-(2-chloropropyl) phosphate and halogenated organic polyphosphates, as well as large quantities of inorganic compounds such as calcium and ammonium phosphates. Zinc and barium salts of boric acid and aluminum trihydrate are also applied. To suppress afterglow of phenolic resins, use is made of compounds such as aluminum chloride, antimony trioxide and organic amides.

A survey of flame retardants used in phenolic resins is given by Sunshine [14, Chapter 4] and Conley [18, Chapter 8]. Melamine and urea resins are covered in the same articles.

5.2.2.3 Engineering Plastics

The group of engineering plastics includes such well-known polymers as polyamides (6 and 6,6), poly(butylene terephthalate) (PBT), poly(ethylene terephthalate) (PET), polycarbonate (PC) and blends (with ABS or PBT or PET) and poly(2,6-dimethyl-1,4-phenylene) oxide (PPO) blended with HIPS. Because of their different chemical structures, they need different flame retarding systems. In recent years, flame retarded engineering plastics have found widespread applications in the electrical and electronic (E&E) industry, where, for example, they are used as housing materials for business machines, and where the fire safety requirements of the UL 94 V test have to be matched.

Commercial flame retarded *polyamides* may contain different flame retardants. Use is made of chlorinated cycloaliphatics such as Dechlorane Plus^{TM}, the Diels-Alder diadduct of hexachlorocyclopentadiene and cyclooctadiene to which antimony trioxide, iron oxide, zinc borate, or zinc oxide may be added [51]. Aromatic bromine compounds such as ethylene bis-tetrabromophthalimide, polybromostyrene, polydibromophenylene oxide, and brominated epoxy resins are also used. Depending on effectiveness, they are applied in amounts between 10 % and 20 %.

The search of polymer producers for nonhalogen flame retardants for polyamide has resulted in several developments based mainly on red phosphorus, melamine, and magnesium hydroxide. Red phosphorus has been known for more than 30 years as a highly efficient flame retardant for polyamide. With amounts of 5–10 %, a V-0 rating is achievable. Problems with handling and stability of the phosphorus could be reduced by special formulations. A disadvantage is the color, which allows only red, gray or black products. Polyamide containing approx. 50 % of magnesium hydroxide meets UL 94 V-0 and provides an extremely low smoke density, however, at the expense of mechanical properties (i.e., very poor impact strength). A suitable flame retardant for unreinforced polyamide with high impact strength is melamine cyanurate. Because of the low loading level (5–10 %), negative effects on mechanical properties are low. A combination of melamine cyanurate and zinc sulfide is claimed to reach a UL 94 V-0 rating in glass fiber reinforced and unreinforced material [52]. Commercial polyamide grades of lower flame retardancy (glow wire test) for E&E applications are also available.

Polybutylene terephthalate (PBT) is a versatile engineering thermoplastic with good toughness, excellent electrical properties, and good molding characteristic. Polyethylene terephthalate (PET) shows high rigidity and dimensional stability and good performance at elevated temperatures. Both polymers are often mineral- or glass-filled.

Usual flame retardants for commercially available PBT are bromine compounds such as decabromodiphenyl ether, oligocarbonates based on tetrabromobisphenol A, ethylene bis-tetrabromophthalimide, brominated epoxy resins, and brominated PS. Commercial blends with up to 10 % bromine compound, antimony trioxide, and dripping inhibitors meet UL 94 V-0. A good compromise of impact strength, processability, blooming, and surface quality properties is obtained with brominated triaryl phosphate [53]. Owing to bromine/phosphorus synergy, no antimony trioxide is necessary for nonfilled or mineral-filled PBT. In glass-filled PBT, melt viscosity is reduced by the brominated triaryl phosphate, thereby improving processability. However, owing to odor problems, brominated triaryl phosphate has not been introduced on a commercial basis.

In commercial flame retardant *polyethylene terephthalate*, generally the same bromine compounds as in PBT combined with an antimony synergist and an antidripping additive may be utilized. However, sodium antimonate is used instead of antimony trioxide, which degrades the polymer.

Polycarbonates are high-impact polymers with better inherent flame retardancy than most other engineering thermoplastics. They exhibit an oxygen index of about 25 and a UL 94 V-2 rating. Frequently, polycarbonates are used in transparent applications, owing to the very good clarity of the basic resin. To obtain UL 94 V-0, brominated or phosphorus compounds or sulfonate salts are added as flame retardants.

Bromine-containing flame retardants mainly include tetrabromobisphenol A oligocarbonates. In commercial products, amounts up to 8 % are necessary to obtain transparent materials with a V-0 rating at 1.6 mm (0.06 in.) thickness. If small amounts of polytetrafluoroethylene (PTFE) powder are added as antidrip, less flame retardant (3 %) is needed to pass UL 94 V-0. The resulting products are translucent. In general, antimony oxide is not used as a synergist, because it may degrade the polymer matrix. The small amounts of PTFE are particularly effective when added as masterbatches, because the latter ensure an optimal dispersion in the matrix.

Aromatic phosphates such as triphenyl phosphate or oligophosphates such as resorcinol bis-(diphenyl phosphate) have proved as effective flame retardants for polycarbonates [54]. More recently, alkyl-substituted oligophosphates have appeared and are commercially available.

A specific method for flame retarding commercial polycarbonates is the incorporation of sodium or potassium salts of organic sulfonates [55]. Owing to the high effectiveness of these salts, only catalytic concentrations of less than 0.1 % are necessary to require a V-0 rating at 1.6 mm thickness. These salts catalyze the decomposition of the polymer resulting in the formation of carbon dioxide and in cross-linking of the residual polymer (Fries rearrangement).

Polycarbonate is recognized as a preferred starting material for blending with other polymers. Blending partners are PET, PBT, and, with most commercial importance, ABS. A *polycarbonate-ABS* blend combines heat distortion temperature, impact resistance, and dimensional stability of polycarbonate with low temperature impact resistance, chemical resistance, processability, and good cost performance of ABS. Because of their balanced properties, flame retarded PC/ABS blends have found widespread applications as housing materials in electrical and electronic industry, where the UL 94 V grades are the most important fire safety specifications.

Preferred bromine carriers are polycarbonates or oligocarbonates based on tetrabromobisphenol A, which are compatible with bisphenol A polycarbonate and show no tendency to bloom.

Until the end of the 1980s, the state of the art was to flame retard PC/ABS blends with bromine compounds and antimony oxide as a synergist. As a result of the ecological discussions in Europe, especially in Germany and the Nordic countries, PC/ABS blends with chlorine- and bromine-free flame retardant systems have been commercialized in the 1990s. Many global producers of computers and related equipment are seeking to obtain the so-called "Eco-Labels" such as Swedish TCO 99 and German Blue Angel Marks, which require bromine- and chlorine-free material for the housings. Triaryl phosphates such as triphenyl phosphate (TPP) and oligophosphates such as resorcinol bis-(diphenyl phosphate) (RDP) and bisphenol A bis-(diphenyl phosphate) (BDP) are utilized as effective flame retardants. In addition, small amounts of PTFE (< 0.5 %) are incorporated in the form of masterbatches as antidripping agents.

The amounts of arylphosphate necessary to reach a V-0 rating at 1.6 mm thickness depend on the PC/ABS ratio. Owing to the different inherent flame retardancy of polycarbonate (PC) and ABS, the PC/ABS ratio determines the amount of phosphate flame retardant. The higher the PC/ABS ratio, the lower the phosphorus concentration required for a V-0 rating (Fig. 5.15).

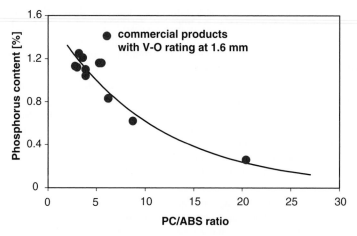

Figure 5.15   Phosphorus content as a function of the PC/ABS ratio

In commercial flame retardant PC/ABS blends, the ratio can vary between 20:1 and 3:1, corresponding to a phosphate level between 3 % and 14 % to reach a V-0 rating. TPP is the most commonly used flame retardant. Because of its high phosphorus content of 9.5 % and its volatility, it is highly effective. Under certain processing conditions TPP shows a tendency to plate out. To overcome these disadvantages, the low volatile but more expensive RDP is utilized. For certain applications, where an improved hydrolytic stability is required, BDP is the more suitable flame retardant. Mixtures of TPP and oligophosphates are also effective and advantageous for certain applications [56].

To meet UL 94 V-0 rating at a thinner thickness (< 1.6 mm) and to follow the trend to thinner molded articles, certain inorganic materials (i.e., talc, aluminum oxides) are added to improve the effectiveness of the phosphates [57].

Blends of *poly(2,6-dimethyl-1,4-phenylene) oxide (PPO) with high-impact polystyrene (HIPS)* are commonly used as housing materials for electrical or electronic devices, for example, as interior parts of printers and copiers. Because of the inherent flame retardancy of PPO, halogen-free flame retardants such as phosphate esters are suitable. The most important additives in commercial blends are triaryl phosphates and resorcinol bis-(diphenyl phosphate). Heat distortion temperature and amount of flame retardant necessary to meet the fire safety requirements depend on the PPO/HIPS ratio. In general phosphate amounts up to 13 % are incorporated.

5.2.2.4 Inherently Flame Retardant Polymers

There are various high temperature resistant polymers with highly aromatic or heteroaromatic chemical structure including polyphenylene sulfides, polysulfones, polyether sulfones, aramides, polyimides, polyarylates, and liquid crystal polymers. They all have only a minimal tendency to burn and consequently do not generally require any additional flame retardant treatment (See also Section 4.2.4).

References for Section 5.2

[1] J. Green: *J. Fire Sci.* 12 (1994) 257
[2] C.P. Yang, F.W. Lee: *J. Appl. Polym. Sci.* 32 (1986) 3005
[3] W. Wanzke: Neue Entwicklungen mit Phosphorverbindungen für Technische Kunststoffe, SKZ Tagung Kunststoffe, Brandschutz und Flammschutzmittel, Würzburg, 2–3 Dec. 1998
[4] J. Green: J. *Fire Sci.* 14 (1996) 362–363
[5] J. Green: *J. Fire Sci.* 14 (1996) 353–366
[6] J. Green: J. *Fire Sci.* 14 (1996) 426–442
[7] K. Shen: *Plastics Compound.* Sept./Oct. 1985
[8] K. Shen, D. Ferm: Use of zinc borate in electrical applications, Fire Retardant Chemical Association Fall conference, 1995
[9] P.A. Cusack, A.J. Killmeyer: *Fire Polym.* 14 (1990) 189–210
[10] P.G. Pape, D.J. Romenesko: 626/ANTEC '97/2941, 1997
[11] J. Green: *Fire Mater.* 19 (1995) 197–204
[12] J.W. Lyons: The Chemistry & Uses of Fire Retardants, Wiley-Interscience, New York, 1970
[13] W.C. Kuryla, A.J. Papa (Eds.): Flame Retardancy of Polymeric Materials, Vol. 1, Marcel Dekker, New York, 1973
[14] W.C. Kuryla, A.J. Papa (Eds.): Flame Retardancy of Polymeric Materials, Vol. 2, Marcel Dekker, New York, 1973
[15] W.C. Kuryla, A.J. Papa (Eds.): Flame Retardancy of Polymeric Materials, Vol. 3, Marcel Dekker, New York, 1975
[16] E.W. Flick: Plastics Additives, Section IX: Fire and Flame Retardants, Noyes Publications, Park Ridge, NJ, p. 213, 1986
[17] Anon: Handbook of Flame Retardant Chemicals and Fire Testing Services, Technomic, Lancaster, 1988
[18] M. Lewin, S.M. Atlas, E.M. Pearce (Eds.): Flame Retardant Polymeric Materials, Vol. 1, Plenum, New York, 1975
[19] Kirk-Othmer: Encyclopedia of Chemical Technology, 3rd edit., Vol.10, John Wiley, New York, 1980
J.W. Lyons: An Overview, p. 348
J.M. Avento, I. Touval: Antimony and Other Inorganic Compounds, p. 355
E.R. Larsen: Halogenated Flame Retardants, p. 373
E.D. Weil: Phosphorus Compounds, p. 396
G.L. Drake Jr.: Flame Retardants for Textiles, p. 420
[20] H.S. Katz, J.V. Milewski (Eds): Handbook of Fillers and Reinforcements for Plastic", Van Nostrand Reinhold, New York, 1978
H.H. Waddell, I. Touval: Antimony Oxide, Chapter 13
E.A. Woycheshin, I. Sobolev: Alumina Trihydrate, Chapter 14
J.V. Milewski, H.S. Katz: Miscellaneous Flame Retardants, Chapter 15
[21] R. Reichherzer: *Kunststoffber. Rdsch. Techn.* 20 (1975) 18
[22] R.C. Kidder: *Plast. Engn.* 33 (1972), p. 38
[23] H. Jenkner: *Kunststoffe* 62(2) (1972) 690
[24] H.E. Stepniczka: *Kunststoff-J.* 10(1–2) (1976) 12
[25] R.P. Levek in R.B. Seymour (Ed.): Additives for Plastics, Vol. 1: State of the Art, Academic Press, New York, p. 259, 1978
[26] H. Jenker in R. Gächter, H. Müller (Eds.): Plastics Additives Handbook. 2nd edit. , Carl Hanser, Munich, p. 535, 1987
[27] S. Salman, D. Klempner: *Plast. Engn.* 35(2) (1979) 39
[28] C.F. Cullis, M.M. Hirschler: The Combustion of Organic Polymers, Claredon Press, Oxford, 1981
[29] R.M. Aseeva, G.E. Zaikov: Combustion of Polymer materials, Carl Hanser, Munich, 1985
[30] R.G. Gann, R.A. Dipert, M.J. Drews: Flammability. Encycl. Polym. Sci. Eng., Vol. 7, Wiley, New York, pp. 154–210, 1987
[31] G. Camino, L. Costa: *Polym. Degrad. Stab.* 20(3–4) (1988) 271–94
[32] A. Hofmann: *Kunststoffe* 61 (1971) 811
[33] R. Frye: US Pat. 4,387,176, to General Electric Company (7 June 1983)
[34] S. Miyata: *J. Appl. Polym. Sci.* 25 (1980) 415
[35] W. Wanzke: Flame Retardants '98, Conf. Proc., pp. 195–206, 1998

[36] W. Wanzke: SKZ Tagung Kunststoffe, Brandschutz und Flammschutzmittel, Würzburg, 2–3 Dec. 1998
[37] J.A. Albright, C.J. Kmiec: J. *Appl. Polym. Sci.* 22 (1978) 2451
[38] A. Granzow, C. Savides: *J. Appl. Polym. Sci.* 25 (1980) 2195
[39] V.D. Rumyantsev: *Plasticheskie Massy* 3 (1977) 57 (translated in: Int. Pol. Sci. Techn.4 (1977) 7 T 103
[40] Anon, Flame Retardant Polymeric Materials, Vol. 3, Plenum Press, New York, 1982
[41] H Hentschel: *Kunststoffe-Plastics* 24(7) (1977) 18
[42] P.V. Bonsignore, P.L. Claasen: J. *Vinyl Technol.* 2(2) (1980)114
[43] M. Hirschler: Flame Retardants '98, Conf. Poc., pp. 103–123, 1998
[44] L.S. Burn, K.G. Martin: A Review of Combustion Characteristics of UPV Formulations, Commonwealth Scientific and Industrial Research Organization (CSIRO), Division of BuildingResearch, Highett, Victoria, Australia.
[45] H.K. Felger in Becker/Braun (Eds.): Polyvinylchlorid, Vol. 2/1 in Kunststoff-Handbuch, Carl Hanser, Munich, 1986
[46] B. Plaitin, A. Fonzé, R. Braibant: Flame Retardants '98, Conf. Proc., pp. 139–150, 1998
[47] R. Walz: SKZ Tagung Kunststoffe, Brandschutz und Flammschutzmittel, Würzburg, 9–10 April 1997
[48] E. K. Moss, D.L. Skinner: J. *Cell. Plast.* 13(7–8) (1977) 276
[49] P.A. Cusack: *Fire Mater.* 10 (1986) 41
[50] W. Krolikowski, W. Nowaczek, P. Penczek: *Kunststoffe* 77 (1987) 864
[51] R.L. Markezich: Int. Prog. Fire Safety Papers, Fire Retard. Chem. Assoc., Lancaster, pp. 17–27, 1987
[52] T. Uhlenbroich: SKZ Tagung Kunststoffe, Brandschutz und Flammschutzmittel, Würzburg, 2–3 Dec. 1998
[53] J. Green: *J. Fire Sci.* (1990) 254–265
[54] E.A. Murashko, G.F. Levchik, S.V. Levchik, D.A. Bright, S. Dashevsky: *J. Fire Sci.* 16 (1998) 278–296
[55] A. Ballistreri, G. Montaudo, E. Scamporrino, C. Puglisi, D. Vitalini, S. Cucinella: *J. Polym. Sci. Polym. Chem.* 26(8) (1988) 2113–2127
[56] T. Eckel, D. Wittmann, M. Öller, H. Alberts: US Pat. 5, 672,645, to Bayer AG (30 Sept. 1997)
[57] M. Bödiger, T. Eckel, D. Wittmann, H. Alberts: US Pat. 5, 849,827, to Bayer AG (15 Dec. 1998)

6 Textiles

A.R. HORROCKS AND B.K. KANDOLA

The high fire hazard posed by textiles both in historical times and in the present day is a consequence of the high surface area of the fibers present and the ease of access to atmospheric oxygen. In spite of the combustibility of textiles, however, across the world very few comprehensive fire statistics exist, especially those which attempt to relate deaths and injuries to cause, such as ignition and burning propagation properties of textile materials.

The annual UK Fire Statistics [1] are some of the most comprehensive available and do attempt to provide information perhaps representative of a European country with a population of about 55 million. For instance, up to 1998, these statistics have demonstrated that while approx. 20 % of fires in dwellings are caused by textiles being the first ignited material, more than 50 % of the fatalities caused by these fires. Table 6.1 presents typical data for 1982–1998, although since 1993 such detailed data have not been as freely available.

Table 6.1 UK dwelling total and textile – related fire deaths, 1982 – 1998 [1]

Year	Deaths in UK dwelling fires	Textile-related fatalities in dwellings				
		Total	Clothing	Bedding	Upholstery	Floor coverings
1998[a]	497	217	62	71	69	11
1997[a]	566	253	59	51	119	8
1996[a]	556	251	60	79	108	11
1995[a]	549	274	85	71	108	8
1994[a]	477	231	65	68	86	5
1993	536	269	51	85	105	19
1992	594	322	71	82	134	22
1991	608	293	59	85	127	10
1990	627	346	61	89	157	20
1988	732	456	92	141	195	20
1986	753	485	69	150	219	17
1984	692	396	59	124	167	22
1982	728	426	86	140	152	23

[a] values based on sampling procedure.

This shows that generally deaths from fires in UK dwellings have fluctuated around 700 per annum between 1982 and 1988; since then they have fallen to the 500–600 level. Fatalities from textile-related fires show a similar pattern, and it may be concluded that legislation associated with the mandatory sale of flame retarded upholstered furnishing fabrics into the domestic UK market since 1989 has played a significant factor in these reductions [2]. If these figures are representative of a typical EU community member, then it is likely that deaths from dwelling fires associated with textiles are of the order of 2000 per annum within the European Community.

It is rarely the direct effects of the fire, such as burn severity that are the prime causes of death, however, but rather the effects of the smoke and emitted fire gases that cause disorientation and impede escape initially followed by subsequent incapacitation, asphyxiation and death [1]. Only in clothing related fires are injury and death caused primarily by burns, especially when loose fitting and worn directly over the body such as nightwear and summer dresses.

6.1 Burning Behavior of Textiles

6.1.1 Burning Behavior of Fibers

The burning behavior of fibers is influenced by and often determined by a number of thermal transition temperatures and thermodynamic parameters. Table 6.2 [3] lists the commonly available fibers with their physical glass (T_g) and melting (T_m) transitions, if appropriate, which may be compared with their chemically related transitions of pyrolysis (T_p) and ignition and the onset of flaming combustion (T_c). In addition, typical values of flame temperature and heats of combustion are given. Generally, the lower the respective T_c (and usually T_p) temperature and the hotter the flame, the more flammable is the fiber. This generalization is typified by the natural cellulosic fibers cotton, viscose and flax as well as some synthetic fibers such as acrylics.

In Table 6.2 respective limiting oxygen index (LOI) values are listed, which are measures of the inherent burning character of a material and may be expressed as a percentage or decimal [4]. Fibers having LOI values of 21 % or 0.21 or below ignite easily and burn rapidly in air (containing 20.8 % oxygen). Those with LOI values above 21 % ignite and burn more slowly, and generally when LOI values rise above values of about 26–28 %, fibers and textiles may be considered to be flame retardant and will pass most small flame fabric ignition tests in the horizontal and vertical orientations.

Nearly all flammability tests for textiles, whether based on simple fabric strip tests, composite tests (e.g. BS 5852: 1979, ISO 8191/2, EN 1021 and 597), or more product/hazard related tests (e.g., BS 6307 for carpets, BS 6341 for tents and BS 6357 for molten metal splash), are essentially ignition-resistance tests.

Within the wider community of materials fire science, it is widely recognized that under real fire conditions, it is the rate of heat release that determines burning hazard. While the heats of combustion, ΔH_c, in Table 6.2 indicate that little difference exists between all fibers, and indeed some fibers such as cotton appear to have a low heat of combustion compared to less flammable fibers such as the aramid and oxidized acrylic fibers, it is the speed at which this heat is given out that determines rate of fire spread and severity of burns.

Currently only textiles used in building materials, aircraft, and transport interiors and seating are required to have minimal levels of rate of heat release, which is measured using instruments such as the cone calorimeter [5, 6] and the Ohio State University calorimeter [7] (used to assess aircraft interior textile performance at an incident heat flux of 35 kW/m^2). There are currently very little published heat release data for textiles, however.

Table 6.2   Thermal transitions of the more commonly used fibers [3]

Fiber	T_g (°C) (softens)	T_m (°C) (melts)	T_p (°C) (pyrolysis)	T_c (°C) (ignition)	LOI (%)	ΔH_c (kJ/g)
Wool			245	600	25	27
Cotton			350	350	18.4	19
Viscose			350	420	18.9	19
Nylon 6	50	215	431	450	20–21.5	39
Nylon 6.6	50	265	403	530	20–21.5	32
Polyester	80–90	255	420–447	480	20–21	24
Acrylic	100	> 220	290 (with decompo-sition)	> 250	18.2	32
Polypro-pylene	5	165	470	550	18.6	44
Mod-acrylic	< 80	> 240	273	690	29–30	–
Poly(vinyl-chloride)	< 80	> 180	> 180	450	37–39	21
Oxidized Acrylic	–	–	≥ 640	–	55	–
n-aramid (e.g., Nomex)	275	375	410	> 500	29–30	30
f-aramid (e.g., Kevlar)	340	560	> 590	> 550	29	–
Polybenz-imidazole (PBI)	–	> 400	> 500	> 500	40–42	–

6.1.2 Effect of Fabric and Yarn Structures

The burning behavior of fabrics comprising a given fiber type or blend is influenced by a number of factors including the nature of the igniting source and time of its impingement, the fabric orientation and point of ignition (e.g., at the edge or face of the fabric or top or bottom), the ambient temperature and relative humidity, the velocity of the air, and last but not least fabric structural variables. Fabric orientation, point of ignition source and time, and the atmos-

pheric variables are controlled in any standard test (see below). However, notwithstanding these, and as shown by Backer et al. [8], low fabric area density values and open structures aggravate burning rate and so increase the hazards of burn severity more than heavier and multilayered constructions.

Hendrix et al. [9] have related LOI linearly with respect to area density and logarithmically with air permeability for a series of cotton fabrics, although correlations were poor. Thus fabric flammability is determined not only by the fiber behavior but also by the physical geometry of fibrous arrays in fabrics. Miller et al. [10] considered that an alternative measure of flammability was to determine the oxygen index (OI) at which the burning rate was zero. The resulting intrinsic OI value for cotton is 0.13 and considerably less than the quoted LOI value of 0.18–0.19 (see Table 6.2).

Horrocks et al. [4, 11, 12] defined the extinction oxygen index (EOI) as the oxygen concentration at which the fabric just will not sustain any flame for a finite observable time when subjected to an LOI ignition source at the sample top for a defined ignition time. For simple flammable fabrics such as cotton, nylon, and polyester, respective EOI values decreased with decreasing igniter application time. Extrapolation enabled EOI values at zero time, $[EOI]_0$, to be defined. For a single layer of a typical cotton fabric, a value of 0.14 was derived, which was considered to be independent of igniter variables. Flame retardant cottons with LOI values of about 0.30 have EOI values much less than and close to 0.21 and so may be considered to be intrinsically flame retardant.

The effect of yarn geometry and structure on burning behavior is less clear and has not been studied in depth, although the above referenced works on fabric structure infer that coarser yarns will have a greater resistance to ignition. This assumes that fiber type and area density remain constant (for coarser yarns, the cover factor will decline and the air permeability will increase, which will have the converse effect). Recent work by Garvey et al. [13] has examined the burning behavior of blended yarns comprising modacrylic/flame retardant viscose and wool/flame retardant viscose, where the flame retardant viscose is Visil (Säteri Oy, Finland), produced by both ring-spinning and rotor-spinning methods, having the same nominal linear densities and knitted into panels.

Figure 6.1 shows the LOI results for all blends, indicating that the difference in yarn structure significantly influences the fabric burning behavior. The more flammable rotor spun yarns are believed to be a consequence of the improved fiber component randomization that occurs using this spinning method; in ring-spun yarns, component fiber aggregation is known to be a feature.

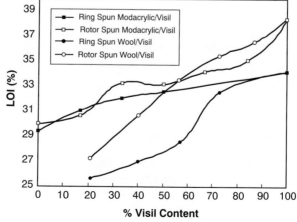

Figure 6.1 LOI values of knitted fabrics from blended yarns

6.1.3 Particularly Hazardous Textiles

The statistical data in Table 6.1 show that upholstered furnishings are the most hazardous in terms of fatality frequency, with smokers' materials being associated with a major ignition source in these fires [1]; legislation in the United Kingdom has, of course, addressed this issue [2]. During the 1980s bedding appeared to be the next most hazardous textile group, however, the need to use ignition-resistant tickings and mattress covers (and combustion-modified foam and fillings if appropriate) was included in the UK furnishings regulations [2]. While cigarettes are a prime cause of smoldering and flaming ignition, a detailed consideration of these is beyond the scope of this review. However, recent interest in both these areas has been revived [14].

The effect of using flame retardant covers and tickings is most probably a factor, as bedding-related fatalities have fallen by more than half during the 15-year period since 1982. At the other end of the spectrum is the relatively low involvement of floor coverings in fire-related deaths. This is not surprising, as carpets are used in relatively low flame-propagating horizontal geometry and they have heavy area densities. Within the United Kingdom the predominance of wool, polyamide, and blends in the middle to upper price ranges ensure a low ignition and flammability hazard. This is not always the case with the popular lower priced polypropylene carpets, which in certain constructions are known to burn quite easily in the horizontal mode. It is not surprising, therefore, that flame retardant additives for polypropylene are targeted at this market, especially in the contract area where fire regulations demand minimum ignition and fire spread criteria.

From Table 6.1 it is evident that the one set of significant statistics which has shown least change is that concerned with clothing-related fatalities, which typically constitute about 10 % of total UK fire deaths; this compares with the similar figures of 8–10 % determined by Weaver [15] for US clothing fire deaths. The earlier study of Tovey and Vickers [16], which in 1976 analyzed 3087 case histories of textile ignition caused fire deaths, showed that in the United States loose-fitting clothes such as shirts, blouses, trousers, and underwear ranked higher as potential hazards than bedding and upholstered furniture; pyjamas, nightgowns, dresses and housecoats presented very similar hazards to these latter two. Although the hazard and severity of burn injury by clothing has been related to sex, age, activity, and accident location [17], major intrinsic textile properties such as time to ignition, high heat release rates, and total heat release [18–20] are still of prime importance. An added factor that can increase heat transfer is whether a burning fabric adheres to the skin and burning continues once in contact; this may, of course, be a problem with thermoplastic fiber containing fabrics, especially in blends with nonthermoplastic components such as cotton.

A recent UK government-sponsored study [21] analyzed clothing burn statistics over the period from 1982–1992 collected from the UK Home Office [1], the UK Consumer Unit's Home Accident Surveillance System (HASS), and UK burns units to identify trends within the 60–80 clothing-related burns fatalities each year (see Table 6.1). The main conclusions agreed with those of other studies in that the very young and old are at greatest risk, with overall fatalities higher for women (55 %) and associated with loose fitting garments, with dresses and nightwear especially posing the highest hazard. With regard to fiber type, natural fibers accounted for 42 % of the accidents, synthetic fibers 42 %, and natural/synthetic blends 16 %; the most commonly mentioned and/or identified materials were cotton (29 %), polyamide (26 %), cotton/polyester blends (13 %), and wool (6 %), although "jeans material", presumably cotton, in the main was separately quantified at 4 %.

6.2 Flammability Testing of Textiles

It is probably true that nearly every country has its own set of textile fire testing standard methods, which are claimed to relate to the special social and technical factors peculiar to each. In addition, test methods are defined by a number of national and international bodies such as air, land, and sea transport authorities; insurance organizations; and governmental departments relating to industry, defense, and health, in particular. A brief overview of the various and many test methods available up to 1989 is given in [4] and a more recent list of tests specifically relating to interior textiles has been published by Trevira GmbH in 1997 [22] with a focus on Europe and North America. Since 1990, within the EU in particular, some degree of rationalization has been underway as "normalization" of EU member standards continues to occur; for detailed information, the reader is referred to respective national and CEN standard indexes. Because of the process of normalization, standards are increasingly serving a number of standards authorities; thus, for example, in the United Kingdom most new British Standards are prefixed by BS EN or BS EN ISO.

The complexity of the burning process for any material such as a textile, which is not only a "thermally thin" material but also has a high specific volume and oxygen accessibility relative to other polymeric materials, proves difficult to quantify and hence rank in terms of its ignition and post-ignition behavior. Most common textile flammability tests are currently based on ease of ignition and/or burning rate behavior, which can be quantified easily for fabrics and composites in varying geometries. Few, however, yield quantitative and fire science related data unlike the often maligned oxygen index methods [4]. LOI, while it proves to be a very effective indicator of ease of ignition, has not achieved the status of an official test within the textile arena. For instance, it is well known that to achieve a degree of fabric flame retardancy sufficient to pass a typical vertical strip test (see below), a LOI value of at least 27–28 % is required which must be measurable in a reproducible fashion. However, because the sample ignition occurs at the top to give a vertically downward burning geometry, this is considered not to be representative of the ignition geometry in the real world. Furthermore, the exact LOI value is influenced by fabric structural variables (see above) for the same fiber type and is not single valued for a given fiber type or blend. However, it finds significant use in development of new flame retardants and optimization of levels of application to fibers and textiles.

As Table 6.3 attempts to show, textile flammability tests may be categorized by various means depending on whether they are ignition/burning parameter or textile structure/composite related.

At the simplest level, most test procedures are a defined standard procedure (e.g., BS 5438 for vertical fabric strips), the use of a standard test to define a specific performance level for a given product (e.g., BS 5722 uses BS 5438 to test and define performance levels for nightwear fabrics [23]), or a combined test and performance-related set of defining criteria (e.g., BS 5852 Parts 1 and 2:1979 and EN 1021 Parts 1 and 2 for testing upholstered furnishing fabric/filling composites against simulated cigarette and match ignition sources). Ideally, all practical tests should be based on quite straightforward principles, which transform into a practically simple and convenient-to-use test method. Observed parameters such as time-to-ignition, post-ignition after-flame times, burning rates, and nature of the damage and debris produced should be reproducibly and repeatably measured with an acceptable and defined degree of accuracy. Figure 6.2 shows a schematic representation of a typical vertical strip test such as BS 5438 and BS EN ISO 6941/2 tests in which a simple vertically oriented fabric may be subjected to a standard igniting flame source either at the edge or on the face of the fabric for a specified time such as 10 s. For flame retarded fabrics, the properties measured after extinction of the ignition source are the damaged (or char) length, size of hole if present, times of after-flame and afterglow, and nature of any debris (e.g., molten drips, etc.). For slow burning fabrics, such as are required in nightwear, a longer fabric strip is used across which cotton trip wires connected to timers are placed. In BS 5722, for example, these are at 300 mm (11.8 in.) and 600 mm (23.6 in.) above the point of ignition and the time taken to cut through each thread enables an average burning rate to be determined. Tests of the type shown are simple and give reproducible and repeatable results. Furthermore, for similarly flame retarded fabrics, the length of the damaged or char length can show semiquantitative relationships with the level of flame retardancy as determined by methods such as LOI.

Table 6.3 Selected test methods for textiles

Nature of test	Textile type	Standard	Ignition source
British Standard based vertical strip method BS 5438	Curtains and drapes	BS 5867:Part 2:1980(1990)	Small flame
	Nightwear	BS 5722:1991	Small flame
	Protective clothing (now withdrawn)	BS 6249:Part 1:1982	Small flame
ISO vertical strip similar to Tests 1 and 2 in BS 5438	Vertical fabrics	BS EN ISO 6940/ 1:1995	Small flame
Small-scale composite test for furnishing fabrics/fillings/ bedding materials	Furnishing fabrics	BS 5852: Pts 1 and 2:1979 (retained pending changes in legislation [2])	Cigarette and simulated match flame (20-s ignition)
	Furnishing fabrics	BS 5852:1990(1998) replaces BS 5852: Pt 2	Small flames and wooden cribs applied to small and full-scale tests
		ISO 8191:Pts 1 and 2 (same as BS 5852:1990)	
		BS EN 1021–1:1994	Cigarette
		BS EN 1021–2:1994	Simulated match flame (15 s ignition)
	Bedding items	BS 6807:1996	Ignitability of mattresses or divans by sources in BS 5852
		BS 7175:1989 (1994)	Ignitability of bed covers and pillows by sources in BS 5852

Table 6.3   (Continuation)

Nature of test	Textile type	Standard	Ignition source
		ISO 12952–1/4:1998	Ignitability of bedding items by cigarette and small flame sources
Cleansing and wetting procedures for use in flammability tests	All fabrics	BS 5651:1989	Not applicable but used on fabrics prior to submitting for standard ignition tests
	Commercial laundering	BS EN ISO 10528: 1995	
	Domestic laundering	BS EN ISO 12138: 1997	
Use of radiant flux	Aircraft seat assemblies, so-called "Boeing" test	ASTM E 906 1983, uses Ohio State University heat release calorimeter	Irradiate under 35 kW/m^2 with small flame igniter
	All fabrics/composites	NF P 92501–7, French "M test"	Irradiate with small burner
Protective clothing	Resistance to radiant heat	BS EN 366:1993 (replaces BS 3791:1970)	Exposure to radiant source
		BS EN 367:1992	Determine heat transfer index
	Resistance to molten metal splash	BS EN 373:1993	Molten metal
	Gloves	BS EN 407:1994	Radiant, convective and molten metal
	Firefighters clothing	BS EN 469:1995	Small flame
	General flame spread	BS EN 532:1994 (replaces BS 5438)	Small flame
	General protection	BS EN 533:1997 (replaces BS 6249)	Small flame
	Contact heat transmission	BS EN 702:1994	Contact temperature 100–500 °C

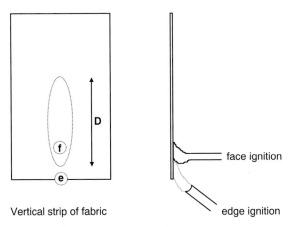

face ignition

edge ignition

Vertical strip of fabric

Figure 6.2   Schematic representation of a simple vertical strip test

With the recognition of the hazards posed by upholstered fabrics, the development of the small-scale composite test BS 5852 (see Table 6.3 and Fig. 6.3) in 1979 represented a milestone in the development of realistic model tests that cheaply and accurately indicate the ignition behavior of full-scale products of complex structure. Again, the test has proved to be a simple to use, cost-effective, and reproducible test, which may be located in the manufacturing environment as well as formal test laboratory environments.

However, in all flammability test procedures, conditions should attempt to replicate real use and so while atmospheric conditions are specified in terms of relative humidity and temperature ranges allowable, fabrics should be tested after having been exposed to defined cleansing and aftercare processes. Table 6.3 lists BS 6561 and its CEN derivatives as being typical here, and these standards define treatments from simple water soaking through dry cleaning and domestic laundering to the more harsh commercial laundering processes used in commercial laundries and hospitals, for instance.

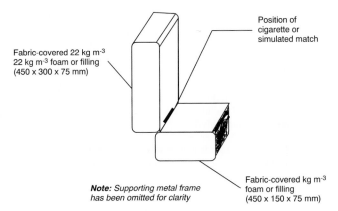

Position of cigarette or simulated match

Fabric-covered 22 kg m^{-3}
22 kg m^{-3} foam or filling
(450 x 300 x 75 mm)

Fabric-covered kg m^{-3}
foam or filling
(450 x 150 x 75 mm)

Note: Supporting metal frame has been omitted for clarity

Figure 6.3   Schematic diagram of BS 5852/EN 1021–1/2 composite test

As textile materials are used in more complex and demanding environments, so the associated test procedures have become more complex. This is especially the case for protective clothing and its components, which not only have to function as a typical textile material but also need to be resistant to a number of agencies including heat and flame. Table 6.3 shows also a set of tests that have recently been developed across the EU to accommodate the different demands of varying types of protective clothing and the hazards —— whether open flame, hot surface, molten metal splash, or indeed a combination — that are addressed. One test not yet standardized is that based on the simulation of a human torso and its reaction to a given fire environment when clothed; the original DuPont "Thermoman" [24] or instrumented manikin provided the means of recording the temperature profile and simulated burn damage sustained by the torso when clothed in defined garments (usually prototype protective garments) during exposure to an intense fire source. This latter is typically a series of gas burners yielding a heat flux of 80 kW/m^2. Sorensen [25] recently reviewed attempts to establish this and related manikin methods as a standard method.

The measurement of ease of ignition under a high heat flux and the associated heat release have been used to define both the ignition and fire propagating of textiles used in commercial aircraft seatings since the late 1980s [7]. The current specification demands that seating composites and all interior textiles shall have peak heat release rates ≤65 kW/m^2 and average rates ≤65 m^{-2} min^{-1} when exposed to a heat flux of 35 kW/m^2. The more recently available cone calorimeter [9] has yet to make a significant impact in the assessment of textile fire behavior apart from applications in the defense and extreme protective clothing related sectors.

6.3 Flame Retardant Textiles

Table 6.4 lists all the commonly available flame retardant and inherently flame-resistant textiles and generally these have LOI values of 27% or greater and will pass most vertical strip tests for all but the lightest weight fabrics.

Table 6.4   Durably-finished and inherently flame retardant fibers in common use

Fiber	Flame retardant structural components	Mode of introduction
Natural:		
Cotton	Organophosphorus and nitrogen-containing monomeric or reactive species, e.g., Proban CC (Rhodia, formerly Albright & Wilson), Pyrovatex CP (Ciba), Aflammit P and KWB (Thor), Flacavon WP (Schill & Seilacher)	F
	Antimony-organohalogenated systems, e.g., Flacavon F12/97 (Schill & Seilacher), Myflam (B F Goodrich, formerly Mydrin)	F
Wool	Zirconium hexafluoride complexes, e.g., Zirpro (IWS); Pyrovatex CP (Ciba), Aflammit ZR (Thor)	F
Regenerated:		
Viscose	Organophosphorus and nitrogen/sulfur-containing species, e.g., Sandoflam 5060 (Clariant, formerly Sandoz) in FR Viscose (Lenzing)	A
	polysilicic acid and complexes, e.g., Visil AP (Säteri)	A

Table 6.4   (Continuation)

Fiber	Flame retardant structural components	Mode of introduction
Inherent synthetic:		
Polyester	Organophosphorus species: phosphinic acidic comonomer, e.g., Trevira CS, (Trevira GmbH, formerly Hoechst); phosphorus-containing additive, Fidion FR (Montefibre)	C/A
Acrylic (modacrylic)	Halogenated comonomer (35–50 % w/w) plus antimony compounds, e.g., Velicren (Montefibre); Kanecaron (Kaneka Corp.)	C
Polypro-pylene	Halo-organic compounds usually as brominated derivatives, e.g., Sandoflam 5072 (Clariant, formerly Sandoz)	A
Poly-haloalkenes	Poly(vinyl chloride), e.g., Clevyl (Rhône-Poulenc) Poly(vinylidene chloride), e.g., Saran (Saran Corp.)	H
High heat and flame resistant (aromatic):		
Polyaramids	Poly(m-phenylene isophthalamide), e.g., Nomex (DuPont), Conex (Teijin)	Ar
	Poly (p-phenylene terephthalamide), e.g., Kevlar (DuPont), Twaron (Enka)	Ar
Poly(aramid-arimid)	e.g., Kermel (Rhône-Poulenc)	Ar
Novoloid	e.g., Kynol (Kynol, Japan)	Ar
Polybenzimi-dazole	e.g., PBI (Hoechst-Celanese)	Ar
Carbonized acrylics (semicarbon)	e.g., Panox (RK Textiles)	

F = chemical finish; A = additive introduced during fiber production; C = copolymeric modifications; H = homopolymer; Ar = aromatic homo- or copolymer

6.3.1 Cellulosics

Flame retardant cellulosic textiles generally fall into three groups based on fiber genus:

- Flame retardant cotton,
- Flame retardant viscose (or regenerated cellulose),
- Blends of flame retardant cellulosic fibers with other fibers, usually synthetic or chemical fibers.

6.3.1.1 Flame Retardant Cottons

It is most important that effective flame retardants are also effective afterglow retardants [26]. All flame retardant cottons are usually produced by chemically aftertreating fabrics as a textile finishing process that, depending on chemical character and cost, yields flame retardant properties having varying degrees of durability to various laundering processes. These may be simple soluble salts to give nondurable finishes (e.g., ammonium phosphates, polyphosphate, and bromide; borate-boric acid mixtures); they may be chemically reactive, usually functional finishes to give durable flame retardancy (e.g., alkylphosphonamide derivatives (Pyrovatex, Ciba; Aflammit KWB, Thor; Flacavon WP, Schill & Seilacher); tetrakis(hydroxymethyl)phosphonium (THP) salt condensates (Proban, Rhodia (Albright & Wilson); Aflammit P, Thor) and back-coatings, which often usually comprise a resin-bonded antimony-bromine flame retardant system. Back-coatings typically contain antimony (III) oxide in combination with decabromodiphenyl ether or hexabromocyclododecane in an Sb/Br molar ratio of 1:3 and suspended in a resin.

6.3.1.2 Flame Retardant Viscose

In these fibers, flame retardant additives are usually incorporated into the spinning dopes during manufacture, which therefore yields durability and reduced levels of environmental hazard as this eliminates the need for a chemical flame retardant finishing process (see Table 6.4). Additives such as Sandoflam 5060 [27] are phosphorus based and so are similar to the majority of flame retardant cotton finishes in terms of their mechanisms of activity (condensed phase), performance, and cost-effectiveness. Again, environmental desirability may be questioned and this issue has been minimized by Säteri Oy (formerly Kemira), Finland, with their polysilicic acid containing Visil flame retardant viscose fiber [27, 28].

Flame Retardant Cellulosic Blends

In principle, flame retardant cellulosic fibers may be blended with any other fiber, whether synthetic or natural. In practice, limitations are dictated by a number of technical limitations including:

- Compatibility of fibers during spinning or fabric formation
- Compatibility of fiber and textile properties during chemical finishing
- Additivity and, preferably synergy, should exist in the flame retardant blend

Consequently, the current rules for the simple flame retarding treatment of blends are either to apply flame retardant only to the majority fiber present or apply halogen-based back-coatings, which are effective on all fibers because of their common flame chemistries.

The prevalence of polyester-cotton blends coupled with the apparent flammability-enhancing interaction in which both components participate (the so-called scaffolding effect, reviewed elsewhere [4, 29]) has promoted greater attention than any other blend. However, because of the observed interaction, only halogen-containing coatings and back-coatings find commercial application to blends, which span the whole blend composition range.

Durable, phosphorus-containing cellulose flame retardants are generally effective only on cellulose-rich blends with polyester. THP-based systems such as Proban CC (Rhodia, formerly Albright & Wilson) are effective on blends containing no less than 55 % cotton if a combination of flame retardation and acceptable handle is required. Application of methylolated phosphonamide finishes (e.g., Pyrovatex CP, Ciba) is effective on blends containing 70 % or less cellulose content. This is because the phosphorus present is less effective on the polyester component than in THP-based finishes [29]. The reasons for this are not clear but are thought to be associated with some vapor phase activity of phosphorus in the latter finish on the polyester component.

6.3.1.3 Flame Retardant Wool and Blends

Of the so-called conventional fibers, wool has the highest inherent flame retardancy, and for some end-uses in which high density of structure and horizontal orientation (e.g., carpets) are required in the product, wool fabrics will often pass the required flame retardancy tests untreated. Table 6.2 shows it to have a relatively high LOI value of about 25 and a low flame temperature of about 680°C. Its similarly high ignition temperature of 570–600°C is a consequence of its higher moisture regain (8–16 % depending on relative humidity), high nitrogen (15–16 %) and sulfur (3–4 %) contents, and low hydrogen (6–7 %) content by weight.

The review by Horrocks [30] comprehensively discusses developments in flame retardants for wool up to 1986 and very little has changed since that time. Of particular note is the well-established durable Zirpro process based on the exhaustion of negatively charged complexes of zirconium or titanium onto positively charged wool fibers under acidic conditions at a relatively low temperature of 60°C. Zirpro treatments can be applied to wool at any processing stage from loose fiber to fabric using exhaustion techniques either during or after dyeing. The relatively low treatment temperature is an advantage because this limits the felting of wool.

Recently the process has come under the critical eye of environmentalists as a consequence of release of heavy metal ions into effluent discharges.

Wool blends pose different challenges, but given the complexity of wool and the position of the Zirpro process as the currently major durable flame retardant treatment, its specificity ensures that little if any transferability occurs to other fibers present. Furthermore, antagonisms between Zirpro and other flame retardant fibers were reported by Benisek in 1981 [31]. In the absence of any back-coating treatment, acceptable flame retardancy of Zirpro-treated blends is obtainable in 85/15 wool-polyester or polyamide combinations. For lower wool contents in blends and without the possibility of using alternative flame retardant treatments, flame retardancy can be maintained only if some of the Zirpro-treated wool is replaced by certain inherently flame retardant fibers, except for Trevira CS polyester [31]. Chlorine-containing fibers such as poly(vinyl chloride) and modacrylics are particularly effective in this respect.

6.3.1.4 Flame Retardant Synthetic Fibers

Inherently Flame Retardant Synthetic Fibers

The conventional synthetic fibers may be rendered flame retardant during production by either incorporation of a flame retardant additive in the polymer melt or solution prior to extrusion or by copolymeric modification. Synthetic fibers produced in these ways are often said to be inherently flame retardant. However, problems of compatibility, especially at the high temperatures used to extrude melt-extruded fibers such as polyamide, polyester, and polypropylene, have ensured that only a few such fibers are commercially available. Table 6.4 lists examples of inherently flame retardant synthetic fibers and the absence of polyamides reflects their high melt reactivities and hence poor flame retardant compatibilities. Flame retardant acrylics are currently not available and modacrylics are used instead. This latter group has been commercially available for 40 years or so, but at present few manufacturers continue to produce them. This is largely because of the success of back-coatings applied to normal acrylic fabrics, which create high levels of flame retardancy more cost-effectively. On the other hand, one group, which continues to be successful is flame retardant polyester typified by the well-established Trevira CS, which contains the phosphinic acid comonomer. Other flame retardant systems, both based on phosphorus-containing additives, are also shown although only the Toyobo GH (and variants) are commercially available. The Rhodia (formerly Albright & Wilson) Antiblaze 1045 additive is the former Mobil Chemical Antiblaze 19 compound, which is available in dimeric form as a melt additive and in monomeric form as a polyester textile finish (Amgard/Antiblaze CU). All three of these flame retardant polyester variants do not promote char but function mainly by reducing the flaming propensity of molten drips normally associated with unmodified polyester. As yet, no char-promoting flame retardants exist for any of the conventional synthetic fibers and this constitutes the real challenge for the next generation of acceptable inherently flame retardant synthetic fibers.

Flame Retardant Finishes for Synthetic Fibers

Polyamide, polyester, polyacrylic, and polypropylene are also candidates for semidurably and durably flame retarding textiles if suitable finishes are available. In the case of acrylics, because of the difficulty in finding an effective flame retardant finish, modacrylic fibers are preferred, unless a back-coating is considered as an acceptable solution, as it would be for finishing fabrics to be tested to BS 5852 : 1979 or EN 1021–1/2. While back-coatings may be similarly effective on other synthetic fiber-containing fabrics and may offer sufficient char-forming character and char coherence to offset fiber thermoplastic and fusion consequences, this is less easily achieved for polypropylene fabrics.

The low melting point, nonfunctionality, and high hydrocarbon fuel content (see Table 6.2) of polypropylene are three factors that have created problems in finding an effective durable flame retardant finish and also pose difficulties in the design of effective back-coatings.

6.4 High Heat and Flame Resistant Synthetic Fibers and Textiles

Inherently flame and heat resistant fibers and textiles, including the inherently flame retardant viscose and synthetic fibers, comprise approx. 20 % of total flame retardant usage.

Table 6.4 includes the main members of the group of high heat and flame-resistant fibers, which have fundamentally combustion-resistant all-aromatic polymeric structures. Table 6.2 shows that most of these decompose above 375 °C or so. Their all-aromatic structures are responsible for their low or nonthermoplasticity and high pyrolysis temperatures. In addition, their high char-forming potentials are responsible for their low flammabilities and, as established by van Krevelen [32], their high LOI values. This group is typified by the polyaramids, poly(aramid-arimids), and polybenzimidazole, which may have end-uses where high levels of heat resistance (HR) are required in addition to flame retardancy. However, also included in Table 6.4 are the novoloid and carbonized acrylic fibers, which, while having poorer fiber and textile physical properties, do have significant char-forming potentials and flame and heat resistance. These fibers tend to be used in nonwoven structures or in blends with other fibers to offset their less desirable textile characteristics. The high cost and high temperature performance of these HR synthetic fibers restrict their use in applications where performance requirements justify the cost. In practice, these fibers find use in high-performance protective clothing and barrier fabrics, particularly in fire-fighting, transport, and defense areas. However, the recently reported intumescent systems developed and described in the next section demonstrate heat barrier characteristics similar to and in some cases superior to those of the HR synthetic fibers.

6.5 Intumescent Application to Textiles

Clearly any enhancement of the char barrier in terms of thickness, strength, and resistance to oxidation will enhance the flame and heat barrier performance of textiles. Generation or addition of intumescent chars as part of the overall flame retardant property will also reduce the smoke and other toxic fire gas emissions. The application of intumescent materials to textile materials has been reviewed [27] and is exemplified in the patent literature by the following fiber-intumescent structures which offer opportunities in textile finishing:

- More conventional, flexible textile fabrics to which an intumescent composition is applied as a coating have been reported [33]. In one example of this patent, the glass-fiber-cored yarns used in the woven or knitted structure complement the flame and heat resistance of the intumescent coating. Presence of sheath fibers of a more conventional generic type ensures that the textile aesthetic properties may be optimized. More recently the Flammentin IST flame retardant from Thor Chemicals, UK, is based on the use of intumescents as replacements for antimony-bromine systems in coating and back-coating formulations. This has been demonstrated to be effective on polyester-based fabrics [34].
- The recent development of a back-coating for technical nonwovens by Schill & Seilacher has been reported and is based on exfoliated graphite [35]. This seems to be particularly effective on polyamides and polyester.
- Recently, Horrocks et al. has patented [36] a novel range of intumescent-treated textiles that derive their unusually high heat barrier properties from the formation of a complex char that has a higher than expected resistance to oxidation. Exposure to heat promotes simultaneous char formation of both intumescent and fiber to give a so-called "char-bonded" structure. This integrated fibrous-intumescent char structure has a physical integrity superior to that of either charred fabric or intumescent alone and, because of reduced oxygen accessibility, demonstrates an unusually high resistance to oxidation when exposed to temperatures above 500 °C and even as high as 1200 °C. Furthermore, these composite structures show significantly reduced rates of heat release when subjected to heat fluxes of 35 kW/m^2, thus demonstrating additional significant fire barrier characteristics [37].
- More recent work has been reviewed elsewhere [38] and has shown that the intumescents, which are based on ammonium and melamine phosphate-containing intumescents applied in a resin binder, can raise the fire barrier properties of flame retarded viscose and cotton fabrics to levels associated with high-performance fibers such as aramids.

Clearly there is an increasing interest in the development and use of intumescent flame retardants across the whole spectrum of flame retardant polymeric materials. This is driven by the need to reduce the concentrations and usage of the common and environmentally questioned Sb-Br formulations [29] coupled with the superior fire barrier and reduced toxic combustion gas properties that they generally confer. In the next few years there will be increased use of these materials in the textile and related sectors.

References

[1] Fire Statistics, United Kingdom,1998, The Home Office, The Government Statistical Office, UK, ISSN 0 143 6384, 1999
[2] Consumer Protection Act (1987), the Furniture and Furnishings (Fire) (Safety) Regulations, 1988, SI1324 (1988), HMSO, London, 1988
[3] A R. Horrocks: *J.Soc. Dyers. Col.* 99 (1983) 191
[4] A R. Horrocks, D. Price, D. Tunc: *Text. Prog.* 18 (1989) 1–205
[5] V. Babrauskas, S. J. Grayson (Eds): Heat Release in Fires, Elsevier, London, 1992
[6| Standard Test Method for Heat and Visible Smoke Release Rates for Materials and Products, ASTM E906 1983
[7] US Federal Aviation Regulation FAR 23:853", Appendix F, Part IV
[8] S. Backer, G.C. Tesoro, T.Y. Toong, N.A. Moussa N.A.: Textile Fabric Flammability, MIT Press, Cambridge, MA, 1976
[9] J.E. Hendrix, G.L. Drake, W.A. Reeves: *J. Fire Flamm.* 3 (1972) 38
[10] B. Miller, B.C. Goswami, R. Turner: *Text. Res. J.* 43 (1973) 61
[11] A.R. Horrocks, M. Ugras: *Fire Mater.* 7 (1983) 119

[12] A.R. Horrocks, D. Price, M. Tunc: in Fundamental Aspects of Polymer Flammability, IOP Short Meeting Series No.4, Bristol, UK, Institute of Physics, p. 165, 1987

[13] S.J. Garvey, S.C. Anand, T. Rowe, A.R. Horrocks in M. Le Bras, G. Camino, S. Bourbigot, and R. Delobel (Eds.): Fire Retardancy of Polymers The Use of Intumescence, Roy. Soc. Chem. Pub., Cambridge, p. 376, 1998

[14] A.J. Dyakonov, D.A. Grider: *J. Fire Sci.* 16 (1998) 297–322

[15] Weaver W.J.: *Text. Chem. Col.* 10(1) (1978) 42

[16] H. Tovey, A. Vickers: *Text. Chem. Col.* 8 (1976) 19

[17] J.M. Laughlin, A.M. Parkhurst, B.M. Reagan, C.M. Janecek: *Text. Res. J.* 55 (1985) 285

[18] B. Miller, C.H. Meiser: *Text. Res. J.,* 48 (1978) 238–243

[19] J.F. Krasny, A.L. Fisher: *Text. Res. J.* 43 (1973) 272–283

[20] J.F. Krasny: *Text. Res. J.* 56 (1986) 287

[21] Clothing Flammability Accidents Survey, Consumer Safety Unit, Department of Trade and Industry, London, 1994

[22] Standards for Testing of Interior Textiles, Trevira (formerly Hoechst) GmbH, Frankfurt, 1997

[23] The Nightwear (Safety) Regulations 1985, SI 1985/2043, HMSO, London, 1985

[24] M.P. Chouinard, D.C. Knodel, H.W. Arnold: *Text. Res. J.* 43 (1973) 166–175

[25] N. Sorensen: *Techn. Text. Int.* (1992 June) 8–12

[26] B. Kandola, A.R. Horrocks, D. Price, G. Coleman: *Revs. Macromol. Chem. Phys.* C36 (1996) 721–794

[27] A.R. Horrocks: *Polym. Deg. Stab.* 54 (1996) 143–154

[28] S. Heidari, A. Kallonen: *Fire Mater.* 17 (1993) 21–24

[29] A.R. Horrocks in D. Heywood (Ed.): Textile Finishing, Vol. 2, Society of Dyers and Colourists, Bradford, UK, 1999

[30] A.R. Horrocks: *Rev. Prog. Colour.* 16 (1986) 62–101

[31] L. Benisek: *Text. Res. J.* 51 (1981) 369

[32] R.W. Van Krevelen: Polymer 16 (1975) 615

[33] T.W. Tolbert, J.S. Dugan, P. Jaco P, J.E. Hendrix: US Pat. 333 174, 4 to Springs Industries, Fire Barrier Fabrics, April 1989

[34] C. Cazé, E. Devaux, G. Testard, T. Reix T in M. Le Bras, G. Camino, S. Bourbigot, R. Delobel (Eds.): Fire Retardancy of Polymers The Use of Intumescence, Roy. Soc. Chem., London, pp. 363–375, 1998

[35] P.J. Wragg in A.R. Horrocks (Ed.): in 2nd Int. Conf. Ecotextile'98 – Sustainable Development, Woodhead, Cambridge, UK, pp. 247–258, 1999

[36] A.R. Horrocks, S.C. Anand, B.J. Hill: UK Pat. GB 2279084B, 20 June 1995

[37] A.R. Horrocks, S.C. Anand, D. Sanderson: Int. Conf. Interflam'93, Interscience, London, pp. 689–698, 1993

[38] A.R. Horrocks, B.K. Kandola in M. Le Bras, G. Camino, S. Bourbigot, R. Delobel (Eds.): Fire Retardancy of Polymers The Use of Intumescence, Roy. Soc. Chem., London, pp. 343–362. , 1998

7 Smoke Development and Suppression

M. Le Bras, D. Price, and S. Bourbigot

Fire hazards are associated with a variety of properties of a material in a particular scenario. They are determined by a combination of factors, including product ignitability, flammability, amount of heat release on burning, flame spread, smoke spread, smoke production, and smoke toxicity [1, 2]. Data analyses of fire "risks" have shown that most fire-related deaths are due to smoke inhalation, rather than to burns by flames and/or heat [3]. A typical example of a smoke effect was the fire inside the Mont Blanc Tunnel linking France and Italy (March 24, 1999; 42 persons killed), in which gas fumes were one of the biggest hindrances to rescue workers and responsible for the death of a fire fighter in a shelter, which turned into a "gas chamber" [4]. Much is already known about how the inhalation of smoke leads directly to the deaths of building occupants [5, 6] and fire fighters, and about the subtle effects of sublethal exposures to smoke [7]. These risks are drastically increased by the general use of polymeric materials in buildings, transport vehicles, and electrical appliences. However, smoke formation does have one advantage as it serves as a spectacular or instrumentally detectable indicator of a fire or a smoldering process (combustion without flame, eventually a pyrolysis process, usually with incandescence and moderate smoke [8, 9]). Early detection of smoke evolution will activate the fire fighting services much sooner than would be the case in the absence of smoke. This more rapid response can be crucial to saving lives and limiting material damage.

The level of smoke development depends on numerous factors such as source of ignition, oxygen availability, constitution and properties of the combustible material [10], regime of the flame [11], buoyancy of the luminous flame environment [12], and location of the fire, in particular within buildings [13]. Smoke development is difficult to describe because it is not an intrinsic material property and may be influenced by the various factors already listed. Moreover, additives such as flame retardants in the polymeric material show an effect on smoke and toxic gas production. Indeed, it has often been noted that a flame retarded polymer can produce a more smoky flame than the corresponding non-flame-retarded polymer, the effect of the retardant on smoke depending on temperature [14].

The amount and properties of smoke generated in a fire depend on the polymeric materials being consumed [10]. For example, poly(vinyl chloride) (PVC), although inherently a good flame retarded material, is a notorious generator of thick smoke in a fire. On the other hand, poly(methyl methacrylate) (PMMA) is highly flammable but gives little if any smoke in a fire. Not surprisingly, PVC is the most extensively investigated polymer system from the viewpoint of identification of the major factors involved in smoke generation. Such studies of PVC and other smoke generating polymer systems are essential to the development of suitable smoke suppressant systems that minimize the smoke hazard. A brief survey of the influence of polymer structure on smoke generation and the mechanisms of smoke suppression is given. A review of the literature concerned with smoke suppressant systems used with the most common types of plastics concludes the chapter. The emphasis is on the chemistry involved rather than on providing an exhaustive list of suppressant systems used. These are numerous and are often based on empirical experience. Reference to the literature, particularly the patent literature, will keep the interested reader informed of current developments.

This chapter is an attempt to present the main points of this complex subject of smoke, which has in fact been investigated only in the last three decades. The present account reflects the literature consulted and tries to give a compilation of pertinent literature references to allow further research, but no claim is made with regard to completeness.

7.1 Constitution of Smoke

7.1.1 Definition of Smoke

ASTM terminology relating to fire Standards (E 176) defines "smoke" as the "airborne solid and liquid particulates and gases evolved when a material undergoes pyrolysis or combustion." According to Gaskill [15] and Hilado [16] or to the ISO working Draft 13943 (1993) [17], this broad definition is used in this chapter, in preference to dictionary ("volatilized products of combustion" [18]), Prado et al. (soot particles, main constituents of smoke [19]), or Mulholland's (aerosol or condensed phase components of the products of combustion [20]) definitions. It permits a clear production account of the whole combustion products from the polymers and thus enables discussion of the whole so-called "secondary fire effects" [10] including eventual synergistic effects (e.g., the combination of carbon monoxide with hydrogen chloride increases mortality above that expected from carbon monoxide alone [5]) which occur in conjunction with the combustion process in fire accidents. According to this definition, smoke results from both pyrolysis and thermooxidative processes during smoldering or ignition, from complete and incomplete oxidation of the degradation products in the flame, and/or from free-radical reactions (radical trapping, condensation, or nucleation) in the flame.

7.1.2 Origin of Smoke from Polymers in Fire

Polymers burn with diffusion flames: decomposition products of the polymer are mixed in the flame with oxygen diffusing from the surrounding air [21]. The chemistry of turbulent, fuel-rich flames, which is what polymer flames tend to be in actuality, is extremely complex. The breakdown of the polymer and the formation of decomposition and combustion products proceeds in long sequences of elementary reactions corresponding to pyrolysis and thermooxidative processes in both the condensed and gas phases [5]. Models to describe the combustion of polymers and to predict product formation are, as a consequence, too complex (e.g., a model of the combustion of butane involves 344 reversible reactions [22]) to solve on the level of elementary reaction mechanisms leading to polycyclic aromatic hydrocarbons or soot components.

Moreover, in building fires, a deep upper layer of "combustion" gases forms in the space adjacent to the burning room during the pre-flashover period of the fire. At the onset of flashover, a high level of incomplete combustion products begins issuing into adjacent spaces as discussed elsewhere [23]. The depth of the oxygen-deficient combustion gas layer then plays an integral role in the degree to which the products of incomplete combustion become oxidized [14].

In addition, a low rate-low temperature (100–250 °C) degradation is observed with almost all polymers [24] in the vicinity of a fire. This degradation may lead to the accumulation of toxic products (monomers, solvents, or oxidized products such as acrolein from polypropylene) in an unventilated room.

A more detailed discussion on the flame mechanisms is beyond the scope of this chapter. Previous considerations were reported only to explain that it is not easy to presume smoke composition from the chemical composition of the polymer. The smoke composition may in fact be very different depending on the flame regime and on the fire localization.

The *gaseous smoke fraction* is, of course, composed of air components but also of pyrolysis products from polymers (hydrogen halides, hydrogen cyanide, CO, CO_2, NH_3, carbonyl fluoride) and hydrocarbons (monomers and oligomers, alkanes, alkenes, alkynes, polyunsaturated and aromatic hydrocarbons) and products of the thermooxidative degradation processes in the condensed and/or gaseous phase (carbon oxides, H_2O, NO_x, SO_2) and oxidized hydrocarbons (alcohols, aldehydes, ketones and organic acids) [24]. It should be noted that the composition of the hydrocarbon fraction in the gaseous mixtures depends obviously on the temperature of smoke and on the partial pressure of the particular chemical substances.

The presence of *aerosols* (liquid and solid suspension in the gaseous fraction) gives smoke, which ranges in color and composition from black, highly graphitic solid particles (specific extinction area higher than $10,000\,m^2/kg$ due to light absorption and scattering), to colorless or "white" smoke that may consist largely of liquid particles (specific extinction area in the vicinity of 2000–$5000\,m^2/kg$ mainly due to scattering) [25]. The *liquid fraction* contains the condensation products of the gaseous phase with droplets of inorganic (HCl, HBr, H_2SO_4, etc.) and/or organic (methanoic acid or acetic acid) acids in solution in water.

Polycyclic aromatic hydrocarbons (PAHs) can be located both in the gas phase and the aerosol fraction. Careful sampling strategies are needed to assess the different locations of PAHs (most in the gas phase [26] or 90 % associated with particles in the breathable range [27]). The occurrence of PAHs in the environment has been intensively studied [28]. It has been proposed that phenanthrene is the most abundant of the PAHs produced by the burning plastics; other aromatics such as stilbene or biphenyl associated with phenyl-substituted PAHs, characteristic of burning polystyrene (PS) smoke, indicate a clear effect from the polymer structure [27]. Moreover, high loading of polymers in composite fuels has been shown to be responsible for the large quantities of PAHs in combustion products, for example, when plastic fuels are added to firewood [26, 27, and references therein].

The burning rate of polymeric materials, oxygen concentration, thermal environment, sample geometry, and additives all influence the generation of smoke particles from burning polymers [28]. *Soot* formation (via nucleation, coagulation, growth, and post-oxidation steps) has been extensively studied. A study of the generation of smoke from underventilated laminar diffusion flames versus their global equivalence ratio (Φ: fuel-to-air ratio normalized by the stoichiometry fuel-to-air ratio equal to 1) has shown that large Φ conditions give a smoke lighter in color with a large liquid fraction than that of smoke observed from low Φ conditions. Moreover, a comparison of transmission electron microscopy images (Fig. 7.1) shows differences in the solid particle fractions collected at different Φ conditions: in both cases the

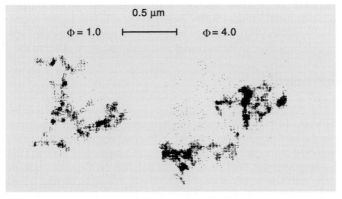

Figure 7.1   TEM micrographs of smoke collected from an ethene flame at $\Phi =1$ and 4 (from Santoro et al. [11] and [26])

material has an agglomerated structure that is clusters of distinct particles. At highest Φ the material has an agglutinated structure (indicating the presence of a liquid-like component [29]) rather than the agglomerated structure with distinct primary spheres usually observed in over-ventilated burning [30, 31]. The repartition of the radii of the soot particles is very large (from 0.2 nm to about 20,000 nm in large agglomerates, that is from 4 to approx. 1015 carbon atoms per particle, as a function of the temperature [32]).

Several authors [33–36] have studied the influence of the flame temperature and propose in particular that the soot surface growth rates in premixed flames can be understood in terms of a first-order thermal decomposition of fuel (modeled using an acetylene flame) on the soot surface. Moreover, they show that the C/H ratio of the soot is important in determining its reactivity and that the surface growth rate increases steeply with increasing equivalence ratio.

Size distribution of smoke particles in flame retarded plastics depends on the nature of the flame retardant. For instance, in the case of phenolic laminates, halogens seem to decelerate the coagulation of smoke while phosphates act as strong accelerators for this coagulation [37]. Size distribution is also affected by the amount of oxygen in the flame [38]. In consequence, smoke from polyurethane or acrylonitrile-butadiene-styrene (ABS) resin comes from the formation of the largest diameter particles by the coagulation of the sooty particles owing to the lack of oxygen for their combustion [37, 38].

Composition and, as a consequence, morphology of the soot particles depends also on the burning polymer. In illustration, comparative studies of the combustion of PVC, polyethylene (PE), and PS show that PS produces the larger number of PAHs (absorbed or constitutive) species in the sooty material and completely different species, such as oxygenated species and PAHs with infused rings [39]. This effect is significant in diffusion flames, but fuel structure (in fact the polymer structure) has little influence in premixed flames [40].

Prediction of the composition of smoke particles is not easy under real fire conditions: combustion of plastic materials occurs often in lively conditions and is sometimes violent and in all cases irregular, with large particle-soot agglomerates and polymer fragments continually liberated in parallel with more homogeneous liquid-based soot [41]. The traditional model of soot is that of polynuclear aromatic compounds with quantities of noncarbon atoms (hydrogen, oxygen, nitrogen, and/or sulfur located in the edges of the structure), which are commensurate with the edge site density [42].

Other *solid particles* may be carried away from the flame zone as ashes. In the condensed phase, cracked high-temperature char residue, mineral fillers such as glass fibers, and solid products of the degradation of additives (such as calcium oxide) may be swept out of the fire zone by gas turbulence. In the gas phase, volatilized additives (Sb_2O_3, SnO_2, zinc stannate, etc.) may condense as solids. Their reactions with halogenated species in the flame can form relatively nonvolatile metal halides ($SnBr_2$, $SnBr_4$, $ZnBr_2$ [43]) or highly volatile salts ($SbCl_3$, $SbBr_3$ and Sb-O-Br species [44]), which condense in low-temperature zones around the flame. These particles also contribute to the optical density and corrosive effect of the smoke.

7.1.3 Smoke Production

The importance of oxidation reactions on the surface of the burning polymer has been emphasized in several works [45–47]. However, a more generally accepted conclusion is that pyrolysis, defined as thermal degradation without oxidation, occurs on the surface [45, 48–50]. Mechanisms involved in thermal degradation have been previously studied [51, 52, and references therein]. Accordingly, in the first step of the polymer combustion cycle, polymers gasify as monomers, oligomers. or as decomposition products, whose molecular weights are close of those reported in vacuum pyrolysis [53]. These volatiles then fuel the flame.

Attempts to propose a model for the prediction of the composition of smoke or at least of the formation of toxic species in flames are recent (typical example in [54]). Treatments have to consider both the condensed and the gas phase to gain fundamental kinetic and mechanistic insight into the corresponding oxidation chemistry [55].

Carbon dioxide is generated roughly in proportion to the mass burning rate of the polymeric material. It is the product of the complete combustion of the material (at its surface under heat or flame or in the whole of the material during a glowing process) and of the fuels (produced by the pyrolysis process) in the flame.

In a diffusion flame, *CO* and *soot* production depend on the chemical structure of fuels produced via the polymer pyrolysis step according to four rules [56]:

- The yields of CO and particulates increase with the molecular weight of the fuel,
- With increase of bond unsaturation (changes in the chemical bonds from single to double, to triple bonds, and to benzene rings), the yields of CO and particulates increase,
- Yields of CO and particulates are the lowest for fuels with carbon-hydrogen-oxygen-based structures, but increase if a benzene ring and, more generally, any rings are present in the structure,
- Introduction of N and S atoms instead of O atoms in the structure results in an increase of the yields of CO and particulates, S atoms showing a higher effect than the N atoms.

Carbon monoxide is both the most important and one of the more complex in its generation. It can be produced directly in a fire plume. Studies of the chemistry and dynamics of diffusion flames indicate that, although there is some effect of atmosphere vitiation (oxygen depletion) on CO yields, the CO yields in a free burn are controlled mainly by the flame temperature [57]. These observed yields themselves are too small to account for the high values (several kg%) observed in post-flashover fires [57, 58].

Research has been conducted therefore in both full-scale and reduced-scale enclosures to determine the factors that can result in high CO yields; it has been shown that CO yields of about 0.2 kg per kg of fuel result from post-flashover fires of a diversity of fuels in several compartment geometries [59]. In addition, very high CO yields can result from anaerobic pyrolysis of oxygen-containing fuels in the upper layer of a compartment after flashover [60]. CO may then be oxidized to CO_2 via a highly exothermic reaction with hydroxyl radical in the Fenimore's radical chain oxidation process [61] proposed to take place in the gas phase mechanism of the ignition of the polymers [44].

Mechanisms responsible for the generation of high CO concentrations are poorly understood. Simple estimating and/or detailed predictive models for a fundamental understanding of the mechanisms of carbon monoxide formation and consumption in flames [62] thus need to be developed with, first, fundamental investigations of the principal chemical and physical mechanisms.

Hydrogen cyanide formation is associated principally with polyurethane, polyacrylonitrile, and polyamide degradation [63]. It has been shown that HCN forms from the high-temperature decomposition of degradation residues (chars and smoke particles) obtained from the lowest temperature polymer degradation process [23, 64]. Moreover, its formation recently has been shown to occur during the combustion of ammonium phosphate based intumescent polymeric formulations [65]. It has been proposed that HCN and other nitrogen-containing products (acetonitrile, acrylonitrile) can be converted in the flame to molecular nitrogen and nitrogen oxides, mainly as nitric oxide [23].

Hydrogen halides from halogen-containing polymers can be assumed to be generated in room fires in roughly the same manner that they are emitted from samples burned in small-scale tests such as the cone calorimeter [59]. HCl results from the depolymerization of PVC, chlorinated plastics, and some flame retardant systems [44]; various other halogen-acid gases are produced by fluoroplastics and some brominated formulations [63].

Nitrogen oxides (NO_x) emission may result from pyrolysis or combustion of N-containing polymers, such as polyurethanes [64], or from char nitrogen conversion [66]. Ammonia production is linked chiefly to the thermal degradation of amino plastics and polyamide [63] and/or to the thermal degradation of additives such as ammonium polyphosphate [67] or melamine [63].

PAHs (naphthalene, acenaphthylene, acenaphthene, fluorene, phenanthrene, anthracene, pyrene, fluoranthene, including the highly carcinogenic benzo(a)pyrene, etc.) are suspected to be the main organic smoke components generated from unventilated combustion and thought to be the precursors to soot formation [68]. For example, the emission of these compounds from pyrolysis and combustion of PVC, PE, and PS, respectively, results from intramolecular cyclization of PVC pyrolysates (polyenes), on evolution of HCl, and intramolecular condensation reactions among the straight-chain pyrolysates of PE and the aromatic rings of PS [69]. The total amounts of PAHs increase with the equivalence ratio Φ, their composition being a function of Φ. Formation of these substances in ethane or acetylene flames agrees nevertheless with the "old" scheme of formation of biphenyl and PAHs which involves radicals free of poly(acetylene) [70].

Soot (i.e., carbon particles) has been assumed to be formed via different routes involving poly(acetylene) and/or PAH intermediates [71]. Several hypotheses for soot formation have been previously proposed (such as Chien and Seader's, Lawson and Kay's or Lahaye and Prado's schemes [72]). The three-step mechanism for processes leading to solid particles in smoke postulated by Pasternak et al. [73] may be considered as a reasonable scheme for soot formation:

(i)

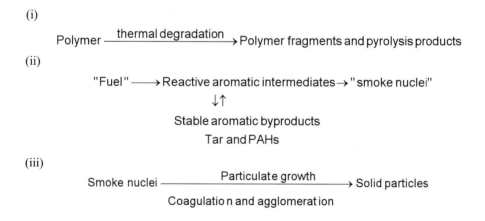

(ii)

(iii)

Large polyaromatic species are generally proposed as nuclei for soot formation. Structure studies show that soot is turbostratic, is lacking an epitaxial relationship between benzenoic graphite-like planes. The axially symmetric "crystallites" can give rise to a spherical particle morphology [74, 75]. Positive and negative ions of the type C_{2n}, in particular, the C_{60}^{+} cation (Fig. 7.2 {1}), has been discovered in a sooting flame [76]. This led Kroto et al. [77] to propose the formation of quasiicosahedral spiral shell soot particles via an initial growth sequence evolved, as shown in Fig. 7.2 (2), from: (a) a corannulene carbon framework through (b) and (c), species in which edge bypass occurs to (d), an embryo in which the second shell is forming.

The resulting solid material shows a high stability that arises from the ability of its graphite-like sheet to close into the spheroidal shell, thereby eliminating its reactive edges [81].

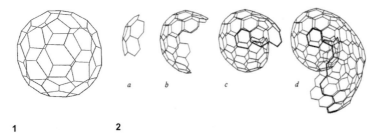

1   **2**

Figure 7.2   Solid particles in soot: 1, structure of the Buckminster-fullerenes (C_{60}) [76] and 2, hypothetical initial growth sequence for the formation of quasiicosahedral spiral shell carbon particles (from L.B. Ebert et al. [78], Herndon [79], and H.W. Kroto et al. [80])

Absorbed, easily extracted compounds are generally characterized in combination with the solid particles. For instance, PVC and PS give mainly absorbed aromatic combustion products (such as naphthalene, phenanthrene, pyrene, and chrysene) [82]. In contrast, PE and polypropylene (PP) form aliphatic and olefinic products with aromatics (including aromatic esters and ketones [84]) and PAHs (such as benzofluoranthene and benzopyrenes [83]); the nature of these PAHs depends on the mode of the combustion: in a nonflaming mode, three- and four-membered ring compounds predominate when, in a flaming mode, again the same species predominate, but many other PAHs of up to six-membered rings are found [85]. Polyamide 6 forms monomeric and dimeric ε-caprolactam [83], which can be absorbed on the surfaces of the solids.

Reactive species consist of radical sites in absorbed PAHs molecules and surfaces of carbonaceous solids, which play important roles in the mechanisms and kinetics of carbon addition (recombination of $CH_3\cdot$, $C_3H_3\cdot$ and other radicals) and oxidation (competitive chemical pathway via reactions with $OH\cdot$ species of the flame) reactions in combustion [86].

Moreover, it has been shown that although volatile halogen-based species (NH_4Cl, HBr, $SbCl_3$, or alkyl bromides and chlorides) have no effect on a methane flame, they do promote visible smoke generation in diffusion flames from burning polymers [87]. In addition, it is known that soot is an effective adsorbent for polymer additives such as flame retardants (chlorinated phosphates), plasticizers (phthalate esters, trialkyl or triaryl phosphates, fatty acid esters, glycerides), antioxidants (hindered phenols), and lubricants (fatty acids) that are distilled from a polymer during combustion [88].

7.2 Smoke Production from the Most Important Polymers

Fire performance is not a material property of plastics, nor are secondary effects such as the production of smoke and toxic gases. Smoke production depends on many variables such as chemical composition, oxygen availability, heat intensity, nature and conditions of the combustion process (e.g., flaming or smoldering), and the structure of the plastic component and the environment in which it is situated. Smoke suppressants are additives used in plastic formulations for the pupose of reducing the amount of smoke evolved if the article is involved in a fire.

Various standard test methods, for example, the NBS smoke chamber for static conditions and the cone calorimeter for dynamic measurements, are used to investigate the influence of chemical composition and structure of a plastic on smoke evolution. Comparative assessments of different materials can be made, thus providing knowledge on which to base development of smoke suppression systems for specific polymer types with a likelihood of effectiveness in real fire situations.

Different polymer systems will be more or less prone to evolve smoke depending on their chemical structure. For example, PMMA burns with little or no smoke evolution, whereas PVC fires are renowned for the large quantities of black smoke that can be produced in the absence of an effective smoke suppressant in the formulation. This is due to the different natures of the chemical species evolved when the different polymer systems are subjected to heat in fire situations. It is widely accepted that aromatic compounds make the largest contribution, in effect by undergoing gas phase condensation processes within the flame to form polyaromatic species, which are precursors to soot and hence smoke. The ability to form smoke decreases roughly in the order:

Naphthalenes > benzenes > diolefins > monoolefins > paraffins

Thermoplastics whose structure contains one or more aromatic groups are more likely to evolve aromatic species and are thus more prone to yield smoke in the event of a fire. The obvious examples are polystyrene, SAN, and ABS, which depolymerize to give styrene monomers and oligomers. In other cases the polymers can decompose to yield species that subsequently cyclize or undergo reactions that yield aromatic structures and thus favor smoke formation. Thus PVC on heating first evolves copious amounts of corrosive HCl, leaving an unsaturated residual carbon chain. Further breakdown of these chains yields polyene sequences, which dehydrogenate and cyclize to aromatic structures on reaching the flame, again a process that favors smoke formation. The large amount of aromatic plasticizers to be found in many PVC formulations only increases the amount of smoke likely to be generated in the event of a fire. There is evidence [89] that the choice of plasticizers is significant if the smoke generating potential of a formulation is an important consideration. Plasticizers that yield benzene rather than naphthenic or higher aromatics will result in relatively lower smoke levels under similar fire conditions. The statistical chain rupture mechanism by which polyolefins pyrolyze yields a wide variety of saturated and unsaturated hydrocarbons. These would yield very little aromatic material in a flame and therefore polyolefins will contribute little to smoke generation if they are involved in a fire.

Oxygen-containing polymers such as polyacrylics and polyacetals form mainly oxygen-containing, nonaromatic products on pyrolysis and thus make little contribution to smoke even under smoldering fire conditions. PMMA pyrolyzes to more than 90 % of methyl methacrylate monomer, which burns with little smoke owing to the presence of the oxygen-containing ester groups. Polyoxymethylene depolymerizes virtually quantitatively to formaldehyde, which contains more than 50 % oxygen and burns with a clean blue flame. Polyamides generate little smoke, as the major gaseous decomposition products are nitrogen- and oxygen-containing compounds such as ammonia, nitriles, amines, cyclic ketones, and esters. Polyurethanes undergo depolycondensation to isocyanates and alcohols. Polyisocyanates would cause much smoke formation if they were to escape from the pyrolysis zone where they are produced. However, the isocyanate is cross-linked to form isocyanate rings, very little of which escape from the pyrolysis zone to the gas phase. Hence, as a consequence, very little smoke is produced via polyisocyanates.

Linear thermoplastic polyesters decompose by statistical chain scission. In the case of poly(ethylene terephthalate) and poly(butylene terephthalate), the main products are aldehydes and terephthalic acid. The latter can cause smoke evolution. Aromatic fragments from bisphenol A, contained in polycarbonates, are responsible for smoke generation from such plastics.

In contrast to thermoplastics, most thermoset polymers yield little smoke when undergoing combustion because their cross-linked structure facilitates char formation under exposure to high temperature with few decomposition products entering the gas phase. Only if the thermoset polymer contains aromatic groups in its structure is there the possibility that there will be significant smoke generation in the event of a fire. Such polymers will include unsaturated polyester resins cross-linked with PS bridges. In a fire, styrene will be liberated into the vapor phase leading to smoke. Epoxy resins based on bisphenol A also contribute to smoke formation. Little smoke is produced via combustion of phenolic resins, which char intensely with only small quantities of aromatics escaping to the vapor phase. Similarly, nitrogen-containing thermosets such as urea and melamine resins also have strong tendencies to char and yield little smoke. This is again the case with high temperature resistant polymers, which, although they are based on ring structures, are reduced to a graphite-like char structure when exposed to high temperatures. However, for example, in the case of poly(phenylene ether), high-impact polystyrene may be included in the formulation to aid processability. Under fire conditions, evolution of aromatic species and hence smoke can be greatly enhanced by the presence of PS. However, very little smoke is evolved when polymers such as poly(phenylene sulfide), and poly(ether sulfones), which char strongly, are combusted. Polyimides decompose exclusively to char and thus evolve virtually no smoke in a fire. Polyphosphazenes, which are based on an inorganic backbone, will contribute to smoke to an extent dependent on the nature of the organic side chains. If these are aromatic, the contribution to smoke can be consider-able.

7.3 Smoke Suppressants

As with flame retardants, smoke suppressant action can occur in both the condensed and gas phases.

Both physical and chemical action can be attributed to suppressant behavior *in the condensed phase*. Suppressants can act physically in a similar fashion to flame retardants, for example, by forming glassy or intumescent coatings or by dilution due to addition of inert fillers, which limit the amount of smoke emitted. Some fillers, for example, calcium carbonate (chalk), act physically as fillers, but also chemically by reacting with acidic pyrolysis products, for example, HCl from PVC, or affecting the cross-linking processes to reduce the extent to which smoke precursor species escape to the gas phase.

There are many chemical approaches to smoke reduction. In thermosets, structural changes can be beneficial, for example, replacement of aromatic cross-linkers such as styrene by nonaromatic ones such as ethyl acetate [90]. An advantageous approach is to encourage the formation of a charred layer on the polymer/plastic surface in the event of a fire. This is the case when organic phosphates are used in unsaturated polyester resins. In PVC and similar halogen-containing resins, iron compounds interact with HCl liberated in a fire, to form the strong Lewis acid iron(III) chloride, which catalyzes char formation. Compounds such as ferrocene cause condensed phase oxidation, visible as glow, leading to enhanced CO and CO_2 formation, as opposed to aromatic compound evolution [91]. Molybdenum oxide is a well known suppressant for the PVC system. It acts by removing benzene formed during PVC degradation, probably via chemisorption, thus preventing this well known smoke precursor from reaching the gas phase [92].

Again both physical and chemical modes of action occur *in the gas phase*. The former occurs mainly due to shielding of the plastic surface against thermal attack by heavy gases. These gases also reduce smoke density by dilution. Chemically, smoke can be reduced either by elimination of any smoke precursors or even smoke particles. Aromatic smoke precursors can be removed by oxidation, particularly in the presence of transition metal complexes. Soot particles can be oxidized by high-energy OH· radicals formed by the catalytic action of metal oxides or hydroxides. Elimination of the ionized nuclei necessary for soot formation is another successful method for smoke suppression. This can be achieved by the presence of metal oxides. Some transition metal oxides are able to cause soot particles to flocculate.

It is rare for a smoke suppressant system to act by affecting just one single parameter, as smoke generation is normally determined by a combination of various parameters. The classic example is that of ferrocene, which acts by oxidizing soot in the gas phase as well as simultaneously facilitating char formation in the condensed phase. It is not possible to say if one process is predominant. Thus no uniform theory as to the mode of action of smoke suppressants has been established. Suffice to say, smoke suppression processes are extremely complex.

7.4 Smoke Suppressants in Plastics

Considerable effort has been extended into the best use of smoke suppressants to eliminate or at least significantly diminish the tendency of some polymeric materials (e.g. PVC, polystyrene, polyurethane) to generate copious quantities of smoke when involved in a fire. Of these, PVC is by far the most extensively researched. The realization that smoke can significantly increase the hazard to people caught up in a fire has led to increasingly stringent fire performance requirements for plastic materials. Resistance to ignition and flame spread is normally achieved by the presence of flame retardant systems. If the latter act in the gas phase by removing the OH·, H· and O· radicals responsible for flame propagation, this action has the additional effect of suppressing oxidation reactions of any smoke precursors and/or particles, which would otherwise occur during combustion. This will increase the extent of smoke generation. Gas phase acting flame retardants are extensively used in thermoplastics, many of which are inherently low smoke producers, for example, PMMA. The problem with the use of a smoke suppressant, which is chemically active in the gas phase, is that the suppressant acts by activating the combustion process, that is, behavior opposite to that of the flame retardant. To some extent, this dilemma can be solved by the use of a suppressant, which acts by a physical mechanism; however, such suppressants have only a limited effectiveness.

The main current approach to solve this dilemma is to develop flame retardant systems that act in the condensed phase, particularly by encouraging char formation. This not only reduces the amount of volatile organic fuel reaching the flame, but also decreases the amount of carbon available to form soot in the gas phase. Intumescent flame retardants are a particularly good example. Such flame retardant systems are particularly important for thermoplastics that have low char forming tendencies. The following provides a limited account of investigating the smoke suppressant action mechanisms for a number of polymers, in particular PVC, which is by far the most widely studied. For more information, the reader is referred to Hilado and Kosola [93], who have reviewed the literature of smoke production from plastics, Brauman et al. [94], who have investigated smoke development from various plastics, and Lawson [95], who details the application of smoke suppressants in plastics.

7.4.1 Smoke Suppressants in PVC

Smoke generated from PVC consists mainly of carbon particles formed from aromatic species. The large quantities of HCl gas evolved at the same time result in formation of clouds of corrosive HCl via interaction with atmospheric moisture. A common approach to diminishing the latter problem is to include additives such as $CaCO_3$ and $MgCO_3$ in the formulation.

The best known suppressants are ferrocene and molybdenum trioxide, both of which act by reducing soot formation. In concentrations below 1 %, ferrocene reduces both smoke density and flammability of PVC. However, it has a number of disadvantages, namely, high vapor pressure, is yellow in color, and is not compatible with plasticized PVC [96]. Ferrocene has also been used for a considerable time in aviation fuels because of its ability to destroy soot particles by promoting the combustion reactions in the gas phase.

In PVC, ferrocene initiates gas phase reactions that lead to the formation of high-energy radicals such as OH·. These subsequently oxidize soot particles to CO via:

$$C_{(solid)} \; + \; OH· \longrightarrow CO + H· \tag{I}$$

Certain metal oxides (Ba, Sr, Ca) are also known to remove soot. In a fire, it is assumed that they catalyze the breakdown to molecular hydrogen and water vapor in the zone where the soot is formed as a combustion product according to:

$$MO \; + \; H_2 \longrightarrow MOH \; + \; H· \tag{II}$$

$$MOH + H_2O \longrightarrow M(OH)_2 + H· \tag{III}$$

$$M(OH)_2 + (X) \longrightarrow MO + H_2O + (X) \tag{IV}$$

$$H· \; + \; H_2O \longrightarrow OH· \; + \; H_2 \tag{V}$$

where M represents the metal. The OH· radicals, thus liberated, remove soot particles as per reaction (I).

This mechanism may also be valid for the smoke suppression mechanism of transition metals and oxidized species formed from ferrocene, for example, α-Fe_2O_3.

Ferrocene and other metal compounds probably interfere with the nucleation reactions leading to soot formation by deactivating the ionized nucleating centers and the growth steps. Attack of the nucleating centers $C_nH_m^+$ occurs either by thermal or catalytic ionization of the metal followed by deactivation of the center or by reaction. Soot precursors, that is, condensed aromatic systems, can be eliminated by oxidation in which case the action of ferrocene or other metal compounds occurs earlier in the combustion process, facilitating the production of OH· radicals. Gas phase formation of benzene and other aromatics would probably be hindered by the presence of ferrocene and other transition metals.

Ferrocene and many other metals act in the condensed phase, where they accelerate the primary dehydrochlorination as well as the cross-linking and charring reactions occurring simultaneously on the PVC surface. Lawson [97] postulated the formation of ferrocenium ions in the presence of HCl and traces of oxygen. These ions then act as a Lewis acid catalyzing dehydrochlorination, cross-linking, and charring of PVC. Descamps et al. [91] have described an additional secondary reaction that occurs when ferrocene and other metal compounds such as V_2O_5 and CuO are used. Such compounds induce "chemical incandescence" and partially convert the decomposition products from PVC to carbon monoxide, thus reducing the amount of smoke emission.

The other commonly used compound for reducing smoke evolution from PVC is molybdenum trioxide, which acts as a weak flame retardant and a smoke suppressant in the condensed phase. The trioxide acts as a Lewis acid accelerating the scission of the C-Cl bonds, thus accelerating dehydrochlorination of PVC. Transpolyene segments, which do not cyclize to benzene or other aromatic systems, are the main products [98]. Thus, the evolution of smoke precursors and hence smoke is reduced. Limited cross-linking also occurs due to the metal oxide. Lum [92] suggests that the reduction in benzene or toluene is due to chemisorption reactions with formation of relatively stable π-aryl complexes with MoO_3.

Over the past 5–10 years, two groups have dominated the literature concerned with PVC and its derivative polymers such as CPVC, namely Starnes and co-workers at the College of William and Mary, Williamsburg, VA, USA, and Carty and White in Newcastle, UK. The former group made extensive studies of the role of metal-containing additives for smoke suppression in PVC. In many cases, their action appears to be a Lewis acid based catalysis of cross-linking processes to enhance char formation [99]. Potential attractive replacements for such Lewis acid acting additives are additives containing transition metals that can promote cross-linking by a process that involves reductive coupling of the polymer chains. Because reductive coupling agents tend to have low acidities, they are not expected to facilitate the cationic cracking of the char [100]. Additives that promote the coupling process are sources of a zero- or low-valent metal on pyrolysis. These include a number of transition metal carbonyls, divalent formiates, or oxalates of the late transition metals, simple Cu(I) containing phosphites or other ligands. Complexes of Cu(I) are especially attractive as reductive coupling agents and thus potential PVC smoke suppressants [101]. A detailed account of this strategy for smoke suppression, also considering the chemical mechanisms involved, has recently been published [102].

Carty and White have extensively investigated the use of various inorganic [103, 104] and organometallic [104] iron compounds as smoke suppressants for PVC and various PVC-containing polymer blends. The very effective smoke suppressing action of hydrated ferric oxide in CPVC-ABS polymer blends was explained to be a consequence of a number of chemical effects, namely:

- Lewis acid cross-linking catalyzed by FeOCl and/or $FeCl_3$ formed *in situ* and producing significant amounts of carbonaceous char
- Formation of chlorinated aromatics by reaction of $FeCl_3$ with benzene formed during the decomposition of PVC and CPVC

Recent work carried out in collaboration with the University of Salford has offered more insight into the chemistry involved [105–107]. An interesting current development is the use of tin compounds to enhance combustion performance. For example, coating $Mg(OH)_2$ and aluminum trihydrate (ATH) with zinc hydroxystannate confers significant increases in flame retardancy (higher limiting oxygen index [LOI] values, reduced heat release rates) and lower levels of smoke evolution, enabling large reductions in additive loadings compared to the unmodified retardant [108, 109]. Synergism between zinc borate and ATH or ATH partially or totally replaced by $Mg(OH)_2$ is also known [110].

A major market for PVC is that of cable insulation. Brown [111] has recently reviewed flame retardant and smoke suppressant additives for cables. As well as PVC, the use of PE and PE copolymers as cable insulation materials is considered. Molybdenum- and boron-based additives have been used to enhance smoke suppression in cable jacket material [112]. With a proper choice and combination of additives and plasticizers, a more than 40% reduction in smoke evolution may be obtained without loss in flame retardant properties. The trend and developments in flame retardants and smoke suppressants for cables has been reviewed [111]. Both inorganic and organic compounds are considered.

7.4.2 Smoke Suppressants in Styrene Polymers

By its very nature, decomposition of PS will inevitably yield aromatic species and thus smoke. The extent of smoke evolution can be limited by incorporating suitable smoke suppressants in the formulation of the polymer material. If organic metallic additives such as ferrocene are used, then, according to Deets and Lee [113], their effectiveness is related to the extent to which the metal is volatilized in the event of a fire. Smoke suppression occurs mainly in the gas phase. Heavy metal salts (Fe, Mn, Cr) of 8-hydroquinoline, copper, and lead phthalocyanines and radical initiators such as lead tetraphenyl [114] suppress smoke generation by reacting with any gaseous aromatic species. In contrast, iron acetonylacetonate is a highly effective smoke suppressant for ABS, acting in the condensed phase. Magnesium and aluminum hydroxides and/or zinc borate are effective suppressants, whereas magnesium carbonate surprisingly is not. A new class of metal hydroxide flame retardants with smoke suppressant abilities has been reported [115]. The mode of action of pyrogenic silica has been discussed [116].

7.4.3 Smoke Suppressants in Unsaturated Polyester Resins

Unsaturated polyester resins are normally cross-linked with short PS bridges that cause high levels of smoke emission in fires. Replacement of the polystyrene bridges with nonaromatic ones, for example, methyl methacrylate or ethyl acrylate, is an obvious approach to reducing smoke levels [90]. Some carboxylic acids such as fumaric, maleic, and succinic can also be effective. Materials that significantly enhance char formation are another approach. For instance, organic phosphates, which dehydrate the resin, also form polyphosphoric acids on heating. The latter form a protective layer inhibiting smoke generation.

Iron compounds, such as Fe_2O_3, which in the presence of a chlorine source such as is found in halogen-containing resin systems based on hexachloroendomethylenetetrahydrophthalic (HET) acid, convert to $FeCl_3$ on heating. This is a strong Lewis acid, which facilitates increased cross-linking and thus enhances char formation. At the same time, the $FeCl_3$ also behaves as a Friedel-Crafts catalyst enabling coupling reactions between the alkyl chlorides and aromatic entities present. The aromatics are thus retained in the condensed phase with the consequential reduction in smoke evolution [90].

Flame retardancy and smoke suppression can be obtained via the synergistic interaction of MoO_3 and halogen-containing compounds such as dibromoneopentyl glycol [117, 118]. Smoke emission from polyester resins reinforced with glass fibers can be reduced by incorporation of combinations such as MoO_3 + Sb_2O_3 + $Al(OH)_3$. If the green color of MoO_3 is a problem, then ammonium molybdates are suitable substitutes [119]. The popular flame retardant aluminum trihydrate acts by cooling the pyrolysis zone due to the large endothermicity of its decomposition and also by evolving large quantities of water. The latter not only removes heat from the flame, but limits the nucleation and agglomeration of soot into large particles. Consequently, the smoke density is reduced and the color of the smoke is white rather than black [90]. Magnesium hydroxide [120] and zinc borate have also been used as smoke suppressants for polyester resin systems. The combination of various zinc stannates with both halogenated and halogen-free polyester resins promises the possibility of significant improvements in respect of smoke suppression [121].

Experimental design with statistical interpretation of the results is a potentially successful approach to optimize the formulation of materials for a particular purpose. An example is its application in the case of "sheet molding compounds-low profile materials," which are widely used in the automotive, transportation and building industries [122].

7.4.4 Smoke Suppressants in Polyurethane Foams

The use of isocyanurates, imides, and carbodiimides increases the extent of cross-linking during production of polyurethane foams. This cross-linked structure facilitates enhanced char formation and thus smoke reduction in the event of a fire. A similar effect can be obtained by the use of solid dicarboxylic acids such as maleic, isophthalic, and HET acids [114, 123]. Alcohols, for example, furfuryl alcohol, have been suggested to be effective smoke suppressants [124]. The mechanism is thought to be via removal of the pyrolysis product polyisocyanates, which are the precursors of the thick variety of smoke. The alcohol is oxidized to aldehyde, which reacts with polyisocyanurate via a Schiff's base and further isocyanate to form a cross-linked structure, which remains in the condensed phase. As is well known, various phosphorus compounds can be used to produce significant charring and thus reduced smoke in the event of a fire. Ferrocene, some metal chelates and potassium or ammonium tetrafluoroborate have the same effect owing to their Lewis acid behavior in the condensed phase.

Smith [125] has investigated the factors affecting smoke generation from textile/polyurethane upholstery. The combustion behavior of polyurethane foam furniture has been extensively investigated for the European Community and a report of this work has been published [126].

7.4.5 Miscellaneous

Flame retardants such as aluminum and magnesium hydroxides, which act via physical mechanisms and evolve water, which causes dilution of the flame region, can in many cases be considered to be general smoke suppressants. There is evidence [127] that adding silane cross-linkable copolymer to PE-metallic hydroxide systems can significantly improve the flame retardant, smoke suppressant, and impact properties of these systems. This approach may be extendable to other polymer types. The use of zinc borate as a smoke suppressant with a range of polymer types, for example, polyolefins, siloxanes, fluoropolymers, and PVC, has been reviewed by Shen and Ferm [128]. Advantageously, in the appropriate form, zinc borate also functions as a flame retardant, an afterglow suppressant, an anti-arcing and anti-tracking agent, and as a char promoter. The prediction of smoke toxicity and the use of toxicant suppressants in materials or products to reduce the formation of toxic smoke has been discussed by Levin [129].

Smoke suppression is a complex topic requiring knowledge as to how smoke formation during the combustion process can be controlled. Innes and Cox [130] have highlighted zinc molybdate smoke suppressant technology. Some possible chemical routes are presented that attempt to account for the reduction in smoke during the burning process as well as the burning process itself. Green [131] has reviewed the mechanisms of both smoke suppression and flame retardancy. Wang [132] is developing the use of X-ray photoelectron spectroscopy to investigate smoke suppression mechanisms. Rothon [133] has provided a brief account of the use of particulate fillers as flame retardants and their influence on smoke and toxic gases.

References

[1]   M.M. Hirschler: *J. Fire Sci.* 5 (1987) 289
[2]   M.A. Barnes, P.J. Briggs, M.M. Hirschler, A.F. Matheson, T.J. O'Neill: *Fire Mater.* 20(1) (1996) 1–16
[3]   P. Bulman, J. Reynolds: NIST Will Lead Study on Dangers of Fire Smoke, NIST 99–11, Gaithersburg, MD, (1999)
[4]   W. Golberine: Tunnel du Mont Blanc: Histoire et photos d'une catastrophe in Paris Match, M2533 (08 April 1999) 112–123
[5]   W.D. Woolley, M. M. Raftery: *J. Hazard. Mater.*, 1 (1976) 215–222
[6]   W.D. Woolley: *J. Macromol. Sci., Chem. A* 17 (1982) 1–33
[7]   G.R. Gann, V. Babrauskas, R.D. Peacock, J.R. Hall Jr.: *Fire Mater.* 18 (1994) 193–199
[8]   K.W. Kuvshinoff, R.M. Fristrom, G.L. Ordway, R.L. Tuve (Eds.): Fire Sciences Dictionary, John Wiley, New York, 1977
[9]   C.J. Hilado, R. M. Murphy in E. E. Smith, T. Z. Harmathy (Eds.): Fire Response of Organic Polymeric Materials (Organic Material in Fire Combustibility) – Design for Fire Safety, ASTM STP 685, Am. Soc. Testing Mater., Philadelphia, pp. 76–105, 1979
[10] J. Troitzsch (Ed.): International Plastics Flammability Handbook, 2nd edit., Carl Hanser, Munich, pp. 63–65, 1990
[11] R.J. Santoro: Fundamental mechanisms for CO and soot formation, Final report, NIST GCR 94–661, NIST, Gaithersburg, MD, 167p, 1994

[12] G.M. Faeth, T.L. Farias, U.O. Koylu, S.S. Krishnan, J.S. Wu: Mixing and radiation properties of buoyant luminous flame environments – II Structure and optical properties of soot, NIST GCR 99–770, NIST, Gaithersburg, MD, 145 p, 1999

[13] E.D. Weil: Fire retardant mechanisms in Life Prop. Prot., Pap., Semi-Annual Meeting Fire Retard. Chem. Assoc., Fire Retardant Chem. Assoc., Westport, CT, p. 104–129, 1977

[14] B.Y. Lattimer, U. Vandsburger, R.J. Roby: in Annu. Conf. Fire Res.: Book of Abstr., NIST, Gaithersburg, MD, NISTIR 5904, pp.121–122, 1996

[15] J. R. Gaskill in: Polymer Conference Series, Flammability Characteristics of Polymeric Materials, University of Utah, Salt Lake City, June 5–10, 1970

[16] C.J. Hilado in: Polymer Conference Series, Flammability Characteristics of Polymeric Materials, University of Detroit, June 8–12, 1970

[17] ISO/IEC WD 13943, Glossary of fire terms and definitions" (Revision of ISO/IEC Guide 52 : Fire Tests – Terminology), ISO/TC92/WG7 N 282, BSI, London, International Organization for Standardization, p. 26, 1993

[18] The College Standard Dictionary of the English Language, Funk and Wagnalls, New York, 1971

[19] G. Prado, J. Jagoda, J. Lahaye: *Fire Res.* 1 (1977/78) 229–235

[20] G.W. Mulholland in: SFPE handbook of fire protection engineering , The National Fire Protection Association, Boston, p. 1–377, 1993

[21] R. Hindersin in H.F. Mark, N.M. Bikales (Eds.): Encyclopedia of Polymers Science and Technology, Supplement, Vol. 2, John Wiley, New York, p. 292, 1977

[22] A. Chakir, M. Cathonnet, J.C. Boettner, F. Gaillard: *Combust. Sci. Technol.* 65 (1989) 207

[23] W.D. Woolley, P.J. Fardell: *Fire Res.* 1 (1977) 11–21

[24] N.H. Margossian: *L'actualité chimique* (May 1980) 25–35

[25] V. Babrauskas, G.W. Mulholland: Smoke and soot data determinations in the cone calorimeter - Mathematical modeling of fires, ASTM STP 983, Am. Soc. Testing Mater., Philadelphia, pp. 83–104, 1987

[26] E. Keränen, A. Leppänen, J.-P. Aittola: Pakkausmuovien Polttaminen, Pakkaustechnologiarihmäry Raportti No 22, Espoo, 1989

[27] M. Elomaa, E. Saharinen: *J. Appl. Polym. Sci.* 42 (1991) 2819–2824

[28] A. Bjorseth (Ed.): Handbook of polycyclic aromatic hydrocarbons, Marcel Dekker, New York, 1983

[29] S. Léonard, G. W. Mulholland, R. Puri, R.J. Santoro: *Combust. Flame* 98 (1994) 20–34

[30] B.A. Benner, N.P. Bryner, S.A. Wise, G.W. Mulholland: *Environ. Sci. and Technol.* 24 (1990) 1418–1427

[31] R.L. Dod, N.J. Brown, F.W. Mowrer, T. Novakov, R.B. Williamson: *Aerosol Sci. Technol.* 10 (1989) 20–27

[32] D.E. Jensen: *Proc. R. Soc. Lond. A* 338 (1974) 375–396

[33] C.J. Dash: *Combust. Flame* 61 (1985) 219–225

[34] J.S. Harris, A.M. Weiner: *Combust. Sci. Technol.* 31 (1983) 155–167

[35] J.S. Harris, A.M. Weiner: *Combust. Sci. Technol.* 32 (1983) 267–275

[36] J. S. Harris, A. M. Weiner: *Ann. Rev. Phys. Chem.* 36 (1985) 31–52

[37] T. Handa, T. Nagashima, Y. Takahashi, K. Suda, N. Ebihara, F. Saito: *Fire Res.* 1 (1977–1978) 265–272

[38] B. Kim, S. Sankar, J.I. Jagoda, B.T. Zinn: Chem. Phys. Proc. Combust., paper 59, p. 4, 1983

[39] T. Panagiotou, Y.A. Levendis, J. Carlson, Y.M. Dunayevsky, P. Vouros: *Combust. Sci. Technol.* 116–117 (1996) 91–128

[40] I. Glassman: 22nd Symp. (Int.) on Combustion, the Combustion Institute (1988) 295–311

[41] M.T. Pinorini, C.J. Lennard, P. Margot, I. Dustin, P. Furrer: *J. Forensic Sci.*, 39(4) (1994) 933–973

[42] L.B. Ebert: *Science* 247 (1990)1468–1471

[43] P.A. Cusack, A.J. Killmeyer in: 52nd Annual Tech. Conf. – Soc. Plast. Eng, Vol. 3, p. 2824–2828, 1994

[44] G. Camino, L. Costa: *Polym. Deg. Stab.* 20 (1988) 271–294

[45] M. Elomaa, L. Sarvaranta, E. Mikkola, R. Kallonen, A. Zitting, C.A.P Zevenhoven, M. Hupa: Critical Reviews in Biochem. Mol. Biol., 27 (1997) 137–197

[46] D.E. Stuetz, Z.F. Diedwardo, B.P. Barnes: *Polym. Flamm. II, J. Polym. Sci., Polym. Chem. Ed.* 18 (1980) 967

[47] D. E. Stuetz, Z. F. Diedwardo, B. P. Barnes: *Polym. Flamm. II, J. Polym. Sci., Polym. Chem. Ed.* 18 (1980) 987

[48] C.P. Fenimore, F.J. Martin in: Proc. 4th Mater. Res. Symp., Gaithersburg, MD, October 26–29 1970, Natl. Bur. Stand. Spec. Publ. 357 (1972) 159–170
[49] K. Seshadri, F.A. Williams: *J. Polym. Sci, Polym. Chem Ed.* 16 (1978) 1755
[50] G.L. Nelson: *Plastic Flamm., Intern. J. Polym. Mater.* 7 (1979) 140
[51] N. Grassie, G. Scott in N. Grassie (Ed.): Polymer degradation and stabilisation, Cambridge University Press, Cambridge, 1985
[52] D.W. Van Krevelen (Ed.) in: Properties of polymers – Correlations with chemical structure, 3rd edit., Elsevier, Amsterdam (1990); ibid., 2nd edit. (1976); ibid., 1st edit. (1972)
[53] S.L. Madorsky (Ed.): Thermal degradation of organic polymers, Interscience, New York, 1964
[54] O.A. Ezekoye, Z. Zhang: *Combust. Flame* 110 (1997) 127–139
[55] R.R. Skaggs, M.P. Tolocka, J.H. Miller: *Combust. Sci. Technol.* 116–117 (1996) 111
[56] A. Tewarson: Smoke point height and fire properties of materials, Report NIST/GCR-88/555, Order No. PB89–141089, NIST, Gaithersburg, MD, 51 pp, 1988
[57] G. Mulholland, M. Janssens, S. Yusa, W. Twilley, V. Babrauskas: in Fire Safety Science – Proc. 3rd Int. Symp., Elsevier, London, p. 585, 1991
[58] J.H. Morehart, E.E. Zukovski, T. Kubota: in Fire Safety Science – Proc. 3rd Int. Symp., Elsevier, London, p. 575, 1991; see also *Fire Safety J.* 19 (1992) 177
[59] R. G. Gann: in Proc. 12th Joint Panel on Fire Research and Safety, Tsukuba, October 27- November 2 1992, Ibaraki and Building Research Institute, Tokyo, 1994
[60] W. M. Pitts: in BFRL/NIST Annu. Conf. Fire Res., Gaithersburg, MD, Sept. 1992, NIST Monograph 179 (June 1994)
[61] C. P. Fenimore, G. W. Jones: *Combust. Flame* 10 (1966), 295
[62] W. M. Pitts: *Fire Safety J.* 23 (1994) 271
[63] A.P. Jain, B.G. Sharma: *Indian J. Environ. Protect.* 7(6) (1987) 438–440
[64] W.J. Potts, T.S. Lederer: *J. Combust. Toxicol.* 4 (1977) 114–162
[65] Y. Claire, E. Gaudin, C. Rossi, A. Périchaud, J. Kalioustan, E. El Watik, H. Zinnedine in M. Le Bras, G. Camino, S. Bourbigot, R. Delobel (Eds.): Fire Retardancy of Polymers: The Use of Intumescence, Roy. Soc. Chem. Pub., Cambridge, p. 437–448, 1998
[66] A. Sarofim: in Proc. Medit. Combust. Symp. – MCS-99", Antalia, June 20–25 1999
[67] R. Delobel, N. Ouassou, M. Le Bras, J.-M. Leroy: *Polym. Deg. Stab.* 23 (1989) 349–357
[68] M. P. Tolocka, J. H. Miller: in Proc. Chem. Phys. Poc. Combust. – Fall Techn. Mtg, October 16–18 1995, Combustion Institute/Eastern States Section, Worcester, pp. 253–256, 1995
[69] T. Panagiotou, Y. A. Levendis, J. Carlson, P. Vouros in: Proc. 26th Symp. (Int.) Combust., Vol. 2, The Combustion Institute, pp. 2421–2430, 1996
[70] B.D. Crittenden, R. Long: in R.I. Freudenthal, P.W. Jones (Eds.) Polynuclear Aromatic Hydrocarbons: Chemistry, Metabolism and Carcinogenesis –Carcinogenesis, Vol. 1, Raven Press, New York, pp.209-223, 1976
[71] S.J. Grayson, R.J.S. Green, J. Hume, S. Kumar: in V.M. Bhatnagar (Ed.) Fire Retard., Eur. Conf. Flammability Fire Retard. – 1st Meeting , Technomic, Westport, CT, pp. 57–62, 1979
[72] J. A. Fritz in Mémoires de la Conférence CEMP – Comportement au feu et sécurité incendie, CEMP, Paris April 16, 1980
[73] M. Pasternak, B.T. Zinn, R.F. Browner: *Combust. Sci. Technol.* 28 (1982) 263–270
[74] J. B. Donnet: *Carbon* 20 (1982) 266–282
[75] L.B. Ebert, J.C. Scanlon, C.A. Clausen: *Energy Fuels* 2 (1988) 438–445
[76] P. Gerhardt, S. Löffler, K.H. Homann: *Chem. Phys. Lett.* 137 (1987) L5
[77] H.W. Kroto, K. McKay: *Nature* 331 (1988) 328–331
[78] L.B. Ebert: *Science* 247 (1990) 1468
[79] W.C. Herndon in L.B. Ebert (Ed.) Polynuclear Aromatic compounds, Am. Chem. Soc. Pub., Washington, pp. 1–12, 1988
[80] H. W. Kroto: *Science* 242 (1988) 1139
[81] Q.L. Zhang, S.C. O'Brien, J.R. Heath, Y. Liu, R.F. Curl, H.W. Kroto, R.E. Smalley: *J. Phys. Chem.* 90 (1986) 525–528
[82] W. Klusmeier, A. Kettrup, K.-H. Ohrbach: *J. Thermal Anal.* 35 (1989) 497–502
[83] M. Elomaa, E. Saharinen: *J. Appl. Polym. Sci.* 42 (1991) 2819–2824
[84] M. Pasternak, B.T. Zinn, R.F. Browner: *Combust. Flame* 41 (1981) 335–337
[85] M. Pasternak, B.T. Zinn, R.F. Browner: in 18th Symp. (Int.) Combust. (5 Proc.) (1981), 91–99, 1980
[86] J. B. Howard in 23rd Symp. (Int.) on Combust., Combustion Institute, p. 1107–1127, 1988
[87] S.K. Brauman, N. Fishman, A.S. Brolly, D.L. Chamberlain: *J. Fire Flammability* 6 (1976) 41–58

[88] K.J. Voorhees in G.L. Nelson (Ed.): *ACS Symp. Ser., Fire Polym. II*, 599 (1995) 553–578
[89] D. Price, G.J. Milnes, P.J. Taylor, J.H. Scrivens, T.G. Blease: *Polym. Deg. Stab*. 25 (1989) 307
[90] J.E. Selley, E. Vaccarella: *Plast. Engn*. 35 (1979) 4
[91] J.M. Descamps, L. Delfosse, M. Lucquin: *Fire Mater*. 4 (1980) 37
[92] R.M. Lum: *J. Appl. Polym. Sci*. 23 (1979) 1247
[93] C.J. Hilado, K.L. Kosola: *J. Fire Flamm*. 8 (1977) 532
[94] S.K. Brauman, N. Fishman, A.S. Brolly, D.L. Chamberlain: *J. Fire Flamm*. 7 (1976) 41
[95] D.F. Lawson in M. Lewin, S.M. Atlas, E.M. Pearce (Eds.) Flame Retardant Polymeric Materials, Vol. 3, Plenum Press, New York, 1982
[96] J.J. Kracklauer, C.J. Sparkes: *Plast. Engn*. 30 (1974) 57
[97] D.F. Lawson: *J. Appl. Polym. Sci*. 20 (1976) 2183
[98] W.H. Starnes, D. Edelson: *Macromolecules,* 12 (1979) 797
[99] R.D. Pike, W.H. Jr. Starnes, S.W. Bryant, J.P. Jeng, P. Kourtesis: *Recent Adv. Flame Retard. Polym. Mater*. 6 (1995) 141
[100] R.D. Pike, W.H. Jr. Starnes, S.W. Bryant, J.P. Jeng, P. Kourtesis, C.W. Adams, S.D. Bunge, Y.M. Kang, A.S. Kim, J.H. Kim, J.A. Macko, C.P. O'Brian: *Macromolecules* 30 (1997) 6957
[101] R.D. Pike, W.H. Jr. Starnes, S.D. Bunge, Y.M. King, A.S. Kim, J.A. Macko: *Recent Adv. Flame Retard. Polym. Mater*. 8 (1997) 44
[102] R.D. Pike, W.H. Jr. Starnes, J.P. Jeng, S.W. Bryant, P. Kourtesis, C.W. Adams, S.D. Bunge, Y.M. Kang, A.S. Kim, J.H. Kim, J.A. Macko, and C.P. O'Brian in S. Al-Malaika, A. Golovoy, C.A. Wilkie (Eds.): Chemistry and Technology of Polymer Additives, Chapter 11, Blackwell Science, Oxford, 1999
[103] P. Carty, S. White: *Polym. Compos*. 6(1) (1998) 33
[104] P. Carty, E. Metcalfe, S. White: *Polymer* 33 (1992) 2704
[105] P. Carty, S. White, D. Price, L F. Lu: *Polym. Deg. Stab*. 63 (1999) 465–468
[106] L.F. Lu, D. Price, P. Carty, S. White: *Polym. Deg. Stab*. 64 (1999) 601
[107] P. Carty, C. Ratcliffe, D. Price, G.J. Milnes, S. White: *J. Fire Sci*. 17 (1999) 483–493
[108] P.R. Hornsby, P.A. Cusack: Ann. Tech. Conf.-Soc. Plast. Eng., 56th (Vol. 3) (1998) 3310
[109] R.G. Baggaley, P.R. Hornsby, R. Yahya, P.A. Cusack, A.W. Monk: *Fire Mater*. 21 (1997) 179
[110] D.J. Ferm: Proc. Int. Conf. Fire Safety 17 (1992) 242
[111] S.C. Brown: Proc. Int. Wire Cable Symp., 47th, p. 727–735, 1997
[112] F.W. Moore, W.J. Kennelly: Proc. Int. Conf. Fire Safety, 16 (1991) 227
[113] G.L. Deets, Y.C. Lee: *J. Fire Flamm*. 10 (1979) 41
[114] D.F. Lawson, E.L. Kay: *JFF/Fire Retardant Chem*. 2 (1975) 132
[115] W.E. Jr. Horn, J.M. Stinson, D.R. Smith: 50th Annu. Tech. Conf. Plast. Eng. (2) (1992) 2020
[116] R. Chalabi, C.F. Cullis, M.M. Hirschler: *Eur. Polym. J*. 19 (1983) 461
[117] G.A. Skinner: *Fire & Mater*. 1 (1976) 154
[118] G.A. Skinner, P.J. Haines: *Fire & Mater*. 10 (1986) 63
[119] D.A. Church, F.W. Moore: *Plast. Engn*. 31(12) (1975) 36
[120] Z. Klosowska-Wolkowicz: *Polimery* (Warsaw), 33 (1988) 428
[121] P.A. Cusack: *Polimery* (Warsaw), 40(11/12) (1995) 650
[122] A. Hernangil, M. Rodriguez, L.M. Leon, J. Ballestero, J.R. Alonso: *J. Compos. Mater*. 32 (1998) 2120
[123] H.P. Doerge, M. Wismer: *J. Cell. Plast*. 8 (1972) 311
[124] K. Ashida, M. Ohtani, T. Yokoyama, S. Ohkubo: *J. Cell. Plast*. 14 (1978) 200 and 256
[125] J.A. Smith: *J. Soc. Dyers Colour*. 108 (1992) 454
[126] B. Sundström (Ed.): The CBUF Report, SP Pub, Sweden; Report EUR 16477 EN (1995), available from Interscience Communications, London
[127] J.T. Yeh, H.M. Yang, S.S. Huang: *Polym. Deg. Stab*. 50 (1995) 229
[128] K.K. Shen, D.J. Ferm: *Proc. Int. Conf. Fire Safety,* 21 (1996) 224
[129] B.C. Levin: *Drug Chem. Toxicol*. 20 (1997) 271
[130] J.D. Innes, A.W. Cox: *J. Fire Sci*. 15 (1997) 227
[131] J. Green: *J. Fire Sci*. 14 (1996) 426
[132] J. Wang: *Macromol. Chem. Symp*. 74 (1992) 101
[133] R. Rothon in R. Rothon (Ed.): Particulate-Filled Polymer Composites, Chapter 6, Longman, Essex (UK) and John Wiley, New York, 1995

National and International Fire Protection Regulations and Test Procedures

8 Methodology of Fire Testing

F.-W. WITTBECKER

8.1 Fire Safety Objectives

A fire safety concept begins by identifying fire safety objectives and acceptable levels of safety. Specifications of the design to be assessed are required for the fire safety performance to be tested and/or modeled. The fire scenarios need to be specified for which the design will be required. For example, for aircraft, three possible fire accident types (on the ground, during the flight, and after a crash) must be considered. For rail vehicles, for example, underground trains, other special hazards, such as a fire during passage through a tunnel, are relevant. Additional assumptions, such as the conditions of the fire environment, have to be specified. A fire hazard assessment needs a specified design to be acceptable under the specified assumptions.

Testing and calculation methods are used to determine whether the objectives will be met by a specified design for a specified fire scenario under the specified assumptions. For the fire hazard assessment procedure to be valid, it is necessary that the calculation methods and the characteristic fire test responses used produce valid estimates of success or failure in achievement of the fire safety objectives. The primary fire safety objective is to ensure that in case of fire the time required to evacuate the fire compartment is shorter than the time for the fire to create untenable conditions in the compartment. The evacuation time includes the time required for the occupants to reach or be transported to a safe location. Consequently, the shorter the interval between ignition and discovery, the faster escape and earlier intervention are possible. This explains why dwelling fires detected by smoke alarms are smaller and result in a reduced number of fatalities.

Tenability is assessed based on fire effects on the occupants, including both direct effects, such as heat, toxic gases, or oxygen deprivation and indirect effects, such as reduced visibility due to smoke obscuration.

Other fire safety objectives are to prevent flashover inside a building and to avoid serious injuries to fire fighters.

With regard to the fire safety objectives, methodology of testing focuses on

- Consideration of the fire risks, that is, ignition and flame spread, in the compartment of origin.
- Consideration of fire hazards, that is, the developing stage and the flashover situation. It is presumed that people do not stay inside the fire compartment; a natural reflex to try to escape is assumed. The hazard considerations are stability of the structure, heat and smoke movement, and occurrence of thresholds. The hazard is associated with the fire environment - design of the enclosure and configuration of the fire load - and with a number of potency values of materials, products, or assemblies including rate of heat release, smoke generation and obscuration, toxicity of fire effluents, and ease of extinguishment.

The investigation of the risk determines whether the assessment will focus on a material, product, or system:

- Is the product likely to be the source of ignition?
- Is the product likely to be the secondary ignited item?
- Is the product a potentially significant fuel source even if not being the first or secondary ignited item?

- What is the potential avenue to contribute to the risk (and hazard)?
- How close are occupants and/or critical equipment to the origin of a fire?

In fire safety engineering guidelines [1–8], the aforementioned decision route may be followed. In international standards, this product-related safety assessment strategy is only partly taken into account. Fire tests vary considerably and focus on different simulated risk situations – walls, roofs, and so forth. Numerous regulations and regional, state, and local codes refer to fire tests and standards. Although numerical material rankings derived from these tests are the most common means available today to compare the various combustion characteristics of products, they are valid only as measurements of the performance under specific controlled test conditions and may not necessarily be representative of their behavior under actual fire conditions. More than one test and possibly intermediate or full-scale tests may be necessary to qualify a (plastic) product for intended or proposed use. This may lead to different classifications depending on the test procedure and end-use conditions. For example, several classification systems require different levels of performance for composite elements: one for the component itself and a second one for the end-used product.

If the product under consideration is a structural element, its fire resistance has to be assessed. Structural safety in fires is defined in terms of specified distances and times within which collapse of structural elements – for example, walls, doors, barriers and ceilings – is prevented. The objective of these requirements is to protect occupants. The different evacuation/escape time requirements are reflected in different demands on individual structural components.

The fire scenario associated with any test method should reproduce the conditions in which the hazard exists.

8.2 Fire Scenarios

Fires can develop in numerous ways, dependent on factors such as enclosure type and size, ventilation conditions, heat and smoke movement. A fire scenario is a detailed description of conditions for one or more of the stages from before ignition to the completion of combustion in an actual fire at a specific location. In Fig. 8.1, a schematic diagram of the course of a fire is shown from the time before ignition to the completion of combustion in an enclosure with sufficiently high fire load to create flashover [9].

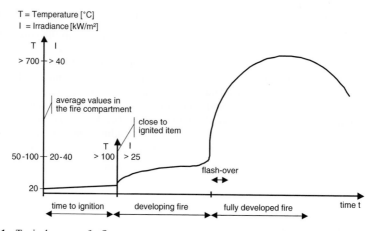

Figure 8.1   Typical course of a fire

Fire parameters between the pre-flashover and the post-flashover conditions are distinctly different. First, it has to be assessed whether ignition is likely to occur. In the second stage of a fire, smoldering or an open flame spread must be considered. The main subject in these stages is the assessment of whether flashover is to be expected or not.

In the pre-flashover phase, reaction-to-fire characteristics of (plastic) products are important, whereas in the post-flashover phase, resistance-to-fire parameters of complete assemblies apply. Almost all fire regulations make a distinction between these two conditions. The most important fire parameters associated with reaction and resistance to fire are:

- Developing fire
  Reaction-to-fire: ignitability, flame spread, heat release
- Fully developed fire
  Resistance to fire: load-bearing, insulation and integrity capacity

Any hazard assessment of an incipient fire will cover the contribution of various factors to surface flame propagation and burning rate, which determine inward spreading of the fire, the latter depending on the effectiveness of protective covers. Material-specific smoke data and temperature increase are dependent on fire duration, the quantity of material burning, and on the fire scenario. In addition, and with regard to the post-flashover phases of an advanced fire, the problems of smoke movement must be considered. As smoke data gained under different test conditions will depend mainly on the area burning, smoke potency data comparison should be based on volume rather than on weight.

A primary fire risk is the potential for initiating fires associated with the ignition source. An ignition source may pose a variety of risks dependent on the associated environmental conditions and on a number of characteristic fire test responses of materials, products, or assemblies. In small rooms, the typical ignition source is small, such as candles, matches, hot electrical wires, and so forth [10]. External irradiance is zero for the first ignited item. The O_2 content of air is almost 21 %. The relevant risk for further assessing the fire is flame spread. Products near the first ignited item are heated by convection and irradiance. The O_2 content in the room air begins to decrease. After a certain time flashover may occur, the room temperature exceeds 500 °C and the overall irradiance exceeds 25 kW/m² (see Fig. 8.1). In such cases, the oxygen content in small rooms is normally not sufficient to allow complete combustion. A typical ignition source for smoldering fires may be a cigarette on a mattress or faulty electrotechnical equipment. Smoldering fires will not significantly increase room temperatures, but may begin to deplete oxygen and cause major smoke development. Another scenario is an open-flame fire caused by primary ignition sources, which ignite wastepaper baskets, curtains, mattresses, and other items acting as secondary ignition sources.

Small ignition sources may cause an accelerated development of the fire, which, in the case of storage of combustible liquids, results in immediate flashover. Heat release may be expressed as the hydrocarbon curve for which, however, relatively high ventilation is necessary. The CO_2/CO ratio is about 100 and low ventilation is likely to lead to temperatures in the range of 600–900 °C.

The development of a fire depends on the size and on the ventilation conditions of the room of fire origin. In large and very large rooms – theatres, open plan offices, warehouses, supermarkets and sports halls – fires are free-ventilated for a long time. Contrary to small rooms, there are hardly any interrelated effects and the fire development is directly dependent on the successive combustion of the burning items. The scenario can be compared with fires in open air for a certain period. Flashover causes rapid decrease of the CO_2/CO ratio.

The course of a real-life fire depends on the nature of combustibles, ignition source, fire load, and conditions in the fire compartment; it is practically impossible to predict a real-life fire taking into account all interactions and boundary conditions. Therefore, it is necessary to define design fires and fire scenarios. Different fire scenarios have been identified in an EC prenormative research program, which was finalized in 1995 [11]. Examples of typical design fire scenarios include:

- Large/medium/small room fires (corner, ceiling, floor and wall),
- Corridor fires,
- Roofing fires,
- Cavity fires,
- Staircase fires,
- Fires in/on facades,
- Single burning item fires (furniture, cable conduit, pipes).

There are two distinctly different methods [2] of determining the design fire for a given scenario. One is based on the knowledge of the amount, type, and distribution of combustible materials in the compartment of fire origin. The other is based on the knowledge of the type of occupancy, where very little is known about the details of the fire load.

Design fires are usually quantified as the heat release rate of the assumed ignition source as a function of time. Once the heat release rate is known, the flame area and height can be estimated. Sources for typical heat release rates of ignition are given in [10]. If the net heat flux from the surface of the actual ignition sources is known, the ignition sources may be simulated by radiant panels. Typical heat fluxes are given in [10]. National and international standards describe a variety of reference ignition sources [12]. Gas burners with different flame heights up to 250 mm (9.75 in.), glow-wires, and fuels are used to test and classify products. When gas burners or radiant panels are used as ignition sources, it has to be considered that the thermal shock created by these regimes might influence especially the charring and melting behavior as well as the burning performance of many plastic products.

Ideally, in structural design for fire safety purposes, determining the thermal impact on a component should reproduce the conditions in the actual fire compartment. Time-temperature fire curves, such as the ISO 834 diagram, can definitively not be taken as models for real-life fires. They consist of arbitrary time-temperature relationships. For almost all of these nominal fire descriptions, the temperature continuously increases. Parametric fires may also be described by time-temperature curves, but theses curves depend, in each particular case, on scalar values representing the main physical parameters that influence a real fire. In a performance-based, risk assessment approach, designers should be permitted, if so desired and subject to appropriate validation, to undertake their own calculations in this regard. Alternatively, advantage can be taken of the design fire concept. In either event, each and every enclosure throughout the construction works concerned must be considered, although in many of the works similar data will apply to a number of the spaces involved.

Many design fires can be represented as having a growth stage in which the rate of heat release is proportional to the square of elapsed time (known as t-squared fires) and, unless controlled or suppressed by the intervention of automatic or manual self-extinguishment, are assumed to grow until restricted by the ventilation limits of the enclosure concerned.

Semiempirical approaches for various model fire loads have been derived using test results and physical principles. These have proved to be relatively successful, particularly with wood cribs. For all fire loads, the release of energy as a function of time can be assumed roughly equivalent to that of wood. Wood is used as the reference material, because it has been the subject of numerous studies, resulting in knowledge about its burning behavior as a function of ventilation.

When fires start and the different proportions of these fires have been ascertained, the probable rates of release of heat, smoke, and toxic and corrosive gases remain to be determined as the main data needed for fire safety engineering.

8.3 Historical Development of Fire Testing

Early awareness of the potential fire hazard in towns of predominantly wooden buildings led people to carry out *ad hoc* fire tests to ascertain the combustibility and hence the fire resistance of buildings. Some of the earliest investigations were made about 1773 and 1790 in Germany and the United Kingdom; 100 years later in the United Kingdom and the United States building structures were tested for fire resistance by subjecting them to wood fires. The present day test procedure for the determination of the fire resistance of building components has developed from this by a process of standardization.

All *ad hoc* investigations and the standardized fire test procedures developed were established to examine natural materials such as wood and other cellulose-based materials. The reaction-to-fire characteristics of these materials are often very different from those of synthetic materials. Thermoplastics especially may pose difficulties for testing; they may melt, drip, and shrink. In addition, other problems are associated with plastics. For example, intumescence occurs with some plastics exposed to radiant heat in tests such as ISO 5657, ISO 5660, and ISO 5659–2. Some products exhibit excessive swelling into the cone radiators or onto the pilot igniters in these tests. Extinguishment of pilot flames with highly flame-retarded plastics can occur with tests such as ISO 5657 and ISO 5659–2 due to the release of vapor phase active flame quenching species. Most pilot burners used with a radiant cone or panel tests have diffusion flames, which are more readily extinguished than pilot flames premixed with air. When vertically oriented thermoplastic sheets are exposed to radiant heat in tests such as ISO 5658–2, the specimens soften and often slump toward the source of radiant heat. Although this effect is realistic under certain fire conditions, the slumping behavior may inhibit fire development, especially flame spread. Other effects are additive evaporation/sublimation, out-gassing, char layer formation, and delamination.

Some applications for plastics that present particular problems for their fire performance assessment and that may require the use of specific fire testing today, are [10]:

- Semifinished products,
- Films and foils,
- Profiled sheets, for example, for roofing or panels,
- Profiles, for example, conduit for electric cables, window-frames, extruded sections,
- Weatherproof glazing,
- Foam pipe sections,
- Pipes, for example, rainwater drainage and discharge pipes,
- Furniture, for example, chairs,
- Air admittance valve ventilating pipes,
- Containers for liquids and solids in general.

Any position change of the specimen surface versus the applied ignition source will lead to a different exposure to heat. For nonplanar products, different parts of the specimen will be heated at different flux levels at any time. To counter these uncertainties, numerous *ad hoc* tests, model fire tests, and laboratory test procedures have been developed and introduced alongside with the classical fire tests. These approaches have helped to build up experience on the fire performance of these products.

Many simple, cheap laboratory test methods were developed principally to test the properties of plastics and to screen materials during product development or for quality control. These tests are only intended to compare one material with another and do not correlate to real-life fires. Because they are easy to run, such tests are widely used and have been introduced as standardized procedures especially in the United States as ASTM, NFPA, and Underwriters Laboratories standards.

Today, fire statistics are an important tool to obtain information on the frequency of certain types of fire, its causes, on the ignition source, and the extent of the fire. They may give an indication of the correlations between the type of a fire, the type of construction, people's habits, and casualty rate. Historically, the United Kingdom and the United States are the countries with the most detailed fire statistics. For example, an important cause of fires is arson, which can be as high as 70 % in schools. Malicious ignition is the lowest in dwellings, with a rate of approx. 20 %. In dwellings, the most frequent item first ignited is cooking appliances; in other types of buildings, this may be different.

Current regulations on fire safety tend to be largely prescriptive in nature and, in many cases, rather conservative. This type of regulation may have a number of adverse consequences, such as the stifling of innovation, the over-design of construction works leading to higher construction costs, and restricting the trading and use of products. However, attempts are being made to develop object-specific fire protection principles that more accurately reflect the actual conditions found in the fire compartments and use these as a basis for assessment. This leads to a better understanding of the actual level of safety and results in structures and construction works that are more economical. Object-specific hazard analyses of this type comprise a number of individual stages, as follows:

- Estimation of the quantity of energy released in the fire compartment as a function of time
- Determination of the gas temperatures in the fire compartment considering boundary conditions
- Calculation of the changes in the reaction of the components over time during the natural fire and during a standard fire
- Determination of the equivalent fire duration by comparing the calculations

In this context, the use of Fire Safety Engineering (FSE) could yield considerable benefits and economies in construction and transportation. The value of such a more fundamental, performance-based approach to fire safety in buildings has already been recognized in the Interpretative Document No. 2, "Safety in Case of Fire," produced in support of the Construction Products Directive CPD (89/106/EEC) in the EU. Similar approaches are currently made in many countries around the world and initiatives on FSE may be considered as some of the longer term measures.

8.4 Terminology

Materials tested according to standards are often classified by terms such as "self-extinguishing" and "difficult to burn." These technical terms refer to a specific test method, for example, "self-extinguishing to ASTM D 1692." However, in practice, the significance and necessity of this is frequently not recognized. If the test procedure is not specified, the term is devalued and the impression is given, for example, that a self-extinguishing material would under all circumstances extinguish itself, thereby implying a nonexistent null risk with possible serious consequences. Even more critical is the widespread use of markings that are not based on any test procedure such as "flame-proof," "nonburning," "fire-proof," and so forth. This confusion in fire terminology is increased by the use of different expressions, for example, in commerce; by science and administration; and in the building, electrical, transport, and mining industries. Problems of terminology occur in many countries and it is almost impossible to translate such terms from one language into another.

The test results shall be given in an objective manner by numbers and/or letter combinations not leading to confusion or false judgment of the fire hazard. A list of subjective and objective markings is given in Table 8.1.

Table 8.1 Subjective and objective fire protection terms

Subjective markings Deprecated	Objective markings Allowed
fireproof flameproof flame-resistant non-ignitable low combustibility low flammability self-extinguishing nonburnable	combustibility flammability ignitability flame flame spread glowing combustion incandescence heat release

A harmonized list of terms and definitions has been published for three languages [13]. For further details on terminology, refs [14–16] and Section 3 of the Appendix, p. 680, may be consulted.

8.5 Test Methods

A product is allowed to be used if it complies with fire safety regulatory requirements. Ignitability, flame spread, and heat release are currently assessed for regulatory purposes via small-scale reaction-to-fire test methods, which differ from country to country and from application to application. Dependent on the test method, the respective importance of the fire parameters being assessed is different. The tests are performed either on the material or on a small or intermediate size composite.

Alongside these, however, there are other tests that in principle do not simulate a fire situation but only compare, for example, the combustibility of certain materials under fixed conditions. Such tests are applied as screening tests for materials during product development or for quality control. They are usually simple in conception and easily carried out. They seldom, if ever, give trustworthy information about the behavior of a material in an actual fire.

The observed phenomena of ignition, a transition region, flame spread, and growth often differ from test to test and may not give any consistent results. This is certainly the reason for the disappointing results obtained in a joint investigation carried out in the mid-1960s by seven laboratories with 24 materials under the auspices of the former Working Group (WG) 4 of ISO/TC 92. When the materials were investigated and classified according to the different national fire test methods, it was found that there was scarcely any correlation between the tests and none at all between the different laboratories.

Because of these poor results it was decided by ISO/TC 92 not to adopt any of the existing intuitive and/or empirical fire test methods but to develop simple, basic tests that concentrate on one of the parameters determining the course of the initiating fire [17, 18].

8.5.1 Design Requirements

Difficulties are often experienced in extrapolating upwards from small- to large-scale performance, because there are always implications to be taken into account in the tradeoff between test reproducibility and test relevance to end-use hazard assessment. Anticipated correlation with a real fire performance is a function of the scale of the simulated fire. The purpose of an intermediate; large-, and full-scale fire test is to generate information in the fire growth stage. For example, the fire hazards of cables for use in atomic power stations are investigated in conditions close to reality. In mining, underground conditions are simulated and the fire performance of materials and finished parts is tested under working conditions in fire galleries or shafts.

Small-scale tests require the entire specimen to be subjected to the exposure condition used in the test. This cannot easily simulate the range of thermal ventilation and physical condition experienced by products during the fire growth phase, where fire boundary conditions will be constantly changing. The effects of thermal deformation, delamination, fixing failure, substrates effects, joints, and so forth, on product performance often require larger specimens to be investigated.

Many large-scale tests require extensive post-combustion product handling facilities to cope with the fire effluents. This makes them expensive and useful mainly as type approval tests. It is therefore desirable to develop flexible intermediate-scale tests that can effectively quantify the relevant parameters.

The relevance of ignition sources depends on the selection of fire scenarios in which the product is to be evaluated. Fundamentally, heat flow from the heat source to the specimen is a major parameter in such an evaluation; this also depends on the relative sizes of specimen and ignition source. The following characteristics of the ignition source should be taken into account:

- Radiant/conductive/convective effects,
- Flaming/nonflaming,
- Impinging/nonimpinging,
- Precision/quantification of measurements,
- Instability of the flame.

Tests should take into account the following considerations [10]:

- The size of the specimen should allow handling in a laboratory.
- The apparatus installation should not be excessively difficult and the test should be designed for efficient testing and easy specimen handling.
- The test should have simple and commonly available instrumentation.
- The ignition source should have a size that can be accommodated in laboratories.
- The ventilation conditions should reflect realistic fire conditions as far as possible.
- The test configuration should allow for open, semi-open or closed conditions.
- Correlation with a larger-scale regulatory test is desirable.

Systematic investigations of the events in the initiating fire have established the parameters that form the basis for estimating the hazard in this phase of the fire. The resulting relatively simple fundamental tests conceived by ISO/TC 92 are now in an advanced stage of development. These "reaction-to-fire" tests allow an estimation of the parameters:

- Combustibility,
- Ignitability,
- Flame spread,
- Heat release,
- Smoke yield.

These requirements arise from the aim of being able to describe a real fire and all its parameters quantitatively and thus obtain simultaneously measurements of as many of these parameters as possible.

Reaction-to-fire tests ascertain whether materials take part in a fire and their contribution to fire growth. According to the ISO concept, the phenomena accompanying the fire (smoke development and toxicity of the fire effluents), which can also be arranged according to "reaction-to-fire," are not to be equated with the parameters mentioned earlier, as they are not independent quantities but depend on the extent to which a material contributes to the fire.

8.5.2 Laboratory-Scale Tests

In a small-scale test, it is not easy to simulate the area, thickness, profile, and orientation of the product in its end-use, or to replicate actual mountings or fixations. This also applies to the joints and air gaps, which may have an influence on real-life product performance. The advantage of small-scale tests is that, if they correlate with large-scale tests, they can provide meaningful data for mathematical modeling. Smaller scale models have already been successfully used in predicting fire performance of wall linings, furniture, cable, and floor coverings in large-scale scenarios.

However, these tests are often carried out using standard substrates for the test specimens; these substrates may not replicate real-scale conditions. Small-scale tests cannot replicate the range of conditions and avenues found in fires (including the size and duration of application of ignition sources).

The individual "reaction-to-fire" parameters are discussed in the following paragraphs. Details of the test rigs and test specifications are given in the subsequent chapters of this handbook.

The test methods that were, and often still are, called "noncombustibility" tests should be designated "*combustibility* tests," as the latter emphasizes the extent of combustion and not the deceptive safety of noncombustibility. These test methods permit the differentiation of combustible (mostly organic) and noncombustible (mostly inorganic) materials. Even so, this does not mean that all materials that are considered inorganic are noncombustible. For example, mineral fiber mats contain small amounts of synthetic resin binders and may not pass a particular test.

Ignition of a material can occur either by convection or irradiation without the influence of a flame *(ignitability)* or by the direct effect of a flame *(flammability)*. Whether and when a material can be ignited depends on whether a flame is present and on radiation intensities. Using a suitable test method, it is possible to simulate the propensity of a material to ignite during various stages of the initiating fire up to flashover. One can thus determine whether the material can cause a fire when exposed only to a low-intensity ignition source without radiation being applied, or whether it causes the fire to grow to flashover under high radiation intensities.

Different *flame spread* tests put different demands on the fire performance of the product to be tested. In the simplest case, a low-intensity flame impinges on the material. Far higher demands are made if, in addition to the ignition source, a heat source reinforces the flame spread. A test of this type can be seen as a special case of a procedure for igniting a material, as the radiation intensity at which the flame front does not advance any further is obtained. This point thus corresponds to the radiation intensity necessary for igniting the material, which could be determined with one test. In the case of the preceding ignitability tests, a series of investigations at different radiation intensities would be necessary. Some flame spread tests are even more rigorous: they enable the effects of convection and radiation from the flame itself to be taken into account by suitable equipment design. The resulting feedback increases the rate of flame spread.

The *heat release* of a material in a fire situation is the heat liberated by the actual material burned and must not be confused with the calorific potential describing the quantity of heat that can theoretically be liberated. The rate at which heat is liberated from a product is especially important because of its influence on the initiating fire. Determination of the quantity of heat evolved per unit time allows an estimation of the potential hazard posed by a product: if the rate of heat release is small, the fire spread can be influenced, but if it is large, this is no longer possible as flashover is quickly reached. At the current stage of fire safety engineering development, heat release is the most important input data for numerical fire modeling.

The parameters discussed in the preceding occur in chronological order in a fire and are cumulative. In the case of flame spread, this means that ignition has already taken place and that a high rate of heat release will lead to a high rate of flame spread. Consequently, the test methods for measuring the aforementioned parameters make ever-increasing demands on the product to be tested.

The secondary fire effects, smoke and toxic fire gases, occur alongside these phenomena, particularly as the rate of flame spread and heat released increases. Together with radiant heat and lack of oxygen, they represent the severest hazard concerning escape and rescue.

To ascertain the smoke hazard, test methods have been developed that try to simulate the effect of smoke on human vision. These tests are based on highly simplified assumptions and must, if possible, be modified and optimized to a realistic simulation. Test and assessment methods for the hazard of toxic fire gases are published in [19], proposing to assess the hazard either by chemical analysis of the fire effluents or by their effect on biological systems; the second method is preferred. Two procedures are recommended:

- First, identification of the materials that cause an unusually high toxic environment,
- Second, development of a standard for classifying all commonly used materials with respect to the relative degree of acute inhalation toxicity that they create.

Of the two possible ways of measuring the toxicity of fire gases, the analytical method is considered less desirable because, *inter alia*, it does not consider the combined effect of several individual components and because a component can show biological activity when present in extraordinarily small (sometimes immeasurably small) amounts. Biological methods enable the total effect of the decomposition gases on the biological system to be investigated and a more realistic estimation of the hazard arising from toxic fire gases to be made. However, to avoid animal tests, efforts are made to develop mathematical models for predicting the toxic potency posed by fire effluents from analytical data [19–21].

It has to be emphasized that laboratory test methods can take into account only single specific test effects (ventilation, heat, geometry, etc.) [22–32]. For validated judgments, additional large- and even full-scale tests must be carried out to allow more realistic assessment of the hazard. One step in this direction is the acceptance of so-called reference scenarios for the development of the European harmonized test methods for building products. The starting point was the link between laboratory tests and the large-scale test defined in [33]. Other reference scenarios include large rooms, ducts, corridors, stairwells, and facades.

8.5.3 Large-/Full-/Real-Scale Tests

In full-scale model fire tests, whole buildings or vehicles (or at least their essential parts such as a furnished room or the compartment of a train or an aircraft) are tested. Large-scale tests are carried out to determine the fire performance of products under more or less normal operating conditions. Such tests may offer the only available realistic assessment of the gross effects of thermal deformation, gravity, and effects of fixations and joints experienced under end-use conditions in real fires. This is especially necessary for new products, for which results obtained in the laboratory must be confirmed in practical application. Conversely, large-scale tests can indicate if an existing laboratory method is suitable for testing a given material or not. This raises questions about the acceptability of small-scale test data on parts of products (or on materials used in products) given that these play a key role in the fire safety of the product in its application scenario. The disadvantages lie in operational hazards (including safety and test environment), cost, and uncertain reproducibility.

Originally, large-scale tests were carried out exclusively in an ad-hoc manner to simulate a real fire without setting down exact conditions. A literature survey is given in [34–36].

Despite the advantages of a large-scale test on the complete object, the enormous costs and massive expenditure of time weigh against such methods. Therefore, there are natural limits set on the size of the objects to be investigated: on economic grounds alone, one would not set fire to a complete tower block or a large passenger jet; the information obtained would in no way justify the expenditure.

An especially satisfactory information/expenditure ratio is achieved with the "compartment size" test [37, 38]. A compartment is defined as a closed spatial element in which a fire develops and in which flashover occurs. This closely mirrors the true relationships, for the initiating fire phase always takes place in such a compartment and people spend much of their time in such compartments (e.g., in a room of a building, inside a car, in a ship's cabin, or in a train compartment).

In addition to the large-scale tests, investigations are also made on a 1:1 scale in which only one type of material or finished part is tested. Such tests have the advantage that they are less expensive and more cost effective; at the same time they allow certain parameters that constitute a special fire hazard to be more exactly described.

Examples of such test methods are the "corner tests" that are often used in the United States for assessing the fire hazard of linings. Here, the ignition source is situated in the corner of the room; this arrangement puts the greatest demands on the material being tested.

Corner tests are applied by Insurance Companies (e.g., Factory Mutual) for estimating the fire hazards of certain materials such as foams, to determine the size of the premium.

Further part tests exist, for example, in the measurement of the fire hazard of rows of seats that are intended for use in passenger aircraft. In the construction field, fire tests are carried out on building elements in various countries.

For the development of assessment methods and fire safety engineering principles, prenormative research using larger-scale tests is still required.

8.5.4 Correlation of Test Methods

Important criteria in assessing the relevance of a test procedure and apparatus to the intended end-use scenario include:

- *Ventilation effects*. Particular care is necessary for interpreting results obtained in closed-box apparatuses, where the ratio of specimen mass to chamber volume is important. The significance of under-/over-ventilation conditions relative to the end-use scenario and relevance to apparatuses such as the cone calorimeter, cable tray tests, and radiant panel flooring test should be taken into account. This raises questions about wind-assisted versus wind-opposed flame-spread regimes.
- *Mechanical effects*. Mechanical restraint on a specimen during thermal expansion can result in specimen warping or fracture. The use of restraining devices such as grids may prevent the formation of a beneficial cohesive char layer on the specimen surface.
- *Thermal effects*. Heat sinks and incident flux variations/gradient over the surface of a specimen should be considered. Every effort should be made to eliminate such effects. If they are unavoidable, they should be taken into account in the final appraisal of the results.

One of the aims in fire protection is to obtain a correlation between the test results and the actual course of a fire so that the fire hazard can be estimated properly. In general, it has been found that the laboratory and pilot-scale methods contained in the regulations allow a certain correlation with large-scale tests and actual fire incidents. Numerous investigations have indicated that products that satisfy the highest demands according to certain standardized tests usually do not contribute to fire spread and fire propagation in large-scale investigations. However, hazards cannot be reliably assessed based on laboratory tests or even large-scale tests because, even in these, the fire situation is only partially reproduced under the given experimental conditions. Experience gained under actual fire conditions must be included as well.

Besides correlating laboratory to large-scale tests and then to real-life fires (scaleup), it is useful to apply the reverse procedure. Starting with the real situation, physical models for large-scale tests can be made, and then by scaling down, retaining the important chemical and thermodynamic properties and the geometry of the system, a correlation with a laboratory test may be sought. In practice, this latter method is used because of financial and time limitations; thus compartment-size tests are replaced by geometrically identical, scaled-down laboratory versions. Similarly, scaled-down corner tests are used for testing only a part of the system for screening materials.

Even today, fire hazard assessment is usually based on procedures that arose intuitively and empirically without knowledge of the parameters that influence the course of the fire.

Systematic investigations of such parameters are still being made (ISO). These parameters must not be regarded, however, as isolated events or else the fire will be even less well described than by the empirical, intuitive methods. As shown in fire safety engineering guidelines [1–8], work is already in progress that no longer regards the events in a fire as isolated, point occurrences but describes the fire as a complete system dynamically and with reference to fundamental parameters and empirical knowledge.

References

[1] ISO 13387-1 Fire safety engineering – Part 1: The application of fire performance concepts to design objectives
[2] ISO 13387-2 Fire safety engineering – Part 2: Design fire scenarios and design fires
[3] ISO 13387-3 Fire safety engineering – Part 3: Assessment and verification of mathematical fire models
[4] ISO 13387-4 Fire safety engineering – Part 4: Initiation and development of fire and generation of fire effluent
[5] ISO 13387-5 Fire safety engineering – Part 5: Movement of fire effluent
[6] ISO 13387-6 Fire safety engineering – Part 6: Structural response and fire spread beyond the enclosure of origin
[7] ISO 13387-7 Fire safety engineering – Part 7: Detection, activation and suppression
[8] ISO 13387-8 Fire safety engineering – Part 8: Life safety: Occupant behaviour, location and condition
[9] ISO/TR 10840: Plastics - Burning behaviour – Guidance for development and use of fire tests
[10] ISO DIS 15791-1: Plastics – Development and use of intermediate scale fire tests for plastics products – Part 1: General guidance
[11] A. Pinney: Update on the European Construction Harmonisation Programme, Flame retardants 96, 23–33 (1996)
[12] ISO 10093: Plastics – Fire tests – Standard ignition sources
[13] ISO 13943: Glossary of fire terms and definitions
[14] J.M. Kingsland, C. Meredith: Fire Tests and Fire Hazards. National Technical Conference, Safety & Health with Plastics, Society of Plastics Engineers, 8–10. 11. 1977, Denver, USA, p. 77
[15] BS-DD 64: 1979. Draft for development: Guidelines for the development and presentation of fire tests and for their use in hazard assessment. British Standards Institution. Replaced by BS 6336: 1982
[16] SAA MP 32 – 1977. SAA Guide for the Presentation, Preparation and Application of Fire Tests. Standards Association of Australia
[17] ISO TR 3814 – 2nd edition 1989. The Development of tests measuring "Reaction to fire" of building materials
[18] V. Babrauskas: *Fire Safety J.* 24, 1995
[19] ISO/IEC/TR 9122 Toxicity testing of fire effluents

[20] ISO 13344 Determination of the lethal toxic potency of fire effluents

[21] B.C. Levin: *Toxicology* 115, 89–106, 1996

[22] R.G. Gann, V. Babrauskas, D. Peacock, J.R. Hall: *Fire Mater.* 18 (1994) 193–199

[23] B.T. Zinn et al.: Analysis of Smoke Produced during the Thermal Degradation of Natural and Synthetic Materials, Int. Symp. Salt Lake City, 1976

[24] R.G. Silversides: Measurement and Control of Smoke in Building Fires, ASTM-STP 22, 1966

[25] R.W. Bukowski: Smoke Measurements in Large and Small Scale Fire Testing, NBSIR-78-1502

[26] C.P. Bankston, R.A. Cassanova, E.A. Powell, B.T. Zinn: Initial Data on the Physical Properties of Smoke Produced by Burning Materials under Different Conditions, J.F.F., 7:165. 1976

[27] H.L. Malhotra: Smoke-Tests - A Critique, QMC Symp. January 1982

[28] D.D. Drysdale et al.: *Fire Safety J.* 15 (1989) 331

[29] D.D. Drysdale et al.: Smoke Production in Fires, Fire Safety Science and Engineering," ASTM-STP 882, p. 285, 1985

[30] P.G. Edgerly, K. Pettett: *J. Fire and Flamm.* 10 (1979)

[31] S.D. Christian: The Performance and Relevance of Smoke Tests for Material Selection for Improved Safety in Fires, Polytechnic of the South Bank Borough Road, London SO1 OAA, 1984

[32] V. Babrauskas: The generation of CO in bench scale fire tests and the prediction for real scale fires, First int. Fire and Material conference, proceedings, Interscience communications, London, 1992

[33] ISO 9705: Fire tests – Reaction-to-fire – full-scale room test for surface products

[34] R. John, P.G. Seeger: Auswertung des in- und ausländischen Schrifttums über die Durchführung von Brandversuchen in bzw. an Gebäuden im Maßstab 1:1, Research report (in German), Forschungsstelle für Brandschutztechnik an der Universität Karlsruhe

[35] P. Shakeshaft: A review of data on active methods of fire protection in residential premises: experimental fires. Building Research Establishment Current Paper CP 21/78, January 1978

[36] C.-D. Sommerfeld, R. Walter, F.-W. Wittbecker: Fire performance of facades and roofs insulated with rigid polyurethane foam – a review of full-scale tests, proceedings, Interflam 1996

[37] EUREFIC European Reaction to Fire Classification, Proceedings, Kopenhagen, 1991

[38] CBUF – Fire safety of upholstered furniture – the final report on CBUF research programme, edited by B. Sundström, EU Commission report EUR 16477 EN, 1995

9 Regulations and Testing

M. MITZLAFF AND J. TROITZSCH

9.1 Sets of Regulations

The aim of fire protection is to minimize the risk of a fire, thus protecting life and possessions. The state, as official custodian of public safety, ensures such protection via relevant legislation, which includes laws and statutory orders. Standards and codes of practice based on recognized technical principles are the means of putting the general requirements of fire protection defined in the legislation into practice. Materials and semifinished and finished products are tested according to methods laid down in the standards and classified according to the test results. Such tests are carried out by officially recognized materials testing institutes. The certificate of the test result and classification provides a basis for the use of the material or product. A test mark is frequently required as evidence of the suitability of a material and its identity with the material tested. This implies a quality check on production either by the manufacturer or an outside body. In the latter case an agreement must be entered into with a state-recognized institution.

Numerous public and private organizations are concerned with the rules and regulations of fire protection. In addition to governmental bodies, these include professional societies and industrial, commercial, technical, and insurance associations.

This multiplicity is necessary to create an appropriate and practical "state of the art" system of fire precautions.

In law the principal requirements are usually laid down in quite general terms. Proof of their fulfillment requires determination of verifiable criteria. This work is performed mostly by standardization organizations where representatives of government, test institutes, professional and industrial societies and insurance associations collaborate and introduce their particular knowledge and experiences. Such organizations, for example, the Deutsches Institut für Normung (DIN) e.V., the British Standards Institution (BSI), the Association Française de Normalisation (AFNOR), or the American Society for Testing and Materials (ASTM), are financially independent (not for profit organizations) and apply profits to promote further activities. In those fields where only few and general legal regulations exist, safety regulations are also drawn up by professional bodies.

Building classifications relating to the fire resistance of structures and the fire performance of materials as well as specific safety regulations formulated by the associations of property insurance are applied in civil law. They are especially significant because they form the basis of fire insurance policies. Some of these regulations cover design, installation, and routine maintenance of fire protection systems and set down testing and approval procedures.

Special requirements in manufacturers' purchasing specifications must, in practice, be complied with because products that do not satisfy them have little chance of commercial success. For example, such requirements are imposed by the aircraft industry with the aim of improving the fire safety of their products in advance of official regulations.

Depending on the field of application, regulations are applied nationally and/or internationally. For example, national codes are encountered almost exclusively in building and mining, because every country possesses its own set of rules and test methods that have developed historically. International harmonization of regulations would in some cases have major economic implications, primarily negative in the eyes of individual states. The safety concepts or fire philosophies of many countries are in many cases diametrically opposed to internationalization in certain areas. Even so, vigorous efforts are being made to harmonize building codes. This has already occurred in the case of the Nordic countries (Denmark, Finland, Iceland, Norway, and Sweden), where fire testing is practically identical in all five countries and test results and test marks are mutually recognized. Unification of national building regulations relating to fire protection is also planned. Details of the situation for building products in the most economically important countries of the world are covered in Chapter 10.

In the European Union (EU), the harmonization of the legislative, regulatory, and administrative provisions of the 15 member states is the responsibility of the EC Commission. The "Construction Products Directive (CPD)" (89/106/EC) deals with types of products and only includes the main requirements. For this purpose, the Interpretative Document ID 2 (Essential Requirement "Safety in Case of Fire") was issued. The technical specifications, to which the products must conform for complying with these requirements, will be defined by European standards or European technical approvals.

European standards are issued by the European Committee for Standardization (CEN) (with members from 19 countries) and the European Committee for Electrotechnical Standards (CENELEC). These standards assist in eliminating technical barriers to trade between member states as well as between these and other European countries (see also Section 10.4 European Union).

The new EN standards are in the course of being transferred into national standards, for example, BS EN, NF EN, or DIN EN. After a definite time ("transition period" ending in 2005 or later) all national fire standards will be withdrawn in favor of the EN standards. This will also be true for the foreseen new EU-members like Poland, Czech Republic, or Hungary.

In other fields, regulations exist that are, in the main, internationally valid. This applies in transportation and in civil aviation in particular. The most important new technologies were developed in the USA and thus the US Federal Aviation Regulations (FAR) of the Federal Aviation Administration (FAA) set the criteria for corresponding internationally accepted regulations. Evidence of airworthiness must be furnished with the aid of these regulations, which have been partially or totally adopted by most countries.

Sets of regulations in international use also exist in other areas of transportation. For example, the International Convention for the Safety of Life at Sea (SOLAS) is accepted in nearly all countries. The regulations, recommendations, and conditions of supply relating to railways are being harmonized within the Union Internationale des Chemins de Fer (UIC) (International Railway Union) and now within the EU by a CEN mandate. Further details on transportation are given in Chapter 11.

International harmonization of standards and mutual recognition of test results are planned in electrical engineering to remove trade barriers. International electrical standards are being developed by the International Electrical Commission (IEC) to achieve this aim. The Commission for Certification of Electrical Equipment (CEE) will base mutual recognition of test results by all member countries on these standards (see Chapter 12).

While the IEC is responsible for international electrical standards, all other technical fields are covered by the International Organization for Standardization (ISO). This organization aims to promote the worldwide development of standards to facilitate the exchange of goods and services and to encourage mutual cooperation in intellectual, scientific and economic activities. ISO member committees are the national standards organizations. Further details on ISO are found in Section 10.23.

9.2 Overview of the Subject

The topics discussed below reflect, in the main, those fields in which fire protection requirements are imposed by regulations. These include building, transportation, and electrical engineering. Building is dealt with in particular detail, as this is the field where regulations, test
methods, and approval procedures are most numerous, and where the greatest variations exist
from country to country. Furthermore, the use of all plastics as combustible building materials
is dealt with here from the point of view of fire protection. This field therefore forms the most
important part of this book. The European Union, 15 European and 6 non-European countries
are covered. In addition, the activities of ISO are dealt with in detail.

In transportation and the electrical industry, regulations and test methods are rather more international so that they are less numerous and varied.

Furniture and furnishings are covered in detail in Chapter 13, as much thought is being given
to the question of fire protection in this area. Fire protection regulations relating to furniture
and furnishings exist at present in the United States and the United Kingdom. The European
Commission, CEN, and ISO are also active in this field.

The various fields covered in the following are dealt with according to a uniform system. This
is intended to help the reader become acquainted with, and to review rapidly, the various regulations, requirements, test methods, and approval procedures for the principal areas of application of products and plastics. After a short introduction to the statutory regulations or codes in
force, the classifications and tests necessary for meeting the requirements are described. The
tests are elucidated according to a uniform scheme in which diagrams illustrating the principles of the methods and tables of the specifications are given. Further details are given where
necessary for clarity. Only those methods contained in regulations or referred to in associated
supporting documents are discussed here. Most of the tests are applications-related and can be
carried out on laboratory or pilot scales. Full-scale tests are seldom required except in certain
countries or in the mining industry. Tests that are not clearly defined such as *ad hoc* investigations, voluntary tests for product development and quality control and methods for determining the fire resistance of building components (which only contain plastics in exceptional
cases) are not considered.

It is not the aim of this book to evaluate the fire performance of various plastics or of semifinished or finished parts on the basis of the test methods discussed in the following. After
describing the test methods, a list of appointed and recognized test laboratories that carry out
fire tests, is given. The ways of obtaining approval and test or quality marks are described
together with the associated inspection and monitoring procedures.

Test or quality marks on a material signify that it has met performance criteria set down in
standards or association regulations. Manufacturers' quality control associations, which have
their own technicians and inspectors, are also frequently approved by the state for testing and
inspecting materials. Tests are carried out at materials testing laboratories, technical organizations or by specially appointed experts. Inspection may be carried out by the manufacturer or
independently by an authorized outside party with whom an inspection agreement has been
entered into.

All manufacturers entitled to use the quality mark are listed in a directory, which is published
periodically and is available to the public. The right to use the mark can be withdrawn if the
inspection regulations are infringed or if several negative test results are obtained. This may
have serious commercial consequences for the manufacturer, who is thus motivated to ensure
that his product meets the requirements.

Short sections on future developments follow each subject and country dealt with in the
following. These cover briefly planned national and international changes or new trends in
regulations, testing and approval.

Secondary fire effects such as smoke development, toxicity and corrosivity of fire effluents are dealt with separately in Chapters 15–17 to present a clear picture of this important subject. A short introduction to each topic is followed by a discussion of the relevant requirements and test methods. The determination of smoke density is dealt with and the techniques for determining the toxicity of fire gases are discussed in the light of the latest knowledge. A general review of the present situation regarding corrosivity of fire gases is given, as few established test methods exist.

The fire protection regulations and test procedures enumerated in Chapters 10–17 cover the principal areas of application of plastics in the economically most important countries but cannot represent this vast subject exhaustively. They are intended to give the reader a comprehensive review of the most important aspects in this field and enable him or her to become rapidly acquainted with the intricate field of fire protection.

9.3 Test Methods

The fire tests discussed below are arranged according to the test criteria on which they are based and their contribution to measuring certain fundamental parameters that affect the initiating fire. These criteria and parameters are, however, presented in a general form because, as mentioned earlier, it is not the aim of this book to evaluate the test methods or the plastics examined by them.

Size is one of the most important factors affecting test procedures. Tests may be carried out on a laboratory scale, pilot scale or full scale. The first two are the most commonly used and are discussed below.

It is usually assumed, particularly with laboratory-scale tests, that the test apparatus can be considered as a closed system, that is, without interaction with its environment. The results are decisively influenced by certain variables in such systems related in particular to the specimen and ignition source. Other variables, which are not directly connected such as the ventilation and volume of the test system, are disregarded.

The principal variables as regards the specimen are:

Type of specimen: for practical reasons, panels are usually preferred. Many methods do in fact allow realistic testing ("end-use condition") as an assembly and with the same thickness as the finished part. However, such finished parts represent only a section of a structure, for example, inner linings, particularly in the field of building. The electrical sector is an exception since actual complete finished parts such as plugs and switches can frequently be tested on the laboratory scale.

The specimen can be positioned in many different ways. The preferred orientations of the specimen in the methods described here are: horizontal and vertical.

The following ignition source parameters play a prime role:

Type and intensity: the main ignition sources have flames of varying intensity, for example, matches, gas burners (small flame burners, Bunsen burners, multiple jet burners, sand bed burners), wood cribs, and so forth. Gas or electric radiators or radiant panels are frequently used, while hot media (e.g., air) are less popular. A pilot flame is also frequently used in addition.

Positioning: horizontal and vertical positions are preferred.

The results are highly dependent on how the ignition source impinges on the specimen (i.e., edge or surface) and on the duration of exposure.

The list of variables above relating to specimen and ignition source illustrate how much individual tests may vary in just two aspects and why test results obtained by different methods seldom agree. Test results can only be correlated empirically with a real fire.

Certain phenomena, which decisively affect the initiating fire, are defined in the various test methods. They are combustibility, ignitability, flame spread, heat release, time to flashover, as well as the various secondary fire effects, that is, smoke development, toxicity, and corrosivity of fire effluents. Some of the above mentioned fire parameters, particularly heat release (HR, rate of HR: RHR or total HR: THR), are now part of indices used for assessing the fire behavior of combustible building products (Fire Growth Rate FIGRA, Smoke Growth Rate SMOGRA, see Section 10.4).

Some test methods are designed to measure only one parameter, while others are designed to measure several simultaneously. An example of the former case is the ISO 1182 combustibility test and of the latter are the French Epiradiateur test and the European SBI test (EN 13 823) in which heat release, ignitability and flame spread are all determined. These parameters are referred to in many national tests under identical names, but, owing to differing test conditions, they are not comparable. Eventually, it is hoped to define these parameters by internationally accepted test procedures within the scope of ISO/TC 92 (see Section 10.23). This problem is discussed further in Chapter 8.

The addresses of national and international standards organizations are given in Sections 4 and 5 of the Appendix.

10 Building

M. MITZLAFF AND J. TROITZSCH

10.1 Introduction

Statutory regulations and provisions relating to fire protection are furthest advanced in the field of building, particularly in the industrialized countries, where comprehensive sets of regulations may differ significantly. Most industrialized countries belong to one or other economic grouping. The regulations are then affected by the attempts at harmonization within such groups to eliminate trade barriers. In Europe, the European Union (EU) now basically harmonizes legal and administrative regulations.

In view of the impending completion of the internal market in the field of building products, on December 21, 1988, the Council of the European Communities (CEC) issued a directive for building products, the "Construction Products Directive" (CPD, 89/106/EC), which is based on the application of a new approach to technical harmonization and standardization. The CPD covers the approximation of laws, regulations, and administrative provisions of the member states relating to construction products and aims to remove technical barriers to trade arising from national laws and regulations in member states. In this approach, legislative harmonization is restricted to the adoption of essential requirements for safety, health, or other requirements for human welfare. The essential requirements in Annex 1 of the CPD do not apply directly to the products themselves, but are broad functional requirements that apply to the construction works in which the products are to be incorporated. The Interpretative Documents (ID) give concrete form to the mentioned essential requirements. The ID No. 2 (safety in case of fire) generally describes actions to be taken in case of fire [1].

The task of defining the technical characteristics of the products suitable to construction works is allocated to bodies competent in the field of standardization, for example, the European Committee for Standardisation (CEN). Draft EC mandates on building products were agreed by the Technical Bureau of CEN and allocated to existing or new Technical Committees (TC).

In the field of fire safety, the work has been performed by CEN/TC 127 since its first meeting in May 1988. TC 127 developed a work program based on the draft CEC mandates. This program concerns both reaction to fire and fire resistance. Many parts are already complete and other parts are still in progress.

The test methods of ISO 1182 and ISO 1716 were adopted for assessing noncombustibility. The lowest level of performance is determined by a test for ignitability by direct flame impingement and under zero irradiance (small flame test to EN ISO 11925-2). The Single Burning Item Test (SBI) to EN 13 823 covers the fire performance of wall and ceiling linings. The radiant panel test according to EN ISO 9239 was introduced to determine the fire performance of floor coverings. The test method prEN 1187, Parts 1–3, is used for evaluating the fire performance of roof coverings and consists of three parts based on the German standard DIN 4102 Part 7, a French and a Nordic Countries test method. With the exception of linear products (such as pipes or cables), which still have to be defined, the European classification system has been published in the Official Journal of the European Communities as a Commission Decision of the CEC in 2000/147/EC [2].

For fire resistance testing, harmonized standards were introduced as Eurocodes 1–9. Eurocodes 1–6 and 9 in particular deal with fire resistance and fire protection calculation.

The countries discussed in the following sections have been selected on the basis of their industrial and commercial importance and their comprehensive regulations on fire precautions. Some states have not been included, as no information was forthcoming and others because fire protection regulations and test methods are still under development.

Each country is discussed according to a uniform scheme:

- Statutory regulations,
- Classification and testing of the fire performance of building materials and components,
- Official approval,
- Future developments.

It is stressed again that fire resistance is not dealt with in this book. A bibliography for each country is given at the end of each section.

10.1.1 Statutory Regulations

Regulations covering fire protection in building are more or less centralized depending on the particular form of government. Countries with centralized regulations include France, Belgium, Italy and Japan (where additional local regulations exist). In principle, the United Kingdom also belongs to this group. However, separate regulations apply in England and Wales, Scotland, and Northern Ireland. The regulations, however, are gradually being brought into line with those in force in England and Wales.

The Nordic countries have centralized national regulations and classifications and test methods largely have been harmonized within the framework of the Nordic Committee for Building Regulations.

In countries with a federal structure, each province or state usually has its own decentralized building regulations, which are generally based on a model building code and framework guidelines in order to maintain some uniformity. Germany, Austria, Switzerland, Canada, Australia, and the United States all have a federal structure. The execution of regulations is the responsibility of the Länder, Cantons, Provinces, or States, although, in certain cases, this function may be transferred to centralized institutions such as the Deutsches Institut für Bautechnik (DIBt) in Germany. The situation is almost chaotic in the United States, where the individual states are theoretically responsible for building regulations. In practice, however, local authorities determine their application via some 20,000 local regulations. A special situation exists in the Netherlands, where responsibility for enforcing building regulations rests largely with local authorities.

10.1.2 Classification and Testing of the Fire Performance of Building Materials and Components

The classification of the fire performance of building materials and components reflects safety concepts in each country. Although national "fire philosophies" differ considerably from one another in certain respects, they are all based on the same concepts. Building materials are usually divided into four levels of contribution to fire:

- Minimal,
- Slight,
- Normal,
- Large.

Within the definition of this safety concept, contribution to fire provides a measure of fire hazard.

Minimal (meaning none or very low) contribution to fire of building materials is usually determined by a noncombustibility test. This is carried out in a furnace at 750 °C (cylindrical specimen) using a procedure based on ISO 1182 and according to ASTM E 136 (cuboid specimen), which is basically similar. Only France does not use this method in the building sector, but tests by a calorimetric bomb technique based on ISO 1716. Such calorimetric measurements are used in various countries including Germany in addition to the furnace method. "Minimal contribution to fire" is subdivided in the EU as well in some countries such as Germany (A1, A2) and Switzerland (6, 6q). This enables totally inorganic building materials to be differentiated from those with low (usually up to 5 %) organic content such as mineral fiber panels containing binders.

The classifications "low," "normal," and "high" contribution to fire of a building material are a measure of its behavior in an initiating fire. Materials that make a low contribution to fire usually start to become involved in the fire at a later stage of the initiating fire. Those that make a normal contribution resist medium- and low-intensity ignition sources for a short time (on the order of minutes) and those that make a high contribution are set on fire by practically any ignition source and are therefore generally not permitted to be used as building materials. Many national classification systems contain intermediate classes to be able to differentiate materials more clearly.

Test methods simulating all the important phases of an initiating fire, from ignition to flashover, are available, and can provide information on the contribution of building materials to each stage. Roughly 50 methods for simulating the initiating fire and 20 methods for describing secondary fire effects (i.e., smoke development and toxicity of fire effluents) exist in the various countries covered in the following. In some cases, an additional test is carried out to determine whether materials give off burning drips. In the European harmonized reaction-to-fire tests dripping, has been introduced as a test criterion.

Fire tests using low-intensity ignition sources (usually small burners) establish the ignitability of a material and allow the elimination of those with a high contribution to fire. Various classes exist based on the demands made on the material. The least stringent case is for a specimen positioned horizontally. Such methods based on the American ASTM D 635 test are used in the United States, the United Kingdom, Australia, and, in modified form, in France. The toughest demands are made on the material if the specimen is positioned vertically. The igniting flame may be applied to the specimen surface (lower demands) or edge (higher demands) as in the DIN 4102-1 or EN ISO 11925-2 small flame test and in similar tests in Austria, Switzerland, and Italy. Tests involving surface flame action only are used in the United Kingdom, while exclusively edging flame action is found in American, Canadian, and Swedish procedures.

The aforementioned methods are sometimes used to establish whether materials have a normal fire contribution and describe the flame spread, which sets in after the material ignites. With a further increase in the rate of flame spread, the instant of flashover is approached and only materials with a small fire contribution match up to these requirements. For this region, some tests take into account heat release, especially the European Union Single Burning Item (SBI) test to EN 13823.

Materials with a normal or low fire contribution are tested with the aid of medium- or high-intensity ignition sources such as naked flames or radiators.

Various test methods involving exposure to flame are used. Normal and low fire contributions are determined in Scandinavia by test method 004 or 002 (the Schlyter test, which is also used in a modified form in Austria), in the United States and Canada by the tunnel test (ASTM E 84), and by the Brandschacht method in Germany. Finally, the Single Burning Item (SBI) test to EN 13823 covers the different grades of fire contribution for combustible building products.

Radiators are used as ignition sources almost exclusively in the determination of normal and low fire contribution. Such methods include the Epiradiateur (F, B, E, P), Brûleur Electrique

(F, E, P), Surface Spread of Flame (UK, NL, B, GR), Vlamoverslag (NL), DS 1058.3 Panel (DK), and Early Fire Hazard (AUS) tests.

In some countries the test methods relate to finished components such as linings and floor coverings rather than to building materials in general. In certain cases, plastics must undergo additional tests if established methods do not provide unambiguous data, such as, for example, with thermoplastics, which can withdraw from the ignition source.

Floor coverings are frequently investigated in tests, either separately or complementary to those for building materials, summarized as follows:

European Union:	Radiant panel – EN ISO 9239-1
Germany:	Small burner – DIN 54 332 and
	Radiant panel – DIN 4102-14
France:	Radiant panel – NF P 92-506
	(corresponds to small radiant panel in BS 476: Part 7)
Austria:	Radiant panel – ÖNORM B 3810
Switzerland:	Small burner – SNV 198 897 (similar to DIN 54 332)
Nordic countries:	NT Fire 007 (INSTA 414)
USA:	Methenamine tablet – ASTM D 2859 or
	Radiant panel – ASTM E 648
Canada:	Modified tunnel test – ULC-S 102.2 (flame applied downwards)

10.1.3 Official Approval

The fire performance of building materials is tested in almost all countries, with the notable exceptions of the United Kingdom, Australia, and the United States, by officially recognized test laboratories. In the United Kingdom, Australia, and the United States, private test institutes can also be chosen. Nevertheless, the building authorities prefer test certificates issued by laboratories, which have been accredited under UKAS (United Kingdom Accreditation Service) in the United Kingdom and by NATA, the National Association of Testing Authorities in Australia. In the United States, large test organizations such as Underwriters Laboratories Inc. (UL) are favored by the authorities. In the European Union, test laboratories notified to the Commission (Notified Bodies, NB) carry out fire testing.

After a test has been carried out, the test laboratory issues a test certificate. This is usually accepted as sufficient evidence by the authorities, although in some cases a committee has to verify and approve it. Some countries require a test mark linked with inspection and quality control for materials subject to stringent fire protection requirements. This applies in Germany, Japan, the Nordic countries, Canada, and the Czech Republic. In a few cases, a quality label indicating quality control and independent inspection is required for certain building materials such as polystyrene foam (e.g., in Austria and the Netherlands). Organizations such as the American Underwriters' Laboratories grant test marks for certain building materials, which require internal and independent monitoring. Further information on test marks and inspection is found in the sections on individual countries.

10.1.4 Present Situation and Future Developments

The following account relates to the situation in 2001/2002. However, it is subject to modifications in light of changes in the regulations covering fire precautions and improvements in test procedures due to advances in technology.

In most countries, building regulations are revised according to a more or less regular cycle to ensure that they reflect the state of the art and take into account new social developments and challenges.

In Europe, the developments are dominated by the efforts of the Commission of the European Communities to harmonize fire test methods and classification criteria within the framework of the Construction Products Directive. Non-EU-member countries such as Switzerland are also influenced by the effects of European harmonization. In the United States, a comprehensive revision of the building codes is underway with the objective to develop, if possible, one unified model building code and to focus on performance-based rather than prescriptive requirements.

The test methods laid down in ISO standards will play a key role in all future developments. One example is the noncombustibility test to ISO 1182, which has already been or will shortly be adopted by numerous countries.

References for Section 10.1

[1] Interpretative Document, Essential Requirement No. 2 Safety in case of fire, February 1994
[2] Commission Decision 2000/147/EC: Classification of the reaction to fire performance of construction products, February 2000

10.2 United States of America

K. A. REIMANN

10.2.1 Statutory Regulations

Building codes in the United States have developed over the years principally by locality and region. Local municipalities can choose to adopt their own building code version. Thousands of such jurisdictions across the country could make this potentially unworkable for material suppliers, designers, architects, and the construction industry. Even today there are virtually no nationally mandated building codes or regulations to cover fire protection and design.

Developing from the many locality-oriented building code interests were some 15 or so code-making organizations in the early part of the last century. These organizations merged or reorganized over the years eventually to become the three Model Building Codes that have been in effect since about 1940. These codes, until recently, have been updated every 2 or 3 years. Their use has been preferred in the following regions:

- *The West:* The Uniform Building Code (UBC) issued by the International Conference of Building Officials (ICBO) [1]. The UBC was last updated in 1997. ICBO was founded in 1922.
- *The Midwest and Northeast:* The BOCA National Building Code issued by Building Officials and Code Administrators International, Inc. [2]. The BOCA code was last updated in 1998. BOCA was founded in 1915.
- *The South:* The Standard Building Code issued by the Southern Building Code Congress International, Inc. (SBCCI) [3]. The Standard Building Code was last updated in 1999. SBCCI was founded in 1940.

These model building codes are favored in the areas where they originate and are adopted in full or in part in state or city building regulations, although this is not mandatory. Local or regional variations in building code acceptance allow for particular concerns of that area; for example, heavy wind resistance is needed along the Gulf Coast and Florida because of the hurricane threat and building codes and regulations have been altered in California because of the likelihood of earthquakes. Localities can adopt a model building code but with specific changes or provisions needed in their particular location. Fire precautions are dealt with comprehensively in these model building codes. Many of the fire standards referenced in the codes are issued by the American Society for Testing and Materials (ASTM) [4].

Many building authorities also use the nationally available NFPA 101 Life Safety Code of the National Fire Protection Association (NFPA) [5], which also covers fire precautions. In large part, NFPA 101 covers fire protection provisions of building construction by occupancy type, a different approach than the other model building codes.

In addition, certain building regulations issued by the federal agencies apply nationwide. One of these regulations is administered by the US Consumer Product Safety Commission (CPSC) [6]. The US CPSC is an independent federal regulatory agency, which protects the public's safety by reducing the risk of injury or death from consumer products. The one flammability standard for plastics enforced by CPSC concerns textile floor coverings, mentioned later. Minimum property standards have also been established by the Department of Housing and Urban Development (HUD) and the Department of Health and Human Services (HHS). To obtain federally sponsored mortgage insurance builders and developers would have to conform to these standards.

10.2.2 New Developments in Statutory regulations

10.2.2.1   International Building Code (IBC)

Even with most communities adopting one of the three main model building codes, there have been problems and challenges with material suppliers. In the past, it was difficult for building industry professionals to move into different regions within the United States, marketing their products on a national level. On December 9, 1994, the International Code Council (ICC) [7] was established as a nonprofit organization dedicated to developing a single set of comprehensive and coordinated national codes, including the IBC and others such as the International Mechanical Code, to promote code uniformity throughout the country. The ICC founders BOCA, ICBO, and SBCCI created the ICC in response to technical disparities among the three sets of model codes in use within the United States.

This single family of codes has received widespread public support from leaders in the building community, including the American Institute of Architects, US Federal Emergency Management Agency, US General Services Administration, Building Owners and Managers Association International, National Association of Home Builders, National Multi-Housing Council/National Apartment Association, Insurance Industry Building Code Coalition, and numerous other national and international stakeholders in the construction industry, citizens groups, and all levels of government in the United States.

The IBC was first issued in 2000. Because at this time it is relatively new, it may take several years for localities to adopt the new building code, although again, it is not mandatory for them to do so. For example, in October 2000, the California Building Standards Commission voted not to use the IBC as part of its codes process. The code, like the other three model codes has one primary chapter on plastics, which includes most of the flammability regulations. It was agreed that standards adopted by the code would be based on consensus processes, for example, ASTM or NFPA.

10.2.2.2 International Residential Code (IRC)

Since 1972, the Council of American Building Officials (CABO) had served as the umbrella organization for BOCA, ICBO, and SBCCI. In November 1997, it was agreed to incorporate CABO into the ICC. The 2000 International Residential Code replaces the International (formerly CABO) One- and Two-Family Dwelling Code. Designed as a companion to the

International Building Code and other International Codes, the IRC concentrates on one- and two-family dwellings, as well as townhouses up to three stories high. Flammability resistance of plastics in residential housing is no more stringent than in other construction and generally must pass either ASTM E 84 for flame spread and smoke or the requirements of the federal standard for textile floor coverings.

All of the aforementioned building codes contain similar structural fire protection measures such as regulations on escape routes, fire resistance of building components, precautions against the spread of fire and smoke, the fire performance of combustible building materials and finished parts, fire alarms, extinguisher systems, and the classification of buildings into different categories of fire hazard (public buildings, cinemas, hospitals, etc.).

10.2.3 Classification and Fire Testing of Building Materials and Components

A good reference for plastics flammability including information on flammability in building codes can be found in [8]. Another useful reference for finding fire tests and standards is the document ASTM D 3814-99, "Standard Guide for Locating Combustion Test Methods for Polymeric Materials." This is a comprehensive listing of fire tests for many areas of plastics uses.

In the United States, in addition to fire tests on building materials and components in general, there are requirements relating to specific materials such as plastics, outlined in the IBC Chapter 26. Other chapters in the IBC, which reference plastics, include 8 for Interior Finishes, 14 for Exterior Walls, and 15 for Roof Assemblies and Rooftop Structures.

As a result of the numerous building regulations, there is no unified system of testing and classification of plastics. In practice, however, most fire test methods have been issued by ASTM. Many of these standard test methods were developed by other organizations, for example, NFPA or UL [9]. Factory Mutual Research Corporation (FMRC) [10] is another organization that has issued frequently used flammability standards for the construction industry.

As in other countries, fire testing and classification are based on certain parameters such as noncombustibility, fire resistance, ignition temperature, flame spread, smoke development, and so forth. The following are test methods referenced in one of the building codes and arranged by general category; however, one test actually may be referenced in more than one area in the building codes. There is duplication of many of these test methods; for example, ASTM E 84 has equivalent tests within the NFPA and UL organizations. The tests may be identical or somewhat different in form but are the same functionally. In these cases, the similar tests are not separately listed but most of the equivalencies are noted. There are also testing standards called out by the Uniform Building Code, for example, UBC 8-1 (similar to ASTM E 84) but are not referenced here as they will no longer be updated in the International Building Code. The date of the most current revision is noted for each test, for ASTM it is the last two digits in the test number. Also, a number of standards are listed more than one time depending on the number of potential use areas.

10.2.3.1 Test Methods Referenced in US Building Codes

Ignition Behavior

ASTM D 1929-96, Determining Ignition Temperature of Plastics

NFPA 268 (1996), Determining Ignitability of Exterior Wall Assemblies Using a Radiant Heat Energy Source

NFPA 259 (1998), Potential Heat of Building Materials

Flame Spread and Smoke Development

ASTM D 2843-99, Density of Smoke from the Burning or Decomposition of Plastics

ASTM E 84-00a [NFPA 255, UL 723], Surface Burning Characteristics of Building Materials

ASTM E 662-97 [NFPA 258], Specific Optical Density of Smoke Generated by Solid Materials

Insulation

ASTM E 84-00a [NFPA 255, UL 723], Surface Burning Characteristics of Building Materials

ASTM E 970-98, Critical Radiant Flux of Exposed Attic Floor Insulation Using a Radiant Heat Energy Source

Walls

ASTM E 84-00a [NFPA 255, UL 723], Surface Burning Characteristics of Building Materials

ASTM E 119-00 [NFPA 251, UL 263], Fire Tests of Building Construction and Materials

FM-4880 (1994) [UL 1040], Building Corner Fire Test Procedure

NFPA 268 (1996), Determining Ignitability of Exterior Wall Assemblies Using a Radiant Heat Energy Source

NFPA 285 (1998), Evaluation of Flammability Characteristics of Exterior Non-Loadbearing Wall Assemblies Containing Combustible Components Using the Intermediate Scale, Multi-Story Test Apparatus

Roofs

ASTM E 108-00 [NFPA 256, UL790], Fire Tests of Roof Coverings

FM-4450 (1989) [UL 1258], Approval Standard for Class I Insulated Steel Roof Decks

FM-4880 (1994) [UL 1040], Building Corner Fire Test Procedure

UL 1256 (2000), Fire Test of Roof Deck Constructions

Doors

ANSI/DASMA 107-1997 [ref. 11], Room Fire Test Standard for Garage Doors Using Foam Plastic Insulation

NFPA 252 (1999) [UL 10B or UL 10C], Fire Tests of Door Assemblies

Floor/Wall Coverings

ASTM D 2859-96 [Carpet and Rug, 16 CFR, Part 1630.4, 1631.4], Ignition Characteristics of Finished Textile Floor Covering Materials

ASTM E 648-99 [NFPA 253], Critical Radiant Flux of Floor Covering Systems Using a Radiant Heat Energy Source

NFPA 265 (1995) [UL 1715], Fire Tests for Evaluation of Room Fire Growth Contribution of Textile Wall Coverings

NFPA 701 (1999), Fire Tests for Flame Resistant Textiles and Films

Light Transmitting/Decorative

ASTM D 635-98 [UL 94], Rate of Burning and/or Extent and Time of Burning of Plastics in a Horizontal Position

ASTM D 1929-96, Determining Ignition Temperature of Plastics

ASTM D 2843-99, Density of Smoke from the Burning or Decomposition of Plastics

UL 1975 (1996), Fire Tests for Foamed Plastics Used for Decorative Purposes

Special Approval

Other testing standards, published or unpublished, may be applied in the Special Approval sections of the building codes. These are examples of frequently called out or used standards:

FM-4880 (1994) [UL 1040], Building Corner Fire Test Procedure

UL 1715 (1997) [NFPA 265], Fire Test of Interior Finish Material (Room Test)

10.2.3.2 Additional Fire Test Standards

In addition to the flammability tests listing in the preceding that are called out in one of the building codes, there are other tests and standards that may be of use in characterizing the flammability performance of plastics used in buildings, listed in the following.

ASTM D 240-97 [UL 44], Heat of Combustion

ASTM D 2863-97, Measuring the Minimum Oxygen Concentration to Support Candle Like Combustion of Plastics (Oxygen Index)

ASTM D 3014-99, Flame Height, Time of Burning, and Loss of Mass of Rigid Thermoset Plastics in a Vertical Position

ASTM E 136-99 [NFPA 251], Behavior of Materials in a Vertical Tube Furnace at 750°C

ASTM E 162-98, Surface Flammability of Materials Using a Radiant Heat Energy Source

ASTM E 662-97 [NFPA 258], Specific Optical Density of Smoke Generated by Solid Materials

ASTM E 1321-97a, Determining Material Ignition and Flame Spread Properties

ASTM E 1354-99, Heat and Visible Smoke Release Rates for Materials and Products Using an Oxygen Consumption Calorimeter

ASTM E 1623-99, Fire and Thermal Properties of Materials, Products and Systems Using an Intermediate Scale Calorimeter (ICAL)

ASTM E 1740-95, Determining the Heat Release Rate and Other Fire Test-Response Characteristics Using a Cone Calorimeter

ASTM E 2074-00, Fire Tests of Door Assemblies, Including Positive Pressure Testing of Side Hinged and Pivoted Swinging Door Assemblies

NFPA 257 (2000), Fire Tests of Window Assemblies

NFPA 272 (1999), Method of Test for Heat and Visible Smoke Release Rates for Materials and Products Using an Oxygen Consumption Calorimeter

NFPA 273, Determining Degrees of Combustibility of Building Materials

It is of interest that ASTM generally uses two statements to caution the user of flammability test standards about the limitations and liability of their use. These statements are:

"This standard should be used to measure and describe the response of materials, products, or assemblies to heat and flame under controlled conditions and should not be used to describe or appraise the fire hazard or fire risk of materials, products, or assemblies under actual fire conditions. However, results of this test may be used as elements of a fire hazard or fire risk assessment which takes into account all of the factors which are pertinent to an assessment of the fire hazard or fire risk of a particular end use."

"This standard does not purport to address all of the safety concerns, if any, associated with its use. It is the responsibility of the user of this standard to establish appropriate safety and health practices and determine the applicability of regulatory limitations prior to use."

10.2.3.3 ASTM D 1929-96 Determining Ignition Temperature of Plastics

The ignition properties of plastics are tested by ASTM D 1929 using a hot air furnace (see Table 10.1 and Fig. 10.1). Both the flash ignition and spontaneous ignition temperatures of plastics are determined. Tests can be useful in comparing the relative ignition characteristics of different materials. Test values represent the lowest air temperature that will cause ignition of the material under the conditions of the test. Normally, this test method is used for light transmitting plastics in the building codes. Other flame test properties are also required for these types of materials.

The test specimens can be in any form but are typically pellets or powder normally supplied for molding. Specimens can also be used as 20 mm x 20 mm (0.79 in. x 0.79 in.) squares [to weigh 3 g (0.1 oz)] or for less dense forms, 20 mm x 20 mm x 5 mm squares.

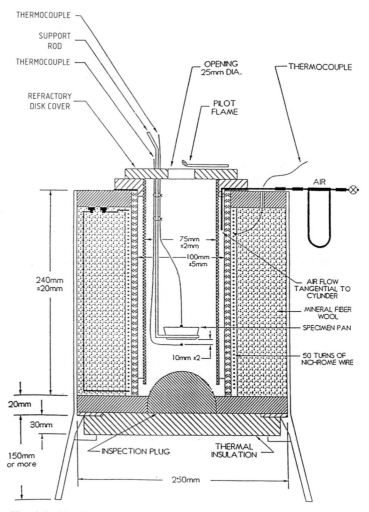

Figure 10.1   Hot air ignition furnace for plastics

Table 10.1 ASTM D 1929-96 Test specifications for flash- and spontaneous-ignition temperatures

Specimens	Two minimum, but several may be required to determine approximate oven initial set temperatures
Specimen position	Pan containing 3 g of specimen suspended in furnace
Ignition source	Electric furnace which can be heated to 750 °C
Test duration	Until no further ignition of the test specimen occurs at the minimum temperature measured, 10 min minimum period
Conclusions	Determination of the flash- and spontaneous-ignition temperatures of plastics

Generally, the procedure uses an initial 400°C oven temperature. The specimen is lowered into the furnace and ignition either is or is not observed. Ignition can be monitored by watching the temperature difference between thermocouples T_1 and T_2. On ignition, T_1 temperature will rise more rapidly than T_2. If ignition has not occurred, raise the oven temperature by 50°C and repeat.

The actual determination of the flash-ignition temperature is made at constant furnace temperature, which should be 10°C below the ignition temperature established in the preliminary tests. If ignition occurs, runs are carried out lowering the temperature each time by 10°C until ignition no longer occurs within 10 min.

The self-ignition temperature is determined as described in the preceding but without the pilot flame.

10.2.3.4 NFPA 268 (1996) Determining Ignitability of Exterior Wall Assemblies Using a Radiant Heat Energy Source

This is a method to determine the propensity of ignition of planar or nearly planar exterior wall assemblies from exposure to a gas radiant heater of 12.5 kW/m^2 and exposed to a pilot burner ignition source. The test specifications are listed in Table 10.2.

Table 10.2 NFPA 268 Test specifications for ignitability of exterior wall assemblies

Specimens	One 1.22 m × 2.44 m × use thickness of exterior wall assembly
Specimen position	Vertical
Ignition source	Propane gas radiant heater, 0.91 m × 0.91 m, 12.5 kW/m^2; pilot burner
Test duration	20 min
Conclusions	Ignition or non-ignition

A conditioned specimen is mounted securely in a trolley assembly and positioned directly in front of the propane-fired radiant panel in a vertical position and at a predetermined distance. The specimen and panel are oriented in a parallel plane configuration. The energy of the panel is measured with a heat flux meter before the test begins. Also, before the start of the test a radiation shield is placed between the sample and radiant panel. A spark igniter/pilot burner is placed on the vertical centerline of the test sample 460 mm (18.1 in.) above its horizontal centerline and about 16 mm (0.63 in.) from its surface and ignited 30 s before the start of the test. The radiation shield is removed and the time to sustained ignition is recorded. If no ignition occurs, the test is terminated at 20 min by placing the radiation shield back in position. The non-ignition or time to ignition of the specimen is recorded.

10.2.3.5 NFPA 259 (1998) Potential Heat of Building Materials

While NFPA 259 is not an ignition behavior test, it is listed here as a matter of convenience. This test method provides a means of determining the potential heat of building materials subjected to a defined, high temperature exposure condition. This method yields a property type measurement of the amount of heat that can potentially be given off by building materials when exposed to a heat source of 750°C. As determined by this test, the potential heat of a material is the difference between the gross heat of combustion per unit mass of a representative specimen and the heat of combustion per unit mass of any residue of the material to a defined heat source using combustion calorimetric techniques. Test specifications are given in Table 10.3.

Table 10.3   NFPA 259 Test specifications for potential heat of building materials

Specimens	One, for oxygen bomb calorimeter, powder, ≥ 10 g Three for electric muffle furnace, ½ in × 1/8 in × ¾ in 1.27 mm × 0.32 mm × 1.9 mm
Specimen position	On wire specimen holder
Ignition source	Oxygen bomb calorimeter Electric muffle furnace
Test duration	2 h
Conclusions	For siding backer board (IBC), < 22.7 MJ/m^2 (2000 Btu/lb)

The oxygen bomb calorimeter (see ASTM D 2015 or ASTM D 3286) measures the total possible heat given off by a specimen and the muffle furnace measures the heat given off under a realistic fire condition (750°C). The reference to "siding backer board" is given above but the test is also called out for defining some wall insulation properties.

10.2.3.6 ASTM D 2843-99 Density of Smoke from the Burning or Decomposition of Plastics

This standard covers a laboratory procedure for measuring and observing the relative amounts of smoke produced by the burning or decomposition of plastics. It is intended to be used for measuring the smoke-producing characteristics of plastics under controlled conditions of combustion or decomposition. Correlation with other fire conditions is not implied. The measurements are made in terms of the loss of light transmission through a collected volume of smoke produced under controlled, standardized conditions. The apparatus is constructed so that the flame and smoke can be observed during the test. See Table 10.4 for test specifications and Fig. 15.7, p.621.

The test specimen is exposed to a flame for the duration of the test, and the smoke is substantially trapped in the chamber in which combustion occurs. A specimen is placed on a supporting metal screen and burned in a lab test chamber under active flame conditions using a propane burner. Specimen thicknesses other than 6 mm (0.24 in.) may be tested but their size must be reported with the smoke density values. The 300 mm x 300 mm x 790 mm (11.8 in. x

11.8 in. x 31.1 in.) test chamber is instrumented with a light source, photoelectric cell, and meter to measure light absorption horizontally across the 300-mm light beam. The chamber is closed during the 4 min test period except for the 25 mm (0.98 in.) high ventilation openings around the bottom. The time until the specimen bursts into flame and the times for flame extinguishment or specimen consumption and obscuration of the EXIT sign in the test chamber are recorded. A light absorption versus time curve is plotted to calculate a smoke rating. The total smoke produced by the area under the curve is divided by the total area of the graph, 0–4 min, 0–100 % light absorption, times 100.

Table 10.4 ASTM D 2843-99 Test specifications for smoke density of burning plastics

Specimens	Three specimens, ~25 mm × 25 mm × 6 mm
Specimen position	Horizontal
Ignition source	Propane burner at 276 kPa
Test duration	Flame extinguishment or specimen consumption (or 4 min)
Conclusions	Passes when smoke rating ≤ 75 (in IBC)

10.2.3.7 ASTM E 84-00a Surface Burning Characteristics of Building Materials

ASTM E 84 is probably the most used or referenced fire test standard and the generally most important in the building codes to characterize flammability of plastics. The test fixture is known as the Steiner Tunnel and was developed by Underwriters Laboratories. The standard is used to assess the comparative surface burning behavior of building materials and is applicable to exposed surfaces such as walls and ceilings, although building codes rely on data for most cellular product insulation materials, even if behind other barriers. The test is conducted with the specimen in the ceiling position with the surface to be evaluated exposed face down to the ignition source. The material, product, or assembly must be capable of being mounted in the proper test position during the test. Thus, the specimen must either be self-supporting by its own structural quality, held in place by added supports along the test surface, or secured from the back side. The purpose of this test method is to determine the relative burning behavior of the material by observing the flame spread along the specimen. Flame spread and smoke developed indices are reported, however, there is not necessarily a relationship between the measurements. Table 10.5 gives test specifications.

Table 10.5 ASTM E 84-00a Test specifications for flame spread and smoke density

Specimens	At least one sample, 0.51 m × 7.32 m × thickness of use up to a maximum
Specimen position	Horizontal
Ignition sources	Two gas burners, 5.3 MJ/min output located 190 mm below the specimen at a distance of 305 mm from and parallel to the fire end of the test chamber
Test duration	10 min
Conclusions	Flame spread index and smoke-developed index (smoke density)

Two preliminary tests are carried out after the apparatus is calibrated. A first run is made with a standard red oak specimen to obtain numerical values of 100 for flame spread and smoke density. In the second preliminary test, an inorganic reinforced cement board specimen is tested in order to obtain zero values for these same parameters.

In the actual main test, the contribution of the material under test to smoke development and flame spread is measured and the material classified on the basis of the results. The Steiner Tunnel is illustrated in Fig. 10.2.

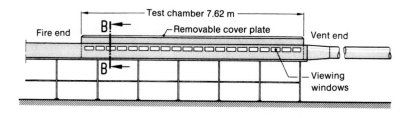

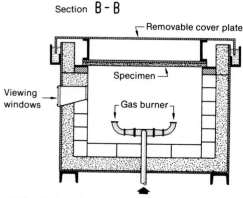

Figure 10.2   The 25-foot Steiner Tunnel

To obtain the Flame Spread Index (FSI) of the material the distance traveled by the flame front (visually observed) is plotted as a function of time. If the total resultant area A_T under the flame spread time-distance curve ≤ 97.5 (ft · min), then

$$FSI = 0.515 \cdot A_T$$

If A_T is greater than 97.5 (ft · min), then

$$FSC = 4900/195\text{-}A_T$$

The smoke density calculation is also derived from the area under a response curve. The area under the curve is divided by the area under the curve for red oak, multiplied by 100, and rounded to the nearest multiple of 5 to establish a numerical smoke-developed index. The performance of the material, as for the flame spread, is compared with that of inorganic reinforced cement board and select grade red oak flooring. For smoke-developed indices of 200 or more, the calculated value is rounded to the nearest 50 points.

In the IBC, Chapter 26, the general requirement for plastic insulation and foam plastic cores of manufactured assemblies, unless otherwise indicated, is not more than a flame spread index of 75 and a smoke-developed index of not more than 450.

Also in the IBC, there are requirements for Interior Finishes in Chapter 8. These finishes must be classified in accordance with ASTM E 84 and are grouped in the following classes according to their flame spread and smoke-developed indices:

Class	Flame spread	Smoke index
A	0–25	0–450
B	26–75	0–450
C	76–200	0–450

10.2.3.8 ASTM E 970-98 Critical Radiant Flux of Exposed Attic Floor Insulation Using a Radiant Heat Energy Source

ASTM E 970 describes a procedure for measuring the critical radiant flux of exposed attic floor insulation subjected to a flaming ignition source in a graded radiant heat energy environment in a test chamber. The specimen can be any attic floor insulation. This test method is not applicable to insulation that melts or shrinks away when exposed to the radiant heat energy environment or the pilot burner. This standard measures the critical radiant flux at the point at which the flame advances the farthest. The test specifications are listed in Table 10.6. It provides a basis for estimating one aspect of fire exposure behavior for exposed attic floor insulation. The imposed radiant flux simulates the thermal radiation levels likely to impinge on the floors of attics whose upper surfaces are heated by the sun through the roof or by flames from an incidental fire in the attic. This standard was developed to simulate an important fire exposure component of fires that may develop in attics, but is not intended for use in estimating flame spread behavior of insulation installed other than on the attic floor.

Table 10.6   ASTM E 970-98 Test specifications for exposed attic floor insulation

Specimens	Min. three specimens, mounted in tray 250 mm × 1000 mm × 50 mm
Specimen position	Horizontal
Ignition source	Air-propane radiant heater, 305 mm × 457 mm with flux to specimen of 1.0– 0.1 W/cm^2, inclined at 30° to specimen and lower edge 140 mm above specimen Pilot flame on pivot, 254 stainless steel tube with perforations
Test duration	2 min for non-ignition, other specimens until all flaming ceases
Conclusions	Critical radiant flux ≥ 0.12 W/cm^2

A horizontally mounted specimen tray is filled with loose fill, batting or board stock insulation and is exposed to a radiant energy panel 305 mm x 457 mm (12 in. x 18 in.) in dimension and inclined to a 30° angle with respect to the specimen. See Fig. 10.3 for a test rig diagram.

The specimens, especially the loose material, must be at the final use density in the test tray. The test is carried out in a test chamber about 1400 mm long x 500 mm deep x 710 mm (55 in. x 19.7 in. x 28 in.) high. The specimen is exposed to a pilot burner used to ignite the material after pre-heating. The distance to the farthest advance of the flaming is measured, converted to kW/m^2 from a prepared radiant flux profile graph, and reported as the critical radiant flux. This standard is referenced in the IBC in the Fire-Resistance-Rated Construction section as an alternative to the ASTM E 84 test.

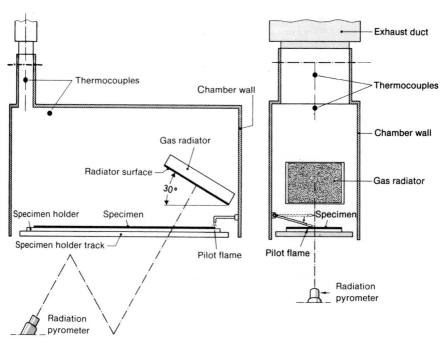

Figure 10.3   Radiant panel test rig side view

10.2.3.9 ASTM E 119-00 Fire Tests of Building Construction and Materials

This standard was first issued by ASTM in 1918 and has been used extensively to evaluate the thermal resistance of materials and assemblies. The test methods described in this standard are applicable to assemblies of masonry units and to composite assemblies of structural materials for buildings, including bearing and other walls and partitions, columns, girders, beams, slabs, and composite slab and beam assemblies for floors and roofs. They are also applicable to other assemblies and structural units that constitute permanent integral parts of a finished building. It is the intent that classifications must register comparative performance to specific fire-test conditions during the period of exposure and must not be construed as having determined suitability for use under other conditions. See specifications in Table 10.7. Test chamber configurations can vary depending on the nature of the material being tested.

These test methods prescribe a fire exposure following a standard time temperature curve (see Fig. 10.4) for comparing the test results of building construction assemblies. The results of these tests are one factor in assessing predicted fire performance of building construction and assemblies. The standard is not used with exposed plastic construction materials but is specified for thermal barriers that are used in combination with foamed plastic insulation. A typical thermal barrier will be 13 mm (0.5 in.) gypsum wallboard but equivalent performance is acceptable.

The test exposes a specimen to a standard fire-controlled period. When required, the fire exposure is followed by the application of a specified standard fire hose stream. The test provides a relative measure of the fire test response of comparable assemblies under these fire exposure conditions. The exposure is not representative of all fire conditions because conditions vary with changes in the amount, nature, and distribution of fire loading, ventilation, compartment size and configuration, and heat sink characteristics of the compartment.

Table 10.7 ASTM E 119-00 Test specifications for flammability of construction materials

Specimens	One
Specimen position	Room configuration
Ignition source	Numerous gas burners sufficient to provide the standard time/temp. curve
Test duration	Varies, but 15 min for most insulated wall assemblies
Conclusions	Average temperature rise <130 °C of unexposed side after 15 min

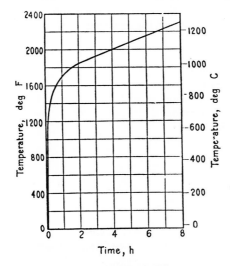

Figure 10.4   Standard time-temperature curve of ASTM E 119

10.2.3.10 FM 4880 (1994) Building Corner Fire Test Procedure

This standard sets performance requirements for insulated wall or wall and roof/ceiling panels, plastic interior finish materials, plastic exterior building panels, wall/ceiling coating systems, and interior or exterior finish systems in wall or wall and roof/ceiling constructions installed to maximum heights of 7.6 m (25 ft) or 15.2 m (50 ft) or unlimited heights when exposed to an ignition source simulating a building fire. The performance of these assemblies depends in part on the fasteners, adhesives or other accessories used in their installation and the substrate over which they are installed. It is necessary to include the components of the construction in the constructions tested. This standard is frequently cited in the Special Approvals section of building codes.

This standard actually encompasses several individual fire tests, which are needed for different types of materials seeking approval. These tests include flammability characterization by a 50 kW scale apparatus, 7.6 m high corner test [allowing a maximum building height of 9.1 m (see Table 10.8)], 15.2 m high corner test, a room test (ISO 9705), or other tests as noted in the standard. Class 1 performance in FM 4880 for roofing materials is referenced as class A in ASTM E 108-00.

Table 10.8   FM 4880 Test specifications for 7.6 m building corner test Class 1 approval

Specimens	One
Specimen position	Vertical for wall materials, horizontal for ceiling/roof assemblies
Ignition source	340 kg (750 lb) wood crib
Test duration	End of burn or when fire reaches structure limits
Conclusions	Must not support self-propagating fire reaching any structure limits

For thermoset plastic foam core panels, the bare thermoset foam may be approved without the 7.6 m corner test (see Fig. 10. 5) if the convective flame spread parameter (FSP_c) is $< 0.39 \, s^{-1/2}$. This is determined through the following:

$L_e = \Delta H_{ch} (50 - q_{cr})/Q_{ch}$ and
$FSP_c = [(\Delta H_c/ L_e) (50 - q''_{cr})]/TRP$ where
FSP_c = flame spread parameter
L_e = effective heat of gasification
ΔH_{ch} = chemical heat of combustion
q_{cr} = critical heat flux for ignition
Q_{ch} = chemical heat release rate
ΔH_c = convective heat of combustion
TRP = thermal response parameter

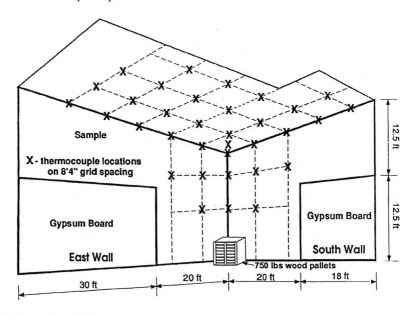

Figure 10.5   7.6 m (25 ft) Corner test structure

All other materials covered by this standard must pass the 7.6m (25 ft) room corner test burn, limiting the self-propagating flaming to less than the limits of the structure. Taller constructions will require the larger corner [15.2m (50 ft)] test be performed.

10.2.3.11 NFPA 285 (1998) Flammability Characteristics of Exterior Wall Using the Intermediate-Scale, Multistory Test Apparatus

This test method determines the flammability characteristics of insulated, exterior, non-load-bearing wall panel assemblies that are required to be of noncombustible construction. The wall panel assemblies will typically have foamed plastic as the insulation. The wall assemblies are evaluated in the intermediate-scale multistory apparatus. The primary performance characteristics to be evaluated are: (1) the capability of the panels to resist vertical flame spread within the combustible insulation core from one story to the next, (2) the capability to resist flame propagation over the exterior face of the panels, (3) the capability to resist vertical flame spread over the interior surface of the panels from one story to the next, and (4) the capability to resist lateral flame spread from the compartment of fire origin to adjacent spaces. The test specifications are listed in Table 10.9.

Table 10.9   Test specifications for exterior walls, intermediate scale multi-story apparatus

Specimens	One, 4.06 m × 5.33 m high
Specimen position	Vertical
Ignition sources	Two gas fired burners, increasing gas flow rate
Test duration	30 min
Conclusions	Flame spread and temperature increases

The test apparatus consists of a two story structure with minimum floor to top dimensions of 4.62m (15.16 ft). Each floor level has one access opening about 1.07m x 2.06m (3.5 ft x 6.76 ft) and the bottom level has a window for the burner that is 76 mm high x 1.98m (3.0 in. x 6.5 ft) long in the test assembly. The window area is the only opening in the first level during the test. Each room has inside dimensions of 3.05 m x 3.05m (10 ft x 10 ft) with a floor-to-ceiling height of 2.13m (7.0 ft). One of the burners is placed inside the first floor room and the second burner inside the window opening of the test wall assembly. Gas flow to the burners is increased in 5 min intervals to a maximum rate equivalent of 904 kW for the room burner and 398 kW for the window burner. The test wall assemblies can be built directly onto the test apparatus or they can be built onto a movable frame system that is itself fastened to the test apparatus, but the assembly must completely close the front face of the test apparatus except for the window area of the first floor.

Acceptance criteria for NFPA 285 include:

- Flame propagation must not occur either vertically or laterally beyond the area of flame impingement on the exterior wall.
- Flame propagation must not occur either vertically or laterally through the core components.
- Flame propagation must not occur laterally beyond the limits of the burn room.
- Temperatures at 25 mm (0.98 in.) from the interior surface of the test wall assembly within the second floor area must not exceed 278°C above the initial ambient temperature.
- Flames must not occur in the second floor room.

10.2.3.12 ASTM E 108-00 Fire Tests of Roof Coverings

This fire test response standard covers the measurement of the relative fire characteristics of roof coverings under simulated fire originating outside the building. It is applicable to roof coverings intended for installation on either combustible or noncombustible decks, when applied as intended for use. The following test methods are included:

- Intermittent flame exposure test,
- Spread of flame test,
- Burning brand test,
- Flying brand test,
- Rain test.

Three classes of fire test exposure are described:

- Class A Tests are applicable to roof coverings that are effective against severe test exposure, afford a high degree of fire protection to the roof deck, do not slip from position, and do not present a flying brand hazard.
- Class B Tests are applicable to roof coverings that are effective against moderate test exposure, afford a moderate degree of fire protection to the roof deck, do not slip from position, and do not present a flying brand hazard.
- Class C Tests are applicable to roof coverings that are effective against light test exposure, afford a light degree of fire protection to the roof deck, do not slip from position, and do not present a flying brand hazard.

The procedures measure the surface spread of flame and the ability of the roof covering material or system to resist fire penetration from the exterior to the underside of a roof deck under the conditions of exposure. The tests can be conducted with a gas burner or with burning brands (cribs) of pine. See Table 10.10 for specifications of a class A roof.

Table 10.10   ASTM E 108-00 Test specifications for Class A burning brand test

Specimens	Four, 1.0 m × 1.3 m
Specimen position	Adjustable angle
Ignition sources	Gas burner (369–387 kWh) or burning wood brands, 305 mm square, 57 mm thick, three layers deep, pine (about 2000 g)
Test duration	Until all evidence of flame, glow and smoke is gone
Conclusions	Flaming < 1.8 m (6 ft) for class A roof, little lateral flame spread

The test roof deck is positioned on the test structure. An air flow passes over the angled roof deck specimen and either the gas burner is applied or the appropriate burning brand is positioned at a location that is considered most vulnerable with respect to ignition of the deck but in no case closer than 100 mm from either side or 305 mm (12 in.) from the top or bottom edge of the deck. Test requirements depend and vary, of course, on the desired classification rating of the assembly. See Fig. 10.6 for a cross section of the test structure set up with a gas burner.

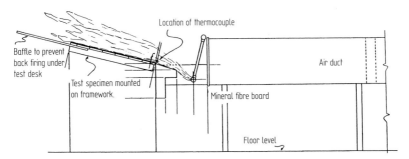

Figure 10.6   Cross section of ASTM E 108 roof deck test structure

10.2.3.13 FM 4450 (1989) Approval Standard for Class I Insulated Steel Roof Decks

While there are other test methods to characterize the flammability properties of roofing materials, FM 4450 and UL 1256 are the two predominant methods used to qualify foam plastic insulation containing roof systems in the building codes. FM 4450 is used specifically to classify insulated steel deck roof assemblies. The standard covers not only fire resistance but also wind uplift, live load resistances, corrosion of metal parts, and fatigue of plastic parts. These roof assemblies are usually made up of a steel deck, the insulation material fastened to the deck followed by weatherproof covering installed on top of the insulation. The final covering can be fastened or ballasted in position. The test specifications are given in Table 10.11. The principle of the test is to measure the fuel contribution rate under the roof assembly in a room test chamber, the Factory Mutual Construction Calorimeter, shown in Fig. 10.7. Only the flame contribution under the specimen and within the test chamber is measured and compared to the contribution from the heptane heat source and an inert specimen. For foam plastic insulation assemblies that pass this test, a thermal barrier is not required underneath the insulation in the building codes.

Table 10.11   FM 4450 Test specifications for Class I insulated steel roof decks

Specimens	One, 1.4 m × 1.5 m (1.2 m × 1.2 m exposed), plus inert control specimen
Specimen position	Horizontal
Ignition source	Three heptane burners (17 l/h), 5000 kW
Test duration	30 min
Conclusions	Average heat release rate <54.0 kW/m^2

The roof assembly test specimen is secured over the opening on top of the test chamber. The edges are sealed with mineral wool/clay cement to prevent heat loss from the calorimeter chamber. The test is run with a specified amount of heptane pumped into the burners and the heat output is measured as a time-temperature curve. This curve represents the combined fuel contribution of the specimen and heptane burners. A second burn is conducted with an inert specimen. The amount of heptane is adjusted in the second run to match the time-temperature curve of the actual specimen and from that the heat output of the specimen is determined.

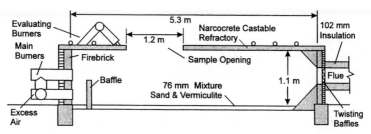

Figure 10.7   Factory Mutual construction materials calorimeter

10.2.3.14 UL 1256 (2000) Fire Test of Roof Deck Constructions

Similar to FM 4450 in evaluation of roof deck constructions for the building codes, UL 1256 is referenced by the codes as an alternate means of approving roof assemblies for use without a thermal barrier. The test approaches, however, are different. There are currently two techniques listed in the UL standard. The first technique referenced in UL 1256, and now listed as Part II, was the one based on the Steiner Tunnel as used in ASTM E 84 (see Fig. 10.2). The other, now listed as Part I, utilizes a large-scale tunnel type structure. This larger structure was originally used by Underwriters Laboratories as a tool to investigate safer roofing assemblies after a very large automotive plant fire in Livonia, Michigan in 1953, which destroyed the 14-hectare plant. A third part of UL 1256 based on an intermediate-scale room fire test for roof decks was withdrawn in March of 2000. The test specifications for Part I of the standard are given in Table 10.12.

Table 10.12   UL 1256 Test specifications for fire tests of roof deck constructions

Specimens	One, 6.1 m × 30.5 m roof deck assembly
Specimen position	Horizontal
Ignition sources	Two heptane atomization burners, heptane flow max. 0.171l/s
Test duration	30 min
Conclusions	Limited flaming propagation, limited distance of flaming droplets, limited thermal degradation of roof components

The roof assembly is fastened to the large test structure measuring 30.5m (100 ft) long, 6.1m (20 ft) wide and 3.1m (10 ft) high. The two heptane burners are on the closed end and the other end is open. Defined steel roof supports are used throughout the length of the structure. Observation windows are placed periodically down the side of the structure and thermocouples are used where called out. Air and atomized heptane are introduced. The heptane flow is increased from time zero at 0.06 l/s to 0.17 l/s at 17 min into the test and that flow is maintained until the end of the 30 min test. The heptane flow was chosen so that flames from it alone do not go past about 6m (19.7 ft) into the tunnel from the burners.

Conditions of acceptance for UL 1256 Part I are:

- The maximum sustained flame front within the structure due to under deck propagation must not exceed 18.3 m (60 ft) from the fire end of the structure during the 30 min test.
- The flaming of any molten material falling from the roof deck must not exceed 18.3 m from the fire end during the 30 min test.

- Intermittent underdeck flaming must not exceed 21.9 m (72 ft) from the fire end during the 30 min test.
- Post-examination of the roof deck must show that combustion damage has decreased at increasing distances from the fire end.
- Thermal degradation of the roof components must not have extended to the extremity of the structure.

Part I of this standard was brought into UL 1256 in the late 1990s as a result of additional testing of some foam plastic roof assemblies. Historically, for foamed plastics, it was viewed that only some thermoset plastic insulation materials, for example, isocyanurate board, could pass this type of test. However, it was found that under the test conditions outlined here, some thermoplastic insulation materials could pass if the roof deck assembly was able to vent combustion products through the steel decking toward the fire end of the structure because of heat warping of the steel deck sections. There are no acceptance criteria for above deck flaming or smoke.

Part II of the standard using the Steiner Tunnel is not discussed in this section.

10.2.3.15 ANSI/DASMA 107-1997 Room Fire Test Standard for Garage Doors Using Foam Plastic Insulation

This standard is designed to evaluate the flammability characteristics of garage doors with foam plastic insulation in the creation of fire hazard under specific fire conditions. The method also assesses the potential for fire spread beyond the room under the particular conditions simulated. The test is conducted in a standard room configuration and the mounted garage door is subjected to a specified flaming ignition source under well-ventilated conditions. The test indicates the maximum extent of fire growth in a room, the rate of heat release, smoke obscuration, flame propagation tendencies, and, if they occur, the time to flashover and the time for flame extension beyond the doorway. Table 10.13 lists the test specifications for garage doors with this method.

The effect of the fire on objects in or near the room, but remote from the ignition source, is evaluated by measurements of the total heat flux incident on the center of the floor, the upper level gas temperature in the room, and the instantaneous peak rate of heat release.

Table 10.13   ANSI/DASMA 107-1997 Test specifications for insulated garage doors

Specimens	One, 2.13 m high × 2.4 m wide foam plastic insulated door
Specimen position	Vertical, in a room configuration
Ignition source	Gas burner, diffusion through sand, 40 kW for 5 min, then 150 kW for 10 min
Test duration	15 min
Conclusions	Max. instantaneous net peak rate of heat release ≤ 250 kW, flames do not propagate the full width of the specimen and smoke $\leq 60 \text{m}^2$ 5 min after test start and $\leq 150 \text{m}^2$ 7.5 min after test start as defined in the standard

The interior room dimensions are 2.44 m x 3.66 m x 2.44 m (8.0 ft x 12.0 ft x 8.0 ft) high. There is an open doorway (0.76 m x 2.30 m) in the center of the wall opposite the garage door specimen. Air flow is provided into the test room from 0.47 m³/s initially increasing to

$3.4\,\mathrm{m^3/s}$, as required to keep the oxygen content above 14 % and to capture all the effluents from the fire room. Appropriate measurements are made on the exhaust gases.

Garage doors not requiring a fire resistance rating may not have to comply with this standard if the door facing has at least 0.8 mm (0.31 in.) aluminum, 0.01 mm (0.0004 in.) steel or 3.2 mm (0.125 in.) wood.

10.2.3.16 NFPA 252 (1999) Fire Tests of Door Assemblies

This standard outlines specific fire and hose stream test procedures for fire door assemblies that are to retard the spread of fire through door openings in fire resistant walls and need a fire protection rating. The standard evaluates the ability of a door to remain in a wall opening during a set period of time, followed by the application of a water hose stream for door integrity. The standard is referenced in the Fire-Resistance-Rated Construction section of the IBC building code and test specifications are given in Table 10.14. The most significant change in this standard in recent years has been the inclusion of a positive pressure element of the test in 1995. Before that time the neutral pressure plane of the test chamber was adjusted to the top of the door. The test chamber behind the door was under negative pressure.

Table 10.14   NFPA Test specifications for fire tests of door assemblies

Specimens	One, normal use size, can be single-, double- or multiple-door assemblies
Specimen position	Vertical
Ignition sources	Gas burners to produce a chamber standard temperature-time curve
Test duration	Varies depending on fire rating desired, typically 30 min–4 h, or to test criteria failure
Conclusions	Door remains in place in test wall and must not develop openings, no sustained flaming during first 30 min, limited flaming allowed

The door or doors are mounted with frames on the front wall of a fire test chamber. Swinging doors are mounted to swing into the furnace chamber. Mounting hardware and door/frame clearances are specified in the standard. The test is started with the ignition of the furnace. For doors not requiring a positive pressure test, the neutral pressure plane is established at the top of the door $\pm$ 25 mm (0.98 in.). For doors requiring a positive pressure test, the neutral pressure plane is established at about 1.0 m (3.28 ft) or less above the bottom of the door. Within 2 min of the end of the fire test, the fire-exposed side of the door assembly is subjected to the impact of a standard hose stream (not required for a 20 min rated door = a smoke and draft control door).

General acceptance criteria for NFPA 252:

- The door assembly must remain in the test wall.
- The door assembly must not develop any openings.
- No flaming can occur on the unexposed surface of the door assembly during the first 30 min of the test (intermittent flames < 152 mm (5.98 in.) or for longer than 10 s are allowed).
- After 30 min of the test some intermittent flames < 152 mm are allowed around the door edges if < 5 min in duration.
- For doors with a test duration > 45 min, flames < 152 mm are allowed on the unexposed surface during the last 15 min of the test if they are within 76 mm (3.0 in.) of the door top.

Some additional requirements are needed for swinging or sliding doors. Side-hinged or pivoted swinging doors require the positive pressure element of the test whereas other types of doors such as swinging elevator doors do not.

10.2.3.17 ASTM D 2859-96 [Carpet and Rug, 16 CFR, Part 1630.4, 1631.4] Ignition Characteristics of Finished Textile Floor Covering Materials

This standard provides a test method to determine the surface flammability of carpets and rugs when exposed to a standard small source of ignition under carefully prescribed draft protected conditions. It is applicable to all types of carpets and rugs used as floor covering materials regardless of their method of fabrication or whether they are made of natural or synthetic fibers or combinations of these. Test specifications are listed in Table 10.15. If the carpet specimens have been treated with flame retardant a washing procedure is called out. The specimens are conditioned, placed in the steel frame, and a methenamine tablet is placed in the center and lighted with a match. The test is ended with self-extinguishment or if burning reaches the edge of the steel frame.

Table 10.15   ASTM D 2859-96 Test specifications for the methenamine pill test

Specimens	Eight specimens, 230 mm × 230 mm
Specimen position	Horizontal, with a covering steel frame having an opening of 205 mm x 205 mm
Ignition source	Methenamine reagent tablet, 0.15 g
Test duration	Until flames extinguish or reach the edge of the steel frame
Conclusions	Passes if charred area is < 25 mm from frame for seven out of eight specimens

10.2.3.18 ASTM E 648-99 Critical Radiant Flux of Floor Covering Systems Using a Radiant Heat Energy Source

This standard describes a procedure for measuring the critical radiant flux of horizontally mounted floor-covering systems exposed to a flaming ignition source in a graded radiant heat energy environment.

The critical radiant flux at flameout is measured by the standard. It provides a basis for estimating one aspect of fire exposure behavior for floor-covering systems. The imposed radiant flux simulates the thermal radiation levels likely to impinge on the floors of a building whose upper surfaces are heated by flames or hot gases, or both, from a fully developed fire in an adjacent room or compartment. The standard was developed to simulate an important fire exposure component of fires that may develop in corridors or exit ways of buildings and is not intended for routine use in estimating flame spread behavior of floor covering in building areas other than corridors or exit ways. Test specifications are given in Table 10.16.

The test chamber is shown in Fig. 10.3 and is the same as used in ASTM E 970 for characterizing insulation on an attic floor. A specimen is mounted over underlayment, a simulated concrete structural floor, bonded to a simulated structural floor, or otherwise mounted in a representative way and may be bolted to the frame if necessary. The test is carried out by exposing the test specimen to the radiation for 5 min and then applying the pilot flame for 5 min to the narrow edge nearest the radiant panel. If the specimen has ignited, the test is continued until the flames extinguish; otherwise it is terminated after the specified time.

Table 10.16   ASTM 648-99 Test specifications for measurement of critical radiant flux of floor covering systems

Specimens	Three specimens, approx. 254 mm × 1070 mm × thickness of use
Specimen position	Horizontal
Ignition source	Air-propane radiant heater, 305 mm × 457 mm with flux to specimen of 1.0–0.1 W/cm², inclined at 30° to specimen and lower edge 140 mm above specimen Pilot flame on pivot, 254 stainless steel tube with perforations
Test duration	5 min preheat, 5 min pilot burner contact, up to 5 min longer if specimen does not propagate flame; otherwise until flames go out
Conclusions	Radiant flux measurement

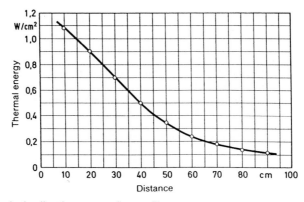

Figure 10.8   Standard radiant heat energy flux profile

The furthest advance of the flame front is measured after the test has ended using the curve shown in Fig. 10.8 obtained with a blank sample. This distance is converted into the critical radiant heat flux in W/cm² required for ignition.

10.2.3.19 NFPA 265 (1995) Fire Tests for Evaluation of Room Fire Growth Contribution of Textile Wall Coverings

For textile wall or ceiling coverings, the International Building Code calls for compliance to a Class A flame spread index according to ASTM E 84 and protecting with automatic sprinklers. In lieu of that, for textile wall coverings, the requirements of NFPA 265 may be met. The standard's method A or B may be used when tested in the manner intended for use. The only difference in running the test is that method A calls for textile specimens to be mounted only on portions of two walls of the test chamber while method B calls for specimens to be mounted on three walls of the chamber, two long walls and one short wall. See Table 10.17 for test specifications of method B.

For method B the specimens are mounted on three full walls of the chamber, which measures 2.44 m x 3.66 m x 2.44 m (8.0 ft x 12.0 ft x 8.0 ft) high and has one doorway in a short wall. The gas diffusion burner is placed in the one corner of the room. The burner is pre-set to deliver a rate of heat release of 40 kW. This level is maintained for 5 min. At that point the burner is adjusted to deliver 150 kW for 10 min.

Table 10.17 NFPA 265 Test specifications for evaluation of fire growth of textile wall coverings, Method B

Specimens	To cover three chamber walls, 2.44 m × 3.66 m (2) and 2.44 m × 2.44 m (1)
Specimen position	Horizontal
Ignition source	Propane gas diffusion burner, "Sand burner", max. capacity of 150 kW
Test duration	15 min
Conclusions	No flaming to the ceiling with 40 kW exposure with 150 kW, no flashover and flames must not spread to the outer extremities

Performance criteria for NFPA 265 Method A:
- Flames do not spread to the ceiling during the 40 kW exposure.
- During the 150 kW exposure:
  - Flames do not spread to the outer extremity of the wall samples.
  - The specimen should not burn to the outer extremity of the 0.6 m (1.97 ft) wide vertically mounted samples.
  - Burning droplets should not be formed and drop to the floor (cannot burn for >30 s).
  - Flashover must not occur.
  - The maximum instantaneous peak heat release rate must not exceed 300 kW.

Performance criteria for NFPA 265 method B:
- Flames do not spread to the ceiling during the 40 kW exposure.
- Flashover must not occur during the 150 kW exposure.

10.2.3.20 NFPA 701 (1999) Fire Tests for Flame Resistant Textiles and Films

NFPA 701 also covers textile applications but includes decorations and trim as well. In some occupancies, this standard is required for curtains, draperies, hangings, and other decorative materials suspended from walls or ceilings. These can include textiles and other plastic materials. There is an additional limitation on the amount of flame-resistant materials covering a certain percentage of wall space depending on occupancy type. There are two test methods in this standard. Test method 1 is called for with single-layer fabrics or multilayer assemblies of curtains, draperies, or window treatments of $< 700 \, g/m^2$ while test method 2 is used for materials of $> 700 \, g/m^2$ or for vinyl-coated fabric blackout linings. Table 10.18 gives the test specifications for test method 1 of this standard.

Table 10.18 NFPA 701 Test specifications for flame resistant textiles and films of test 1

Specimens	Ten, 150 mm × 400 mm
Specimen position	Vertical
Ignition source	Meeker burner with grid top channels, methane gas/air
Test duration	Until flaming ceases
Conclusions	Average weight loss < 40 % for 10 specimens Average burn time of fragments on floor ≤ 2 s

One of 10 specimens is mounted with a pin bar in the test chamber (see Fig. 10.9) and binder clips on the bottom two corners. The burner is adjusted with a methane flow of 1205 ml/min and air flow of 895 ml/min and positioned 25 mm (0.98 in.) from the bottom edge of the specimen. The burner is ignited and allowed to burn 45 s before it is moved away from the specimen and turned off. The after flame time of the specimen is measured as well as the burning time of material that may have fallen to the floor during the test.

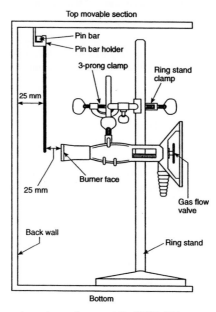

Figure 10.9   Test apparatus and specimen placement for NFPA 701

Performance criteria for NFPA 701 test method 1:

- Average weight loss of 10 specimens must be < 40 %.
- Fragments that fall to the chamber floor must not continue to burn for more than an average of 2 s.
- No individual specimen's mass loss percent can deviate more than 3 standard deviations from the mean.
- When a retest is needed, the percent mass loss of no individual specimen in the second set can deviate from the mean value by more than 3 standard deviations calculated for the second set.

10.2.3.21 ASTM D 635-98 Rate of Burning and/or Extent and Time of Burning of Plastics in a Horizontal Position

The ASTM D 635 standard is used to test rigid plastics in a horizontal position. This test method covers a lab screening procedure for comparing the relative rate of burning or extent and/or time of burning of plastics in the form of bars molded or cut from sheets, plates, or panels. The method is normally used for light transmitting plastics. The apparatus is illustrated in Fig. 10.10 and the test specifications are summarized in Table 10.19.

This test method forms the basis for classifying the burning behavior of plastics in the building codes. In the new International Building Code, a Class CC1 material exhibits a burning extent of 25 mm (0.98 in.) or less [with a 1.5 mm (0.06 in.) thickness]. Class CC2 must have a

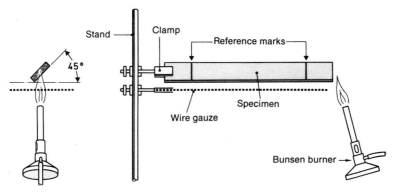

Figure 10.10 Horizontal rate of burning test fixture

Table 10.19 ASTM D 635-98 Test specifications for rate of burning of plastics in a horizontal position

Specimens	Ten specimens, 125 mm × 13 mm × 3 mm with 100 mm reference marks
Specimen position	Horizontal
Ignition source	Bunsen burner with 20 mm high blue flame
Test duration	30 s
Conclusions	Different classifications based on burning time from the 25 mm mark to the 100 mm mark

burning rate of < 1.06 mm/s (0.04 in./s) at the same thickness. The practical differentiation is that more class CC1 material may be used in a given building area than class CC2.

The procedure is that a conditioned bar specimen is supported horizontally at one end. The free end is exposed to a 20 mm (0.79 in.) blue flame for 30 s. The time and extent of burning are measured and reported if the specimen does not burn to the 100 mm (3.9 in.) mark. An average burning rate is reported for a material if it burns to the 100 mm mark.

10.2.3.22 UL 1975 (1996) Fire Tests for Foamed Plastics Used for Decorative Purposes

Certain foamed plastics used for decorative purposes or as part of interior signs must comply with the requirements of UL 1975 although the use of the test method is limited in the building codes. The purpose of the standard is to determine the extent to which these materials resist rapid heat release when subjected to a flaming ignition source. The test method can be applied to construction of open-ceiling, portable exhibit booths, individual decorative objects such as murals or signs, or to some theater, motion picture, or television stage settings. The test specifications for an exhibit booth configuration are given in Table 10.20.

In the case of an exhibit booth construction test, an open room is made up of a 2.4 m x 2.4 m (7.9 ft x 7.9 ft) rear wall and two 1.8 m x 2.4 m (5.9 ft x 7.9 ft) side walls. There is no front wall or ceiling in the test structure. The wood crib is placed in a back corner of the structure, 25 mm (0.98 in.) from the back and side wall. The wood crib and excelsior are ignited with a match and the test samples are allowed to burn until no combustion is observed. The heat release rate is calculated from an oxygen depletion calorimeter. The maximum heat release rate under this test, for example, for interior foamed plastic signs, is 150 kW.

Table 10.20   UL 1975 Test specifications for decorative foamed plastics, exhibit booth

Specimens	One set, 2.4 m × 2.4 m (1) and 1.8 m × 2.4 m (2) wall sections, each 50.8 mm thick
Specimen position	Vertical
Ignition source	Wood crib made of pine sticks, 340 g with 20 g wood excelsior
Test duration	Until all flaming ceases
Conclusion	Heat release rate

10.2.4 Official Approval

In almost all cases in the United States, products and materials manufactured for use in the built environment must meet the provisions of one of the model building codes. However, there are cases from time to time in which a manufacturer has a new product or an existing product for a new application for which the established building code requirements are not met. Each of the three model building codes has set up an Evaluation Service (ES) to handle special requests from manufacturers. Let's look at the ICBO Evaluation Service as an example. The ES is made up of an Evaluation Committee and staff. The Evaluation Committee is made up of nine voting members, each of whom are themselves building code officials. The ES staff is made up of knowledgeable people in various building areas of specialization. The ES issues approved reports or Acceptance Criteria (AC) for different applications, for example, Foam Plastic Insulation – AC12.

The manufacturer of a new product or material would present its case to the ES The Evaluation Committee reviews the request and rationale and may give approval to consider the request. A review is made with the ES staff to determine if an existing AC would apply to the new material or application. The objective of the ES is to see that a new application meets the intent of the building code, but through alternate tests not specifically called out in the code. If no existing AC report can be applied, an attempt will be made to alter an existing AC. In that case, the manufacturer may have to present testing data that supports the safe use of their material. In the case where the application does not fit an existing AC at all, the manufacturer must present compelling evidence that the new application, from a fire perspective, is safe, normally through testing data. New requests are presented in a public forum before a final AC is issued.

Testing data presented to the ICBO Evaluation Service must come from an accredited laboratory. As of 2002, the ES accepts only data generated from ISO 17025 accredited laboratories. The intent of three model codes is to merge into the new International Code Council by about 2002 and the three Evaluation Services may also merge into one. The final makeup and operation of a new organization have not yet been established.

10.2.5 Future Developments

10.2.5.1 NFPA 5000 Building Code

In March of 2000, the Board of Directors of NFPA approved a plan to develop a consensus based building code as part of a full set of consensus codes and standards. There seem to be two main driving forces for this additional national building code. First, NFPA has said they want to develop the best scientifically based code that is possible. Second, they are looking to develop and implement standards and a building code that utilize the NFPA full and open consensus process. Until now in the IBC only code enforcers have been allowed to vote on changes to the building code. This code, which is currently in draft form, is to be a compilation of NFPA 101 Life Safety Code and the EPCOT building code. EPCOT is part of the Disney World complex in the vicinity of Orlando, Florida. The EPCOT building code is comprehensive and has been in existence for more than 30 years. This area's structural fire loss rate is one of the lowest of any community of its size in the world. The code has been known to be adaptable to innovative construction techniques with the many unconventional building structures found in the Disney World complex. The first draft of the NFPA 5000 building code is now completed and is scheduled for issue in late 2002. It is unclear at this point whether the IBC or NFPA national (they reference themselves as international) building code will be the prevalent one adopted by most US communities in the future.

10.2.5.2 Performance-Based Building Codes

Whereas some other countries in the world have mandated performance, design objective based building codes, in the United States, this approach has been slow to arrive. There are specific areas in which a performance rather than prescriptive approach can be taken but most US construction is still directed through prescriptive language. Performance-based construction can be handled through the current "Special Approval" sections of the three older model building codes. Prescriptive code enforcement is certainly easier for the code official, but as new construction techniques or innovative building styles are found, performance-based code language may be the preferred approach. Current fire and structural models may have to be updated to adequately assess performance based construction. Now, within the draft NFPA 5000 building code, there is a chapter on performance-based code approaches that may be used in construction. This is certainly a step in the right direction in allowing methods as flexible as possible in modern construction.

The International Code Council has also addressed performance codes. In August of 2000, the ICC completed the final draft of its own Performance Code for Buildings and Facilities. This code is not to completely replace its prescriptive codes, however. The performance code provides procedures to help address design and review questions or concerns associated with new or alternate building materials and techniques of construction. In the end, performance-based codes may allow for increased opportunities and competitiveness in the global market.

References for Section 10.2

[1] Uniform Building Code, 1997 Edition, International Conference of Building Officials (ICBO). 5360 Workman Mill Road, Whittier, California 90601-2298. www.icbo.org

[2] The BOCA National Building Code, 1998 Edition, Building Officials & Code Administrators International, Inc. 4051 West Flossmoor Road, Country Club Hills, Illinois 60477-5795. www.bocai.org

[3] Standard Building Code, 1999 Edition, Southern Building Code Congress International (SBCCI), Inc. 900 Montclair Road, Birmingham, Alabama 35213-1206. www.sbcci.org

[4] ASTM, 100 Barr Harbor Drive, West Conshohocken, PA 19428-2959. www.astm.org

[5] NFPA 101 Life Safety Code, National Fire Protection Association, 1 Batterymarch Park, Quincy, MA 02269-9101 www.nfpa.org

[6] U.S. CPSC, 4330 East-West Highway, Bethesda, MD 20814-4408. www.cpsc.gov

[7] International Code Council. 5203 Leesburg Pike, Suite 600, Falls Church, VA 22041. www.intl-code.org

[8] C.J. Hilado: Flammability Handbook for Plastics, 5th edit., Technomic, Lancaster, PA, 1998

[9] Underwriters Laboratories Inc., 333 Pfingsten Road, Northbrook, IL 60062-2096. www.ul.com

[10] Factory Mutual Research Corporation, 1151 Boston-Providence Turnpike, P.O. Box 9102, Norwood, MA 02062 www.factorymutual.com

[11] American National Standards Institute, 11 West 42nd Street, New York, NY 10036. www.ansi.org

10.3 Canada

J. MEHAFFEY

10.3.1 Statutory Regulations

The regulation of buildings in Canada is the responsibility of provincial and territorial govern-ments. Each of the 10 Provinces and 3 Territories enacts its own building code. All of the provincial and territorial building codes, however, are based on a single model code, the National Building Code of Canada (NBC) [1]. Although the NBC is intended to establish a minimum standard of fire safety for the construction of new buildings or the renovation of existing buildings, several provinces make amendments to the NBC that render their codes somewhat more demanding.

10.3.2 Classification and Testing of the Fire Performance of Building Materials and Components

For housing and small buildings, the fire safety requirements for building materials, products, and assemblies are contained in Part 9 of the NBC. For all other buildings, the fire safety requirements are contained in Part 3, "Fire Protection, Occupant Safety and Accessibility." The explicit NBC fire safety requirements depend on the intended use and the size of a building. Separate sections address fire resistance ratings required of building assemblies and the fire performance requirements of combustible building products such as interior finish.

The fire test methods cited in the NBC were all developed by Underwriters Laboratories of Canada (ULC) and are listed in Part 2 of the NBC.

The following standards relate to the fire resistance of building assemblies [2–10]:

CAN/ULC-S101-M89   Standard Methods of Fire Endurance Tests of Building Construction and Materials

CAN4-S104-M80   Standard Method for Fire Tests of Door Assemblies

CAN4-S105-M85   Standard Specification for Fire Door Frames Meeting the Performance Required by CAN4-S104

CAN4-S106-M80   Standard Method for Fire Tests of Window and Glass Block Assem-blies

CAN/ULC-S107-M87   Standard Methods of Fire Tests of Roof Coverings

CAN/ULC-S110-M86   Standard Methods of Test for Air Ducts

ULC-S111–95   Standard Method of Fire Tests for Air Filter Units

CAN/ULC-S112-M90   Standard Method of Fire Test of Fire-Damper Assemblies

CAN4-S113–79   Standard Specification for Wood Core Doors Meeting the Perform-ance Required by CAN4-S104–77 for Twenty Minute Fire Rated Closure Assemblies

Fire resistance test methods are not dealt with further in this chapter.

Whether a building is classified as combustible or noncombustible is determined by subjecting it to CAN4-S114 [11], which differs somewhat from ISO 1182.

Several methods for testing the fire performance of combustible materials are prescribed in the NBC. The most important method used for testing interior finishes is CAN/ULC-S102 [12], which is almost identical to ASTM E 84. Two further methods for evaluating floor coverings, CAN/ULC-S102.2 [13], and light diffusers, ULC-S102.3 [14], are derived from CAN/ULC-

S102. They have different mounting systems (floor or ceiling, respectively). An additional test for nonmelting building materials, ULC-S127 [15], is used if other tests give ambiguous results. In certain cases, the NBC requires fabrics and films to be flame tested to ULC-S109 [16].

All of the standards mentioned in the preceding [2–16] can be obtained from Underwriters Laboratories of Canada, 7 Crouse Road, Scarborough, Ontario, Canada M1R 3A9.

The test methods listed in the preceding are discussed in detail in the paragraphs that follow, and, where appropriate, the classifications employed in the NBC are given.

10.3.2.1 CAN4-S114. Determination of the Noncombustibility of Building Materials

The noncombustibility of building materials is determined employing CAN4-S114 [11]. The Canadian method differs somewhat from ASTM E 136 (see Section 10.2), so the apparatus is shown in Fig. 10.11 and the test specifications summarized in Table 10.21.

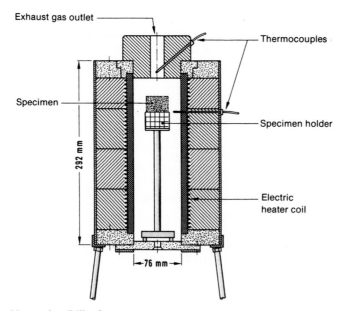

Figure 10.11   Noncombustibility furnace

Table 10.21   Test specifications for determining the noncombustibility of building materials to CAN4-S114

Specimens	Three specimens 38 mm × 38 mm × 51 mm
Specimen position	vertical
Ignition source	electric furnace heated to 750 °C
Test duration	15 min (max.)
Conclusions	passed if • the three specimens do not contribute on average more than 36 °C to the temperature rise in the exhaust gas outlet, • no specimen burns 30 s after the start of the test, • no specimen exceeds a maximum weight loss of 20 %.

10.3.2.2 CAN/ULC-S102. Test for Surface Burning Characteristics of Building Materials

The surface burning characteristics of building materials are tested by CAN/ULC-S102 [12], which is similar to ASTM E 84. The apparatus and test specifications correspond to those given in Fig. 10.2 and Table 10.5, respectively, in Section 10.2 USA. The experimental results, however, are evaluated differently. The results from at least three runs are used to determine the Flame Spread Classification as FSC_1 or FSC_2. The distance traveled by the flame front is plotted graphically as a function of time and the total area A_T under the flame spread time-distance curve is calculated. The Flame Spread Classification FSC_1 is calculated in terms of A_T as follows:

if $A_T \leq 29.7$ (m · min), then $FSC_1 = 1.85 \cdot A_T$
if $A_T > 29.7$ (m · min), then $FSC_1 = 1640 / (59.4 - A_T)$

During the testing of some materials, particularly those of low thermal inertia, the flame front may advance rapidly during the initial stages of the test and subsequently slow down or even fail to reach the end of the specimen. In such cases the Flame Spread Classification FSC_2 is calculated according to the following equation:

$FSC_2 = 92.5 \, d/t$

where t is the time in minutes required by the flame front to travel d meters where a marked slowdown of the flame front occurs.

If the flame spread behavior is such that it is difficult to ascertain the point at which the speed of the flame front starts to decrease, then FSC_2 is determined by consideration of test results obtained according to ULC-S127 (see later).

The Flame Spread Classification is the greater of FSC_1 or FSC_2.

Smoke Developed Classifications are also determined as in ASTM E 84.

10.3.2.3 CAN/ULC-S102.2. Test for Surface Burning Characteristics of Floor Coverings and Miscellaneous Materials

The surface burning characteristics of floorings are tested according to CAN/ULC-S102.2 [13]. The method is also used for materials that cannot be tested when affixed to the ceiling such as thermoplastic and loose fill materials.

The test rig is identical to that for CAN/ULC-S102, but the test specimen lies on the floor and the burners are directed downwards at an angle of 45° to the specimen. Diagrams are shown in Fig. 10.12.

The test specifications are similar to those described in CAN/ULC-S102. The specimen, however, is placed on the floor of the tunnel and is therefore narrower with dimensions of 0.44 m (1.44 ft) x 7.32 m (24 ft) x usual thickness.

The values obtained from tests are used to compute the Flame Spread Classification and Smoke Developed Classifications.

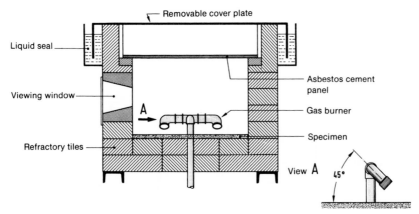

Figure 10.12 Modified tunnel for testing floor coverings and miscellaneous materials

10.3.2.4 ULC-S102.3. Fire Test for Light Diffusers and Lenses

The tendency of light diffusers and lenses to become loose and fall out of their holders when exposed to fire is assessed by ULC-S102.3 [14] using the tunnel described in CAN/ULC-S102 and ASTM E 84.

The test procedure is similar to CAN/ULC-S102 except that two tests are carried out on specimens half the length [3.66 m (12 ft)]. In one case, the specimen is simply placed on the ledges of the furnace and in the second it is clamped in position. Note is made of whether and when the specimens fall from position and whether or not ignition occurred before the material fell.

10.3.2.5 ULC-S127. Corner Wall Test for Flammability Characteristics of Nonmelting Building Materials

The ULC-S127 corner wall test method [15] is used in addition to the CAN/ULC-S102 and CAN/ULC-S102.2 to calculate FSC_2 when these two methods do not give clear data on the advance of the flame front. The test rig is illustrated in Figs. 10.13 and 10.14 and the test specifications are summarized in Table 10.22.

Table 10.22 ULC-S127 corner wall test specifications

Specimens	Test material lines ceiling, two walls, and canopy of fire test chamber.
Fire chamber	Cube-shaped with internal dimension of 1.3 m, canopy measures 0.52 m from top to bottom; 47.5 cm × 15 cm cutout in front canopy
Ignition source	Cylindrical tray, diameter 22 cm, depth 14 cm, filled with sand and ceramic beads; in lower part of bead layer, natural gas with a heat output of 2 MJ/min flows through a copper tube with 48 holes; gas is ignited by a small burner located above sand bed; the tray is located on the floor in the corner formed by the two walls
Test duration	Until flames issue beneath canopy or cotton thread stretched 2.54 cm beneath it breaks; max. 5 min after lighting ignition source
Conclusions	Measure time t until flames issue beneath canopy or cotton thread breaks; determine Flame Spread Classification FSC_2.

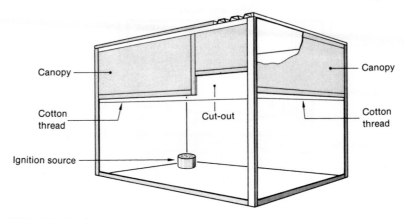

Figure 10.13   Fire chamber

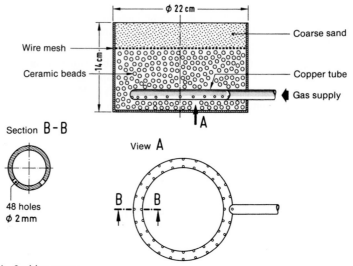

Figure 10.14   Ignition source

The time t at which flames appear from beneath the canopy or the cotton thread breaks is recorded and is converted to the Flame Spread Classification FSC_2 with the aid of the following equation:

$$FSC_2 = 51.47 \, t^{1.215}$$

10.3.2.6 NBC Requirements for Interior Finish Materials

Subsection 3.1.12 of the NBC requires that the flame-spread rating (Flame Spread Classification) and smoke-developed classification of an interior finish material be assessed by CAN/ULC-S102 or CAN/ULC-S102.2.

Requirements for the flame-spread ratings and smoke-developed classifications of the interior finish of walls, ceilings and floors in buildings are specified in Subsection 3.1.13 of the NBC. The requirements are expressed in terms of the implicit classes defined in Table 10.23.

Subsection 3.1.13 also includes requirements for the performance of light diffusers and lenses as assessed by ULC-S102.3.

Flame-spread ratings and smoke developed classifications of some common interior finish materials are presented in Section D-3 of Appendix D of the NBC.

Table 10.23   Implicit grades of flame spread and smoke development

Flame-spread rating	Range	Smoke-developed classification	Range
25	0–25	50	0–50
75	26–75	100	51–100
150	76–150	300	101–300
X[a]	>150	X[a]	>300
X[b]	>300	X[b]	>500

[a] For walls and ceilings
[b] For floors

10.3.2.7 ULC- S109. Flame Tests of Flame Resistant Fabrics and Films

The ULC-S109 [16] method for testing of flame-resistant fabrics and films (equivalent to US method NFPA 701) involves two tests with ignition sources of differing intensity (small and large flame). In both cases the flame is applied vertically to the specimen with a Bunsen burner.

The apparatus for the small-flame test is shown in Fig. 10.15. The test specifications are summarized in Table 10.24.

Table 10.24   Small flame test specifications

Specimens	Ten specimens: 70 mm × 254 mm × usual thickness; 5 specimens in direction of warp and 5 in direction of weft; specimens are clamped in a metal frame so that an area of 50 mm × 254 mm is tested
Specimen position	Vertical in combustion chamber 305 mm × 305 mm × 750 mm; lower end of specimen 19 mm above tip of burner
Ignition source	Bunsen burner with luminous 38 mm long flame with air supply shut off; burner located vertically beneath specimen or at an angle of 25° from vertical if dripping expected
Test duration	Flame applied for 12 s; afterflame time and charring noted
Conclusions	Passed if • Afterflame time ≤ 2 s, • No burning drops or material falling, • Length of char due to vertical flame spread and afterglow does not exceed specified limits (see below).

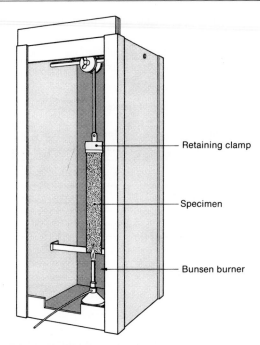

Retaining clamp

Specimen

Bunsen burner

Figure 10.15   Flammability test for fabrics and films

Charring of the specimen caused by vertical flame spread and afterglow must not exceed the values given in Table 10.25. The length of char is determined by tearing.

Table 10.25   Classification of fabrics and films by their extent of destruction

Weight of material $(g \cdot m^{-2})$	Mean tear length (10 specimens) (mm)	Maximum tear length (individual specimens) (mm)
> 340	90	115
≤ 340–> 200	115	140
≤ 200	140	165

The large flame test is similar to the small-flame test but employs flat and folded specimens, up to 2 m (6.5 ft) long. Normally, however, flat specimens need not be longer than 760 mm (30 in.) and folded specimens not longer than 1065 mm (42 in.) in order to establish compliance with the test requirements. The maximum lengths are necessary when it is desirable to ascertain the extent of destruction following a failed test. The specifications for the large flame test are given in Table 10.26.

Table 10.26   Large flame test specifications

Specimens	• Flat: 10 specimens 127 mm × 762 mm; up to 2134 mm if necessary • Folded: 4 specimens 635 mm × 762 mm; up to 2134 mm if necessary; each specimen is folded lengthwise four times so that folds are approx. 127 mm apart
Test stack	305 mm × 305 mm × 2134 mm on 305 mm high supports, open at top and bottom
Specimen position	Vertical, bottom of specimen 100 mm above tip of burner; folded specimens set up with folds approx. 127 mm apart
Ignition source	Bunsen burner with 280 mm long oxidizing flame; burner vertical under specimen or at an angle of 25° to vertical if dripping expected
Test duration	Flame applied for 2 min; afterflame time and charring noted
Conclusions	Passed if • Afterflame time ≤ 2 s, • No burning drops or material falling, • Length of char due to vertical flame spread and afterglow does not exceed 254 mm.

10.3.3 Official Approval

The following Canadian laboratories are accredited by the Standards Council Canada to carry out fire tests on building materials and components:

Underwriters Laboratories of Canada (ULC), Crouse Road, Scarborough, Ontario, Canada M1R 3A9

Bodycote Ortech Inc., 2395 Speakman Drive, Mississauga, Ontario, Canada L5K 1B3

Intertek Testing Services NA Ltd. (ITS), 211 Schoolhouse Street, Coquitlam, B.C., Canada V3K 4X9

ULC and ITS provide testing and certification services related to the fire performance of building products. Products that have been certified by ULC are identified by the ULC Mark (which consists of ULC in a circle) or by a ULC Certificate. Products certified by ITS are identified by the Warnock Hersey Listed Mark. Both ULC and ITS publish directories of listed products.

10.3.4 Future Developments

The NBC is normally published on a 5-year cycle; however, the last edition was published in 1995 and the next will not be published until 2003. This delay is a result of the fact that the NBC is being completely recast into an objective-based format. In the new format, the fundamental "objectives" of the NBC will be stated up front. Based on these objectives, a number of more specific, but qualitative "functional requirements" will be identified. The existing (1995) requirements will be considered "acceptable solutions"; that is, one way to meet the objectives and functional requirements. Where possible "quantitative performance criteria" will be provided to foster the evaluation of alternatives to the acceptable solutions. Future revisions of the NBC will increasingly encourage performance-based design.

In 1992, a new Canadian test method was approved, CAN/ULC-S135 [17], to determine the degrees of combustibility of building materials. A new test method was deemed necessary to replace the existing test CAN4-S114 [11] that determines whether a material is combustible or noncombustible. In CAN/ULC-S135, the heat release rate of materials exposed to specified conditions in the cone calorimeter (ASTM E1354 and ISO 5660) is used to define degrees of combustibility of building materials. A Task Group is preparing new sections of the NBC that would cite CAN/ULC-S135 rather than CAN4-S114. It is possible that the new test method will be cited in the 2003 edition of the NBC.

References for Section 10.3

[1] National Building Code of Canada 1995. Eleventh Edition. Canadian Commission on Building and Fire Codes, National Research Council Canada, Ottawa, NRCC 38 726

[2] CAN/ULC-S101-M89. Standard Methods of Fire Endurance Tests of Building Construction and Materials

[3] CAN4-S104-M80. Standard Method for Fire Tests of Door Assemblies

[4] CAN4-S105-M85. Standard Specification for Fire Door Frames Meeting the Performance Required by CAN4-S104

[5] CAN4-S106-M80. Standard Method for Fire Tests of Window and Glass Block Assemblies

[6] CAN/ULC-S107-M87. Standard Methods of Fire Tests of Roof Coverings

[7] CAN/ULC-S110-M86. Standard Methods of Test for Air Ducts

[8] ULC-S111-95. Standard Method of Fire Tests for Air Filter Units

[9] CAN/ULC-S112-M90. Standard Method of Fire Test of Fire-Damper Assemblies

[10] CAN4-S113-79. Standard Specification for Wood Core Doors Meeting the Performance Required by CAN4-S104-77 for Twenty Minute Fire Rated Closure Assemblies

[11] CAN4-S114-M80. Standard Method of Test for Determination of Non-Combustibility in Building Materials

[12] CAN/ULC-S102-M88. Standard Method of Test for Surface Burning Characteristics of Building Materials and Assemblies

[13] CAN/ULC-S102.2-M88. Standard Method of Test for Surface Burning Characteristics of Flooring, Floor Covering, and Miscellaneous Materials and Assemblies

[14] ULC-S102.3-M1982. Standard Method of Fire Test of Light Diffusers and Lenses

[15] ULC-S127-1988. Standard Corner Wall Method of Test for Flammability Characteristics of Non-Melting Building Materials

[16] ULC-S109-M87. Standard for Flame Tests of Flame-Resistant Fabrics and Films

[17] CAN/ULC-S135-1992. Standard Method of Test for Determination of Degrees of Combustibility of Building Materials Using an Oxygen Consumption Calorimeter (Cone Calorimeter)

10.4 European Union

B. SUNDSTRÖM

10.4.1 The Construction Products Directive

The European Commission published the construction products directive (89/106/EEG) in 1989. A major reason was to promote free trade of construction products. The directive contains six essential requirements that apply to the building itself. One of the requirements is safety in case of fire. Therefore, construction products must have a fire classification based on the same standards throughout Europe. A member state that regulates for a certain safety level will be able to identify the fire properties of a construction product corresponding to that level. Products complying with the essential requirements of the directive are labeled with the CE Mark.

The function of the construction products directive relies on a number of specifications. In the fire area, a definition of European fire classes, harmonized test standards, and rules for attestation of conformity are such important specifications. The European fire classes and the rules for attestation of conformity are published by the European Commission. The reaction-to-fire standards are published by CEN.

10.4.2 The Role of Regulators, Notified Bodies, and Standardization Bodies

The European standards for reaction-to-fire are published by the European Committee for Standardization (CEN). CEN works out the test standards based on specifications given in a mandate from the European Commission. An expert group set up by the European Commission, the Fire Regulators Group (FRG), worked out the basis for the European system of classification and the request for test methods. The representation in FRG is one regulator and one technical expert per country. In addition, liaison is maintained with industry and CEN. In particular, the work of defining the Euroclasses was performed by FRG.

Once the system is working, there is a continuous need for quality assurance work. This can include interpretation of test procedures, extended application of test data, technical cooperation between test laboratories, agreements of practice between certification bodies, and so forth. A major role for keeping the system in work is expected to be held by the Fire Sector Group. This group consists of the notified bodies for testing and certification throughout Europe. A notified body is a body, for example, a test laboratory or a certification organization, that a member state has indicated to the European Commission is suitable for performing testing/certification under the European system.

Technical work such as development of good technical practice in testing is expected to rely heavily on EGOLF, the European Group of Official Fire Laboratories, and European industrial organizations.

10.4.3 The Euroclasses

The European Commission published the Euroclasses on February 8, 2000. Reaction-to-fire testing will be done following a new concept compared to existing procedures in Europe. Seven main classes are introduced, the Euroclasses. These are A1, A2, B, C, D, E, and F. A1 and A2 represent different degrees of limited combustibility. For linings, Euroclasses B-E represent products that may go to flashover in a room and at certain times. F means that no performance is determined. Thus, there are seven classes for linings and seven classes for floor coverings. Additional classes of smoke and any occurrence of burning droplets are also given, see Table 10.27 and 10.28.

Table 10.27 Classes of reaction-to-fire performance for construction products excluding floorings (*)

Class	Test method(s)	Classification criteria	Additional classification
A1	EN ISO 1182[a]); and	$\Delta T \leq 30\,°C$; *and* $\Delta m \leq 50\,\%$; *and* $t_f = 0$ (i.e. no sustained flaming)	–
	EN ISO 1716	$PCS \leq 2.0\ MJ.kg^{-1}$ [a)]; *and* $PCS \leq 2.0\ MJ.kg^{-1}$ [b, c)]; *and* $PCS \leq 1.4\ MJ.m^{-2}$ [d)]; *and* $PCS \leq 2.0\ MJ.kg^{-1}$ [e)]	–

Table 10.27 (Continuation)

Class	Test method(s)	Classification criteria	Additional classification
A2	EN ISO 1182[a]; or	$\Delta T \leq 50\,°C$; *and* $\Delta m \leq 50\,\%$; *and* $t_f \leq 20\,s$	–
	EN ISO 1716; and	$PCS \leq 3.0\,MJ.kg^{-1\ a)}$; *and* $PCS \leq 4.0\,MJ.m^{-2\ )b)}$; *and* $PCS \leq 4.0\,MJ.m^{-2\ d)}$; *and* $PCS \leq 3.0\,MJ.kg^{-1\ e)}$	–
	EN 13823 (SBI)	$FIGRA \leq 120\,W.s^{-1}$; *and* LFS < edge of specimen; *and* $THR_{600s} \leq 7.5\,MJ$	Smoke production[f]; *and* Flaming droplets/particle s[g]
B	EN 13823 (SBI); and	$FIGRA \leq 120\,W.s^{-1}$; *and* LFS < edge of specimen; *and* $THR_{600s} \leq 7.5\,MJ$	Smoke production); *and* Flaming droplets/particles[g]
	EN ISO 11925-2[i]: Exposure = 30 s	$Fs \leq 150\,mm$ within 60 s	
C	EN 13823 (SBI); and	$FIGRA \leq 250\,W.s^{-1}$; *and* LFS < edge of specimen; *and* $THR_{600s} \leq 15\,MJ$	Smoke production[f] ; *and* Flaming droplets/particles[g]
	EN ISO 11925-2[i]: Exposure = 30 s	$Fs \leq 150\,mm$ within 60 s	
D	EN 13823 (SBI); and	$FIGRA \leq 750\,W.s^{-1}$	Smoke production[f]; *and* Flaming droplets/particles[g]
	EN ISO 11925-2[i]: Exposure = 30 s	$Fs \leq 150\,mm$ within 60 s	
E	EN ISO 11925-2[i]: Exposure = 15 s	$Fs \leq 150\,mm$ within 20 s	Flaming droplets/particles[h]
F	No performance determined		

(*) The treatment of some families of products, e.g., linear products (pipes, ducts, cables, etc.), is still under review and may necessitate an amendment to this decision.

a) For homogeneous products and substantial components of nonhomogeneous products

b) For any external nonsubstantial component of nonhomogeneous products

c) Alternatively, any external nonsubstantial component having a PCS $\leq$ 2.0 MJ.m^{-2}, provided that the product satisfies the following criteria of EN 13823(SBI) : FIGRA $\leq$ 20 W.s^{-1}; *and* LFS < edge of specimen; *and* THR$_{600s} \leq$ 4.0 MJ; *and* s1; *and* d0

d) For any internal nonsubstantial component of nonhomogeneous products

e) For the product as a whole

f) s1 = SMOGRA $\leq$ 30 m^2.s^{-2} *and* TSP$_{600s} \leq$ 50 m^2 ; s2 = SMOGRA $\leq$ 180 m^2.s^{-2} *and* TSP$_{600s}$ $\leq$ 200 m^2; s3 = not s1 or s2.

g) d0 = No flaming droplets/ particles in EN 13823 (SBI) within 600 s; d1 = no flaming droplets/ particles persisting longer than 10 s in EN 13823 (SBI) within 600 s; d2 = not d0 or d1; ignition of the paper in EN ISO 11925-2 results in a d2 classification.

h) Pass = no ignition of the paper (no classification); Fail = ignition of the paper (d2 classification)

i) Under conditions of surface flame attack and, if appropriate to end-use application of product, edge flame attack

Table 10.28 Classes of reaction-to-fire performance for floorings

Class	Test method(s)	Classification criteria	Additional classification
$A1_{fl}$	EN ISO 1182[a)]; and	$\Delta T \leq 30\,°C$; *and* $\Delta m \leq 50\,\%$; *and* $t_f = 0$ (i.e., no sustained flaming)	–
	EN ISO 1716	$PCS \leq 2.0\ MJ.kg^{-1\ a)}$; *and* $PCS \leq 2.0\ MJ.kg^{-1\ b)}$; *and* $PCS \leq 1.4\ MJ.m^{-2\ c)}$; *and* $PCS \leq 2.0\ MJ.kg^{-1\ d}$	–
$A2_{fl}$	EN ISO 1182[a)]; or	$\Delta T \leq 50\,°C$; *and* $\Delta m \leq 50\,\%$; *and* $t_f \leq 20\,s$	–
	EN ISO 1716; and	$PCS \leq 3.0\ MJ.kg^{-1\ a)}$; *and* $PCS \leq 4.0\ MJ.m^{-2\ b)}$; *and* $PCS \leq 4.0\ MJ.m^{-2\ c)}$; *and* $PCS \leq 3.0\ MJ.kg^{-1\ d)}$	–
	EN ISO 9239-1[e)]	Critical flux[f)] $\geq 8.0\ kW.m^{-2}$	Smoke production[g]
B_{fl}	EN ISO 9239-1[e)] and	Critical flux[f)] $\geq 8.0\ kW.m^{-2}$	Smoke production[g)]
	EN ISO 11925-2[h)]: Exposure = 15 s	Fs $\leq 150\,mm$ within 20 s	
C_{fl}	EN ISO 9239-1[e)] and	Critical flux[f)] $\geq 4.5\ kW.m^{-2}$	Smoke production[)g]
	EN ISO 11925-2[h)]: Exposure = 15 s	Fs $\leq 150\,mm$ within 20 s	
D_{fl}	EN ISO 9239-1[e)] and	Critical flux[f)] $\geq 3.0\ kW.m^{-2}$	Smoke production[g)]
	EN ISO 11925-2[h)]: Exposure = 15 s	Fs $\leq 150\,mm$ within 20 s	
E_{fl}	EN ISO 11925-2[h)]: Exposure = 15 s	Fs $\leq 150\,mm$ within 20 s	–
F_{fl}	No performance determined		

a) For homogeneous products and substantial components of nonhomogeneous products
b) For any external nonsubstantial component of nonhomogeneous products
c) For any internal nonsubstantial component of nonhomogeneous products
d) For the product as a whole
e) Test duration = 30 min
f) Critical flux is defined as the radiant flux at which the flame extinguishes or the radiant flux after a test period of 30 min, whichever is the lower (i.e., the flux corresponding with the furthest extent of spread of flame)
g) s1 = Smoke $\leq 750\,\%.min$; s2 = not s1

h) Under conditions of surface flame attack and, if appropriate to the end-use application of the product, edge flame attack.

Abbreviations: The characteristics are defined with respect to the appropriate test method:

ΔT	= temperature rise;
Δm	= mass loss;
t_f	= duration of flaming;
PCS	= gross calorific potential;
FIGRA	= fire growth rate;
THR_{600s}	= total heat release;
LFS	= lateral flame spread;
SMOGRA	= smoke growth rate;
TSP_{600s}	= total smoke production;
Fs	= flame spread.

10.4.4 Definitions

Material: A single basic substance or uniformly dispersed mixture of substances, for example, metal, stone, timber, concrete, mineral wool with uniformly dispersed binder, polymers.

Homogeneous product: A product consisting of a single material, of uniform density and composition throughout the product.

Nonhomogeneous product: A product that does not satisfy the requirements of a homogeneous product. It is a product composed of one or more components, substantial and/or nonsubstantial.

Substantial component: A material that constitutes a significant part of a nonhomogeneous product. A layer with a mass per unit area ≥ 1.0 kg/m^2 or a thickness ≥ 1.0 mm is considered to be a substantial component.

Nonsubstantial component: A material that does not constitute a significant part of a nonhomogeneous product. A layer with a mass per unit area < 1.0 kg/m^2 and a thickness < 1.0 mm is considered to be a nonsubstantial component.

Table 10.29   New system of Euroclasses for linings

A1		
A2s1d0	A2s1d1	A2s1d2
A2s2d0	A2s2d1	A2s2d2
A2s3d0	A2s3d1	A2s3d2
Bs1d0	Bs1d1	Bs1d2
Bs2d0	Bs2d1	Bs2d2
Bs3d0	Bs3d1	Bs3d2
Cs1d0	Cs1d1	Cs1d2
Cs2d0	Cs2d1	Cs2d2
Cs3d0	Cs3d1	Cs3d2
Ds1d0	Ds1d1	Ds1d2
Ds2d0	Ds2d1	Ds2d2
Ds3d0	Ds3d1	Ds3d2
E		Ed2
F		

Two or more nonsubstantial layers that are adjacent to each other (i.e., with no substantial component(s) in between the layers) are regarded as one nonsubstantial component and therefore must altogether comply with the requirements for a layer being a nonsubstantial component.

For nonsubstantial components, distinction is made between internal nonsubstantial components and external nonsubstantial components, as follows:

Internal nonsubstantial component: A nonsubstantial component that is covered on both sides by at least one substantial component.

External nonsubstantial component: A nonsubstantial component that is not covered on one side by a substantial component.

A Euroclass is intended to be declared as, for example, *Bs2d1* where *B* stands for the main class, *s2* stands for smoke class 2, and *d1* stands for droplets/particles class 1. This gives theoretically a total of about 40 classes of linings (see Table 10.29) and 11 classes of floor coverings to choose from. However, each country is only expected to use a very small fraction of the possible combinations.

In many cases, the test methods used come from the ISO. They are well known and some of them have been in use in various countries throughout the world for many years. In liaison with CEN, ISO/TC92/SC1 has actively been involved in development of European standards. These standards are called EN ISO to indicate that they are worldwide as well as European.

10.4.5 Classification and Fire Testing of Building Products

10.4.5.1 EN 13501 Fire Classification and EN 13238 Standard Substrates

EN 13501 [1] basically repeats the classification criteria given in Tables 10.27 and 10.28. It also gives general requirements, provides a model for reporting, and gives background information of the testing and classification system. Classification reports on products will be given based on EN 13501.

Prior to all testing, product samples shall be prepared, conditioned, and mounted in accordance with the relevant test methods and product standards. If relevant, aging and washing procedures are carried out in accordance with the actual product standard.

EN 13501 allows for two additional tests to increase accuracy of classification under certain circumstances. This rule applies to all of the tests described in the following subsections.

EN 13238 [2] recommends standard substrates on which the product sample can be attached before testing. The standard substrates represent various end-use conditions. Thus the test results become more general and the amount of testing can be kept down.

10.4.5.2 EN ISO 1716 Calorific Potential

EN ISO 1716 [3] determines the potential maximum total heat release of a product when completely burning, regardless of its end use. The test is relevant for classes A1, A2, $A1_{fl}$, and $A2_{fl}$.

The calorific potential of a material is measured in a bomb calorimeter. The powdered material is completely burned under high pressure in a pure oxygen atmosphere.

10.4.5.3 EN ISO 1182 Noncombustibility Test

EN ISO 1182 [4] identifies products that will not, or will not significantly, contribute to a fire, regardless of their end use. The test is relevant for classes A1, A2, $A1_{fl}$, and $A2_{fl}$.

EN ISO 1182 is a pure material test and a product cannot be tested in end-use conditions. Therefore only homogeneous building products or homogeneous components of a product are tested. EN ISO 1182 was first published by ISO during the 1970s and is well known. The EN ISO version is shown in Fig. 10.16 and the test specifications are given in Table 10.30.

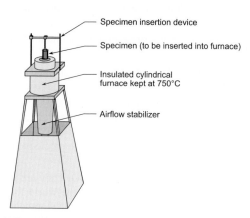

Specimen insertion device

Specimen (to be inserted into furnace)

Insulated cylindrical furnace kept at 750°C

Airflow stabilizer

Figure 10.16   EN ISO 1182 test for noncombustibility

Table 10.30   EN ISO 1182 test specifications

Specimens	Five cylindrical samples, diameter 45 mm, height 50 mm
Specimen position	Vertical in specimen holder in the center of the furnace
Heat source	Electrical cylindrical furnace at 750 °C (measured by the furnace thermocouple)
Test duration	Depends on temperature stabilization
Conclusions	Classification is based on temperature rise as measured by the furnace thermocouple, duration of flaming and mass loss of the sample Details are given in Tables 10.27 and 10.28

10.4.5.4 ISO 9705 Room/Corner Test

To find limit values for the Euroclasses a reference scenario, a room fire test, was selected for the SBI (see 10.4.5.5). Test data from the reference scenario could then be used to determine the Euroclass of the product and to "position" EN 13823, the SBI test procedure. The reference scenario chosen is the international standard Room/Corner Test, ISO 9705 [5], see Fig. 10.17. The test specifications are summarized in Table 10.31.

The Room/Corner Test was first published by ASTM in 1982 [6] and then by NORDTEST in 1986 [7]. The international standard, ISO 9705, was published in 1993.

The Room/Corner Test is a large-scale test method for measurement of the burning behavior of construction products (linings) in a room scenario. The principal output is the occurrence and time to flashover. Results from the test also include a direct measure of fire growth [heat release rate (HRR)] and light obscuring smoke [smoke production rate (SPR)].

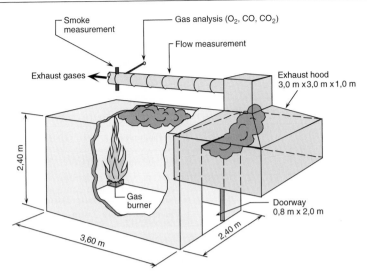

Figure 10.17   The Room/Corner Test to ISO 9705

Table 10.31   The Room/Corner Test, ISO 9705, test specifications

Specimens	Sample material enough to cover three walls and the ceiling of the test room. The wall containing the doorway is not covered
Specimen position	Forms a room lining
Ignition source	Gas burner placed in one of the room corners. The burner heat output is 100 kW for the first 10 min and then 300 kW for another 10 min
Test duration	20 min or until flashover
Conclusions	A number of parameters relating to a room fire such as temperatures of the gas layers, flame spread, and heat fluxes can be measured. However, the most important outputs are HRR, SPR, and time to or occurrence of flashover

The product is mounted on three walls and on the ceiling of a small compartment. A door opening ventilates the room.

Experience on testing products has been gained during more than 10 years of work with the Room/Corner Test. A considerable amount of information on product burning behavior in this method is available and the thermal conditions during a test fire has been carefully mapped [8–10].

The development of EN 13823, the SBI test, included testing of 30 construction products across Europe. The same products were tested according to the reference scenario, the Room/Corner Test. The subsequent analysis then resulted in a correlation between FIGRA (ISO 9705) and FIGRA (SBI) [11]. The correlation between the tests is also relating flashover in ISO 9705 to certain Euroclasses (see Table 10.32).

Table 10.32 Description of the Euroclasses as the tendency of the actual product to reach flashover in the Room/Corner Test, ISO 9 705

Euroclass	Limit value FIGRA (SBI) (W/s)	Expected burning behavior in the Room/Corner Test
A2	120	No flashover
B	120	No flashover
C	250	No flashover at 100 kW
D	750	No flashover before 2 min at 100 kW
E	>750	Flashover before 2 min

It is clear that a strict correlation in all cases between the SBI results and the occurrence of flashover in a room test will not be true. However, it is seen that a concept of safety was one of the reasons behind selecting the specific class limits.

10.4.5.5 EN 13823 Single Burning Item (SBI) Test

EN 13 823 SBI [12] evaluates the potential contribution of a product to the development of a fire, under a fire situation simulating a single burning item in a room corner near to that product. The test is relevant for classes A1, A2, B, C, and D

The SBI is the major test procedure for classification of linings (see Fig. 10.18). The test specifications are summarized in Table 10.33.

The SBI is of intermediate scale size. Two test samples, 0.5m x 1.5m (1.65 ft x 4.9 ft) and 1.0m x 1.5m (3.28 ft x 4.9 ft) are mounted in a corner configuration where they are exposed to a gas flame ignition source. As for ISO 9 705, a direct measure of fire growth (HRR) and light obscuring smoke (SPR) are principal results from a test. Other properties such as the occurrence of burning droplets/particles and maximum flame spread are observed.

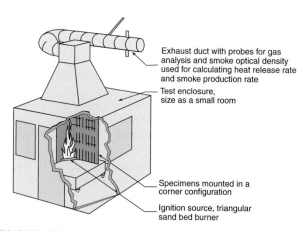

Exhaust duct with probes for gas analysis and smoke optical density used for calculating heat release rate and smoke production rate

Test enclosure, size as a small room

Specimens mounted in a corner configuration

Ignition source, triangular sand bed burner

Figure 10.18   EN 13 823, SBI, the Single Burning Item Test

Table 10.33   EN 13 823 SBI test specifications

Specimens	Samples for three tests. Each test requiring one sample of 0.5 m × 1.5 m and one sample of 1.0 m × 1.5 m
Specimen position	Forms a vertical corner
Ignition source	Gas burner of 30 kW heat output placed in corner
Test duration	20 min
Conclusions	Classification is based on FIGRA, THR_{600s}, and maximum flame spread. Additional classification is based on SMOGRA, TSP_{600s}, and droplets/particles. Details are given in Table 10.27

The index FIGRA, FIre Growth RAte, is used to determine the Euroclass. The concept is to classify the product based on its tendency to support fire growth. Thus, FIGRA is a measure of the biggest growth rate of the fire during a SBI test as seen from the test start. FIGRA is calculated as the maximum value of the function (heat release rate)/(elapsed test time), in units of W/s. A graphical presentation is shown in Fig. 10.19.

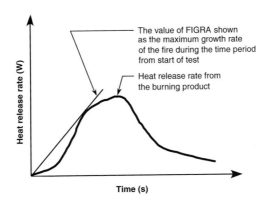

Figure 10.19   Graphical representation of the FIGRA index

To minimize noise, the HRR data is calculated as a 30 s running average. In addition, certain threshold values of HRR and the total heat release rate must first be reached before FIGRA is calculated.

The additional classification for smoke is based on the index SMOGRA, SMOke Growth RAte. This index is based on principles similar to those for FIGRA. SMOGRA is calculated as the maximum value of the function (smoke production rate)/(elapsed test time) multiplied by 10,000. The data on smoke production rate, SPR, is calculated as a 60 s running average to minimize noise. In addition, certain threshold values of SPR and integral values of SPR must first be reached before SMOGRA is calculated.

The detailed definitions of FIGRA and SMOGRA can be found in EN 13 823 (SBI).

10.4.5.6 EN ISO 11925-2 Small-Flame Test

EN ISO 11925-2 [13] evaluates the ignitability of a product under exposure to a small flame. The test is relevant for classes B, C, D, E, B_{fl}, C_{fl}, D_{fl}, and E_{fl}.

The small-flame test is quite similar to the DIN 4102-1 test used for the German class B2. Variants of this procedure are also found in other EU member states regulations. Test rig and specifications are shown in Fig. 10.20 and Table 10.34.

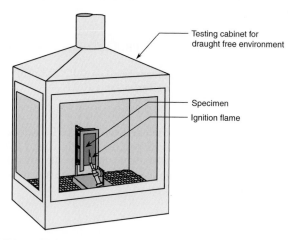

Testing cabinet for draught free environment

Specimen
Ignition flame

Figure 10.20   EN ISO 11925-2 Small-flame test

Table 10.34   EN ISO 11925-2 Small-flame test. Specifications

Specimens	250 mm long, 90 mm wide, max. thickness 60 mm
Specimen position	Vertical
Ignition source	Small burner. Flame inclined 45° and impinging either on the edge or the surface of the specimen
Flame application	30 s for Euroclass B, C, and D; 15 s for Euroclass E
Conclusions	Classification is based on the time for flames to spread 150 mm and occurrence of droplets/particles. Details are given in Tables 10.27 and 10.28

10.4.5.7 EN ISO 9239-1 Floor Covering Test

EN ISO 9239-1 [14] evaluates the critical radiant flux below which flames no longer spread over a horizontal flooring surface. The test is relevant for classes $A2_{fl}$, B_{fl}, C_{fl}, and D_{fl}. The test apparatus is shown in Fig. 10.21 and the test specifications are given in Table 10.35.

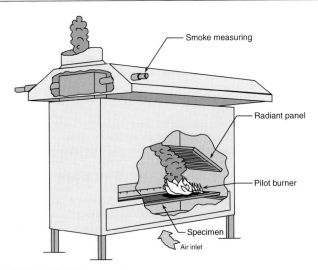

Figure 10.21   EN ISO 9239-1 Floor covering test

Table 10.35   EN ISO 9239-1 Floor covering test. Specifications

Specimens	1050 mm long × 230 mm wide
Specimen position	Horizontal
Ignition source	A gas fired radiant panel that gives a heat flux to the specimen. The maximum heat flux is 11 kW/m² that drops to 1 kW/m² at the end of the specimen. A pilot flame is impinging on the surface of the hot end of the specimen to initiate any flame spread
Test duration	Until the flames extinguish or max. 30 min
Conclusions	Classification is based on the critical heat flux below which flame spread is not occurring. Additional classification is based on smoke density. Details are given in Table 10.28

10.4.6 Future Developments

There is a footnote in Table 10.27 referring to linear products. These are cables, pipes, and pipe insulation. Like floor coverings, they will have their own classification system, but there is no final decision yet on how these products will be tested.

The role of the reference scenario, ISO 9705, is not yet decided. As it was used to identify the limit values in the SBI test for the Euroclasses, ISO 9705 could be used in special cases for direct classification. It was recognized by the FRG that such needs may occur for products or product groups that for technical reasons cannot be tested in the SBI. Recently, the European Commission has given a mandate to CEN to make ISO 9705 a European standard. Once an EN, the test can be used for reference tests for the European system also in the future.

References for Section 10.4

[1] EN 13 501: 2002 Fire classification of construction products and building elements Part 1 Classification using test data from reaction-to-fire tests

[2] EN 13 238: 2002 Reaction-to-fire tests for building products – Conditioning procedures and general rules for selection of substrates

[3] EN ISO 1716: 2002 Reaction-to-fire tests for building products – Determination of the calorific value

[4] EN ISO 1182: 2002 Reaction-to-fire tests for building products – Noncombustibility

[5] ISO 9705: 1993 Fire tests – Full-scale room test for surface products

[6] 1982 Annual book of ASTM standards, Part 18 – Proposed Method for Room Fire Test of Wall and Ceiling Materials and Assemblies

[7] Surface Products: Room Fire Test in Full Scale, NT FIRE 025, Helsinki 1986.

[8] B. Sundström, Full-Scale Fire Testing of Surface Materials. Measurements of Heat Release and Productions of Smoke and Gas Species, Technical Report SP-RAPP 1986:45, Borås 1986

[9] EUREFIC Seminar Proceedings, Interscience Communications Ltd, London, Sep. 1991, ISBN 0 9516320 19

[10] B. Östman, R. Nussbaum, National Standard Fire Tests in Small Scale Compared with the Full-Scale ISO Room Test, Träteknikcentrum Rapport I 870217

[11] B. Sundström, P. van Hees, P. Thureson, Results and Analysis from Fire Tests of Building Products in ISO 9705, the Room/Corner Test, The SBI Research Programme, SP REPORT 1998:11

[12] EN 13 823: 2002 Reaction-to-fire tests for building products – Single burning item test

[13] EN ISO 11 925-2: 2002 Reaction-to-fire tests for building products – Ignitability

[14] EN ISO 9239-1: 2002 Reaction-to-fire tests for floor coverings – Determination of the burning behaviour using a radiant heat source

10.5 Austria

C. PÖHN

10.5.1 Statutory Regulations

In the nine Austrian provinces, structural fire protection is governed by the provincial building codes. These building codes are based principally on that in force in Vienna contained in legislation originally passed in 1883 following the Ring Theatre fire and that has existed as a separate, frequently revised, building code for Vienna since 1930 [1].

Efforts are being made to unify the various Austrian building codes by incorporating existing urban regulations in the relevant provincial building codes. A model building code was issued in 1961 with the intention of achieving further unification of the provincial codes. In practice the Viennese building code still has a certain pilot function for most provincial regulations. In the past few years many building codes were newly issued with lower levels in the field of resistance to fire of building products. In particular, the political pressure to open up opportunities for the use of wood and wood products led to this development.

For the use of combustible construction materials the following essential documents are of special interest:

- The product standards for combustible materials specify reaction-to-fire levels. For example, expanded polystyrene EPS has to be "schwer brennbar B1" for complying with the product standard ÖNORM B 6050 [2].
- For products without a special product standard, a guideline for combustible materials (TRVB B 109) is issued by the Austrian Institute for Construction.

- If there exists no product standard for a product and this product is not considered in the guideline mentioned above, the technical guideline about fire safety – TRVB B 109 needs to be used. In this guideline, the lowest class concerning reaction-to-fire for every application can be found. This guideline has been adapted to the Euroclasses and published as ÖNORM B 3 806 [3].

The Austrian Institute for Construction (OIB) is currently pushing the unification of the technical parts of the nine building codes.

10.5.2 Classification and Testing of the Fire Performance of Building Materials and Components

Classification and testing of the fire performance of building materials are described in ÖNORM B 3 800. This standard is arranged in four parts, Part 1 of which deals with fire performance of building materials, while Parts 2–4 cover fire performance of building components [4–7].

- ÖNORM B 3 800-1 Behavior of building materials and components in fire – Building materials: Requirements and tests (VORNORM; 1988-12-01)
- ÖNORM B 3 800-2 Behavior of building materials and components in fire – Components: Definitions, requirements, and tests (1997-03-01)
- ÖNORM B 3 800-3 Behavior of building materials and components in fire – Special components: Definitions, requirements and tests (1995-12-01) [9]
- ÖNORM B 3 800-4 Behavior of building materials and components in fire – Components: Classification of fire resistance (2000-05-01)
- ÖNORM B 3 810 Fire behavior of floor coverings (1986-11-01)
- ÖNORM B 3 836 Behavior of building components in fire – Cable penetration seals (1984-12-01)
- ÖNORM B 3 850 Fire resisting doors – Hung doors and swing doors – Single- and double-sided design (2001-06-01)
- ÖNORM B 3 852 Fire resisting doors and gates – Vertical lifting, sectional overhead, tilting, rolling, sliding, and folding doors and gates (1997-08-01; new issue under development)
- ÖNORM B 3 860 Closures for roof voids (attics) with or without stair (1987-01-01)
- ÖNORM DIN 4 102-12 Fire behavior of building materials and building components – Part 12: Circuit integrity maintenance of electric cable systems – Requirements and testing (2000-02-01)
- ÖNORM M 7 625 Ventilation equipment; fire dampers; requirements, testing, marking of conformity (1985-11-01; new issue in development)
- ÖNORM M 7 626 Ventilation; ventilation ducts with requirements for fire resistance (1980-11-01; new issue in development)

All Austrian standards are issued by the Österreichisches Normungsinstitut, Leopoldsgasse 4, A-1021 Wien 2.

ÖNORM B 3 800-1 was issued as a provisional standard in November 1979. It was revised in December 1988 but remained a provisional standard (VORNORM). Fire performance as defined in ÖNORM B 3 800-1 includes characteristics such as combustibility, smoke production, and droplet formation. As the European standards EN 13 501-1 for classification and the European reaction-to-fire standards on noncombustibility, gross calorific potential, the SBI and small-flame tests (see Section 10.4) will supersede the existing national Austrian standards; the latter will be withdrawn in the next few years.

According to ÖNORM B 3 800-1, combustibility, smoke generation, and droplet formation during the burning of building materials are divided as shown in Table 10.36.

Table 10.36 Classification of the fire performance of building materials according to ÖNORM B 3800-1

Class of building product	Designation
Combustibility	
Combustibility class A	Noncombustible
Combustibility class B	Combustible
Combustibility class B1	Low combustibility
Combustibility class B2	Moderately combustible
Combustibility class B3	Highly combustible
Smoke production during burning	
Smoke production class Q1	Low smoke production
Smoke production class Q2	Moderate smoke production
Smoke production class Q3	Strong smoke production
Formation of drops during burning	
Drop formation class Tr1	Nondripping
Drop formation class Tr2	Dripping
Drop formation class Tr3	Flaming drops

Combustibility, smoke generation, and drop formation are measured by various methods that enable building materials to be classified in various classes as shown in Table 10.36.

Smoke generation is determined only by an ancillary test. Drop formation is evaluated in connection with the combustibility tests.

Although the formation of toxic gases in fire incidents is considered to be of major importance, standard tests or evaluation criteria are yet to be established. Consequently no binding requirements relating to toxicity exist.

10.5.2.1 Noncombustibility Test to ÖNORM B 3800 Part 1

The noncombustibility test according to the supplement to ÖNORM B 3800 Part 1 corresponds partially to ISO 1182. The test device is shown in Fig. 10.22. The test specifications

Figure 10.22 Noncombustibility test according to ÖNORM B 3800-1

are restricted to temperature rise in the furnace and mass loss. As all Austrian fire testing laboratories have been accredited, the noncombustibility test is now fully in accordance with ISO 1182 (temperature rise T, mass loss m and duration of flaming t_f). Therefore, in the future, there will be no problems to adapt the results to the new Euroclasses. According to ÖNORM B 3800 Part 1, a building material is considered to be noncombustible if it cannot be made to burn or made to ash in air at 750 °C; furthermore, there should be no or only slight smoke development (smoke generation class Q1). Inorganic materials such as sand, earthenware, plaster, steel, and so forth are considered to be noncombustible without any special proof (see ÖNORM B 3800 Part 1, Section 3.1.1). A special definition of non-combustibility exists for some special products. For example, the coating of external insulation systems has such a high share of organic components that the duration of flaming always exceeds 20s. Thus, the duration of flaming is not reported when such products are tested according to the ÖNORM B 6135. The two other test results (T and m) are reported and have the same limits as ISO 1182.

10.5.2.2 Combustibility Class B1: Low Combustibility According to the Schlyter Test

The Schlyter test is generally be used to provide evidence that building materials are of low combustibility; however a few building materials such as certain wood products, plaster board, and special grades of plastics (see ÖNORM B 3800 Part 1, Section 3.1.2) are considered to fulfill this description without proof. The test equipment is illustrated in Fig. 10.23 and the test specifications are summarized in Table 10.37.

Figure 10.23   Schlyter test to ÖNORM B 3800-1

10.5.2.3 Combustibility Class B2: Moderately Combustible According to the Small Burner Test

Various building materials such as certain wood products and plastics listed in ÖNORM B 3800 Part 1, Section 3.1.3 are considered to be moderately combustible without testing. All other materials are required to pass the so-called small burner test (see Fig. 10.24), which is practically identical to the German test DIN 4102-1 for class B2 building materials. The equipment and test specifications are almost the same as those given in Section 10.9. As, in addition, this test method is also virtually identical to EN ISO 11925-2, all building products meeting combustibility class B2 to ÖNORM B 3800-1 fulfill Euroclass E to EN 13501-1.

Table 10.37   Schlyter test specifications

Specimens	Three × two specimens 800 mm × 300 mm
Specimen position	Two vertical specimens parallel to and at a distance of 50 mm from each other (surface to surface), one specimen displaced 50 mm downwards
Ignition source	Row of six gas jets at 30 mm intervals; flames impinge on lower specimen at an angle of 30° to horizontal; distance of jets from flamed specimen: 40 mm; flame length in vertical position: 120 mm; gas supply 3.5 parts N_2 to 1 part propane (21 MJ/m³)
Test duration	15 min
Results	Combustibility class B1 achieved if: • Specimen that is not flamed does not ignite, • After removal of ignition source, flamed specimen continues to burn ≤ 1 min and to glow ≤ 5 min, • Unburned length of flamed specimen ≥ 400 mm.

Figure 10.24   Small flame test to ÖNORM B 3800-1 for B2 classification

10.5.2.4 Combustibility Class B3: Highly Combustible

Building materials are considered to be highly combustible if they cannot be classified in classes A, B1, or B2.

Smoke Generation

Building materials are classified in three smoke generation categories as shown in Table 10.36. The test apparatus is shown in Fig. 10.25. The test criteria are given in Section 10.17 Switzerland and the smoke classification shown in Table 10.38.

Figure 10.25   Smoke density test (Lüscher test) to ÖNORM B 3 800-1

Table 10.38   Smoke density test specifications

Smoke production class	Maximum light absorption
Q1	0–50 %
Q2	> 50–90 %
Q3	> 90 %

Drop Formation

Drop formation during burning of building materials is classified as shown in Tables 10.39 and 10.40. The Schlyter test for combustibility class B1 and the small burner test for combustibility class B2 are used with a basket containing a filter paper placed under the test rig.

Table 10.39   Dripping criteria for class B1 and B2 building products

Dripping after start of flaming	Time
Class B1 materials	≤ 20 min
Class B2 materials	≤ 20 s

Table 10.40   Criteria for drop formation classes

No dripping	Tr1 – nondripping
Filter paper not ignited	Tr2 – dripping
Filter paper ignited	Tr3 – flaming drops

Certain materials listed in Section 3.3 of ÖNORM B 3800-1 are considered to be nondripping without special proof.

10.5.2.5 ÖNORM B 3810. Fire Behavior of Floor Coverings

The test described in ÖNORM B 3810 is similar to ASTM E 648 (NBSIR 75-950). Essential disparities concern the smaller dimensions of the test apparatus and the angle between the radiant panel and the horizontal line which is altered to 60°, thus giving a different profile of the radiant flux along the sample. The test device is shown in Fig. 10.26 and the specifications are given in Table 10.41.

Figure 10.26   floor coverings test ÖNORM B 3810

Table 10.41   Test specifications floor covering test

Specimens	Three specimens 200 mm × 800 mm
Specimen position	Horizontal
Ignition source	Gas heated radiator 300 mm × 360 mm. Inclined at 60° to the horizontal, radiation of specimen 1.7–0.15 W/cm² swivelling pilot flame, 40–60 mm long, acting on the specimen 2 min after start of the test for 2 min
Test duration	Until flames on the specimen extinguish, ≤ 20 min
Conclusions	If the maximum extent burnt corresponds to a heat flux of > 0.4 W/cm² B1 0.4–0.3 W/cm² B2 < 0.3 W/cm² B3

10.5.3 Official Approval

Since 1994, accreditation according to ÖNORM EN 45001 and EN 45004 is necessary for test and inspection institutes. In Austria there are two official accreditation offices:

Federal Ministry for Labor and Economy (Bundesministerium für Arbeit und Wirtschaft – BMWA)
(for the execution of the federal accreditation law – AkkG)
phone: +43 1 71100 8248
fax: +43 1 7143582
e-mail: guenter.friers@bmwa.gv.at
Web site: www.bmwa.gv.at/service/akkservice_fs.htm

Austrian Institute of Construction Engineering – OIB
(for the execution of the accreditation laws of the provinces – e.g. WBAG)
phone: +43 1 5336550
fax: +43 1 5336423
e-mail: mail@oib.or.at
Web site: oib.or.at

This organization, OIB, represents Austria at EOTA (European Organisation for Technical Approvals).

Two institutes in Austria are accredited for the whole range of test methods – fire resistance as well as reaction-to-fire tests – based on Austrian standards:

Versuchs- und Forschungsanstalt der Stadt Wien – MA 39
A-1110 Vienna, Rinnböckstraße 15
phone: +43 1 79514 8039
fax: +43 1 79514 99 8039
e-mail: post@m39.magwien.gv.at

Institut für Brandschutztechnik und Sicherheitsforschung
A-4020, Linz, Petzoldstraße 45/47
phone: +43 732 76170
fax: +43 732 7617 29
e-mail: office@ibs-austria.at

Both are members of EGOLF, the European Group of Official Laboratories for Fire Testing.

Some institutes are accredited for special methods, especially for testing the reaction-to-fire of specific products:

Österreichisches Kunststoffinstitut
A-1030 Vienna, Arsenal, Objekt 213

Versuchsanstalt für Kunststofftechnik am Technologischen Gewerbemuseum
A-1200 Vienna, Wexstr. 19–23

Österreichisches Textilforschungsinstitut
A-1050 Wien, Spengergasse 20

After the first complete test with positive results, the test laboratory provides a test report giving the test results and the classification of the building material according to the applied test and classification standard. This document forms the basis for the use of currently known and proved products according to the specifications given in [8] or for obtaining approval from the building authorities in the various Austrian provinces for the use of new products. Approval of some building materials remains valid only for a limited period. The validity of the test report is restricted to a first period of 4 years and may be prolonged three times for 2 years according to ÖNORM B 3800.

Marking and inspection of building materials do not take place except where material standards for certain building materials require goods to be marked "ÖNORM ... geprüft" (Austrian standard ... tested). This is usually coupled with independent inspection. A list of products that have the right to carry the test mark is given in a register published by the Austrian Standards Institute [9].

10.5.4 Future Developments

The use of the harmonized European standards is under consideration: the above mentioned ÖNORM B 3806 was prepared by the Technical Committee (TC) FNA 006. Since 2001, this TC has been working on the ÖNORM B 3807 transposition document for integrating the Euroclasses dealing with resistance to fire.

References for Section 10.5

[1] R. Moritz: Bauordnung für Wien, Manzsche Kurzkommentare, Wien 1997
[2] ÖNORM B 6050 Materials for thermal and/or acoustic insulation in building construction – Expanded polystyrene particle foam EPS – Marking of conformity
[3] Proposal for ÖNORM B 3806 Requirements for fire behavior of building products (building materials)
[4] ÖNORM B 3800-1 Behavior of building materials and components in fire – Building materials: Requirements and tests (VORNORM; 1988-12-01)
[5] ÖNORM B 3800-2 Behavior of building materials and components in fire – Components: Definitions
[6] ÖNORM B 3800-3 Behavior of building materials and components in fire – Special components: Definitions, requirements and tests (1995-12-01)
[7] ÖNORM B 3800-4 Behavior of building materials and components in fire – Components: Classification of fire resistance (2000-05-01)
[8] Bundesländerausschuss für die Beurteilung neuer Baustoffe und Bauweisen (Bauarten)
[9] Österreichisches Normungsinstitut, Register ÖNORM ... geprüft.

10.6 Belgium

P. VAN HESS

10.6.1 Statutory Regulations

10.6.1.1 Overview of Authorities

Subsequent to the third phase of state reform in 1988 [1], there are now five different authority levels in Belgium with respect to fire safety regulations in buildings as shown in Table 10.42.

The highest level of responsibility is the federal government, which has authority for general legislation concerning safety. With respect to fire safety, this means it establishes the minimum requirements that have to be applied generally for new buildings. The authorities at a lower level can add specific, additional requirements.

Table 10.42 Overview of authority levels

Governmental level		Authority concerning fire safety requirements
Federal government		Three different authorities: • General legislation to which all buildings have to satisfy • Specific legislation for buildings for which the federal government has responsibility (hospitals, stadiums etc.), • Legislation concerning the safety at work (ARAB Article 52).
3 Communities	• The Flemish community • The French-speaking community • The German-speaking community	Legislation concerning buildings for public and cultural activities Old people's homes Hotels Schools
3 Regions	• The Flemish Region • The Walloon Region • Brussels-Capital Region	Legislation related to regional activities Environmental legislation Urban law (town and country planning law)
Provinces	10 provinces	Additional requirements concerning fire safety based on province laws and the urban law
Local communes	589 Local communes	Additional requirements concerning fire safety based on municipal law and the urban law

10.6.1.2 The Federal Level

The minimal requirements a building has to satisfy are adressed here. They have been published as a Royal Approval (KB) dated July 7, 1994 resulting in the Basic Standards for the prevention of fire and explosion. New buildings are required to satisfy these Basic Standards. This KB has been amended by a KB dated December 19, 1997 with changes in the annexes [2].

The federal government can also decide on specific regulations for those buildings for which it has responsibility. An example of this is the fire safety requirements for hospitals formulated by a KB dated 06/11/1979 [Official Journal (Moniteur Belge, Belgisch Staatsblad BS) publication date 11/01/1980]. Stadiums and football stadiums are also the responsibility of the federal government within the KB of 17/07/1989 (B.S. 20/08/1989), amended by the KB of 08/09/1990 (BS 15/10/1990).

The last responsibility is the fire safety requirements within the workplace safety laws. These are listed in the ARAB regulations Article 52.

More recent fire protection and fire prevention requirements are described in refs. [3–5].

10.6.1.3 Regions and Communities

The communities and regions have been established to take into account respectively the cultural and economical differences in Belgium. They have each their own responsibilities. The communities are, for example, responsible for matters related to persons and culture. For fire safety, this means that the communities are responsible for specific requirements for buildings that have a function toward persons. An example is the fire requirements for retirement homes, hotels, and schools (see Table 10.43).

Table 10.43   Overview of legislation in the regions and Communities

Specific building	Community	Legislation
Old people's homes	Flemish	K.B. 15/03/1989 (B.S. 23/03/1989)
	French-speaking	K.B. 23/12/1998 (B.S. 27/01/1997)
	German-speaking	K.B. 20/02/1995 (B.S. 10/10/1995)
Hotels	Flemish	K.B. 27/01/1988 (B.S. 27/05/1988)
	French-speaking	K.B. 24/12/1990 (B.S. 21/06/1991)
	German-speaking	They use the standard NBN S21-205
Schools	Flemish	The Basic Standards ("Basisnormen") are the basis and the standard NBN S21-204 is an extension to it
	French-speaking	The Basic Standards ("Basisnormen") are the basis and the standard NBN S21-204 is an extension to it. In addition there is a K.B. dated 04/11/1996 (B.S. 20/03/1997). This KB has requirements with respect to evacuation and stability
	German-speaking	The Basic standards ("Basisnormen") are the basis and the standard NBN S21-204 is an extension to it.

At the regional level it is possible that additional requirements are established with respect to the environment and the urban laws.

An example of such legislation can be found in the Flemish environmental laws, called VLAREM. This law was established on June 1, 1995 (B.S. 31/07/1995). In Part 5, a number of supplementary fire safety requirements are given for specific buildings such as swimming pools, cinemas, theatres, and so forth.

The Flemish decree of May 18, 1999 can be given as an example of urban laws. In Article 54, the Flemish region receives the authority to establish town and country planning orders, which, among others, can lead to additional requirements for fire safety.

10.6.1.4 Provinces and Communes

Provinces and communes can establish additional fire safety requirements based on the province law and the municipal law. Additional requirements can be decided on within Article 55 of the decree of urban law for those buildings under the responsibility of the provinces and communes.

10.6.1.5 Belgian Standards

Apart from the legislation, a number of Belgian standards include rules for fire safety. These standards have no regulatory status and include so-called rules for good practice. They concern specific buildings and include specific requirements for these buildings. For this reason, it is recommended that they be followed. Examples are found in Table 10.44. They are an obligation only if a law or KB includes them or refers to them.

The Belgian standards are issued by BIN-IBN, Avenue de la Brabançonne 29, B-1040 Brussels.

Table 10.44 Overview of Belgian standards

Reference	Application area	Confirmation by K.B.
NBN S21-204	Fire Safety of buildings – School buildings – General requirements for reaction-to-fire	K.B. of 06/12/1982 (B.S. 08/03/1983)
NBN S21-205	Fire Safety of buildings – Hotels and similar buildings – General requirements	K.B. of 25/10/1993 (B.S. 20/11/1993)
NBN S21-207	Fire Safety of buildings – High rise buildings – Heating and ventilation equipment	K.B. of 05/01/1988 (B.S. 28/01/1988)
NBN S21-208-1	Fire Safety of buildings – Design and calculation of installation for smoke and heat extractions Part 1: Large nondivided internal spaces with one level of construction	K.B. of 25/04/1996 (B.S. 19/06/1996)

10.6.2 Classification and Testing of the Fire Performance of Building Materials and Components

Classification and testing of the fire performance of building materials are described in the Basic Standards (KB 7/7/1994), modified by KB 19/12/1997.

Prescriptions about resistance to fire are given in refs. [6–8]. In these references there is a link to NBN 713-020 [9].

The fire performance of a building material is defined as "the totality of the properties of a building material with regard to the initiation and development of a fire." Details are given in reference [10] and are summarized below.

A material is considered noncombustible "if during the course of a standardised test, in which it is subjected to prescribed heating, no external indication of a recognisable development of heat is manifested." Determination of noncombustibility is carried out in accordance with the

ISO noncombustibility test, the former recommendation ISO/R 1182-1970 being superseded by ISO 1182-1975. A diagram of the non-combustibility furnace and the test specifications are given in Section 10.23.

If the material fails the ISO 1182 test, its combustibility is then investigated. Flammability according to the French decrees of 9.12.1957 and 10.7.1965 and rate of flame spread to "British Standard 476: 1953 Part 1, Section 2" were used as criteria for determining combustibility and as basis for classification in NBN 713-010. NBN S 21-203 now relates to ISO 1182-1975, to the French standards NF P 92-501 and 92-504 (1975), the latter for melting materials, and to the British Standard BS 476 Part 7: 1971. Building materials are divided into five classes, A0 to A4, as specified in KB of 19/12/1997 annex 5 and summarized in Table 10.45.

Table 10.45   Classification of building materials in Belgium

Class	Test method	Requirements
A0	ISO 1182	No temperature rise $> 50\,°C$, sustained flaming not $> 20\,s$, weight loss not $> 50\,\%$
A1	NF P 92-501 NF P 92-504 BS 476 Part 7	$s = 0; i = 0; h = 0; c < 1$ No afterflame, no burning droplets As for British class 1
A2	NF P 92-501 NF P 92-504 BS 476 Part 7	$s < 0.2; i$ to choice; $h < 1; c < 1$ or $s < 1; i < 1; h < 1; c < 1$ no classification possible as for British class 2
A3	NF P 92-501 NF P 92 504 BS 476 Part 7	$0.2 < s < 1; i$ to choice; $h < 1.5; c < 1$ or $1 < s < 5; i < 2;$ $h < 2.5; c < 2.5$ No burning up to the second reference mark, no burning droplets As for British class 3
A4		Not complying with one of the preceding classes

i = flammability index; s = flame spread index; h = index of maximum flame length; c = combustibility index

Fire performance is determined by the French Epiradiateur test and for melting materials by the French rate of spread of flame test. Diagrams of the test apparatus, test specifications, calculation of the various indices and classification of building materials are given in Section 10.8.

Flame spread is determined by the British "Surface Spread of Flame Test" to BS 476 described in Part 1 (1953) (old version) and in Part 7 (1971) (new version). A diagram of the apparatus, test specifications, and classification of building materials according to test performance are given in Section 10.18.

10.6.3 Official Approval

The tests described in the preceding must be carried out to classify the fire performance of building materials and to obtain official recognition. The principal fire test laboratories for testing of building materials in Belgium are:

Universiteit Gent Laboratorium voor Aanwending der Brandstoffen en Warmte-overdracht,
Ottergemsesteenweg 711
B-9000 Gent
Belgium
Phone + 32 09-243 77 50
Fax + 32 09-243 77 51
E-mail: paul.vandevelde@rug.ac.be
Web site: http://www.allserv.rug.ac.be/~gblanche

Centre d'étude des matériaux plastiques, Université de Liège
Quai Banning 6
B-4000 Liège
Belgium
Phone + 32 43 66 93 39
Fax + 32 43 66 95 34
E-mail: a.bruls@ulg.ac.be

Centexbel
Technologiepark 7
B-9052 Zwijnaarde
Belgium

ISSEP
Rue du Chéra, 200
B-4000 Liège
Belgium
Phone + 32 (0)4 22 98 203
Fax + 32 (0)4 25 24 665
E-mail: fire@issep.be

The test results and classification achieved are recorded in a test report that meets the requirements of the authorities. Marking and inspection are not carried out.

10.6.4 Future Developments

In the future, Article 52 of ARAB shall be included in a new legislation concerning safety at work, namely the CODEX3, which is based on the European rules for safety at work. Two appendices will also soon be added to the Basic Standards with respect to fire safety of industrial buildings. In addition, work is being performed to adapt the requirements to the newly developed European standards.

References for Section 10.6

[1] Belgium, a federal State, Information from the federal government, www.fgov.be
[2] Basis normen voor brandveiligheid, Royal approval (KB) 7/7/94 and 19/12/1997
[3] P. Vandevelde, A. Brüls: Brandveiligheid in gebouwen – deel 1 Passieve beveiliging, 2000, Gent – Luik, ISIB
[4] J. Debyser, W. Ghysel, A. Clymans: Codex Openbare Hulpverlening – Preventie I & II & III, aanvulling nr. 33 april 1999, 1999, Brugge, Die Keure
[5] J. Debyser, W. Ghysel, A. Clymans: Codex des Service de Secours – Prevention I & II & III, compl. n° 27 maart 1999, 1999, Brugge, Die Keure
[6] NBN 713-010 (1971) Protection contre l'incendie dans les bâtiments – Bâtiments élevés – Conditions générales
[7] NBN S 21-201 Protection contre l'incendie dans les bâtiments – Terminologie (1980)

[8] NBN S 21-202 Protection contre l'incendie dans les bâtiments – Bâtiments élevés et batiments moyens – Conditions générales (1980 and Addendum I 1984)

[9] NBN 713-020 (1968) Protection contre l'incendie – Comportement au feu des matériaux et éléments de construction – Résistance au feu des éléments de construction., with Addendum 1 doors (1982), fire resistance shutters (1985), fire resistance of electrical cables (1994)

[10] NBN S 21-203 Protection contre l'incendie dans les bâtiments – Prescriptions relatives à la réaction au feu des matériaux – Bâtiments élevés et bâtiments moyens (1980)

10.7 Czech Republic

R. Zoufal

10.7.1 Statutory Regulations

Health and safety at work as well as fire protection in the Czech Republic are governed by a range of laws, government decrees, proclamations, and other mandatory directives issued by certain authorities and institutions.

The requirements of fire protection in buildings are described in a comprehensive series of legally binding standards that deal with the regulations, classification, and tests concerning the fire performance of building materials and components. The standards contain requirements relating to fire-resistant structures, escape routes, separation of buildings, division of buildings into fire compartments, and so forth. Preventive measures such as sprinklers, smoke and heat extractors, and fire alarms are also described. These requirements form a unified concept for the fire protection of buildings.

CSN 73 0802, "Fire protection of buildings, nonindustrial buildings," and CSN 73 0804 "Fire protection of buildings, industrial buildings," are the fundamental standards for fire precautions and contain a summary of all the important relevant requirements. These standards are supplemented by others covering various types of building such as dwellings, hospitals, storage rooms, buildings for agricultural production, and so forth [1].

CSN 73 0810, "Fire protection of buildings," has a special status. It is the basis for adapting the Czech project standards for building structures to the requirements laid down in the Interpretative Document No. 2 of the European Commission and thus enables the introduction of the European fire testing standards.

10.7.2 Classification and Testing of the Fire Performance of Building Materials and Components

The Czech fire protection standards are divided into five groups:

• Project standards [1],
• Test standards [2],
• Evaluation standards [3],
• Object standards [4],
• Special standards outside the 73 08 series.

The project standards concern structural fire protection requirements and have already been mentioned in the introduction. The test and evaluation standards are discussed in more detail in the following subsections, while the object standards listed under [4] are mentioned for the sake of completeness.

The various standards concerning testing and evaluation of the fire resistance of building components lie outside the scope of this book and are not dealt with.

CSN 73 0823 defines five fire classes for building materials (Table 10.46).

Table 10.46   CSN 73 0823 classes for the combustibility of building materials

Fire class	Designation
A	Noncombustible
B	Not easily combustible
C1	Low combustibility
C2	Moderately combustible
C3	Easily combustible

Classification is made on the basis of performance in tests laid down in CSN 73 0861 and 73 0862 (see later). CSN 73 0823 also contains a list of materials that can be classified without testing. These include inorganic materials such as plaster, concrete, and so forth in class A; polystyrene-concrete in class B; wood and plastics building materials conforming to certain CSN standards in classes C1 and C2; and nonstandardized combustible building materials in class C3.

10.7.2.1 CSN 73 0861. Noncombustibility Test for Building Materials

The CSN 73 0861 noncombustibility test for building materials is identical to ISO 1182. A diagram of the noncombustibility furnace and test specifications are given in Section 10.23 ISO.

10.7.2.2 CSN 73 0862. Determination of Flammability of Building Materials

CSN 73 0862 classifies building materials on the basis of temperature rise of flammable combustion products. The aim of the fire test is to determine if and how the particular building material contributes to fire intensity; the test method, which is equivalent to the British Fire Propagation Test, is based on heat release in relation to time.

The test chamber is illustrated in Section 10.18 United Kingdom. The test specifications are summarized in Table 10.47.

The differences between the temperatures of the test specimen and of the calibration test are decisive for the test results. The calibration test is carried out in a closed testing device with a calcium silicate test specimen as follows: The gas burner is ignited. The gas flow must ensure a continuous temperature rise (measured in the exhaust chimney) to $90 \pm 5\,°C$ at the end of the third minute. After 3 min the gas burner is shut off. The device cools down for 2 min. At the end of 5 min, the exhaust chimney temperature must be $55 \pm 5\,°C$. At 6 min, the burner is reignited and the heating elements are switched on at the same time. The exhaust chimney temperature must rise steadily to $255–275\,°C$ by the end of 20 min. The surface temperature of the calcium silicate specimen during the test must reach at least $600\,°C$. The test procedure is the same as for calibration. The following parameters are determined for each of the five specimens as shown by Fig. 10.27:

- The difference between the temperature at 3 min in the specimen and calibration tests (ΔT_3).

- The same at 5 min (ΔT_5).
- The maximum temperature difference between the specimen and the calibration tests (ΔT_{max}).
- The time at which the maximum temperature was reached (t_{max}).
- The specific weight (m) of the specimen.

Table 10.47   Test specifications for the flammability of building products

Test Unit	0.4 m × 0.4 m × 0.75 m high
Specimens	Five specimens 220 mm × 195 mm
Specimen thickness	Material density ≥ 100 kg/m³: 20–30 mm Material density 100–700 kg/m³: 5–20 mm Material density > 700 kg/m³: 3–5 mm
Specimen position	Vertical
Ignition Source	• Two 100 W electric elements with variable output, distance from specimen: 31 mm • Gas pipe burner of 20 mm diameter • Gas supply 4 l/min propane/butane distance from specimen: 15 mm
Test duration	20 min
Conclusion	Classification in combustibility classes A–C3 depending on test performance

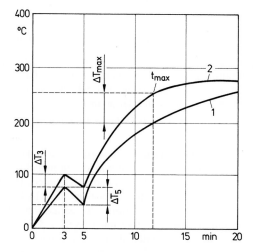

Figure 10.27   Calibration (1) and specimen (2) test temperature curves

The average values (T_i, t_{max}) are calculated from the measured values and a Q-value obtained from the equation

$$Q = \left(\frac{\overline{\Delta T}_5}{\overline{\Delta T}_3 \cdot m}\right)^{1/2} \cdot \overline{\Delta T}_{max} \cdot \frac{20}{t_{max}}$$

The Q-value determines the classification of the material as indicated in Table 10.48.

Table 10.48 Classification of materials according to CSN 73 0862

Combustibility grade		Q-value
A	Noncombustible	< 50
B	Not easily combustible	$\geq 50 - 150$
C1	Low combustibility	$\geq 150 - 300$
C2	Moderately combustible	$\geq 300 - 600$
C3	Easily combustible	> 600

10.7.3 Official Approval

The Czech Republic applies the law No. 22/1997 digest on technical requirements for construction products and the Governmental Decree No. 178/1998 digest, which determines the technical requirements for building products and refers to the EU Directive 106/1989 for basic requirements. It contains a list of products and materials designed to be certified and assessed from the conformity point of view. This is done by notified bodies with the rights for product certification to CSN EN 45011, for quality system certification to CSN EN 45012, and accredited testing laboratories to CSN EN 45001. The authorization is granted by the Czech Office for Standards, Metrology, and Testing UNMZ.

For building products having to fulfil fire safety requirements, the notified body is:

Fire Research Institute Praha Inc.
(PAVUS Praha, a.s.)
Prazska ul. 16
102 45 Praha 10 – Hostivar
Czech Republic
Phone + 420 2 8101 7344
Fax + 420 2 8101 7365
E-mail: vanis@pavus.cz

PAVUS Praha, Inc. has an accredited laboratory for testing fire technical properties of products at Vesell nad Luznici and both above mentioned accredited certification bodies in Prague.

10.7.4 Future Developments

Since April 1, 1997, the Czech Republic has been a full member of CEN. In the field of fire safety of buildings, it will adopt the European standards prepared by CEN TC 127 both for fire resistance and reaction-to-fire. These standards will replace the existing national Czech standards.

References for Section 10.7

[1] CSN (Czech State Standard) Fire protection of buildings
      CSN 73 0802 Non-industrial buildings 2/1995
      CSN 73 0804 Industrial buildings 10/1995
      CSN 73 0810 Requirements in civil engineering 9/1996
      CSN 73 0831 Assembly rooms 5/1979
      CSN 73 0833 Buildings for dwelling and lodging 1/1996
      CSN 73 0834 Changes of buildings 6/1995
      CSN 73 0835 Buildings for sanitary matters 4/1996
      CSN 73 0842 Agricultural premises 4/1996
      CSN 73 0843 Premises for telecommunications 8/1978
      CSN 73 0845 Storage rooms 2/1997
[2] CSN EN 1363-1 Fire resistance tests – Part 1: General requirements 7/2000
      CSN EN 1363-2 Fire resistance tests – Part 2: Alternative and additional procedures 7/2000
      CSN PrEN 1363-3 Fire resistance tests – Part 3: Verification of furnace performance 7/2000
      CSN EN 1364-1 Fire resistance tests for non-loadbearing elements – Part 1: Walls 7/2000
      CSN EN 1364-2 Fire resistance tests for non-loadbearing elements – Part 2: Ceilings 7/2000
      CSN EN 1365-1 Fire resistance tests for loadbearing elements – Part 1: Walls 7/2000
      CSN EN 1365-2 Fire resistance tests for loadbearing elements – Part 2: Floors and roofs 7/2000
      CSN EN 1365-3 Fire resistance tests for loadbearing elements – Part 3: Beams 7/2000
      CSN EN 1365-4 Fire resistance tests for loadbearing elements – Part 4: Columns 7/2000
      CSN EN 1366-1 Fire resistance tests for service installations – Part 1: Ducts 7/2000
      CSN EN 1366-2 Fire resistance tests for service installations – Part 2: Fire dampers 7/2000
      CSN EN 1634-1 Fire resistance tests for door and shutter assemblies – Part 1: Fire doors and shutters 10/2000
      CSN 73 0861 Combustibility testing of building materials – non combustible materials 5/1979
      CSN 73 0862 Determination of flammability of building materials 6/1980
      CSN 73 0863 Determination of the flame propagation along the surface of building materials 2/1991
      CSN 73 0864 Determination of the heating values of combustible solids under fire conditions 3/1988
      CSN 73 0865 Determination of materials dripping from ceilings and roofs in fire 2/1987
      CSN 73 0866 Determination of burning rate of substances in silos, warehouses and reservoirs 2/1987
[3] CSN 73 0818 Person/surface rate in buildings 7/1997
      CSN 73 0821 Fire resistance of structures 2/1973
      CSN 73 0822 Flame propagation along the surface of building materials 9/1986
      CSN 73 0823 Flammability of building materials 5/1983
      CSN 73 0824 Calorific value of combustible materials 12/1992
[4] CSN 73 0872 Protecting buildings from the spread of fire via ventilation systems 1/1996
      CSN 73 0873 Equipment for fire-extinguishing water supply 10/1995
      CSN 73 0875 Design of electrical fire alarm signal systems 3/1991

10.8 France

T. BONNAIRE AND G. TOUCHAIS

10.8.1 Statutory Regulations

French building regulations for the public building sector differ from those for the private sector. French fire regulations relate mainly to high-rise buildings, buildings open to the public, residential buildings, workplaces, and classified installations. The regulations are contained in brochures entitled "Sécurité contre l'Incendie" (Fire Safety) [1] as follows:

- Regulations for buildings open to the public [établissements reçevant du public (ERP)] laid down in Sections R.123-1-R.123-55 of the building and housing code (code de la construction et de l'habitation ([1] No. 1477-I). The implementation of these safety regulations is covered in the Arrêté (decree) of 25.6.80 and several modifying and supplementing decrees ([1] No. 1477-I-XI).
- Regulations for high rise buildings [immeubles de grande hauteur (IGH) contained in sections R.122-1 to 122-29 and R.152-1-52-3 of the building and housing code ([1], No. 1536) and in the decree of 18.10.1977 modified by the decree of 22.10.82 regarding the construction of high rise buildings and antifire and antipanic measures ([1], No. 1536).
- The decree of 31.1.86 modified by the decree of 18.8.86 [1] Official Journal of March 16 and September 20, 1986 regarding fire protection of residential buildings.

10.8.2 Classification and Testing of the Fire Performance of Building Materials and Components

The classification of building materials and components with regard to fire hazard is laid down in Sections R.121-1–121-13 of the building and housing code ([1] No. 1540-III).

The terms "réaction au feu" (reaction-to-fire or fire performance) and "résistance au feu" (fire resistance) are defined in Article 2. The French view "réaction au feu" as including the supply to the fire and fire development, while the "résistance au feu" is defined as "the time during which the building components can fulfill their intended function in spite of the effects of fire."

The classification characteristics laid down in Article 3 for the "réaction au feu" are "the heat evolved during combustion and the presence or absence of combustible gases. The classification should thus specify the noncombustible or combustible nature of the building material and in the latter case its degree of combustibility".

The methods used to determine reaction to fire performance are described in the decree of 30.6.1983 modified by the decree of 28.8.91 ([1] No. 1540-II). These methods are also the subject of a series of standards issued by the Association Française de Normalisation (AFNOR) in Dec. 1985, namely NF P 92-501/503/504/505 (Dec 95), 92-506 (Dec 85), 92-508 (Dec 85), 92-510 (Sep 96) – 92-512 (Dec 85) and fascicule de documentation P 92-507 (Sep 97) [2], which correspond to the regulations of the decree, but are expressed in a more easily understood form. These standards are issued by AFNOR, Tour Europe, Cedex 7, F-92 080 Paris la Défense.

The fire performance of building materials is divided into the five classes M0 (no contribution to fire), M1, M2, M3, and M4.

Only the abstract symbols M0–M4 are used to characterize the results of fire tests. Terms such as nonflammable, and so forth used previously were frequently regarded as properties of the material and often led to incorrect interpretation.

Building materials are divided into two groups for test purposes:

- Flexible materials up to 5 mm (0.2 in.) thick,
- Flexible materials more than 5 mm thick and rigid materials of any thickness.

A primary test is carried out in both cases to ascertain whether the material can be classified in categories M1–M3. Complementary tests are carried out for classification in class M4 or, in certain cases, to confirm classification in classes M1–M3. If there are doubts as to which group the material belongs, both types of test are carried out and the least favorable result evaluated. Where possible, materials are tested as used in buildings. For example, wall coverings are stuck on fiber-cement panels and tested as rigid materials.

The radiators used as heating sources in the following tests are calibrated in accordance with NF P 92-508 (electrical burner test to NF P 92-503) and NF P 92-509 (épiradiateur test to NF P 92-501 and drip test to NF P 92-505) [3]. NF P 92-510 [4] describes the determination of the upper calorific potential. NF P 92-511 and 92-512 [5] define special test conditions for composites and aging tests.

10.8.2.1 NF P 92-501. Epiradiateur Test

The Epiradiateur test is the primary method of determining the fire performance of rigid samples of any thickness and of flexible samples more than 5 mm (0.2 in.) thick. The apparatus is illustrated in Fig. 10.28 and the test specifications are summarized in Table 10.49.

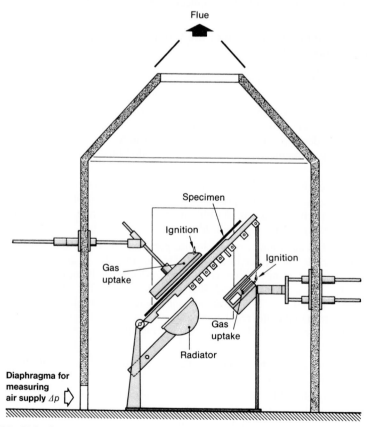

Figure 10.28   Epiradiateur test cabin

Table 10.49   Epiradiateur test specifications

Specimens	Four specimens 300 mm × 400 mm × max. 120 mm
Specimen position	Inclined at 45°
Ignition sources	• Electric radiator (inclined at 45°) 500 W, radiation falling on specimen (distance from radiator: 30 mm): 3 W/cm² • Two butane pilot flames for igniting the combustible decomposition gases above and below the specimen
Test duration	20 min
Conclusion	Classification according to flammability in classes M1–M4

During the test the following must be recorded:
- The time elapsed until flaming first occurs for longer than 5 s on one or both sides of the specimen. The pilot flames are removed when flaming appears,
- The maximum flame height every 30 s period,
- The mean temperature of the thermocouples recorded continuously by chart recorder or every 15 s manually,
- Further primary and secondary fire effects such as smoke generation, burning droplets, after-flaming, glowing, and so forth.

One combined index, calculated from the preceding data, serves as a basis for classification in classes M1–M4:

Classification Index

$$q = \frac{100 \Sigma h}{t_i \sqrt{\Delta t}}$$

where

Σh = sum of the maximal flame lengths,

t_i = time to ignition from specimen positioning to first sustained ignition (> 5 s),

Δt = total combustion duration in seconds calculated for each specimen.

Classification

$\bar{q}$ mean value from four specimens

$\bar{q}$ < 2.5 M1

$\bar{q}$ < 15 M2

$\bar{q}$ < 50 M3

$\bar{q}$ ≥ 50 M4

10.8.2.2 NF P 92-503. Electrical Burner Test

The electrical burner test (essai au brûleur électrique) is used for determining the fire performance of flexible materials. A diagram and test specifications are given in Fig. 10.29 and Table 10.50, respectively.

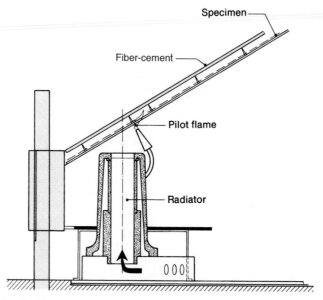

Figure 10.29   Electrical burner

Table 10.50   Electrical burner test specifications

Specimens	Four specimens 600 mm × 180 mm
Specimen position	Inclined at 30° to horizontal
Ignition sources	• Electric radiator 500 W, 30 mm from specimen • Butane gas pilot flame (orifice 20 mm × 0.5 mm, flame 30 mm high)
Test duration	Five min for radiation; if the specimen continues to burn, until extinction of specimen
Conclusion	Classification according to flammability in classes M1–M4

The following must be noted during the test, which lasts until the specimen extinguishes completely:

• Instant and duration of flaming,
• Burning drippings,
• Appearance of the damaged specimen,
• Effect of complete or partial spread of flame, afterglow.

After the test, the maximum damaged length and width of the four specimens is determined and the arithmetic mean of each calculated.

10.8.2.3 NF P 92-504. Rate of Spread of Flame Test

The rate of spread of flame test (essai de vitesse de propagation de la flamme) serves as a complementary test to the primary tests and enables the criteria that cannot be determined by the latter to be established.

This test is used to determine the contribution to flame spread to enable a classification in categories M1–M4 to be made if a material melts rapidly without burning in the vicinity of the radiator in the primary test and if a material does not achieve classes M1–M3. In the first case afterflame, nonpropagation of flame and falling off of burning or nonburning droplets is observed, in the second, flame spread rate is measured.

The test is performed with a horizontal specimen and the German small burner initially described in DIN 50051 is used. Details of the test specifications are summarized in Table 10.51.

Table 10.51   Rate of spread of flame test specifications

Specimens	Four specimens 400 mm × 35 mm Reference marks at 50 mm and 300 mm parallel to the free end of the specimen
Specimen position	Long side (400 mm) horizontal, short side (35 mm) vertical
Ignition source	Small burner with 20 mm flame is moved under the long side of the specimen
Test procedure	• For afterflame and nonpropagation of flame, the flame is applied to the free end of the specimen 10 times for 5 s and time for afterflame is measured • For flame spread rate the flame is applied on the free end of the specimen for 30 s and time for flame travel between the reference marks is measured
Conclusion	Classification in classes M1–M4 depending on course of test

The rate of spread of flame, v, is computed from $v = 250/t$ where t is the time in s required to reach the second reference mark. Otherwise duration of afterflame and falling off of burning or nonburning droplets after application of the flame are noted.

10.8.2.4 NF P 92-505. Dripping Test

The dripping test (essai de goutte) is a further complementary test to determine burning drops, a phenomenon that cannot be clearly assessed in the primary tests. It is carried out if, during the relevant primary test, nonburning drops are observed or if the material withdraws very rapidly or without burning from the vicinity of the radiator. The material is classified in classes M1–M4 depending on test performance. A diagram of the test rig and the test specifications are presented in Fig. 10.30 and Table 10.52, respectively.

If the geometry of the specimen alters during the test, the distance of the radiator from the specimen must be readjusted to 30 mm (1.2 in.). If the specimen ignites during the first 5 min, the radiator is removed 3 s after ignition; irradiation is continued as soon as the specimen extinguishes. During the second 5 min, irradiation is maintained throughout regardless of whether the specimen burns. Note is made of any dripping, whether the cotton wool ignites, the level of smoke development, appearance, and amount of residues. Ignition of the specimen is noted if it is longer than 3 s.

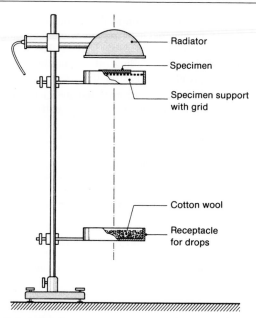

Figure 10.30   Test rig for burning drops

Table 10.52   Dripping test specifications

Specimens	Four specimens 70 mm × 70 mm, minimum weight 2 g
Specimen position	Horizontal on a grid
Ignition source	• Horizontal electric radiator 500 W, radiation intensity on specimen (30 mm from radiator): 3 W/cm²
Receptacle for catching droplets	contains cotton wool and is located 300 mm below the grid
Test duration	10 min
Conclusion	If cotton wool ignites, material is classified in class M4.

10.8.2.5 NF P 92-506. Radiant Panel Test for Floor Coverings

The radiant panel test (essai au panneau radiant) is a test used exclusively for determining the fire performance of floor coverings if the material does not achieve class M1 or M2 in the primary test. The apparatus is illustrated in Fig. 10.31 and the test specifications are summarized in Table 10.53.

The specimen is subjected to the temperature of the radiant panel (850 °C) and the effects of the pilot flame until it ignites or for a maximum of 60 s. If the specimen extinguishes, the pilot flame is reapplied for 10 s at 2, 3, 4, 6, and 8 min. The distance traveled by the flame front at 1 min, as well as the maximum spread of flame are noted. The arithmetic mean is computed from the results of three tests.

This method corresponds in principle to the small-scale surface spread of flame test described in the British Standard BS 476: Part 7.

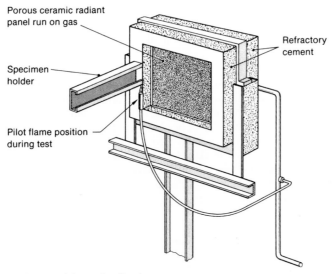

Figure 10.31   Radiant panel for testing floorings

Table 10.53   Radiant panel test specifications for floorings

Specimens	Three specimens 400 mm × 95 mm × max. 55 mm
Specimen position	vertical, long side (400 mm) at 90° to radiator surface
Ignition sources	• Radiant panel 305 mm × 305 mm run on gas, surface temperature 850 °C • Gas pilot flame (flame 40 mm high) below the specimen on the same side as the radiator
Test duration	10 min for radiation; if the specimen continues to burn, until extinction of specimen
Conclusion	Classification in classes M3 or M4 depending on test performance

10.8.2.6 NF P 92-510. Determination of Calorific Potential

The calorific potential (pouvoir calorifique) [4] is determined to provide proof that a building material is noncombustible. Classification M0 applies if the upper calorific potential does not exceed 2500 kJ/kg and if the requirements for M1 are satisfied (with flaming duration < 5 s). Most class M0 building materials are inorganic and usually contain organic components only as binders.

The test apparatus consists of a bomb calorimeter in which the powdered material is placed. At least three determinations are made and the arithmetic mean computed.

This test method is basically similar to the ISO 1716 high-pressure bomb calorimeter method (see also Section 10.23 ISO).

10.8.3 Classification of Building Materials

Class M0 (Noncombustible)

Materials are placed in class M0 if the requirements for M1 are satisfied (with flaming duration < 5 s) and the upper calorific potential, determined by NF P 92-510, does not exceed 2500 kJ/kg. This applies to flexible as well as rigid materials.

Classification of Flexible Materials ≤ 5 mm Thick

Classification in classes M1–M3 is carried out with the electrical burner test (NF P 92-503). If classification in M1, M2, or M3 is not achieved in the primary test, the rate of flame spread test (NF P 92-504) must be carried out. If the rate of flame spread is ≤ 2 mm/s the material is classified in class M4; otherwise it cannot be classified. If droplets occur during the primary test, the NF P 92-505 dripping test must be carried out. If the cotton wool is ignited by the droplets, the material is classified in class M4. If ignition does not occur the classification achieved in the primary test is retained if the droplets in the primary test were nonburning. If there were burning droplets the material is placed in a lower performance class.

The classification criteria are shown in Table 10.54.

Table 10.54 Criteria for classifying flexible materials

Test	Classification criteria				
Dripping test		No cotton ignition	No cotton ignition	Cotton ignition	Cotton ignition
Electrical burner test	No drips	Nonburning drips	Burning drips or debris	Nonburning drips	Burning drips or debris
Flames ≤ 5 s	M1	M1	M2	M4	M4
Flames > 5 s and mean of destroyed lengths < 350 mm	M2	M2	M3	M4	M4
Flames > 5 s and mean of destroyed widths < 90 mm betw. 450/500 mm	M3	M3	M4	M4	M4
Flame spread test < 2 mm/s			M4	M4	M4

Classification of Rigid Materials of Any Thickness and Flexible Materials > 5 mm Thick

Classification in classes M1–M3 is carried out with the Epiradiateur test (NF P 92-501). If classes M1, M2, or M3 are not achieved in the primary test, the rate of flame spread is determined to NF P 92-504 to ascertain whether the material can be classified in class M4 (rate of flame spread ≤ 2 mm/s) or is not classifiable (> 2 mm/s).

The classification criteria are given in Table 10.55.

Table 10.55 Classification criteria for rigid materials and flexible materials > 5 mm thick

Epiradiateur test		Flame spread test		Calorific potential	
Classification criteria		< 2 mm/s	> 2 mm/s	< 2.5 MJ/kg	> 2.5 MJ/kg
No flame	M0			M0	M1
Q < 2.5	M1				
Q < 15	M2				
Q < 50	M3				
Q ≥ 50		M4	No class		

For materials with a particular behavior, that is, that melt without igniting in the vicinity of the radiant panel, the dripping test to NF P 92-505 and the rate of flame spread test to NF P 92-504 are carried out. The classification criteria are then the same as those in Table 10.56.

Table 10.56 Classification for materials with a particular behavior

Test	Classification criteria				
Dripping test		No cotton ignition	No cotton ignition	Cotton ignition	Cotton ignition
Flame spread test	No drips	Nonburning drips	Burning drips or debris	Nonburning drips	Burning drips or debris
No flames	M1	M1	M2	M4	M4
Flames ≤ 5 s	M2	M2	M3	M4	M4
Flames > 5 s and flame spread < 2 mm/s	M3	M3	M4	M4	M4

The NF P 92-506 radiant panel test is intended to classify floor coverings in classes M3–M4 if they do not achieve M1 or M2 classification in the Epiradiateur test. M3 is achieved if the mean destroyed length is ≤ 300 mm. Materials are classified in class M4 if the mean destroyed length is ≤ 100 mm after 1 min and exceeds 300 mm at the end of the test. If the material exceeds 100 mm after 1 min, it is not classified.

10.8.4 The Use of Synthetic Materials

The problem of secondary fire effects and, in particular, of fire gases has achieved prominence due to various spectacular fire incidents. This has led to official controls on the use of synthetic materials in buildings open to the public.

The regulations governing the use of such materials in buildings open to the public laid down in the Arrêté of 4.11.1975 and its revision of 1.12.1976 [1], No. 1540-III specify that the total amounts of nitrogen (N) and chlorine (Cl) contained in synthetic materials that can be liberated as HCN or HCl must not exceed 5 g and 25 g respectively per m^3 of enclosed space.

Plastics, man-made fibers and textiles, elastomers, paints, varnishes, and adhesives are considered as synthetic materials. Materials that achieve classes M0 and M1 are not taken into account. Certain materials, such as foams, and special applications are rated with other fire risk factors as shown in Table 10.57.

Table 10.57　Special provisions concerning chlorine and nitrogen levels

Building material classification	Application		
	ceiling	floor	miscellaneous
M0 or M1	0	0	0
M2 or M3 Density $\geq 0.02\,g/cm^3$	$\dfrac{4 \cdot bP^{a)}}{3}$	0	bP
Density $< 0.02\,g/cm^3$	$\dfrac{16 \cdot bP}{9}$	0	$\dfrac{4 \cdot bP}{3}$
M4 Density $\geq 0.02\,g/cm^3$	–	$\dfrac{bP}{5}$	bP
Density $< 0.02\,g/cm^3$	–	–	$\dfrac{4 \cdot bP}{3}$

a) P = weight in kg of the nitrogen- or chlorine-containing product
　b = nitrogen or chlorine content in % by weight

10.8.5 Official Approval

The tests described in the preceding must be carried out to classify building materials and obtain official approval. Five test institutes in France are agreed by the Ministry of the Interior to test the reaction to fire performance of building materials (réaction au feu des matériaux de construction), CSTB being the pilot laboratory:

Centre scientifique et technique du bâtiment (C.S.T.B.)
Laboratoire de réaction au feu.
84, avenue Jean-Jaurès, Champs-sur-Marne
F-77428 Marne la Vallée Cédex 2
Tel. + 33 1 64 68 84 12
Fax. + 33 1 64 68 84 79

reaction@cstb.fr
http://feu.cstb.fr

Laboratoire central de la préfecture de police (L.C.P.P.)
39 bis, rue de Dantzig
F-75015 Paris
Phone + 33 1 64 68 84 12
Fax + 33 1 64 68 84 79
jean-claude.labarthe@interieur.gouv.fr

Laboratoire national d'essais (L.N.E.)
CEMAT
29, Avenue Roger Hennequin
F-78197 Trappes
Phone + 33 1 30 69 10 73
Fax + 33 1 30 69 12 34
alain.sainrat@lne.fr
http://www.lne.fr

Société nationale des poudres et explosifs (S.N.P.E.)
Centre de recherches du Bouchet
B.P. 2
F-91710 Vert-le-Petit
Phone + 33 1 64 99 14 82
Fax + 33 1 64 99 14 14
m.mauny@propulsion.snpe.com
http://www.snpe.com

Institut textile de France (I.T.F.)
Avenue Guy de Collongue
F-69130 Ecully
Phone + 33 4 72 86 16 21
Fax + 33 4 78 43 39 66
http://www.itf.fr

Materials should be submitted directly to one of the above addresses. Following the tests, the laboratory issues so-called procès-verbaux (classification documents) valid for 5 years and describing the test performance and resultant classification. The "demande d'homologation de classement" (application for official approval) enclosed with the procès-verbal should be sent to the Ministry of the Interior, Dept. of Civil Safety at the following address:

Ministère de l'Intérieur
Direction de la défense et de la sécurité civiles (D.D.S.C.)
1, Place Beauvau
F-75800 Paris Cédex

In practice the "homologation" is used only for publication in the Journal Officiel as official recognition comes into force de facto with the issue of the procès-verbal by the test institute.

In cases of dispute, a committee of experts, the comité d'étude et de classification des matériaux et éléments de construction par rapport au danger d'incendie (CECMI) (committee for study and classification of building materials and components with regard to fire hazard), can order further fire tests, not laid down in the regulations, to be carried out.

10.8.6 Future Developments

Like the other member countries of the European Union, France will adopt the European Standards and the national fire test methods described here will be withdrawn.

References for Section 10.8

[1] Journal Officiel de la République Française
Sécurité contre l'Incendie:
N° 1536I.G.H. (sécurité contre l'incendie dans les immeubles de grande hauteur)
N° 1540-I. Résistance au feu des éléments de construction. Essais de ventilateurs de désenfumage
N° 1540-II. Réaction au feu
N° 1540-III. Textes généraux
Règles de sécurité contre l'incendie dans les E.R.P.:
N° 1477-I. Dispositions générales. Instructions techniques
N° 1477-II. Dispositions particulières (Type M): Magasins de vente. Centres commerciaux
N° 1477-IV. Dispositions particulières (Types R et X): Etablissements d'enseignement. Colonies de vacances. Etablissements sportifs couverts
N° 1477-V. Dispositions particulières (Types N et O): Hotels et pensions de famille. Restaurants et débits de boissons
N° 1477-VI. Dispositions particulières (Type L): Salles d'audition, de conférences, de réunions, de spectacles ou à usage multiple
N° 1477-VII. Dispositions particulières (Types PA, SG): Etablissements de plein air. Structures gonflables
N° 1477-VIII. Dispositions particulières (Types V et W): Etablissements de culte. Administrations. Banques. Bureaux
N° 1477-IX. Dispositions particulières (Type P): Salles de danse et salles de jeux
N° 1477-X. Application aux personnes de droit public
N° 1477-XI. Etablissements spéciaux (Types CTS): chapiteaux, tentes et structures itinérants
Journaux Officiels, 26, rue Desaix, 75 727 Paris CEDEX 15. (These brochures are kept up to date by supplements available from the Journaux Officiels)
[2] Bâtiment. Essais de réaction au feu des matériaux. Normes NF P 92-501, P 92-503 to P 92-506 and fascicule de documentation P 92-507. Association Française de Normalisation (AFNOR), Tour Europe, CEDEX 7, 92 080 Paris la Défense
NF P 92-501: Essai par rayonnement applicable aux matériaux rigides ou rendus tels (matériaux de revêtement) de toute épaisseur et aux matériaux souples d'épaisseur supérieure à 5 mm. December 1995
NF P 92-503: Essai au brûleur électrique applicable aux matériaux souples d'une épaisseur inférieure ou égale à 5 mm. December 1995
NF P 92-504: Essai de propagation de la flamme applicable aux matériaux non destinés à etre collés sur un subjectile (essai complémentaire). December 1995
NF P 92-505: Essai de goutte, au radiateur, applicable aux matériaux fusibles (essai complémentaire). December 1995
NF P 92-506: Essai au panneau radiant pour revêtements de sol. December 1985
FD P 92-507: Matériaux de construction et d'aménagement. Classement selon leur réaction au feu. September 1997
[3] NF P 92-508: Réglage du brûleur électrique. December 1985
NF P 92-509: Réglage du radiateur. December 1985
[4] NF P 92-510: Détermination du pouvoir calorifique supérieur. September 1996
[5] NF P 92-511: Détermination des essais à réaliser suivant la nature de l'utilisation des matériaux – Supports-types – Modèles de fiches d'information. December 1985
NF P 92-512: Détermination de la durabilité des classements de réaction au feu des matériaux – Essais. December 1985

10.9 Germany

M. MITZLAFF

10.9.1 Statutory Regulations

The regulations relating to building inspection in Germany are derived from the Muster-bauordnung (MBO, Model Building Code) which forms the basis of all the Landesbauord-nungen (LBO, State Building Codes) [1]. The building inspectorate is responsible for averting hazards that threaten the life, health, and property of the individual. It is backed by a comprehensive range of legislation, directives, and standards (Table 10.58). The MBO and also the LBO [latest version 1997] take into account the European Construction Products Directive CPD 89/106/EC (see Chapter 10.4).

Table 10.58   Building inspection regulations and structural fire protection provisions

Laws	"Länder" building codes
Statutory orders	Implementing order Furnace order Garage order Business premises order Places of assembly order Industrial premises order Hospital building order Restaurant building order Test mark and quality order
Recognized practice in buildings	DIN 4102 – Fire performance of building materials and components DIN 18082 – Fire barriers DIN 18090 ff – Lift doors DIN 18160 – Chimneys VDE 0100 – Power installations
Building inspections provisions	DIN 18230 – Structural fire protection in industrial premises DIN 18231 – Structural Fire Protection: total structures DIN 18232 – Structural fire protection: smoke and heat control installations

Structural fire precautions are covered in the state building regulations in line with MBO § 17 as follows:

- Structural installations are to be laid out, erected, and maintained in such a way that the incidence and spread of fire and smoke are prevented and that in the event of a fire, effective measures exist for extinguishing it and rescuing humans and animals.
- Building materials that are highly flammable must not be used in the erection and installation of structures; this does not apply to building materials if they are not highly flammable when used with other building materials.

The fire performance requirements for building materials and components are laid down in further sections of the state building regulations, associated ordinances, and additional statutory orders and administrative directives.

10.9.2 Classification and Testing of the Fire Performance of Building Materials and Components

Compliance/conformity with building inspection regulations can be substantiated with the aid of the standards generally recognized as standard building practice.

The DIN 4102 Standard – Fire performance of building materials and components – defines in tangible terms the terminology of fire protection (e.g., combustible, non-combustible) employed in the rules and regulations covering building inspection and fire protection. The original version dates back to 1934. The complete version of this standard includes 20 parts of which now 19 have been issued. Part 20 is in preparation. The status is summarized in Table 10.59.

Table 10.59   DIN 4102 (Parts 1–20)

DIN 4102-	Fire behavior of building materials and building components
	1. Building materials, terminology, requirements, and tests (1998-05)
	2. Buildings components (1977-09)
	3. Fire walls and non-load-bearing external walls (1977-09)
	4. Synopsis and application of classified building materials, components, and special components (1994-03)
	5. Fire barriers, barriers in lift wells and glazing resistant against fire (1977-09)
	6. Ventilation ducts (1977-09)
	7. Roofing (1998-07)
	8. Small-scale test furnace (1986-05)
	9. Seals for cable penetrations (1990-05)
	11. Pipe encasements, pipe bushings, service shafts and ducts, and barriers across inspection openings (1985-12)
	12. Circuit integrity maintenance of electric cable systems (1998-11)
	13. Fire-resistant glazing (1990-05)
	14. Determination of the burning behavior of floor covering systems using a radiant heat source (1990-05)
	15. "Brandschacht" (1990-05)
	16. "Brandschacht" tests (1998-05)
	17. Determination of melting point of mineral fiber insulating materials (1990-12)
	18. Fire barriers, verification of automatic closure (continuous performance test) (1991-03)
	19. Surface products on walls and under floors inside rooms, test room for additional assessments[a] (1998-12)
	20. Special proof of the burning behavior outside wall facing[b]

[a] Draft
[b] Draft in preparation

The following section deals only with the fire performance of building materials (DIN 4102-1); building components lie outside the scope of this book.

The list and application of classified building materials is contained in DIN 4102-4. DIN standards are issued by DIN (Deutsches Institut für Normung, D-10772 Berlin).

10.9.2.1 DIN 4102-1 Building Materials

Building materials may be divided into various classes using this standard and the fire tests described therein. These classes and the prescribed test methods are summarized in Table 10.60.

Table 10.60   DIN 4102-1 classification and test methods for building materials

Building material class	Building inspection designation	Test method
A A1[a] A2	**Noncombustible**	• Furnace test 750 °C • Brandschacht • Smoke density • Toxicity • Calorific potential to DIN 51900-1 and heat release to DIN 4102-8 or furnace test 750 °C
B B1[b]	**Combustible** Low flammability	• Brandschacht (including smoke density and burning drops) and small burner test • Special case floor coverings: radiant panel test to DIN 4102-14
B2	Moderately flammable	• Small burner test • Special case textile floor coverings: small burner test to DIN 54332
B3	Highly flammable	• No tests, no compliance with B2

[a] Class A2 requirements must also be satisfied.
[b] Class B2 requirements must also be satisfied.

Building materials can be classified in classes A (noncombustible) or B (combustible). Owing to their organic structures, plastics usually achieve only class B. In certain cases, for example, as composites with inorganic materials, they achieve class A2.

The test methods for noncombustible and combustible building materials are described in DIN 4102-1, Sections 5 and 6, DIN 4102-14 and DIN 4102-16. Additional requirements for building materials subject to mandatory test marking are laid down in specific guidelines [2, 3].

Noncombustible Building Materials

Class A1 Building Materials

Noncombustible building materials satisfy the prerequisites for inclusion in class A1 if they pass the 750 °C furnace test and fulfill the additional requirements of class A2. At present, the Deutsches Institut für Bautechnik (DIBt, German Institute for Building Technology) does not require a test mark in the case of building materials with an organic content of < 1 %; the official tests for class A1 suffice. Test marks are mandatory for materials with an organic content > 1 %.

A diagram and specifications for the furnace test are given in Fig. 10.32 and Table 10.61, respectively.

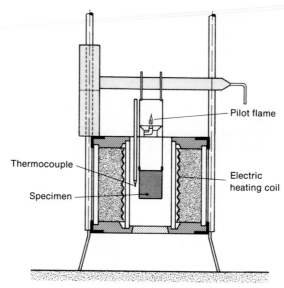

Figure 10.32   Electric furnace

Table 10.61   Test specifications for class A1 electrically heated furnace

Specimens	Five specimens 40 mm × 40 mm × 50 mm
Specimen position	Vertical
Ignition source	Electric furnace: heated to 750 °C Pilot flame: height 20 mm
Test duration	15–30 min
Conclusions	Passed if none of the five samples give rise to • Flaming, • A temperature rise of more than 50 °C in the furnace.

Flaming is considered to have occurred if flames are observed in the furnace or if the sample glows or if the enlarged pilot flame exceeds a height of 45 mm (1.8 in.) or fills the opening in the furnace lid.

Class A2 Building Materials

To achieve class A2, building materials must satisfy the following:

- Brandschacht test,
- Smoke density tests,
- Toxicity test (no longer required in all cases),
- Calorific potential and heat development test or the furnace test.

A diagram and specifications for the Brandschacht test are given in Fig. 10.34 (see under combustible building materials below) and Table 10.62 respectively.

Table 10.62   Test specifications for class A2 Brandschacht

Specimens	Three sets of four specimens 190 mm × 1000 mm × original thickness (max. 80 mm)
Specimen position	Vertical, specimens at right angles to one another
Ignition source	Ring burner
Test duration	10 min
Conclusions	Passed if • The mean residual length is at least 350 mm; • No specimen < 200 mm; • Mean smoke gas temperature does not exceed 125 °C; • The back of any sample does not flame; • No other reservations exist
Observation	Smoke development during test time: < 400 % × min or exact value

The determination of smoke density and toxicity of fire gases is described in the Annexes A–C of DIN 4102-1. The measurement is based on ASTM D 2843-70 (XP2 apparatus) and DIN E 53 436/37 (see Chapter 15).

The calorific potential H_u is determined according to DIN 51 900-2 [4] using an isothermal bomb calorimeter. To satisfy the test, the calorific value H_u must not exceed 4200 kJ/kg.

Heat release is measured on a small-scale test rig to DIN 4102-8. The test is passed if the heat liberated determined from the calorific value H_u and the weight per unit area before and after the test does not exceed 16,800 kJ/m².

Tests for calorific potential and heat release can be dispensed with if the furnace test (Fig. 10.33) is carried out under the conditions for A2 rather than A1 materials and is passed. A 15 min test suffices and flaming for a total of 20 s is allowed.

Combustible Building Materials

Class B1 Building Materials

Combustible building materials are placed in class B1 if they pass the Brandschacht test and satisfy class B2 requirements.

A diagram of the Brandschacht equipment is shown in Fig. 10.33 and the test specifications are listed in Table 10.63.

Fire performance testing of floor coverings is a special case. Classification in class B1 is not based on successful completion of the Brandschacht test but of the flooring radiant panel test (DIN 4102-14). Further evidence that class B2 requirements are satisfied is necessary. The flooring radiant panel test is illustrated in Fig. 10.34 while the test specifications are summarized in Table 10.64.

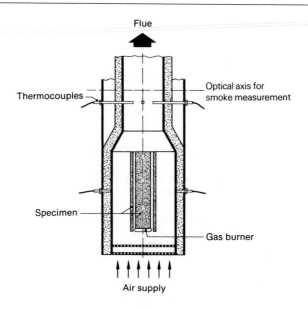

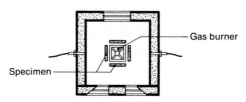

Figure 10.33   Brandschacht

Table 10.63   Class B1 Brandschacht test specifications

Specimens	Three sets of four specimens 190 mm × 1000 mm × original thickness (max. 80 mm)
Specimen position	Vertical, specimens at right angles to one another
Ignition source	Ring burner
Test duration	10 min
Conclusions	Passed if • The mean residual length is at least 150 mm; • Residual length must not be 0 mm for any specimen, • Mean smoke gas temperature does not exceed 200 °C, • No other reservations exist.
Observation	Smoke development during test time: < 400 % × min or exact value

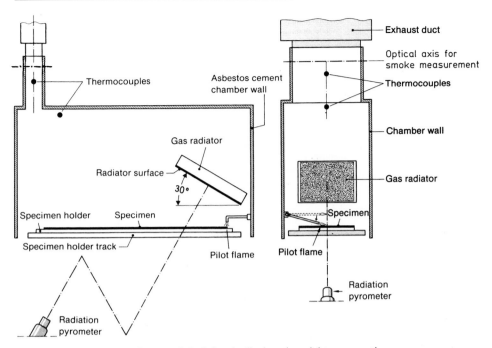

Figure 10.34   Flooring radiant panel. Left: longitudinal section; right: cross section

Table 10.64   Flooring radiant panel test specifications

Specimens	Three specimens 230 mm × 1050 mm × usual thickness
Specimen position	Horizontal
Ignition source	• Gas heated radiation panel 305 mm × 457 mm. Operation temperature up to 815 °C, inclined at 30° to the horizontal; lower edge 140 mm above the specimen; radiation on specimen 1.1–0.1 W/cm², • Swiveling propane pilot flame, inner blue flame cone 13 mm long. Flame impinges perpendicularly to the longitudinal axis on the middle of the narrow edge on the radiant panel side; ignition can be pivoted up parallel to the specimen at a height of 50 mm.
Test duration	10 min of flaming and irradiation; if no ignition occurs, swing flame up and irradiate for additional 10 min; if ignition occurs, continue until flame extinguishes, up to a max. of 30 min
Conclusions	Passed if • The average of the burnt length corresponds to a radiation flux of at least 0.45 W/cm², • Smoke development does not exceed 750 % × min.

Class B2 Building Materials

Building materials achieve class B2 if they satisfy the test requirements given in Table 10.65. Diagrams of the test rigs are shown in Figs. 10.35–10.37.

Table 10.65 Test specifications for class B2 building materials

Specimens	• Edge application of flame: five specimens 90 mm × 190 mm × original thickness (max. 80 mm), reference mark 150 mm from lower edge • Surface application of flame: five specimens 90 mm × 230 mm × original thickness (max. 80 mm), reference marks 40 mm and 150 mm from lower edge
Specimen position	Vertical
Ignition source	Small flame burner, inclined at 45°, flame height 20 mm
Flame application	15 s
Test duration	20 s
Conclusions	Passed if the tip of the flame does not reach the reference marks within 20 s on any sample for: • Edge application of flame (not applicable for materials used with protected edges), and for • Surface application of flame, if failure is expected with this test. If the filter paper under the sample ignites within 20 s after flaming, the material is judged to burn with flaming droplets.

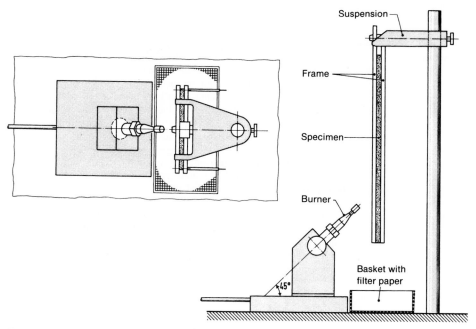

Figure 10.35 Small burner for testing homogeneous building materials (class B2)

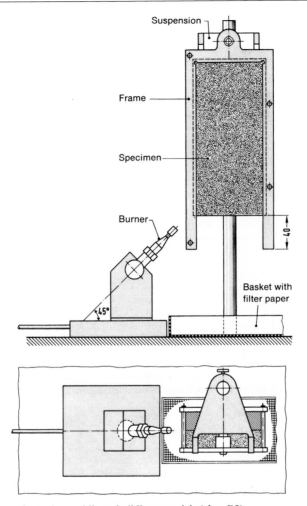

Figure 10.36   Set up for testing multilayer building materials (class B2)

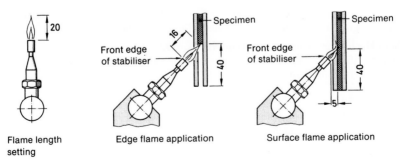

Figure 10.37   Various burner settings

These specifications are contained in the revised version of DIN 4102-1 published in May 1998 and apply to single and multilayer building materials. For the latter, the flame impinges on the least favorable point on the front edge of the specimen as shown in Fig. 10.38.

Textile floor coverings can be classified in class B2 if they are tested according to DIN 54 332 [5] and meet at least the requirements of burning class T-b to DIN 66 081.

The DIN 54332 small burner test corresponds in principle to the apparatus for testing class B2 building materials, although with larger specimens and specimen holder and without the basket and filter paper for testing burning drops. A diagram of the test equipment is shown in Fig. 10.36. The test specifications are listed in Table 10.66, the flame is applied for 15 s and alternatively for 5 s, in contrast to DIN 4102.

Table 10.66   Small burner test specifications for textile floor coverings

Specimens	Five specimens 340 mm × 104 mm × original thickness on a fiber-cement board, reference mark 40 mm from lower edge, cotton thread drawn in front of sample reaching 250 mm above the ignition point
Specimen position	Vertical
Ignition source	Small flame burner, inclined at 45°, flame height 20 mm
Flame application	15 s, resp. 5 s
Test duration	20 s
Conclusions	Passed if the cotton thread has not burnt through in any of the 5 samples before 25 s at a flame application of 15 s or before 30 s at a flame application of 5 s

Class B3 Building Materials

Combustible building materials that cannot be classified in class B1 or B2 are placed in class B3.

10.9.3 Official Approval

In Germany, fire performance and secondary fire effects of building materials are tested by the following recognized institutions ([6] and status December 2001):

Brandhaus Höchst der Siemens Axiva GmbH & Co. KG [1, 2]
Industriepark Höchst C 369
D-65926 Frankfurt am Main
Tel.: +49-69 305 3476
Fax: +49-69 305 17071
brandhaus@siemens.com
www.siemens-axiva.com/po_monomere

Forschungs- und Materialprüfungsanstalt Baden-Württemberg (FMPA)
Otto-Graf-Institut [2]
Pfaffenwaldring 4
D-70569 Stuttgart
Tel. +49-711 685 2713
Fax: +49-711 685 6829
stefan.lehner@po.uni-stuttgart.de
www.fmpa.de

Gesellschaft für Materialforschung und Prüfanstalt für Bauwesen Leipzig mbH
-MFPA Leipzig GmbH-
Hans-Weigel-Str. 2B
D-04319 Leipzig
Tel. +49-341 6582 134
Fax: +49-341 65 82 197
Brandschutz@mfpa-leipzig.de
www.mfpa-leipzig.de

Institut für Baustoffe, Massivbau und Brandschutz der Technischen Universität Braunschweig
Amtliche Materialprüfanstalt [1, 2]
Beethovenstrasse 52
D-38106 Braunschweig
Tel. +49-531 391 5466
Fax: +49-531 391 8159
p.nause@tu-bs.de
www.ibmb.bau.tu.bs.de

Institut für Baustoffkunde und Materialprüfung der Universität Hannover
Amtliche Materialprüfanstalt
Nienburger Strasse 3
D-30167 Hannover
Tel. +49-511 762 2240
Fax: +49-511 762 4001
bernd.restorff@mpa-bau-hannover.niedersachsen.de
www.mpa-bau-de

Institut für Holzforschung (HFM)[1]
Universität München
Winzererstrasse 45
D-80797 Munich
Tel. +49-89 2180 6420
Fax: +49-89 2180 6429
Ehr@holz.forst.tu-muenchen.de
www.holz.forst.tu-muenchen.de

MPA Dresden
Referat Brandschutz Freiberg
Fuchsmühlenweg 6F7
D-09599 Freiberg
Tel. +49-3731 34850
Fax: +49-3731 34842
thomas.huebler@mpa.smwa.sachsen.de
www.smwpa.sachsen.de/mpa/brandschutz

Materialprüfungsamt Nordrhein-Westfalen (MPA NRW)[1, 2]
Außenstelle Erwitte
Auf den Thränen 2
D-59597Erwitte
Tel. +49-2943 89711
Fax: +49-2943 89733
mpaerwitte@aol.de
www.mpanrw.de

Prüfinstitut Hoch für das Brandverhalten von Bauprodukten
Lerchenweg 1
D-97650 Fladungen
Tel. +49-9778 740 163
Fax: +49-9778 740 164
Hoch.fladungen@t-online.de
www.brandverhalten.de

For toxicity:

Rheinisch-Westfälische Technische Hochschule
Institut für Hygiene und Umweltmedizin
Pauwelsstr. 30
D-52074 Aachen
Tel. +49-241 80 88385
Fax: +49-241 80 8888587
Wolfgang.dott@post.rwth-aachen.de
www.ukaachen.de

For testing class B2 building materials only:

Bundesanstalt für Materialforschung und -prüfung (BAM)[1]
Unter den Eichen 87
D-12205 Berlin
Tel. +49-30 8104 4238
Fax: +49-30-8112 029
Manfred.korzen@bam.de
www.bam.de

Forschungsinstitut für Wärmeschutz e. V. München (FIW)
Lochhamer Schlag 4
D-82166 Gräfelfing
Tel. +49- 89858000
Fax: +49- 89858040
Info@fiw-muenchen.de
www.fiw-muenchen.de

For smoke: those institutions marked[1]
For testing floor coverings with radiant panel: those institutions marked[2].

Proof of compliance (Conformity Certification) with established building regulations suffices for most materials utilized in the building industry. Special proof of usability is required for new and certain specifically defined materials. By the middle of the year 2000, the German § 20 MBO system was adapted to the European Construction Products Directive (CPD) 89/106/EC. The German legal equivalent to the CPD is the Bauproduktengesetz (BauPG of August 1992).

Approvals obtained from the previous system are still valid for a maximum period of 5 years. In the past, they were obtained as follows:

- General building inspection approval,
- Test mark,
- Permission for use in individual cases [7].

Because all building products need the CE mark for handling within the European Community and also in Germany, the German Institute for Building Technology DIBt has issued Lists A, B, and C for building products (Bauregelliste A, B, and C) [8, 9]. In accordance with the LBOs, List A contains building products and constructions for which regulations exist (List A Part 1, Regulated Building Products [geregelte Bauprodukte]). List A Parts 2 and 3 contain building products for which no regulations exist and whose use can be assessed by generally recognized test procedures (List A Part 2, Nonregulated Building Products [Nicht geregelte Bauprodukte], and List A Part 3, Nonregulated Building Constructions [Nicht geregelte Bauarten]). List B contains all building products that are CE mark labeled according to European Directives. Building products for which neither technical building regulations nor established technical rules exist are contained in List C [other building products ("sonstige Bauprodukte")].

The institutions recognized by the DIBt with respect to the LBOs as testing laboratories, inspection, and certification bodies according to List A are contained in [6]. The products covered in Lists A, B, and C are summarized in Table 10.67.

List A

The use of List A building products requires a test mark (Ü-Zeichen), documenting the conformity with the technical rules and regulations contained in the list. This can be achieved in three ways:

- Declaration of conformity by the manufacturer (ÜH),
- Declaration of conformity by the manufacturer after product testing in a generally recognized test laboratory (ÜHP),
- Conformity certificate by a recognized certification body (ÜZ).

If products of *List A Part 1* differ in major points from the required technical rules, the producers need a proof of compliance via a:

- General building authority test report (P), or a
- General building authority approval (Z).

List A Part 2 deals with nonregulated building products

- Whose use does not serve essential requirements for the safety of building construction, and
- For which generally recognized technical rules do not exist,

or

- For which neither technical directives nor generally acknowledged technical rules for all requirements exist and
- That can be assessed by generally recognized test procedures with regard to these requirements.

These products need only a general building authority test report (P) or permission for use in individual cases. The proof of compliance (Ü-Zeichen, Conformity Certification) corresponds to the general building authority test report (P).

List A Part 3 deals with nonregulated building constructions

- Whose use does not serve essential requirements for the safety of the building construction, and
- For which no recognized technical rules exist,

or

- For which neither technical directives nor technical rules for all requirements exist, and
- Of which an opinion can be formed by recognized test procedures with regard to those requirements.

These constructions need only a general building inspection test report (P). The Ü-Zeichen corresponds to the general building inspection test report (P). The user of these building constructions has to declare that the used building construction and building products correspond to the general building inspection test report (P).

Table 10.67   Lists A–C for German building rules (Bauregelliste A bis C)

List A Part 1	Content
Section 1	Building products for concrete and steel construction
2	Building products for stonework and brick construction
3	Building products for timber construction
4	Building products for metal construction
5	Insulating products for thermal and acoustic protection
6	Doors
7	Storage rooms
8	Special constructions
9	Building products for wall and ceiling linings
10	Building products for draught proofing of building structure and roof
11	Glass building products
12	Building products for the drainage of plots of land
13	Facilities for waste water treatment
14	Heating installations
15	Building products for stationary used facilities for stocking and handling of dangerous substances for water
16	Building products for erection of scaffolding
Part 2	
Sections 1.1–1.10	Building products with no technical regulations or generally recognized technical rules and with no applications serving essential requirements for building construction safety
Sections 2.1–2.33	Building products with no technical regulations or generally recognized technical rules or without rules for all requirements, and of which can be formed an opinion by generally recognized test procedures with regard to these requirements
Part 3	
Sections 1–10	Building constructions differing in main points from technical regulations or with no generally recognized technical rules or with no applications serving essential requirements for building construction safety of which can be formed an opinion by generally recognized test procedures with regard to these requirements

List B Part 1	No entry until beginning of 2001
Part 2	
Section 1 2 3	Technical building equipment Building products for stationary used facilities for stocking and handling Fire safety equipment
List C	Building products which have to meet class B2 (DIN 4102-1 B2) according to building regulations
Section 1 2 3 4 5 6	Building products for the shell Building products for (constructions works) completion of the building Building products for domestic installation Building products for stationary used facilities for stocking and handling of dangerous substances for water Other building products Building products for dumping sites

List B

List B contains all building products that are in compliance with European Directives for Construction Products and which are labeled with the CE mark.

In 2002, only a few building products were listed, since only few new harmonized product standards were available so far.

List C

Building products that are required to meet DIN 4102-1 Class B2 are contained in List C.

The most common proof of usability is the Ü-Zeichen test mark. Regulations exist in Germany specifying which building materials must be testmarked within the framework of fire protection laid down in Lists A, B, and C.

These are:

- Building materials that must be noncombustible but have combustible components
- Building materials and textiles that must be of low flammability according to class B1,
- Fire retardants for building materials and textiles that must be of low, class B1 flammability.

Details of fire testing procedures (number of specimens, ageing tests, reserve samples, composites, etc.) are described in the standards DIN 4102-1, -14, and -16.

DIN 4102-16 also contains a short section on burning drops. Building materials are considered to give rise to burning drops if during at least two tests for class B1, drops or parts of the specimen continue to burn for 20 s or longer on the sieve beneath the burner or, if during class B2 test, burning drops are detected. Burning drops are not otherwise evaluated in building inspection techniques.

10.9.4 Future Developments

For an as yet undefined transition period, the DIN 4102 standards series will still be nationally valid in parallel to the new European standards series.

References for Section 10.9

[1] Introductory decrees for DIN 4102 of the German Federal States
[2] Richtlinien zum Übereinstimmungsnachweis nichtbrennbarer Baustoffe (Baustoffklasse DIN 4102-A) nach allgemeiner bauaufsichtlicher Zulassung – Version Oct. 1996 – Mitt. DIBt 28 (1997) 42–44
[3] Richtlinien zum Übereinstimmungsnachweis schwerentflammbarer Baustoffe (Baustoffklsse DIN 4102-B1) nach allgemeiner bauaufsichtlicher Zulassung – Version Oct. 1996 – Mitt. DIBt 28 (1997) 39–41
[4] DIN 51900-2, 1977-08. Testing of Solid and Liquid Fuels; Determination of the Gross Calorific Value by the Bomb Calorimeter and Calculation of the Net Calorific Value; Method Using Isothermal Water Jacket
[5] DIN 54332, 1975-02. Testing of Textiles; Determination of the Burning Behaviour of Textile Floor Coverings
[6] Verzeichnis der Prüf-, Überwachungs- und Zertifizierstellen nach den Landesbauordnungen DIBt Mitt. Sonderheft 25 (2001)
[7] G. Lichtenauer: Schadensprisma (1995) 53–57
[8] Bauregelliste A, Bauregelliste B und Liste C – Version 2001/1 – Mitt. DIBt Sonderheft 24 (2001) 1–151
[9] Änderung der Bauregelliste A, Bauregelliste B und Liste C – Version 2001/2 – Mitt. DIBt 32 (2001) 178–192

10.10 Hungary

T. Banky

10.10.1 Statutory Regulations

The statutory regulations for structural fire protection in Hungary are contained in Standard-series MSZ 595, "Fire protection of buildings" [1].

MSZ 595-2:1994 "Fire protection of buildings. Classification of building materials to combustibility, smoke generation, and burning drips" covers the classification of building materials, lists the prescribed test standards (see later) and contains a summary of materials classified with the aid of these tests.

MSZ 595 Part 3 contains requirements for the fire endurance of building elements and structures in relation to their application, the height and number of stories of the building, and category of stability against fire of the building which depends on the fire hazard of usage of building. It also details the classification methods of building elements according to combustibility. Parts 5–9 contain the regulations covering permitted use of building materials according to their fire performance, the necessary preconditions for intervention by the fire brigade, division into fire compartments, size of escape routes, calculation of the required fire endurance based on fire load, as well as smoke and heat extraction and size of ventilation apertures.

The Hungarian Fire Code, the "Országos Tűzvédelmi Szabályzat" OTSZ (State Regulation for Fire Protection), details, in conjunction with MSZ 595, general regulations for erecting and operating structural works. Further standards cover fire precautions in computercenters, theatres and cinemas, and so forth.

10.10.2 Classification and Testing of the Fire Performance of Building Materials and Components

According to MSZ 595-2:1994 building materials are divided into the following groups:

Combustibility groups:

"nem éghető" (noncombustible):

Class A1

Class A2

"éghető" (combustible):

Class B1, "nehezen éghető" (low combustibility)

Class B2, "közepesen éghető" (moderately combustible)

Class B3, "könnyen éghető" (highly combustible)

Smoke development classes:

Class F0, "füstöt nem kibocsátó" (non-smoke-producing)

Class F1, "mérsékelt füstfejlesztő" (moderately smoke-producing)

Class F2, "fokozott füstfejlesztő" (highly smoke-producing)

Burning-dripping classes:

Class C0, "olvadékot nem képző" (no drips)

Class C1, "gyulladást okozó olvadékot nem képző" (no burning drips)

Class C2, "égve csepegő" (burning drips)

Classification of building materials and components is made on the basis of tests conforming to MSZ 14799 [2] and 14800 Parts 1–16 [3].

MSZ 14799, 14800-1, MSZ 14800-5, and MSZ 14800-7, which cover the fire resistance of building components, are not dealt with.

10.10.2.1 MSZ 14800-2:1994 Noncombustibility Test

The MSZ 14800-2:1994 noncombustibility test is basically similar to EN ISO 1182. The noncombustibility furnace and test specification are discussed in Section 10.23 ISO.

10.10.2.2 MSZ 14800-3:1982 and MSZ 14800-4:1984. Testing Building Materials for "Low Combustibility" and "Moderate Combustibility"

These tests largely resemble the DIN 4102 Part 16 "Brandschacht test" shown in Section 10.9 Germany. The test specifications that are summarized in Table 10.68 differ somewhat from the German version.

Combustible floorings are tested to *MSZ 14800-9:1985* and classified. The test corresponds to the flooring radiant panel test to ASTM E 648 and DIN 4102-14 (see Section 10.9) and lasts 10 min. The surface spread of flame L_T in cm is measured and the critical irradiation intensity I is determined. The following limiting values are used for classification:

Low spread of flame:	$L_T \leq 40\,\text{cm}$	$I \geq 0.50\,\text{W/cm}^2$
Moderate spread of flame:	$40\,\text{cm} < L_T \leq 60\,\text{cm}$	$0.5\,\text{W/cm}^2 > I \geq 0.25\,\text{W/cm}^2$
High spread of flame:	$L_T > 60\,\text{cm}$	$I < 0.25\,\text{W/cm}^2$

Table 10.68 Specifications for testing low and moderate combustibility of building materials

Specimens	Three sets of four specimens 1000 mm × 190 mm × usual thickness
Specimen position	Vertical, perpendicular to each other
Ignition source	Quadratic gas burner with 32 jets, gas supply with 88.0 ± 2.2 MJ/h
Test duration	10 min (low combustibility) 2 min (moderate combustibility)
Conclusions	Classification "low combustibility" (after 10 min) or "a moderate combustibility" (after 2 min) achieved if (mean values for the test set): • The consumed length of the specimen does not exceed 85 %, • The mean exhaust-gas temperature does not exceed 235 °C, • The weight loss of the specimen is max. 80 %, • Sustainable flaming does not exceed 30 s, • The individual values must not exceed 90 %, 250 °C, 85 % and 60 s.

10.10.2.3 MSZ 14800-6:1980. Fire Propagation Test for Facades

In this standard, vertical and horizontal fire propagation and their limiting values on light-weight facades, combustible wall claddings, curtain walls, and facades with large areas of glazing are determined by means of a model fire test using a defined test fire. The standard applies to wall structures for buildings of at least two stories. The layout of the test rig (test building) is shown in Figs. 10.38 and 10.39. The test specifications are summarized in Table 10.69.

Table 10.69 Specifications for testing fire propagation on facades

Specimen	One test component 9.2 m high × 8.5 m wide × usual thickness
Specimen position	Vertical
Ignition source	Wood crib (fire load depends on type of building and contents, see below), ignition with 10 kg of fuel oil
Test duration	Until maximum fire propagation or until complete combustion of the crib
Conclusions	Fire propagation limits specified in hours

The fire tests are carried out in a three-story test building shown above with the test component built into the front wall. A wood crib is located on the ground floor representing a fire load calculated according to MSZ 595-7:1994.

The wood crib is placed 30–50 cm (11.8 in.–19.7 in.) away from the wall under test and must cover at least 50 % but not more than 60 % of the floor. The crib is set alight with the help of 10 kg of fuel oil contained in a trough slid under it. The temperature in the fire room is measured with five thermocouples and that in the upper story with two thermocouples. The test is continued until maximum fire spread or until complete combustion of the crib.

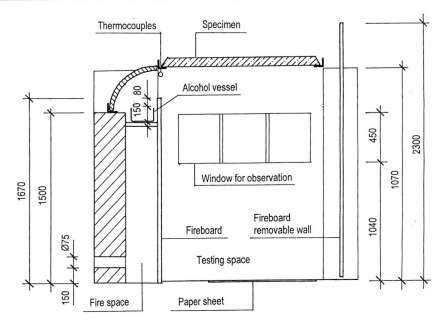

Figure 10.38 Vertical section of test rig

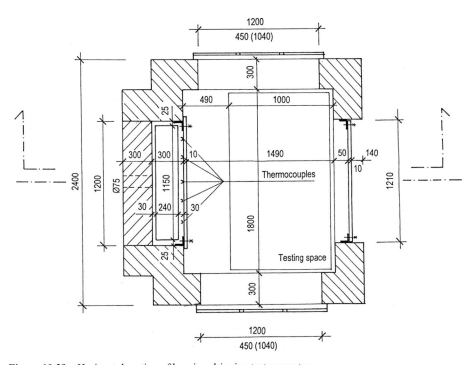

Figure 10.39 Horizontal section of burning-dripping test apparatus

The fire spread limit is the time elapsed until one of the following events occurs:

- Flames or hot smoke gases penetrate between the facade and ceiling.
- The temperature rise on the inner face of the facade averages 150°C or is 190°C at one measuring point.
- Flames or hot gases penetrate the test wall in the second floor.
- A curtain made of combustible material located in the first floor and the second floor is set on fire by the glazed surface breaking.
- The fire on the surface of the front wall spreads to the ceiling of the second floor.
- The fire on the surface of the front wall spreads sideways to 1.5 m (4.9 ft) or more.
- The test wall or its holder is destroyed.

10.10.2.4 MSZ 14800-8:1994. Test for Burning Dripping of Plastic Building Materials

The test specifications are summarized in Table 10.70.

Table 10.70   Test for burning drips

Specimens	Two specimens: 2000 mm × 1500 mm × usual thickness
Specimen position	Horizontal
Ignition source	$8000 \, cm^3$ ethanol
Test duration	20 ± 5 min
Conclusions	Observation of burning and nonburning droplets and ignition of paper on the floor

10.10.3 Official Approval

In Hungary, fire performance tests on building materials and structures are carried out by the official test institute for the building industry:

Építésügyi Minőségellenőrző Innovációs Kht. (ÉMI)
(Institute for Quality Control and Innovation in Building)
H-1113 Budapest, Diószegi u. 37
Hungary
Tel. + 36 1 361 36 72
Fax. + 36 1 361 36 72
E-mail: tbanky@mail

ÉMI is in possession of a fire testing laboratory for building industry accredited according to EN 45001 (EN 17025) by the Hungarian Accreditation Board (NAT), and also performs product certification according to prescriptions of EN 45011 as an accredited institute.

In Hungary, in accordance with the regulations in force, conformity attestation of the marketed building materials, products, structures as well as the applied construction technologies is compulsory.

The attestation should be based on Hungarian Standards related to the material, product, and so forth, or in the absence of Hungarian Standards on Construction Technical Approvals. For marketing and applying "new-type" products it is compulsory to obtain the relating Construction Technical Approvals.

ÉMI is the only designated approval body for the development and issuing of Construction Technical Approvals. Test reports, certificates, and technical approvals issued by ÉMI are fully accepted by Hungarian authorities.

The Fire Protection Laboratory of ÉMI is a member with full powers of the European Group of Official Laboratories for Fire Testing (EGOLF), which is the relevant organization of this field.

10.10.4 Future Developments

Hungary intends to join the European Union as soon as possible. Hungarian experts are participating in the work within CEN/TC 127 ("Fire Safety in Buildings") on development of resistance-to-fire and reaction-to-fire test methods. There is a bilateral contract existing between the EU and Hungary: – among others – all CEN testing and classification standards will be adopted by Hungary after their official acceptance and publication.

Meanwhile the ÉMI Fire Laboratory is continuously expanding its testing repertory with the accepted EN methods.

References for Section 10.10

[1] MSZ 595-1:1986 Fire protection of buildings. Definitions
     MSZ 595-2:1994 Fire protection of buildings. Classification of building materials to combustibility, smoke generation and burning drips
     MSZ 595-3:1986(M1987) Fire protection of buildings. Fire resistance requirements of building structures
     MSZ 595-4:1986 Fire protection of buildings. Medium-high and high-rise buildings
     MSZ 595-5:1987 Fire protection of buildings. Fire compartments
     MSZ 595-6:1980 Fire protection of buildings. Evacuation of buildings
     MSZ 595-7:1994 Fire protection of buildings. Method for determination of calculated fire load and calculation of authoritative fire resistance requirements
     MSZ 595-8:1994 Fire protection of buildings. Heat and smoke exhausts for hall-type buildings
     MSZ 595-9:1994 Fire protection of buildings. Explosion overpressure relief vent and rupture diaphragms
[2] MSZ 14799:1988 Fire resistance requirements, laboratory testing and qualification of chimney structures
[3] MSZ 14800-1:1989 Resistance-to-fire tests. Fire resistance limit test of building structures
     MSZ 14800-2:1994 Reaction-to-fire tests. Test for "noncombustibility" of building materials
     MSZ 14800-3:1982 Reaction-to-fire tests. Determination of "low combustibility" of combustible building materials
     MSZ 14800-4:1984 Reaction-to-fire tests. Determination of "moderate combustibility" of combustible building materials
     MSZ 14800-5:1994 Resistance-to-fire tests. Determination of fire resistance for doors and inner glazing systems
     MSZ 14800-6:1980 Resistance-to-fire tests. Fire propagation test for building facades
     MSZ 14800-7:1983 Resistance-to-fire tests. Determination of fire protective ability for suspended ceilings
     MSZ 14800-8:1994 Reaction-to-fire tests. Test for burning dripping of plastic building materials
     MSZ 14800-9:1985 Reaction-to-fire tests. Test for the flame propagation on combustible floor coverings and classification
     MSZ 14800-10:1987 Reaction-to-fire tests. Test for smoke development characteristics of solid combustible materials
     MSZ 14800-11:1991 Reaction-to-fire tests. Test for fire propagation on roof structures.
     MSZ 14800-14:1990 Resistance-to-fire tests. Test method for determination of fire resistance limit of cable penetration
     MSZ 14800-15:1990 Resistance-to-fire tests. Test for fire dampers and ventilation ducts
     MSZ 14800-16:1992 Reaction-to-fire tests. Determination of ignition temperature of solid materials

10.11 Italy

S. MESSA

10.11.1 Statutory Regulations

The state building law (Legge Urbanistica) was enacted in 1942, brought up to date in 1967 by further legislation (Legge Ponte), and regulates the principles for the exploitation of building sites and the planning of buildings. Fire precautions must be supervised by the fire brigade.

In 1982 the decree of the President of the Republic No. 577 introduced the concept that all regulations on fire safety are governed by official laws and not just by instructions. Subsequently a revision of all fire prevention regulations was carried out. The new regulations in the "Decretos" of the Ministry of the Interior apply to theaters, garages, hotels, and gymnasiums and will be completed by codes relating to schools, hospitals, high-rise buildings, and department stores. These regulations concern compartments, exits, safety routes, and all other measures for active and passive fire precautions. Existing buildings require a provisional approval (nullaosta provvisorio) according to Law 818/84.

The classification of the fire behavior and the certification of materials for the purpose of fire prevention is regulated by the Decreto of the Ministry of the Interior issued on June 26, 1984 [1].

10.11.2 Classification and Testing of the Fire Performance of Building Materials and Components

The Decree of June 26, 1984 contains test procedures enabling the fire performance of building materials to be determined and classified. These test methods have been adopted by the Italian Standards Organization, UNI (Ente Nazionale Italiano di Unificazione) and other organizations, such as Uniplast, Unitex, and so forth.

The methods contained in the Decree of June 26, 1984 are the noncombustibility test for building materials based on ISO 1182, the CSE RF 1/75/A (edge application of flame), and CSE RF 2/75/A (surface flame application) small burner ignitability tests for combustible materials, the CSE RF 3/77 test for the contribution of a material to spread of flame and the CSE RF 4/83 test for upholstered furniture.

According to Italian opinion, these procedures define the characteristic parameters determining the fire hazard of building materials in the early and advanced stages of an initiating fire. The results of these tests enable building materials to be classified in various fire classes that are also specified in the Decree of June 26, 1984.

The standardization situation has radically been evolving in the year 2000. For several years, it has been attempted to line up the situation of our country to the "new approach" philosophy: in practice, the technical rules – standards – are the ones produced by the standardization body (in Italy UNI), while the regulator has to fix the safety levels and the technical rules for the uses of products.

All Italian fire safety standards were established by the Ministry of the Interior, which is also the Ministry responsible for fire prevention and fire brigades. Representatives from all industrial sectors nominated by an interdisciplinary Ministerial Committee, several members of Ministry, fire testing laboratories, representatives of professional associations, public administrations, and so forth contributed to their finalization.

The text of each standard has been published in the Official Gazette, as CSE methods, and the Ministry has applied –through regulations – the results of tests in different situations.

A UNI version of all the standards used in regulations has been prepared, with several substantial changes, derived from experience and changing laws. The Ministry has prepared a Decree establishing that:

- The UNI standards will be adopted in the most up-to-date version, including the various amendments already published or to be published,
- The standards published in the Official Gazette as CSE methods are to be cancelled.

The versions of the standards included in the following are based on the latest UNI versions that will go into effect with the publication of this Decree, approved by our Ministry and sent to Brussels for the necessary procedure. The preparation of the decree took much longer than expected, but is expected to be published soon.

Standards are issued by the Ente Nazionale Italiano di Unificazione (UNI) Via Battistotti Sassi 11/b, I-20133 Milano.

10.11.2.1 UNI ISO 1182 Noncombustibility Test

Test method UNI ISO 1182 of December 1995 is equivalent to the noncombustibility test of building materials described in Section 10.23.

10.11.2.2 UNI 8456 Small Burner Test (Edge Application of Flame)

Method UNI 8456 [2] is used to determine afterflame time, afterglow time, the extent of damage of materials, and flaming droplets such as curtains and awnings that may be subjected to the effects of a flame from both sides. Droplets are defined as all burning debris in addition of drops. The method is basically similar to the EN ISO 11925-2 small-flame test (see Section 10.4 European Union). The test specifications are somewhat different and summarized in Table 10.71.

The flame is applied to the specimen for 12s. The afterflame time, afterglow time, extent of damage, and flaming droplets are observed and recorded. The four parameters above are divided into three grades as shown in Table 10.72.

Table 10.71   UNI 8456 small burner test specifications

Specimens	340 mm × 104 mm Two series of 10 specimens (5 in the direction of warp and 5 in the direction of weft)
Specimen position	Vertical
Ignition source	Small propane gas burner inclined at 45°, flame length 40 mm, edge application of flame
Flame application	12 s
Test duration	10 min
Conclusions	Depending on test performance classification into various categories taking into consideration afterflame time, afterglow, extent of damage, and flaming droplets

To establish the category that serves as the basis for classifying the building material, the grades of the four parameters are multiplied by weighting factors (Table 10.73). Materials that are consumed in < 17s (12s for flame application and 5s for afterflame) are placed in the worst category, IV; if they burn for more than 10 min they are classified in grade 3 for extent of damage.

Table 10.72   Grading of building materials to UNI 8456

Grade	Afterflame time (s)	Afterglow time (s)	Extent of damage (mm)	Time for drippings to extinguish (s)
1	≤ 5	≤ 10	≤ 150	non burning
2	> 5 – ≤ 60	> 10 – ≤ 60	> 150 – ≤ 200	≤ 3
3	> 60	> 60	> 200	> 3

Table 10.73   Determination of building material categories

Parameter	Weighting factor
Afterflame time	2
Afterglow time	1
Extent of damage	2
Dripping	1
Category	Weighted sum-of grades (grade × weighting)
I	6–8
II	9–12
III	13–15
IV	16–18

10.11.2.3 UNI 8457 Small Burner Test (Surface Application of Flame)

Method UNI 8457 [3], with the update UNI 8457/A1 (May 1996), is used to determine afterflame time, afterglow time, the extent of damage of materials, and flaming droplets and hence fire performance of linings and materials for floors, walls, and ceilings by surface application of flame from one side.

This test method closely resembles the EN ISO 11925-2 small-flame test (see Section 10.4 European Union). The Italian version, however, does not include a basket with a filter paper to test for burning droplets. The test specifications are summarized in Table 10.74.

The flame is applied to the specimen for 30s. The four parameters, afterflame time, afterglow time, extent of damage, and flaming droplets are observed and recorded. These parameters are divided into three grades. The evaluation criteria are identical to those in Table 10.72 except for extent of damage which is placed in grade 3 if ≥ 200 mm (7.9 in.).

The procedure for establishing the building material category is the same as that given for UNI 8456 in Table 10.73. Similarly, materials that are consumed within 35 s are placed in category IV and those that burn for longer than 10 min are placed in grade 3 with regard to the extent of damage.

Table 10.74   UNI 8457 small burner test specifications

Specimens	340 mm × 104 mm Two series of 10 specimens (5 in the direction of warp and 5 in the direction of weft)
Specimen position	Vertical
Ignition source	Small propane burner, inclined at 45°, flame height 20 mm (flame applied to surface of specimen 40 mm above lower edge)
Flame application	30 s
Test duration	10 min
Conclusion	Depending on test performance, classification into various categories taking into account afterflame and afterglow times, extent of damage, and flaming droplets

10.11.2.4 UNI 9174 Spread of Flame Test

The UNI 9174 spread of flame test for building materials [4], with the update UNI 9174/A1 (May 1996), is carried out with a small flame and a radiant panel. The intensity of this ignition source system corresponds to the advanced phase in the initiating fire. By varying the position of the specimen, materials and finished components can be tested under conditions simulating use in floors, walls, and ceilings.

This test method is based on and practically identical with ISO/TC 92 draft proposal DP 5658 (Spread of flame). A diagram of the spread of flame test apparatus and the test specifications are given in Fig. 10.40 and Table 10.75.

The specimen is exposed to the radiant panel and pilot flame. Note is made of whether the specimen self-ignites and of the rate of spread of flame, extent of damage, afterglow, and of flaming droplets. The four parameters are each divided into three grades (Table 10.76). The rate of spread of flame is not measurable if the flame front does not reach the 150 mm (5.9 in.) mark.

The building material category on which classification is based is arrived at by multiplying the above levels by weighting factors that, in the case of flaming droplets, distinguish between floor, wall, and ceiling use (Table 10.77).

If a material burns for longer than 60 min, it is placed in extent of damage grade 3. If the rate of flame spread is ≥ 200 mm/min and the extent of damage ≥ 650 mm, the material is classified in the worst category, IV. If the rate of flame spread is ≥ 30 mm/min and the extent of damage ≤ 200 mm the material usually attains grade 2 for the rate of flame spread.

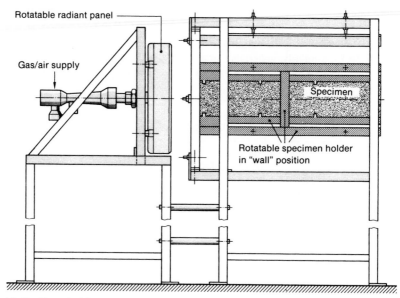

Figure 10.40   Spread of flame apparatus

Table 10.75   UNI 9174 spread of flame specifications

Specimens	Three specimens 800 mm × 155 mm for each orientation (wall, floor, ceiling)
Specimen position	End of exposed area nearest radiator at a distance of 100 mm • *Wall:* 450 mm side of radiant panel horizontal, specimen vertical, long (800 mm) horizontal side at 45° to radiant panel • *Floor:* 450 mm side of radiant panel vertical, specimen horizontal and located centrally in the plane of the bottom of the radiant panel, long (800 mm) side at 90° to radiant panel • *Ceiling:* 450 mm side of radiant panel vertical, specimen horizontal and located centrally in the plane of the top of the radiant panel, long (800 mm) side at 90° to radiant panel
Ignition sources	• Vertical variable propane radiant panel 300 mm × 450 mm, radiation intensity 62 kW/m², surface temperature 750 °C • Variable propane pilot flame, length 80 mm, impinges on specimen 20 mm from edge nearest to radiator
Test duration	Test terminated when flame front stops or spreads to the end of the specimen
Conclusions	The maximum distance in millimeters of flame travel and the time to extinction if the end of the specimen is not reached are recorded.

Table 10.76 Grading of building materials according to UNI 9174

Grade	Spread of flame rate (mm/min)	Extent of damage, maximum length (mm)	Afterglow time (s)	Time for drippings to extinguish (s)
1	Not measurable	≤ 300	≤ 180	nonburning
2	≤ 30	≥ 350 – ≤ 600	> 180 – ≤ 360	≤ 3
3	> 30	≥ 650	> 360	> 3

Table 10.77 Determination of building material categories

Parameter	Weighting		
Rate of spread of flame	2		
Extent of damage	2		
Afterglow	1		
Dripping: floor wall ceiling	0 1 2		
Category	Weighted sum of grades (grade × weighting)		
	Floor	Wall	Ceiling
I	5–7	6–8	7–9
II	8–10	9–12	10–13
III	11–13	13–15	14–17
IV	14–15	16–18	18–21

10.11.3 Classification of Building Materials

The categories obtained from the tests laid down in the old and new Decree to be published form the basis for classifying building materials as shown in Table 10.78.

Building materials are classified by testing according to UNI-ISO 1182, UNI 8456, UNI 8457 and update, as well as UNI 9175 and update. Two, five and seven equivalent combinations of the categories obtained are available for classification in classes 2, 3, and 4, respectively. Fire hazard increases from class 0 (lowest hazard) to class 5 (highest hazard).

Table 10.78   Classification of building materials according to UNI-ISO 1182, UNI 8456, UNI 8457 and UNI 9174

Test method	Category	Class
UNI-ISO 1182	as described in Section 10.23 ISO	0
	compliance with categories to choice	1
UNI 8456 or UNI 8457	I	
UNI 9174	I	
UNI 8456 or UNI 8457	II I	2
UNI 9174	I II	
UNI 8456 or UNI 8457	III II I III II	3
UNI 9174	II III III I II	
UNI 8456 or UNI 8457	IV III III IV II IV I	4
UNI 9174	III IV III II IV I IV	
UNI 8456 or UNI 8457	IV	5
UNI 9174	IV	

10.11.3.1 UNI 9175. Fire Reaction of Upholstered Furniture Under Small-Flame Action

The UNI 9175 spread of flame test for upholstered furniture, with the update UNI 9175/FA1 (July 1994), is carried out with a small flame. The test device is shown in Fig. 10.41 and the specifications are summarized in Table 10.79.

Table 10.79   UNI 9175 Upholstered furniture reaction-to-fire specifications

Specimens	• Seat: 450 mm × 150 mm × 75 mm (five series) • Back: 450 mm × 300 mm × 75 mm (five series) • Fabric: 800 mm × 650 mm (three series)
Specimen position	• Seat: horizontal • Back: vertical
Ignition source	Small propane gas burner, flame length 40 mm
Flame application	20 s – 80 s – 140 s
Test duration	280 s for first part, 600 s for second part
Conclusions	Depending on test performance, classification into various categories based on flame duration

First, the test is carried out only on the filling (two tests). The flame application time is 20 s. If one of the two tests is negative, no more tests are made and no classification given.

Second, if positive results are obtained for filling, the test is performed on a complete specimen with filling, interliner (if present) and fabric with three flame applications (20 s, 80 s and 140 s) in three tests. If the first application is negative, no classification is given.

The test result is negative if the specimen does not stop to burn within 120 s after flame removal or if it is completely burnt within 120 s. The obtainable classifications are listed in Table 10.80.

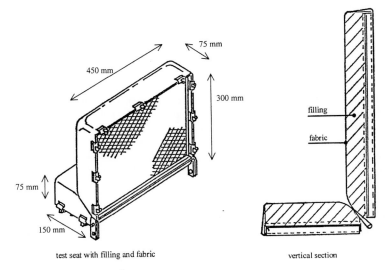

test seat with filling and fabric   vertical section

Figure 10.41   UNI 9 175 test specimen

Table 10.80   Grading of materials according to UNI 9 175

Category	Positive results for
1.IM	First, second, and third flame application (20/80/140 s)
2.IM	First and second flame application (20/80 s)
3.IM	First flame application only (20 s)

10.11.3.2 Specimen Preparation Before Testing

The specimen preparation before testing is done according to UNI 9176 (second edition, January 1998).

Method A: For textile materials susceptible to the effects of flames from both sides, to be tested to UNI 8456 and UNI 9174

Wash five times with a standard washing machine. Each time the procedure is: prewash (35–40 °C, detergent 0.5 g/l), wash (40 °C, detergent 1.5 g/l), dry in stove (60 °C, 2 h) and iron with a steam press or standard iron.

Method C: For all materials to be tested to UNI 8457 and for non-textile materials to be tested to UNI 8456

Procedure for specimens: Brush 200 times for materials for walls and ceiling and for materials susceptible to be attacked by both sides, 5000 times for flooring. The material is put into a detergent solution.

Method D: For nontextile materials to be tested to UNI 9174, for textile materials subjected to the effects of flames from one side to be tested to UNI 9174 and for all covering materials to be tested to UNI 9175

Procedure for specimens: Brush 200 times for materials for walls and ceiling and for materials susceptible to be attacked by both sides, 5000 times for flooring. The material is put into a detergent solution.

10.11.4 Official Approval

Classification and certification of the fire behavior and the official attestation (omologazione) of materials are governed by the old and the new decree to be published. Whereas classification and certification is performed by officially recognized or authorized laboratories and test institutes, the "omologazione" for the production of the material tested is granted by the Ministry of the Interior. The material must then be marked or delivered with an attestation of conformity. The tests for classification of fire behavior can be carried out and certificated by:

State test institute:
C.S.E. (Centro Studi ad Esperienze dei Vigili del Fuoco)
I-00178 Capanelle (Roma)

Authorized laboratories:

Laboratories joint to A.L.I.F. (fire testing officially authorized Italian laboratories association) and E.G.O.L.F. (European Group of Official Laboratories for Fire Testing)

C.S.I. (Servizi dei laboratori di analisi del comportamento al fuoco)
Viale Lombardia, 20
I-20021 Bollate (MI)

LAPI (Laboratorio di Prevenzione Incendi)
Via Della Quercia, 11
I-59100 Prato (PO)

L.S.F. (Laboratorio di Studi e ricerche sul Fuoco)
Via Garibaldi, 28/A
I-22070 Montano Lucino (CO)

I.T.L. (Istituto per la Tecnologia del Legno)
Via Biasi, 75
I-38010 S.Michele all'Adige (TN)

LSF SUD (Laboratorio di Studi e ricerche sul Fuoco)
Via della Bonifica, 4
I-64010 Controguerra (TE)

ISRIM (Istituto Superiore di Ricerca e formazione sui Materiali speciali per tecnologie avanzate)
Loc. Pentima Bassa, 21
I-05100 Terni (TR)

Istituto Giordano S.p.A. (Centro Politecnico di ricerche e certificazioni)
Via Rossini, 2
I-47041 Bellaria (RN)
Other authorized laboratories:
Istituto Sperimentale delle Ferrovie dello Stato

Istituto Masini

Labortec

CATAS

10.11.5 Future Developments

Like the other member countries of the European Union, Italy will adopt the European Standards and the national fire test methods described here will be withdrawn after a not yet defined coexistence period.

References for Section 10.11

[1] Ministero dell'Interno: Decreto Ministeriale 26 giugno 1984: Classificazione di reazione al fuoco ed omologazione dei materiali ai fini della prevenzione incendi
[2] UNI 8456. Reazione al fuoco dei materiali sospesi e suscettibili di essere investiti da una piccola fiamma su entrambe le facce. Oct. 1987
[3] UNI 8457. Reazione al fuoco dei materiali che possono essere investiti da una piccola fiamma su una sola faccia. Oct. 1987
[4] UNI 9174. Reazione al fuoco dei materiali sottoposti all'azione di una fiamma d'innesco in presenza di calore radiante. Oct. 1987

10.12 The Netherlands

P. VAN HEES

10.12.1 Statutory Regulations

The Dutch building regulations are based on the "Woningwet" (the housing law), which refers to the "Bouwbesluit" (building decree) for all technical building requirements. The requirements are differentiated according to the type of building (residential, offices, hotels, schools, industry buildings, etc.) and whether the building is new or existing. In addition to this, the local authorities use a "Gemeentelijke bouwverordening" (municipal building regulation) to control the safe use of buildings.

The Building Decree contains functional (nonquantified) requirements for all building types. These requirements are further specified quantitatively for residential buildings, office buildings, and hotels only, in (mainly) performance-based requirements. A new building decree containing performance-based requirements for all types of buildings is under development.

The requirements in the building decree refer to assessment methods given in technical NEN standards. These standards are issued by the "Nederlands Normalisatie-Instituut" (Dutch Standards Body), Vlinderweg 6, 2623 AX Delft, Netherlands, www.nen.nl.

10.12.2 Classification and Testing of the Fire Performance of Building Products and Components

10.12.2.1 Test Methods for Reaction to Fire

The classification and testing of the reaction to fire performance of building products are described in four standards: NEN 6064 (noncombustibility), NEN 1775 and NEN 6065 (contribution to fire propagation of floorings and of other building products, respectively), and NEN 6066 (smoke production). Both fire propagation tests (NEN 1775 and NEN 6065) contain two tests.

The determination of noncombustibility of building products is described in NEN 6064 (1997) [1] and corresponds in principle to ISO 1182 [2].

Determination of the combustibility of building products excluding floorings is laid down in NEN 6065 (1997) [3]. The measured "bijdrage tot brandvoortplanting" (contribution to fire propagation) is a combination of the parameters "vlamuitbreiding" (flame spread) and "bijdrage tot vlamoverslag" (contribution to flashover).

The *testing of flame spread* is based on the British surface spread of flame test to BS 476 Part 7 (1971). A diagram of the test apparatus is given in Section 10.18 UK. The test specifications differ slightly and are shown in Table 10.81.

Table 10.81   Test specifications for vlamuitbreiding (flame spread)

Specimens	Six specimens 1000 mm × 300 mm × max. 60 mm or 40 mm for melting materials
Specimen position	Vertical, long side (1000 mm), at approx. 90° to radiator surface
Ignition Source	• Gas fired radiant panel, surface temperature approx. 900 °C • Gas pilot flame (height: 180 mm) below the test specimen on the same side as the radiant panel
Test duration	10 min
Conclusion	Classification in classes 1–5 depending on test performance

The test involves subjecting the specimen to a position dependent radiant level, decreasing from 32 kW/m^2 at one end to 5 kW/m^2 at the other end in combination with a pilot flame at the 32 kW/m^2 position. In contrast to BS 476 Part 7, the pilot flame burns throughout the test (only 1 min in the British test). Considerable differences in classification can thus occur with certain building products. The maximum spread of the flame front in the middle of the specimen is noted at 1.5 and 10 min.

The *contribution to flashover* is determined with the apparatus shown in Fig. 10.42. The test specifications are presented in Table 10.82. One of the two test specimens in the apparatus is flamed by the nine flames of the pipe burner. The filaments, which are located exactly midway between the specimens, are run at different levels between 190 and 1875 W. The flashover time (time period until the second specimen burns for more than 5 s) is determined. At least three tests are carried out running the filaments at different heat outputs. The occurrence of flashover within 5 or 15 min at the different power levels form the basis for classifying products in classes 1–4.

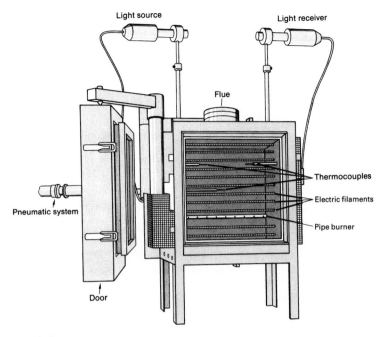

Figure 10.42   Flashover test apparatus

Table 10.82   Test specifications for vlamoverslag (flashover)

Specimens	16 specimens 295 mm × 295 mm × max. 75 mm or 50 mm for melting materials
Specimen position	Two specimens vertical and parallel to each other, separated by 160 mm
Ignition Source	• 12 electric filaments with variable output (190 W – 2250 W); distance from specimen 80 mm • Pipe burner (diameter 10 mm) with nine holes at intervals of 30 mm, flame length 20 mm, gas supply 500 W, distance from first test specimen 25 mm
Test duration	Until flashover occurs
Conclusion	Classification into classes 1–4 depending on the energy supply to the filament required for flashover at exactly 15 or 5 min

The determination of the combustibility of floorings is laid down in NEN 1775 (1997) [4]. The measured "bijdrage tot brandvoortplanting van vloeren" (contribution to fire propagation of floorings) is also a combination of two parameters: "vlamuitbreiding" (flame spread) and "ontvlambaarheid" (ignitability by a small flame).

The *testing of flame spread* for floorings is identical to the German flooring radiant panel test to DIN 4102-14 (1990) described in Section 10.9, the only difference being the absence of an integrated smoke production measurement.

The test involves subjecting the specimen to a position-dependent radiation level, decreasing from 11 kW/m^2 at one end to 1 kW/m^2 at the other end in combination with a pilot flame at the 11 kW/m^2 position. The performance is reported as a critical radiant flux which is the radiation flux level at the position of maximum flame spread.

The testing of ignitability is similar to the German small burner test to DIN 4102-1 (1998) described in Section 10.9, with the main difference being the flame application time. The determining parameter is the period of time in which the flame reaches a mark at 150 mm (5.9 in.) above the flame application position.

The "Rookproduktie bij brand van bouwmaterialen" (Smoke production during fire of building products) is determined with NEN 6066 [5]. This standard corresponds in principle to the ISO dual chamber smoke box method (ISO TR 5924, see Section 15.3). The smoke test is of the static type with a conical heater attacking the specimen at radiant levels of 20, 30, 40, and 50 kW/m². The maximum smoke production of those measured at the different radiation levels is reported as a light attenuation factor, no classes are used.

10.12.2.2 Other Important Fire-Related Test Methods

Roofs are tested on their external fire performance in NEN 6063 "Brandgevaarlijkheid van daken" (Fire behavior of roofs on external exposure to flying brands" [6], nearly identical to the German DIN 4102-7 (1998). This test involves the assessment of damaged surface area, damaged area of inner layers, and fire penetration aspects when the roof is exposed to burning brands. The result is reported as pass or fail.

Resistance to fire aspects are covered by test and calculation methods, the main test standard being NEN 6069 "Experimentele bepaling van de brandwerendheid van bouwdelen" (Determination of fire resistance of elements of building constructions) [7]. This method is based on ISO 834. Standardized calculation methods are available for resistance to fire of concrete, steel, and wood constructions (in NEN 6071-6072-6073).

10.12.2.3 Classification

The noncombustibility of building products is reported as a pass or fail.

The contribution to fire propagation according to NEN 6065, a combination of the aspects flame spread and contribution to flashover is expressed as the classification in one of five classes 1–5 (Table 10.83). The aspect flame spread is divided into five classes as shown in Table 10.84, the contribution to flashover into four classes as shown in Table 10.85. If a building product falls into different classes for flame spread and flashover, the least favorable is taken for its classification for contribution to fire propagation.

Table 10.83 Classification of contribution of building products to fire propagation

Class	Contribution to fire propagation
1	Very weak
2	Weak
3	Moderate
4	Strong, satisfies basic requirements to NEN 6065
5	Strong, does not satisfy basic requirements to NEN 6065

Table 10.84   Division of flame spread into classes

Flame spread [mm]				Flame spread class
During the first 1.5 min		During 10 min		
Class limit	Max. permitted upper deviation for 1 sample	Class limit	Max. permitted upper deviation for 1 sample	
175	25	175	25	1
250	50	550	50	2
350	50	750	100	3
500	50	If not classes 1 to 3		4
If not classes 1 to 4				5

Table 10.85   Division of flashover intensity into classes

E_{15}	E_5	Flashover intensity class
$\geq 1500\,W$	$\geq 1875\,W$	1
	$< 1875\,W$	2
$< 1500\,W$	$\geq 1125\,W$	
$\geq 750\,W$	$< 1125\,W$	3
$< 750\,W$	$\geq 565\,W$	
$\geq 190\,W$	$< 565\,W$	4
$< 190\,W$	–	

The contribution to fire propagation of floorings according to NEN 1775 is expressed as the classification in one of three classes: T1, T2, or T3. Classes T2 and T3 are based on the ignitability test only. T2 products that also meet the flame spread requirement (the critical radiant flux being $\leq 4.5\ kW/m^2$) are classified as T1.

The contribution to smoke production is reported in terms of the level of light attenuation. No classes exist.

10.12.2.4 Requirements

The reaction to fire requirements for building products may be summarized (and simplified) as follows:

- Near fireplaces: noncombustibility according to NEN 6064,
- On high facades: fire propagation class 2 according to NEN 6065, lowest 2.5 m: class 1 according to NEN 6065,

- In escape routes:
  - Floorings: fire propagation class T1 according to NEN 1775,
  - Other building products: fire propagation class 2 according to NEN 6065 and smoke production ≤ 2.2 m^{-1} according to NEN 6066 (in case of class 1 products, a smoke production of ≤ 5.4 m^{-1} according to NEN 6066 is accepted).
- Other cases / basic requirement:
  - Floorings: fire propagation class T3 according to NEN 1775 and smoke production ≤ 10 m^{-1} according to NEN 6066,
  - Other building products: fire propagation class 4 according to NEN 6065 and smoke production ≤ 10 m^{-1} according to NEN 6066,
  - No smoke requirement is used outside buildings.

For completeness, a special aspect of the Dutch fire resistance requirements is mentioned here. The requirements are given as a level of resistance to a fire going from one room to the other, independent of the route of fire propagation, either inside or outside the building. Fire resistance of constructions or spatial separation of façade openings along that route may be summed to satisfy the requirement.

10.12.3 Official Approval

The tests described above and the resultant classifications provide the basis for official approval by the local authority. In the Netherlands, fire performance testing of building products is carried out by a single test institute:

Centrum voor Brandveiligheid
TNO Bouw
Lange Kleiweg 5
NL-2288 GH Rijswijk ZH or
Postbus 49
NL-2600 AA Delft
Tel. + 31 15 284 2307
Fax. + 31 15 284 3955
E-mail: l.twilt@bouw.tno.nl

The test results and classification achieved are recorded in a test report issued by TNO. This document substantiates the fire performance of the building products or constructions for officials and the fire brigade. Marking and inspection are not compulsory.

10.12.4 Future Developments

In the Netherlands, a new building decree is to be introduced in 2003. This new decree will contain quantified requirements for all building types. The assessment methods and the requirements will be brought in line with the current developments under the European Construction Product Directive.

References for Section 10.12

[1] NEN 6064 (1997) Bepaling van de onbrandbaarheid van bouwmaterialen
[2] ISO/R 1182 – 1970 Building materials – Non-combustibility test, see also ISO/DIS 1182 Aug. 1973
[3] NEN 6065 (1997) Bepaling van de bijdrage tot brandvoortplanting van bouwmateriaal(combinaties)
[4] NEN 1775 (1997) Bepaling van de bijdrage tot brandvoortplanting van vloeren
[5] NEN 6066 (1997) Bepaling van de rookproduktie bij brand van bouwmateriaal (combinaties)
[6] NEN 6063 (1997) Bepaling van de brandgevaarlijkheid van daken
[7] NEN 6069 (1997) Experimentele bepaling van de brandwerendheid van bouwdelen.

10.13 Nordic Countries

B. Sundström

10.13.1 Statutory Regulations

The Nordic countries – Denmark, Finland, Iceland, Norway, and Sweden – have a long tradition of cooperation in the area of building regulations. The results are fire regulations, testing, and classification procedures that are much the same in all of the Nordic countries. The relevant national building regulations are:

- Denmark: Bygningsreglement 1995 [1] and for small houses, Bygningsreglement 1998 [1]
- Finland: Suomen Rakentamismääräyskokoelma [2]
- Iceland: Byggingerreglugerd nr 441/1998 and Lög um brunvarnir og brunemål nr 41/1992 [3]
- Norway: Plan og bygningslov av 14 juni 1985 nr. 77. Last update 1 June 1999 [4]
- Sweden: Boverkets byggregler, BBR [5]

On a Nordic basis the "Nordiska Kommitten för byggbestämmelser" (NKB) (Nordic committee for building regulations) had the task of harmonizing the Nordic building regulations including those relating to fire protection. NKB issued recommendations, which can be incorporated voluntarily in the national regulations. NKB has published guidelines, "product rules," on testing and classification of product groups that have had an influence on the Nordic regulations. For reaction to fire the relevant documents are:

NKB report no. 51 [6]Nordic guidelines for mutual acceptance of centrally approved building products and official control measures

NKB Product rules no. 6 [7] Floor coverings

NKB product rules no. 7 [8] Roof coverings

NKB Product rules no. 14 [9] Linings

NKB product rules no. 15 [10] Fire protecting coverings

10.13.2 Nordtest Methods

Nordic building regulations make reference to common Nordic test methods to a very large extent. Nordtest [11] has published approx. 50 fire test methods. They cover reaction to fire and resistance to fire in a variety of areas. Relevant Nordtest methods for reaction to fire in the building sector are listed below:

NT Fire 001/ISO 1182:1990 (1991–09)[1] Fire tests – Building materials – Noncombustibility test NT Fire 002. (1985–11) Building products: Ignitability

NT Fire 003. (1985–11) Coverings: Fire protection ability NT Fire 004. (1985–11) Building products: Heat release and smoke generation

NT Fire 006. (1985–11) Roofings: Fire spread

NT Fire 007. (1985–11) Floorings: Fire spread and smoke generation

NT Fire 012/ASTM E 662–97. (1999–02) Specific optical density of smoke generated by solid materials

NT Fire 013/ISO 4589–1, 2, and 3: 1996. (1998–02) Plastics – Determination of burning behaviour by oxygen index Parts 1, 2 and 3

---

1   Date when adopted by Nordtest.

NT Fire 025/ISO 9705:1993. (1993–11) Fire tests – Full-scale room tests for surface products

NT Fire 030. (1987–02) Buildings products: Fire spread and smoke production – full-scale test

NT Fire 033/ISO 5657:1997. (1998–02) Fire tests – Reaction to fire – Ignitability of building products

NT Fire 035. (1988–02) Building products: Flammability and smouldering resistance of loose-fill thermal insulation

NT Fire 036. (1988–02) Pipe insulation: Fire spread and smoke production – full-scale test

NT Fire 038. (1989–09) Building materials: Combustible content

NT FIRE 047. (1993–05) Combustible products: Smoke gas concentrations, continuous FTIR analysis

NT FIRE 048/ISO 5660–1:1993 (1993–05) Fire tests-Reaction to fire-Part 1: Rate of heat release from building products (Cone calorimeter method)

NT FIRE 050 (1995–05) Heat flux meters: Calibration

The procedures for testing the fire performance of building products are described below.

10.13.2.1 NT Fire 001 Noncombustibility Test

The combustibility or noncombustibility of building products is assessed by Nordtest method NT Fire 001 which is identical to ISO 1182. An illustration of the noncombustibility furnace and test specifications are thus given in Section 10.23 ISO.

10.13.2.2 NT Fire 002 Ignitability Test

The ignitability of building products, particularly, of coverings and linings, is tested according to Nordtest method NT Fire 002 also known as the Schlyter test. The apparatus is illustrated in Figure 10.43 and the test specifications are summarized in Table 10.86.

Table 10.86   Schlyter test specifications

Specimens	Three sets of two specimens 800 mm × 300 mm, max thickness 70 mm
Specimen position	Two specimens vertical at intervals of 50 mm parallel to each other, one specimen displaced 125 mm upwards, both mounted into frames of angle iron
Ignition source	Propane burner 25 mm below the shielded edge of the upper specimen. The burner is at right angles to and 60 mm away from the lower specimen
Test duration	The flame is applied to the lower specimen until both specimens burn on their own. If the specimens do not ignite, the test duration is 30 min
Conclusion	Classification is based on time to ignition of the specimen not exposed to the ignition flame

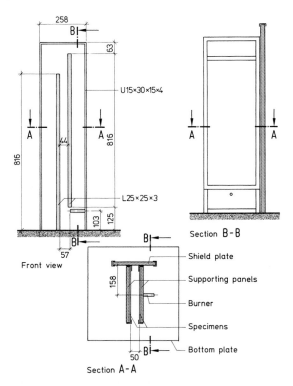

Figure 10.43   Ignitability apparatus, the Schlyter test, for linings and coverings

The use of NT Fire 002 has almost disappeared in the Nordic building regulations. It now only appears as an alternative to NT Fire 033 to determine ignitability of linings in the Finnish building regulations. Sometimes NT Fire 002 is also used as an ad-hoc test for special cases.

10.13.2.3 NT Fire 004 Heat Release and Smoke Generation Test

The tendency of exposed surfaces of building products to release heat and generate smoke is tested by Nordtest method NT Fire 004. A diagram of the apparatus and the test specifications are given in Figure 10.44 and Table 10.87, respectively.

Table 10.87   Specifications for the heat release and smoke generation test

Specimens	Three sets of four specimens, square with sides 228 ± 4 mm and a thickness of 12 ± 3 mm
Specimen position	Four specimens fixed to the rear wall, side walls, and top of the fire chamber
Ignition source	Propane burner premixed with gas and air
Test duration	5 or 10 min
Conclusion	Classification is based on the time-temperature curve from the fire inside the box as well as the smoke density

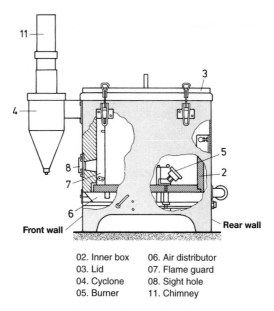

02. Inner box	06. Air distributor
03. Lid	07. Flame guard
04. Cyclone	08. Sight hole
05. Burner	11. Chimney

Figure 10.44   Apparatus for determining heat release and smoke generation

The test specimens in principle form a room lining inside the box. The propane burner then acts as the ignitions source for the "model room fire." The time – temperature relation of the smoke gases leaving the box through the chimney reflects the fire growth and is used for classification. In addition the optical density of the smoke is included.

Classes are defined using limiting curves; see Figure 10.45.

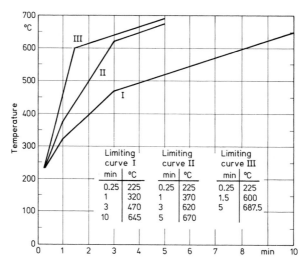

Limiting curve I		Limiting curve II		Limiting curve III	
min	°C	min	°C	min	°C
0.25	225	0.25	225	0.25	225
1	320	1	370	1.5	600
3	470	3	620	5	687.5
10	645	5	670		

Figure 10.45   Limiting smoke gas temperature curves for NT Fire 004

The different product categories and the terminology are defined by NKB as:

- A product is defined as having a *low* heat release, if in three tests in accordance with NT FIRE 004, over a test period of 10 min, it is shown that the temperature curve for the smoke, as an average of the three tests, does not exceed the limiting curve *I* for a total period > 30 s, and that the area thus enclosed between the mean temperature curve and the limiting curve is not > 15 min x °C.
- A product is defined as having a *moderate* heat release, if in three tests in accordance with NT FIRE 004, over a test period of 5 min, it is shown that the temperature curve for the smoke, as an average of the three tests, does not exceed the limiting curve II for a total period > 30 s, and that the area thus enclosed between the mean temperature curve and the limiting curve is not > 15 min x °C.
- A product is defined as having a *normal* heat release, if in three tests in accordance with NT FIRE 004, over a test period of 5 min, it is shown that the temperature curve for the smoke, as an average of the three tests, does not exceed the limiting curve III for a total period > 30 s, and that the area thus enclosed between the mean temperature curve and the limiting curve is not > 15 min x °C.
- A product is defined as having a *low* smoke production, if in three tests in accordance with NT FIRE 004, over a test period of 10 min, it is shown that the area below the mean density curve for the smoke, determined as an average over the three tests, is not > 50 min x density %.
- A product is defined as having a *normal* smoke production, if in three tests in accordance with NT FIRE 004, over a test period of 5 min, it is shown that the area below the mean density curve for the smoke, determined as an average over the three tests, is not > 150 min x density %.

The Nordic countries make use of these requirements to define national classes, see Table 10.91.

10.13.2.4 NT Fire 006 Test for Roofings: Fire Spread

Fire spread on roof coverings is tested by Nordtest method NT Fire 006. The apparatus is shown in Figure 10.46 and Figure 10.47 while the test specifications are given in Table 10.88.

Table 10.88   Specifications for testing the fire spread on roof coverings

Specimens	Six specimens 1000 mm × 400 mm (three specimens for each wind speed)
Specimen position	Inclined at 30° to horizontal between two air ducts, upper duct with suction fan, lower duct with blower
Ignition source	Wood crib 100 mm × 100 mm consisting of eight pieces of pine wood 10 mm × 10 mm × 100 mm, ignited by a special gas burner, and placed on the test specimen on the center line 100 mm from the lower edge
Test duration	15 min or when the flame front has reached the upper end of the specimen
Conclusion	Classification is based on the fire damage on the surface and on the substrate.

The roof covering is attached to a substrate and mounted at an angel of 30° to the horizontal in the test apparatus. The blower and suction fan are switched on giving an airflow of 2 m/s (6.5 ft/s) or 4 m/s (13 ft/s). A flame is applied for 30 s to the wood crib by the separate pilot burner shown in Figure 10.47. Then the air supply to the lower fan is shut off and the burning crib is placed on the specimen. After 15 s the air supply to the lower fan is opened again. A note is made of the time elapsed until the specimen ignites and the flames and glow die out. The extent of damage to both the roofing and substrate is measured.

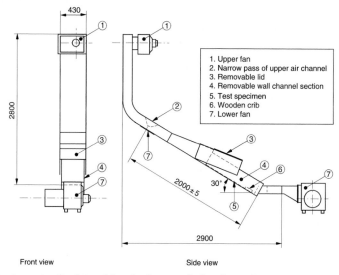

1. Upper fan
2. Narrow pass of upper air channel
3. Removable lid
4. Removable wall channel section
5. Test specimen
6. Wooden crib
7. Lower fan

Front view   Side view

Figure 10.46   Apparatus for determining the fire spread of roof coverings

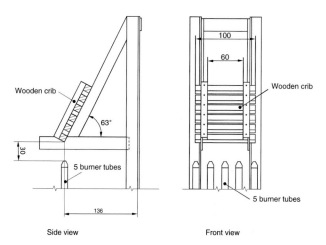

Side view   Front view

Figure 10.47   Crib ignition stand with the wood crib in position for ignition

10.13.2.5 NT fire 007 Test for Floorings: Fire Spread and Smoke Generation

The ability of floor coverings to resist fire spread and smoke development is tested by Nordtest method NT Fire 007. In principle the same apparatus and test specifications apply as for testing roof coverings (see Figs. 10.47 and 10.48 as well as Table 10.88, respectively). In addition to the NT Fire 006 specifications, the upper air duct is equipped with a device for measuring light transmission through smoke.

The actual test is identical to that for roof coverings except that only a single flow rate of 2 m/s is used and that smoke density is measured. Asymmetrically woven floor coverings must be tested both warp- and weftwise.

10.13.2.6 NT Fire 025/ISO 9705:1993 Fire tests – Full Scale Room Tests for Surface Products

ISO 9705, also called the Room/Corner Test, is used as reference test for NT Fire 004. An illustration of the test room and test specifications is given in Section 10.23 ISO.

10.13.2.7 NT Fire 033/ISO 5657:1997 Fire Tests – Reaction to Fire – Ignitability of Building Products

ISO 5657 is described in Section 10.23 ISO. The Nordic countries have defined two levels of ignitability used for classification purposes. The different categories and the terminology are defined by NKB as:

- A product is defined as having a *low* ignitability if in five tests in accordance with ISO 5657 at a radiation intensity of 40 kW/m², and in five tests in accordance with ISO 5657 at a radiation intensity of 30 kW/m², it is shown that the time to sustained ignition on the surface of the test specimen, as an average of the five tests, is
- At least 1 min at 40 kW/m², and
- At least 3 min at 30 kW/m²
- A product is defined as having a *normal* ignitability if in five tests in accordance with ISO 5657 at radiation intensity of 30 kW/m², and in five tests in accordance with ISO 5657 at a radiation intensity of 20 kW/m² it is shown that the time to sustained ignition on the surface of the test specimen, as an average of the five tests, is
  – At least 1 min at 30 kW/m², and
  – At least 3 min at 20 kW/m².

10.13.2.8 NT Fire 036 Pipe Insulation: Fire Spread and Smoke Production – Full-Scale Test

NT Fire 036 test equipment is identical to ISO 9705, the Room/Corner Test. However, pipe insulation cannot be tested as a lining owing to the geometrical dimensions and the specific mounting conditions. Therefore ISO 9705 test procedure was adapted to reflect more the end use conditions for pipe insulation, see Fig. 10.48.

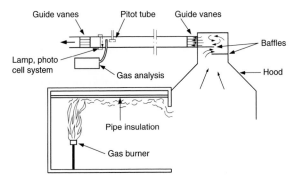

Figure 10.48   NT Fire 036. The test ISO 9705 is adapted to be useful for testing pipe insulation

10.13.3 Equivalents of the Nordtest Methods in the Individual Nordic Countries

The Nordtest methods NT Fire 002, 003, 004, 006, and 007 were developed by the Nordic fire laboratories and are unique to the Nordic countries. They have also been transposed without modification into national standards; see Table 10.89. The ISO standards mentioned that also have a Nordtest number can generally be identified nationally by the ISO designation preceded by the prefix of the national standardization organization, for example, SS ISO 5660.

Table 10.89   National equivalents to Nordtest NT Fire methods

Nordtest method	Denmark	Finland	Norway	Sweden
NT Fire 002. (1985-11) Building products: Ignitability	DS/INSTA 410	SFS 4190	NS-INSTA 410	SS 024821
NT Fire 003. (1985-11) Coverings: Fire protection ability	DS/INSTA 411	SFS 4191:E	NS-INSTA 411	SS 024822
NT Fire 004. (1985-11) Building products: Heat release and smoke generation	DS/INSTA 412	SFS 4192:E	NS-INSTA 412	SS 024823
NT Fire 006. (1985-11) Roofings: Fire spread	DS/INSTA 413	SFS 4194:E	NS-INSTA 413	SS 024824
NT Fire 007. (1985-11) Floorings: Fire spread and smoke generation	DS/INSTA 414	SFS 4195:E	NS-INSTA 414	SS 024825

10.13.4 Classification of Building Products

In general, classification of building products is quite similar in the Nordic countries. When differences appear they tend to be minor. In Denmark and Norway the relevant classification is found in standards while the classes in Finland and Sweden are given in guidelines related to the building regulations. The references to relevant classification criteria are given in Table 10.90.

In addition, guidelines for products that in Denmark may be considered deemed to satisfy without testing are published by DBI [12].

Table 10.90 Documents containing classification criteria

Product group	Denmark	Finland	Norway	Sweden
Noncombustible materials	DS 1057.1	National building code "Structural Fire Safety"	NS 3919	Guidelines for type approval, fire protection. Boverket Allmänna råd, 1993:2 utgåva 2
Linings/ Building materials	DS 1065-1 and MK 6.00/005		NS 3919	
Roof coverings	DS 1063.1		NS 3919	
Floor coverings	DS 1063.2		NS 3919	
Pipe insulation				
Fabrics used for tents	MK 6.00/016			

10.13.4.1 Linings

In Table 10.91 the classes for linings are summarized.

NT Fire 004 is used for a large variety of linings. However, there are cases when NT Fire 004 may give uncertain test data due to exotic fire behavior of the product, for example excessive melting. In those cases it may be possible to run a large-scale test. There are two options, NT Fire 030 and NT Fire 025/ISO 9705.

NT Fire 030 is a two-room building. The tested product is lined to both rooms. Classification criteria are basically fire spread, smoke production, and melting/dripping/falling of the product.

ISO 9705, the Room/Corner Test may be used on a similar basis as NT Fire 030. The product is lined to the walls and the ceiling of the test room according to the standard mounting configuration given in ISO 9705. The standard ignition source of 100/300 kW is used.

Denmark [13], Norway [14] and Sweden [15] have published classification criteria for NT Fire 030 and ISO 9705. The criteria for ISO 9705 are shown in Table 10.92.

Finland

In Finland NT Fire 002 for ignitability may be used instead of ISO 5657. Different degrees of ignitability (class 1 and class 2) may be combined with heat release (class I and class II) according to NT Fire 004 to form more classes. According to the Finnish rules for type approval regarding structural fire safety, "Brandtekniskt godkännande av byggnadsprodukter", combustible building materials are classified as Surface Class 1/I, 1/II, 1/-, 2/I, 2/II, 2/-, -/I, -/II or -/-.

Table 10.91   Terminology and classes used in conjunction with the NT Fire 004 test

Description used by NKB	Test method	Equivalent class used in the Nordic countries				
		Sweden	Denmark	Norway		Finland
Low heat release	NT Fire 004	Class I	Class A	In 1[a]	Ut 1	Class 1/I
Low smoke production	NT Fire 004				–	
Low ignitability	ISO 5657	–			Ut 1	
Moderate heat release	NT Fire 004	Class II	–	–		Class 2/II
Normal smoke production	NT Fire 004		–	–		
Normal ignitability	ISO 5657	–	–	–		
Normal heat release	NT Fire 004	Class III	Class B	In 2	Ut 2	–
Normal smoke production	NT Fire 004				–	–
Normal ignitability	ISO 5657	–			Ut 2	–

[a] The criterion for smoke density is 100 min × density %.

Norway

The surface classes called "In" refer to interior linings. However, in Norway there are also the corresponding classes for exterior linings, "Ut." The classification criteria are the same as for interior linings except that the smoke requirements are dropped.

Table 10.92   Swedish, Norwegian and Danish classification criteria for linings when tested to ISO 9705

Class	Test time (min)	HRR$_{ave}$[a] excl. burner (kW)	HRR$_{peak}$ excl. burner (kW)	SPR$_{ave}$[a] (m²/s)	SPR$_{peak}$ (m²/s)	Flame spread	Burning droplets/ particles
Class A/ class I	20	50	300	0.7	2.3	Max 0.5 m from floor along the walls, except area inside 1.2 m from ignition corner	Only occasionally and only inside the area of 1.2 m from ignition corner
Class B/ class III	2	–	900	4.6	16.1	As class A/ class I	–

[a] Calculated over the entire test time
HRR = heat release rate, SPR = smoke production rate. For detailed definitions, see ISO 9705

10.13.4.2 Roof Coverings

According to the Nordic product requirements [8] a roof covering is considered to belong to class T if it meets the following requirements in tests according to NT FIRE 006:

- The average damaged length of the four specimens may not exceed 550 mm.
- Maximum damaged length in the roof covering and in the underlay may not exceed 800 mm for any specimen.

The Nordic countries except Finland use class T (called *Ta* in Norway) as defined by NKB.

Finland

Finland uses two classes of roof coverings. Products are classified in classes K 1 "does not burn and protects its substrate from igniting," K 2 "burns but does not assist in spreading a fire and protects to some extent its substrate from igniting," and others (-) if the conditions for classes K 1 or K 2 are not met. Class K1 includes criteria of no flame spread and very limited damage of the substrate. Class K2 corresponds to class T.

10.13.4.3 Floor Coverings

According to the Nordic product requirements [7] a floor covering is considered to belong to class G if it meets the following requirements in tests according to NT FIRE 007:

- The average damaged length of the four specimens may not exceed 550 mm.
- Maximum damaged length in the floor covering and in the underlay may not exceed 800 mm for any specimen.
- The average value of the four maximum smoke densities measured during the first 5 min does not exceed the scale value 30%. The smoke density is given in percentages of total non-transparency.
- The average value of the four maximum smoke densities measured during the last 10 min does not exceed the scale value 10%.

Fire classified floor coverings are called "L" in Finland. In the other Nordic countries, this class is denoted with "G".

10.13.4.4 Pipe Insulation

Sweden and Norway

Pipe insulation is tested according to NT Fire 036 and classified directly based on the large-scale results. There are three fire classes of pipe insulation parallel to the three classes for linings. The classification criteria are given in Table 10.93.

Table 10.93   Swedish classes for pipe insulation

Class	Test time (min)	Classification criteria
PI	15	Peak HRR excluding burner max. 650 kW, Peak SPR max. $2.3\,m^2/s$ and no burning droplets/particles
PII	10	Peak HRR excluding burner max. 650 kW, Peak SPR max. $18.4\,m^2/s$ and no burning droplets/particles
PIII	5	Peak HRR excluding burner max. 650 kW, Peak SPR max. $18.4\,m^2/s$

10.13.4.5 Fabrics Used for Tents

Large constructions made of fabric, for example, covered tennis courts, are considered a building and fall under the building regulations. Denmark, Norway, and Sweden have specified procedures for testing fabrics for such constructions.

Swedish standard SIS 65 00 82 [16], ignitability of fabrics, is used. The test apparatus is shown in Figure 10.49 and the test specifications for building products < 3 mm thick in Table 10.94.

Table 10.94   Specifications for testing ignitability of building products < 3 mm thick

Specimens	Six specimens 50 mm × 300 mm (if necessary three specimens each in warp and weft directions)
Specimen position	Vertical in a cabinet 0.3 m × 0.3 m × 0.76 m
Ignition source	Bunsen burner, diffusion flame 38 mm long. Edge ignition
Test duration	The flame is applied for 12 s. The test stops when flaming and/or glowing has ceased
Conclusion	Classified as difficult to ignite if five of six specimens • Burn on average ≤ 2 s and in no case > 3 s, • Average burn length ≤ 90 mm and in no case > 115 mm.

Large-scale reference tests of constructions were made by SP. The tests showed that the simple ignitability test gave a sufficient level of safety. SP also has a test procedure documented for large-scale testing of constructions made out of fabric.

Sweden

Building products required to be difficult to ignite and that are thinner than 3 mm (0.12 in.) are tested according to SIS 65 00 82. This means that the test has a broader use than for fabrics.

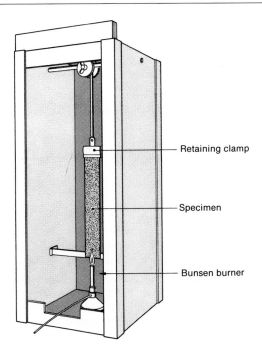

Retaining clamp

Specimen

Bunsen burner

Figure 10.49   SIS 65 00 82 Apparatus for testing ignitability of building products <3 mm thick

10.13.4.6 Tests in Conjunction with Production Control

Keeping a type approval valid normally requires annual retesting of samples taken at the site of the manufacturer. If the type test were based on large-scale investigations the cost of verification would be very high. For that reason small-scale tests are often used in regular production control. Reference data are taken when the type test is done and that value is then used for control. Typical test methods used in conjunction with production control are NT Fire 012, 013, 038, and 048.

10.13.4.7 Modeling Fire Growth

ISO 5660, the cone calorimeter (see Section 10.23), is a modeling tool for fire growth. Test data can be used to predict the fire growth in the Room/Corner Test, ISO 9705, for a building product. This may be helpful to reduce testing and for extended application of test data.

10.13.5 Official Approval

The tests described in the preceding for determining the fire performance of building products can be carried out by the officially accredited test institutes in each country with the exception of Iceland owing to the lack of a suitable establishment. The major institutes are:

Dansk Brandteknisk Institut (DBI)
Jernholmen 12
DK-2650 Hvidovre
Denmark
www.brandteknisk-institut.dk

Valtion teknillinen tutkimuskeskus (VTT)
Bygg och transport
PB 1803
FIN-02044 VTT
Finland
www.vtt.fi

Norges Branntekniske Laboratorium AS (NBL), SINTEF
N-7465 Trondheim, Norway
Visiting address: Tiller Bru, Trondheim
www.sintef.no/units/civil/nbl

SP Sveriges Provnings- och Forskningsinstitut
Box 857
SE-501 15 Borås
Sweden
www.sp.se

The test organization issues test reports that are recognized in all the Nordic countries. The product is classified on the basis of the test report according to national regulations.

Organizations accredited for certification in each Nordic country issue type approvals based on the test reports. However, in Finland, the ministry of the environment issues type approvals. The approval, which includes rules for marking and inspection of the product, is not mandatory and is issued only on application. In practice, a type approval is almost always requested, as it is stipulated by most local authorities and is required by the market.

A type approval is obtained by submitting the test report to the certifying body. All Nordic countries require the existence of an inspection agreement, production control, with the relevant recognized test institute before a type approval can be issued. The test laboratories mentioned in the preceding are all recognized for production control. Details for each country are given below.

10.13.5.1 Denmark

Applications for type approval should be made to

ETA-Danmark A/S
Venlighetsvej 6
DK-2970 Hörsholm
Denmark

Type approvals are normally valid for 3 years and a list is published annually with quarterly supplements, which are sent to all building authorities.

The underlying procedure for type approval is given in a Building Ministry Circular [17]. Detailed information on how to apply for a type approval for different products is given in circulars "MK Prövnings- og godkendelsebetingelser." These circulars are very detailed and contain all information needed including prices for approvals. The relevant circulars are listed in Table 10.95.

Table 10.95   Danish testing and approval requirements for different product groups

Product group covered by the circular	Designation
Materials of class A and class B	MK 6.00/005 5.udgave Juli 1999
Fire protecting coverings of class 1 and class 2	MK 6.00/006 5.udgave Juli 1999
Floor coverings of class G	MK 6.00/001 5.udgave Juli 1999
Fabrics for tents	MK 6.00/016 3.udgave Juli 1999
Roof coverings of class T	MK 6.00/002 5.udgave Juli 1999

10.13.5.2 Finland

Applications for type approval should be made to

Ympäristöministeriö
Kaavoitus- ja rakennusosasto
(Ministry of the Environment, Dept. of Building and Planning)
PL 380
SF-00131 Helsinki
Finland

Type approvals are usually valid for 5 years or less and a list is published annually. Specifications for type approval are given in these publications:

- Typgodkännanden i byggbranschen. Miljöhandledning 12, augusti 1996. (Type approvals in the building field),
- Brandtekniskt godkännande av byggnadsprodukter. Miljöhandledning 35, januari 1998. (Fire technical approval of building products).

10.13.5.3 Iceland

The type approvals of the other Nordic countries, and in particular of Denmark, form the basis for inspection in Iceland since no fire testing is carried out there. The body responsible for compliance with the regulations is the State Authority for Fire Protection:

Brunamalastofnun rikisins
Laugaveg 59
IS-101 Reykjavik
Iceland

10.13.5.4 Norway

Application for type approval should be made to

Nemko Certification Service AS (NCS)
Postboks 10 Skøyen
NO-0212 Oslo
Norway

NCS issues type approvals based on national standards. In the case when there is no standard available, "product documentation" can be issued. The "product documentation" then works as a type approval. The "product documentation" is given by the Norwegian Fire Research Laboratory, SINTEF, in Trondheim.

10.13.5.5 Sweden

Application for type approval should be made to

SITAC AB
Box 553
371 23 Karlskrona
Sweden

Type approvals are generally valid for 5 years, and are listed in "Approval List B, type approved, fire protection" published annually.

General guidelines for type approvals are given in "Boverkets Författningssamling Bfs 2000:27 Typ 2, August 29, 2000".

Guidelines for inspection and production control procedures in the fire area are given in "Boverkets allmänna råd om tillverkningskontroll, 1996:2, Boverket."

10.13.6 Future Developments

The building regulations in all of the Nordic countries are being adapted to the European harmonization in this area. As the product standards become available and CE marking becomes possible, the national regulations covering that specific area will disappear. The rate of change depends on the European development and will vary from product group to product group.

Nordic building codes are being developed to specify performance criteria instead of being prescriptive. This opens possibilities for fire safety design based on calculations. Some of the tests described in this chapter are also useful for fire modeling, for example, the cone calorimeter. The concept of large fire scenarios is also seen to be useful in performance-based design.

References for Section 10.13

[1] Bygningsreglement 1995, February 13 and Bygningsreglement for småhuse 1998, June 25. Issued by the Bygge- og Boligstyrelsen, Stormgade 10 , DK-1470 Köpenhamn, Denmark
[2] Suomen rakentamismääräyskokoelma Issued by the Ministry of the Environment, P.O. Box 380, 00121 Helsinki, Finland, www.vyh.fi
[3] Brunamalastofnun rikisins, Laugaveg 59, IS-101 Reykjavik, Iceland, www. brs.is
[4] Plan og bygningslov av 14. 06. 1985 Nr. 77 and forskrift om krav til byggverk, 22.01.1997 nr 33. Kommunal- og regionaldepartementet, Oslo, www.be.no
[5] Boverkets byggregler, BBR, BFS 1998:38. Issued by Boverket, publikationsservice, Box 534, 371 23 Karlskrona, Sweden, www.boverket.se
[6] Nordiska riktlinjer för ömsesidigt accepterande av centralt godkända byggprodukter och officiell kontrollordning, NKB-rapport nr 51, januari 1984
[7] Produktregler for Brandmæssigt egnede Gulvbelægninger. NKB Produktregler 6. July 1988.
[8] Produktregler for Brandmæssigt egnede Tagdækninger. NKB produktregler 7. January 1989.
[9] Produktregler for Brandmæssigt egnede Overfladelag (Ytskikt). NKB produktregler 14. January 1990.
[10] Produktregler for Brandmæssigt egnede Beklædninger. NKB produktregler 15. January 1990.
[11] Nordtest – P.O.Box 116 – FIN-02151 ESPOO – Finland, www.vtt.fi/nordtest/
[12] Brandtekniske eksempler, Brandteknisk vejledning 30, 2. udgave september 2000, Dansk Brandteknisk Institut
[13] MK 6.00/005 5.udgave Juli 1999, ETA-Danmark A/S
[14] Melding HO-1/94. Plast i bygninger. Statens bygningstekniske etat, www.be.no/beweb/regler/meldinger/plast.html
[15] Boverkets Allmänna råd 1993:2, utgåva 2 "Riktlinjer för typgodkännande brandskydd"
[16] Svensk Standard SIS 650082. 1966. Textilvaror. Bestämning av brandhärdighet hos vävnader.
[17] Byggestyrelsens cirkulär 20 augusti 1986

10.14 Poland

J. FANGRAT

10.14.1 Statutory Regulations

There are two basic documents regarding fire safety of buildings in Poland:
- Decree of Minister of the Interior of 3 November 1992, on fire safety of buildings, other building structures and sites [1],
- Decree of Urban Planning and Building Minister from 14 December 1994, on the technical requirements for buildings [2].

According to [1] buildings, their parts as well as rooms are divided on the basis of their function into the following five categories of life hazard:

- ZL I Public buildings or their parts, where more than 50 persons can be present,
- ZL II Buildings or their parts intended for use by people with low mobility,
- ZL III Schools, office buildings, dormitories, hostels, hotels, health care centers, ambulatories, nursing homes, shops, stores and restaurants where over 50 persons can be present, barracks, computer rooms, prisons, and so forth.
- ZL IV Apartments, dwellings, and private homes,
- ZL V Registers, museums, and libraries.

In addition, the following groups of buildings are distinguished according to their height:

Low (N)	up to 12 m (39.4 ft) over ground level or domestic up to four stories
Medium height (SW)	from 12 m to 25 m (82 ft) over ground level or domestic up to nine stories
High (W)	from 25 m to 50 m (164 ft) over ground level
High rise (WW)	more than 55 m (180 ft) over ground level

The basic technical requirements on fire safety of buildings are given in document [2], Part VI: Fire Safety. In Chapter 1, general rules are given, one of them stating that the requirements regarding fire protection of buildings and their parts depend on the function and use of the building, explosion hazard, and fire load and are settled on the basis of the Decree, fire safety and explosion regulations, as well as Polish Standards.

In addition to the above documents, the lists of the obligatory Polish Standards were issued by these Ministers.

According to document [2], there are five fire classes of buildings (A–E) depending on fire load, category of life hazard, and building height (Table 10.96). In addition, the following general requirements are given:

- Building construction must fulfill the load bearing criteria for a certain period of time.
- Fire initiation as well as fire and smoke spread in the building is limited within and beyond the room of origin.
- Fire spread to the neighboring buildings is limited.
- Occupants must be able to leave the building or otherwise to be rescued.
- The safety of the rescue team is ensured.

Building parts of certain fire class shall meet basic requirements listed in Table 10.97. Under certain conditions described in this document, these requirements may differ from the given one.

It is required, that noncombustible materials are used in escape routes. Wood and other combustibles are allowed in:

- Low buildings (< 12 m) of D and E category of fire resistance,
- Temporary buildings,
- Two-story domestic buildings.

In buildings with the category of life hazard, the use of finished materials, which generate highly toxic effluents and heavy smoke, is forbidden.

The use of readily ignitable materials is not allowed in escape routes.

The use of readily ignitable floor covering is not allowed:

- Inside building of ZL II category,
- In rooms with more than 50 people present,
- In production and storage rooms.

In conference rooms, restaurants, and production rooms, the use of partitions and coatings made of readily ignitable materials is forbidden.

Table 10.96 Fire classes of buildings

Class		A	B	C	D	E
Fire load density per fire zone (MJ/m²)	Production and storage buildings	>4000	(a) 2000–4000 (b) For buildings of type W and WW: 1000–2000	(a) For buildings of type SW: <2000 (b) For buildings of type N: 1000–2000	Buildings of type N: <1000	Single floor buildings <500
	ZL category buildings	–	(a) For buildings over two floors of ZL I, ZL II, ZL V categories (b) For buildings of type W and WW of category ZL III (c) For buildings of type WW of category ZL IV	(a) For 2 floor buildings of category: ZL I, ZL II, ZL V (b) For buildings of more than two floors type N and SW of ZL III category (c) For buildings of more than three floors type N, SW and W of ZL IV category	(a) For one floor buildings of ZL II category (b) For buildings of up to two floors of ZL III category (c) For three floor buildings of ZL IV category	(a) Single floor buildings with nonspreading fire elements of ZL I and ZL V categories (b) up to two floors of ZL IV category

Table 10.97 Building elements and sections of different fire category

Building Fire category	Building elements							
	Main construction (walls, columns, floors)		Ceilings		Partition and curtain walls		Roofs, terraces, bearing construction of roof	
	Minimum fire resistance (min)	Fire spread	Minimum fire resistance (min)	Fire spread	Minimum fire resistance (min)	Fire spread	Minimum fire resistance (min)	Fire spread
A	240	NRO	120	NRO	60	NRO	30	NRO
B	120	NRO	60	NRO	30	NRO	30	NRO
C	60	NRO	60	NRO	15	NRO	15	NRO
D	30	NRO	30	NRO	(–)	SRO[a]	(–)	SRO[a]
E	(–)	SRO	(–)	SRO	(–)	SRO	(–)	SRO

NRO = non-spreading fire, SRO = hardly spreading fire, (–) = no requirements. [a] For ZL II category buildings, NRO is required

In multiple dwellings and public buildings where more than 50 persons are present, fixed finished materials must be made of at least hardly ignitable materials.

Ceiling coverings and suspended ceilings must be made of noncombustible or non-ignitable materials that will not fall down in case of fire (not valid for dwellings).

In category ZL I rooms limited to 100 persons, the use of suspended ceilings made of hardly ignitable, nondripping and nonfalling materials is permissible (not valid for dwellings).

The free space between suspended ceiling and floor shall be subdivided into areas not exceeding 1000 m^2. Corridors must be divided by noncombustible barriers placed every 50 m (164 ft).

In bathrooms and saunas with gas or electric heaters, the use of combustible wall coverings is permissible if the distance from apparatus to the covering is at least 0.3 m (1.18 ft).

The use of wall coverings made of readily ignitable materials is forbidden in bathrooms and saunas with solid fuel furnaces.

In the case of explosion hazard, use of windows and doors that do not give off sparks while opening or closing is required.

10.14.2 Classification and Testing of the Fire Performance of Building Materials and Components

The methods of testing material fire properties that are subject to the Polish Standards are presented in the Table 10.98.

As can be seen from Table 10.98, building products are divided into noncombustible and combustible materials on the basis of ISO 1182 [3] and of the determination of the gross calorific value to ISO 1716 [4].

Combustible materials are tested according to one of the following reaction-to-fire test methods listed in Table 10.98. Ignitability and flame spread are determined depending on the anticipated application of the material in building. Flammable building products are divided into the three classes

- Nonignitable,
- Hardly ignitable,
- Readily ignitable.

For flooring materials there are only two groups:

- Hardly ignitable, and
- Readily ignitable.

Different methods as well as criteria for product evaluation apply to the aforementioned groups. The main testing methods of interest are described below.

10.14.2.1 Noncombustibility of Building Products to PN-B-02862

This standard method is based on ISO 1182. Apparatus and test specifications are described in Section 10.23 ISO.

10.14.2.2 Gross Calorific Value by Oxygen Bomb Calorimeter to ISO 1716

This measurement is made as specified in ISO 1716 and described in Section 10.23 ISO.

10.14.2.3 Fire Performance of Building Materials to PN-B-02874

The test method is based on the French Standard NF P 92-501:1985 and describes a procedure for determining the ignitability factor i and the combustibility factor c. Apparatus and test specifications are similar to the Epiradiateur test covered in Section 10.8 France.

The classification of building products is based on the following requirements:

I　　rate of combustibility (non ignitable material):　　$i = 0$ and $c < 1$

II　　rate of combustibility (hardly ignitable material):　　$i \leq 1$ and $c \leq 1$

III　　rate of combustibility (readily ignitable material):　　$i > 1$ or $c > 1$

Table 10.98　Polish standard methods for testing reaction to fire of building products

Building material/product/element	Fire property	Standard method of testing in Poland	Foreign equivalent
All building products	Noncombustibility	PN-B-02862 [3]	ISO 1182:1990
All building products	Heat of combustion	–	EN ISO 1716 [4]
Combustible wall and ceiling coverings, suspended ceilings, combustible insulation of ventilation ducts	Ignitability	PN-B-02874 [5]	NF P 92–501:1985
Thin, flexible products	Ignitability Flame spread	PN EN ISO 6940 [6] PN EN ISO 6941 [7]	
Floor coverings, all layers of multilayered floor systems (tested subsequently in one test)	Flame spread	PN-B-02854 [8]	ASTM E 648 DIN 4102 Part 14[a)]
Internal and external walls	Fire spread	PN-B-02867 [9]	–
Roofs	Fire spread, Fire penetration	PN-B-02872 [10]	DIN 4102 Part 7
Ventilation ducts	Fire spread	PN-B-02873 [11]	DIN 4102 Part 6
All building materials	Toxicity	PN-89/B-02855 [12]	–
All building materials	Smoke generation	PN-89/B-02856 [13]	–

[a)] In this German standard a different substrate material (concrete board) is used

10.14.2.4 Flame Propagation of Floorings to PN-B-02854

The method is based on ASTM E 648-78 and German DIN 4102-14 and describes a procedure for measuring the critical radiant flux of horizontally mounted floor covering systems exposed to a flaming ignition source (pilot flame) in a graded radiant heat energy environment in a combustion chamber. Apparatus and test specifications are shown in Section 10.9 Germany.

The test method was taken over with minor changes by the European Union as EN ISO 9239-1.

The following classification criteria apply:

- CRF ≥ 4.5 kW/m^2 – hardly ignitable material
- CRF < 4.5 kW/m^2 – readily ignitable material

10.14.2.5 Testing of Fire Spread Rate of Building Elements

The tests are described in [9–11]. Samples representative for classified building parts are exposed to standard fire under conditions corresponding to the initial period of fire development. Flame propagation range, temperature rise, and visual observations are registered during the test. External walls are exposed to fire from both sides: indoor and outdoor. Asymmetric internal walls are tested from both sides. The testing conditions are as follows:

Temperature $20 \pm 10°C$

No rain, ice-rain, and frost

Wind speed:
- Tested walls exposed to indoor fire up to 1 m/s,
- Tested wall exposed to outdoor fire 2 ± 0.5 m/s,
- Tested roof 2 ± 0.5 m/s.

An electric ventilator is used to obtain requested wind speed.

Testing period:
- Standard fire – wood stock – 30 min (15 min fire exposure and 15 min watch),
- Standard fire – wood chips and diesel oil – 10 min (5 min fire exposure and 5 min watch).

Various testing devices and standard fires are used. The sample dimensions of building parts are shown in Table 10.99. Samples are conditioned to $15 \pm 10°C$ prior to the test. A temperature above 450°C in specified points is interpreted as burning. Proper visual observation is required and photographs are taken during and after the tests. The classification criteria based on the test results are summarized in Table 10.100.

Table 10.99 Sample dimensions of building parts (in meters) for fire spread testing according to [9–11]

Internal and external wall exposed to indoor fire [9]	2.1×1.8
External wall exposed to outdoor fire [9]	2.3×1.8[a]
Internal veneer wall on noncombustible substrate [9]	2.1×1.8
External veneer wall on noncombustible substrate [9]	2.3×1.8[a]
Roof part with support [10]	1.5×2.5[a]
Ventilation ducts [11]	$2 \times 1.5 + 1 \times 1.0$

[a] Greater dimension in supposed flame spread direction

Table 10.100 Classification criteria of products tested to fire spread according to [9–11]

Type	Temperatures measured in specified points		Combustion during test at certain zones	Combustion after test period	Burning drips
Nonspreading fire	< 450	< 450	No	No	No
Hardly spreading fire	No limits	< 450	Yes	Yes	No
Readily spreading fire	No limits				

10.14.2.6 Evaluation of Toxicity to PN-88/B-02855

The method is based on the analysis of toxic gases in fire effluents and based on the German DIN 53436 tube furnace test (see Chapter Toxicity, Section 16.5.2). Gas chromatography is used for analysis of CO and CO_2, and other methods for the analysis of NO_2, HCN, TDI, HF, HCl, and SO_2. The tests are conducted in a quartz tube heated by a ring furnace. The test conditions are:

- Furnace velocity: 20 mm/min,
- Air flux (in countercurrent to direction of furnace motion): 100 dm^3/h,
- Temperatures of decomposition: 500°C, 550°C, 750°C,
- Quartz tube length: 1000 mm,
- Quartz tube diameter: 30 or 40 mm,
- Sample mass: 5.4 g.

The analyzed toxic components are defined as the mass of combustion product generated from the mass unit of the sample. The toxic gas components formed are evaluated based on critical concentrations for each component called W_{LC50}.

Lethal concentrations of various toxic components after 30 min of exposure are:

$$CO:\ 3.75\,g/m^3 \qquad NO_2:\ 0.205\,g/m^3$$
$$CO_2:\ 196\,g/m^3 \qquad HCN:\ 0.16\,g/m^3$$
$$HCl:\ 1.0\,g/m^3$$

Toxic hazard is assessed on the basis of the arithmetic mean of factors W_{LC50M} for the temperatures 500°C, 550°C, and 750°C (W_{LC50SM}). This factor enables to:

- Determine toxic dominant combustion products for each material,
- Classify materials into groups according to their toxic products.

On the basis of the W_{LC50SM} values, materials are divided into three groups: very toxic, toxic, and merely toxic (Table 10.101).

Table 10.101   Exemplary classification of materials on the basis of toxic hazard to PN-88/B-02855

W_{LC50SM} (g/m^3)	Category of toxic hazard in fire	Material
0–15	Very toxic	PVC, rubber, EVA, PA, PAN
15–40	Toxic	Viscose rayon, PU, ABS, wool, PE, PP, fiber board
> 40	Merely toxic	Cotton, hard board, plywood, paper, latex foam, oak, beech wood, pine, silicone rubber, styrofoam, linen, PMMA, PS, poly(ethylene terephthalate)

10.14.2.7 Smoke Density Testing to PN-89/B-02856

The test method is based on the modified ASTM E 662-83, "Test method for specific optical density of smoke generated by solid materials," for determining smoke density and smoke evolution rate in the NBS smoke chamber. This modified method works with the concept of two parameters: the contrast weakening factor Y and its derivative $\dot{Y}$ are applied; these parameters are based on the light scattering, light absorption, and visibility as a function of contrast in smoke clouds.

The new parameters have some theoretical justification and are much less sensitive to sample thickness and the density of foamed materials than the widely used specific optical density.

Both final values and the time dependency of Y are of interest. Both flaming and nonflaming modes of material degradation are treated. Specimens are tested in a vertical orientation at heat flux level 70 kW/m^2.

10.14.3 Official Approval

The Building Research Institute at Warsaw is an accredited governmental institution to issue certificates and technical approvals for building products. Enquiries should be made to the:

Building Research Institute
Fire Research Department,
Ksawerów 21, 02-656 Warsaw
Poland
Tel. + 48 22 848 34 27
Fax. + 48 22 847 23 11
E-mail: fire@itb.pl

All necessary tests can be performed in its accredited laboratories. The fire laboratory of Building Research Institute is able to carry out all standard tests currently required by law in Poland.

10.14.4 Future Developments

The system of Euroclasses [14] will be gradually implemented in Poland within the next few years during a transition period in which the national standards will still be valid. This process was initiated in the year 2000.

References for Section 10.14

[1] Rozporządzenie Ministra Spraw Wewnętrznych z dnia 3 listopada 1992r w sprawie ochrony przeciwpożarowej budynków, innych obiektów budowlanych i terenów, Dziennik Ustaw Rzeczypospolitej Polskiej Nr 92, Warszawa, 10 grudnia 1992

[2] Rozporządzenie Ministra Gospodarki Przestrzennej i Budownictwa z dnia 14 grudnia 1994r w sprawie warunków technicznych, jakim powinny odpowiadać budynki i ich usytuowanie, Dziennik Ustaw RP Nr 15 z dnia 25 lutego 1999

[3] PN-B-02862 Ochrona przeciwpożarowa budynków. Metoda badania niepalnści materiaów budowlanych

[4] prEN ISO 1716 Reaction to fire tests for building materials – Determination of gross calorific value

[5] PN-B-02874 Ochrona przeciwpożarowa budynków. Metoda badania stopnia palnoci materiaów budowlanych

[6] PN-EN ISO 6940 Płaskie wyroby włókiennicze. Zachowanie się podczas palenia. Wyznaczanie zapalności pionowo umieszczonych próbek

[7] PN-EN ISO 6941 Płaskie wyroby włókiennicze. Zachowanie się podczas palenia. Pomiar właciwości rozprzestrzeniania się płomienia na pionowo umieszczonych próbkach

[8] PN-B-02854 Ochrona przeciwpożarowa budynków. Metoda badania rozprzestrzeniania pomieni po posadzkach podłogowych

[9] PN-B-02867 Ochrona przeciwpożarowa budynków. Metoda badania rozprzestrzeniania ognia przez ściany

[10] PN-B-02872 Ochrona przeciwpożarowa budynków. Metoda badania odporności dachów na ogie zewnętrzny

[11] PN-B- 02873 Ochrona przeciwpożarowa budynków. Metoda badania stopnia rozprzestrzeniania ognia po instalacjach rurowych i przewodach wentylacyjnych

[12] PN-88/B-02855 Ochrona przeciwpożarowa budynków. Metoda badania wydzielania toksycznych produktów rozkładu i spalania (Fire protection of buildings. Method for testing emission of toxic products of decomposition and combustion of materials)

[13] PN-89/B-02856 Ochrona przeciwpożarowa budynków. Metoda badania właściwości dymotwórczych materiałów (Fire protection of buildings. Method for testing smoke generation of materials)
[14] Commission Decision of 8 February 2000 implementing Council Directive 89/106/EEC as regards the classification of the reaction to fire performance of construction products

10.15 Spain

S. GARCIA ALBA

10.15.1 Statutory Regulations

The main Spanish regulations on fire performance of building products are:

NBE-CPI: Norma Basica de la Edificacion – Condiciones de Protección contra Incendio en los Edificios (Basic Building Regulations – Fire Safety Conditions)

In Spain, there are the most generally applicable prescriptive regulations for building fire protection. NBE-CPI is part of a wider family of building codes used in the Spanish construction activity.

It is issued by the Building Ministry (Ministerio de Fomento) and introduced with the Real Decreto 2177/1996 [1]. Only the general regulations are normative. Annexes are informative. This rule covers building design features (prescriptive occupational density, egress ways dimensions, allowed sector sizes, lift shafts, signals, etc.), reaction-to-fire of materials and fire resistance of elements levels, general service installations, special risk rooms, and active protection means. Industrial buildings are excluded from its scope.

Proyecto de Reglamento de Seguridad contra Incendios de Edificios Industriales (Fire Safety of Industrial Buildings Rule (Draft)

Industrial buildings are going to be covered by this regulation, which is issued by the Ministry of Technology and Science, the former Industry Ministry (Ministerio de Ciencia y Tecnologia), and is in a final draft stage today. This regulation includes all information from the NBE-CPI and contains a risk assessment procedure limiting design allowances.

Reglamento General de Policia de Espectaculos Públicos y Actividades Recreativas (General Regulation for Public Places of Assembly and Leisure Activities)

Mainly places of assembly, cinemas, theatres, sports facilities, and so forth are contained in the aforementioned regulation. It is issued by the Ministry of the Interior (Ministerio del Interior), and introduced with the Real Decreto 2816/1982 of August 27 [2]. In 1984, several additional comments especially relevant to upholstered furniture were issued. At present, this regulation with all the requirements about fire safety building practice is part of NBE-CPI-96.

Ordenanzas municipales de Prevención de Incendios. (Local Fire Safety Regulations)

Specific local building regulations apply in larger towns, through Fire Safety Regulations issued by the local City Council, which must be based on the prescriptions contained in NBE-CPI-96 or may even be more severe.

10.15.2 Classification and Testing of the Fire Performance of Building Materials and Components

According to NBE-CPI and other regulations, building materials are classified according to the effects of their reaction-to-fire as laid down in the classification standard:

UNE 23-727-90 Ensayos de reacción al fuego de los materiales de construcción. Clasificación de los materiales utilizados en la construcción.

Also to be noted is an additional standard, UNE 23-730-90, related to special classification cases, product information sheets and standards substrates.

NBE-CPI gives mandatory rules to check the durability of flame retardancy in textile and fabric materials by endurance procedures before fire testing.

In UNE 23-727, the reaction-to-fire of building materials is divided into five classes: M-0, M-1, M-2, M-3, and M-4. Materials showing a high flame spread and not meeting class M-4 are "Not Classifiable" (NC). Classification depends on the results of the tests laid down in the following standards:

UNE 23-102 Determination of noncombustibility of building materials

UNE 23-721 Radiation test for building materials [rigid materials of any thickness, materials to be used in rigid substrates and flexible materials more than 5 mm (0.2 in.) thick] (main test)

UNE 23-723 Electrical burner test for building materials (flexible materials up to 5 mm thick) (main test)

UNE 23-724 Flame propagation and persistency test for building materials (complementary test)

UNE 23-725 Dripping test for building materials (complementary test)

UNE 23-726 Radiant panel test for floor coverings (complementary test)

UNE 23-735 Part 1 Quick wear and ageing procedures: general requirements

UNE 23-735 Part 2 Quick wear and ageing procedures: textile materials not exposed to weathering

NOTE: The alcohol frame test to UNE 23-722 was withdrawn in 1990.

The standards are issued by AENOR (Asociación Española de Normalización) – Genova, 6 – E-28004- Madrid. The responsible Technical Committee is CTN 23.

UNE 23-102 corresponds to ISO 1182-1983 and A-1 Amendment of 1989. All other standards partially correspond to French standards NF P 92-501 to 92-506 in the 1985 edition.

NOTE: Partial correspondence is due to several interpretative changes introduced by AELAF (Spanish Fire Laboratories Association) for the practical application of the standards in Spain. These changes have led to several differences to the French standards and therefore do not result in the same classification levels when the same material is tested according to French or Spanish methods.

The ISO 1182 noncombustibility test procedure is described in Section 10.23 ISO.

UNE 23-721, 23-723, 23-724, 23-725 and 23-726 are based on the French tests NF P 92-501, 92-503, 92-504, 92-505 and 92-506 described in Section 10.8 France. However, Spain did not take over the modifications on equipment and classification criteria introduced in 1991 by the French authorities (the Spanish requirements remain unchanged, i.e., the same as stated in the French "arrêté" of 1983).

The Spanish regulations specify different test procedures for certain classes of material:

- Flexible materials up to 5 mm thick are tested to UNE 23-723 or, if necessary, to UNE 23-724 and UNE 23-725.
- Flexible materials > 5 mm thick and rigid materials of any thickness are tested to UNE 23-721 and, if necessary, to UNE 23-724 and UNE 23-725.
- Floor coverings are tested to UNE 23-721 and, if necessary, to UNE 23-726.

The requirements that must be met in these tests to achieve the various classifications are summarized in Tables 10.102–10.106.

The main method for classifying rigid materials, materials on rigid substrate or flexible materials > 5 mm thick is the test to UNE 23-721. It partially corresponds to the test to NF P 92-501 described in Section 10.8. Four indices are calculated from the test data:

- Flammability index, i
- Flame spread index, s
- Index of maximum flame length, h
- Heat release evolution index, c

The classification criteria are summarized in Table 10.103.

Table 10.102   Classification of noncombustible building materials

Tests	Requirements
Radiation test UNE 23-721	Proofing class M-1 prior to noncombustibility test (see Table 10.103)
Noncombustibility test UNE 23-102	Testing five specimens. Mean values are used to classify: • increase in temperature compared to initial furnace temperature for the thermocouple in the furnace and inside the specimen ≤ 50 °C, • duration of flames lasting more than $5\,\mathrm{s} \leq 20\,\mathrm{s}$, • mean mass loss $\leq 50\,\%$.
	M-0

Table 10.103   Classification of combustible rigid building materials and flexible materials above 5 mm thick

Tests	Requirements/Classification				
Radiation test [a] UNE 23-721	$s = 0$	$s < 0.2$	$s < 1$	$s < 1$	$s < 5$
	$h = 0$	$h < 1$	$h < 1$	$h < 1.5$	$h < 2.5$
	$c < 1$	$c < 1$	$c < 1$	$c < 1$	$c < 2.5$
	$i = 0$	any i	$i < 1$	any i	$i < 2$
	M-1	M-2		M-3	
Flame spread test UNE 23-724	Materials out of mentioned classes and flame spread speed < 2 mm/s			Materials out of mentioned classes and flame spread speed > 2 mm/s	
	M-4			NC	

[a] If the material presents a special behavior (melting, shrinkage, disappearing), see table 10.105

The other main method for classifying flexible materials < 5 mm thick is the UNE 23-723 test. It partially corresponds to the test to NF P 92-503 described in Section 10.8. The following criteria are considered from the test data:

- Ignition duration,
- Flammable dripping,
- Destroyed material surface (length and width).

Materials that shrink or disappear rapidly on exposure to thermal radiation must, in addition, be tested to UNE 23-724 (becoming main test) and UNE 23-725. Classification is made on the basis of the requirements shown in Table 10.105.

For flooring materials, there is a special requirement, indicated in Table 10.106. The radiant panel test only gives a distinction between M-3 and M-4 levels.

Table 10.104   Classification of combustible flexible building materials up to 5 mm thick

Tests	Requirements/Classification				
Dripping test UNE 23-725		No ignition of cotton pad	No ignition of cotton pad	Ignition of cotton pad	Ignition of cotton pad
Electrical burner test UNE 23-723	No dripping	Nonflammable dripping	Nonflammable dripping	Nonflammable dripping	Flammable dripping
Flaming ≤ 5 s[a)]	M-1	M-1	M-1	M-4	M-4
Destroyed length < 350 mm	M-2	M-2	M-3	M-4	M-4
If destroyed length < 600 mm, destroyed width < 90 mm	M-3	M-3	M-4	M-4	M-4
Fire spread test UNE 23-724	Materials out of mentioned classes and flame spread speed < 2 mm/s	Materials out of mentioned classes and flame spread speed > 2 mm/s			
	M-4	NC			

[a)] If the material presents a special behavior (melting, shrinkage, disappearing); see Table 10.105.

Table 10.105   Classification of material with melting, shrinking, disappearing, and so forth behavior

Tests	Requirements/Classification				
Dripping test UNE 23-725		No ignition of cotton pad	No ignition of cotton pad	Ignition of cotton pad	Ignition of cotton pad
Persistent flame test UNE 23-724	No dripping	Non flammable dripping	Non flammable dripping	Non flammable dripping	Flammable dripping
No persistency	M-1	M-1	–	M-4	M-4
Persistency ≤ 5 s	M-2	M-2	M-3	M-4	M-4
Persistency > 5 ≤ 2 s	M-3	M-3	M-4	M-4	M-4

Note: Persistency is the average value of the maximum persistency value obtained from each specimen in the same serial test.

Several interpretative agreements especially dealing with the UNE 23-724 test were issued by AELAF.

Table 10.106   Classification of floor coverings materials

Test Radiant panel test to UNE 23-726		Final classification according to UNE 23-721 and UNE 23-726 results
Destroyed length	Destroyed length at 1 min test	
		M-1 or M-2
≤ 300 mm	–	M-3
> 300 mm	≤ 100 mm	M-4
> 300 mm	> 100 mm	NC

Note: Persistency is the average value of the maximum persistency value obtained from each specimen in the same serial test

10.15.3 Official Approval

NBE-CPI also allows justifying the compliance of building materials to the reaction-to-fire requirements through certification marks, but to date no officially approved mark is available (CE marking is awaited).

Today, to obtain a classification for a building material according to NBE-CPI, the manufacturer must provide evidence of the classification by means of positive test certificates on representative samples, issued by an officially acknowledged laboratory. This is not necessary for M-0 recognized products such as stone, glass, and metallic and ceramic materials.

The list of Recognized Laboratories is continuously updated and requires an EN 45001 quality assurance system approved by ENAC (Official Accreditation Body) and a positive Building Ministry acknowledgement.

The following laboratories are officially recognized for issuing reaction-to-fire test certificates:

Laboratorio de Investigacion y Control del Fuego de la Asociación para el Fomento de la Investigación y Tecnologia contra Incendios (AFITI-LICOF)
Ctra. Valencia Km 23,400
E-28500 Arganda del Rey
Madrid
Spain
Tel. + 34 91 871 35 24
Fax + 34 91 871 20 05
E-mail: licof@afiti.com
Laboratori General d´Assaigs i Investigaçións de la Generalitat de Cataluña (LGAI)
P.O.Box 18
E 08193 Bellaterra (Barcelona)
Spain
Tel. + 34 93 691 92 11
Fax + 34 93 691 59 11
E-mail: xescriche@lgai.es

Laboratorio del Fuego del Instituto Tecnologico Textil (AITEX),
Alicante
Spain

No arrangements have yet been made regarding homologation of classifications achieved in tests, as the implementation of the CPD 89/106/CEE is awaited.

Nonmandated third-party certification is available for several products families including reaction-to-fire characteristics (e.g., building insulation materials) by nonmandatory marks, mainly those created by AENOR.

10.15.4 Future Developments

10.15.4.1 Regulations

The recently issued Ley de Ordenamiento de la Edificación [3] provides a general frame for developing and modifying the building activities at national level.

A fundamental step toward European harmonization is the mandatory new and unified technical building code Codigo Técnico de la Edificación. This new code is expected to be performance-based and strongly connected to the essential requirements of the Construction Products Directive 89/106/CEE (CPD) and will be used instead of the NBEs. All building regulations will merge in the new code, although regional and local administration levels will be free to regulate fire matters.

As in others European countries, there will be a transitional period with a double regulatory input from European test methods based on the CPD and the aforementioned national test methods.

10.15.4.2 Standardization

Like the other European countries, Spain is awaiting the implementation of the European reaction-to-fire standards and classification to Euroclasses of building products, derived from the European decision 2000/147/CEE.

CTN 23 remains connected to European technical fora and several Spanish laboratories are almost fully equipped to carry out such tests.

10.15.4.3 Certification

There are an increasing number of construction products in a certification procedure, mainly using AENOR product marks. After the introduction of the CE Mark, widespread certification procedures will take place, especially for those involved in the certification of conformity procedures based on the CPD.

References for Section 10.15

[1] B.O.E. Num. 261 of 29 Oct. 1996
[2] B.O.E. Num. 267 of 06 Nov. 1982
[3] B.O.E. Num. 266 of 06 Nov. 1999

10.16 Slovak Republic

O. GREXA

10.16.1 Statutory Regulations

The basic fire regulation in the Slovak Republic is given by The Law of Fire Safety No 126/ 1985. The fire safety of buildings is regulated by the Slovak Technical Standard (STN) series STN 7308xx which are also called "The Code of Fire Standards." This code includes design standards [1], test standards [2], value standards [3], object standards [4] and special standards.

General requirements for the passive and active fire measures are defined in two standards: STN 730802: Fire protection of buildings – Common regulations, and STN 730804: Fire protection of buildings – Industrial buildings. These standards define the fire risk and requirements for the fire safety degree; fire compartments; fire resistance; and flammability of constructions, escape routes, active fire measures, and so forth. Specific standards exist for various buildings (e.g., buildings for dwelling and lodging, hospitals, garages, storage rooms, assembly rooms, etc.) within the design standards. These specialized standards further define the requirements for the specific buildings and activities in conjunction with STN 730802 and STN 730804.

10.16.2 Classification and Testing of the Fire Performance of Building Materials and Components

Fire characteristics needed for the evaluation of the fire safety of buildings are divided into two main categories – fire resistance and reaction-to-fire.

The fire resistance of constructions is tested according to the following standards: STN 730851 – Determination of fire resistance of building constructions; STN 730852 – Determination of fire resistance of fire closures; and STN 730855 – Fire resistance test of external walls. In principle, the tested construction is subject to the thermal exposure defined by the standard temperature curve. In the test, the time is measured during which one of the failure criteria is reached (increase of temperature of 160°C above the ambient temperature at the unexposed side, failure of the load bearing structures and loss of integrity).

The main reaction-to-fire parameter is flammability measured according to STN 730862. The apparatus used in this test method is almost identical to the one defined in BS 476, Part 6 (see Section 10.18 UK). All building materials are classified into one of the five fire classes based on the test results as stated in Table 10.107.

Table 10.107   STN 73 0862 flammability classes for building materials

Fire class	Designation
A	Noncombustible
B	Not easily combustible
C1	Low combustibility
C2	Moderately combustible
C3	Easily combustible

Flame spread on the surface of linings and finishes is measured according to STN 730863. This test method determines the flame spread index i_s. In some applications, the material must also pass the STN 730861 noncombustibility test, which is identical to the ISO 1182 test method (see Section 10.23 ISO).

The fire load density in buildings is calculated using the heat of combustion of materials and products used in the interior furnishing. The heat of combustion of solids is measured according to STN 441352.

To measure the fire behavior of plastic materials, other standard test methods are also used in the Slovak Republic (series STN 64xx xx) [5]. These test methods characterize parameters such as ignitability, optical density, reaction to a small flame exposure, flammability of plastics in the form of films, and so forth. However, these test methods are not required by the building fire regulations in Slovakia. All materials used for interiors of road vehicles, tractors, and machinery must pass the test according to STN ISO 3795.

Further details regarding the tests mentioned here are described in Section 10.7 Czech Republic.

10.16.3 Official Approval

The certification of building products in the Slovak Republic is regulated by the Law of Building Products No 90/1998. In this law, the basic requirements for construction products expressed in the EU Directive 89/106 are incorporated.

For all construction products, conformity with legislation and technical specifications must be documented.

The certification of construction products is done by the certification body. The authorization is granted by the Ministry of Architecture of the Slovak Republic.

10.16.4 Future Developments

The Slovak Republic has a status of observer within the CEN, and it takes part in the CEN TC 127 meetings. The official policy of the Slovak Republic is to harmonize its regulations and legislation with those of the European Union. In the area of fire safety, the new fire standards as well as the Governmental Decree within the Law of Fire Safety No 126/1985 are being prepared. These regulations incorporate the requirements of the EU Construction Products Directive 89/106 and its interpretative documents as well as the Eurocodes. The Eurocodes are in the process of adoption into the Slovak legal system. The new fire standards and Governmental Decree will replace the existing fire standards and the system of fire safety in the Slovak Republic.

References for Section 10.16

Slovak Technical Standards in the Area of Fire Protection of Buildings and Fire Behavior of Plastics

[1] STN 73 0802 Nonindustrial buildings 6/1992
    STN 73 0804 Industrial buildings 5/1991
    STN 73 0831 Assembly rooms 5/1979
    STN 73 0833 Buildings for dwelling and lodging 1/1993
    STN 73 0834 Changes of buildings 3/1987
    STN 73 0835 Buildings for sanitary matters 6/1980
    STN 73 0837 Small Garages 8/1977

STN 73 0838 Multi-storey and underground garages 8/1977
STN 73 0842 Buildings for agricultural production 7/1989
STN 73 0843 Buildings for telecommunication service 8/1978
STN 73 0844 Storage rooms 6/1986
[2] STN 73 0850 Technical demands on fire furnaces 11/1986
STN 73 0851 Determination of fire resistance of building constructions 5/1984
STN 73 0852 Determination of fire resistance of fire closures 2/1974
STN 73 0855 Fire resistance test of external walls 6/1980
STN 73 0856 Determination of fire resistance of suspended ceilings 11/1986
STN 73 0857 Determination of fire resistance of ventilation ducts 2/1991
STN 73 0861 Testing of combustibility of building materials – noncombustible materials 5/1979
STN 73 0862 Determination of flammability of building materials 6/1980
STN 73 0863 Determination of flame propagation along the surface of building materials 2/1991
STN 73 0864 Determination of heating values of combustible solids under fire conditions 3/1988
STN 73 0865 Evaluation of materials drainage of the soffits of ceilings and roofs 2/1987
STN 73 0866 Determination of burning rate of substances in silos, stocks and reservoirs 2/1987
[3] STN 73 0818 Building occupation 1/1982
STN 73 0821 Fire resistance of engineering structures 2/1973
STN 73 0822 Flame propagation along the surface of building materials 9/1986
STN 73 0823 Flammability of building materials 5/1983
STN 73 0824 Heating value of flammable substances 12/1992
[4] STN 73 0872 Protection of buildings to extension of fire by air-distributing equipment 8/1978
STN 73 0873 Equipment for fire-water supply 10/1995
STN 73 0875 Design of electric fire-alarm signal systems 3/1991
[5] STN 64 0146 Heat stability and fire resistance during the hot mandrel test 3/1984
STN 64 0150 Determination of smoke optical density 11/1989
STN 64 0756 Plastics. Determination of flammability by oxygen index 6/1989
STN 64 0757 Determination of flammability of plastics in the form of film 10/1977
STN ISO 1210   Plastics. Determination of the burning behavior of horizontal and vertical speci-mens in contact with a small flame ignition source 1992
STN ISO 9772   Cellular plastics. Determination of horizontal burning characteristics of small spec-imens subjected to a small flame 1994
STN ISO 3795   Road vehicles, and tractors and machinery for agriculture and forestry. Determina-tion of burning behavior of interior materials 1989
STN ISO 871   Plastics Determination of ignition temperature using hot-air furnace 1996

10.17 Switzerland

R. HOFFMANN

10.17.1 Statutory Regulations

The cantons are responsible for matters regarding the public fire authorities according to the federal constitution of Switzerland. These cantons enact their own laws and regulations. For enacting these laws, most of the cantons use as a basis the standard form of regulations for fire protection issued by the Vereinigung Kantonaler Feuerversicherungen VKF (Association of Cantonal Fire Insurers). These regulations can be subdivided into a fire protection standard (rule of law) as the basic paper and the guidelines for fire protection as the implementing provisions as well as into the regulations for the testing of material and equipment.

As early as in 1962, the fire behavior of building materials was considered in the fire protection guidelines on the use of combustible building materials in buildings. These guidelines were elaborated by the Technical Commission of the VKF in close cooperation with the

Eidgenössische Materialprüfungs- und Forschungsanstalt EMPA (Federal Materials Testing and Research Institute), the Schweizerische Ingenieur- und Architekten-Verein SIA (Swiss Engineers and Architects Association) and the Schweizerische Institut zur Förderung der Sicherheit (Sicherheitsinstitut) (Swiss Institute of Safety and Security) and were revised and brought up to date several times in the course of years.

10.17.2 Classification and Testing of the Fire performance of Building Materials and Components

The fire performance is assessed according to the combustion and smoke developing behavior of the material and is classified by a fire coding. The fire coding is established by standardized tests.

The classification of combustible building materials is described in the "Building material and components" guideline [1]. This guideline deals with the combustion behavior, the smoke grades, and the classification of combustible building material.

10.17.2.1 Fire behavior

In the sense of this assessment the combustibility of a material is defined by its flammability and the burning rate and is substantiated by testing. The classification of building materials is based on combustibility grades 3–6. Materials of combustibility grades 1 and 2 are not approved as building materials. Details are shown in Table 10.108.

Table 10.108 Classification of building materials

Combusti-bility grade	Fire behavior	Example
1	Extremely highly flammable and extremely fast burning	Nitrocellulose
2	Highly flammable and fast burning	Celluloid
3	Highly combustible Highly flammable building material, burning spontaneously and fast without additional heat supply	Plastic foams without flame retardants
4	Average combustibility Average combustibility building material, continuing to burn spontaneously for a longer period of time without additional heat supply	Conditioned white wood
5	Low combustibility Low combustibility building material, continuing to burn slowly or carbonizing only with additional heat supply. After removal of the heat source the flames must go out within a short time interval and afterglowing must cease	Plastics containing flame retardants
5 (200 °C)	Low combustibility at 200 °C Building material fulfilling the requirements of combustibility grade 5 even under the effect of an elevated ambient temperature of 200 °C	Rigid PVC

Table 10.108 (Continuation)

Combusti-bility grade	Fire behavior	Example
6q	Quasi non-combustible Building material containing a low content of combustible components, which are classified as nonflammable for noncombustible application purposes	Mineral wool
6	Noncombustible Building material without combustible components, which does not ignite, carbonize or incinerate	Concrete

10.17.2.2 Smoke Developing Behavior

The smoke developing behavior of building materials is classified as in Table 10.109:

Table 10.109 Classification of the smoke developing behavior of building materials

Smoke grade	Smoke developing behavior	Example
1	Heavy formation of smoke	Polystyrene
2	Average formation of smoke	Polyethylene
3	Low formation of smoke	Poly(methyl methacrylate)

10.17.2.3 Classification

The combustibility and smoke grades established on the basis of the test results are expressed as "fire coding," which is a combination of the combustibility and the smoke developing classification.

A fire coding of 4.1 means, for example, that the building material has an average combustibility and develops heavy smoke when burnt.

10.17.2.4 Application of Combustible Building Materials in Buildings

The VKF "Use of combustible building materials" fire protection guideline [2] offers guidance for the cantons in terms of combustible building materials. It applies to materials such as slabs, linings, floorings, pipe work, and insulating materials used for the construction of a building.

This guideline establishes the following principles:

- Building materials of combustibility grades 1 and 2 may not be used.
- Highly combustible building materials (combustibility grade 3) are allowed in exceptional cases only. Such materials must be covered on both sides by F30 fire resistant noncombustible materials with no hollow spaces in between.
- Heavy smoke developing building materials may not be used without linings on the room end.
- Restrictions in terms of surface apply to transparent elements consisting of combustible materials.

- Use of building materials developing toxic, strongly irritating, panic-inducing fire gases or vapors may lead to increased exposure in highly frequented rooms.
- Noncombustible materials must be used or sufficient safety distances must be observed in areas with existing or anticipated sources of ignition.
- In certain application cases, building materials of combustibility grade 5 are required to be of low combustibility up to an ambient temperature of 200 °C.

The admission of combustible building materials is regulated in the guideline according to the type of application as follows:

- External walls,
- Internal walls and ceilings,
- Flooring material,
- Pipe work and insulation,
- Roofing.

Testing of building materials and components

The "Building Materials and Components, Part B: Test Regulations" fire protection guideline describes the test methods for establishing the combustibility and smoke grades of building materials and the tests for components [3].

For further properties, relevant in case of fire, such as toxicity or corrosivity of fire gases, reference is made to the classification according to SIA [4].

Determination of Combustibility Grade

The application of the combustibility grade test is normal practice. Special regulations exist for certain materials such as flooring and textile materials.

The test is conducted using a standardized test apparatus shown in Fig. 10.50.

The test criteria are summarized in Table 10.110.

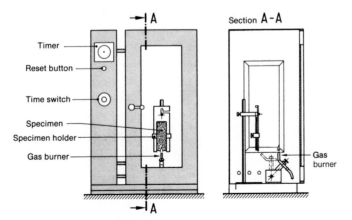

Figure 10.50 Combustibility test apparatus

Test Procedure at Room Temperature

A minimum of three tests are conducted. If these three tests do not result in one and the same classification, the number of tests is increased to six, deleting always the highest and the lowest result. The remaining worst result is the one deciding on the classification.

A conditioned specimen of the building material is mounted in vertical position on the test rig and a standardized ignition source is applied in the center of the lower front edge.

Test Procedure at an Ambient Temperature of 200 °C

In a test apparatus that can be heated, the temperature is increased to 200 °C until constant conditions have been reached. The specimen is clamped into the specimen holder. After the specimen has been heated up for 5 min, the tests are conducted as described in the preceding.

Classification

The decisive criterion for the classification is the time elapsed from the start of the flame application until the tip of the flame reaches the upper part of the specimen holder [150 mm (5.9 in.) from the lower edge of specimen] (time) or until the flame extinguishes (burning time).

If rising of the tip of the flame to the upper part of the specimen holder is not unambiguously observed, a cotton thread according to SN 198 898 standard must be tensioned at this level for measuring the time that it takes to burn the thread. For classification purposes the test using the cotton thread has priority over a visual observation. Details of classification and requirements are summarized in Table 10.111.

Table 10.110   Combustibility test criteria

Specimens	Six specimens Compact materials: 160 mm × 60 mm × 4 mm Foams: 160 mm × 60 mm × 6 mm
Specimen position	Vertical
Ignition source	Propane-operated displaceable burner, flame tip temperature approx. 900 °C, flame length 20 mm, inclined at 45° to horizontal
Test duration	Flame application 15 s on lower front edge until the flame has reached the upper part of the specimen holder or until extinction
Conclusion	According to test result, classification into combustibility classes 3–5 as specified below

Table 10.111   Classification of building materials based on combustibility test requirements

Classification	Requirements
Combustibility grade 3	Time: 5–20 s
Combustibility grade 4	Time, burning time: > 20 s
Combustibility grade 5	The flame does not reach the upper part of the specimen holder (150 mm) Burning time: ≤ 20 s
Combustibility grade 5 (200 °C)	The flame does not reach the upper part of the specimen holder (150 mm) Burning time at temperature of 200 °C ≤ 20 s
Combustibility grades 6q and 6	No ignition, incineration or carbonizing and noncombustibility test

Noncombustibility Test

Noncombustibility is tested according to DIN 4102, Part 1 (1981), Chapter 5. For details, see Section 10.9 Germany.

Classification

The decisive criteria for the classification are the flame duration and the temperature increase in the noncombustibility furnace and/or the level of the lower calorific value of the tested building material. Details are shown in Table 10.112.

Table 10.112   Classification of noncombustible building materials

Classification	Requirements
Combustibility grade 6q	Flaming ≤ 20 s and temperature increase (ΔT) ≤ 50 K or calorific value, lower (H_u) ≤ 4200 kJ/kg
Combustibility grade 6	no flaming Temperature increase (ΔT) ≤ 50 K

Radiant Panel Test for Floor Coverings

The combustibility of floor coverings is tested by the radiant panel test. Test apparatus and specifications are based on DIN 4102-14 and shown in Section 10.9 Germany.

The test chamber has a temperature of $180\,°C \pm 5\,°C$. The air throughput rate of the chamber is approx. $170\ \text{m}^3/\text{h}$. The incident heat flux radiated from the following distances by the radiant panel onto the plane of the specimen must range between:

at 200 mm: $0.87 - 0.95\,\text{W/cm}^2$

at 400 mm: $0.48 - 0.52\,\text{W/cm}^2$

at 600 mm: $0.22 - 0.26\,\text{W/cm}^2$

All values from 100 to 900 mm measured for the heat flux – plotted as a function of the distance – result in the heat flux profile required for the assignment of heat flux densities (W/cm^2).

Classification

To specimens that do not ignite or burn < 10 cm wide, a heat flow density of $\geq 1.1\,\text{W/cm}^2$ is assigned.

Specimens burning more than 90 cm wide have a lesser heat flux density compared to the calibration value at 90 cm. In all the other cases, a heat flux density corresponding to the burning distance is assigned to the specimens on the basis of the heat flux density profile.

The value critical for classification is found by averaging the heat flux densities of three specimens. The classification criteria are summarized in Table 10.113.

Smoke Density Test

The test is conducted in a standardized test box based on the American XP2 chamber to ASTM D 2843. Test apparatus and specifications are described in Chapter 15.3.2.2. Six specimens are tested with the following dimensions listed in Table 10.114.

Table 10.113   Classification criteria for floor coverings

Classification	Requirements
Combustibility grade 3	Heat flux density $< 0.25\,\text{W/cm}^2$
Combustibility grade 4	Heat flux density $0.25\text{--}0.49\,\text{W/cm}^2$
Combustibility grade 5	Heat flux density $\geq 0.5\,\text{W/cm}^2$

Table 10.114   Dimensions of test specimens for smoke test

Dimensions in mm $\pm\,10\,\%$ tolerance	Compact materials	Foams	Composite flooring materials
Length	30	60	30
Width	30	60	30
Thickness	4	25	Original thickness

Three tests are conducted. The specimen is placed on a defined wire grid and is burnt by means of a flame of 150 mm (5.9 in.) length. Any melting material is flamed in a metal sheet cup according to DIN 4102, Part 1, Item 5.1.2.2 (Version 1977). Flaming is continued until complete combustion of the specimen.

Classification

The decisive criterion for classification is the maximum light absorption. The requirements are shown in Table 10.115.

Table 10.115   Classification criteria smoke test

Classification	Requirements
Smoke grade 1	Maximum light absorption $> 90\,\%$
Smoke grade 2	Maximum light absorption $> 50\text{--}90\,\%$
Smoke grade 3	Maximum light absorption $0\text{--}50\,\%$

10.17.3 Official Approval

The classification of the fire behavior of building materials according to fire coding and their official approval are based on the test methods described in the preceding. In Switzerland, fire tests are conducted by the following official testing bodies:

Testing body	Services offered
Eidgenössische Materialprüfungs- und Forschungsanstalt Überlandstr. 129–133 CH-8600 Dübendorf	Reaction-to-fire and fire resistance of building materials and components
Eidgenössische Materialprüfungs- und Forschungsanstalt Lerchenfeldstr. 5 CH-9014 St.Gallen Homepage: www.empa.ch	Reaction-to-fire of textile and resilient flooring materials
Sicherheitsinstitut (Swiss Institute of Safety and Security) Klybeckstr. K-32.302 Postfach 4002 Basel CH-4057 Basel BS Homepage: www.swissi.ch	Reaction-to-fire of building materials and certification

After completion of the fire tests, the testing body issues a test report that serves as a basis for a certificate. The results of this test report and of this certificate form the basis for the decision of the cantonal fire authorities on whether or not to admit this material in their cantons.

For the purpose of simplifying this procedure the cantonal fire authorities contracted the

Vereinigung Kantonaler Feuerversicherungen (VKF)
(Association of Cantonal Fire Insurers)
Bundesgasse 20
Postfach 3001 Bern
CH-3011 Bern

and more specifically the VKF "Fachkommission Bautechnik" (Building Technology Technical Commission) for conducting the approval procedure. The following institutions are represented in the VKF Building Technology Technical Commission:

- Eidgenössische Materialprüfungs- und Forschungsanstalt (EMPA) (Federal Materials Testing and Research Institute),
- Schweizerisches Institut zur Förderung der Sicherheit (Sicherheitsinstitut) (Swiss Institute of Safety and Security),
- Vereinigung Kantonaler Feuerversicherungen (VKF) (Association of Cantonal Fire Insurers),
- Representatives of the cantonal fire authorities.

The approvals were published by VKF in a directory that is brought up-to-date annually, entitled "Swiss Fire Protection Register" [5] and in the BVD sheet [6].

10.17.4 Future Developments

Switzerland is actively involved in the European harmonization of test procedures with the intention of introducing these procedures in due time. Up until completion of these activities in Europe, the above described Swiss procedures will be applied in Switzerland.

References for Section 10.17

[1] Brandschutzrichtlinie Baustoffe und Bauteile, Klassierungen (Fire protection guideline: Building materials and components, Classification) available through VKF, Bern
[2] Brandschutzrichtlinie Verwendung brennbarer Baustoffe (Fire protection guideline: Use of combustible building materials) available through: VKF, Bern
[3] Wegleitung für Feuerpolizeivorschriften Baustoffe und Bauteile, Teil B: Prüfbestimmungen mit Nachträgen 1990, 1994 und 1995 (Guideline for Fire Authorities "Building materials and components, Part B: Test rules" including Amendments 1990, 1994 and 1995) available through: VKF, Bern
[4] SIA Empfehlung 183, Ausgabe 1996, Baulicher Brandschutz, (SIA-Swiss Engineers and Architects Association Recommendation 183, Edition 1996, Structural Fire Protection, available through SIA, Selnaustraße 16, Postfach CH-8039, Zurich)
[5] Schweizerisches Brandschutzregister / Verzeichnis der Schweizerischen Brandschutz-Zulassungen (Swiss Fire Protection Register/ Directory of Swiss Fire Protection Approvals) available through : VKF, Bern
[6] Brandschutztechnische Klassierung und Verzeichnis von Stoffen und Waren, BVD-Blatt SW1. (Classification in terms of fire behavior and directory of substances and commodities, BVD sheet SW1) available through: Swiss Institute of Safety and Security, Zurich

10.18 United Kingdom

J. Murrell

10.18.1 Statutory Regulations

The following building regulations apply in the various parts of the United Kingdom:
- England and Wales: The Building Regulations 1991 [1]
- Scotland: The Building Standards (Scotland) Regulations 1990–1997 [2]
- Northern Ireland: The Building Regulations (Northern Ireland) 1991 [3]

Technical provisions for use and fire performance of building materials and components are given in the supporting documents to the Building Regulations, for example, under Approved Document B for England and Wales. The technical provisions for England and Wales, and Northern Ireland, are virtually identical. Those for Scotland differ in minor aspects only. All the technical provisions are based on the same test methods specified in British Standards.

The provisions given in Approved Document B are given as guidance and provide one method of showing compliance with the Regulations. There is no obligation to follow that method provided compliance with the Regulations can be satisfactorily demonstrated.

10.18.2 Classification and Testing of the Fire Performance of Building Materials and Components

Compliance with the fire behavior of building materials and components identified in the Building Regulations and supporting documents is related to performance under certain standard tests. The methods are described in British Standards BS 476 Parts 3–33 and, for certain cases, in BS 2782 Method 120A and Method 508A, and in BS 5867 Part 2 and BS 5438 Test 2.

These Standards are issued by and available from the British Standards Institution 389, Chiswick High Road, London, W4 4AL.

Standard BS 476 "Fire Tests on Building Materials and Structures" consists of many parts covering fire behavior and fire resistance [4]. BS 476 Parts 3 and 20–24 describe the testing of roofs and elements of construction and are not covered in this book. BS 476 Parts 10, 12, 13, 31.1, 32, and 33 are not referred to in the Building Regulations or their supporting documents and are therefore also omitted.

Separation into combustible and noncombustible materials is carried out according to BS 476 Part 4 and "Materials of limited combustibility" are determined using BS 476 Part 11. With the aid of the test methods described in BS 476 Parts 6 and 7, combustible materials can be graded into classes 0–4 with increasing fire hazard.

The performance of certain plastics that are difficult to test to BS 476 Parts 6 and 7 can be investigated with the quality control tests in BS 2782, Methods 120 A and 508 A, together with BS 5867 Part 2 when tested in accordance with BS 5438 Test 2.

Thermoplastic materials that cannot be assessed to BS 476 Part 7, can be classified TP(a) rigid, TP(a) flexible, or TP(b) according to a series of methods or based on generic acceptance.

Limitations on the use of combustible building materials are made in connection with external walls, compartment walls or floors, separating walls, fire protecting suspended ceilings, internal wall and ceiling linings, and window openings.

No requirements exist at present for testing combustible floor coverings and for determining smoke evolution or the toxicity of fire gases.

10.18.2.1 BS 476 Part 4 Noncombustibility test

The noncombustibility of building materials is tested using BS 476 Part 4 (1970). The test methodology detailed in this Standard has been used as the basis for BS 476 Part 11, which is substantially similar to ISO EN 1182. A diagram of the furnace is shown in Section 10.23. Test specifications are given in Table 10.116.

Table 10.116   Noncombustibility test specifications

Test unit	0.2 m × 0.2 m × 1.425 m high
Specimens	Three specimens 40 mm × 40 mm × 50 mm
Specimen position	Loose laid in bottom of specimen basket
Ignition source	cylindrical furnace preheated to 750 °C
Test duration	20 min
Conclusions	Noncombustible if none of the three specimens either: • Causes the temperature reading from either of the two thermocouples (furnace and specimen) to rise by 50 °C or more above initial furnace temperature, or • Is observed to flame continuously for 10 s or more inside the furnace.

10.18.2.2 BS 476 Part 6 Fire Propagation Test

This method of test, the result being expressed as a fire propagation index, provides a comparative measure of the contribution to the growth of fire made by an essentially flat material, composite, or assembly. It is used primarily for the assessment of the performance of wall and ceiling linings. A diagram of the equipment and test specifications are reproduced in Fig. 10.51 and Table 10.117, respectively.

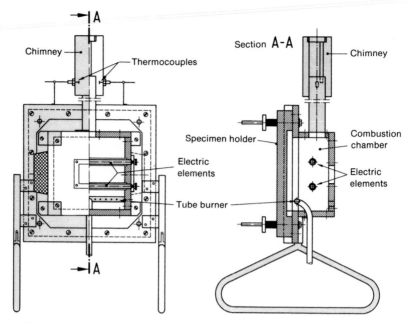

Figure 10.51   Fire propagation test apparatus

Table 10.117   Fire propagation test specifications

Test unit	0.4 m × 0.4 m × 0.75 m high
Specimens	Minimum of three and maximum of five specimens 225 mm × 225 mm × max.50 mm
Specimen position	Vertical
Ignition source	• Two 100 W electric elements with variable output (1800 W after 2 min 45 s and 1500 W after 5 min); distance from specimen: 45 mm • Gas pipe burners (internal diameter 9 mm) with 14 holes (internal diameter 1.5 mm) at 12.5 mm centers, distance from specimen 3 mm, • Flame applied at 25 mm above the bottom of the exposed face of the specimen
Test duration	20 min
Conclusion	Results are reported as subindices derived using the equations given in the text and a total fire propagation index (I)

Fire propagation is determined by exposing the specimen to the 14 jets of a gas pipe burner at a distance of 3 mm (0.12 in.). The heat evolved is 530 J/s. The two electric elements with a total output of 1800 W are switched on after 2 min 45 s. Their output is reduced to 1500 W after 5 min and maintained constant until the end of the test (20 min).

To evaluate a material, the difference between ambient temperature and that in the chimney is recorded continuously using thermocouples and compared with a calibration curve derived in a similar manner using a noncombustible board of a specific density. The two curves are evaluated using the temperatures from the calibration curve and test curve at 30 s intervals from the start of the test to 3 min, at 1 min intervals from 4 to 10 min and a 2 min interval from 12 to 20 min.

The individual indices are calculated as follows from these values:

$$i_1 = \sum_{1/2}^{3} \frac{\theta_m - \theta_c}{10t}, \quad i_2 = \sum_{4}^{10} \frac{\theta_m - \theta_c}{10t}, \quad i_3 = \sum_{12}^{20} \frac{\theta_m - \theta_c}{10t}$$

The total index $I = i_1 + i_2 + i_3$

where i_1, i_2 and i_3 are the subindices at 3 min, 10 min, and 20 min, respectively,

t = time in min from the origin at which the reading is taken,

θ_m = temperature in °C of the test curve at time t,

θ_c = temperature in °C of the calibration curve at time t.

If four or five specimens are needed to obtain three valid results, a suffix R is added.

The fire propagation test serves mainly to investigate whether materials that have achieved class 1 according to BS 476 Part 7, conform to class 0. To achieve class 0, a material must be of limited combustibility to BS 476 Part 11 or must meet the requirements of class 1 to BS 476 Part 7 and have a subindex $i_1 \leq 6$ and a total index $I \leq 12$ using BS 476 Part 6.

The requirements of class 0 make increased demands on the fire performance of class 1 building products and differentiate between them with regard to fire hazard. This test method is also used to examine the suitability of building materials for certain applications. For example, external cladding to walls 1m (3.3 ft) or more from the boundary on buildings exceeding 20m (66 ft) in height, must have a total index I of not more than 20 where located less than 20m above the ground. Any cladding above this height must be class 0.

10.18.2.3 BS 476 Part 7 Surface Spread of Flame Test

The lateral spread of flame along the surface of a specimen of a product orientated in a vertical plane, is determined by this test method using the apparatus shown in Fig. 10.52. The test specifications are summarized in Table 10.118. The test provides data suitable for comparing the performances of essentially flat materials, used primarily as the exposed surface of walls or ceilings.

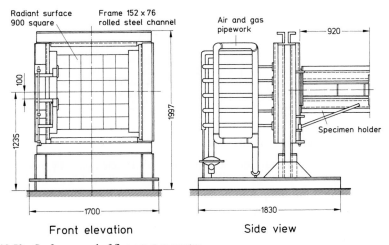

Front elevation   Side view

Figure 10.52   Surface spread of flame test apparatus

Table 10.118   Surface spread of flame test specifications

Test unit	1.7 m × 1.83 m × 2 m high
Specimens	Minimum of six and maximum of nine specimens 885 mm × 263 mm × maximum 50 mm
Specimen position	Vertical, longitudinal axis (885 mm) perpendicular to the radiant panel
Ignition source	• Gas fired radiant panel, radiation intensity 75 mm from radiant panel surface: 32.5 kW/m² • Gas pilot flame (height: 75–100 mm) impinging on the specimen on the same side as the radiant panel
Test duration	10 min
Conclusion	Classification in classes 1–4 depending on test performance

The specimen is mounted in a watercooled holder and is exposed to a radiant panel for 10 min. In addition, a pilot flame is applied to the bottom corner of the specimen during the first minute of the test. The time required for the flame front to reach reference marks on the specimen is noted, together with the extent of flame spread at 1½ min and at the end of the test. Observations are also made of the burning behavior and a suffix Y is added to the result, if softening or slumping occurs. If more than six specimens are required to obtain a result, a Suffix R is added, and if the product is tested in a modified form, a prefix D is added.

Materials are classified according to test performance as shown in Table 10.119. Class 4 materials are considered high risk and their use is not permitted.

Table 10.119   Spread of flame classification

Classification	Spread of flame at 1.5 min		Final spread of flame	
	Limit (mm)	Tolerance for one specimen (mm)	Limit (mm)	Tolerance for one specimen (mm)
Class 1	165	25	165	25
Class 2	215	25	455	45
Class 3	265	25	710	75
Class 4	Exceeding the limits for class 3			

10.18.2.4 BS 476 Part 11 Heat Emission Test

The assessment of heat emission from building materials is determined by this test method. The equipment used is substantially similar to ISO EN 1182 and BS 476 Part 4. A diagram of the furnace is shown in Section 10.23 ISO.

The method is applicable to simple materials that are reasonably homogeneous. Nonhomogeneous materials can be tested provided irregularities within the material are small. It is not normally suitable for assessing composites or assemblies.

Specimens are exposed for a maximum of 120 min in a preheated (750°C) cylindrical furnace and note is taken of any rise in temperature of both the furnace and specimen, and of any sustained flaming. From these results are calculated: the mean furnace temperature rise (°C), and the mean duration of sustained flaming(s), for each specimen tested. Calculation of the temperature rise is based on the maximum temperature achieved minus the equilibrium temperature reached which determines the end of the test. The test specifications are given in Table 10.120.

Table 10.120   Heat emission test specifications

Test unit	$0.4\,m \times 0.4\,m \times 1.25\,m$ high
Specimens	Five cylindrical specimens 45 mm diameter $\times$ 50 mm height
Specimen position	Vertical in specimen basket
Ignition source	Cylindrical furnace preheated to 750 °C
Test duration	Maximum of 120 min or until final temperature equilibrium is established
Conclusion	Results are reported as furnace and specimen temperature rises, duration of flaming, and mass loss.

This test is used to classify "materials of limited combustibility" [5], which are required for certain end use applications, for example, external walls situated within 1 m of the boundary and compartment walls of 1 h or more fire resistance.

Typical requirements are:

- *For structural frames of the above walls:*
  the material must not flame and there is no rise in temperature on either thermocouple.
- *For linings to such frames:*
  the material, having a minimum density of 300 kg/m^3, must not flame, and the rise in temperature on the furnace thermocouple is not more than 20°C.
- *For cavity insulation within such walls:*
  the material, of less than 300 kg/m^3, must not flame for more than 10 s and the rise in temperature on the specimen thermocouple is not more than 35°C and on the furnace thermocouple is not more than 25°C.

Note: Materials classified as noncombustible under BS 476 Part 4 are acceptable in all situations.

10.18.3 Test Methods for Thermoplastics

The methods to test fire performance referred to in the technical provisions of the Building Regulations, in general, are for all types of building materials and not specific groups. However, where the usual test methods have been found inappropriate, for example, with some thermoplastics, use is made of quality control test methods to provide information on plastics performance to allow for their use.

The use of such materials is, however, subject to restrictions on size, thickness, and separation from each other depending on their application.

The procedures identified in connection with the Building Regulations for testing thermoplastics are BS 2782, Methods 508 A [6] and 120 A [7]. BS 5438 Test 2 [8] is also used, in compliance with the requirements given for Type C in BS 5867 Part 2 for performance criteria [9].

Thermoplastics can be classified TP(a) rigid, TP(a) flexible, or TP(b); these classifications being specified for certain use areas such as windows, rooflights, and lighting diffusers.

TP(a) rigid applies to any solid poly(vinyl chloride) (PVC), and solid polycarbonate of 3 mm thickness and above, and multiskinned PVC or polycarbonate that is class 1, or any other thermoplastic that meets the requirements of BS 2782 Method 508 A 1970.

TP(a) flexible applies to products not more than 1 mm thick, which comply with Type C requirements in BS 5867 Part 2 as tested to BS 5438 Test 2.

TP(b) applies to all other PVC or polycarbonate products or other thermoplastics that have a rate of burning of less than 50 mm/min when tested to BS 2782 Method 508 A.

Method 120 A of BS 2782 Part 1, which is used for thermoplastics, does not constitute a fire test and is therefore not discussed below. Plastics with a softening point lower than 145°C cannot be classified as Class 0 although there are exceptions, which are detailed in Appendix B of Approved Document B [5].

10.18.3.1 BS 2782 Method 508A, Rate of Burning

The rate of burning (for production control purposes only) is determined by BS 2782 Part 5 – Method 508 A 1970, using the apparatus illustrated in Fig. 10.53. The test specifications are summarized in Table 10.121.

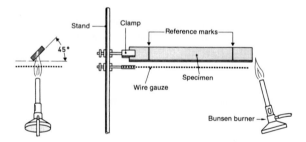

Figure 10.53   Rate of burning apparatus. Left: Front view, right: Side view

Table 10.121   Rate of burning test specimens

Specimens	Three specimens 150 mm × 13 mm × 1.5 mm reference lines at 25 mm and 125 mm
Specimen position	parallel to the free end of the specimen Longitudinal axis (150 mm), horizontal, short edge (13 mm) inclined at 45° to horizontal
Ignition source	Alcohol or Bunsen burner (flame length 13 mm to 19 mm)
Test duration	Flame applied to free end of specimen 10 s, specimen monitored until flame extinguishes
Conclusion	Performance criteria in connection with Building Regulations are: • Burnt length ≤ 25 mm and residual flaming ≤ 5 s [TP(a) rigid] • Rate of burning ≤ 50 mm/min [TP(b)]

A nonluminous alcohol or Bunsen flame is applied for 10s to the free end of the specimen. The times at which the first and second reference lines are reached by the flame front are noted and the rate of burning computed in millimeters per minute. If the flame does not reach the first line, the duration of residual flaming or afterglow is measured.

10.18.3.2 BS 5438 Flammability of Textile Fabrics

The extent of melting away or flaming is determined by BS 5438 Test 2A, using the apparatus illustrated in Fig. 10.54 and is limited by the requirements of BS 5867 Part 2, Type C. The test specifications are summarized in Table 10.122.

Table 10.122   Extent of burning specifications

Test unit	0.6m × 0.6m × 1.05m high
Specimens	24 specimens 200 mm long x 160 mm wide 12 with the longer dimension in the machine direction 12 with the shorter dimension in the machine direction
Specimen position	Vertical
Ignition source	Butane flame inclined at 30° (flame length 40 ± 2 mm)
Test duration	Flame applied for 5, 15, 20, or 30s Specimen monitored for a further 5s for afterflame or afterglow
Conclusion	Performance criteria in connection with Building Regulations are: • No flame or hole to any edge, • No mean afterflame or afterglow time exceeding 2.5s, • No flaming debris.

The small flame is applied to the surface of a vertically held specimen. The duration of flaming, afterglow and damaged length are measured. Spread of flame to any edge, the formation and extent of holes, and the occurrence of flaming debris are also noted.

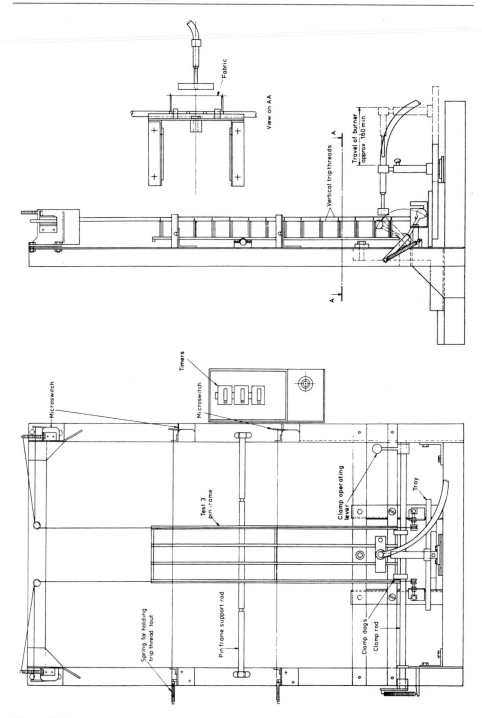

Figure 10.54   Extent of burning test

10.18.4 Official Approval

The tests described in the preceding are used to classify the fire performance and to obtain official approval of building materials. In principle the fire performance of building materials can be assessed by any test laboratory in the United Kingdom. There are no state test laboratories. In general, the tests are carried out by private test institutes that participate in a National Accreditation Scheme run by UKAS (United Kingdom Accreditation Service).

The principal fire test laboratories are listed below:

Building Test Centre
East Leake
Loughborough
Leicestershire LE12 6JQ
Tests: BS 476 Parts 4 to 11, 20, and 22

Chiltern Fire
Hughendon Valley
High Wycombe
Buckinghamshire, HP14 4ND
Tel. + 44 01494 569 818
Fax + 44 01494 564 895
E-mail: kdtowler@chilternfire.co.uk
Web site: http://www.chilternfire.co.uk
Tests: BS 476, Parts 20 to 24

Fire and Risk Sciences (FRS)
Building Research Establishment
Garston
Watford WD25 9XX
Tel. + 44 01923 66 40 00
Fax + 44 01923 66 49 10
E-mail: smithda@bre.co.uk
Web site: http://www.bre.co.uk/frs
Tests: BS 476 Parts 4, 6, 11

Warrington Fire Research Centre
Holmesfield Road
Warrington
Cheshire, WA1 2DS
Tel. + 44 (0)1925 655116
Fax + 44 (0)1925 646622
E-mail: geoff.deakin@wfrc.co.uk
Web site: http://www.wfrc.co.uk
Tests: BS 476 Parts 3–33
BS 2782 Method 508A
BS 5438 and BS 5867

After the test, the laboratory issues a report giving the test results and, where appropriate, the classification of the building material. This document provides sufficient confirmation of the fire performance of the building materials. Building materials are not marked or inspected by the test laboratory.

10.18.5 Future Developments

The United Kingdom is participating in work within the EEC on the harmonization of fire test methods. When these methods are adopted as European Standards they will subsequently be dual numbered as British Standards and the test methods described here will be withdrawn.

References for Section 10.18

[1] The Building Regulations, 1991 London HMSO, 1992
[2] The Building Standards (Scotland) Regulations, 1990–1997 London HMSO
[3] The Building Regulations (Northern Ireland), Belfast HMSO, 1991
[4] BS 476. Fire tests on Building Materials and Structures
      Part 3:1975. External Fire Exposure Roof Test (Note, the Building Regulations tests are still based on the 1958 version)
      Part 4:1970. Non-Combustibility Test for Materials
      Part 6:1989. Method of Test for Fire Propagation for Products
      Part 7:1997. Method for the Classification of the Surface Spread of Flame of Products
      Part 10:1983. Guide to the Principles and Application of Fire Testing
      Part 11:1982. Method for Assessing the Heat Emission from Building Materials
      Part 12:1991. Method of Test for Ignitability of Products by Direct Flame Impingement
      Part 13:1987. Method for Measuring the Ignitability of Products Subjected to Thermal Irradiance
      Part 15:1993. Method for Measuring the Rate of Heat Release of Products
      Part 20:1987. Method for Determination of the Fire Resistance of Elements of Construction (General Principles)
      Part 21:1987. Methods for Determination of the Fire Resistance of Loadbearing Elements of Construction
      Part 22:1987. Methods for Determination of the Fire Resistance of Non-loadbearing Elements of Construction
      Part 23:1987. Methods for Determination of the Contribution of Components to the Fire Resistance of a Structure
      Part 24:1987. Method for the Determination of the Fire Resistance of Ventilation Ducts
      Part 31:1983. Method for Measuring Smoke Penetration through Doorsets and Shutter Assemblies. Section 31.1:1983 Method of Measurement under Ambient Temperature Conditions
      Part 32:1989. Guide to Full Scale Tests within Buildings
      Part 33:1993. Full-scale Room Test for Surface Products
[5] Department of the Environment and the Welsh Office. The Building Regulations 1991. Approved Document B. Fire Safety. London HMSO 1992
[6] BS 2782:1970 Methods of Testing Plastics: Method 508A: Rate of Burning (Laboratory Method)
[7] BS 2782 Methods of Testing Plastics. Part 1 Thermal Properties
      Method 120A:1976 Determination of the Vicat Softening Temperature of Thermoplastics
[8] BS 5438:1989. Flammability of Textile Fabrics When Subjected to a Small Igniting Flame Applied to the Face or Bottom Edge of Vertically Oriented Specimens
[9] BS 5867 Specification for Fabrics for Curtains and Drapes. Part 2:1980 Flammability Requirements

10.19 Australia

V. DOWLING

10.19.1 Statutory Regulations

The Commonwealth of Australia consists of six atates – New South Wales, Queensland, South Australia, Tasmania, Victoria, and Western Australia – and two Territories – the Australian Capital Territory where the capital, Canberra, is situated, and the Northern Territory.

Building regulations are the responsibility of the states and territories. All the State and Territory Building Regulations call up the Building Code of Australia (BCA). The individual building regulations are [1–8]:

- Building Regulations (Australian Capital Territory)
- Building Regulations (Northern Territory)
- Building Regulations (Victoria)
- Local Government Approvals Regulation (New South Wales)
- Building Regulations (Queensland)
- Building Regulations (Western Australia)
- Development Regulations (South Australia)
- Building Regulations (Tasmania)

The BCA is published by the Australian Building Codes Board (ABCB), a body formed by the Federal Government with representatives from the states and territories. The BCA is performance based. It contains Objectives, Functional Statements, and Performance Requirements. The Performance Requirement for materials is that they must resist the spread of fire and limit the generation of smoke, heat, and toxic gases to a degree appropriate to the building type. The BCA also contains Building Solutions that are deemed to satisfy the Performance Requirements.

10.19.2 Classification and Testing of the Fire Performance of Building Materials and Components

There are two paths to meeting the requirements of the BCA. To use the performance path for materials, it is necessary to demonstrate that a particular material or system achieves the same level of fire safety in use as a material or system that meets the deemed-to-satisfy provisions. This can be achieved by carrying out comparative tests in any agreed large-scale or small-scale test.

Alternatively, the deemed-to-satisfy, or prescriptive, path can be used. The test methods used to ascertain whether building materials and components meet the deemed-to-satisfy provisions for fire safety are described in Australian Standards AS 1530.1 [9], AS 1530.2 [10], AS/NZS 1530.3 [11], and AS 1530.4 [12]. AS/NZS 1530.3 is a joint Australian-New Zealand Standard. The standards are available from Standards Australia, GPO Box 5420, Sydney, NSW 2000, Australia. Full contact details are available on their web site, www.standards.com.au.

AS 1530.1 only enables an opinion on combustibility to be made. In AS 1530.2 and AS/NZS 1530.3 various indices enable fire performance to be gradually differentiated. AS 1530.2 applies to thin flexible materials such as roof sheeting, curtains, or wall coverings while AS/NZS 1530.3 covers all other building materials.

AS 1530.4 is a test for determining the fire resistance of building components and is not dealt with in this book. In three specific cases, however, it is of importance for combustible building materials.

- The BCA permits the use of materials that have not met the deemed-to-satisfy provisions when tested according to AS/NZS 1530.3, provided they are protected on all sides by material that is not combustible and the composite passes a 10 min furnace test. In this test, 1m^2 specimens of such composite systems are heated for 10 min up to approx. 700 °C according to the AS 1530.4 time-temperature curve. This is not a determination of fire resistance, but a demonstration that a combustible core of a composite system will not be involved in a fire for 10 min.
- Electrical cables that are required to maintain circuit integrity in the event of a fire must be assessed in a test, AS/NZS 3013 [13], which is based on AS 1530.4. The cables, with all connections, junction boxes, etc., are fixed to a concrete slab and exposed to the furnace conditions, where their ability to maintain circuit integrity is assessed.
- Fire-separating building elements that have been penetrated by plastic pipes or conduits are required to be protected with a penetration system that would reinstate the fire resistance level of the penetrated building elements. The performance of a prototype of the building element with the penetration system is assessed in a fire resistance test, AS 4072.1 [14], which is based on AS 1530.4. In AS 4072.1 it is noted that the material properties of the various plastics impact on the performance of penetration protection devices. The Standard therefore requires all pipe sizes and pipe types to be individually tested.

Ductwork must comply with AS 4254 [15], which calls up AS/NZS 1530.3 and UL 181 for the fire testing.

10.19.2.1 AS 1530.1 Combustibility Test for Materials

The AS 1530.1 test apparatus and procedure for combustibility is essentially the same as ISO 1182 (discussed in Section 10.23). However, there are differences in terminology and the criteria used for "combustibility":

- The standard specifies criteria of failure for assessment of combustibility not specified in ISO 1182. These are based upon three out of the five criteria recommended in ISO 1182.
- However, the other two criteria, namely the specimen center thermocouple rise and the mass loss, are required to be reported in the standard to maintain compatibility with ISO 1182.
- The term "non-combustible" is not mentioned in the title or text of the standard. This is to avoid any possible interpretation of the test results that may imply that a material that passes the test is in all circumstances "non-combustible" or is "inert" in every sense of the term, including the strictest scientific sense.

10.19.2.2 AS 1530.2 Test for Flammability of Materials

The flammability of thin flexible building materials – in this case fabrics and films – is tested by AS 1530.2. A diagram of the apparatus and the test specifications are given in Fig. 10.55 and Table 10.123, respectively.

If the material has an asymmetric weave, six specimens are tested in the warp and six in the weft direction. If both sides have differing surface structure then two sets of six specimens are tested with the different surfaces facing outwards.

The material stretched over the frame is ignited by burning 0.1 l of pure ethanol in a copper container 13 mm (0.5 in.) below the lower edge of the specimen. The temperature of the fire gases is measured by thermocouples positioned 570 mm (22.4 in.) above the top edge of the specimen in a flue. Observations are made for a maximum of 160 s after ignition of the specimen. If the flame does not reach the highest (21st) mark, the maximum flame height is noted. The temperature of the fire gases is measured at least every 5 s over a period 180 s from ignition of the specimen. The following are also recorded:

- The time taken for the flame to reach the 21st mark if this occurs before 160 s have elapsed,
- The area between the recorded temperature curve for the combustion gases and the ambient temperature curve over the 180-s period.

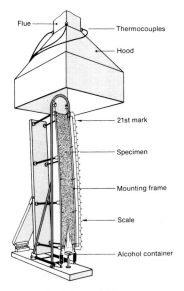

Flue — Thermocouples
Hood
21st mark
Specimen
Mounting frame
Scale
Alcohol container

Figure 10.55   Flammability apparatus for fabrics and films

Table 10.123   Flammability test specifications

Specimens	Six specimens 535 mm × 75 mm
Specimen position	Stretched over a slightly convex frame at 3°–4° to the vertical
Ignition source	Alcohol flame (0.11 pure alcohol in copper container) 13 mm below lower edge of specimen
Test duration	Max. 160 s
Conclusion	Flammability index computed from a "speed" factor, heat factor, and spread factor

Factors calculated from the test data are used to compute the flammability index and thus classify the material. These are: speed factor, heat factor, and spread factor and are calculated as follows:

- *Speed factor*: The speed factor S is expressed as: $S = (60 - 3t/8)$,

where t is the time in s taken by the flame to reach the 21st mark. The mean value from six tests is used in the equation. If the flame does not reach the 21st mark on any one specimen the speed factor is recorded as 0.

- *Heat factor*: The heat factor H is obtained from: $H = 0.24 \cdot A$

where A is the mean value determined from six tests of the area between the temperature curve of the combustion gas and the ambient temperature curve over the 180-s test period.

- *Spread factor*: The spread factor E is calculated only if the speed factor S equals 0,

$E = 20/9\,(D - 3)$, where D is the mean scale mark (0–21) reached by the flame determined for six specimens.

The flammability index (I) is obtained via one of the following equations:

$$I = H + E \text{ or}$$
$$I = H + S$$

The expression "$H + E$" is used if the flame does not reach the 21st mark and "$H + S$" if it does reach this mark.

10.19.2.3 AS/NZS 1530.3 Simultaneous determination of ignitability, flame propagation, heat release, and smoke release (known as the Early Fire Hazard Test)

The fire behavior of building materials is assessed by AS/NZS 1530.3. The apparatus is illustrated in Fig. 10.56 and the test specifications are given in Table 10.124. Normally six specimens are tested.

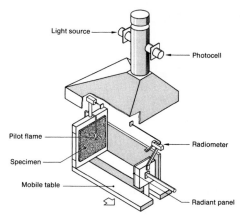

Figure 10.56   Early Fire Hazard Test apparatus

Table 10.124   AS/NZS 1530.3 specifications

Specimens	Nine specimens 600 mm × 450 mm × usual thickness
Specimen position	Vertical mobile specimen holder, positioned parallel to radiant panel, moving during test from 850 mm to 175 mm from radiant panel
Ignition source	• Vertical gas radiator 300 mm x 300 mm, radiation intensity 850 mm from radiator: 2.4 kW/m² • Two gas pilot flames at a distance of 15 mm from specimen, length of luminous flame portion 12 mm; first pilot flame 50 mm above center of test specimen, second variable for igniting decomposition gases
Test duration	Variable, depending on ignition time and emitted radiation increase rate
Conclusion	Four indices – ignitability, spread of flame, heat evolved, smoke developed

In the test, a specimen is vertically positioned on a mobile holder opposite a vertical gas-fired radiant panel. The holder and specimen are advanced, every 30 s, toward the radiant panel until ignition, or the 12 min mark of the test, when the movement ceases. Two pilot flames are placed near the surface of the specimen to ignite pyrolysis gases produced by the heated specimen. The specimen holder has side shields placed on either side of the specimen to restrict the effects of draughts on the burning behavior.

A radiometer viewing the heated face of the specimen records the radiation produced by ignition of the specimen. The radiometer moves with the specimen holder so that a set distance is maintained between the two. The smoke is collected by a hood, located over the radiant panel and specimen, and rises through a vertical duct where its optical density is recorded.

A light source and photoelectric cell are mounted on the flue above the test apparatus such that a 305 mm (12 in.) long light beam is directed through the flue. The sensitivity of the photocell corresponds to that of the human eye (greatest sensitivity in the range 500–600 nm; only 50 % of maximum sensitivity in the range < 400 nm and > 700 nm). The attenuation of the light beam due to smoke gases is recorded every 3 s during the test.

The test results are used to calculate the maximum smoke density for any 1 min period using the following formula:

$$D = \frac{1}{L} \log_{10} \frac{100}{100 - R}$$

where L is the length of the light path (305 mm) and R is the maximum value of the average reduction in percentage transmission of light for any 1 min period of the test. The mean of the optical densities so calculated is used to determine the "*Smoke Developed*" Index (see later).

The time elapsed from the start of the test until ignition is measured. Ignition is considered to have occurred if the flame continues for 10 s. Four indices are calculated from the test results. These are: Ignitability, Spread of Flame, Heat Evolved, and Smoke Developed.

- *Ignitability*: The index for the ignitability is 20 minus the time in min until ignition of the material. If fewer than three specimens ignite, the Ignitability Index is 0.
- *Spread of Flame*: If the radiation intensity increases by more than 1.4 kW/m² in < 203 s from the time of ignition, this time is used to determine the Spread of Flame Index (Table 10.125). If the ignitability index is zero or the radiation intensity increases by < 1.4 kW/m² within 203 s, the Spread of Flame Index is 0.
- *Heat Evolved*: The radiant heat evolved for each sample is calculated from the integral for a period of 2 min after ignition of the difference between the instantaneous radiation intensity and that just before ignition. The index is determined from the mean radiation intensity integral as shown in Table 10.126. If five or more specimens fail to ignite, the Heat Evolved Index is 0.
- *Smoke Developed*: The mean of the calculated optical densities is used to determine the Smoke Developed Index as shown in Table 10.127.

If some of the specimens do not ignite, the Index is calculated for each of the cases of ignition and non-ignition.

A report listing the results and the indices obtained based on test performance is issued. The ignitability indices run from 0 to 20 while the other indices run from 0 to 10. In practice only the Smoke Developed and Spread of Flame indices are considered in the BCA for the approval of building materials.

Table 10.125   Spread of flame index numbers

Index number	1.33 × mean time for radiation intensity to increase by 1.4 kW/m² (s)
0	≥ 270
1	$240 - <270$
2	$210 - <240$
3	$180 - <210$
4	$150 - <180$
5	$120 - <150$
6	$90 - <120$
7	$60 - <90$
8	$30 - <60$
9	$10 - <30$
10	<10

Table 10.126   Heat evolved index

Index number	Value of radiation intensity integral $(kJ \cdot m^{-2})$
0	<25
1	$25 - <50$
2	$50 - <75$
3	$75 - <100$
4	$100 - <125$
5	$125 - <150$
6	$150 - <175$
7	$175 - <200$
8	$200 - <225$
9	$225 - <250$
10	≥ 250

Table 10.127   Smoke developed index

Index Number	Mean optical density [m^{-1}]	
0	$< K^*$	< 0.0082
1	$K - <2K$	$0.0082 - <0.0164$
2	$2K - <2^2 K$	$0.0164 - <0.0328$
3	$2^2 K - <2^3 K$	$0.0328 - <0.0656$
4	$2^3 K - <2^4 K$	$0.0656 - <0.131$
5	$2^4 K - <2^5 K$	$0.131 - <0.262$
6	$2^5 K - <2^6 K$	$0.262 - <0.525$
7	$2^6 K - <2^7 K$	$0.525 - <1.05$
8	$2^7 K - <2^8 K$	$1.05 - <2.10$
9	$2^8 K - <2^9 K$	$2.10 - <4.20$
10	$\geq 2^9 K$	≥ 4.20

$* K = 0.0082$

10.19.3 Official Approval

According to the deemed-to-satisfy provisions of the BCA, one or more of the tests described in the preceding must be carried out to ascertain the fire performance of building materials. The tests can be performed by the CSIRO BCE Fire Science and Technology Laboratory (FSTL) or any laboratory approved by the National Association of Testing Authorities, Australia (NATA). Approval involves regular inspection of the test laboratories by NATA experts. A NATA directory [16] listing recognized test laboratories is published annually. Laboratories that test fire performance of plastics are not listed separately. By agreement between NATA and International Accreditation of New Zealand, tests performed by laboratories in New Zealand are also acceptable (see Section 10.20).

The test laboratory issues a NATA endorsed Report containing the test results. This document is sufficient proof of performance of the building materials. Building materials are not marked or inspected in Australia. Alternatively, the ABCB or a State or Territory accreditation authority can issue a certificate stating that the properties and performance of a building material fulfill specific requirements of the BCA.

Plastics for electrical use must also conform to the "Wiring Rules" [17], and are generally required to pass appropriate IEC tests or equivalent.

Two other Standards are used in the fire testing and specification of plastics materials but not called up in the BCA and therefore not mandatory. Both methods are based on research carried out at CSIRO Building, Construction and Engineering. The first, AS 2122.1 [18], assesses the flame propagation properties of rigid and flexible cellular plastics, in the form of small bars. The method is similar to ASTM D 3014, in that the specimen is examined in a vertical orientation in a metal "chimney" with a Bunsen burner type ignition source. The second, AS 2122.2

[19], assesses the minimum oxygen concentration [limiting oxygen index (LOI)] required for flame propagation on small specimens of plastics bars and sheets and textile fabrics. The method is similar to the ISO version of the oxygen index test; except that an incandescent electrically heated igniter is used, instead of a flame, for ignition of the specimen.

Building fire tests for plastics carried out by laboratories in Australia are listed below. All are NATA approved for the tests required by the BCA. The list of tests performed includes some required by building authorities outside Australia.

Australian Wool Testing Authority (AWTA)
Textiles Testing Division
26 Robertson Street, Kensington, Victoria 3031
Phone +613 9371 2126
Fax +613 9371 2102
E-mail: textiles@awta.com.au
Tests: AS 1530.2; AS/NZS 1530.3; AS 2111.18/ISO 6925; AS 2122.1; AS 2122.2

CSIRO Building, Construction and Engineering
Fire Science and Technology Laboratory (FSTL)
PO Box 310, North Ryde, NSW 1670
Phone +612 9490 5666
Fax +612 9490 5777
E-mail: information@dbce.csiro.au
Tests: AS 1530.1; AS 1530.2; AS/NZS 1530.3; AS 1530.4; AS/NZS 3013; AS/NZS 3837/ISO 5660; AS 4072.1; BS 476.6; BS 476.7; ISO 9705; UL 181

Warrington Fire Research (Aust) Pty Ltd
PO Box 4282, Dandenong South, Victoria 3164
Phone +613 9793-0088
Fax +613 9793-0111
E-mail: testing@wfra.com.au
Tests: AS 1530.4; AS 4072.1; AS/NZS 3013

10.19.4 Future Fevelopments

In response to the perceived need to reform Australian building regulations, the Fire Code Reform Centre (FCRC), a consortium of industry, research institutions and government, represented primarily by the ABCB, was formed in 1993. One FCRC project looked at the level of control on interior linings, and at the fire tests used to set these controls. International portability of data was one of the aspects considered.

The cone calorimeter (AS/NZS 3837 [20], technically identical to ISO 5660, except with smoke measurement, and the room fire test (ISO 9705) were recommended by researchers for possible use with wall and ceiling linings. The flooring radiant panel ASTM E 648 and the cone calorimeter were recommended for floorings. Both sets of recommendations are under consideration for possible inclusion in the deemed-to-satisfy provisions of the BCA. The report on wall and ceiling linings [21] is available from Fire Code Reform Centre Ltd, Suite 1201, Level 12, 66 King St, Sydney, New South Wales 2000.

References for Section 10.19

[1] ACT Building Regulations, Publications no 4, (reprinted 1999)
[2] NT Building Regulations 1993 (reprinted 1997)
[3] Victoria Building Regulations 1994 (reprinted 1999)
[4] Local Government Approvals Regulation NSW (1999)

[5] Building Regulations QLD 1991 (reprinted 1998)
[6] Building Regulations WA 1989 (reprinted 1999) – 4 amendments
[7] Development Regulations SA (1993)
[8] Building Regulations Tasmania (1994) – 3 amendments since
[9] AS 1530.1-1994. Methods for fire tests on building materials, components and structures – Combustibility test for materials
[10] AS 1530.2-1993. Methods for fire tests on building materials, components and structures – Test for flammability of materials
[11] AS/NZS 1530.3-1999. Methods for fire tests on building materials, components and structures – Simultaneous determination of ignitability, flame propagation, heat release and smoke release
[12] AS 1530.4-1997. Methods for fire tests on building materials, components and structures – Fire-resistance tests of elements of building construction
[13] AS/NZS 3013-1995. Electrical installations – Classification of the fire and mechanical performance of wiring systems
[14] AS 4072.1-1992. Components for the protection of openings in fire-resistant separating elements – Service penetrations and control joints
[15] AS 4254-1995. Ductwork for air-handling systems in buildings
[16] NATA Directory (Current Issue) available from the National Association of Testing Authorities, Australia, 7 Leeds Street, Rhodes NSW 2138
[17] AS/NZS 3000-2000. Electrical installations (known as the Australian/New Zealand Wiring Rules)
[18] AS 2122.1-1993. Combustion characteristics of plastics – Determination of flame propagation – Surface ignition of vertically oriented specimens of cellular plastics
[19] AS/NZS 2122.2-1999. Methods of test for determining combustion propagation characteristics of plastics – Determination of minimum oxygen concentration for flame propagation following top surface ignition of vertically orientated specimens
[20] AS/NZS 3837-1998. Method of test for heat and smoke release rates for materials and products using an oxygen consumption calorimeter
[21] *V.P.* Dowling, J.M. Blackmore: 1998: Fire Performance of Wall and Ceiling Linings, Final Report, Project 2 Stage A, FCRC PR 98-02, Fire Code Reform Centre Ltd, Sydney, 1998

10.20 New Zealand

V. DOWLING

10.20.1 Statutory Regulations

The Building Act [1] established a three-part framework of building controls:

- The Building Act 1991 sets down the law for building work in New Zealand.
- The Building Regulations 1992 contain the mandatory New Zealand Building Code and details about the processing of building approvals.
- The Approved Documents are (nonmandatory) documents written by the Building Industry Authority to assist people to comply with the Building Code.

The New Zealand Building Code is a performance-based code. It sets out objectives to be achieved rather than prescribing construction methods. The emphasis is on how a building and its components must perform as opposed to how the building must be designed and constructed.

10.20.2 Classification and Testing of the Fire Performance of Building Materials and Components

Some test methods for materials are contained in the Approved Documents as acceptable standard tests. They are:

- Floor coverings – BS 4790,
- Fabrics – AS 1530.2 (see Section 10.19),
- Lining materials – AS/NZS 1530.3 (see Section 10.19),
- Noncombustibility – AS 1530.1 (see Section 10.19) or NZS/BS 476.4,
- Tents or marquees – NFPA 701.

10.20.3 Official Approval

One or more of the tests described in the preceding must be carried out on building materials if conformity to the Approved Documents is to be demonstrated. The tests can be performed by any laboratory approved by International Accreditation of New Zealand (IANZ). Approval involves regular inspection of the test laboratories by IANZ experts. An IANZ directory [2] listing recognized test laboratories is published regularly. Laboratories that test fire performance of plastics are not listed separately. By agreement between IANZ and NATA tests performed by laboratories in Australia are also acceptable (see Section 10.19).

The test laboratory issues an IANZ endorsed Report containing the test results. This document is sufficient proof of performance of the building materials. Building materials are not marked or inspected in New Zealand.

Building fire tests for plastics carried out by laboratories in New Zealand are listed below. All are IANZ approved for the tests accepted in the Approved Documents unless otherwise indicated. The list includes some tests required by building authorities outside New Zealand.

Applied Physics Laboratory (APL)
24 Umere Crescent Ellerslie Auckland 1131, New Zealand
Phone +649 579 3912
Tests: AS/NZS 1530.3

Building Research Association of New Zealand (BRANZ)
Private Bag 50 908, Porirua City 6220 (Moonshine Road, Judgeford)
New Zealand
Phone +644 235 7600
Fax +644 235 6070
E-mail: branz@branz.org.nz
Tests: AS 1530.2; AS/NZS 1530.3; AS 1530.4; BS 4790; ISO 5660; NFPA 701

Firelab Pacific Ltd
PO Box 12-907, Penrose (31 Titi Street, Otahuhu) Auckland
New Zealand
Phone +649 276 5006
Fax +649 276 5008
E-mail: director@firelab.co.nz
Tests: AS 1530.4; AS 4072.1

Wool Research Organisation of New Zealand (WRONZ)
Private Bag 4749, Christchurch, (cnr Springs Rd & Gerald St, Lincoln), New Zealand
Phone +643 325 2421
Fax +643 325 2717
greer@wronz.org.nz
Tests: AS 1530.1 (not IANZ accredited); AS/NZS 1530.3 (agent); BS 4790

10.20.4 Future Developments

The Approved Documents are continuously updated to reflect advances in technology and changes in standards.

References for Section 10.20

[1] The Building Regulations 1992, Building Industry Authority, Wellington
[2] New Zealand Accreditation Directory (Current Issue), International Accreditation of New Zealand, Private Bag 28908, Remuera, Auckland 1136, New Zealand

10.21 People's Republic of China

QIAN JIAN-MIN

10.21.1 Statutory Regulations

The use of building materials in China basically conforms to the "Fire Law of the People's Republic of China" (issued and implemented in 1984). In practical applications, the performance requirements and the application fields for materials are specified more clearly in the relevant laws and regulations, such as the

- Fire Protection Code for Building Design (GBJ 16-87),
- Fire Protection Code for Civil High Rise Building Design (GB 50045-95),
- Fire Protection Code for Building Internal Decorations (GB 50222-95) and others.

To define the fire performance of building materials, the China State Bureau of Quality and Technical Supervision has issued the standard "Classification of the Burning Behavior for Building Materials" (GB 8624-1997), in which the classification of fire properties of building materials are definitely specified and the technical parameters and the test methods that determine whether the materials reach the relevant class are also described in detail.

10.21.2 Classification and Testing of the Fire Performance of Building Materials and Components

GB 8624-1997, which is the amendment of GB 8624-88, offers the methods to classify and evaluate the fire performance of building materials that cover the structural and decorative materials used in all kinds of industrial and public buildings.

Compared with GB 8624-88, GB 8624-1997 has more new provisions that specify the class A composite materials (sandwich panels) and some materials for specific use, such as floor coverings, textiles for curtains, plastic conduit pipes for wire and cable, and insulating cellular plastics for pipes. However, it should be pointed out that the testing and classification of the materials for specific use mentioned above should still be in line with the specifications of the materials for nonspecific use, when they are used as wall coverings or ceilings.

The classes, designations, and test methods of fire performance of building materials specified in GB 8624-1997 are given in Table 10.128.

Table 10.128 Classes, designations and test methods of the fire performance of building materials

Class	Designation	Test Method							
		Materials for nonspecific use		Materials for specific use					
		Homogeneous materials	Composite materials (Sandwich panel)	Floor coverings	Textiles for curtains	Plastic conduit pipes for wires and cables		Insulating cellular plastics for pipes	
						Thermoplastics	Thermosets		
A	Non-combustible	GB/T5464-1999	GB/T8625-88 GB/T8627-88 GB/T14402-93 GB/T14403-93 GA 132–1996	—	—	—	—	—	
B1	Low flammability	GB/T8625-88 GB/T8626-88 GB/T8627-88		GB/T11785-89	GB/T5454-1997 GB/T5455-1997	GB/T2406-93 GB/T2408-1996 (horizontal method) GB/T8627-88	GB/T2406-93 GB/T2408-1996 (vertical method) GB/T8627-88	GB/T2406-93 GB/T8333-87 GB/T8627-88	
B2	Moderately flammable	GB/T8626-88				GB/T2406-93 GB/T2408-1996 (horizontal method)	GB/T2406-93 GB/T2408-1996 (vertical method)	GB/T2406-93 GB/T8332-87	
B3	Highly flammable	Materials that have not reached the grades above							

10.21.3 Generally Used Tests and Classifications for Building Materials

10.21.3.1 GB/T 5464-1999 Noncombustibility of Building Materials

This standard is identical to ISO 1182. A diagram of the apparatus is shown in Section 10.23 ISO. The tests specifications are summarized in Table 10.129.

Table 10.129 Non-combustibility test for building materials

Test specifications	• Specimens: five cylindrical samples, external diameter (45 ± 2) mm, height (50 ± 2) mm • Furnace temperature: (750 ± 5) °C • Test duration: 30 min
Pass criteria	Classified as class A homogeneous materials if • The average furnace temperature rise does not exceed 50 °C, • The mean duration of sustained flaming does not exceed 20 s, • The average mass loss rate is not more than 50 %.

10.21.3.2 GB/T 8625-88 Low flammability of Building Materials

This standard is based on the German Brandschacht test in the version of DIN 4102 Part 1 of May 1981. The test apparatus is shown under Section 10.9 Germany. The test specifications are summarized in Table 10.130.

Table 10.130 Low flammability class A and B1 test for building materials

Test specifications	Specimens: three groups of four samples, 1000 mm × 190 mm × max. thickness 80 mm Gas flow: methane 35 (± 0.5) l/min, air 17.5 (± 0.2) l/min Test duration: 10 min
Pass criteria	Classified as class A composite materials (sandwich panel) if • The mean residual length is at least 350 mm; no specimens 200 mm, • Smoke gas temperature does not exceed 125 °C, • The back of any samples does not flame. Classified as class B1 if • The mean residual length is at least 150 mm; no specimens 0 mm, • Smoke gas temperature does not exceed 200 °C.

10.21.3.3 GB/T 8626-88 Moderate Flammability of Building Materials

This standard is based on the German small burner test. The testing device is illustrated in Section 10.9 Germany. The test specifications are summarized in Table 10.131. The small burner test must be passed before testing to the low flammability test.

Table 10.131   Moderate flammability test specifications

Test specifications	Specimens: edge flame application: five specimens, 190 mm × 90 mm × max. thickness 80 mm Surface flame application: five specimens, 230 mm × 90 mm × max. thickness 80 mm Ignition source: inclined at 45°, flame height 20 mm Test duration: 20 s (including 15 s for flame application)
Pass criteria	• Class B1 achieved only if the specimens pass class B2 • Classified as class B2 if none of the five specimens reach the reference mark of 150 mm within 20 s after ignition

10.21.3.4 GB/T 8627-88 Smoke Density of Building Materials

This standard is based on ASTM D 2843. The test design is shown in Section 15 Smoke under 15.3.2. The test specifications are summarized in Table 10.132.

Table 10.132   Smoke density test

Test specifications	Specimens: three specimens, 30 mm × 30 mm Thickness: • 6 mm if density $\geq$ 1000 kg/m^3 • 10 mm if 100 kg/m^3 < density < 1000 kg/m^3 • 25 mm if density <100 kg/m^3 • or in accordance with its original thickness in practical use Test apparatus: the same as ASTM D 2843 Test duration: 4 min
Pass criteria	• Class A composite materials (sandwich panel) determined if the smoke density rating (SDR) $\leq$ 15 • Class B1 achieved if the smoke density rating (SDR) $\leq$ 75 • Class B1 textiles for curtains achieved if the smoke density rating (SDR) $\leq$ 75 • Determination of class B1 plastic conduit pipes for wires and cables (including thermoplastics and thermosetting plastics) based on the smoke density rating $\leq$ 75 • Determination of class B1 insulating cellular plastics for pipes based on the smoke density rating $\leq$ 75

10.21.4 Specific Tests Used for Plastics, Textiles, and Floor Coverings and Further Tests to Determine Noncombustibility and Smoke Toxicity of Building Materials

In the following, additional test methods used in China for determining the flammability of pipe insulation, plastic conduit for wire and cable, plastic foams and floor coverings are dealt with. In addition, further tests for determining the calorific potential and heat emission of class A building materials, as well as the smoke toxicity of fire effluents and their pass criteria are described.

10.21.4.1 GB/T 2406-88 Flammability of Pipe Insulation

This standard is based on ISO 4589, the limiting oxygen index (LOI). The apparatus is shown in Fig. 10.57 and the specifications are summarized in Table 10.133.

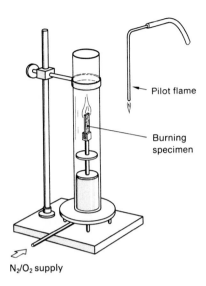

Pilot flame

Burning specimen

N₂/O₂ supply

Figure 10.57 LOI apparatus

Table 10.133 Determination of burning behavior of plastics by the oxygen index test

Test specifications	Specimens: 15 specimens, (70–150) mm × (6.5–52) mm × thickness (for self-supporting materials: thickness 3–10.5 mm; for non-self-supporting materials: thickness > 10.5 mm) Test apparatus: the same as ISO 4589 Test method: up and down method to determine the critical oxygen concentration (by oxygen index)
Pass criteria	• Classified as class B1 plastic conduit pipes for wires and cables (including thermoplastics and thermosetting plastics) if oxygen index ≥ 32 • Classified as class B2 plastic conduit pipes for wires and cables (including thermoplastics and thermosetting plastics) if oxygen index ≥ 26 • Class B1 insulating cellular plastics for pipes achieved if oxygen index ≥ 32 • Class B2 insulating cellular plastics for pipes achieved if oxygen index ≥ 26

10.21.4.2 GB/T 2408-1996 Flammability of Plastic Conduit Pipes

This standard is equivalent to ISO 1210, which is a Bunsen burner test derived from the UL 94 horizontal burning HB and vertical V tests (see Chapter Electrical engineering under 12.2.1). The test criteria are shown in Table 10.134.

Table 10.134   Determination of flammability characteristics of plastics – horizontal and vertical test

Test specifications	Specimens: (125 ± 5) mm $\times$ (13.0 ± 0.3) mm $\times$ thickness (3.0 ± 0.2) to ≤ 13 mm Horizontal test: three samples; vertical test: 5 samples Test apparatus: the same as ISO 1210 Flame application time: horizontal test 30 s; vertical test 10 s
Pass criteria	Horizontal test: • Class B1 plastic conduit pipes for wires and cables (thermoplastics) based on FH-1 • Class B2 plastic conduit pipes for wires and cables (thermoplastics) based on FH-2 Vertical test: • Class B1 plastic conduit pipes for wires and cables (thermosetting plastics) based on FV-0 • Class B2 plastic conduit pipes for wires and cables (thermosetting plastics) based on FV-1

10.21.4.3 GB/T 5454-1997 Flammability of Textiles by the Oxygen Index Test

The burning behavior of textiles is measured with the LOI test. The standard is based on ISO 4589. The apparatus is shown in Fig. 10.58 and the specifications summarized in Table 10.135.

Table 10.135   Determination of burning behavior of textiles by the oxygen index test

Test specifications	Specimens: 150 mm $\times$ 58 mm (in warp direction: 15 samples; in weft direction: 15 samples) Test apparatus: the same as ISO 4589 Test method: up and down method to determine the critical oxygen concentration (by oxygen index)
Pass criteria	• Class B1 textiles for curtains satisfied if oxygen index ≥ 32 • Class B2 textiles for curtains satisfied if oxygen index ≥ 26

10.21.4.4 GB/T 5455-1997 Flammability of Textiles by Using a Vertical Bunsen Burner Test

The standard is based on the Japanese JIS 1091 test and the specifications are shown in Table 10.136

Table 10.136   Determination of burning behavior of textiles – vertical test

Test specifications	Specimens: 300 mm $\times$ 80 mm (in warp direction: five specimens; in weft direction: 5 specimens) Test apparatus: the same as JIS 1091 Flame application time: 12 s
Pass criteria	• Classified as class B1 textiles for curtains if destroyed length ≤ 150 mm, afterflame time ≤ 5 s and afterglow time ≤ 5 s • Classified as class B2 textiles for curtains if destroyed length ≤ 200 mm, afterflame time ≤ 15 s and afterglow time ≤ 10 s

10.21.4.5 GB/T 8332-87 Flammability of Cellular Plastics – Horizontal Burning

This test is a Bunsen burner test equivalent to ISO 3582. The test specifications are shown in Table 10.137.

Table 10.137   Flammability test for cellular plastics – horizontal burning test

Test specifications	Specimens: 10 samples, 150 (± 1) mm × 50 (± 1) mm × 13 (± 1) mm, or original thickness of materials (5 mm < thickness < 13 mm) with surface removed Test apparatus: the same as ISO 3582 Flame application time: 60 s
Pass criteria	Classified as class B2 insulating cellular plastics for pipes if mean burning time ≤ 90 s, mean burning length ≤ 50 mm

10.21.4.6 GB/T 8333-87 Flammability of Cellular Plastics – Vertical Burning

This test is a Bunsen burner test with vertical sample. The test specifications are shown in Table 10.138.

Table 10.138   Flammability test for cellular plastics – Vertical burning test

Test specifications	Specimens: six samples, 250 (± 1) mm × 20 (± 1) mm × 20 (± 1) mm Test apparatus: height of inner flame cone: 25–30 mm temperature of top inner flame cone: 960 (± 2) °C Flame application time: 10 s
Pass criteria	Classified as class B1 insulating cellular plastics for pipes if mean burning time ≤ 30 s, mean burning height ≤ 250 mm

10.21.4.7 GB/T 11785-89 Critical Radiant Flux of Floor Coverings

This test is derived from the radiant flooring panel test described in ISO/DIS 9239. The test apparatus is illustrated in Section 10.9 Germany. The test specifications are shown in Table 10.139.

Table 10.139   Determination of critical radiant flux of floor coverings using a radiant heat source

Test specifications	Specimens: three samples, 1050 (± 5) mm × 250 (± 5) mm Test apparatus: the same as ISO/DIS 9239 Flame application time: 10 min
Pass criteria	• Class B1 achieved if the critical radiation flux ≤ 0.45 W/cm^2 • Class B2 achieved if the critical radiation flux ≤ 0.22 W/cm^2

10.21.4.8 GB/T 14402-93 Calorific Potential of Building Materials

This test is derived from ISO 1716 described under Section 10.23 ISO. The test specifications are shown in Table 10.140.

Table 10.140   Determination of calorific potential of building materials

Test specifications	Specimens: homogeneous powder 1.0–1.2 g Test apparatus: the same as ISO 1716 Test duration: 5 min after the water temperature of inner water jacket reaches the highest point t_m
Pass criteria	Class A composite materials (sandwich panel) achieved if the calorific potential ≤ 4.2 MJ/kg

10.21.4.9 GB/T 14403-93 Heat Emission of Building Materials

This test is equivalent to the DIN 4102 furnace test for class A1 building materials described in Section 10.9 Germany. The test specifications are shown in Table 10.141.

Table 10.141   Determination of heat emission by combustion for building materials

Test specifications	Specimens: three samples, 40 mm × 40 mm × 50 mm Test apparatus: the same as DIN 4102 Part 1 furnace Test duration: 30 min
Pass criteria	Class A composite materials (sandwich panel) achieved if the heat release ≤ 16.8 MJ/m^2

10.21.4.10 GA 132-1996 Smoke Toxicity of Fire Effluents

The test method is derived from the standards ISO TR 9122-2 and 9122-4, DIN 53436 and JIS A 1321. The test specifications are shown in Table 10.142.

Table 10.142   Classification of the smoke toxicity of fire effluents

Test specifications	Specimens: three to five samples max. length 400 mm, determination of the mass of the specimens based on the designed smoke concentration Test apparatus: the smoke generating apparatus identical to DIN 53436, mice rotary cage identical to JIS A 1321, the methods of test and evaluation identical to ISO TR 9122-2 and the terms identical to ISO TR 9122-4 Test duration: 30 min exposure to toxicity, 14 d observation after toxicity infection
Pass criteria	Classified as class A composite materials (sandwich panel) if the smoke concentration causing no death $LC_0 \leq 25$ mg/l

10.21.4.11 Special Requirements for Composite Materials and Surface Coatings

No matter what thickness the composite building materials have or what kind of procedure they are produced by, their burning behavior should be tested as a whole and be given a overall evaluation.

If the thickness of coatings or decorative coverings, which are added to the surface of the substrates by spraying, sticking, or other means, is not more than 0.6 mm (0.02 in.) or the mass of unit area of the coverings is not more than 300g/m^2, the influence of the coverings on burning behavior or classification of the substrates can be neglected. If the thickness of decorative coverings exceeds 0.6 mm or the decorative covers have obvious influence on the fire performance of substrates, the burning behavior tests may be performed with the specimens made of the coverings and the substrates together.

10.21.5 Official Approval

The official institution for product quality supervision and testing is the

National Center for Quality Supervision and Testing
of Fire Building Materials (NCFM)
266 Waibei Street, Dujiangyan,
Sichuan, 611830
China
Tel: + 86 28 7123825, 7132037
Fax: +86 28 7123813, 7132051

As an official institution, NCFM was set up under the approval of the China State Bureau of Quality and Technical Supervision (CSBTS).

In 1998, NCFM received laboratory accreditation in accordance with ISO/IEC Guide 25. Being a notified body, NCFM undertakes the supervision and fire testing of various building materials, fire resistant building components (elements) and fire retardant coatings, and the monitoring of burning behavior classification of building materials. NCFM is also engaged in drafting the national and professional standards (GB, GA) concerning fire protection materials and testing methods on burning behavior of building materials, and in researching and developing the testing technology and test apparatus.

NCFM offers a follow-up service for manufacturers to ensure the quality of their products, and permits the application of the registered mark "NCFM®" to the monitored products.

10.21.6 Future Developments

NCFM keeps in contact with ISO, IEC, CEN, and other developed countries through ISO/TC92, CIB/W14, participates in international technical exchanges and cooperations, and pays close attention to the developing trends of ISO, IEC, and CEN in the fields of standard drafting and testing technology developing.

NCFM's testing laboratory has been approved by the China National Committee of Testing Laboratory Accreditation in accordance with Guide 25 of ISO/IEC, which has laid a base for NCFM to eliminate international trade barriers related to the fire performance of building materials.

10.22 Japan

E. ANTONATUS

10.22.1 Statutory Regulations

The Building Standards Law came into effect on November 16, 1950. The aim of this legislation is "to protect the lives, health, and wealth of citizens, and thus contribute to the prosperity of the community, by laying down guidelines and standards for plots of land, building design, furnishing and use."

In 1998, a revision of the main parts of the BSL (Building Standard Law), including the fire safety design systems, was published and came into effect on June 1, 2000 [1]. This revision has started a process of changing from a specification-based to a performance-oriented design.

The revision is defining:

Basic requirements – definition of categories of building parts and materials
* Fireproof, fire preventive construction, noncombustible materials, etc. (BSL Article 2),
* Quasi noncombustible materials, fire retardant materials, and so forth (Enforcement Order Article 1).

Performance criteria required for defined building parts and materials
* Fireproof, quasi-fireproof, fire preventive construction (Enforcement Order Article 107,107-2,108),
* Noncombustible, quasi-noncombustible, fire retardant materials (Enforcement Order Article 1,108-2).

Approval of building parts and materials with the required performance
* Approval for "performance evaluation report" of tested materials (based on specific test methods and technical criteria) are submitted by designated examination bodies: (BSL Article 68–26),
* Among the specification-based materials listed in the previous notifications, those that proved to satisfy the new fire performance requirements are presented in the new regulation system

Details are contained in the Notifications No. 1399 for fireproof, No. 1358 for quasi-fireproof, No. 1359 for fire preventive construction, No. 1400 for noncombustible, No. 1401 for quasi-noncombustible and No. 1402 for fire retardant materials.

10.22.2 Classification and Testing of the Fire Performance of Building Materials and Components

In Table 10.143, the test methods and criteria for reaction to fire of building products are summarized. The role of the cone calorimeter in fire testing of building products in Japan is described in [2].

Building materials are tested for *noncombustibility* with an electric furnace operating at 750 °C. Apparatus and test method are based on ISO 1182 and described in Section 10.23 ISO. The maximum temperature rise allowed is 20 K, and the mass loss of the specimen may not be more than 30 %.

Alternatively, these materials can be tested in the cone calorimeter, based on ISO 5660-1 [3]. Details of the apparatus and specifications are given in Section 10.23 ISO. The test is carried out for 20 min at a radiant heat level of 50 kW/m². Within this time, the total heat release (THR) is not allowed to exceed 8 MJ and the maximum rate of heat release (RHR) may not exceed 200 kW/m².

Table 10.143 Fire performance of building materials, test methods and criteria

Classification	Required performance criteria	Duration of test	Performance evaluation	Test method and criteria [3]	(E.O. 68–26)
	Performance		Fire performance		Fire gas toxicity[a]
			Test method: (a) or (b)	Performance criteria	
Noncombustible materials (BL)		20 min (E.O. 108–2)	(a) Non-combustibility test (based on ISO 1182) (b) Heat release test (based on ISO 5660-1, radiant heater 50 kW/m²)	(a) Temperature increase ≤ 20K Mass loss $\leq 30\%$ (b) THR ≤ 8 MJ/m² MHR ≤ 200 kW/m² (≥ 10s)	Test method: Previous Notification No.1231 or equivalent
Quasi noncombustible materials (E.O. 1)	1. No burning 2. No deformation, melting, cracking, or other damage detrimental to fire prevention 3. No development of smoke or gas hampering fire protection (E.O. 108–2)	10 min (E.O. 1)	(a) Heat release test (based on ISO 5660-1, radiant heater 50 kW/m²) (b) Box heat test (based on ISO DIS 17431, propane burner 40 kW)	(a) THR ≤ 8 MJ/m² MHR ≤ 200 kW/m² (≥ 10s) (b) THR ≤ 30 MJ MHR ≤ 140 kW (≥ 10s) No through crack or hole	Test method: Previous Notification No.1231 or equivalent
Fire retardant materials (E.O. 1)		5 min (E.O. 1)	(a) Heat release test (based on ISO 5660-1, radiant heater 50 kW/m²) (b) Box heat test (based on ISO DIS 17431, propane burner 40 kW)	(a) THR ≤ 8 MJ/m² MHR ≤ 200 kW/m² (≥ 10s) (b) THR ≤ 30 MJ MHR ≤ 140 kW (≥ 10s) No through crack or hole	Test method: Previous Notification No.1231 or equivalent

BSL: Building Standard Law, E.O. = Enforcement Order, THR = Total Heat Release, MHR = Max. Rate of Heat Release
[a] Materials with a higher content of organic components than specified shall pass the fire gas toxicity test
[3] Applied by Japan Testing Center for Construction Material

Quasi non-combustible materials are tested in the cone calorimeter or in the box heat test to ISO 17341. The box heat test is derived from the former Japanese "Model Box Test". Details of the box heat test are given below.

The pass criteria in the cone calorimeter test are the same as for noncombustible materials, but for a shorter test period of 10 min. If the box heat test is used, a THR of 30 MJ and a RHR of 140 kW may not be exceeded within the 10 min of the test and no burn-through is accepted.

Fire retardant materials are also tested in the cone calorimeter or in the box heat test. The duration of the test for these materials is 5 min; the criteria are the same as above for the period of 5 min.

For fire protective materials, additional toxicity tests are required. A test method based on the incapacitation time of mice (previous notification No. 1231) is used, but results from other test methods are often accepted, if the organic content of the product to be tested does not exceed $400\,g/m^2$.

The box heat test (similar to Draft ISO Standard 17 431) for determining the maximum value of heat release rate and the total value of heat released employs a box-shaped specimen with a gas burner as fire source. The product to be tested shall be cut to make four panels to cover the ceiling, two side walls, and the end wall of the inside surfaces of the combustion chamber. In the standard specimen configuration, three walls and the ceiling of the combustion chamber are covered with the panels.

The dimensions of the panels shall be so determined that, when the panels are assembled and inserted to the combustion chamber, the combustion chamber has the following inner dimensions:

Width: 0.84 m
Length: 1.72 m
Height: 0.92 m

If the product to be tested is a board type, the normal width, length, and thickness of the board shall be used as far as practicable. Thin surface products such as paints and varnishes or thermoplastic products which may melt during the test shall, depending on their end-use, be applied to a substrate.

The product or product with substrate shall be nailed to panels made of C-shaped steel frames [nominal section dimension of 40–45 mm x 40–45 mm (1.57–1.77 in. x 1.57–1.77 in.)] and steel sheet [nominal thickness 0.27 mm (0.01 in.)] with steel nails at intervals of 0.15 m.

The ignition source is a propane gas burner having a 0.17 m x 0.17 m (square top surface layer of a porous, inert material, for example, sand. The heat output of the burner (net heat of combustion of the fuel) is 40 kW during the test period.

At the front door opening, a system for collecting the combustion products is installed that shall have a capacity and be designed in such a way that all of the combustion products leaving the combustion chamber through the opening during a test are collected. The system shall not disturb the fire-induced flow in the opening. The exhaust capacity has to be at least $2.0\,m^3 \cdot s^{-1}$ at normal pressure and a temperature of $25°C$. An example of the design of hood and exhaust duct is given in Annex D of ISO 9705. The measurement instrumentation and method is described in ISO 9705 (see Section 10.23 ISO).

Specimens are judged to have passed the box heat test if in each of two tests the following criteria are met throughout the defined test duration:

- The maximum value of heat release rate during the test shall not exceed 140 kW.
- The total value of heat released during the test shall not exceed 30 MJ.
- The box specimen shall not exhibit combustion, which would present a significant problem in terms of fire protection. No holes are allowed to burn through the walls and no significant cracks are allowed to occur.

10.22.3 Official Approval

10.22.3.1 Approval for the Reaction-to-Fire of Building Materials

The fire performance of building materials is classified with the aid of the above tests performed by officially recognized test institutes to whom materials should be submitted. Some of them are listed below.

Building Research Institute, Ministry of Construction
Tatehara 1, Tsukuba-shi
Ibaraki-Ken, 305 Japan

Forestry and Forest Products Institute, Ministry of Agriculture, Forestry and Fisheries
1 Matsunosato, Kukizaki-Mura,
Ibaraki-Ken, 300–12 Japan

Japan Testing Center for Construction Materials
1804 Inari-cho, Soka-Shi, Saitama-Ken, 304 Japan

General Building Research Corporation
Fujishirodai 5–125, Suita-shi, Osaka, 565 Japan

Hokkaido Building Research Institute
8. 1-Chome, 4-Jo 24-Ken, Nishi-Ku,
Sapporo-Shi, Hokkaido, Japan

Additional official testing institutes have recently been licensed, for example, Center for Better Living, Building Center of Japan and Japan Housing and Wood Technology Center.

A test report and an evaluation of the test results are issued by the test laboratory. The client then brings the results and the evaluation to the Ministry of Land and Infrastructure Management to obtain the approval for the product.

10.22.3.2 Approval for Fire Resistance

The procedures and criteria for fire resistance including fire doors have been changed according to the ISO 834 series. There are only small differences left between ISO and Japanese standards.

10.22.4 Future Developments

Japan is participating actively in the development of ISO standards and trying to cover their regulatory needs based on these standards. The development of a performance-based system will be continued and additional methods introduced. For roofs, a new method for "leaping flames" is under preparation and a new method for stairs is planned (both methods are under development at ISO).

References for Section 10.22

[1] Law for Amendment of the Building Standard Law 98 (Law No.100, June 12th,1988), edited and published by The Building Center of Japan, 30 Mori Bldg. 3–2-2, Toranomon, Minatoku, Tokyo 105–8438 Japan
[2] S. Sugahara, M. Yoshida, K. Ueda: A study on the structure of cone calorimeter as the authorized apparatus. Interflam 2001 Conference, UK
[3] Application form of examination on fire performance evaluation of building construction parts and building materials

10.23 International Organization for Standardization (ISO)

B. SUNDSTRÖM

10.23.1 Introduction

ISO's principal activity is to develop technical standards required by the market. The work is carried out by experts from the industrial, technical, and business sectors that have asked for the standards and that will put them into use. Member bodies are standards organizations from 132 countries (end of 1999). Of these, 90 member bodies are entitled to participate and exercise full voting rights within ISO. The most active ISO countries in terms of holding secretariats for technical committees and subcommittees are ANSI USA (140 secretariats), DIN Germany (130 secretariats), and BSI United Kingdom (113 secretariats). ISO activities in figures are shown in Table 10.144 [1].

Table 10.144   ISO standardization in figures

Period	Production of international standards	Production of pages
1999	961	12,524
Total output	12,524	356,427

Financially, the operational expenditure for ISO's work is estimated at 150 million Swiss Francs, of which about 30 million Swiss Francs finances the Central secretariat in Geneva (1999).

The structure of ISO is shown in Fig. 10.58. The general assembly and the council have the final control of ISO. The technical management board (TMB) has the overall management of technical committee structure and subcommittee structure. TMB establishes and dissolves technical committees and provides outlines of their scopes. The central secretariat supports the work in the technical committees, publishes standards and publications and so forth.

A recent strategic initiative is to focus on increasing the ISO's market relevance. Important goals are a better understanding of market needs and increased involvement of industry, consumers, and other stakeholders. Therefore, business plans are seen as important tools in this process. Each technical committee will have a business plan, which also covers the activities of its subcommittees. The business plans will analyze the conditions and trends in the market sector served by the technical committee and will be required explicitly to link work programs and sector needs. Thus, priorities for which standards are needed can be set.

The technical work in ISO is carried out via technical committees (TCs). A TC may have assigned Subcommittees (SCs) with working groups (WGs) to cover certain areas of work. The actual standardization work is taking place in the WGs and SCs. Further general information of ISO can be found in refs. [2] and [3].

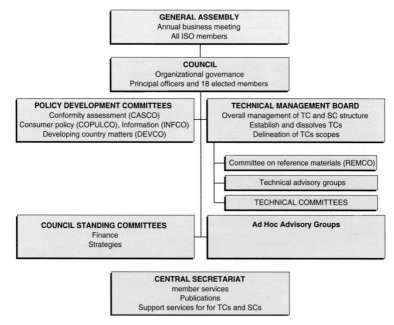

```
┌─────────────────────────────────┐
│        GENERAL ASSEMBLY         │
│     Annual business meeting     │
│        All ISO members          │
└─────────────────────────────────┘

┌─────────────────────────────────┐
│            COUNCIL              │
│    Organizational governance    │
│ Principal officers and 18 elected members │
└─────────────────────────────────┘
```

POLICY DEVELOPMENT COMMITTEES	TECHNICAL MANAGEMENT BOARD
Conformity assessment (CASCO)	Overall management of TC and SC structure
Consumer policy (COPULCO), Information (INFCO)	Establish and dissolves TCs
Developing country matters (DEVCO)	Delineation of TCs scopes

Committee on reference materials (REMCO)

Technical advisory groups

TECHNICAL COMMITTEES

COUNCIL STANDING COMMITTEES	Ad Hoc Advisory Groups
Finance	
Strategies	

```
┌─────────────────────────────────┐
│       CENTRAL SECRETARIAT       │
│         member services         │
│          Publications           │
│   Support services for for TCs and SCs │
└─────────────────────────────────┘
```

Figure 10.58   ISO's structure

10.23.2 ISO/TC 92 Fire Related Activities

Fire issues appear in more than one TC, but only TC 92 is dedicated solely to the fire field. The scope of TC 92 is Fire Safety and therefore its activities are intended to cover a broad spectrum of standardization issues in the fire area. Liaison with other ISO TCs and IEC is maintained with:

ISO/TC 21   Equipment for fire protection and fire fighting
ISO/TC 38/SC 19 Burning behaviour of textiles and textile products
ISO/TC 45   Rubber and rubber products
ISO/TC 59   Building construction
ISO/TC 61/SC 4   Burning behaviour of plastics
ISO/TC 77   Products in fibre reinforced cement
ISO/TC 85   Nuclear energy
ISO/TC 162   Doors and windows
IEC/TC 20   Electric cables
IEC/TC 89   Fire hazard testing

ISO/TC 92 also has liaison with organizations outside of ISO and is internally built up in a network. The structure is shown in Fig. 10.59.

TC 92 has recently been reorganized. It has now an objective to produce standards in the field of fire safety engineering, at the same time supporting the standards used for prescriptive purposes. The major activities of TC 92 are given in Table 10.145.

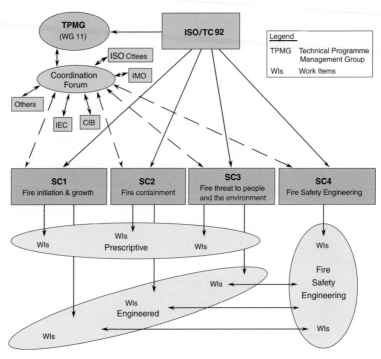

Figure 10.59   The structure of ISO/TC 92 [4]

Table 10.145   Major activities in ISO TC 92

Unit	Chairman	Major activities
TC 92 Fire safety	G. Cox, UK	General management of the TCs activities
SC1 Fire initiation and growth	B. Sundström, Sweden	Measurement of fire initiation and growth and standards relating to fire scenarios and characteristic fire growth of products
SC 2 Fire containment	D. N. Priest, USA	Measurements of fire resistance and integration of fire resistance tests and fire safety engineering
SC 3 Fire threat to people and environment		Effects on people from toxic gases, smoke and heat. Environmental effects of fires
SC 4 Fire safety engineering	J. Kruppa, France	Develop and maintain a set of ISO documents on the use of fire hazard and risk assessment models

Test methods for reaction to fire as well as their application to fire safety engineering are covered by SC1.

10.23.2.1 TC 92/ SC1 Fire Initiation and Growth

The defined objectives of SC 1 are:

Fire safety engineering (FSE)
- Test protocols, measuring techniques and procedures for securing data of fundamental fire properties,
- Test protocols, measuring techniques and procedures for input data to FSE models,
- Standards relating to fire scenarios and characteristic fire growth of products.

Performance codes
- Test protocols for reference scenarios,
- Test protocols, measuring techniques and procedures for fire calorimetry.

Prescriptive codes
- Updating tests already in use.

Test validation
- Protocols for determining the precision of fire test procedures,
- Test protocols for validation of fire growth predictions.

Instrumentation
- Protocols for measurement technologies used in fire test procedures.

10.23.2.2 The Present Package of Standards and Technical Reports from ISO TC 92/SC1

The activities of SC1 have led to a number of different standards for different purposes. Some are intended for prescriptive codes. The use of ISO standards in the European harmonization (see Section 10.4) and the use of ISO standards in IMO, see Section 11.4, are good examples. Other standards find a use for fire safety engineering like the cone calorimeter. Others are used as reference scenarios, for example, the Room/Corner Test, ISO 9705.

The standards and technical reports issued by SC1 are in many cases accompanied by guidance documents. The intention is to provide the user of a certain standard with background information, technical details to support quality assurance in testing, and scientific background for that test. In addition, bibliographies are provided. Three standards are of a horizontal nature and give advice on how to use the test data from the SC1 package of standards for mathematical modeling and fire hazard analysis. The overall intention is to leave the user of a SC1 standard not only with a test procedure but also with a deeper understanding of underpinning technical work as well as how the test data can be used to mitigate hazard.

ISO/TR[1] 3814 Tests for measuring reaction to fire of building materials – their development and application

and

ISO/CD TR 11696-1 Use of reaction to fire tests – Part 1: Application of results to predict performance of building products by mathematical modelling

and

---

1 ISO means that the standard is a published complete ISO standard; EN ISO means that the standard is also a CEN standard; ISO/TR means that the document is a technical report and therefore could be a background document; ISO/CD means the standard is a committee draft and still under preparation; ISO working documents not yet CD are not mentioned.

ISO/TR 11696-2 Use of reaction to fire tests – Part 2: Guide to the use of test results in fire hazard analysis of building products

This series of standards contains the principles of the SC1 tests for fire growth measurements, theoretical modeling of fire processes relating to growth, and hazard analysis. Any user of the standards enumerated and described below should study this package. The benefit would be the maximum use of a given test standard for its purpose.

ISO/TR 14697 Reaction to fire tests – Guidance on the choice of substrates for building products

This standard recommends substrates on which the product sample can be attached before testing. The standard substrates represent various end use conditions. Thus, the test results become more general and the amount of testing can be kept down.

EN ISO 1716 Reaction to fire tests for building products – Determination of the gross calorific value

EN ISO 1716 determines the potential maximum total heat release of a product when completely burning. The calorific potential of a material is measured in a bomb calorimeter. The powdered material is completely burned under high pressure in a pure oxygen atmosphere.

This standard is also used by CEN (see Section 10.4 European Union).

EN ISO 1182 Reaction to fire tests for building products – Noncombustibility

EN ISO 1182 identifies products that will not, or significantly not, contribute to a fire, regardless of their end use. EN ISO 1182 was first published by ISO during the 70ies and is well known. It is used in various building codes and by IMO. This standard is also used by CEN, see Section 10.4 European Union. The EN ISO version is shown in Fig. 10.60 and the test specifications are given in Table 10.146.

ISO 5657 Reaction to fire tests – Ignitability of building products using a radiant heat source

The ISO 5657 apparatus measures the time-to-ignition of a sample exposed to different levels of radiation. The data can be used in fire modeling or for prescriptive purposes. The test apparatus is shown in Fig. 10.61 and the test specifications are given in Table 10.147.

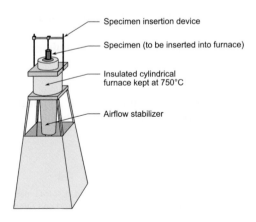

Figure 10.60   EN ISO 1182 test for noncombustibility

Table 10.146 EN ISO 1182 test specifications

Specimens	Five cylindrical samples, diameter 45 mm, height 50 mm
Specimen position	Vertical in specimen holder in the centre of the furnace
Heat source	Electrical cylindrical furnace at 750°C (measured by the furnace thermocouple)
Test duration	Depends on temperature stabilization
Conclusions	Classification is based on temperature rise as measured by the furnace thermocouple, duration of flaming and mass loss of the sample

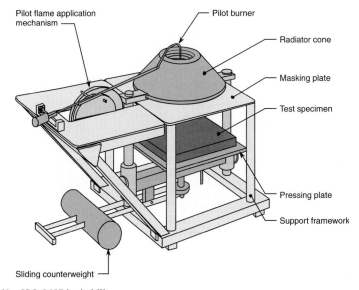

Figure 10.61   ISO 5657 ignitability

Table 10.147   ISO 5657 test specifications

Specimens	Five specimens 165 mm square
Specimen position	Horizontal
Heat source	Electrical conical heater giving a heat flux to the specimen of 50 kW/m^2 or other levels as required
Test duration	Until ignition occurs or max. 15 min
Conclusions	Time to ignition for actual heat flux

ISO/TR 5 658-1 Reaction to fire tests – Spread of flame – Part 1: Guidance on flame spread

and

ISO 5 658-2 Reaction to fire tests – Spread of flame – Part 2: Lateral spread on building products in vertical configuration

The spread of flame standard has two parts. Part 1 gives advice to the use of the test and its results. Part 2 contains the description of the hardware and the actual test procedure to be followed. The lateral flame spread as a function of time is measured on a vertical 800 mm (31.5 in.) long sample that is exposed to heat flux starting at 50 kW/m^2 and then decreasing to 1.5 kW/m^2 along the length axis. The ISO spread of flame test is similar to an IMO test procedure used for approval of marine products to be used on board ships. The IMO version is also equipped with a system for measuring heat release rate with a thermal method. The test apparatus is shown in Fig. 10.62 and the test specifications are given in Table 10.148.

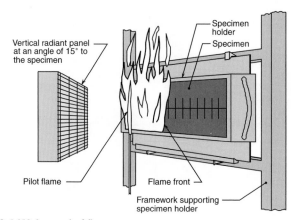

Figure 10.62   ISO 5 658-2 spread of flame

Table 10.148   ISO 5 658-2 test specifications

Specimens	Three specimens 800 mm × 155 mm
Specimen position	Vertical in specimen holder
Heat source	Gas-fired radiant panel
Test duration	30 min or earlier if flame front stops or if the flames spread to the end of the specimen
Conclusions	Average heat for sustained burning[a] Critical heat flux at extinguishments

[a] Sustained burning = time from test start x nominal heat flux at actual flame front position. Units are MJ/m^2

ISO 5 658-4 Reaction to fire tests – Spread of flame – Part 4: Intermediate scale test of vertical spread of flame with vertically oriented specimen

A large vertical sample is exposed to heat flux from a gas-fired radiant panel. Flame may spread in all vertical directions along the sample surface and the flame spread rate and distance are measured. Owing to the size of the sample, flame spread is easily observed and measured. This test may also be equipped with thermocouples for measurement of heat output from the fire.

ISO 5660-1 Fire tests – Reaction to fire- Part 1: Rate of heat release from building products – (Cone calorimeter method)

The cone calorimeter, originally developed by V. Babrauskas [5], is widely used as a tool for fire safety engineering, by industry for product development, and as a product classification tool (IMO). It has been proven to predict large-scale test results in the Room/Corner Test [6], fire growth in upholstered furniture [7], flame spread on cables [8], and in a number of other situations. The cone calorimeter is one of the most important tests developed by SC1 over its entire period of activity.

In the cone calorimeter, ISO 5660-1, specimens of 0.1 m x 0.1 m are exposed to controlled levels of radiant heating. The specimen surface is therefore heated up and an external spark igniter ignites the pyrolysis gases from the specimen. The gases are collected by a hood and extracted by an exhaust fan. The heat release rate (HRR) is determined by measurements of the oxygen consumption derived from the oxygen concentration and the flow rate in the exhaust duct. The specimen is placed on a load cell during testing. The cone calorimeter standard has been revised and a three-part edition will be published. Part 1 is the traditional test for measurement of HRR; Part 2 is intended for smoke measurements and Part 3 is the guidance standard. In addition, a very simple version of the cone calorimeter for measurement only of mass loss rate from the burning specimen will be published.

The test apparatus is shown in Fig. 10.63 and the test specifications are given in Table 10.149.

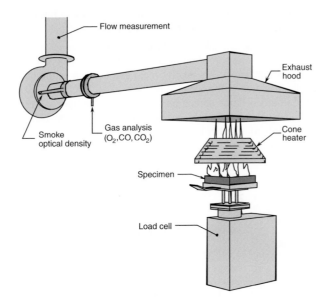

Figure 10.63   ISO 5660-1 and ISO 5660-2 cone calorimeter

Table 10.149   ISO 5660-1 and ISO 5660-2 test specifications

Specimens	Three specimens 100 mm × 100 mm
Specimen position	Horizontal in specimen holder
Heat source	Conical shaped electrical heater giving a heat flux in the range of 0–100 kW/m². If no specifications are given, tests at 25, 35 and 50 kW/m² are recommended
Test duration	32 min unless the burning practically has stopped for 10 min. If there is no ignition, test is stopped after 30 min
Conclusions	*HRR data*: Time to sustained flaming, curve of HRR versus time, 180 s and 300 s average of the HRR, peak HRR and possibly other HRR data *Smoke data*: Graph of smoke production rate per unit area of exposed specimen as well as discrete values for nonflaming and flaming phases. As a supplementary information, it is also possibly to report the yield of smoke per unit mass loss of the specimen, the specific extinction area, σ (units m²/kg)

EN ISO 9239-1 Reaction to fire tests for floor coverings – Determination of the burning behaviour using a radiant heat source

and

ISO 9239-2 Reaction to fire tests – Horizontal surface spread of flame on floor covering systems – Part 2: Flame spread at higher heat flux levels

These standard tests evaluate the critical radiant flux below which flames no longer spread over a horizontal flooring surface. Flame spread on the floor surface of a corridor exposed to heat flux from a hot gas layer or from flames in a nearby room can be seen as the model case. EN ISO 9239-1 is used for classification in Europe (see Section 10.4). EN ISO 9239-1 is considered to model the case of wind-aided flame spread on the surface of a floor covering. In the case of cocurrent airflow, the heat flux levels used in EN ISO 9239-1 are considered too low [9]. ISO 9239-2 exposes the sample to a higher heat flux and would be more versatile as it also covers the cocurrent airflow case. The test apparatus is shown in Fig. 10.64 and the test specifications are given in Table 10.150.

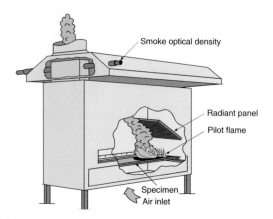

Figure 10.64   EN ISO 9239-1 and ISO 9239-2 floor covering test

Table 10.150   EN ISO 9239-1 and ISO 9239-2 floor covering test specifications

Specimens	1050 mm long × 230 mm wide
Specimen position	Horizontal
Ignition source	A gas fired radiant panel that gives a heat flux to the specimen. A pilot flame is impinging on the surface of the hot end of the specimen to initiate any flame spread EN ISO 9239-1: The maximum heat flux is 11 kW/m^2 that drops to 1 kW/m^2 at the end of the specimen ISO 9239-2: The maximum heat flux is 25 kW/m^2 that drops to 2.6 kW/m^2 at the end of the specimen
Test duration	Until the flames extinguish or maximum 30 min
Conclusions	Classification is based on the critical heat flux below which flame spread is not occurring

ISO 9705 Fire tests – Full scale room test for surface products

and

ISO/TR 9705-2 Reaction to fire tests – Full scale room tests for surface products – Part 2: Technical background and guidance

and

ISO 13784-1 Reaction to fire tests for sandwich panel building systems – Part 1: Test method for small rooms

These standards all make use of the same test equipment, the so-called room/corner test, shown in Fig. 10.65.

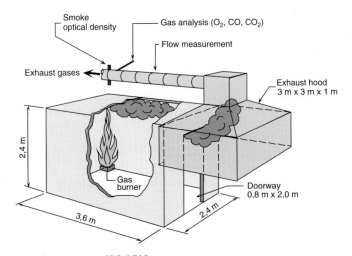

Figure 10.65   Room/corner test to ISO 9705

The room/corner test was first published by ASTM in 1982 [10] and then by NORDTEST in 1986 [11]. The international standard, ISO 9705, was published in 1993.

The room/corner test is a large-scale test method for measurement of the burning behavior of building products (linings) in a room scenario. The principal output is the occurrence and time to flashover. A direct measure of fire growth (heat release rate, HRR) and light obscuring smoke (smoke production rate, SPR) are also results from a test.

The product is mounted on three walls and on the ceiling of a small compartment. A door opening ventilates the room.

Experience on testing products has been gained during more than 10 years of work with the room/corner test. A considerable amount of information on product burning behavior in this method is available and the thermal conditions during a test fire have been carefully mapped [7, 12, 13].

ISO/TR 9705-2 is a valuable source of background data and theory for the test. The complete heat balance of the test room is described. Equations for mass flow are given as well as a mapping of the heat flux from the burner. ISO/TR 9705-2 also contains a large biography of technical papers related to the room/corner test.

ISO 13784-1 is a variant where the room/corner test is used for the testing of sandwich panels. These panels are generally well insulated and self-supporting. Therefore, they are themselves used as construction materials for the test room. In all other aspects, ISO 13784-1 testing is performed in the same way as an ordinary ISO 9705 test. The test specifications are shown in Table 10.151.

Table 10.151   Room/corner test to ISO 9705 and ISO 13784 test specifications

Specimens	Sample material enough to cover three walls and the ceiling of the test room. The wall containing the doorway is not covered except for the ISO 13784-1 method
Specimen position	Lined inside the room (ISO 9705) or forms the room structure itself (ISO 13784-1)
Ignition source	Gas burner placed in one of the room corners. The burner heat output is 100 kW for the first 10 min and then 300 kW for another 10 min
Test duration	20 min or until flashover
Conclusions	A number of parameters relating to a room fire such as temperatures of the gas layers, flame spread and heat fluxes can be measured. However, the most important outputs are HRR, SPR, and time to or occurrence of flashover

ISO 13784-2 Reaction to fire tests for sandwich panel building systems – Part 2: Test method for large rooms

Constructions made of sandwich panels may reach a considerable size. A large-scale test therefore is also needed. ISO 13784-2 test setup is a large open configuration (see Fig. 10.66). The test specification is given in Table 10.152.

perspective

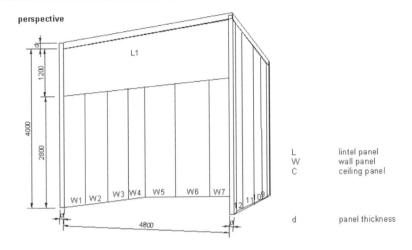

L   lintel panel
W   wall panel
C   ceiling panel

d   panel thickness

top view with burner position

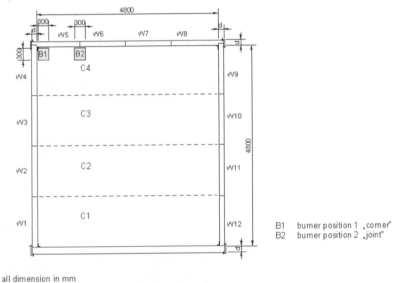

B1   burner position 1 „corner"
B2   burner position 2 „joint"

all dimension in mm

Figure 10.66 ISO 13 784-2 large-scale test for sandwich panels

Table 10.152 ISO 13784-2 test specifications

Specimens for constructing the test room	Test room made up from sandwich panels. Consists of four walls at right angles and a ceiling located on a rigid, non-combustible floor surface. Room inner dimensions: • length: (4.8 ± 0.05) m • width: (4.8 ± 0.05) m • height: (4.0 ± 0.05) m Doorway provided in the center of one wall, no other wall with any openings that allow ventilation. Doorway dimensions: • width: (4.8 ± 0.05) m • height: (2.8 ± 0.05) m Room may be located indoors or outdoors
Ignition source	Burner placed on the floor in a corner directly opposite the doorway wall in contact with the specimens. Heat output from the burner: 100 kW for first 5 min of the test, 300 kW for the subsequent 5 min, increasing to 600 kW for the remaining 5 min test duration, if ignition and sustained burning of the test specimens has not occurred
Test duration	Until flashover, or if the structure collapses, 15 min
Conclusions	A number of parameters relating to a room fire such as temperatures, flame spread, and structural failure

ISO 13785-1 reaction to fire tests for façades – Part 1: Intermediate-scale tests

ISO 13785-2 reaction to fire tests for façades – Part 2: Large-scale tests

These two standards address the problem of fire spread along the surface of a façade. The reference fire for the large-scale test, Part 2, is a flashover inside a compartment in a building followed by a large flame breaking through a window opening and subsequently impinging on the façade. The parameters measured are flame spread and breaking and falling pieces of the façade material. The intermediate-scale test is smaller and intended to measure the flame spread only.

ISO 5659-2 Plastics – Smoke generation – Part 2: Determination of optical density by a single-chamber test

A specimen is exposed to heat flux with or without an impinging pilot flame. Thus smoke is generated both under smoldering and flaming conditions. The smoke produced from the sample is accumulated in an airtight box. The optical density through the smoke is measured by using a lamp and a photocell. IMO is using this standard also for measurement of toxic gas species. There are classification criteria for products to be used on board ships. The apparatus is shown in Fig. 10.67 and the test specifications are summarized in Table 10.153.

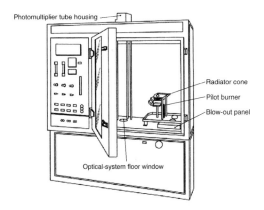

Figure 10.67   ISO 5659 Part 2 smoke density chamber

Table 10.153   ISO 5659 Part 2 test specifications

Specimens	9 specimens 75 mm square
Specimen position	Horizontal in a sample holder exposing 65 × 65 mm
Ignition source	Electrical conical radiator giving the following exposure: • Three specimens at 25 kW/m² with impinging pilot flame • Three specimens at 25 kW/m² without impinging pilot flame • Three specimens at 50 kW/m² without impinging pilot flame
Test duration	Normally 10 min
Conclusions	IMO requirements are: • Average maximum optical density, $D_m \leq 200, 400,$ or 500 depending on product type and use, • Maximum measured gas concentrations of CO 1450 ppm, HCl 600 ppm, HF 600 ppm, NO_x 350 ppm, HBr 600 ppm, HCN 140 ppm, and SO_2 120 ppm

ISO/TR 11 925 Reaction to fire tests – Ignitability of building products subjected to direct impingement of flame – Part 1: Guidance on ignitability

and

EN ISO 11 925-2 Reaction to fire tests – Ignitability of building products subjected to direct impingement of flame – Part 2: Single flame source test

and

ISO 11 925-3 Reaction to fire tests – Ignitability of building products subjected to direct impingement of flame – Part 3: Multisource test

ISO 11 925 series evaluates the ignitability of a product under exposure to a flame. This series of standards covers most cases where a flame test for ignitability determination would be required. Part 1 gives guidance on the theory and use of the ignitability tests. Part 2 is a small flame test that also appears in the European regulations for Euroclass (see Section 10.4). Part 3 is the multisource test where various flame sources from small flames to very large flames are covered. Each flame source is described in terms of the real case it intends to model, for example, a fire in a sauce pan, a wastepaper basket or a candlelike flame.

ISO/TR 14696 Reaction to fire tests – Determination of fire parameters of materials, products and assemblies using an intermediate-scale heat release calorimeter (ICAL)

A vertical sample is exposed to a constant, precisely defined heat flux from a radiant panel. The heat release rate is measured by means of the oxygen-consumption calorimetry. The test is similar in technology to the cone calorimeter. However, the sample in the ICAL test is larger and vertically oriented.

ISO/TS 14934-1 Reaction to fire tests – Calibration and use of radiometers and heat flux meters – Part 1: General principles

A worldwide standard for the calibration of heat flux meters is urgently required. Theories of ignition and subsequent flame spread all rely on data for a certain heat flux. National calibration procedures all show small deviations in results compared to each other. The coming ISO standard is intended to solve this situation by creating a worldwide reference from which secondary national references can be deduced.

10.23.3 Future Developments

ISO/TC 92 direction is now toward fire safety engineering. The work of SC1 is now emphasizing the development of standards for fire safety engineering, reference scenarios and different tools describing the fire growth process. Details of the scope are given here under "TC 92/ SC1 Fire initiation and growth."

References for 10.23

[1] ISO Annual report 1999
[2] ISO website www.ISO.ch
[3] ISO Bulletin, ISO Central Secretariat, 1 rue de Varembé, CH-1211 Genève 20, Switzerland
[4] J. Kruppa.: International Standardisation on Fire Safety Engineering, 3rd International Conference on Performance –Based Codes and Fire Safety Design Methods, 15–17 June 2000 Lund University, Lund Sweden.
[5] V. Babrauskas: The Cone Calorimeter (Section 3/Chapter 3), pp. 3–37 to 3–52, The SFPE Handbook of Fire Protection Engineering, second edition, National Fire Protection association, Quincy, MA (1995)
[6] EUREFIC – European Reaction to Fire Classification, International Seminar, Inter Science Communications Limited, London, September 1991, ISBN 0 9516320 1 9.
[7] CBUF Fire Safety of Upholstered Furniture-the final report of the CBUF research programme. Edited by B. Sundström, SP Fire Technology. EU project, Report EUR 16477 EN
[8] FIPEC, Fire Performance of Electric Cables, EU project
[9] P. van Hees: Wind-aided flame spread of floor coverings. Development and validation of large scale and small scale test methods, University of Gent (in Dutch), 1995
[10] 1982 Annual book of ASTM standards, part 18 – Proposed Method for Room Fire Test of Wall and Ceiling Materials and Assemblies
[11] Surface Products: Room Fire Test in Full Scale, NT FIRE 025, Helsinki 1986
[12] B. Sundström: Full-Scale Fire Testing of Surface Materials. Measurements of Heat Release and Productions of Smoke and Gas Species. Technical Report SP-RAPP 1986:45, BORÅS 1986
[13] B. Östman, R. Nussbaum: National Standard Fire Tests in Small Scale Compared with the Full-Scale ISO Room Test, Träteknikcentrum Rapport I 870217

11　Transportation

E. ANTONATUS AND J. TROITZSCH

At the present time plastics are considered as essential materials in all vehicles. In the past, they were used mainly because of their production advantages over metallic materials, for example, rapid, cost saving manufacture of complex moldings and reduced assembly costs, and by consumer demands for more comfort and better design. Today, the significant weight savings required for energy conservation, more stringent industrial and traffic safety requirements, and low maintenance and running costs can frequently be achieved only by the increased use of plastics.

11.1 Motor Vehicles

E. ANTONATUS

Approximately 10 to 15 % by weight – 100 to 150 kg (220–330 lb) – of a modern medium size car consists of plastics. The quest for lower fuel consumption and the desire for safety, corrosion resistance, and comfort will lead to increased application of these materials in the automotive field. The interior of cars, including seats, but also an increasing amount of technical functional parts are made of plastics today. Modern technology enables also reinforced thermoplastics and thermosets to be used on a wider scale for mass produced body parts. Following more than two decades of experience, high-density polyethylene (HDPE) fuel tanks have been fitted in more than 10 million vehicles and have proved their advantages all over the world.

11.1.1 Statutory Regulations

The object of safety legislation is to guarantee the user a minimum level of protection. Performance requirements are generally laid down in safety standards to prevent development being impeded by construction regulations. These standards reflect the state of the art and can be relatively easily brought into line with technical advances in safety.

Legal provisions often refer to relevant standards such as those of ISO, DIN and so forth. Derived from these standards, legislation has been set up on a national as well as on an international basis.

In Germany, for example, the Strassenverkehrsgesetz (StVG) (Road Traffic Act), Strassenverkehrsordnung (Road Traffic Order), and Strassenverkehrs-Zulassungs-Ordnung (StVZO) (Road Traffic Licensing Regulations) [1] form an important part of traffic legislation. The StVZO lays down the conditions for vehicles to be used on roads. Besides the licensing procedure for motor vehicles and trailers, Part III contains the construction and regulations for use, § 30 of which states:

Vehicles must be so constructed and equipped so that

- Their normal operation neither harms nor endangers, hinders, or disturbs anyone more than necessary,

- The occupants are protected as far as possible from injury, particularly in accidents and the extent and consequence of injuries should remain as slight as possible.

These fundamental requirements, which of course also include fire safety, were the basis for defining technical requirements for type approval of vehicles on a national basis. Similar developments occurred in other countries.

In the last years, international harmonization has resulted in an increasing transfer of responsibilities to international committees and in particular to those of the European Union (EU) and the United Nations Economic Commission for Europe (ECE).

The aim of harmonizing technical regulations within the EU is to remove technical trade barriers between member states. EU Directives are established by the General Directorate III (GD III) for the Internal Market of the European Commission (EC). The adoption of approval directives in national legislation must normally occur within 18 months. EC type approval will be granted for complete vehicles but since some directives have not yet been finalized, partial type approvals are being issued in the interim.

Within ECE, regulations finalized by the Working Party are implemented if two states declare their use to the UN Secretary General. It is significant that such provisions are adopted not only by the parties to the agreement but also by other countries. The technical provisions of a particular ECE agreement are then declared as authoritative in the national legislation.

At present more than 100 regulations have been issued by the UN Secretary General in New York including No. 34 "Uniform provisions concerning the approval of vehicles with regard to the prevention of fire risks" [2], Annex 5 of which contains the test regulations for plastic fuel tanks (now also adopted by the EC with directive 2000/8/EC [3]).

Besides regulations, ECE working parties can also draw up "Recommendations" for example flammability requirements for vehicle interiors [4].

The US Federal Motor Vehicle Safety Standards (FMVSS) based on the US Safety Act occupy a special position in the regulations of other countries owing to the importance of European and Japanese motor industry exports to the United States. The process of issuing a new safety standard commences with an announcement in the Federal Register (FR). Interested parties thus have the opportunity to express their opinion, raise objections, or suggest additions which are then dealt with in a "Notice of Proposed Rule Making." The intended deadline is fixed at the start.

The US regulations not only cover vehicles and components that are already in use but also accelerate new developments that are thought to be particularly important for vehicle safety. This dynamic legislation enables rapid changes and amendments to be made so that new technical trends can be put into practice relatively rapidly.

Besides the FMVSS, the Motor Carrier Safety Regulations (MCSR) for motor vehicles has limited validity for interstate and cross-border traffic.

11.1.2 Technical Fire Protection Requirements and Tests

All safety requirements including those in the field of fire protection are intended to be orientated toward real accident situations. Statistical and phenomenological analyses are used to establish the required levels of performance. Sensible performance standards should be laid down in safety regulations taking into account technical and financial limitations.

Fire incidents are relatively rare in motor vehicles compared to the large number of road accidents. According to a survey of accidents carried out by the German HUK-Verband (Association of Liability Insurance, Accident and Traffic Underwriters) involving 28,936 vehicles and 40,464 casualties, only 64 fires were reported [5]. This represents 0.24 % of the accidents investigated. It is nevertheless significant that the number of deaths in fire incidents is 10 times greater than in accidents that did not result in fires.

The cause of fires is usually difficult to ascertain. It can be assumed, however, that in the incidents included here, leaking fuel ignited by short circuits, sparks, or hot exhausts was a prime source. The survey does not include fire incidents that did not involve personal injuries. According to Trisko [6], these are usually due to carburetor fires, short circuits, or to careless handling of smokers' requisites and naked flames. The increasing incidence of arson is also noteworthy.

Fire spreads relatively slowly in a vehicle as long as no large quantities of fuel escape. Thus the Allianz-Zentrum für Technik, Ismaning near Munich (Allianz Insurance Company, Technology Center) reported flashover in the interior of a car only some 5–10 minutes after leaking fuel had ignited in the engine compartment [7].

To reduce these fire risks in motor vehicles, the following measures have been introduced:

- Materials and components in the passenger cabin may not be ignited by small ignition sources such as cigarettes and should not contribute to rapid fire propagation.
- The passenger compartment must protect the occupants for an adequate period from external flames and heat radiation.
- Electrical wires must be safely routed and batteries and components must be firmly anchored.
- The fuel system must remain leak free even if it is distorted in an accident. In particular, no fuel should penetrate the passenger compartment. Also, if a turnover occurs, no fuel may leak out of the fuel tank.

11.1.2.1 Materials and Components for Use in Car Interiors

Specific requirements for materials and components in the interiors of cars, trucks, and buses were first established by the US National Highway Transport Safety Administration (NHTSA) and brought into force as Federal Motor Vehicle Safety Standard (FMVSS) 302 in 1972 [8].

Specimens taken from the passenger compartment are clamped horizontally and subjected for 15 s to a Bunsen burner flame. The rate of flame spread, measured over a distance of 254 mm (10 in.), should not exceed 101.6 mm/min (4 in./min) for any of the specimens (Fig. 11.1). The specimen thickness must correspond to that of the component and should not exceed 12.7 mm (0.5 in.). The test specifications are summarized in Table 11.1.

The test procedure has been adopted by some countries with a significant automotive industry as national standards (DIN 75 200 [9], BS AU 169, JIS D 1201[10]). Furthermore, on the basis of this method, ISO Standard 3 795 has been established by ISO/TC 22/SC 16 to provide international unified test conditions for such components [11]. ISO 3 795 (as well as, e.g., German Standard DIN 75 200 and BS AU 169) does not contain any explicit requirements for the materials tested. These are laid down in legal provisions or conditions for delivery of car producers.

In Germany, for example, according to the "Richtlinien zu § 30 StVZO über die Verwendung schwerentflammbarer Werkstoffe im Kraftfahrzeugbau" (§ 30 StVZO Guidelines for the use of low flammability materials in motor vehicle construction), the maximum permitted rate of flame spread for interior components subject to mandatory testing is 110 mm/min (4.4 in./min) [12]. Proof must be furnished when a new vehicle is tested for type approval by officially recognized experts.

A number of car producing companies have defined several classes of materials by means of different limits for flame spread rate, burning time, and extent of flame spread, using the aforementioned test procedures. The manufacturers' requirements for the products used in vehicles therefore often are even more severe than those defined by the legislators.

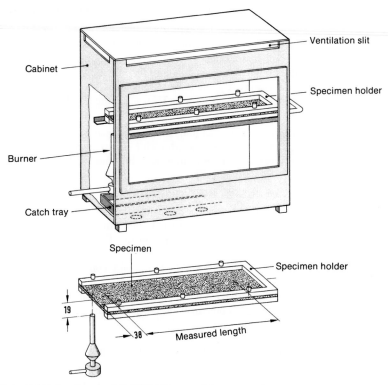

Figure 11.1   US Safety Standard FMVSS 302 rate of flame spread test for motor vehicle interior components

Table 11.1   FMVSS 302 test specifications for flammability of materials for vehicle

Specimens	Five specimens, 356 mm × 100 mm × thickness in use reference marks 38 mm, 292 mm, measured length 254 mm
Specimen position	Horizontal
Ignition source	Bunsen burner, 9 mm diameter
Duration of flame application	15 s
Result	rate of flame spread over measured length, max. permitted rate of flame spread 101.6 mm/min (4 in/min)

11.1.2.2 Buses and Coaches

In France, specific technical fire protection requirements for motor coaches for more than 16 passengers were established by the French minister of transport as a result of a severe road accident in 1982 near Paris. The "Specification Technique ST 18–502" [13] specifies the testing of interior materials in accordance with ISO 3795. Materials used for curtains and blinds must be tested in vertical position to ISO 6940 [14]. Furthermore, no burning droplets are allowed from materials used for head linings in coaches when tested according to the French standard NF P 92–505 [15].

In 1995, the European Union released the directive 95/28/EC [16] defining requirements for the fire behavior of materials used in the internal fittings of vehicles of category M3 buses and coaches (not applicable for buses for less than 22 passengers, or only for urban use).

Appendix IV describes a horizontal test that is similar to the FMVSS 302 test method and allows a maximum flame spread rate of 100 mm/min (3.9 in./min).

This has to be proven for:

- Materials used for the upholstery of any seat and its accessories (including the driver's seat),
- Materials used for the interior lining of the roof,
- Materials used for the interior lining of the side and rear walls, including separation walls,
- Materials with thermal and/or acoustic function,
- Materials used for the interior lining of the floor,
- Materials used for the interior lining of luggage racks, heating and ventilation pipes,
- Materials used for the light fittings.

Annex V is used to determine whether in case of fire a material produces burning drips. The test is similar to NF P 92-505. It uses the same test apparatus as this French standard, but the heat flux at the sample surface is calibrated with a heat flux meter to 3 W/cm².

This procedure applies for

- Materials used for the interior lining of the roof,
- Materials used for the interior lining of the luggage racks, heating and ventilation pipes situated in the roof,
- Materials used for the lights situated in the luggage-racks and/or roof.

Annex VI describes a test for the determination of the vertical burn speed of materials. This test is identical to ISO 6941. It is applied for curtains and blinds (and/or other hanging materials).

Materials that are not required to undergo the tests described in Annexes IV–VI are:

- Parts made of metal or glass,
- Each individual seat accessory with a mass of non-metallic material < 200 g. If the total mass of these accessories exceeds 400 g of nonmetallic material per seat, then each material must be tested,
- Elements of which the surface area or the volume does not exceed respectively:
  - 100 cm² or 40 cm³ for the elements that are connected to an individual seating place;
  - 300 cm² or 120 cm³ per seat row and, at a maximum, per linear meter of the interior of the passenger compartment for these,
- Elements that are distributed in the vehicle and that are not connected to an individual seating place,
- Electric cables,
- Elements for which it is not possible to extract a sample in the prescribed dimensions.

11.1.2.3 Bodywork

Only in Germany are fire protection requirements laid down for bodywork and body components made of plastics. According to the § 30 StVZO Guidelines relating to the use of low flammability materials in vehicle construction, sheet components of vehicle assemblies enclosing the passenger compartment must be tested to DIN 53 438 Part 3 [17]. Specimens of representative thickness must satisfy "Evaluation Class F1." The test is also intended to ensure that vehicles with plastic bodies cannot be set alight by small ignition sources such as matches and lighters. Proof that materials conform to the requirements must be furnished when applying for type approval.

11.1.2.4 Fuel Tanks

The use of plastic fuel tanks raises the question of their approval and evaluation from the aspect of fire safety. In the seventies, the Federal German Minister of Transport appointed the "Fachausschuss Kraftfahrzeugtechnik" (FKT) (Motor Vehicle Technology Committee) to develop safety requirements and tests for such tanks [18–20]. This task was performed by the "Sonderausschuss Feuersicherheit" (Fire Safety Committee) of the FKT in close cooperation with the plastics industry, and the results have been published in the "Prüfvorschriften des FKT-SA Feuersicherheit" (FKT-SA fire safety test specifications) [21]. These test conditions have been incorporated in ECE Regulation No. 34 [2], which is currently applied by all European countries with a significant automotive industry. The test has also been included into Council Directive 2000/8/EG [3] which includes a turnover test for fuel tanks added to the requirements of ECE.

The principal tests and requirements for plastic fuel tanks according to Council directive 2000/8/EG are summarized in Table 11.2. In addition, plastic tanks, like all tanks have to pass a

Table 11.2   Special requirements and tests for plastic fuel tanks according to Directive 2000/8/EC

Parameter	Test conditions				Minimum requirement
	Type and quantity of contents	Temperature (°C)	Time/pressure	Other	
Impact resistance	Water-glycol mixture 100 %	– 40 °C	–/–	Energy of impact: As close as possible to 30 Nm and 30 Nm	No leakage allowed
Mechanical strength	Water 100 %	+ 53 °C	5h / 0.3 bar		No leakage and no cracks allowed Deformation allowed
Fuel permeability	Petrol (or diesel if car is designed for use of diesel) 50 %	40 °C If criteria are not met 23 °C	Storage to constant weight loss/per time After that: refill and test for 24h		Quantity lost: ≤ 20 g/day ≤ 10 g/day
Resistance to fuel	Requirements regarding impact resistance and mechanical strength must still be met after test of fuel permeability				
Resistance to fire	Petrol (or diesel if car is designed for use of diesel) Different quantities for the three tests		60 s preheating 60 s exposure to flames of freely burning fuel 60 s reduced flame	Actual installation conditions simulated	No leakage allowed
Resistance to high temperature	Water 50 %	95 °C	1 h	Ventilation system open	No leakage and no serious deformation

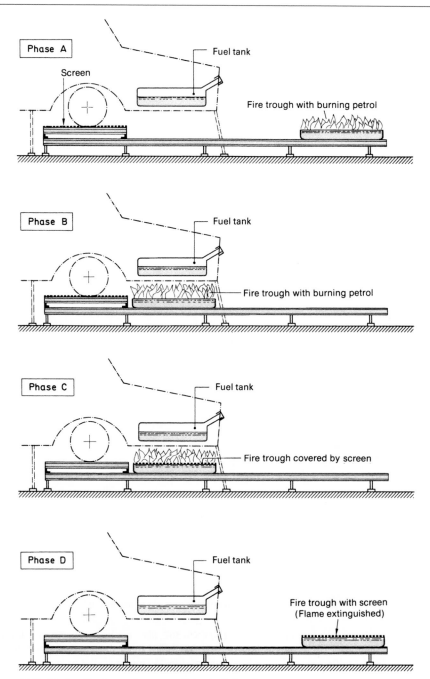

Figure 11.2 Testing the fire performance of plastic fuel tanks

hydraulic test and an overturn test. All these requirements, except the overturn test also are defined in ECE Regulation No. 34.

The fire performance test is illustrated schematically in Fig. 11.2. The fuel tank is filled with fuel to different percentages of its nominal capacity and installed in a test rig or in the vehicle assembly, that should include any parts which may affect the course of the fire in any way. The filler pipe is closed and the venting system should be operative. The flame is applied to the fuel tank in four phases from an open fire trough filled with petrol:

- Phase A (Preheating): The fuel in the fire trough is ignited and burns for 60 s.
- Phase B (Direct flame exposure): The fuel tank is subjected to the flames of the fully developed fuel fire for 60 s.
- Phase C (Indirect flame exposure): Immediately after completing Phase B, a defined screen is placed between the fire trough and the tank. The latter is subjected to this reduced fire exposure for a further 60 s.
- Phase D (End of test): The fire trough covered by the screen is brought to the initial position. The tank and fire trough are extinguished.

The requirements of the test are met if, at the end of the test, the tank is still in its mount and does not leak.

11.1.2.5 Plastic Windows

In Germany, the flammability requirements for plastic windows are laid down in the "Technische Anforderungen an Fahrzeugteile bei der Bauartprüfung nach § 22 a StVZO" [22] (Technical requirements for vehicle components in the type test according to § 22 a StVZO). Technical fire requirements are given in Part II No. 29, "Safety glass made of glasslike materials (rigid and flexible plastics)" (Fig. 11.3 and Table 11.3).

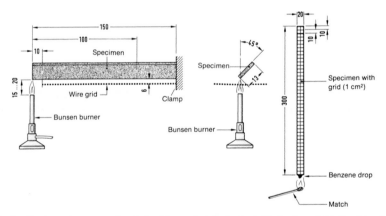

Figure 11.3   Testing of panes made of glasslike materials. Left: rigid plastics, right: flexible plastics

In the United States, the requirements of FMVSS 302 also apply to plastic windows in convertible hoods. In the European Directive 92/22/EC [23] the same test is applied, but the maximum acceptable flame spread rate has been fixed to 250 mm/min (9.8 in./min).

Table 11.3 Test specifications for windows of glasslike materials

Rigid plastics:	
Specimens	Three specimens, 150 mm × 13 mm × usual thickness Calibration marks at 10 mm and 100 mm
Specimen position	Longitudinal axis horizontal, Transverse axis inclined at 45°
Ignition source	Bunsen burner, flame height 15–20 mm
Duration of flame application	30 s
Conclusion	Rate of flame spread over measured length, maximum permitted rate of flame spread: 90 mm/min

Flexible plastics:		
Specimen	One specimen 300 mm × 20 mm usual thickness	
Specimen position	Vertical	
Ignition source	Benzene (C_6H_6) drop	
Conclusion	Rate of burning, maximum permitted values:	
	thickness of film (mm)	area rate of burning (cm^2/s)
	0.1–0.25	≤ 6.0
	0.25–0.5	≤ 3.0
	0.5–1.5	≤ 1.5

11.1.3 Future Developments

The continuing development of fire safety specifications will be based on the harmonization of regulations within the framework of the UN and of the European Communities. For fuel tanks, a new issue of the ECE regulation is expected, achieving full accordance with Council Directive 70/221/EEC.

Limitation of smoke and toxic gas emissions has been investigated by ECE Report Group GRSG 1978 and is still not considered as significant.

References for Section 11.1

[1] Strassenverkehrs-Zulassungs-Ordnung (StVZO) as in the Announcement of 15.11.1974, BGBl.I, p. 3193
[2] United Nations agreement concerning the adoption of uniform conditions of approval and reciprocal recognition of approval for motor vehicle equipment and parts – Uniform provisions concerning the approval of vehicles with regard to the prevention of fire risks

[3] Council directive 2000/8/EC, March 2000

[4] ECE Recommendation: Flammability of materials intended to be used inside road vehicles. Trans/ SC 1/WP 29/78, Annex 4

[5] Innere Sicherheit im Auto – Das Unfallgeschehen und seine Folgen. Issued by: HUK Verband der Haftpflicht-, Unfall- und Kraftverkehrsversicherer e. V., Hamburg, 1975

[6] E.M. Trisko: *Fire J.* 3 (1975) p. 19

[7] M. Danner, D. Anselm: *Der Verkehrsunfall* 12 (1977) p. 229

[8] Title 49 – Transportation § 571.302. Standard 302: Flammability of interior materials (valid Sep. 1972). US-Federal Register Vol. 36, No. 232

[9] DIN 75 200 (Sep. 1980): Bestimmung des Brennverhaltens von Werkstoffen der Kraftfahrzeuginnenausstattung

[10] Japanese Industrial Standard JIS D 1201 – 1998: Test method for flammability of organic interior materials for automobiles

[11] International Standard ISO 3795 (1989): Road vehicles – Determination of burning behavior of interior materials for motor vehicles

[12] FKT-Sonderausschuss "Feuersicherheit": Richtlinien zu § 30 StVZO über die Verwendung schwerentflammbarer Werkstoffe im Kraftfahrzeugbau – April 1977

[13] Union Technique de l'Automobile du Motocycle et du Cycle (U.T.A.C.): Spécification Technique ST 18–502: "Tenue au feu des matériaux utilisés dans l'aménagement intérieur des véhicules"

[14] ISO 6940 (09.84) Textile fabrics – burning behaviour – determination of ease of ignition of vertically oriented specimens

[15] NF-P 92–505 (12.95): Essai de goutte, au radiateur, applicable aux matériaux fusibles (so-called "Dripping test")

[16] Council directive 95/28 EC (10.95)

[17] DIN 53 438 Part 3 (06.84): Prüfung von brennbaren Werkstoffen. Verhalten beim Beflammen mit einem Brenner, Flächenbeflammung

[18] H. Wolf, G. Stecklina: *Materialprüfung* 17 (1975) 6, p. 205

[19] W. Becker, H.J. Bönold, M. Egresi: *Materialprüfung 17* (1975) 6, p. 203

[20] W. Becker, H.J. Bönold, M. Egresi: *Fire Prevent. Sci. Technol.* 21 (1979) 3

[21] FKT-Sonderausschuss "Feuersicherheit": Untersuchungen zur Beurteilung der Gebrauchsfähigkeit von Kraftstoffbehältern aus Kunststoffen – Prüfvorschriften – April 1977

[22] Technische Anforderung an Fahrzeugteilen bei der Bauartprüfung nach § 22a StVZO, 5.7.1973 (VkBl p. 559), Ergänzungen 1999

[23] Council directive 92/22/EC (03.92)

11.2 Rail Vehicles

A. EBENAU

Decisive factors for lightweight construction in modern track-guided transport systems by increased usage of light materials such as plastics or metals such as aluminum or magnesium alloys are the requirements for higher payload and for more efficient use of energy, low maintenance costs, corrosion resistance, and improved comfort. This results in higher fire loads in passenger train vehicles.

The frequency of fires in railways decreased during the last decade [1]. In most cases, the fire can be extinguished very soon or the burning parts self-extinguish. Only few fires cause greater damage, some of which involve loss of life.

The origins of fires in railways can be divided into different groups. Fires caused by technical defects generally occur in the traction unit. The main causes are excess oil and grease mixed with dirt and dust ignited by sparks from short circuits or overheated bearings and explosion of oil switches and high-voltage step selectors. These fires normally break out when the train is not in service; therefore the losses can be high but people are not affected.

Other causes are passenger-induced hazards. Matches or cigarettes carelessly thrown away in combination with apertures, slits, or hidden spaces and draughts, rubbish, dust, and paper can result in a smoldering fire that can remain undetected and flare up only after several hours.

Arson as a fire hazard is being found increasingly more often. Usually, the arsonist uses materials that he can find in the vehicle, such as newspapers left on the seat or litter under the seats. The ignited items often are seats with the ignition source on or under the seat base, sometimes wall linings or luggage in luggage racks.

Fire risks in rail vehicles are determined by different factors such as the design, ventilation, the materials used for interior equipment and their interaction, as well as by train operating conditions. A tram that stops every few minutes and in which passengers are under the supervision of staff represents a lower level of hazard compared to an automatically driven underground vehicle with regard to necessary time for evacuation, time needed for fire detection, and the chance of the fire fighters to be on the spot within a few minutes. Systems for automatic fire detection and suppression play an increasing role and must be considered in the assessment of fire risks.

11.2.1 Statutory Regulations

Technical legal provisions for rail vehicles vary from country to country. Generally, the type of vehicle or the operation mode is taken into account. In addition, operators frequently stipulate that various specifications laid down in conditions for delivery must be met.

Unification of regulations to meet the requirements of international rail transports is important, particularly for those countries connected by a railway network. European and Asian railway administrations are represented in the "Union Internationale des Chemins de Fer" (UIC). They have established the UIC Code covering regulations, recommendations, and uniform conditions of supply for rail vehicles [2].

With the coalescence of the countries of the European Union, the harmonization of national regulations gains more and more significance. A joint working group (JWG) compiled from members of CEN/TC 256 und CENELEC 9X has been working in the formulation of a seven-part standard EN 45545 "Railway applications – Fire protection on railway vehicles" [3] since 1991. A finalization of this standard is expected soon; the implementation in the national regulations will then be done during a transitional period of several years.

11.2.2 Fire Protection Requirements, Classification and Tests on Railway Vehicles in Selected Countries

The fire tests referred to in the next sections were mainly taken over from already existing national and international tests used in building and electrical engineering. Details on test equipment and specifications can be found in Chapters 10 Building (Sections 10.2 USA, 10.4 European Union, 10.8 France, 10.9 Germany, 10.18 UK, 10.23 ISO), 12 Electrical Engineering, 15 Smoke, and 16 Toxicity.

11.2.2.1 Germany

In Germany, the fire protection requirements for rolling stock are given in the German standard series DIN 5510 [4].

The regulations given in these standards are valid for vehicles, which are built under the order of the "Eisenbahn Bau- und Betriebsordnung" (EBO) (Railway Construction and Operation Decree), the "Eisenbahn Bau- und Betriebsordnung für Schmalspurbahnen" (ESBO) (Narrow gauge Railway Construction and Operation Decree), the "Magnetschwebebahn- Bau- und

Betriebsordnung" (MbBO) (Maglev Train Construction and Operation Decree) [5], and the "Strassenbahn-Bau und Betriebsordnung" (BOStrab) (Tramway Construction and Operation Decree) [6]. National and local technical supervisory authorities (federal, state, city) supervise the execution and construction of operating facilities and rolling stock and issue acceptance certificates.

The DIN 5510 standard series "Preventive fire protection in railway vehicles" contains the following parts:

Part 1: Principles, levels of protection, fire prevention measures, and justifiable evidence
Part 2: Requirements and test procedures
Part 4: Vehicle design
Part 5: Electrical equipment
Part 6: Auxiliary measures, emergency brake operation, information systems, fire alarms, fire fighting equipment.

The definitions for preventive fire protection in DIN 5510 are based on three protection objectives. The first objective is the prevention of fires caused by arson in customer areas, where ignition sources equivalent to $100\,g$ (3.6 oz) of paper are regarded as a typical initial fire load. The second objective is the prevention of fires caused by technical defects and finally the postponement of spread of fire to ensure a safe evacuation of the passengers, if the first two objectives fail.

Four operation categories are defined with different requirement levels for the fire performance and the vehicle design, as shown in Table 11.4.

Table 11.4   Operation categories

Operation category	Characteristics
1	Surface operation with no substantial tunnel operation periods
2	Substantial underground operation with a distance between emergency stops of > 2000 m
3	Substantial underground operation with a distance between regular stops of > 2000 m
4	Operation on tracks without escape possibilities

DIN 5510 Part 2 defines test methods and a classification system for materials and parts regarding

- Fire behavior,
- Smoke development,
- Flaming droplets or parts,

and requirements for the different operation categories.

The most important test method is a burning cabinet defined in DIN 54 837 [7] (Table 11.5 and Fig. 11.4). Specimens can be materials and material compounds and should be tested in end use condition.

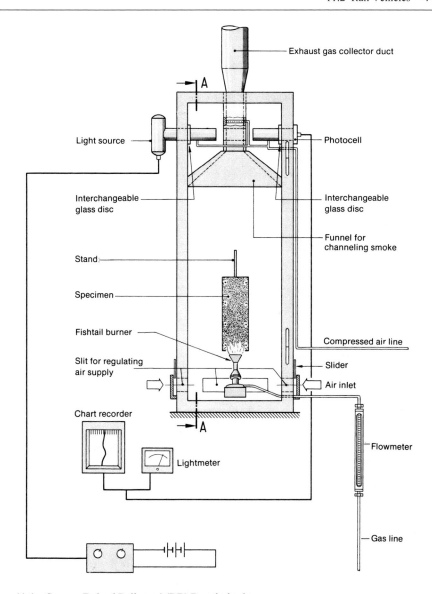

Figure 11.4 German Federal Railways' (DB) Brandschacht

Table 11.5 Test specifications for materials and components in German Federal Railways "rolling stock and installations" according to DIN 54 837

Specimen size	190 mm × 500 mm × thickness; if not possible, original size or flat model with 190 mm × 500 mm 500 mm length for profiles
Number of specimens	5
Specimen position	Vertical
Ignition source	Bunsen burner with fishtail jet, flame temperature 850 °C, inclination 45°
Application	Flame applied to surface, 20 mm from lower edge of specimen. Application time 3 min
Conclusions	Combustibility grades S2–S5 Smoke development grades SR1, SR2 Dripping grades ST1, ST2

Fire Behavior

Specimens are classified into combustibility grades as shown in Table 11.6.

Table 11.6 Classification in combustibility grades

Grade	Test method	Requirements	Comments
S1	DIN 53 438 [8], parts 1 to 3 DIN 4102-1 [9]	max. flame height < 15 cm within observation period of 20 s or K1; F1; K2; F2 Class B2	Only for small parts. Surface or edge flame attack depending on situation in service
S2	DIN 54 837	Destroyed length of specimen < 30 cm	
S3	DIN 54 837	Destroyed length of specimen < 25 cm Afterburning time < 100 s	
S4	DIN 54 837 or DIN 4102-1	Destroyed length of specimen < 20 cm Afterburning time < 10 s Class B1	
S5	DIN 54 837 or DIN 4102-1	Destroyed length of specimen 0 cm No afterburning Class A1 or A2	

For floorings, a combined combustibility and smoke generation classification is defined using the radiant panel test DIN 4102 Part 14 [10]. The criteria are shown in Table 11.7.

Table 11.7   Classification of flooring products for combustibility and smoke development

Grade	Test method	Requirements
SF1	DIN 4102-14	Critical radiant flux > 0.25 W/cm^2
SF2		Critical radiant flux > 0.25 W/cm^2 Integral of light obscuration < 2500 % · min
SF3		Critical radiant flux > 0.45 W/cm^2 Integral of light obscuration < 750 % · min

Smoke Development

For parts other than flooring, light obscuration during combustion is determined by a measuring device for light obscuration incorporated in the exhaust gas collector duct of the DIN 54 837 cabinet. The percentage light attenuation is recorded for classifying specimens in smoke development grades as shown in Table 11.8.

Table 11.8   Classification in smoke development grades

Grade	Test method	Requirements
Not SR1	DIN 54 837	Integral of light obscuration > 100 % · min
SR1		Integral of light obscuration < 100 % · min
SR2		Integral of light obscuration < 50 % · min

Dripping

The occurrence of flaming or non-flaming droplets is evaluated as shown in Table 11.9.

Table 11.9   Classification in dripping grades

Grade	Test method	Requirements
ST1	DIN 54 837	Occurrence of flaming droplets or parts
ST2		No occurrence of flaming droplets or parts (a maximum burning time of droplets or parts of 20s is admissible)

Combustibility of Seats

The fire performance of seats is tested using original complete seats to DIN 54341 Part 1 [11]. The test specifications are shown in Table 11.10. It should be noted that the conditions of delivery of the Federal German Railways specify that upholstered seats must be self-extinguishing, even if the seat covering is slit open.

Table 11.10   Test specifications for seats

Specimen	Original completed seat
Number of specimens	3 + 1 + 1
Conditions	End use condition: seat in a corner of noncombustible material with a second seat beneath
Ignition source and application	100 g paper cushion positioned on the seat base (three tests) centrally below a seat (one test) between two seats below the seats (one test)
Requirements	Test on the seat base: – Maximum flame height 1000 mm • Self-extinguishing within 15 min • The seat edges may not be destroyed Tests under the seat: • Self-extinguishing within 10 min, • Maximum 50 mm of the seat edges may be destroyed. For operation categories 3 and 4, the tests have to be carried out additionally on seats with a slitted base

Requirements

The requirements for fire performance and secondary fire effects of materials and components are laid down in Table 5 of DIN 5510–2 depending on the application and the operation category of the vehicle.

11.2.2.2 Great Britain

In Great Britain, guidelines on the design and construction of railway passenger rolling stock relating to fire are given in BS 6853 [12]. This code of practice for fire precautions in the design and construction of passenger-carrying trains covers new vehicles and changes to existing vehicles. It contains design considerations and demands concerning fire detection and suppression systems and the function of essential and emergency systems. Three categories of vehicles have been defined, as shown in Table 11.11.

Table 11.11   Operation categories

Category	Characteristics
Ia	Substantial operation in single-track tunnels with side exits to a walkaway and escape shafts; sleeper vehicles with substantial underground operation
1b	Substantial operation in multitrack tunnels or tunnels with side exits to a walkaway and escape shafts
2	Surface stock with no substantial operation periods in tunnels

Test compliance criteria regarding flame spread, smoke development, and toxicity of fire gases for products used in passenger vehicles are laid down in BS 6853. For the classification of products, three different surface orientations are defined (see Table 11.12) with reduced demands to surfaces of limited extent and with minimum distances to other similar surfaces.

Table 11.12 Definition of surface orientations in BS 6853

Surface orientation	Description
Vertical (V) surface	"Wall like" surface within 45° of the vertical
Horizontal prone (HP) surface	"Ceiling like" downward facing surface within 45° of the horizontal
Horizontal supine (HS) surface	"Floor like" upward facing surface within 45° of the horizontal

Fire Behavior

The essential test methods used for the classification of the fire behavior are

- BS 476–7 [13] radiant panel surface flame spread test,
- BS 476–6 [14] fire propagation test,
- BS ISO 9239–1 [15] radiant panel test for floorings.

Minor use materials of a mass between 100 g (3.6 oz) and 500 g (18 oz) are classified with the methods

- BS EN ISO 4589–3 [16] Oxygen index at elevated temperature,
- BS ISO 4589–2 [17] Oxygen index.

For most of the applications the requirements listed in Table 11.13 are defined.

Table 11.13 Requirements for surfaces depending on operation category

Surface orientation	Category 1a	Category 1b	Category 2
Vertical (V) surface	Class 0	Class 0	Class 1
Limited vertical (V) surface	Class 1	Class 1	Class 2
Horizontal prone (HP) surface	Class 0	Class 0	Class 1
Horizontal prone (HP) surface	Class 1	Class 1	Class 2
Horizontal supine (HS) surface	Class 1	Class 1	Class 2

Smoke Development

Smoke emission testing is carried out in a 3 m (9.8 ft) cube chamber using two fire sources:

- 1000 ml of 94 % industrial methylated spirit contained in a tray made of sheet steel
- 0.5 kg (1.1 lb) of English soft wood charcoal soaked in the alcohol used as source 1.

Five types of test methods are defined:

- Small-scale screening test,
- 60° panel test,
- Seating test,
- Flooring test,
- Cable test.

The measured absorbance A_m is calculated in accordance with the Beer-Lambert law

$$A_m = \log_{10}(l_0/l_t)$$

l_0 = initial luminous intensity,
l_t = transmitted luminous intensity.

The standard absorbance is

$$A_0 = A_m \cdot V/l \cdot n$$

V = Volume in the cube,
l = length of the optical path,
n = number of units of material constituting the test piece.

Toxicity of Fire Gases

The assessment of toxic potency of fire gases is carried out by determination of a weighted summation index R. Concentrations of eight gas components are measured using the mass based French test method NF X 70–100 by preference [18].

The concentrations c_x for the different gases are weighted with reference values f_x and summed up to the index R.

$$R = \Sigma \, (c_x/f_x)$$

If the index does not meet the requirements, an alternative time based analysis may also allow a technical sound hazard assessment.

11.2.2.3 France

Requirements on materials and parts used in French railway vehicles are laid down in the standards NF F 16–101 [19] (Choice of materials), NF F 16–102 [20] (Electrotechnical equipment), and NF F 16–201 [21] (Fire behavior of seats). NF F 16–101 defines criteria for the fire behavior, smoke development, and toxicity of fire gases for three categories of vehicles (Table 11.14).

Table 11.14   Operation categories

Category	Characteristics
A1	Substantial operation in tunnels
A2	Urban and suburban traffic, sleeper vehicles
B	Inter city operation with no substantial periods in tunnels

Fire Behavior

Different systems for the determination of the fire behavior of materials are defined. The majority of materials is classified using the "M-Classement" according to NF P 92–507 [22]. Small parts are tested with the oxygen index test NF T 51–701 [23] and the glow wire test NF C 20–921 (new version of NF C 20–455) [24]; cables have to satisfy NF C 32–070 [25]. Details are shown in Table 11.15.

Table 11.15 Fire performance classes and test methods

Classification	Classes					Test and classification methods	Application
M	M0	M1	M2	M3	M4	NF P 92-507 (using test methods NF P 92-501, -503, -504, and -505)	Every material with dimensions sufficient for taking specimens
I	I0	I1	I2	I3	I4	NF T 51-701 Oxygen index NF C 20-455 Glow-wire test	Small parts
–	–	A	B	C	D	NF C 32-070	Cables

Seats have to be tested additionally as complete assembly according to NF F 16–201 with a paper cushion of 100 g (3.6 oz) as ignition source and a maximum flame duration time of 10 min.

Smoke Development and Toxicity of Fire Gases

The smoke development of materials is determined using NF X 10–702 [26]. Square specimens of 76 mm x 76 mm (3.0 in. x 3.0 in.) are tested either in the flaming mode or the nonflaming mode with a radiant source of 25 kW/m^2; the values measured are the maximum smoke density D_m and the VOF_4 value with

$$VOF_4 = \frac{1}{2} (D_0 + 2D_1 + 2D_2 + 2D_3 + D_4)$$

The results of that test mode resulting in a higher VOF_4 value are used for classification.

For evaluation of fire gas toxicity, the material is tested to NF X 70–100. The gas concentrations t_i in mg/g are divided by 'critical concentrations' CC in mg/m^3, which cause no irreversible harm within 15 min and the results are summed up to get the toxicity index (ITC).

$$ITC = 100 (t_i/CC_i)$$

Both, smoke development and toxicity evaluation are combined to a smoke index IF.

$$IF = D_m/100 + VOF_4/30 + ITC/2$$

and classified in smoke classes F

Smoke class	F0	F1	F2	F3	F4	F5
Smoke index IF	≤ 5	≤ 20	≤ 40	≤ 80	≤ 120	> 120

Requirements

The requirements for rolling stock materials are listed as a combination of M and F classes and depend on the operation classes. Because many of these requirements cannot be covered by actual materials, grids have been defined with admissible and inadmissible combinations of M and F classes and combinations that require an individual approval by the authority. An example is shown in Table 11.16.

Table 11.16   Example for requirements for wall covering materials in category A vehicles

According to NF F 16-101, Table 7: M1F1
According to NF F 16-101, Table 8: Grid 3

Grid 3

	M0	M1	M2	M3	M4	NC
F0	Admissible	Admissible	Individual approval	Inadmissible	Inadmissible	Inadmissible
F1	Admissible	Admissible	Individual approval	Inadmissible	Inadmissible	Inadmissible
F2	Admissible	Admissible	Individual approval	Inadmissible	Inadmissible	Inadmissible
F3	Admissible	Admissible	Individual approval	Inadmissible	Inadmissible	Inadmissible
F4	Admissible	Admissible	Inadmissible	Inadmissible	Inadmissible	Inadmissible
F5	Admissible	Inadmissible	Inadmissible	Inadmissible	Inadmissible	Inadmissible

Legend: □ Admissible   ⊠ Individual approval   ▨ Inadmissible

11.2.2.4 Other European Countries

The fire test procedures for approval of materials and parts of other European countries often are attached to the systems existing in the larger countries described in the preceding. The testing and classification procedures of Spain, Italy, or Belgium are based on the French system. Other countries do not have autonomous systems. They take pattern from another national system (i.e., the Netherlands) or accept tests according to any proved system. For detailed information, the national authorities should be contacted.

11.2.2.5 United States of America

In the United States, comprehensive safety standards are issued by the Federal Railroad Administration (FRA). Recommendations for test methods and performance criteria for the flammability and smoke emission characteristics of categories and functions of materials to be used in the construction of new or rebuilt rail passenger equipment were developed in the 1970s by the Volpe Center and are published in the Code of Federal Regulations 49 CFR Part 238 [27]. The latest update, which is valid from July 12, 1999 (see Federal Register Vol. 64 No.91, p. 25539–25703) comprises small adjustments of performance criteria and an update of the test methods. An alternative method for the evaluation of seat assemblies and some exceptions for small component parts have been introduced.

The most important test methods for flammability are

- ASTM E 162–98 [28] radiant panel test for all vehicle components except flexible cellular foams, floor coverings and some other items,

- ASTM D 3675–95 [29] radiant panel test for flexible cellular foams,
- ASTM E 648–97 [30] radiant panel test for floor covering.

The determination of smoke emission characteristics uses ASTM E 662–97 [31] smoke box test in either the flaming or nonflaming mode depending on the higher smoke generation.

Alternatively to the methods described above, complete seat assemblies may be tested according to ASTM E 1537–98 [32] using the pass/fail criteria of California Technical Bulletin (CTB) 133 [33]. In addition, a fire hazard analysis must be conducted to consider the operating environment and the risk of vandalism, puncture, cutting, or other acts exposing the seat assembly or components. The risk assessment procedure will be standardized in an ASTM guideline "Guide for Fire Hazard Assessment of Rail Transportation Vehicles" which is in preparation. As a basis for the risk assessment, cone calorimeter test results have to be obtained from all components under consideration.

For structural parts the fire resistance must be shown according to ASTM E 119–98 [34].

The performance criteria for flammability and smoke emission of materials used in passenger cars and locomotive cabs are listed in the Code of Federal Regulations 49 CFR Part 238.103, Appendix B [27].

11.2.3 UIC Regulations

The UIC Code is covering regulations, recommendations, and uniform conditions for delivery for rail vehicles [2]. Sheet 564.2 gives test procedures and performance criteria for flammability and smoke production, but also indications concerning design and fire detection and suppression equipment.

The test procedures concerning fire behavior and smoke development are summarized in Annexes 4–15 of UIC 564.2. Materials used in vehicles for transnational traffic have to fulfill class A or B requirements or national requirements with equivalent classification. The details are summarized in Table 11.17.

Table 11.17 Test procedures concerning the fire behavior and smoke development of materials in annexes 4 to 15 of UIC 564.2

Annex	Test description	Measures	Application
4	Specimen size: 160 mm x 400 mm, inclination 45° Ignition source 4 ml of ethanol	Destroyed surface, afterflame time	Non-thermo-plastic solid materials
5	Specimen size: 190 mm x 320 mm, vertical Bunsen burner, flame height 40 mm, inclination 45°, 30 s at edge	Destroyed surface, afterflame time dripping	Textiles
6	Specimen length 320 mm Ignition conditions as in Annex 5	Destroyed length, afterflame time	Rubber profiles
7	Oxygen index to ISO 4589		
8	Specimen size and flame application time as ISO 3582 [35] Fishtail burner with 48 mm width (ISO) or 42 mm width (DIN 53438)	Destroyed surface, afterflame time dripping	Foams

Table 11.17   (Continuation)

Annex	Test description	Measures	Application
9	Specimen length 600 mm Test to UIC 895	Destroyed length, afterflame time	Cables
10	Specimen size: 220 mm x 20 mm, cut from length and transverse direction, inclination 70° Bunsen burner, flame height 40 mm, increasing flame application times 5 s–30 s, 5 s steps	Destroyed surface, afterflame time dripping	Rubber bellows
11	Specimen size: 160 mm x 300 mm Burner as in Annex 8 Flame application time 3 min	Destroyed surface, afterflame time dripping	Thermoplastic materials
12	Specimen size: 160 mm x 300 mm Burner as in Annex 8 Flame application time 3 min	Destroyed surface, afterflame time	Floorings
13	Seat (complete assembly) Ignition source 100 g paper cushion	afterflame time, falling parts, droplets	Seats
14	Real-scale test, complete compartment built in a 3 m x 3 m x 3 m cube Ignition source 100 g paper cushion		–
15	Smoke development test in a small chamber	Light obscuration	All

11.2.4 Future Developments

In Europe, the development of a harmonized seven part standard for fire protection in railways started in 1991. The draft standard EN 45545 Railway applications – Fire protection on railway vehicles [3] has the following parts:

Part 1: General
Part 2: Requirements for fire behavior of materials and components
Part 3: Fire resistance requirements for fire barriers and partitions
Part 4: Fire safety requirements for rolling stock design
Part 5: Fire safety requirements for electrical equipment including that of trolley buses, track guided buses, and magnetic levitation vehicles
Part 6: Fire control and management systems
Part 7: Fire safety requirements for flammable liquid and flammable gas installations

In Part 1, the objectives of the standard are given as minimization of the risks of a fire starting by accident, arson, or by technical defects and allowing for the safety of staff and passengers if the first objectives fail. Four different design categories as well as four operation categories are defined. Fire hazard levels resulting from different dwell times for passengers and staff are related to combinations of the operation and design categories.

Test methods and a classification system were evaluated in an EC founded research project Firestarr (Fire Standardization Research for Railways). The classification system will comprehend the fire characteristics ignition, flame spread and heat release, smoke development, and fire gas toxicity. The test methods will be chosen from existing European or international methods. For the evaluation of the system, small-scale fire tests are carried out, which will be verified by suitable large-scale and real-scale tests

In the United States, the FRA regulations were updated in 1989. After that, a research program was started to find alternative methods to evaluate the fire performance of rolling stock materials in a more realistic way. The outcome of this project reported in 1993 was an alternative evaluation approach based on heat release rate data. A follow-up program funded by FRA and evaluated by NIST deals with the prediction of fire performance of materials and components of passenger cars by using computer-aided simulation methods, respecting the limitations of small-scale test methods regarding geometric aspects and interactions between different materials. Phase 1 of this program found a good correlation for the ranking of materials tested with classical small-scale test methods described in the FRA regulations compared with that resulting from heat release data [36]. In phase 2, results from heat release measurements are used as input for a fire model for evaluation of fire hazard analyses with different fire models. These were compared with passenger escape times in different car designs and used for evaluation of fire detection and suppression systems. Phase 3 covers the verification of fire hazard analysis with selected real-scale tests.

The outcome of this project will be considered for the next revision of FRA guidelines. An expanded use of fire hazard analysis for parts as an alternative to the conventional small-scale methods can be expected.

References for Section 11.2

[1] European Firestarr research project, Workpackage 1-report, (1998)
[2] UIC Code Vol. V – Vehicles. Union Internationale des Chemins de Fer. Secrétariat Général, Paris.
[3] EN 45545 Railway applications – Fire protection on railway vehicles
    Part 1. General
    Part 2. Requirements for fire behaviour of materials and components
    Part 3. Fire resistance requirements for fire barriers and partitions
    Part 4. Fire safety requirements for rolling stock design
    Part 5. Fire safety requirements for electrical equipment including that of trolley buses, track guided buses and magnetic levitation vehicles
    Part 6. Fire control and management systems
    Part 7. Fire safety requirements for flammable liquid and flammable gas installations
[4] DIN 5510. Preventive fire protection in railway vehicles
    Part 1 (10/1988). Principles, levels of protection, fire preventive measures and justifiable evidence
    Part 2 (E 02/1996). Requirements and test procedures
    Part 4 (10/1988). Vehicle design
    Part 5 (10/1988). Electrical equipment
    Part 6 (10/1988). Auxiliary measures, emergency brake operation, information systems, fire alarms, fire fighting equipment
[5] Magnetschwebebahn- Bau- und Betriebsordnung (MbBO), veröffentlicht im Bundesgesetzblatt I Nr. 64 vom 25.09.1997
[6] Straßenbahn-Bau und Betriebsordnung (BOStrab) 31.8.1965, BGBl. I, p. 1513
[7] DIN 54 837 (E 01/1991). Prüfung von Werkstoffen, Kleinbauteilen und Bauteilabschnitten für Schienenfahrzeuge; Bestimmung des Brennverhaltens mit einem Gasbrenner
[8] DIN 53 438. Verhalten beim Beflammen mit einem Brenner
    Teil 1 (06/1984). Allgemeine Angaben
    Teil 2 (06/1984). Kantenbeflammung
    Teil 3 (06/1984). Flächenbeflammung

[9] DIN 4102. Brandverhalten von Baustoffen und Bauteilen
Teil 1 (05/1998), Baustoffe; Begriffe, Anforderungen und Prüfungen

[10] DIN 4102 Teil 14 (05/1990). Bodenbeläge und Bodenbeschichtungen; Bestimmung der Flamme-nausbreitung bei Beanspruchung mit einem Wärmestrahler

[11] DIN 54341. Prüfung von Sitzen für Schienenfahrzeuge des öffentlichen Personenverkehrs
Teil 1 (01/1988). Bestimmung des Brennverhaltens mit einem Papierkissen

[12] BS 6853 (01/1999), Code of practice for fire precautions in the design and construction of railway

[13] BS 476. Fire tests on building materials and structures
Part 7 (1997). Method of test to determine the classification of the surface spread of flame of products

[14] BS 476 Part 6 (1989). Method of test for fire propagation for products

[15] BS ISO 9239 (03/1998). Reaction to fire tests for floor coverings Determination of the burning behaviour using a radiant heat source

[16] BS EN ISO 4589. Plastics – Determination of burning behaviour by oxygen index
Part 3 (08/1996). Elevated-temperature test

[17] BS EN ISO 4589. Plastics – Determination of burning behaviour by oxygen index
Part 2 (07/1996), Ambient-temperature test

[18] NF X 70 – 100 (06 /86), Essais de comportement au feu, analyse des gaz de pyrolyse et de combustion, méthode au four tubulaire

[19] NF F 16 – 101 (10/1988), Comportement au feu; Choix des matériaux

[20] NF F 16 – 102 (10/1988), Comportement au feu; Choix des matériaux, application aux équipements électriques

[21] NF F 16 – 201 (03/1990), Matériel roulant ferroviaire; Essai de tenue au feu des sièges

[22] NF P 92–507 (12/1985), Building materials – Reaction to fire tests – Classification

[23] NF T 51–071 (11/1985), Plastics –Determination of flammability by oxygen index

[24] NF C 20 – 921/1 (05/97), Essais relatives aux risques de feu part 2-Méthodes d'essais section 1/ feuille 1: essais au fil incandescent sur produits finis et guide, replaces NF C 20 – 455

[25] NF C 32 – 070 (06/1979), Essais de classification des conducteurs et câbles du point de vue de leur comportement au feu

[26] NF X 10–702 Détermination de l'opacité des fumées en atmosphère non renouvelée
Part 1 (11/1995) Description du dispositif et méthode de vérification et de réglage du dispositif d'essai

[27] Code of Federal Regulations, Title 49 – Transportation, Subtitle B – Other Regulations Relating to Transportation, Chapter II – Federal Railroad Administration, Part 238 – Passenger Equipment safety standards, (01–01–1999 Edition)

[28] ASTM E 162–98. Standard Test Method for Surface Flammability of Materials Using a Radiant Heat Energy Source

[29] ASTM D 3675–95. Standard Test Method for Surface Flammability of Flexible Cellular Materials Using A Radiant Heat Energy Source

[30] ASTM E 648–98. Standard Test Method for Critical Radiant Flux of Floor-Covering Systems Using a Radiant Heat Energy Source

[31] ASTM E 662–97. Standard Test Method for Specific Optical Density of Smoke Generated by Solid Materials

[32] ASTM E 1537–98. Test Method for Fire Testing of Real Scale Upholstered Furniture Items

[33] California Technical Bulletin (CTB) 133 (04/1988) Flammability Test Procedure for Seating Furniture for Use in Public Occupancies

[34] ASTM E 119–98. Standard Test Methods for Fire Tests of Building Construction and Materials

[35] ISO 3582 (11/78). Cellular plastics and cellular rubber Materials – Laboratory assessment of horizontal burning characteristics of small specimens subjected to a small flame

[36] R.D. Peacock, et al.: Fire Safety of Passenger trains Phase I: Material Evaluation (Cone Calorimeter) NIST- IR 6132

11.3 Aircraft

E. ANTONATUS

A modern wide-bodied jet such as the Boeing 747 contains some 4000 kg (8800 lb) of plastics. Approximately 2000 kg (4400 lb) of these are used in structural applications, mainly in the form of glass- or carbon fiber-reinforced types, and about the same quantity in compartments occupied by passengers and crew. In addition, up to 8600 kg (18,920 lb) of combustible materials including maintenance materials, provisions and hand baggage are stored in the cabins. To this must be added some 39,000 kg (85,800 lb) of freight, usually in metal containers and approx. 100,000 l of fuel in wing and fuselage tanks.

A survey of transport aircraft accidents indicated that between 1955 and 1974 53,000 persons were involved, including some 11,000 fatalities, 2650 of which were caused by fire. Approximately 75 % of accidents occur on the runway or within a radius of 3000 m (about 10,000 ft) of the airport. By far the largest number of fatalities occurs as a consequence of crash landings in this area, when planes land short of or overshoot the runway on takeoff.

Large quantities of fuel can be released during crashes and are ignited by short circuits or heat. Such fires can develop fully within 1–3 min because of the low fire resistance of the aircraft fuselage; occupants have only a few minutes in which to escape.

In-flight fires occur mainly in the galleys as a result of electrical faults or overheating. It is noteworthy that smokers' materials are a frequent cause of cabin fires. Other commonly cited causes include faults in the electrical and oxygen supply systems.

Fires started by small ignition sources in the cockpit or cabin can generally be successfully tackled with on-board equipment. Exceptions include fires that develop unnoticed in inaccessible compartments or toilets. In 1973, for example, a fire in a toilet in an aircraft at Paris-Orly resulted in the death of 119 passengers. In 1998, a Swissair plane crashed because a fire in the insulation between the interior linings and the fuselage occurred and developed large quantities of smoke.

Fires also occur in aircraft while on the ground, mainly during maintenance work on the electrical and oxygen supply systems. Such incidents usually do not result in casualties and can, as a rule, be brought under control relatively easily.

Since 1980, the authorities started to improve regulations regarding the fire behavior of materials and parts used in aircraft and achieved a significant contribution to risk reduction.

In addition, many airlines no longer allow smoking on board their aircrafts.

In view of the frequency of fires occurring on impact, much attention should be devoted to fuel storage. Mechanically resistant fuel tanks with foam fillings as well as modification of the physical properties of the fuel can limit the formation of ignitable fuel-air mixtures and the spillage of combustible liquids to a certain extent. The US Army has achieved considerable success in this respect. High-strength fuel tanks with self-sealing connections (dry breakaway fuel fittings) have been used for years in helicopters and have led to a significant reduction in the incidence of fire following impact.

In fires, passengers should be protected as long as necessary from heat, smoke, and toxic gases by suitable measures such as division into fire compartments and the use of fire-resistant partitions in front of concealed fuselage areas and spaces where a high fire risk exists. These measures should be supplemented by fire alarm systems and effective automatic extinguishing equipment.

The ignitability, flammability, and secondary fire effects of materials, fire resistance of structural components, and measures for detecting and fighting on-board fires regrettably are often considered as separate problems. The only effective way of reducing fire risk, particularly in air transport, appears to be the introduction of safety systems that take adequate account of all aspects of fire protection. Large-scale fire tests under operating conditions are a prerequisite for more effective fire regulations in aviation. They must take into account the types of accident that pose a particular hazard for occupants. The majority of tests involve small ignition sources simulating the start of a fire during flight. As a rule, they do not enable any predictions to be made on the behavior of structures exposed to large ignition sources.

The Special Aviation Fire and Explosion Reduction (SAFER) advisory Committee and the aircraft industry have spent much time examining the factors affecting the ability of aircraft cabin occupants to survive in a post-crash fire. Many full-scale fire tests were conducted by the Federal Aviation Administration-National Aviation Facilities Experimental Center and by industry. They concluded that major accumulation of hazards occurs in the cabin. If an external fuel fire creates the hazards, ambient wind determines the amount of hazards entering a cabin due to such a fire.

11.3.1 Statutory Regulations

Besides specifications agreed for operational reasons between manufacturers and purchasers of aircraft, airplanes and the materials used in their construction must comply with national and international regulations. The national air transport authority of the country in which the aircraft is registered monitors compliance. Most countries have adopted wholly or in relevant parts the Federal Aviation Regulations (FAR) [1] of the US Federal Aviation Administration (FAA), as the basis for ensuring airworthiness. From time to time the Regulations are amended to reflect the latest technical developments. Experts from the US Aerospace Industries Association (AIA), the US Air Transport Association (ATA), the International Civil Aviation Organization (ICAO) and the International Air Transport Association (IATA) advise the FAA.

The following FAR regulations apply in Germany until further notice [2]:

FAR Part 23 – Airworthiness Standards: Normal, utility, and acrobatic category airplanes
FAR Part 25 – Airworthiness Standards: Transport category airplanes
FAR Part 27 – Airworthiness Standards: Normal category rotorcraft
FAR Part 29 – Airworthiness Standards: Transport category rotorcraft
FAR Part 33 – Airworthiness Standards: Aircraft engines

The regulations cover the airworthiness tests required for type approval by the "Luftfahrt-Bundesamt Braunschweig" (federal aviation agency), which grants a "Verkehrszulassung" (traffic permit) and issues an airworthiness certificate.

11.3.2 Technical Fire Protection Requirements and Tests

The terminology of technical fire protection is defined in FAR Part 1. It should be noted however that the terms employed, such as "fire-proof," "fire resistant," "self-extinguishing," and "flash resistant" often have different meanings in other standards and regulations, a situation that can cause confusion.

The methods used to test the materials and components in the cabins and holds of transport aircraft are described in Appendix F to FAR Part 25.

The requirements for military aircraft are partly similar. In Germany, they are covered by Airworthiness Regulation LTV 1500–850 [3], which makes considerable use of test rigs laid down in German Standards (DIN).

Airline and manufacturers' specifications that contain fire performance requirements based mainly on those of the FAR must also be complied with. They often contain, however, additional requirements particularly relating to smoke emission and toxicity of combustion products.

Requirements for materials and parts used in crew and passenger compartments in transport aircraft are given in FAR 25.853. They also apply to other aircraft where appropriate.

The test procedures for proving compliance with the requirements are given in Appendix F to FAR Part 25. These tests are generally carried out on test specimens taken from or simulating the components used. Composites must be tested as such.

11.3.2.1 Bunsen Burner-Type Tests

These test methods apply for most parts used in the interior. Depending on the type and area of application, additional tests are required. Windows, small parts, and electrical components are tested differently.

Vertical Test

The test for demonstrating that materials are "self-extinguishing" is illustrated schematically in Fig. 11.5. The test specifications and requirements are summarized in Table 11.18. The respective requirements are determined by size, orientation, and application of the tested parts.

The test is carried out in a draught-free cabinet according to Federal Test Method Standard 191, Method 5903.2. The average of three results is evaluated.

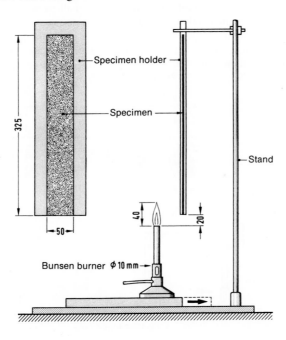

Figure 11.5   FAR Part 25 testing in vertical position

Table 11.18 FAR Part 25 specifications for vertical test

Specimens	Three specimens, 75 mm × 305 mm Thickness: Minimum thickness in use, foamed parts cut to 13 mm		
Specimen orientation	Vertical		
Ignition source	Bunsen or Tirill Burner, inner diameter 10 mm, methane flame 38 mm high at 843 °C		
Requirements	FAR 25, Appendix F Part I	(a)(1) (i)	(a)(1) (ii)
	Flame application Burn length Afterflame time Flame time of drippings	12 s ≤ 152 mm (6 in.) ≤ 15 s ≤ 3 s	60 s ≤ 203 mm (8 in.) ≤ 15 s ≤ 5 s

Horizontal Test

Windows, signs, parts made from elastomeric materials, small parts, and a number of other parts that are not installed permanently in the aircraft, are tested in a horizontal position. Proof must be obtained that a maximum permitted rate of flame spread is not exceeded (Fig. 11.6 and Table 11.19). The test is carried out in a draught-free cabinet according to Federal Test Method Standard 191, Method 5906. The mean value of three measurements is evaluated. The test corresponds to ISO 3795, DIN 75200, and to FMVSS 302.

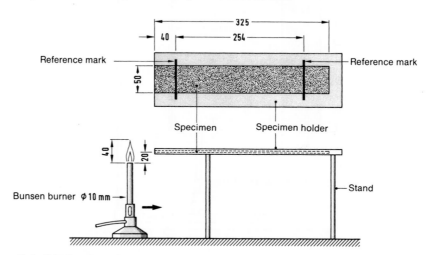

Figure 11.6 FAR Part 25 testing in a horizontal position

Table 11.19 FAR Part 25 specifications for horizontal test

Specimens	Three specimens 76 mm × 305 mm Thickness: Minimum thickness in use, maximum 3 mm Reference marks at 38 mm and 292 mm from the edge where the flame is applied		
Specimen orientation	Horizontal		
Ignition source	Bunsen or Tirill Burner, inner diameter 10 mm, methane flame 38 mm high at 843 °C		
Flame application	15 s		
Class Requirements	FAR 25, Appendix F Part I	(iv)	(v)
	Permitted burn rate	≤ 62.5 mm/min (≤ 2.5 in./min)	≤ 100 mm/min (≤ 4.0 in./min)

Forty-Five Degree Test

Liners for cargo and baggage compartments (FAR § 25.855) must be subjected to a 45° test (Fig. 11.7 and Table 11.20), in addition to the vertical test. This test has to proof that a fire starting in the cargo compartment will not easily pass through the walls of this compartment. For bigger airplanes this is applied only to the floor panels, and an additional test is carried out for walls and ceilings (see Fig. 11.10). The 45° test is carried out in a draught-free cabinet and the average of three measurements is evaluated.

Table 11.20 FAR Part 25 specifications for forty-five degree test

Specimens	Three specimens, 254 mm × 254 mm (exposed area: 203 mm × 203 mm) Thickness: Minimum thickness in use Reference marks at 38 mm and 292 mm from the edge where the flame is applied	
Specimen orientation	45°	
Ignition source	Bunsen or Tirill Burner, inner diameter 10 mm, methane flame 38 mm high at 843 °C	
Flame application	30 s	
Requirements	Flame penetration	not allowed
	Afterflame time	≤ 15 s
	Glow time	≤ 10 s

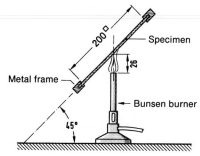

Figure 11.7   FAR Part 25 testing at 45°

Sixty-Degree test

Insulation, electric conductors, and cables must be "self-extinguishing." The test is illustrated in Fig. 11.8 and the test specifications and requirements are summarized in Table 11.21. The test is carried out in a combustion cabinet according to Federal Test Method Standard 191 (with the cabinet door open) or in a test chamber approx. 300 mm x 300 mm x 600 mm high (11.8 in. x 11.8 in. x 23.6 in.) with open top and side.

Table 11.21   FAR Part 25 specifications for sixty-degree test

Specimens	Three specimens, stretched by a weight, exposed length 610 mm Thickness: Minimum thickness in use Reference marks at 38 mm and 292 mm from the edge where the flame is applied	
Specimen orientation	60°	
Ignition source	Bunsen or Tirill Burner, inner diameter 10 mm, methane flame 38 mm high at 843 °C	
Flame application	30 s	
Requirements	Burn length	≤ 76 mm (3 in.)
	Afterflame time	≤ 3 s
	Drip extinguishing time	≤ 30 s
	Wire breakage	Not allowed

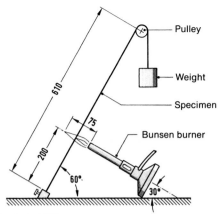

Figure 11.8   FAR Part 25 testing at 60°

11.3.2.2 Fire Behavior and Smoke Tests

For many years the requirements described above were the only measures taken by the authorities to ensure a reasonable fire behavior of interior materials for aircraft. Some major
accidents initiated research programs and subsequent developments, and resulted in amendments of the Federal Aviation Regulations. In 1984, the seat cushion kerosene burner test [4]
was introduced and, in the following years, the fire penetration test for cargo compartment
linings [5], the heat release measurement for large interior parts [6] and the smoke release test
were established [7]. In 2000, as a consequence of a fatal accident of a Swissair aircraft that
occurred in 1998, a NPRM (Notice of proposed rulemaking) was published [8], which
proposes to add to the rules additional requirements for insulation materials. The requirements
regarding heat release and smoke have been developed for transport category airplanes (more
than 19 passengers) and are not applied for other types of aircraft; the other requirements
mentioned above are applicable for all types of aircraft.

Seat Cushion Fire Blocking Test

The seat cushion fire blocking test limits the contribution of aircraft seats to fire spread and
smoke emission (Fig. 11.9 AND Table 11.22). The test is performed with a kerosene burner,
simulating a post-crash fire burning through the fuselage. By limiting the allowed burn length
on the seat, fire spread is limited. Furthermore, the contribution of the seat to heat development and smoke emission is restricted, as only 10 % of the mass of the seat is allowed to be
burnt. Positive results can be achieved by using a fire blocking layer, which protects the seat
foam from the flame attack.

Table 11.22   Seat cushion fire blocking test

Specimens	Three sets of seat bottom and seat back cushion specimens • Bottom: 508 mm × 457 mm × 102 mm • Back: 635 mm × 457 mm × 51 mm
Specimen orientation	Model seat mounted on a steel angle
Ignition source	Kerosene burner horizontally mounted delivering nominal 2 gallons/h
Flame application	2 min to the side of the seat bottom
Requirements	The average percentage weight loss must not exceed 10 %.

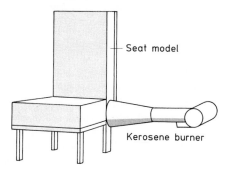

Figure 11.9   Seat cushion fire blocking test

Flame Penetration Resistance Test Of Cargo Compartment Liners

This test has been developed to prevent a fire originating in a cargo compartment to break through the linings and to spread further through the aircraft. Integrity of the linings is also important, because a fire in a cargo compartment can be successfully extinguished only when the concentration of the extinguishing agent can be maintained for a certain time, and no additional oxygen can come into the area of the origin of the fire. Apparatus and specifications are shown in Fig. 11.10 and Table 11.23.

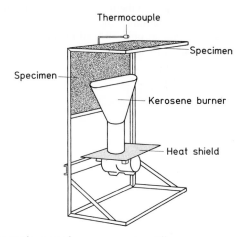

Figure 11.10   Flame penetration test of cargo compartment liners

Table 11.23   Flame penetration test of cargo compartment liners

Specimens	Three sets of ceiling and sidewall panels fixed to steel angles • Ceiling: 406 mm × 602 mm • Sidewall: 406 mm × 602 mm
Specimen orientation	Vertically and horizontally mounted on a steel angle
Ignition source	Kerosene burner, vertically mounted delivering nominal 2 gallons/h
Flame application	5 min to the ceiling panel
Requirements	No flame penetration of any specimen, the peak temperature measured 102 mm above the panel must not exceed 204 °C.

Heat release test

For large parts inside the passenger cabin of aircraft a heat release test was developed at the Ohio State University (OSU). It was introduced in 1990 and resulted in major material developments, because most of the materials used before were not able to pass these requirements. Table 11.24 summarizes the test requirements and the test apparatus is shown in Fig. 11.11.

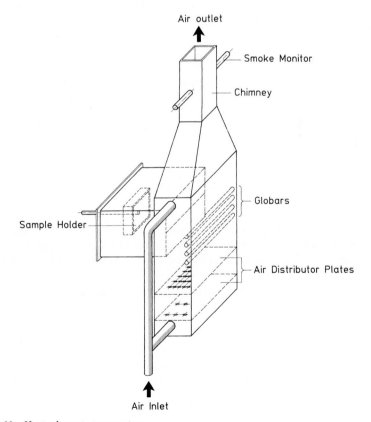

Figure 11.11   Heat release test apparatus

Table 11.24   OSU Heat release test

Specimens	Three specimens, surface 150 mm × 150 mm
Specimen position	Vertical
Ignition source	35 kW/m² radiant heat attack, pilot flame on lower end of specimen
Test duration	5 min
Requirements	• Maximum heat release rate (HRR) within 5 min: 65 kW/m² • Total heat release (HR) during the first 2 min: 65 kW min/m²

Smoke Emission Characteristics of Cabin Materials

For many years, the major aircraft manufacturers (Boeing, Airbus) have fixed requirements regarding optical density of the smoke emitted in case of fire (see Section 11.3.2.3). In 1990, aircraft authorities decided to introduce an FAR amendment with requirements on smoke emission characteristics of cabin materials [7].

The test is done according to ASTM F 814–83. The instrument and the method described in this standard are similar to ASTM E 662–97 (see Chapter 15, Section 15.3.2 for testing device and specifications). The test is performed on three samples, only in the flaming mode, and the specific optical density is measured within 4 min. During this period of time, which is regarded as sufficient for evacuation of passengers, the specific optical density D_s may not exceed 200.

New Requirements for Insulation Materials

As a consequence of the 1998 Swissair accident, in which the spread of fire within the insulation was assumed to be the main reason for the fatal consequences, the FAA developed new tests for insulation materials. These have been published as Notice of Proposed Rulemaking in September 2000 [8], and are likely to be adopted soon in the FAR as statutory requirements.

The proposed amendment was developed with the aim to limit flame propagation and to prevent entry of an external fire into the airplane (burn-through) in the case of a post-crash fire. It is planned to introduce the flame propagation test for all airplanes and, in addition, a requirement for flame penetration resistance for airplanes with a passenger capacity of 20 or more.

The test for flammability and flame propagation is performed in a test apparatus similar to ASTM E 648 (see Section 10.2 USA, flooring radiant panel test). The sample may be shorter (still under discussion) and the heat flux will be higher (18 kW/m² instead of 12 kW/m² at the hot end of the sample). The pilot burner is applied for 15 s and is then removed. The proposed criteria are that:

• No flaming occurs beyond 51 mm (2 in.) to the left of the centerline of the pilot flame application point,
• Not more than one out of the three specimens tested shows an afterflame for more than 3 s after the removal of the pilot flame.

For determining the burn-through resistance, an oil burner similar to that shown in Fig. 11.9 for seat testing is used. The fuel flow is significantly higher than for testing of seats. The specimen is exposed for 4 min to the burner flame. The sample may not burn through, and the heat flux measured 305 mm (12 in.) behind the sample may not exceed 25 kW/m².

"Fireproof" and "Fire-Resistant" Tests

Main flight controls, engine assemblies, and other parts of the airframe that lie in fire risk zones or adjacent areas as well as fire bulkheads, claddings, and shielding subject to heat must be made of fireproof materials or be fire resistant. The terms are defined in FAR Part 1 as follows:

- "Fireproof" with respect to materials and parts used to confine fire in a designated fire zone means the capacity to withstand at least as well as steel in dimensions appropriate for the purpose for which they are used, the heat produced when there is a severe fire of extended duration in that zone.
- "Fire resistant"
  - With respect to sheet or structural members means the capacity to withstand the heat associated with fire at least as well as aluminum alloys in dimensions appropriate for the purpose for which they are used.
  - With respect to fluid-carrying lines, fluid system parts, wiring, air ducts, fittings, and power plant controls means the capacity to perform the intended functions under the heat and other conditions likely to occur when there is a fire at the place concerned.

The basis of the fireproof and fire resistant test is described in FAR § 23.1191 (fire bulkheads). Further details particularly as regards test rigs are not given. The layout is illustrated in Fig. 11.12 and the test specifications and requirements are summarized in Table 11.25.

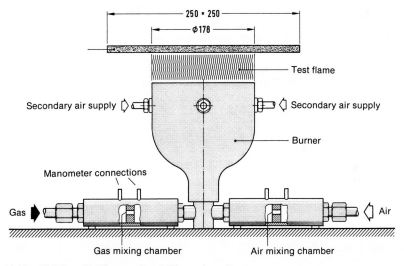

Figure 11.12   FAR Part 23 "fireproof" and "fire-resistant" tests

Certain mandatory fire equipment is tested by *ad hoc* methods as well as by the methods described in the preceding. This applies particularly to the effectiveness against smoke of bulkheads between passenger and crew compartments and cargo and baggage compartments as well as to smoke detectors, fire alarms, and automatic extinguishers.

Table 11.25   FAR Part 23 "fireproof" and "fire resistant" test specifications

Specimens	250 mm × 250 mm for flat specimens or 5 × outside diameter for hoses
Specimen position	Horizontal
Ignition source	Burner applied to specimen over an area of approx. 125 mm × 125 mm, temperature 1093 °C ± 28 °C, with hose pipes, the flame must be approx. as wide as the outside diameter of the hose
Flame application	• "Fireproof": 15 min • "Fire resistant": 5 min Maximum heat release rate (HRR) within 5 min: 65 kW/m²
Requirement	Tested part must be in working condition after test.

11.3.2.3 Technical Fire Requirements in the Conditions of Delivery of the Aviation Industry – Additional Requirements Relating to Smoke Density and Toxicity of Combustion Products

Technical fire requirements are also included in the material specifications of the airlines and aircraft manufacturers. As a rule, they correspond to those of FAR. Compliance must be demonstrated with the same or similar test methods and equipment. Lufthansa has compiled the applicable test methods and acceptance criteria in Specification S-25–98/016 [9].

Additional requirements are often specified, or evidence from standard or modified tests is required to evaluate secondary fire effects. Airbus Industrie, for example, specifies an additional test for smoke density and toxicity of combustion products for materials used in all Airbus models [10]. The basic layout of the test chamber used corresponds to the NBS smoke chamber, specified, for example, in ASTM E 662 (see Section 15.3.2). The specific optical density D_s occurring within 4 min is determined and, in the case of electrical wiring, within 16 min. For all parts in the pressurized part of the fuselage, the following limiting values shown in Table 11.26 should not be exceeded:

Table 11.26   Limits for specific optical density specified in Airbus Directive ABD 0031

Component parts to be tested in acc. with AITM 2.0007	D_m (flaming)	D_m (nonflaming)

Major interior panels
• Ceiling and sidewall panels
• Dado panels (without textile coverings)
• Door and door frame linings
• Partitions
• Cabin walls, e.g., lavatories
• Overhead passenger service units
• Stowage compartments
  (other than underseat stowage compartments and
  compartments for stowing small items)
	150	Not required

• Cargo liners	100	100

Component parts to be tested in acc. with AITM 2.0007	D_m (flaming)	D_m (nonflaming)
• Textile covered panels • Cabin floor panels	200	150
Textile components • Upholstery • Drapery	200	150
• Carpets	250	
Other components • Air-ducting • Thermal and acoustic insulation • Insulation coverings	100	100
• Seat cushion • Components, e.g., transparencies, elastomeric (used within the passenger cabin and for air-ducting) and thermoplastic parts, • Nontextile floor covering • Interior equipment parts (except electrical wire/cable)	200	150
Nonmetallic structural parts		Not required

The toxicity of all parts used in the pressurized area of the fuselage must be tested in the NBS smoke chamber equipped with three gas-sampling probes. These pass through the top of the chamber and reach halfway down the smoke chamber (Fig. 11.13).

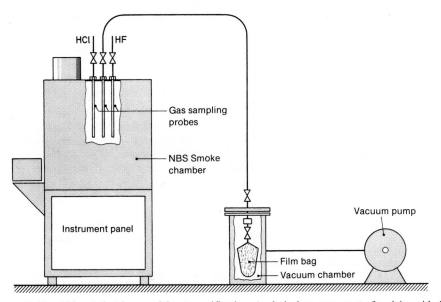

Figure 11.13 Airbus Industrie material test specification: Analytical measurement of toxicity with the NBS smoke chamber

The samples required for determining toxicity should be taken at the same time as the smoke density is measured. "Dräger Tubes" are used for analysis. Analysis for hydrogen chloride (HCl) and hydrogen fluoride (HF) is done directly during the test and other toxic components are determined from a gas sample taken from a film bag filled with the help of a vacuum pump during the test. Analysis of thermal decomposition products resulting from flaming and nonflaming conditions is carried out. The measured values should not exceed the concentrations summarized in Table 11.27.

Table 11.27 Limiting values of toxic smoke gas components within the test duration (16 min for electrical wires and cables, 4 min for all other materials) as specified in Airbus Industrie material test specification

Smoke gas components	Maximum concentration [ppm]
Hydrogen fluoride (HF)	100
Hydrogen chloride (HCl)	150
Hydrogen cyanide (HCN)	150
Sulfur dioxide/Hydrogen sulfide (SO_2/H_2S)	100
Carbon monoxide (CO)	1000
Nitrous gases (NO/NO_2)	100

11.3.3 Future developments

The topic of fires in inaccessible areas is still under consideration. After the development of the proposed rule for insulation materials, work has been started by authorities and aircraft manufacturers, regarding stricter requirements for electrical wiring and ducts in aircraft.

Furthermore, requirements regarding flammability will be extended to accessories such as blankets for passengers. For aircraft seating, until now, the flammability of the seat cushions has been the main criterion. In the future, the seat lining materials and tables will also be regulated more strictly.

Major changes of the regulations for reaction to fire of materials used in airplanes are not expected in the near future. The work of the last 20 years has resulted in a considerable limitation of ignitability, heat release, and flame spread, and consequently of smoke and gas development. A satisfactory level of safety has been reached.

References for Section 11.3

[1] Federal Aviation Regulation (FAR). Airworthiness Standards. Department of Transportation, Federal Aviation Administration
[2] Bundesanzeiger No. 90, 14 May 1965 and NFL B 75/65
[3] LTV 1500–850. Luftfahrt-Tauglichkeits-Vorschrift: Brandverhalten von Luftfahrt-Werkstoffen. Bundesamt für Wehrtechnik und Beschaffung, Jan. 1974
[4] FAR Part 25. Amendment No. 25–59, Nov. 1984

[5] FAR Part 25. Amendment No. 25–60, June 1986
[6] FAR Part 25. Amendment No. 25–61, Aug. 1988 and 1990
[7] FAR Part 25. Amendment No. 25–66, Aug. 1990
[8] Notice of Proposed Rulemaking NPRM – improved flammability standards for thermal/acoustics insulation in transport category airplanes, US DOT/FAA Elizabeth Erickson, 09/08/2000
[9] Lufthansa Technik Specification S-25–98/016. Determination and Testing of Flammability properties for Aircraft Interior Materials
[10] Airbus Industrie: Material Test Specification No. ATS-1000.001, 1996

11.4 Ships

K. YOSHIDA

International safety standards for merchant ships became a worldwide concern after the tragedy of the passenger ship Titanic. Consequently the first version of International Convention for the Safety of Life at Sea (SOLAS) [1] was adopted in 1914 by an intergovernmental conference. The SOLAS was revised in 1929 and 1948.

In 1948, an international conference convened by the United Nations established the Inter-Governmental Maritime Consultative Organization (IMCO). The name of the organization was changed to International Maritime Organization (IMO) in 1982. The function of IMO was, since its establishment as IMCO, to develop international regulations and standards to realize safer ships and cleaner oceans. One of the major works of IMO has been the development and adoption of revisions and amendments to SOLAS. Total revisions of SOLAS were adopted in 1960 and 1974. Since 1974 SOLAS has been amended many times until today to enhance the safety of merchant ships (see http://www.imo.org).

Fire safety of ships engaged in international trade has been regulated by Chapter II-2 of 1974 SOLAS. This chapter has been revised several times when major fire disasters of ships happened and additional fire safety measures became necessary to avoid such disasters. These piled-up amendments made the chapter complex. Therefore, a demand for making the chapter user-friendly has arisen. The Fire Protection Sub-Committee (FP) of IMO has worked on a comprehensive review of SOLAS Chapter II. This resulting new draft Chapter II-2 was approved by the IMO Maritime Safety Committee (MSC) in May 2000, was adopted in November 2000 and entered into effect on July 1st, 2001.

11.4.1 Statistics of Fire on Board Passenger Ships

From 1978 to 1995, a large number of fire casualties occured on board passenger ships and passenger ferries. The statistics in Table 11.28 [2] show that after grounding, fire and/or explosion are the most important type of casualties, causing loss of life and property. Many fire casualties have happened on board of cargo ships. Therefore, fire safety of ships is still a major item for safety of life at sea.

Table 11.28   Casualty types of passenger ships/ferries and number of lives lost

Type of casualty	Number	Lives lost	Lives missing	Lives lost and missing
Collision	136	896	109	1005
Contact	102	401	0	401
Grounding	264	132	372	504
Flooding	92	750	1392	2142
Fire/Explosion	209	525	511	1036
War	20	11	1	12
Mechanical Failure	129	17	17	34
Others	5	0	0	0
Total	957	2732	2402	5134

11.4.2 SOLAS Chapter II-2: International Regulations for Fire Safety of Ships

11.4.2.1 Existing SOLAS Fire Safety Regulations and Revision Activities

Safety of international trading ships has been regulated by the International Convention for the Safety of Life at Sea (SOLAS), adopted in 1974. SOLAS includes a set of fire safety regulations in its chapter II-2. Table 11.29 shows the history of amendments to SOLAS Chapter II-2.

Table 11.29   SOLAS Chapter II-2, history of amendments

Year of adoption	Amendments	Date of enforcement
1978	Enhancing tanker safety	May 1, 1981
1981	Enhancing fire protection of construction	Sept. 1, 1984
1983	Enhancing means of escape and tanker safety	July 1, 1986
1989	Introducing new fire safety measures	Feb. 1, 1992
1992	Enhancing passenger ship safety (after a fire on the passenger ferry "Scandinavian Star")	Oct. 1, 1994
1994	Enhancing protection of fuel lines	July 1, 1998
1995	Enhancing passenger ship safety (after foundering of the passenger ferry "Estonia")	July 1, 1997
1996	Introducing Fire Test Procedures Code (FTP Code) and interpretation to vague expressions	July 1, 1998

The existing SOLAS Chapter II-2 contains many prescriptive requirements for construction taking into account fire protection, fire fighting equipment, limitation of use of combustible materials, arrangement of escape routes, and so forth. However, no basic fire safety philosophy is described.

Technology for enhancing fire safety and fire protection equipment was developed extensively in the 1980s and 1990s. For example, sprinkler systems using water mist and fire-extinguishing systems, new types of high expansion foam fire-extinguishing systems, and newly computerized fire detection and fire alarm systems are now available. In the meantime, a new ship design has been introduced for passenger ships, passenger ferries and many types of cargo ships. Because SOLAS Chapter II-2 requirements are prescriptive, this new technology and design cannot be dealt with by these requirements. Therefore, a demand for making the chapter user-friendly has arisen.

For facilitating the demand, IMO FP has worked on a comprehensive review and prepared a new draft of SOLAS Chapter II-2. To accommodate new fire safety technology, FP agreed that SOLAS Chapter II-2 should specify clearly the fundamental fire safety requirements in a particular regulation and the functional requirements in every regulation, specifying precisely fire safety measures. FP also agreed that precise requirements and technical standards for each type of fire protection system and equipment, such as sprinkler systems and CO_2 fire extinguishing systems, should be moved from SOLAS to the Fire Systems Code, which has been newly developed. This subject is also reviewed in [3].

Fundamental Principles for Fire Protection on Board of Ships

Fundamental objectives for fire safety of ships have been extracted:

- To prevent the occurrence of fire and explosion,
- To reduce the risk to life caused by fire,
- To reduce the risk of damage caused by fire to the ship, its cargo, and the environment,
- To contain, control, and suppress fire and explosion in their compartment of origin,
- To provide adequate and readily accessible means of escape for passengers and crew.

To achieve these fire safety objectives, the following functional requirements have been derived and are embodied in the regulations of SOLAS Chapter II-2:

- Division of the ship into main vertical or horizontal zones by thermal and structural boundaries,
- Separation of accommodation spaces from the remainder of the ship by thermal and structural boundaries,
- Restricted use of combustible materials,
- Detection of any fire in the zone of origin,
- Containment and extinction of any fire in the space of origin,
- Protection of means of escape or access for fire fighting,
- Ready availability of fire-extinguishing appliances,
- Minimisation of possibilities of ignition of flammable cargo vapors.

These objectives and functional requirements are specified in the new regulation 2 of SOLAS Chapter II-2.

11.4.2.2 Structure of New Chapter II-2 of SOLAS

The new Chapter II-2 of SOLAS has been constructed based on the following principles:

- Level of fire safety of the new Chapter II-2 shall be the same as that of the existing Chapter II-2.
- Each regulation shall clearly state functional requirements.

- Each regulation shall have a clear relationship to regulation 2 "Fire safety objectives and functional requirements."
- Amendments already agreed shall be incorporated, but new ideas for amendments not yet agreed shall not be incorporated.
- Vague expressions in existing regulations shall be clarified.
- Precise requirements and technical standards for each type of fire protection system and equipment shall be moved from SOLAS to the Fire Safety Systems Code.
- A new regulation for evaluation of alternative fire safety design, arrangements and solutions shall be developed.

Table 11.30 shows the construction of the new Chapter II-2. Each regulation specifies its functional requirement at the beginning of the regulation. Table 11.31 shows examples of such functional requirements. Table 11.32 shows the contents of the newly developed Fire Safety Systems Code (FSS Code).

Table 11.30   Structure of new Chapter II-2

Part	Regulations
Part A: General	Regulation 1: Application Regulation 2: Fire Safety Objectives and General Requirements Regulation 3: Definitions
Part B: Prevention of Fire and Explosion	Regulation 4: Probability of Ignition Regulation 5: Fire Growth Potential Regulation 6: Smoke Generation Potential
Part C: Suppression of Fire and Explosion	Regulation 7: Detection and Alarm Regulation 8: Control of Smoke Spread Regulation 9: Containment of Fire Regulation 10: Fire Fighting Regulation 11: Structural Integrity
Part D: Means of Escape	Regulation 12: Notification of Crew and Passengers Regulation 13: Means of Escape
Part E: Operational Requirements	Regulation 14: Operational Readiness, Maintenance and Casualty Records Regulation 15: Instruction, Onboard Training and Drills Regulation 16: Operation
Part F: Alternative Design and Arrangements	Regulation 17: Verification of Compliance
Part G: Special Requirements	Regulation 18: Helicopter Facilities Regulation 19: Carriage of Dangerous Goods Regulation 20: Protection of Vehicle Spaces and Ro-ro Spaces

Table 11.31   Examples of functional requirements

> **Regulation 4: Probability of ignition**
>
> The purpose of this regulation is to prevent the ignition of combustible materials or flammable liquids. For this purpose, the following functional requirements shall be met:
> - Means shall be provided to control leakage of flammable liquids.
> - Means shall be provided to limit the accumulation of flammable vapors.
> - The ignitability of combustible materials shall be restricted.
> - Ignition sources shall be restricted.
> - Ignition sources shall be separated from combustible materials and flammable liquids.
> - The atmosphere in cargo tanks shall be maintained out of the explosive range.
>
> **Regulation 5: Fire growth potential**
>
> The purpose of this regulation is to limit the fire growth potential in every space of the ship. For this purpose, the following functional requirements shall be met:
> - Means of control for the air supply to the space shall be provided.
> - Means of control for flammable liquids in the space shall be provided.
> - The use of combustible materials shall be restricted.

Table 11.32   Contents of the fire safety systems code (FSS Code)

Chapter	Contents
1	General, Definitions, etc.
2	International Shore Connections
3	Personal Protection
4	Fire Extinguishers
5	Fixed Gas Fire-extinguishing Systems
6	Fixed Foam-extinguishing Systems
7	Fixed Pressure Water-spray and Water-mist Fire-extinguishing Systems
8	Automatic Sprinkler Systems
9	Fixed Fire Detection and Fire Alarm Systems
10	Sample Extraction Smoke Detection Systems
11	Low Location Lighting for Escape Routes
12	Fixed Emergency Fire Pumps
13	Arrangement of Means of Escape on Passenger Ships
14	Fixed Deck Foam Systems
15	Inert Gas Systems

Alternative Design and Arrangements for Fire Safety

Each regulation in Parts B, C, D, E, and G of the drafted new Chapter II-2 contains prescriptive requirements. However, designers and/or owners of ships may wish to use a new fire safety design and/or arrangements that are not explained in the regulations but may have a fire safety level equal to or better to that specified by the prescriptive requirements. Therefore, FP has developed a regulation (new Chapter II-2 Part F Regulation 17) for evaluation of the level of fire safety provided by alternative design and arrangements. The principle is specified in the draft Regulation 17 as follows:

• The designer may deviate from the prescriptive requirements in the regulations and design the fire safety measures differently to meet the fire safety objectives defined in Regulation 2 or other regulations in Parts B, C, D, E, and G of Chapter II-2.
• The designers, when deviating from the prescriptive requirements, shall provide proof that the alternative design or arrangements have a level of safety equivalent to that achieved by a design in accordance with the prescriptive requirements. To prove the safety level, an engineering analysis shall be carried out.

The fundamentals of such an engineering analysis are also specified in the draft regulation 17. It is required that:

• The proposal states, which prescriptive requirements are not used.
• The proposal specifies fire safety performance and its criteria addressed by the prescriptive requirements.
• Details of alternative design and arrangements, and the assumptions used in the design, are described that may lead to operational restrictions of the ship (such as limiting operation under certain conditions).
• Proof is given that the alternative design and arrangements meet the specified fire safety performance and its criteria.

To specify details of the method of proof, draft "Guidelines for proving equivalence of alternative design and arrangement to SOLAS Chapter II-2" are being developed by CG. These guidelines will specify a detailed method for the aforementioned steps. One of the key issues is that an established technology such as ISO/TR 13387 [4] or SFPE Engineering Guide [5] may be used for this purpose.

11.4.3 Fire Test Procedures Code (FTP code)

In 1996, IMO FP developed the Fire Test procedures Code (FTP Code) [6] which contains fire test procedures for fire safe constructions and materials used on board of ships. This Code has become a mandatory instrument by regulations in SOLAS Chapter II-2 revised in 1996 and put into effect on July 1, 1998. This test system and related regulations are moved into the new Chapter II-2 without any change. Table 11.33 shows the contents of the FTP Code.

Some of the Parts describe the test procedures to be used. However, many of the Parts refer to ISO standards or IMO test standards that have been already established. Table 11.34 shows the referred standards in conjunction with each Part of the FTP Code.

Table 11.33   Contents of the FTP Code

- Scope
- Application
- Definitions
- Testing
- Approval
- Products that may be installed without testing and/or approval
- Use of equivalents and modern technology
- Period of grace for other test procedures
- List of references

Annex 1: Fire test procedures
Part 1 – Non-combustibility test
Part 2 – Smoke and toxicity test
Part 3 – Test for "A", "B", and "F" class divisions
Part 4 – Test for fire door control systems
Part 5 – Test for surface flammability
Part 6 – Test for primary deck coverings
Part 7 – Test for vertically supported textiles and films
Part 8 – Test for upholstered furniture
Part 9 – Test for bedding components

Annex 2: Products that may be installed without testing and/or approval
Annex 3: Use of other test procedures

Table 11.34   Test Procedures of FTP Code

FTP Code	Type of test	Referred test method	Similar test method
Part 1	Noncombustibility Test	ISO 1182:1990	–
Part 2	Smoke and Toxicity Test	ISO 5659-2	–
Part 3	Fire Resistance Test for Fire Resistant Divisions	IMO A.754(18)	ISO 834–1
Part 4	Fire Resistance Test for Fire Door Closing Mechanisms	–	–
Part 5	Surface Flammability Test	IMO A.653(16) IMO A.687(17)	ISO 5658-2
Part 6	Test for primary Deck Coverings	IMO A.653(16)	ISO 5658-2
Part 7	Flammability Tests for Curtains and Vertically Suspended Textiles and Films	IMO A.471(XII) IMO A.563(14)	–
Part 8	Test for Upholstered Furniture	IMO A.652(16)	–
Part 9	Test for Bedding Components	IMO A.688(17)	–

Noncombustibility Test (FTP Code Part 1)

Materials used for fire resistant divisions of ships must be noncombustible. According to Annex 2 of the FTP Code, products made only of glass, concrete, ceramic products, natural stone, masonry units, common metals and metal alloys are considered noncombustible and may be installed in ships without testing and approval. Other materials that need to be classified as noncombustible shall be tested according to Part 1 of FTP Code which refers to ISO 1182/1990 as test procedure. The pass-fail criteria for marine use are different from those specified in ISO 1182. The criteria are:

- The average temperature rise of the furnace thermocouples shall not exceed 30 K.
- The average temperature rise of the surface thermocouples shall not exceed 30 K.
- The mean duration of sustained flaming shall not exceed 10 s.
- The average mass loss shall not exceed 50 %.

Further details of the test apparatus and test specifications are available under Section 10.23 ISO.

Fig. 11.14 shows a view of the ISO 1182 noncombustibility test.

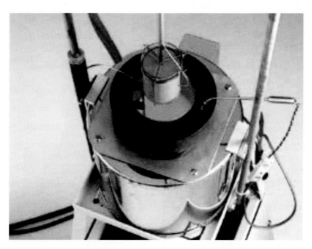

Figure 11.14   ISO 1182 Noncombustibility test

Smoke and Toxicity Test (FTP Code Part 2)

According to the requirements of SOLAS Chapter II-2, interior surface (deck, wall, and ceiling) materials shall not be capable of producing excessive quantities of smoke and toxic effluents. For this purpose, products to be used in interior of ships shall be tested and approved according to the smoke and toxicity test of Part 2 of the FTP Code. The specimen shall be tested according to ISO 5659–2/1994 and the specific optical density shall be measured under three test conditions:

- Irradiance of 25 kW/m^2 in the presence of a pilot flame,
- Irradiance of 25 kW/m^2 in the absence of a pilot flame,
- Irradiance of 50 kW/m^2 in the absence of a pilot flame.

When the optical smoke density reaches the maximum value, gases in the smoke chamber shall be sampled and the concentration of defined gas components measured. Measurement techniques for the gas concentrations are not specified, but the procedure shall be calibrated with a traceable method. Table 11.35 shows the requirements.

Table 11.35 Requirements regarding smoke and toxic gas generation

Product application	Smoke generation	Toxicity	
		Gas component	max. acceptable concentration
	Average of max. optical density not to exceed D_s	CO	1450 ppm
Surface of wall and ceiling	200	HBr	600 ppm
		HCl	600 ppm
Surface of floor	500	HCN	140 ppm
		HF	600 ppm
Primary deck covering (the first layer applied to the steel deck plate)	400	SO_2	120 ppm
		NO_x	350 ppm

Recognizing that the evaluation of toxicity using this test is not fully based on fire safety engineering and that such techniques are under development within ISO/TC 92, the test method and the criteria have been adopted as an interim solution for SOLAS. In the future, the test method and the criteria will be reviewed. Figure 11.15 shows a test apparatus according to ISO 5659–2, which is referred to in Part 2 of FTP Code. Further details are given in this book in Chapter 15.

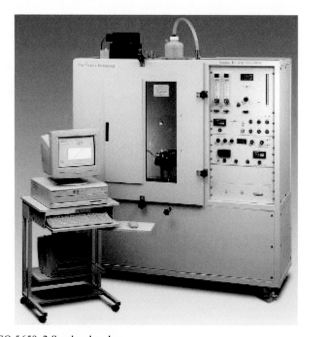

Figure 11.15 ISO 5659–2 Smoke chamber

Test for Surface Flammability (FTP Code Part 5)

Fire safety regulations in SOLAS require interior surface materials to show "low flame spread." This needs to be verified in accordance with Part 5 of the FTP Code, which has a reference to IMO Assembly Resolution A.653(16) as a test method.

The test method in IMO A.653 is almost identical to ISO 5658–2. The difference is that IMO A.653 requires an additional measurement of heat release using thermocouples. A rectangular box type stack is fitted above the specimen position. Combustion fume flows into the stack. Five thermocouples are fitted at the outlet of the stack and measure the temperature of the exhaust gases. The signal of the thermocouples is converted into heat release rate.

The specimen [155 mm x 800 mm (6.1 in. x 31.5 in.)] is supported vertically with the longer edges in horizontal orientation and is exposed to a heat flux generated by a gas-fueled radiant panel heater. The maximum heat flux imposed to one end of the specimen is approx. 50 kW/m². The heat flux decreases along the specimen.

Flame spread time along the specimen is measured in 50 mm (1.97 in.) intervals. Heat for sustained burning (kJ/m²) is the average of the flame spread time (s) multiplied by the incident heat flux (kW/m²) at the respective position on the specimen.

The maximum flame spread distance is used to determine the Critical Heat Flux at Extinguishment (CFE) which is the irradiance at the point of the specimen where the flame front ceased its progress.

The principle of the apparatus is shown in Fig. 11.16. Table 11.36 shows the criteria for interior surface materials. Part 5 of FTP Code provides interpretations by which peculiar phenomena of specimen can be dealt with. Figure 11.17 shows an example of a test in IMO A.653.

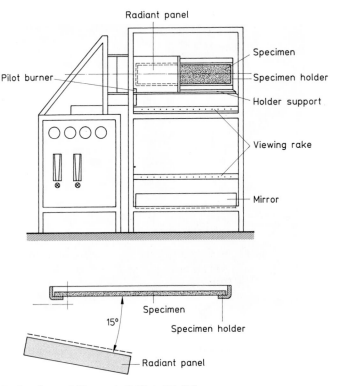

Figure 11.16   Surface flammability test in IMO A.653 (16)

Table 11.36   Criteria for surface flammability

Parameter	Ceiling and wall surface	Deck surface
Critical Heat Flux at Extinguishment (kW/m^2)	≥ 20.0	≥ 7.0
Heat for Sustained Burning (MJ/m^2)	≥ 1.5	≥ 0.25
Maximum Heat Release Rate (kW)	≤ 4.0	≤ 10.0
Total Heat Release (MJ)	≤ 0.7	≤ 2.0

Figure 11.17   IMO A.653 Test (PVC wall surface)

FTP Code Part 5 also specifies that the test method in ISO 1716 is recommended for determining the gross calorific value if there is a requirement regarding the maximum gross calorific value for surface materials (see Section 10.23 ISO).

Test for Primary Deck Coverings (FTP Code Part 6)

Steel deck plates usually have many welding beads and the surface is uneven. "Primary deck covering" is a coating system that is applied directly on the steel deck plate to make the surface flat enough to install floor coverings such as floor tiles, flooring coating, and carpets.

A primary deck covering is required, by the fire safety regulations of SOLAS, to meet the criteria of Part 6 of the FTP Code that specifies that the primary deck covering shall be tested in accordance with IMO A.653(16) and has to satisfy the criteria for floor surfaces of Part 5 of the FTP Code. Specimens shall be made in a way that the primary deck covering is applied to a steel plate with a thickness of 3.0 mm $\pm$ 0.3 mm (0.12 in. $\pm$ 0.012 in.).

Test for Vertically Supported Textiles and Films (FTP Code part 7)

Draperies, curtains, and other suspended textile materials in rooms of restricted fire risk shall behave better or equal to a wool fabric of a mass/area of 0.8 kg/m^2 regarding propagation of flame. To verify this, Part 7 of the FTP Code is used.

Part 6 of FTP Code refers to IMO Assembly Resolution A.471(XII) and A.563(14), which specify a flammability test method for vertically supported textiles and films. The heat source for this test is the German small burner specified in Section 10.9 Germany. The test procedure is slightly different from the DIN 4102 Part 1 test and described in ISO 6940 and ISO 6941.

The classification is based on the following criteria:

- Afterflame time (duration of flaming of specimen after removal of the burner) ($\leq 5\,$s),
- Occurrence of surface flash or flame propagating $> 100\,$mm (3.9 in.),
- Length of charring of the specimen $\leq 150\,$mm (5.9 in.),
- Ignition of cotton wool laid below the specimen.

Test for Upholstered Furniture (FTP Code Part 8)

Upholstered furniture and seating in rooms of restricted fire risk and in high-speed craft have to meet the requirements of Part 8 of the FTP Code regarding resistance to the ignition and propagation of flames. Part 8 of the FTP Code refers to IMO Assembly Resolution A.652(16) for the test method and criteria. The test method in Part 8 of FTP Code was developed based upon BS 5852.

Surface fabric and filler are attached to a test rig 450 mm (17.7 in.) wide, 150 mm (5.9 in.) deep and 300 mm (11.8 in.) high. An ignited cigarette and a small flame of a propane gas burner are applied to the junction of seating part and back part of the test setup. No smoldering or flaming shall propagate beyond the vicinity of the applied ignition source. Details of test rig and specifications are shown in Chapter 13, Section 13.3.2.

Test for Bedding Components (FTP Code Part 9)

Bedding components, such as mattresses, pillows, and blankets, in rooms of restricted fire risk shall meet the requirements of Part 9 of the FTP Code regarding resistance to the ignition and propagation of flames. Part 9 of the FTP Code refers to IMO Assembly Resolution A.688 (17) for the test method and the criteria.

The specimen size is 450 mm x 450 mm x thickness as used. A mockup of the mattress shall have the actual structure except reduced length and width. An actual pillow can be tested. Blankets should be cut to the size of 450 mm x 450 mm. Test rig and detailed specifications are described in Chapter 13, Section 13.3.2.

An ignited cigarette and a small flame from a propane gas burner are applied to the top surface. The cigarette shall be covered by a piece of cotton wool. No smoldering or flaming shall propagate beyond the vicinity of the applied ignition source.

Test for "A," "B," and "F" Class Divisions (FTP Code Part 3)

One of the fundamental requirements regarding fire safety of ships in SOLAS is "containment and extinction of any fire in the space of origin." To meet this requirement, the interior of a ship shall be divided into "Main Vertical Zones" every 40 m (131.2 ft) or less by steel walls (bulkhead). Most of the decks shall be constructed by steel. These decks and bulkheads shall have a fire resistance of 60 min. Materials or systems other than steel may be used for bulkheads and decks, if the material or system is made of noncombustible materials and has the same degree of fire resistance as a steel construction. Other boundaries such as walls and ceilings of cabins, rooms, and corridors shall have, in general, a resistance to fire of 30 min.

The grade of fire resistance of these bulkheads, decks, ceilings, and walls shall be verified by a standard fire resistance test specified in FTP Code Part 3 which refers to IMO Assembly Resolution A.754. This test method was developed in the early 1990s based on ISO 834. The major difference between IMO A.754 and ISO 834 is the specimen size. In IMO A.754, the dimensions of wall and bulkhead specimen are about 2.5 m x 2.5 m (8.2 ft x 8.2 ft) and the dimensions of deck and ceiling specimen are about 2.5 m x 3.0m (8.2 ft x 9.8 ft) The specimen shall be installed in a rigid frame and attached to a test furnace. The time-temperature curve of IMO A.754 is the same as that in ISO 834 and shown in Fig. 11.18. The furnace chamber is illustrated in Fig. 11.19.

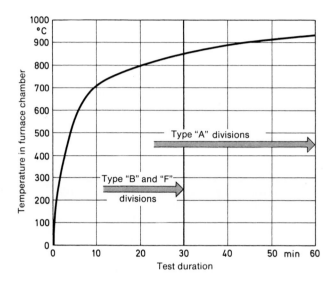

Figure 11.18   IMO standard time-temperature curve

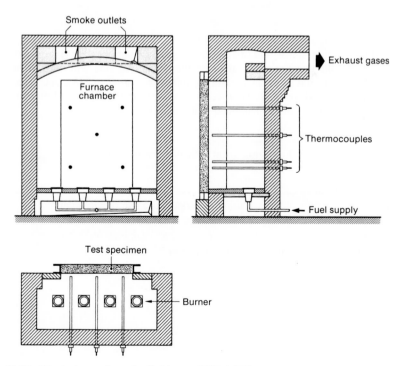

Figure 11.19   Fire test procedures for divisions to IMO A.754

Doors, windows, ducts (including dampers), and other penetrations such as cable transits and pipe trunks shall have the same fire resistance as the boundaries to which these penetrations are fitted.

IMO A.754 specifies details of type and positions of thermocouples to be applied to the specimen, as well as criteria for fire resistance classification. In general, constructions with a fire resistance of 60 min are defined as class A. If the fire resistance can be shown for 30 min, the classification is class B.

Class A divisions shall be capable of preventing the passage of flame and smoke for at least 1 h in the fire resistance test. In addition, for class A divisions, the average temperature of the unexposed side may not rise more than 140K on average, nor may the temperature at any measuring point rise more than 180K above the original temperature. This is normally achieved by insulating the construction with approved non-combustible materials. For all class A constructions (structural integrity against smoke and flames for one hour), the following classes have been defined, using the criteria on temperature rise of the unexposed side of the test specimen:

Class	Time where temperatures do not rise above limit
A-60	60 min
A-30	30 min
A-15	15 min
A-0	0 min (no temperature rise requirement)

Similarly, class B divisions shall be capable of preventing the passage of flame and smoke up to the end of half an hour during the fire resistance test. In addition, class B divisions shall be insulated such that the average temperature on the unexposed side will not rise more than 140K, nor will the temperature at any measuring point rise more than 225K above the original temperature. Class B divisions are classified in relation to the capability of maintaining the temperature rise within the above mentioned limits as follows:

Class	Time where temperatures do not rise above limit
B-15	15 min
B-0	0 min (no temperature rise requirement)

Fig. 11.20 shows an example of the fire resistance test.

Figure 11.20   Fire resistance test for a class B door (IMO A.754)

Test for Fire Door Control Systems

Doors in main vertical zone boundaries shall meet the criteria for class A. Sometimes the door is large and operated pneumatically or hydraulically. A fire safety regulation in SOLAS requires that the door operating system shall be operable up to a temperature of 200°C and shall be capable of keeping the door closed at a temperature above 200°C. Part 4 of the FTP Code specifies the test method for fire door operating systems. The fire test is conducted in a furnace for fire resistance tests.

11.4.4 Other Test Methods for High-Speed Craft

High-speed passenger crafts have been operated in commercial service routes since the 1980s. Hover crafts and hydrofoil crafts were the first generation of high-speed passenger craft. In the early 1990s, there was a demand to develop larger high-speed passenger crafts that could also carry cars (high-speed passenger ferry). The service speed exceeded 40 knots, and the capacity exceeded 500 passengers. The safety of such large high-speed passenger ferries became a matter of public concern. Consequently, IMO developed an international code for safety of high-speed craft (HSC Code) and introduced it as mandatory instrument alternative to SOLAS in 1996. Lightweight construction is the most important design goal for high-speed craft. Fire safety construction is also required to be lightweight. Therefore, the HSC Code allows using fire resistant boundaries that may be constructed from materials, which are noncombustible and restrict spread of fire by two ways:

- The construction has to meet the requirements of FTP Code Part 3, and
- The materials used do not create a flashover in case of fire.

The latter characteristics are verified by IMO Maritime Safety Committee Resolution MSC.40(64), which specifies that the ISO 9705 room corner test shall be used as test method (for details, see Section 10.23 ISO) . The classification criteria are defined as follows:

- The average heat release rate (HRR) (excluding the HRR from the ignition source) does not exceed 100 kW.

- The maximum HRR averaged over any 30 s period of time during the test (excluding the HRR from the ignition source) does not exceed 500 kW.
- The average smoke production rate does not exceed $1.4\,m^2/s$.
- The maximum value of the smoke production rate averaged over any 60 s period of time during the test does not exceed $8.3\,m^2/s$.
- Flame spread must not reach any further down the walls of the test room than 0.5 m (1.6 ft) from the floor excluding the area, which is within 1.2 m (3.9 ft) from the corner where the ignition source is located.
- No flaming drops or debris of the test sample may reach the floor of the test room outside an area which is within 1.2 m (3.9 ft) from the corner where the ignition source is located.

In addition, seats in high-speed crafts governed by HSC Code shall comply with FTP Code Part 8.

References for Section 11.4

[1] 1974 International Convention for the Safety of Life at Sea (74 SOLAS) and amendments thereto
[2] Kaneko, F., Ohta, S., Dai, X., Fukumoto, M.: Probabilistic Safety Assessment Method of Ships: Part 2: A Risk Evaluation Methodology for FSA, No. 6–3, Vol. 186 Journal of the Society of Naval Architects of Japan 1999
[3] Yoshida, K.: New Fire Safety Code for Shipping, Flame Retardants 2000 Conference, London, February 2000
[4] ISO/TR 13387-1 through 13387-8 Fire safety engineering. 1999, ISO/TC92/SC4
[5] SFPE Engineering Guide to Performance-Based Fire Protection Analysis and Design of Building, Society of Fire Protection Engineers and National Fire Protection Association, USA, 1999
[6] International Code for Application of Fire Test Procedures (FTP Code), IMO, 1996

12 Electrical Engineering

B. MÜLLER

Significant technological advances in electrical engineering would not have been possible without the contribution of plastics. Plastics are well known for their durability, corrosion resistance, strength, and outstanding insulation properties. They are therefore the material of choice for many applications in electrical and electronic engineering such as appliances, consumer and office equipment, electrical parts, and cables. The diversity of polymers available as well as their aesthetic design qualities give the development engineer, designer, and manufacturer of electrotechnical products tailor made solutions to meet a variety of needs from function and aesthetics through safety to technical and environmental performance.

In 1999 the world total production of plastics in the electrical machinery / electronics industry amounted to 2.1 billion US $ [1]. In Western Europe, the electrical and electronic sector accounted for 9 % (2.196 million tons) of plastics consumed in 1996 [2].

Although familiar products such as televisions, household appliances, personal computers, and telephones account for 59 % of plastics consumption, there is a less visible yet vital world of wires and cables that accounts for the remaining 4 %. The use of plastics in the electrical and electronics (E&E) sector is forecast to grow faster than that of any other material.

It cannot be denied, however, that the use of plastics in electrotechnical products and cables entrains special fire precautions, because fire hazard is inherent in any energized circuit. The primary objective of component circuit and equipment design and the choice of materials are to minimize the risk of electrically induced ignition, even in the event of foreseeable abnormal use, malfunction, or failure. If, despite precautions, ignition and fire do occur, that fire should be controlled within the bonds of the equipment's enclosure. Potential ignition sources within electrical equipment result primarily from overheated current carrying parts, electrically over-loaded parts or components, overheated terminals due to loosening and faulty installation of connections (bad connections), sparks, and arcs. Like all other organic substances, plastics are combustible. Therefore, plastics are a potential fire hazard in electrical equipment. Flames may develop and impinge on adjacent combustible materials, if, for example, a fault current flows over a tracking path on the surface of the insulation or a bad connection causes insulation to ignite. Special fire tests described in the following paragraphs have been developed to simulate as closely as possible actual ignition risks.

In cases where surfaces of E&E products such as large enclosures, cables, and cable management systems are exposed to an external fire, it has to be made certain that they do not contribute more to fire growth than other combustibles, building components, or structures in the immediately surrounding areas.

Electrical equipment can be operated only if it complies with regulatory requirements. The same is true for plastics used in electrical insulation or equipment housings. In many applications, the polymers used and/or the final component made from them have to meet one or more flammability standards. Historically, the elaboration of standards for fire testing of electrical equipment was the responsibility of national organizations or regional standardization commissions. Numerous fire test methods have been developed, which often differed from country to country. Depending on the test method, the relevant importance of the fire parameters being assessed is different. In standardization, be it on a national or international level, there are two philosophies of testing.

Side by side with hazard-oriented end-product testing, which is based on the view of fire hazard technologists that only the as-operated assembly in the as-installed state can provide a valid hazard assessment, preselection testing of standard specimens is being demanded. These materials tests are often small, bench-scale tests, which examine the reaction-to-fire of standardized specimens under defined conditions. Mostly, they are used to provide data on combustion characteristics, such as flammability, ignitability, and flame spread rate. The results of preselection tests on materials are generally quoted as numerical values or as classifications in flammability classes. Flammability classes based on the results of preselection tests on materials are often cited in materials specifications. Data of preselection tests alone are insufficient for a prediction of the performance of the plastics material in the final end-product, unless that small-scale test has been validated by a fire hazard test, which reflects that real fire scenario. The interaction of all related material and design factors can be investigated only by fire simulation tests on assemblies or final electrotechnical products or parts taken out of them.

However, combustion characteristics tests can be quite useful in product development and quality control.

The data from these tests can help the plastics industry in comparing the performance of different materials, particularly in research or material development or assist the development engineer to select materials in the designing process of an electrotechnical product. In quality control, material tests serve to evaluate the uniformity of a particular property of a product, in this case, the flammability.

A great deal of work has been done in the last two decades by international standard bodies, the International Electrotechnical Commission (IEC) or the European Electrotechnical Standardization Commission (CENELEC), to develop small- and large-scale tests for electrotechnical products and cables that more accurately model actual fire conditions. The real conditions of use including the surroundings of an electrotechnical product are simulated as closely as possible, and the design of the test procedure is related to the actual risk. Special test ignition sources have been developed to simulate the conditions of a fault, improper use, or exposure to an external fire. This concept of fire hazard testing is well accepted by the international and national standardization committees. Special fire testing facilities have been set up by plastics manufacturers, and testing and certification institutes to carry out large-scale tests without polluting the environment [3–5]. Owing to the increased need for a worldwide free movement of electrotechnical equipment, national standards and classification systems for fire hazard assessment in the E&E sector will be increasingly replaced by harmonized systems. Advances in the understanding and use of fire safety engineering will have a positive impact on the standardization process and can help to integrate the two philosophies of preselection testing on materials and hazard oriented end-product testing into an overall system approach to fire hazard assessment.

12.1 International Fire Regulations and Standards (IEC)

12.1.1 Objectives and organization

Harmonization diminishes trade barriers, promotes safety, allows interoperability of products, systems and services, and promotes common technical understanding. The development of international standards is the task of such organizations as the International Organization of Standardization (ISO) and the International Electrotechnical Commission (IEC). ISO is responsible for all technical fields with the exception of electrical technology. International electrical standards are established by the IEC which was founded in 1906 [6, 7].

The objective of the IEC is to promote international cooperation on all questions of standardization and related matters in the fields of electrical and electronic engineering. The IEC is composed of 60 national committees (members and associate members) representing all the industrial countries in the world. The preparation of standards, technical specifications, and technical reports is entrusted to technical committees (TCs).

To promote international understanding, IEC National Committees undertake to apply IEC International Standards as often as possible instead of national standards. IEC cooperates with numerous other international organizations, particularly with ISO, and in Europe with CENELEC. The supreme authority of the IEC is the Council. The council delegates the management of technical work to the Committee of Action. The Advisory Committee on Safety (ACOS) supervises the coordination of IEC safety standards and their coherence. It assigns to certain TCs a "safety pilot function" concerning safety standards or a "safety group function" (preparation of safety rules to a given sector). More than half of the 3,500 IEC publications are safety related. The Advisory Committee on Environmental Aspects (ACEA) coordinates and focuses the IEC efforts to ensure that IEC product standards do not result in any harm to the environment.

Publicly Available Specifications (PAS) and Industry Technical Agreements (ITA) are products of IEC. PAS is an IEC tool for quickly introducing to the market *de facto* standards in a rapidly growing area of technology. The ITA is a new low-cost procedure to cover fast-moving technologies. ITAs are intended to be used by industry for launching products on the market, when business and trade in high technology products and services do not need consensus-based International Standards. IEC publications are known as International Standards, Technical Specifications and Technical Reports. Full information on published standards is available on the IEC Web site http://www.iec.ch [8].The most important technical committees concerned with fire are summarized in Appendix 6 of this book.

12.1.2 Fire Hazard Testing, IEC/TC 89

IEC Technical Committee (TC) No. 89, Fire hazard testing, was formed in 1988 by a decision of the IEC Council to change subcommittee 50D into a full technical committee. TC 89 is a horizontal committee with a pilot safety function within the IEC to give guidance and develop standards related to fire hazards for use by other IEC Product Committees responsible for finished components or electrotechnical products [9].

The scope of TC 89 is to prepare international standards, technical specifications, and technical reports in the areas of:

- Fire hazard assessment, fire safety engineering, and terminology as related to electrotechnical products,
- Measurement of fire effluents (e.g., smoke, corrosivity, toxic gases, and abnormal heat), and the review of the state of the art of current test methods as related to electrotechnical products,
- Widely applicable small-scale test methods for use in product standards and by manufacturers and regulators.

At present, the work of TC 89 is executed through three working groups (WGs) consisting of experts from 14 participating countries:

WG 10 – General guidance, hazard assessment, fire safety engineering and terminology,
WG 11 – Fire effluents (smoke, heat, corrosive and toxic gases),
WG 12 – Test flames and resistance to heat. Small-scale heat and flame test methods.

Standardization is expensive and time-consuming. Wherever possible, IEC works with other standards organizations. To support and advise IEC/TC 89, various active liaisons to other IEC and ISO Technical Committees exist. These liaisons are listed in Table 12.1.

Table 12.1 IEC/TC 89 liaisons

IEC	ISO	Title
TC 10	–	Fluids for electrotechnical applications
TC 20	–	Electric cables
SC 23 A	–	Cable management systems
TC 61/WG 4	–	Safety of household and similar electrical appliances
TC 99	–	System engineering and erection of electric power installations in systems with normal voltages above 1 kV A. C. and 1.5 kV D. C. particularly considering safety aspects
–	TC 61/SC 4	Plastics – Burning behavior
–	TC 92	Fire safety
–	TC 92/SC 1	Fire initiation and growth
–	TC 92/SC 3	Threats to people and environment
–	TC 92/SC 4	Fire safety engineering

12.1.2.1 Fire Hazard Assessment

The risk of fire needs to be considered in any electrical circuit. With regard to this risk, the objective of component circuit and equipment design and the choice of material are to reduce the likelihood of fire even in the case of foreseeable abnormal use, malfunction, or failure.

IEC/TC 89, in its basic guidance document IEC 60695-1-1, provides an overall framework for assessing the hazard that an electrotechnical product poses to life, property, and the environment in a fire situation [10]. It is intended to provide guidance to IEC product committees. This standard makes use of the many elements of fire safety engineering principles outlined in ISO 13387, Parts 1–8 [11] and outlines a process to identify appropriate fire test methods and performance criteria for electrotechnical products based on fire hazard.

In the context of the standard, electrotechnical products are materials, components and subassemblies as well as the complete, fully assembled end-use products.

The basis of the process in IEC 60695-1-1 is to identify the class or range of the product to which the assessment applies, the kind of fire scenarios that will be associated with the product, and to define the fire safety objectives and acceptable levels of safety. The standard includes guidance on how to incorporate preselection tests on materials, and tests on end-products, service conditions, the fire scenario and the environment in the overall hazard assessment.

In IEC 60695-1-1 Annex B, an example of hazard assessment on a hypothetical installation of rigid plastic conduit is demonstrated to illustrate the hazard assessment process and provide guidance on how to use results of materials tests and fire growth models for assessing whether products are likely to provide any significant contribution to the overall fire hazard.

A concept of determining the electrical product fire hazard, including the principles of IEC 60695-1-1, and the assessment of probabilistic and deterministic indices characterizing the resistance of products to an exposure to fire hazardous electrotechnical factors as well as their stochastic trends has been worked out by Smelkow and Pekhotikov [12].

Details of a fire hazard assessment of light diffuser panels carried out according to the principles developd in ISO/TR 11696, Part 2 [13] are given in [14].

Electricity is a potential fire hazard. Fire statistics demonstrate that one of the primary root causes of fire is electrical. In particular, the benefits of using plastics in E&E appliances go hand in hand with the need to manage the risk of fire. Fires can result from faulty product design, but are often the result of misuse or abuse of the appliance. Educating the consumer to, for instance, remove the dust from the appliance backplate, not to place candles or smoking materials on the appliance, and to avoid carrying out do-it-yourself repairs is a long-term process. The increased use of plastics in E&E appliances over the years has become possible only by the parallel development and application of flame retardants. Plastics containing flame retardant materials or specific, inherently flame retardant polymers can meet all fire safety requirements foreseen for combustible materials.

One basic safety objective is to interfere with the course of a fire. The course of a fire always follows the same pattern consisting of the initiating, fully-developed and decreasing fire stages (see Chapter 3 of this book).

The main "reaction-to-fire" parameters and secondary fire effects determining the course of a fire and, particularly, the phase of the initiating fire, are:

Reaction-to-fire parameters	Secondary fire effects
Combustibility Ignitability Spread of flame Heat release	Smoke development Toxic potential of fire effluents Corrosive potential of fire effluents

Fire protection measures for combustible materials and electrotechnical products originating from them can apply only to the early stage of a fire, That is, the initiating fire phase, by preventing ignition first as a fire safety objective, and if ignition occurs, by preventing or reducing fire propagation and minimizing heat release. Common ignition phenomena in electrotechnical products are described in Table 12.2.

According to UL 746C Section 3.30 (see Section 12.2.1 this chapter), a risk of fire is considered to exist at any two points in a circuit where:

- The open circuit voltage is > 42.4 V peak and the energy available in the circuit under any condition on load, including short circuit, results in a current of 8 A or more after 1 min of operation, or
- A power of more than 15 W can be delivered into an external resistor connected between the two points.

Certain electrotechnical products such as large enclosures, insulated cables, and conduits may cover large portions of surfaces and finishing materials of building construction or may penetrate fire-resisting walls. In these circumstances, electrotechnical products, when exposed to an external fire, shall be evaluated from the standpoint of their contribution to the fire hazard in comparison to the building materials or structures without the installation of those electrotechnical products. Therefore, the secondary fire safety objective is that in those cases, electrotechnical products do not contribute to the fire growth to a greater extent than the building products, items, or structures in the immediately surrounding areas.

The fire hazard of an electrotechnical product depends on its characteristic service conditions and the environment in which it is used. Therefore, a fire hazard assessment procedure for a particular product shall describe the product, its conditions of operation, and its environment.

Whether the focus of assessment is a material, product, or system is determined by an investigation in terms of the contribution to the assumed scenario:

- Is the product likely to be the source of ignition?
- Is the product likely to be the secondary ignited item?
- Is the product a potentially significant fuel source, even if it is not the first item ignited?
- What is the potential avenue to contribute to the risk (and hazard)?
- How close are occupants and/or critical equipment to the origin of the fire?

Table 12.2   IEC 60695-1-1: Common ignition phenomena in electrotechnical products

Phenomenon	Origin
Abnormal temperature rises	• Overcurrent in a conductor • Bad connections • Leakage currents • Failure of a component, an internal part or an associated system (e.g., ventilation) • Mechanical distortions that modify electrical contacts or the insulation system • Premature thermal aging
Short-circuit	• Direct contact of conducting live parts at different potentials (loosening of terminals, disengaged conductors, ingress of conducting foreign bodies, etc.) • Gradual degradation of some components causing changes in their insulation impedances • After sudden failure of component or internal part
Accidental sparks and arcs	• Cause external to the equipment (e.g., overvoltage to the system network) • Internal cause (on-off switching with gradual degradation of some components and ingress of moisture) • After sudden failure of a component or an internal part

Source: IEC 60695-1-1, Table 1.

Table 12.3   Fire parameters related to classification and testing

Developing fire	Reaction to fire	Ignitability Flame spread Heat release	**Smoke**
Fully developed fire	Resistance to fire	Load-bearing insulation and integrity capacity, e.g., fire resistance of protective systems for cables; circuit integrity of cables	**Toxicity** **Corrosivity**

Testing and calculations (see Chapter 8 of this book for fire safety engineering) are used to determine whether the fire safety objectives will be met. For the fire hazard assessment procedure to be valid, it is necessary that the characteristic fire test responses used produce valid estimates of success or failure in achievement of the fire safety objectives given in the specified fire scenario (see Table 12.3). A schematic representation of the principles of a fire hazard assessment procedure of IEC 60695-1-1 is shown in Fig. 12.1. In fire safety engineering guidelines such as IEC 60695-1-1, this flowchart may be followed; in national or international fire test standards for electrotechnical products this strategy is at the time being only partly taken into account.

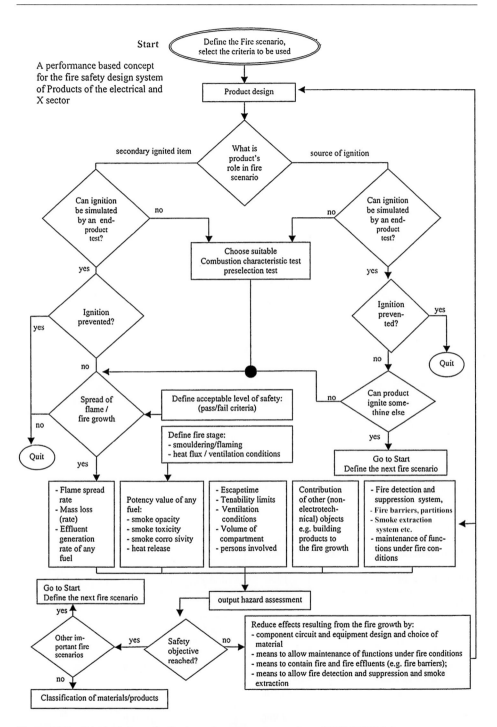

A performance based concept for the fire safety design system of Products of the electrical and X sector

Figure 12.1   Principal steps of a fire hazard assessment procedure of IEC 60695-1-1

12.1.2.2 Fire Statistics

Only if detailed fire statistics are kept can a reliable picture of fire hazards and their effects emerge and result in specific safety measures taken to minimize them. Some countries, especially the United Kingdom and the United States, compile detailed data on both fatal and nonfatal fires. However, the fire statistics vary in terms of detail. It is extremely important that the statistics from the United States and different European countries are studied in detail to determine where they match and where they differ. Fire statistics can be used as an important tool for standardization bodies to obtain information about the cause and sources of ignition, the extent of fire, and the frequency of certain fire types. They can provide interpretation of trends in intentionally set and accidental fires and the interdependence between the type of the fire, the type of building, people's practices, and the casualty rate.

An overview of structural fires reported to fire departments in the United States during the period 1983–1987 that were coded as caused by electrical failure is given in Chapter 2.22, Table 2–3 E of [15]. The reported 90,430 fires represent 11 % of all structure fires in the same period. The statistical analyses shows that nearly all fires involving electrically powered equipment stem from foreseeable or avoidable human error, such as improper installation, lack of maintenance, improper use, carelessness or oversight.

A general view of possible fire hazards in the electrical sector can be gained from the UK Home Office Fire Statistics 1995 [16]. Table 12.4 indicates the causes of all fires in dwellings and other buildings.

Table 12.4   Fire in dwellings and other buildings by cause, United Kingdom, 1995

Cause	Dwelling		Other buildings	
	1995	Percentage	1995	Percentage
Misuse of equipment or appliances	16300	31.7	4600	18.5
Chip/fat pan fires	10100	19.6	600	2.4
Faulty appliances and leads	7500	14.6	5700	23.0
Careless handling of fire or hot substances	5700	11.1	3400	13.7
Placing articles too close to heat	3900	7.6	2000	8.1
Faulty fuel supplies	1900	3.7	2000	8.1
Playing with fire	1100	2.1	300	1.2
Other accidental/unspecified cause	4900	9.5	6200	25.0
Total of accidental fires	**51,500**	**100**	**24,800**	**100**
Malicious fires	**13,500**		**21,400**	
Total	**64,900**		**46,100**	

The main cause of accidental fires in dwellings was the misuse of equipment and appliances, remaining at approx. 32 % of all accidental fires. Chip/fat pan fires are the second highest cause of accidental fires in the home. In other buildings, other factors play a more significant role, such as faulty appliances and cables.

Table 12.5 shows the main sources of ignition for accidental fires in dwellings along with fatal casualties and nonfatal casualties. Cooking appliance fires accounted for more than half of those accidental fires in dwellings and half the nonfatal casualties in accidental dwelling fires.

Table 12.5 Main sources of ignition for accidental dwelling fires, with casualties, U.K., 1995

Source of ignition	Fires	Percentage	Fatal casualties	Non-fatal casualties
Smokers' materials	4886	9.5	145	1607
Cigarette lighters	589	1.1	9	309
Matches	1229	2.4	33	432
Candle	1226	2.4	9	452
Taper, lighted paper, or naked light	292	0.6	6	72
Cooking appliances	27523	53.5	50	5472
Electric cookers	18126	35.2	23	3465
Gas cookers	8107	15.7	27	1855
Space-heating appliances	2653	5.2	64	743
Electric space heating	859	1.7	25	273
Gas space heater	734	1.4	23	217
LPG space heater	235	0.5	4	78
Solid fuel: Fire in grade	530	1.0	12	120
Electrical distribution	2225	4.3	15	290
Plugs	64	0.1	0	19
Sockets and switches	192	0.4	1	33
Leads to appliances	311	0.6	3	58
Wire and cable	1658	3.2	11	180
Other electrical appliances[a]	6155	12.0	29	864
Washing machine	2130	4.1	0	87
Blanket, bedwarmer	784	1.5	17	249
Computer/VDU	11	0.02	0	7
Total accidental	**51,479**	**100**	**446**	**10,855**

[a] These appliances include washing machines, refrigerators, electrical blankets, televisions, tumble dryers, and so forth.

Other electrical appliances, including such items as washing machines, refrigerators, electric blankets, televisions, and tumble dryers, and so forth, accounted for 12 % of the accidental dwelling fires. Smokers' materials were the most frequent source of ignition causing fire deaths. They accounted for 35 % of accidental fire deaths in dwellings.

Other important sources of ignition were smokers' materials (9.5 % of accidental dwelling fires), space heating appliances (5.2 % of accidental dwelling fires) and electrical distribution with 4.3 % of the accidental dwelling fires.

A review of European and US television fire statistics has been carried out by De Poortere, Schonbach and Simonson [17]. The objective of this survey was to identify the relative importance of fires from internal sources due to electrical faults and external ignition sources such as Christmas decorations or due to consumer misuse or lack of maintenance. The statistics suggest that about one third of all television fires are due to external ignition sources.

The number of television fires in Europe is estimated as 100 fires per million televisions per year. Proceeding from the number of television fires in the United States, which is at least an order of magnitude less, the authors recommend to take measures to improve the fire safety of televisions by carrying out a fire hazard assessment taking into account all elements that contribute to fire safety, that is, material fire performance, design, the environment, and consumer education.

An example of the use of fire statistics used to determine the effect of the inclusion of a flame retardant on the size and number of fires associated with a television is discussed in [18]. According to an analysis of fire statistics in 1996, Russia witnessed 293,507 fires, of which electrical fires accounted for 20.5 % [12]. Table 12.6 summarizes the data on fires caused by various types of electrical products and their percentage in terms of overall electrical fires.

Examples of the main causes of fires, fire risks and damages to electrical appliances and installations are given by Arlt [19], Urbig [20] and in [21].

Based on the statistics of the Ministry of the Interior of North Rhine Westphalia in Germany for the period 1994–1996, electrical fires (without lightning) accounted for 5.3 % of the total number of 47,578 fires. In the same analysis, based on 1 million fires for the period 1979–1983, it is shown that, depending on the type of fire, electrical fires accounted for 15 % to 23 % of all fires. The exact estimation of the frequency of electrically induced fires related to the total number of fires is very difficult, because in most cases traces of fire origin and sources are destroyed by the fire.

In 1991, the losses caused by fires in Germany accounted for 6.02 billion DM [22]. According to the German Federal Statistical Agency, in Germany, for the period of 1990–1996, more than 600 persons per year died in fires [23].

Table 12.6   Distribution of fires as to types of products in Russia, 1996

Product type	Number of fires	Percentage
Cable, wire	35,247	58.6
Electrical fire place	6776	11.3
Television	4033	6.7
Electrical hot plate	2054	3.4
Switchboard	3009	5.0

Product type	Number of fires	Percentage
Electrical lamp	1337	2.2
Switch	2160	3.6
Refrigerator	1228	2.0
Electrical bell	725	1.3
Transformer	957	1.6
Tape recorder, receiver	567	0.9
Electrical iron	460	0.8
Automatic switch	405	0.7
Electrical washing device	301	0.5
Other products	825	1.4
Total of electrical fires	**60,134**	**100**
Total of fires in Russia	**293,507**	

12.1.2.3 Fire Hazard Assessment Tests for Electrotechnical Products

The characteristic fire test responses for electrotechnical products or their components, or the materials that may be required for such an assessment, are ignition and flame spread, currently investigated in a set of small-scale fire tests. Realistic ignition sources are used for the fire scenario to be investigated, and are applicable to a variety of potential fire scenarios, to ignite a specimen.

The IEC/TC 89 guidance publications and their equivalent EN Standards are summarized in Table 12.7.

Table 12.8 contains the relevant standards for the fire hazard assessment tests for end-products and the preselection tests for materials.

Table 12.7   IEC/TC 89   Guidance publications

Reference IEC 60695-	Reference EN 60695-	Description	Subject	End-product test	Preselection test for materials
1-1	1-1	General guidance	Fire hazard	X	X
1-2	–	Electronic components		X	
1-30	1-30	Use of preselection procedures			X
1-3[a]	–	Preselection procedures			X
2-10	2-10	Glow-wire apparatus	Flammability	X	X
5-1	5-1	General guidance	Corrosion damage	X	X
5-2	–	Damaging effects of fire effluent – Summary and relevance of test methods		X	X
6-1	6-1	General guidance	Smoke obscuration	X	X
6-1	6-1	Summary and relevance of test methods		X	X
6-2	–			X	X
7-1	7-1	General guidance	Toxic hazard	X	X
7-3	–	Use and interpretation of results		X	X
7-4[a]	–	Guidance on unusual toxic effects		X	X
7-51	–	Guidance on the calculation and interpretation of test results			X
8-1	8-1	General guidance	Heat release		X

Reference IEC 60695-	Reference EN 60695-	Description	Subject	End-product test	Preselection test for materials
8-2	8-2	Guidance on test methods	Heat release	X	X
9-1	9-1	General guidance	Spread of flame	X	X
9-2	–	Guidance on test methods		X	X
1-40	–	Insulating liquids	Fire hazard of insulating liquids	X	X
-4	–	Terminology concerning fire tests			

a) To be withdrawn

Table 12.8 IEC/TC 89 Fire hazard assessment tests for end-products and preselection tests for materials

Reference IEC 60695-	Reference EN 60695-	Description	Subject	End-product test	Preselection test for materials
2-11	2-11	Glow-wire end-product test and guidance	Flammability	X	
2-12	2-12	Glow-wire flammability test on materials			X
2-13	2-13	Glow-wire ignitability test on materials	Ignitability		X
2-2	2-2	Needle-flame test	Flammability	X	
2-3	–	Bad-connection test with heaters		X	
2-4/0	2-4/0	Diffusion type and premixed type flame test methods		X	

Table 12.8 (Continuation)

Reference IEC 60695-	Reference EN 60695-	Description	Subject	End-product test	Preselection test for materials
2-4/1	2-4/1	1 kW nominal premixed test flame and guidance	Flammability	X	
2-20	–	Hot wire coil ignitability test on materials	Ignitability		X
6-30	–	Small-scale static test: Apparatus	Smoke obscuration		X
6-31	–	Smoke obscuration – small-scale static method: Materials			X
7-50	–	Toxicity of fire effluent – toxic potency: Apparatus and test method	Toxic potency		X
10-2	10-2	Method for testing products made from nonmetallic materials for resistance to heat using the ball pressure test	Resistance to abnormal heat	X	
11-3	–	Test flames – 500 W flames – apparatus and confirmational test methods		X	X
11-4	–	Test flames – 50 W flames – apparatus and confirmational test methods		X	X
11-10	11-10	50W horizontal and vertical flame test methods	Flammability		X
11-20	11-20	500W flame test methods			X
IEC 60707	60707[a]	Flammability of solid nonmetallic materials when exposed to flame sources; list of test methods			X

[a] VDE 0304–3, 12.99

Ball Pressure Test, IEC 60695-10-2

The properties of insulating materials in electrotechnical appliances or parts may change when exposed to heat from internal sources, or by an external heat source such as the environment in which the device is used. As a consequence, the insulating material may be exposed to heat sources such as heat producing components or the physical properties may be changed, which could impair the mechanical or electrical suitability of the material. The IEC 60695-10-2 ball pressure test is a method for testing parts of nonmetallic materials for resistance to heat [24]. A steel ball of 5 mm (0.2 in.) diameter is pressed at 20 ($\pm$ 0.2) N for 1 h against the surface of the test specimen at 80 ($\pm$ 3) °C or 125 ($\pm$ 5) °C or a temperature that is 40 ($\pm$ 2) °C in excess of the temperature rise of the relevant part, whichever is higher. The test specimen has to be taken from the electrotechnical equipment, its subassemblies and components or, if this is not practical, a plaque of identical material having a thickness of 3.0 ($\pm$ 0.5) mm (0.12 in.) and at least a square with 10 mm (0.39 in.) sides, or a minimum diameter of 10 mm may be used.

After the ball is removed, the test specimen is cooled by immersion in room temperature water within 10 s and allowed to cool to room temperature. The diameter of the indentation is measured. The result is expressed as a pass, if the diameter caused by the ball does not exceed 2.0 mm (0.08 in.). Considerable work was done to establish a correlation coefficient so that the results obtained from Vicat measurements can be converted into ball pressure results. However, at present, it is not considered possible to produce one correlation coefficient that will work with all plastic materials and the many fillers and additives used by industry.

Glow-Wire Test on End-Product, IEC 60695-2-11

IEC 60695-2-10 (EN 60695-2-10) specifies the glow-wire test apparatus and the test procedure to simulate the effect of thermal stresses that may be produced by heat sources such as glowing wires or overloaded resistors, for short periods, in order to assess, by simulation technique, the fire hazard [25]. The glow-wire test for end-products is prescribed in IEC 60695-2-11 [26]. The test is carried out to ensure that, under defined conditions, the glow-wire does not cause ignition of parts, and that a part if ignited has a limited duration of burning without spreading fire by flames or by burning or with glowing particles falling from the test specimen. If possible, the test specimen should be complete electrotechnical equipment or its subassembly. A typical version of the equipment used is illustrated in Fig. 12.2. The glow-wire itself consists of a loop of nickel/chromium (80/20) wire 4 mm (0.16 in.) in diameter. A sheathed fine-wire thermocouple (NiCr/NiAl) having a nominal overall diameter of 1.0 mm (0.04 in.) is used for measuring the temperature of the glow wire. The test apparatus positions the electrically heated glow-wire in a horizontal plane while applying a force of 1.0 $\pm$ (0.2) N to the test specimen. The force is maintained when the glow wire is moved horizontally toward the test specimen. The depth of penetration of the glow wire is limited to 7 mm. The glow wire is heated by a simple electric circuit and, once the temperature is stable, no further adjustment may be made to the current. The glow wire is applied to the test specimen for a period of 30 ($\pm$ 1) s. To evaluate the possibility of spread of flame, for example, by burning or glowing particles falling from the test specimen, a tissue-covered wooden board or the material or components normally surrounding or situated underneath the test specimen are placed underneath the test specimen. The test specifications are summarized in Table 12.9. Information on a fully automatic system for the flame height analysis during the glow-wire test can be found in [27].

The testing of an electrical socket is shown in Fig. 12.3 as an example of this method. The appropriate test temperature in the range 550–960 °C should be specified by the Product Committees responsible depending on the risk situation for the respective product.

The informative Annex A of IEC 60695-2-11 gives guidance for the selection of the test temperature. The suggestions for the Product Committees are summarized in Table 12.10.

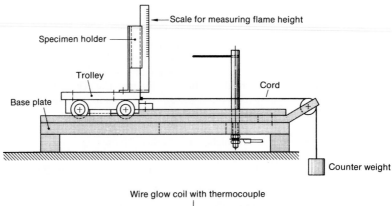

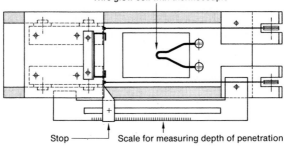

Figure 12.2   IEC 60695-2-10 glow-wire test apparatus

Table 12.9   IEC 60695-2-11 Glow-wire flammability test method for end-products

Test specimen	Complete electrotechnical equipment, its subassemblies and components, solid electrical insulating materials, other combustible materials. As a rule only one test is carried out. If ambiguous data are obtained, two further specimens must be tested
Test layout	The position of the test specimen should correspond to the least favorable case in use, glow-wire coil horizontal, penetration depth limited to 7 mm, pressing force 1 N
Ignition source	Glow-wire coil made of chrome-nickel alloy with built-in miniature thermocouple. Preferred test temperatures:
Suggested test severities	550 °C ± 10 K 650 °C ± 10 K 750 °C ± 10 K 850 °C ± 15 K 960 °C ± 15 K
Application time of ignition source	30 (± 1) s
Conclusions	Unless otherwise specified in the relevant specification, the test specimen is considered to have passed the glow-wire test if there is no flaming or glowing, or if all of the following situations apply: • If flames or glowing of the test specimen extinguish within 30 s after removal of the glow wire, i.e., $t_e \leq t_a + 30\,s$; and • When the specified layer of wrapping tissue is used, there shall be no ignition of the wrapping tissue

Figure 12.3   Glow-wire test of an electrical socket according to IEC 60695-2-11

Table 12.10   IEC 60695-2-11, Annex A: Guidance for glow-wire test

Parts of insulation material		
Temperature	In contact with, or retaining in position, current-carrying parts	For enclosures and covers not retaining current-carrying parts in position
550 °C	To ensure a minimum level of resistance to ignition of and/or spread of fire by parts liable to contribute to a fire hazard, and that are not subjected to other tests in this respect	
650 °C	Equipment for attended use	
	–	Fixed accessories in installations
750 °C	Equipment for attended use but under more stringent conditions	
	Fixed accessories in installations	Equipment intended to be used near the central supply point of a building
	Equipment for unattended use but under less stringent conditions	
850 °C	Equipment for unattended use continuously loaded	
960 °C	Equipment for unattended use, continuously loaded but under more stringent conditions	
	Equipment intended to be used near the central supply of a building	–

Table 12.11 shows examples for glow-wire tests specified in standards for electrotechnical products.

The glow-wire test has also been introduced for the application to test specimens of solid electrical insulation materials or other solid combustible materials for flammability (IEC 60695-2-12) or ignitability (IEC 60695-2-13) testing (see Section 12.1.2.4).

Table 12.11 Glow-wire flammability test methods in standards for electrotechnical products

Reference	Description	Test object	Test temperature (°C)
IEC 60998-1: 1990-04	DIN/EN 60998-1, VDE 0613-1, 1994-04	Connecting devices for low-voltage circuits for household and similar purposes; general requirements	Parts of insulating material in contact with current-carrying parts and for enclosures retaining in position only earthing clamping units	650 [a]
			Parts of insulation material retaining current-carrying parts and parts of the earthing circuit in position	850 [a]
IEC 400: 1996	DIN/EN 60400, VDE 0616-3, 1997-06	Lamp-holders for tabular fluorescent lamps and starter-holders	External parts of nonmetallic material	650
IEC 730-1: 1993	DIN/EN 60730-1, VDE 0631-1, 1996-01	Automatic electrical controls for household and similar use; general requirements	External parts of nonmetallic material; parts of insulating material supporting connections in position and parts in contact with them; enclosures retaining in position only earthing clamping units	550 Category A: 550 Category B: 550 Category C: 750 Category D: 850
IEC 669-1: 1993	DIN/EN 60669-1 VDE 0632-1, 1996-04	Switches for household and similar fixed electrical installations; general requirements	Parts of insulating material retaining current carrying parts and parts of the earthing circuit in position	850
IEC 439-3: 1990	DIN/VDE 0660, Part 504, 1992-02	Switchgear and control-gear; particular requirement for low-voltage switchgear, distribution boards	Parts of insulating materials retaining current-carrying parts	960
IEC 335-1: 1991-04	EN 60335-1: 1994	Safety of household and similar electrical appliances; general requirements	External parts of nonmetallic material; parts of insulating material supporting connections in position; $I \geq 0.5A$; operated while unattended	550 IEC: 750 EN: 850

[a] Application of the glow-wire: 5 s

Bad Connection Test IEC 60695-2-3 (Applies to Screw Terminals Only)

Connections may produce, under certain conditions, for example, loosening, insufficient mechanical pressure, or faulty installation, high heat dissipation due to wattage losses depending on the increased contact resistance and the actual current passing through the connection. After a long period of time, this can result in overheating and glowing. Such "glowing contacts" can ignite the insulating material in contact with the connection. The bad connection test with heaters is a test intended to simulate such an overheated connection to assess the fire hazard by a simulation technique.

This phenomenon has been investigated in depth by the VDE test station and results reported by Schwarz [28]. It was discovered that the point-shaped contact of the parts under electrical tension first leads to low energy ("cold") sparks, which form a thin insulating oxide layer on the conductor. With continuing making and breaking of the contact point, for example, due to vibration or thermal expansion, the semiconducting oxide layer increases in thickness until the unstable "blue spark process" undergoes sudden transition to a stable and continuous "red spark process". An incandescent transition zone is formed between the two parts of the contact. Considerable amounts of heat are liberated, which can result in strong heating of the contact parts and their substrate. The semiconducting layer consists of a mixture of various copper oxides. After extensive preparations, agreement has been reached that the simulation of such glowing contacts should be performed with electrical test heaters as shown in Fig. 12.4. The shape, diameter, and length of the resistance wire depend on the design and the size of the terminal taking into account the test power in relation to the actual current carried by the terminal or termination. A comprehensive description of these processes has been given by Holm [29].

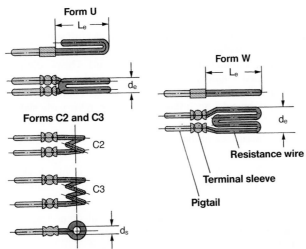

Figure 12.4   Bad connection test according to IEC 60695-2-3: Electrical test heaters

For the test according to IEC 60695-2-3 [30] the test specimen, complete electrotechnical equipment, and its subassemblies and components, is arranged in its most unfavorable position of normal use. A test heater is inserted into the terminal or termination. The heat is generated as close as possible to the area where a bad connection is likely to occur. The duration of the application of the test power is 30 min.

The test is applicable to screw connections having a rated current not exceeding 63 A. The test power values for the bad connection test are shown in Table 12.12. Burning of the test specimen or the parts surrounding it or of the layer placed below it, and the maximum height of any flame, but disregarding the start of the ignition, which may produce a high flame for a period of approx. 1 s, are noted.

Discussions have taken place within TC 89 to expand the bad connection test to include modern connections. Unfortunately, due to lack of interest and resources, this publication will not be the subject of a revision. The publication will remain in the IEC catalogue until 2005.

Table 12.12   Test power values for the bad connection test

Current (A)				Test power (W)
	0.2		0.5	2.0
	0.5		1.0	4.5
	1.5		2.0	9.0
	2.5		3.0	13.0
	4.0		5.0	19.0
Over	6.0	Up to and including	8.0	25.0
	10.0		13.0	33.0
	13.0		16.0	37.0
	16.0		20.0	42.0
	32.0		40.0	60.0
	50.0		63.0	73.0

From Table 1 of IEC 60695-2-3

Flame tests – General

In general, flames are not the primary cause of fires in electrotechnical equipment, but may occur due to fault conditions within the equipment, for example, a fault current flowing over a tracking path. Such flames should not cause a safety hazard. The needle-flame test according to IEC 60695-2-2 is intended to simulate such flames [31]. Flames may also occur in the environment of electrotechnical products and may impinge on combustible parts, for example, cables or enclosures, from outside. Such flames occurring in the early stage of fire should not unduly affect the electrotechnical product.

In these circumstances, the surface of the electrotechnical product needs to be assessed by the use of flame tests.

The 1 kW premixed test flame according to IEC 60695-2-4/1 is intended to simulate the effect of bigger flames, which may arise from other ignited items in the vicinity of the electrotechnical product [33].

IEC standard 60695-2-4/0 gives guidance on the design of flame test methods and general requirements for the test apparatus to produce a series of test flames and the correlated confirmatory tests [32].

Needle-Flame Test, IEC 60695-2-2 (Under Revision)

IEC 60695-2-2 is applicable to a complete electrotechnical equipment, its subassemblies and components, and to solid electrical insulating materials or other combustible materials [31]. The needle-flame test is applied to ensure that under defined conditions, the test flame does not cause ignition of parts, or that a combustible part ignited by the test flame has a limited extent of burning without spreading fire by flames or burning or glowing particles falling from the test specimen. The test specifications are summarized in Table 12.13.

The burner to produce the flame consists of a tube, for example, prepared from a hypodermic needle, and is supplied with butane gas. The preferred duration of the application of the flame is in the range 5–120 s. The actual duration time should be specified by the Product Committees dependent on the risk situation for the respective product. Examples of test arrangements are shown in Fig. 12.5.

Table 12.13   IEC 60695-2-2: Needle-flame test specifications

Test specimen	Complete electrotechnical equipment, its sub-assemblies and components. As a rule three tests are carried out
Test specimen position	It should correspond to the least favorable case in use. Application of the flame at a place on the surface of the test specimen, at which flames resulting from a fault condition can be expected
Ignition source	Needle-flame burner: • tube length: ≥ 35 mm • inner diameter: $0.5 (\pm 1)$ mm • outer diameter: ≤ 0.9 mm • length of the flame: $12 (\pm 1)$ mm • fuel gas: butane with a purity of at least 95 %
Preferred application time of flame	5, 10, 20, 30, 60, and 120 s
Conclusions	Unless otherwise prescribed in the relevant specification, the test specimen is considered to have satisfactorily withstood the needle-flame test if one of the following situations applies: • There is no flame and no glowing of the test specimen and no ignition of the wrapping tissue or scorching of the wooden board when these are used • Flames or glowing of the test specimen, the surroundings, and the layer below extinguish within 30 s after the removal of the needle-flame, that is $t_b < 30$ s, the surrounding parts and the layer below have not burnt away completely, and there has been no ignition of the wrapping tissue or scorching of the wooden board when these have been used

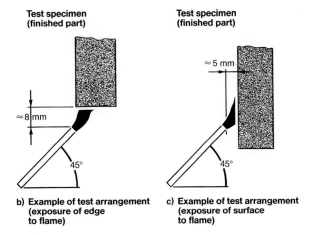

b) **Example of test arrangement (exposure of edge to flame)**

c) **Example of test arrangement (exposure of surface to flame)**

Figure 12.5   Needle-flame test according to IEC 60695-2-2: Examples of test arrangement

1 kW Nominal Premixed-Flame Test, IEC 60695-2-4/1 (Under Revision)

The detailed requirements for the production of the 1-kW nominal, propane based premixed type test flame with an approx. overall flame height of 175 mm (6.9 in.) and guidance for the test arrangements are given in IEC 60695-2-4/1 [33] (Table 12.4).

The standard specifies the hardware for the 1-kW propane burner and the related confirmatory test. The test specimen should be a complete electrotechnical product. Care should be taken to ensure that the test conditions are not significantly different from those occurring in normal use with regard to shape, ventilation, effect of possible flames occurring or burning, or glowing particles falling in the vicinity of the test specimen. The recommended distance from the top of the burner tube to the point on the surface of the test specimen is 100 mm (3.9 in.). Examples of the test arrangements are given in the IEC 60695-2-4/1 Annex A.

Table 12.14   IEC 60695-2-4/1: 1 kW nominal premixed test flame

Test specimen	Complete electrotechnical product (equipment, subassembly, or component). If the test cannot be made on a complete product, a suitable test specimen may be cut or removed from it.
Test specimen position	The test specimen shall be arranged in its most unfavorable position of normal use. 1 kW nominal premixed test flame according to IEC 60695-2-4/1
Ignition source	Blue cone height: 50–60 mm Overall height: 170–190 mm
Preferred application time of test flame	5 s, 10 s, 20 s, 30 s, 1 min, 2 min, 5 min, 10 min, 20 min, 30 min The required time should be specified in the relevant specification.
Conclusions	Unless otherwise specified in the relevant specification, the test specimen is considered to have withstood the test if one of the three following situations applies: • If the test specimen does not ignite • If flames or burning or glowing particles falling from the test specimen do not spread fire to the surrounding parts or to the layer placed below the test specimen, and there is no flame or glowing of the test specimen at the end of the test flame application, or at the end of an additional period as permitted by the relevant specification, or • If the duration or extent of burning specified in the relevant specification has not been exceeded.

Table 12.15 contains examples for fire hazard tests with test flames in standards for electrotechnical products.

Table 12.15 Fire hazard test with test flames in standards for electrotechnical products

Reference	Description	Test object	Requirements
IEC 60998-1: 1996 DIN/EN 60598-1, VDE 0711-1	Luminaries Part 1: General requirements and tests	Parts of insulating material supporting electrical connections in position	Needle flame test[a] Application time: 10s After burning time: ≤ 30s (no burning droplets)
IEC 730-1: 1993 DIN/EN 60730-1, VDE 0631-1, 1996-01	Automatic electrical controls for household and similar use; Part 1: General requirements	Parts in the vicinity of 50 mm from current-carrying parts	Needle flame test[a] Application time: 30s After burning time: ≤ 30s (no burning droplets)
IEC 400: 1996 DIN/EN 60400, VDE 0616-3, 1997-06	Lamp-holders for tubular fluorescent lamps and starter-holders	Parts of insulating material supporting live parts, other than connections, in position	Needle flame test[a] Application time: 10s After burning time: ≤ 30s (no burning droplets)
no equivalent DIN/VDE 0611-20, 1987-12	Modular terminal blocks for connection of copper conductors up to 100V A.C. and up to 1200V A.C.; test for flammability and flame propagation	Insulating material Modular terminal block (end-product test)	Material classified as V-2 according to IEC 60695-11-10 Needle flame test[a] Application time: 5s to 10s After burning time: ≤ 30s (no burning droplets)
DIN/EN 50086-1, VDE 0605-1, 1994-05	Conduit systems for electrical installations; Part 1: General requirements	Conduit systems (end-product test)	1 kW flame according to IEC 60695-2-4/1 Application time depending on the thickness of the material from 15s−500s After burning time: ≤ 30s (no burning droplets) limited burning length

[a] Needle flame test according to IEC 60695-2-2

12.1.2.4 Combustion Characteristic Tests for Preselection of Materials, IEC 60695-1-30 (Under Revision)

The selection of the right material contributes toward the appearance, functionality, and serviceability of an item of equipment. Production costs and overall costs are similarly influenced to a considerable extent by the choice of material, together with its design and process potential.

The basic question, as far as the use of the techniques outlined in IEC 60695-1-30 is also enhancing the value of materials tests to the producers of finished electrotechnical products and parts made from these materials, can be answered only on the basis of test programs that demonstrate the correlation between the results of these materials tests with the performance of the finished products.

The next revision of IEC 60695-1-30 [34] will absorb much of the text contained in IEC 60695-3-1:1987, Combustion characteristics test survey [35]. It is intended to provide procedures for relating preselection fire tests of subassemblies, components, and materials to the specific function of the electrotechnical products. General guidance for the IEC product committees is given for assessing the significance, relevance, and limitations of those IEC and ISO fire tests for preselection compared to fire tests for hazard assessment. From an economic point of view, combustion characteristics tests often are small in scale, so that they can be used as low cost tests for routine laboratory testing. In that context, they can serve as a valuable tool for quality testing, production control, or for screening processes in research and development.

The next section contains detailed information about a selection of IEC tests listed in IEC 60695-1-30, which are most common today, to compare the various combustion characteristics of materials. They are valid only as measurements of the performance under specifically controlled test conditions and not considered indicators of their behavior under actual fire conditions. Single-material assessment can lead to misunderstanding the evaluation of fire risks; the aim should be to simulate the end-use conditions of an electrotechnical product as closely as possible.

Glow-Wire Flammability Test Method on Materials, IEC 60695-2-12

IEC 60695-2-12 specifies the details of the glow-wire test, when applied to test specimens of solid electrotechnical insulating materials or other solid combustible materials for flammability testing [36]. The test results make it possible to provide a relative comparison of various materials according to their ability to extinguish flames on removal of the heated glow wire and their ability not to produce flaming or glowing particles capable of spreading fire to a tissue-covered wooden board.

The test apparatus is identical to that described in IEC 60695-2-10 and is illustrated in Fig. 12.2. The test specimen is fixed on the trolley in a vertical position. The tip of the glow wire is brought into contact with the test specimen for 30 ($\pm$ 1) s. Nine test temperatures ranging from 550 °C to 960 °C can be used. Duration of burning of the test specimen and ignition of the wrapping tissue are noted. The test specifications are summarized in Table 12.16.

The GWFI (glow-wire flammability index) is the highest temperature of the glow wire at which, during three subsequent tests, flames or glowing of the test specimen extinguish within 30s after removal of the glow wire. Ignition of the tissue-covered wooden board by burning droplets or particles is not allowed. The GWFI will usually vary depending on the thickness of the material tested. The preferred values are 0.75 ($\pm$ 0.1) mm; 1.5 ($\pm$ 0.1) mm or 3.0 ($\pm$ 0.2) mm (0.03 in., 0.06 in., or 0.12 in.).

Table 12.16   IEC 60695-2-12: Glow-wire flammability test method for materials

Test specimen	Test specimens having a sufficiently large plane section with fixed dimensions: • Length: ≥ 60 mm • Width (inside clamps): ≥ 60 mm • Thickness: preferred values 0.75 (± 0.1) mm; 1.5 (± 0.1) mm, or 3.0 (± 0.1) mm
Test specimen position	Vertical according to this test
Test layout	Test specimens may be any shape, provided there is a minimum test area of 60 mm in diameter. A set of 10 test specimens will in general be adequate to evaluate their flammability
Ignition source	The description of the test apparatus is given in IEC 60695-2-1/0. Glow-wire coil is made of chrome-nickel alloy with built-in thermocouple.
Test temperatures	550 °C ± 10 K 600 °C ± 10 K 650 °C ± 10 K 700 °C ± 10 K 750 °C ± 10 K 800 °C ± 15 K 850 °C ± 15 K 900 °C ± 15 K 960 °C ± 15 K
Application time of ignition source	30 (± 1) s
Conclusions	The test specimen is considered to have satisfactorily withstood the test if there is no ignition of the test specimen or if both of the following conditions apply: • If flames or glowing of the test specimen extinguish within 30 s after removal of the glow-wire, and • There is no ignition of the wrapping tissue

Glow-Wire Ignitability Test Method on Materials, IEC 60695-2-13

IEC 60695-2-13 describes the glow-wire test when applied to test specimens of solid electrical insulating materials or other solid combustible materials for ignitability testing [37] (Table 12.7).

The test results make it possible to provide a relative comparison of various materials according to the temperature at which the test specimen ignites during the application of the heated glow wire. The test apparatus is identical to that described in IEC 60695-2-10 and is illustrated in Fig. 12.2. The ignitability will usually vary depending on the thickness of the material tested. The preferred thicknesses are 0.75 (± 0.1) mm; 1.5 (± 0.1) mm, or 3.0 (± 0.1) mm. The test specimen is arranged so that its free plane surface is vertical. The tip of the glow wire is brought into contact with the surface of the test specimen for 30 s. By repeated tests with different test temperatures of the glow wire, using a new specimen each time, the glow wire ignition temperature (GWIT) of the material is the test temperature, which is 25 K higher than the maximum temperature that does not cause ignition. The test specifications are shown in Table 12.17.

Table 12.17   IEC 60695-2-13: Glow-wire ignitability test method for materials

Test specimen	Test specimens having a sufficiently large plane section with fixed dimensions: • Length: ≥ 60 mm • Width (inside clamps): ≥ 60 mm • Thickness: preferred values: 0.75 ($\pm$ 0.1) mm; 1.5 ($\pm$ 0.1) mm or 3.0 ($\pm$ 0.1) mm Specimens may be any shape provided there is a minimum test area of 60 mm in diameter. A set of 10 test specimens will in general be adequate to evaluate the ignitability test
Test specimen position	Test specimens position is vertical according to this test
Test layout	The description of the test apparatus is given in IEC 60695-2-1/0
Ignition source	glow-wire coil made of chrome-nickel alloy with built-in thermocouple
Test temperatures	500 °C $\pm$ 10 K 550 °C $\pm$ 10 K 600 °C $\pm$ 10 K 650 °C $\pm$ 10 K 700 °C $\pm$ 10 KK 750 °C $\pm$ 15 K 800 °C $\pm$ 15 K 850 °C $\pm$ 15 K 900 °C $\pm$ 15 K 960 °C $\pm$ 15 K
Application time of ignition source	30 ($\pm$ 1) s
Ignition criteria	The tip of the glow wire is applied to the center of the plane area of the surface for the purpose of this test; ignition means that a flame is visible for more than 5 s; the ignition of the test specimen during the period of application of the glow wire shall be determined
Evaluation of test results	The test temperature, which is 25 K higher than the maximum temperature not causing ignition during three subsequent tests shall be reported as the GWIT

Hot-Wire Coil Ignitability Test on Materials, IEC 60695-2-20 (Under Revision)

IEC 60695-2-20 is intended to provide a relative comparison of the behavior of various solid electrical insulating materials according to the time taken to ignite the test specimen during application of heat from an electrically heated coil used as an ignition source [38]. Hot-wire ignition tests are carried out on five bar-shaped test specimens of dimensions 125 ($\pm$ 5) mm (4.9 in.) long, 13.0 ($\pm$ 0.3) mm (0.5 in.) wide, and 3.0 ($\pm$ 0.1) mm (0.12 in.) thick. Each test specimen is wrapped with five complete turns of 0.5 mm (0.02 in.) diameter nickel/chromium (80/20) wire of approximate length 250 mm (9.8 in.) and with a nominal cold resistance of 5.28 Ω/m spaced 6.35 ($\pm$ 0.5) mm (0.25 in.) between turns.

The test apparatus and ignition source are shown in Fig. 12.6. The test specimen is tested in a horizontal position by heating the wire electrically with a linear power density of 0.26 ($\pm$ 0.01) W/mm. Ignitability is identified by the time required to start ignition of the test specimen under test. The ignitability will usually vary depending on the thickness of the test specimen. IEC 60695-2-20 therefore recommends obtaining, in addition to the standard thickness specified, results for thickness of about 0.8 mm (0.03 in.), 1.6 mm (0.06 in.), and 6.0 mm (0.24 in.).

The test is equal to the HWI Test carried out by the Underwriters Laboratories described in Section 12.2.1.2.

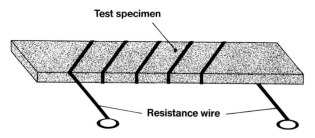

Figure 12.6   Hot wire coil ignitability test on materials according to IEC 60695-2-20

Flammability of Solid Nonmetallic Materials

Flame Sources

One aim of the work of TC 89 is to make available an appropriate series of standardized test flames, covering a range of powers and test methods for the use of all product committees requiring test flames. Wherever possible, these test flames have been based on existing types and methods, but are now available with improved performance and specifications. IEC/TC 89, in cooperation with ISO/TC 61, have harmonized the technical content of IEC 60707 Ed.1 [39], ISO 1210 (withdrawn) [40], ISO 10351 (withdrawn) [41], and the UL 94 of the Underwriters Laboratories [42] to publish IEC 60695-11-10 and IEC 60695-11-20 in 1999. The methods FH and FV flammability tests described in IEC 60707 Ed.1 are specified in IEC 60695-11-10 as the horizontal and vertical burning tests. The method LF flammability test described in IEC 60707 Ed.1 is specified in IEC 60695-11-20. The method BH incandescent bar flammability test described in IEC 60707 Ed.1 has been withdrawn. The actual standard IEC 60707, 1999 [43] lists flame test methods to check the constancy of the flammability characteristics of a material and provides an indication of the process in the development of materials and a relative comparison and classification of various materials (Table 12.18).

Table 12.18   IEC 60707 (EN 60707): Flammability classifications; List of test methods

Test method	Standard	Flammability classifications	Flammability classification of the technically equivalent UL 94 test method
50-W horizontal burning test	IEC 60695-11-10 (EN 60695-11-10)	HB 40, HB 75	HB
50-W vertical burning test		V-0, V-1, V-2	
500-W burning test	IEC 60695-11-20 (EN 60695-11-20)	5VA, 5VB	94-5VA, 94-5VB
50-W vertical burning test for flexible specimens	ISO 9773	VTM-0, VTM-1, VTM-2	
Horizontal burning test on cellular plastics	ISO 9772	FH-1, FH-2, FH-3	HBF HF-1, HF-2

IEC/TC 89 is in the process of additional harmonization to include the UL 94-HB classification criteria and some specifications for the consideration of the range of colors, densities, melt flows, or reinforcement of the test specimens into IEC 60695-11 and IEC 60695-11-20. The amendments to both standards will result in one harmonized worldwide flammability standard for the plastics industry. It is considered essential by the fire testing community because it satisfies the needs of all parties, for example, material suppliers, end-product designers, and safety regulators.

Once this harmonization is achieved, UL plans to withdraw the UL 94 standard in favor of IEC 60695-11-10 and IEC 60695-11-20.

50-W Horizontal Flame Test Method, IEC 60695-11-10

IEC 60695-11-10 [44] describes the application of a 20 (± 1) mm (0.79 in.) test flame for 30 s to a specific part of a horizontally mounted test specimen. The behavior during and after is observed. Successful completion of this test leads to the classifications in HB 75 or HB 40. The test layout is illustrated in Fig. 12.7. The classification depends on the burning rate of the test specimen. The test specifications are summarized in Table 12.19. It is similar to UL 94 with the exception of the classification for the HB test.

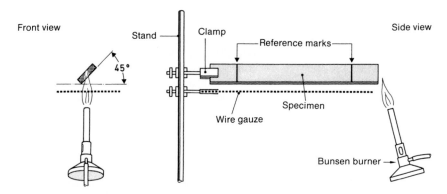

Figure 12.7   Test layout for the classes HB 40 and HB 75 according to IEC 60695-11-10

50-W Vertical Flame Test Method, IEC 60695-11-10

The vertical flame test method according to IEC 60695-11-10 is well known as UL 94 and has been used for more than 40 years in North America. The test classifies solid plastic test specimens as UL 94V-0, UL 94V-1, and UL 94V-2. It is a small-scale laboratory screening procedure for comparing the burning behavior of vertically oriented test specimens exposed to a small flame source. The test layout is illustrated in Fig. 12.8. The test uses the same test specimens as the HB test. The test flame is applied twice to the lower end of the test specimen for 10 s.

Burning and/or glowing times, when dripping occurs, and whether or not the cotton beneath is ignited are all recorded. The best rating, 94V-0, is achieved if the mean afterflame time of all five test specimens is ≤ 50 s. The material is rated 94V-1 if the mean afterflame time is ≤ 250 s. If flaming particles ignite the cotton indicator pad the material is classified as 94V-2.

The test specifications are summarized in Table 12.20.

Table 12.19 IEC 60695-11-10: 50 W horizontal flame test method

Test specimen	Three bar-shaped test specimens 125 (± 5) mm × 13.0 (± 0.5) mm with two reference marks 25 (± 1) mm and 100 (± 1) mm from the end that is to be ignited. Maximum thickness is 13 mm. Depending on test performance, it may be necessary to test a further set of three samples.
Test specimen conditioning	48 h minimum at 23 (± 2) °C and 50 (± 5) % relative humidity
Test specimen position	Longitudinal axis of test specimen approximately horizontal, traverse axis inclined 45 (± 2)° to horizontal
Ignition source	Laboratory burner according to IEC 60695-11-4, flames A, B, or C 20 (± 2) mm nonluminous flame [ISO 10093 describes the burner as ignition source P / PF 2 (50 W)]. The longitudinal axis of the burner is tilted at 45° toward the horizontal during application of flame.
Application time of flame	30 s. If the flame front reaches the 25 mm mark on the test specimen before 30 s, the burner is removed.
Conclusion	Classification in class HB 40 if: • There is no visible burning with a flame after removal of the flame. • The test specimen extinguishes before the 100 mm mark is reached. • The burning rate between the reference marks does not exceed 40 mm/min. Classification in class HB 75 if: • The burning rate between the reference marks does not exceed 75 mm/min.

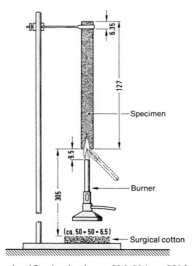

Figure 12.8 Test layout for the classification in classes V-0, V-1, and V-2 according to IEC 60695-11-10

Table 12.20   IEC 60695-11-10: 50 W vertical flame test method

Test specimen	Two sets of five bar-shaped test specimens 125 (± 5) mm × 13.0 (± 0.5) mm × maximum 13.0 mm; two sets of test specimens of material of the maximum and minimum thickness normally supplied should be used. Depending on test performance, it may be necessary to test additional sets of test specimens.
Test specimen conditioning	• Storage of one set of test specimens for at least 48 h at 23 (± 2) °C and 50 (± 5) % relative humidity, • Storage of a second set of test specimens for 168 (± 2) h in an air-circulating oven at 70 (± 2) °C followed by a 4 h cooling to room temperature in a desiccator.
Test specimen position	Test specimens are suspended with longitudinal axis vertical. A horizontal layer of cotton (m ≤ 0.08 g) approx. 50 mm × 50 mm × 6 mm is placed 300 (± 10) mm below the lower edge of the specimen.
Ignition source	Laboratory burner according to IEC 60695-11-4, flames A, B, or C, 20 (± 2) mm high nonluminous flame
Application time of flame	Two applications for 10 (± 0.5) s for each test specimen. The second application starts as soon as the test specimen, ignited by the first application, extinguishes.
Conclusions	Classification in class 94 V-0 if: • Individual afterflame time is ≤ 10 s, the total set afterflame time for any conditioning does not exceed 50 s, • No ignition of the cotton indicator by flaming particles or drops, • Test specimens do not burn up completely, • Individual test specimen afterflame plus afterglow time not more than 30 s after removal of ignition source. Classification in class 94 V-1 if: • Afterflame time ≤ 30 s after removal of ignition source, • The total set afterflame time for any conditioning does not exceed 250 s individual test, specimen afterflame plus afterglow time not more than 60 s after removal of the ignition source, • Other criteria as for class 94 V-0. Classification in class 94 V-2 if: • Ignition of cotton indicator pad by burning drops. Other criteria as for class 94 V-1.

500-W Flame Test Method, IEC 60695-11-20

IEC 60695-11-20 specifies a small-scale laboratory screening procedure for comparing the relative burning behavior of test specimens made from plastics as well as their resistance to burn through when exposed to a flame ignition source of 500 W nominal power [45]. This method requires the use of two test specimen configurations to characterize material performance. Rectangular bar-shaped test specimens 125 (± 5) mm (4.9 in.) long x 13 (± 0.5) mm (0.50 in.) wide in the minimum supplied thickness are used to assess ignitability and burning time. Square plate test specimens 150 (± 5) mm (5.9 in.) x 150 (± 5) mm (5.9 in.) in the minimum supplied thickness are used to assess the resistance of the material to burn through (Fig. 12.9 and Table 12.21).

Table 12.21   IEC 60695-11-20: 500 W flame test methods

Test specimen	• Bar test specimens: Two sets of five bar-shaped test specimens 125 ($\pm$ 5) mm $\times$ 13.0 ($\pm$ 0.5) mm $\times$ minimum thickness normally supplied. The thickness shall not exceed 13 mm. • Plates: Two sets of three plates 150 ($\pm$ 5) mm $\times$ 150 ($\pm$ 5) mm $\times$ minimum thickness normally supplied. Depending on test performance, it may be necessary to test further sets of test specimens.
Test specimen conditioning	• Storage of one set of test specimens for at least 48 h at 23 ($\pm$ 2) °C and 50 ($\pm$ 5) % relative humidity • Storage of second set of test specimens for 168 ($\pm$ 2) h in the air-circulating oven at 70 ($\pm$ 2) °C followed by 4 h cooling to room temperature in a desiccator.
Test specimen position	• Bar test specimen: To be suspended vertically with the flame of the burner applied, at an angle of 20 $\pm$ (5)° from the vertical, centrally to the lower front corner. Plates: To be supported horizontally. Burner flame to be applied, at an angle of 20 $\pm$ (5)° from the vertical, to the approximate center of the bottom surface of the plate.
Ignition source	Laboratory burner according to IEC 60695-11-3. Flames A, B, C or D [ISO 10093 describes the burner as ignition source P/PF 2 and P/PF 4 (500 W)] with axis inclined at 20° to vertical; flame length 125 ($\pm$ 10) mm with 40 ($\pm$ 2) mm blue inner cone
Application time of flame	Five times 5 ($\pm$ 0.5) s with 5 ($\pm$ 0.5) s intervals
Conclusions	Burning category 5 VA: • Individual afterflame time is $\leq$ 60 s after the fifth flame application • No ignition of the cotton indicator by flaming particles or drops Burning category 5 VB: • The flame did penetrate (burn through) any of the plates, other criteria as for 5 VA, • Materials classified either 5 VA or 5 VB shall also conform to the criteria for materials classified either V-0, V-1, or V-2 in the same bar test specimen thickness.

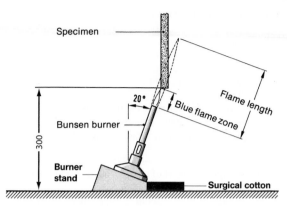

Figure 12.9   Test layout for the classification in class 5 VA (method A) according to IEC 60695-11-20

Confirmational Test Methods

The IEC/TS 60695-11 [46] test flame series of Technical Specifications represents a significant enhancement to previous existing Bunsen burner methodologies through the application of state-of-the-art flame calibration procedures.

IEC/TS 60695-11-3 gives the detailed requirements for the production of a nominal 500 W, pre-mixed type test flame, with an overall height of approx. 125 mm (4.9 in.). These test flames are used, for example, for the IEC 60695-11-20 small-scale laboratory screening procedure or the UL 94-5 V-tests. Four methods are given: flame A and D are produced using only methane, flame B is produced using only propane, and flame C is produced using either methane or propane.

IEC/TS 60695-11-4 provides detailed requirements for the production of a 50-W nominal, premixed type test flame, with an overall height of approx. 20 mm. Three methods – A, B, and C – are given: flame A is produced by methane; flames B and C are produced with either methane or propane. These flames are used, for example, for the 50-W horizontal and vertical flame test methods according to IEC 60695-11-10 or for the technically equivalent UL-94 tests.

Both technical specifications prescribe confirmatory tests based on copper block calorimetry. These confirmatory tests require a thermocouple to be placed in a copper block under which the flame is located at a specified distance.

The flame is considered to be confirmed if the time taken for the copper block to increase from 100 °C to 700 °C is within the time required.

Surface Spread of Flame, IEC 60695-9-1

IEC 60695-9-1 provides guidance on the assessment of surface spread of flame for electrotechnical products and the materials from which they are formed [47].

IEC/TS 60695-9-2 presents a summary and relevance of test methods that are used to determine the surface spread of flame of electrotechnical products and the materials from which they are formed [48]. It reflects the current state of the art and includes special observations on their relevance and use where available. Details on the assessment of surface spread of flame tests for cables are reported in Section 12.5.

12.1.2.5 Side Effects of a Fire

Electrotechnical products, when involved in a fire, contribute to the fire hazard due to the release of heat and smoke. Smoke may cause loss of vision and/or disorientation which could impede escape from the building or fire fighting.

The production of toxic effluents may be a significant contributing factor to the overall fire hazard. Fire effluents from burning plastics are to some degree corrosive and may present a risk of corrosion damage to electrotechnical equipment and systems and/or building structures. In the last few years, there has been an increasing demand for testing and assessment of such subsidiary phenomena of combustion.

Heat release, smoke density, and toxicity, as well as corrosivity, require fire safety engineering methods for the hazard assessment with regard to temperature, visibility, toxic, and corrosive effects. The standardized test method (the fire model) chosen for decomposing materials or products must allow a comprehensive simulation of the many different fire situations.

ISO has published a general classification of fire stages in ISO/TR 9122-1, shown in Section 3 [49]. Conditions for use in the fire model can be derived from this table to correspond, as far as possible, to real-scale fires. The important factors affecting the heat release rate and the smoke production are oxygen concentration and irradiance/temperature.

Heat Release

Fire calorimetry is the technology used to measure heat release rate from a fire in a product or a whole system. Oxygen consumption calorimetry is the usual method used to determine the heat release rate [50]. This testing ranges from small-scale calorimeter test in laboratory scale, to intermediate-scale tests, full-scale room and open calorimeter tests. This technology finds use in a quickly growing field. Such measurements provide basic fire characteristics of materials and products that enable the fire scientist to get a better understanding of the performance of the material or the electrotechnical equipment in actual fire situations. Heat release characteristics are used as an important factor in the determination of fire hazard and also used as one of the parameters in fire safety engineering calculations in IEC 60695-1-1.

IEC 60695-8-1 provides guidance in the assessment of heat release from electrotechnical products and materials from which they are constructed [51]. The IEC/TS 60695-8-2 provides a summary and relevance of test methods that are used to determine heat release for electrotechnical products and the materials from which they are formed. It represents the current state of the art of the test methods and includes special observations on their relevance and use [52].

One further example for fire calorimetry in an open product calorimeter test is the method for determining the fire performance response of discrete products and their accessories in air-handling spaces according to UL 2043 [53]. The purpose of this test is to determine the rate of heat release and the rate of smoke release of the burning product samples in accordance to the requirements of the NEC (NFPA 70) [118]. The test rig and the test specifications are shown in Fig. 12.10 and Table 12.22, respectively. Further details of fire calorimetry and an introduction to this field are given in review articles by [54–56].

The burning characteristics of potential ignition sources of room fires, for example, televisions, wastepaper baskets, curtains, chairs and Christmas trees have been described by Ahonen et al. [57]. In the experiments, the rate of heat release, rate of smoke production, and gas temperatures were measured in a full-scale calorimeter. The most intense fire was found in one of the Christmas tree experiments with a maximum heat release rate of 0.65 MW and a maximum temperature above the burning tree as high as 1300 °C.

Fire tests of packaged and palletized warehouse computer products have been carried out by Hasegawa et al. The test focused on obtaining heat release rate data to be used as input for enclosure fire models and to predict fire spread between commodity arrays [58].

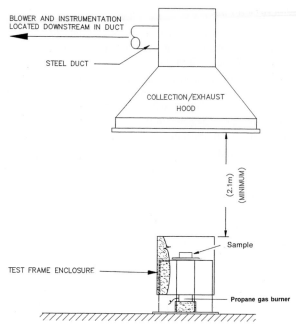

Figure 12.10   Product calorimeter to UL-2043

Table 12.22   Test specifications for determining the heat and visible smoke release by UL 2043

Test specimen	Discrete products (electrical equipment) installed in air-handling spaces. The test specimens are to be representative of the construction being investigated with regard to components and design. *Examples:* Loudspeakers, loudspeaker assemblies and their accessories, ceiling mounted products, suspended in air-handling space products
Test specimen conditioning	At least 24 h in 23/50 standard atmosphere
Test layout	The finished products are mounted on a test frame enclosure and are subjected to an open flame ignition source and evaluated using a product calorimeter. All combustion products released by the test specimen are collected in the hood (face dimensions at least 2.4 m by 2.4 m). Nominal exhaust flow rate: $0.71 \, \text{m}^3/\text{s}$
Ignition source	Propane gas diffusion burner; the burner is 0.318 m wide, has a square surface that is 0.203 m above the bottom of the test frame enclosure
Application time of flame	10 min
Conclusion	Passed if: • The peak rate of heat release (HRR) measured during each test does not exceed 100 kW. • The peak normalized optical density measured during each test does not exceed 0.5. • The average optical density (10 min test duration) does not exceed 0.15.

Smoke Obscuration to IEC 60695-6-1 and Opacity to IEC 60695-6-2

IEC 60695-6-1 [59] gives guidance on the optical measurement of smoke obscuration, the general aspects of optical smoke test methods, and the relevance of optical smoke data to hazard assessment. IEC/TS 60695-6-2 gives a summary and relevance of the test methods that are used in the assessment of smoke opacity [60]. It presents a brief summary of static and dynamic test methods in common use. It includes special observations on the relevance, for electrotechnical products and their materials, to real fire scenarios and gives guidance on their use. It is intended to support product committees wishing to incorporate test methods for smoke obscuration in product standards.

Small Static Test Method for Materials, IEC 60695-6-30 and IEC 60695-6-31

IEC/TS 60695-6-30 [61] and IEC/TS 60695-6-31 [62] describe a small-scale static test method, which is well known as the NBS smoke chamber. Table 12.23 gives an overview of smoke test methods.

For more information about the fundamental principles and the techniques for measuring smoke density and future developments see Chapter 15 of this book.

Toxic Hazard Assessment

Fire does not discriminate between electrotechnical and nonelectrotechnical products. TC 89 is working closely with the fire experts in ISO/TC 92 as a basis for publishing guidance for toxic hazards in fires involving electrotechnical products. The basic philosophy developd by ISO/TC 92 is that toxic hazard cannot be taken in isolation, but forms part of an assessment of the total fire hazard. Other hazards can include smoke, heat and structural endurance. The total combination of all hazards leading to lethality or incapacitation must be considered as the fire grows and spreads. One task of TC 89 was to translate this guidance into terms, which would be useful for IEC product groups. Within the electrotechnical industry, in 1990, the cable industry was the product group most interested in the assessment of toxic hazards in fire. Chapter 16 of this book comprehensively reviews the toxicity of fire effluents.

Toxicity is a function of the exposure dose and the toxic potency of fire effluents. The main questions concerning the reduction in toxic hazards from fires in which electrotechnical products have been involved are:

- Potency: How toxic is the fire effluent?
- Dose: How much of the fire effluent is inhaled?
- Escape: How is escape impeded? Does the time available to escape exceed the time needed to escape?

IEC/TC 89 addresses all these questions, which have resulted in the list of publications summarized in Table 12.24.

IEC 60695-7-1 [63] gives general guidance for IEC product committees.

Table 12.23 IEC 60095-6-2 Smoke opacity – Summary and relevance of test methods[1]

Type of test method	Section	Test method reference	Limitations on test specimen	Relevance to stage of fire						Limitations on use for regulatory purposes	Suitability of data format for input to fire safety engineering
				1(a)	1(b)	1(c)	2	3(a)	3(b)		
Static	6.1	Determination of opacity of smoke in a NBS chamber	Essentially flat only, and less than 75 mm square. Not suitable for liquids or some thermoplastics	No	Yes	No	Yes	No	No	Not recommended	No
	6.2	Determination of opacity of smoke by a single chamber test	Essentially flat only, and < 75 mm square. Possibly suitable for products – see clause 6.2	No	Yes	No	Yes	Yes	No	Only for products with suitable geometry, and for appropriate fire stage	Not as currently reported
	6.3	Determination of opacity of smoke in a "three meter cube" smoke chamber	Self-supporting products only, approx. 1 m long	No	No	No	No	No	No	Only for self-supporting products, and for appropriate fire stage	
Dynamic	7.1	Determination of the opacity of smoke generated by electric cables mounted on a horizontal ladder	Essentially flat building products, or cables	No	No	No	No	No	Yes	Only for products with suitable geometry, and for appropriate fire stage	Yes

Type of test method	Test method reference		Limitations on test specimen	Relevance to stage of fire						Limitations on use for regulatory purposes	Suitability of data format for input to fire safety engineering
	Section			1(a)	1(b)	1(c)	2	3(a)	3(b)		
	7.2	Determination of the opacity of smoke generated by electrical cables mounted on a vertical ladder	Self-supporting products only	No	No	No	Yes	No	Yes	Only for self-supporting products, and for appropriate fire stage	
Dynamic	7.3	Determination of the opacity of smoke using the cone calorimeter	Intended for essentially flat materials. Possibly suitable for products – see 7.3	No	Yes	No	Yes	No	Yes	Only for products with suitable geometry, and for appropriate fire stage	Yes

[1] The suitability of data for fire safety engineering purposes is limited to applications in which the test method is relevant to the appropriate stage of fire

Table 12.24 IEC list of publications concerning toxicity of fire effluent

Publications		Title
IEC 60695-7-1		Guidance on the minimization of toxic hazards due to fires involving electrotechnical products – General
IEC/TS 60695-7-2	In preparation	Toxicity of fire effluent – Summary and relevance of test methods
IEC/TS 60695-7-3		Toxicity of fire effluent – Use and interpretation of test results
IEC 60695-7-4	To be withdrawn in 2003	Guidance on the minimization of toxic hazards due to fires involving electrotechnical products – Unusual toxic effects in fires
IEC/TS 60695-7-50	In preparation	Toxicity of fire effluent – Estimation of toxic potency: Apparatus and test method
IEC/TS 60695-7-51	In preparation	Toxicity of fire effluent – Estimation of toxic potency: Calculation and interpretation of test results

Key messages to IEC product committees incorporating requirements for the assessment of toxic hazard in product standards are:

- Toxic potency data should not be used directly in product specifications to give individual pass/fail criteria and be used only in an integrated system analysis, within a total fire hazard assessment.
- Toxic potency should not be confused with toxic hazard.
- Toxic hazards from fires can best be minimized by the control of ignition, rates of flame spread, fire growth, and smoke development.
- Carbon monoxide is by far the most significant agent contributing to toxic hazard. Other agents of major significance are hydrogen cyanide and carbon dioxide as well as irritants, heat and hypoxia caused by oxygen depletion.

Prediction of toxic potency is based on mathematical models taking into account analytical results of fire effluents for acute toxicity by inhalation in animal tests. This requires the generation of thermal decomposition products and analytical determination of the concentration of those components of effluents responsible for toxicity.

IEC/TS 60695-7-2 provides a summary and relevance of test methods that are in common use [64]. It includes special observations on the relevance, for electrotechnical products and their materials, to real fire scenarios and gives guidance on their use. It is intended to support product committees wishing to incorporate test methods for toxicity in product standards. Details are shown in Table 12.25.

IEC/TR 60695-7-3 provides guidance on the use and interpretation of results from laboratory tests for the determination of toxic potency in assessing the toxic hazard as part of the total fire hazard [65]. It discusses currently available approaches to toxic hazard assessment consistent with the approach of ISO/TC 92/SC 3, as described in ISO/TR 9122 [66]. The detailed methodology is directly applicable to data produced by tests measuring the lethal effects of fire effluents.

IEC/TS 60695-7-50 describes a method for the generation of fire effluents and the identification and measurement of its combustion products [67]. This method uses a moving sample in a tube furnace using different temperatures and air flow rates. This test is based on the widely used DIN 53436 tube furnace method [68]. The DIN method has been found to be a reasonable model for decomposition conditions in various fire stages, and is inexpensive, simple to use, and validated by animal tests [69, 70]. A schematic of the furnace system is shown in Fig. 16.11 in Chapter 16.

IEC/TS 60695-7-51 [71] describes the calculation procedure for converting the data generated in IEC/TS 60695-7-50 or other equivalent methods into estimated toxic potency data for the materials tested under defined conditions. The purpose is to estimate the contribution of the fire effluent of the tested electrotechnical product or material, to the toxic threat to life of the total fire effluent resulting from a fire scenario in which the electrotechnical product or material is involved. The principle of the method is to calculate the fractional effective dose (FED) of toxic products accumulated by exposure to the effluent of a fire in that given scenario. The fractional exposure dose (FED) is the ratio of the sum of the c-t product (concentration-exposure time) of the individual effluent components to the total of the time-concentration products, which cause death of 50 % of the test animals in each case. A retrospective analysis of predicted and observed LC_{50} values is given by Pauluhn [72] (See also Chapter Toxicity under 16.6). A comprehensive description of the scenario dependent judgment of the toxicity of fire gases is given by Wittbecker, Klingsch and Bansemer [73].

The concept of lethal toxic potency was derived originally from experiments on animals. The lethal concentration (LC_{50}) expresses the ratio between the mass or volume concentration of the test specimen and the amount of air used when 50 % of the test animals will die under specific test conditions. This is known as the relative acute toxicity by inhalation LC_{50} as an indicator for toxic potency.

Corrosion Damage to IEC 60695-5 Series (Under Review)

About 40 years ago, the corrosive behavior of smoke began to be investigated systematically and concentrated mainly on determining the effects of chlorine. Recent fires worldwide in power stations, telecommunication facilities, ships, underground railways, and so forth, showed that exposure of sensitive electronic or mechanical equipment to fire effluent has the potential to cause extensive damage. In combination with cleaning, replacement of equipment, and the temporary loss of operation large financial losses may be caused. Work continues in ISO/TC 61 and IEC/TC 89 aimed at the development of universally valid testing and evaluation criteria and the preparation of guidance documents.

It is generally agreed that the problems can be solved only through cooperation by specialists engaged in such separate fields as fire technology, electrochemistry, electrical engineering, and metallurgy. IEC 60695-5-1 [74] gives guidance on the assessment of corrosion damage to electrotechnical equipment, systems, and building structures from fire effluent emitted from electrotechnical equipment and systems.

IEC/TS 60695-5-2 [75] provides a summary and relevance of test methods and advises which are suitable for use in fire hazard assessment and fire hazard engineering. It includes special observations on the relevance, for electrotechnical products and their materials, to real fire scenarios and gives guidance on their use. It is intended to support product committees wishing to incorporate test methods for corrosion damage in product standards. The corrosion test methods and specifications are summarized in Table 12.26.

Table 12.25 Summary of test conditions and relevance of toxicity test methods

Test method	Reference	Static	Dynamic	Heat source	Test temperature (°C)	Irradiance (kW/m²)	Ventilation (l/min)
Airbus Directive ABD 0031	[156]	Yes	No	Electrical heater	n.a.	25 without/with pilot ignition flame	n.a.
NES 713	[157]			Methane Bunsen burner		n.a.	
NF C20-454	[158]	No	Yes	Electrical tube furnace	400; 600; 800	n.a.	2
CEI 20-37-7	[159]			Electrical tube furnace	800		
IEC 60695-7-50	[67]				350 650 825		1.1 2.6 2.7
Cone calorimeter ISO 5660-1	[79]			Electrical heater	n.a.	Up to 100 ignition source possible: electric spark	well ventilated conditions
IMO M 5.41 (64)	[160]	Yes	No			25 without/with pilot flame; 50 without pilot flame	n.a.
DIN 53436	[68]	No	Yes	Moving electrical tube furnace	Variable 200 to 900 (depends on fire scenario)		variable e.g. 100 to 300 (depends on fire scenario)
NBS-Cup	[161]	Yes	No	Electrical tube furnace	25 below and 25 above the test specimens ignition temperature	n.a.	n.a.
UPitt-box furnace	[162] [163]	No	Yes	Electrical furnace	Furnace temperature increased at 20 °C/min up to 1100 °C		11
ASTM E 1678-97 (NIST Radiant panel)	[164]	Yes	No	Radiantly heated combustion cell	n.a.	50 ignition source: high energy spark plug	n.a.

n.a. = not applicable   [a] Validation by animal experiment possible

Fire model		Chemical analysis system	Animal exposure system	Volume of exposure/analysis system (l)	Method	
Test specimen	Corresponds to ISO/TR 9122 fire stages				Provides toxic potency data	Could be adapted to provide toxic potency data
76.2 mm × 76.2 mm × intended installation thickness (up to 25 mm)	Difficult to characterize	No	No	n.a.		No
Strip-like specimen					No	
Piece of material mass dependent on sample density 0.5 g–1 g						Yes
Piece of material ; mass typically 1 g						
Strip-like sample; 10 g spread over 400 mm	1 b 2 3 a				Yes	
100 mm × 100 mm × thickness (up to 50 mm)	Well ventilated conditions				Yes (limited fire model)	
76.2 mm × 76.2 mm × thickness (up to 25 mm)	Difficult to characterize	Yes			No	
Strip-like sample; sample length: 400 mm; specimen width: 15 mm; mass per length of plastics typically 25 mg/mm	Yes			Typically 10		No
Mass typically 1 to 8 g	According to test protocol: flaming and non-flaming mode		Yes[a]	200	Yes	
Mass typically 1 to 10 g	Difficult to characterize			4		
Test specimen up to 76 mm × 127 mm × 51 mm (thickness); mass typically up to 8 g	Well ventilated conditions			200		

Table 12.26   Summary of test conditions of corrosivity test methods

Test method	Fire model — Heat source: Dynamic	Fire model — Heat source: Static	Test temperature (°C)	Ventilation (l/min)	Test specimen	Exposure model / Corrosion target — Corresponds to ISO/TR 9122 fire stages	Product testing possible	Simulated product testing	Indirect assessment (aqueous solutions of combustion effluents) — pH	Indirect assessment (aqueous solutions of combustion effluents) — Electrical conductivity
IEC 754-2		Fixed electrical furnace	950 (±50)	15–30	Mass: 1000 (±5) mg					X
CAN/CSA 22.2 No. 03 (1992)			800 (±10)	6 approx.	Mass: 500 (±50) mg					–
NF C20-453				15 to 30	Mass: 500 (±1) mg	Difficult to characterize	–	–	X	
VDE 0472 Part 813 (1992)			750 to 800	10 (±3)	Mass: 1000 mg					X
Copper mirror test as per ASTM D 2671-98 93.1; Procedure A		Combustion tube immersed in oil bath	Defined in applicable specification sheet	–	Test is applied to heat-shrinkable insulating tubing, area: 150 mm^2			Copper glass mirror		
ASTM D 5485-94a	conical electrical heater, heat flux[c] up to 100 kW/m^2		–	1440	Material, compound[b] or finished product[b], max. size: 100 mm × 100 mm, max. thickness: 6.3 (±0.5) mm	Well ventilated conditions	X	Typically "Rohrback-cosasco probes"[1]	–	–

Test method	Fire model					Exposure model / Corrosion target				
	Heat source		Test temperature (°C)	Ventilation (l/min)	Test specimen	Corresponds to ISO/TR 9122 fire stages	Product testing possible	Simulated product testing	Indirect assessment (aqueous solutions of combustion effluents)	
	Dynamic	Static							pH	electrical conductivity
ISO 11907-2		Small furnace with Ni-Cr electrical heater	Difficult to characterize	–	mass: 600 (±1) mg of the material to be tested, mixed with 100 (±1) mg polyethylene[d]	difficult to characterize	–	Copper[c] printed wiring board		
ISO 11907-3	Traveling electrical furnace		200–900 Depending on fire scenario	100 to 300	strip-like test specimen: length: 400 mm; width: 15 mm. Mass per length of plastics typically 25 mg/mm	yes	X	Copper[c] printed wiring board; sheets of different metals	–	–
ISO 11907-4	Conical electrical heater, heat flux[a] up to 100 kW/m²		–	1440	Material, compound[b] or finished product[b] (50 mm), max. size: 100 mm x 100 mm, max. thickness of material. test specimen 12 mm, max. thickness of finished products: 50 mm	Well ventilated conditions		Typically "Rohr-back-cosasco probes"[c]		

[a] 25 kW/m² and 50 kW/m² hat fluxes are recommended.
[b] Maximum thickness of compounds or finished products: 50 mm.
[c] This method measures the increase in electrical resistance.
[d] Non-flame-retardant polyethylene is added as co-combusting agent.

Prescribed are three approaches to testing differentiated by the type of corrosion target.

- Product testing: The corrosion target shall be a manufactured product, for example, printed wiring board, switchboards.
- Simulated product testing: The corrosion target shall be a reference material simulating a product.
  Effects can be assessed e.g. by mass loss or physical or electrical characteristics.
- Indirect assessment: No corrosion target is used, but characteristics of the gases and vapors evolved are measured, for example, the pH value and the conductivity of a solution in which the effluents have been dissolved.

The fire effluent shall be generated according to the relevant procedures proposed by IEC or ISO. Information on the applicability of the corrosivity tests to electrotechnical products can be found in [76–78].

12.1.2.6 Fire Safety Engineering

ISO/TR 13387 [11] gives the definition: Fire safety engineering is the application of engineering principles, rules, and expert judgment based on a scientific appreciation of the fire phenomena, of the effects of fire, and of the reaction and behavior of people, in order to:

- Save life, protect property and preserve the environment and heritage,
- Quantify the hazards and risk of fire and its effects,
- Evaluate analytically the optimum protective and preventative measures necessary to limit, within prescribed levels, the consequences of fires.

In the field of fire hazard assessment in electrical engineering, fire safety engineering (FSE) is still a relatively new part of the "tool box" TC 89 has made available to the Technical committees by IEC 60695-1-1. Up to now, most fire safety requirements in electrical engineering were based on prescriptive measures, that is, not scenario-dependent normative testing and assessment of products. These measures imply a level of fire safety acceptable to the IEC Technical Committees responsible for finished components and electrotechnical products. They have to fix the type of fire test and the strength of requirements. However, that fire safety level is rarely quantified in numerical terms and depends on external pressure rather than on technical considerations. Many existing fire tests in the field of electrical engineering cannot be used for input to FSE because they are pass-fail tests or their results cannot be expressed in engineering units. In most cases, it is not possible to translate test performance into "end-use" behavior.

There are already some tests that can be used for FSE purposes, for example, the product calorimeter according to UL 2043 [53] or the combined spread of flame/heat release tests for cables (see Section 12.5.6) or the cone calorimeter according to ISO 5660-1 [79]. However, more work is needed to develop these tests further and to standardize new procedures that are especially dedicated to FSE in the field of electrical engineering. To evaluate the applicability of fire modeling and hazard analysis when applied to the design of electrotechnical equipment, an appropriate input parameter must be obtained, fire modeling and hazard analysis conducted, and the results of the methodology tested against real-scale fire simulations designed to verify the predicted outcome. Special attention has to be paid to the side effects of a fire. Heat release, smoke density, and toxicity as well as corrosivity are potency values, which require FSE methods for the hazard assessment in electrical engineering with regard to temperature, visibility, toxic, and corrosive effects. The amount of decomposed or burned polymer, the volume of the room, the ventilation effects, and the stage of fire have to be taken into account. Further details on an introduction to the scenario-dependent assessment and judgment of the fire technological material behavior are given by Wittbecker [80]. Guidance will be given to IEC/TC 89 by ISO/TC 92/SC 4, which is responsible for fire safety engineering.

12.1.3 International Certification

The CB Scheme

Companies wishing to enter the global market have to overcome certain hurdles. Some years ago, manufacturers had to obtain national safety approvals for every country in which they sold their products in. To overcome these problems, the Certified Body (CB) Scheme has been established by IEC. The system is based on the principle of mutual recognition by means of certification through the use of internationally accepted standards. By virtually eliminating duplicate testing, the CB Scheme facilitates the international exchange and acceptance of product safety test results. The International Electrotechnical Committee for Conformity Testing to Standards for Electrical Equipment (IECEE) is a global network of National Certification Bodies (NCBs) that has agreed to mutual acceptance of CB test certificates and reports. The Scheme is based on the use of international (IEC) Standards. If some national standards of a member country adhering to the CB Scheme are not yet completely harmonized to IEC-Standards, national differences are permitted, if clearly declared to all other members. Currently there are more than 35 member countries in the IECEE. Further details about the 52 participating NCBs, the 121 CB Testing Laboratories, and an introduction in the CB Scheme are given in [82].

The CB Scheme applies to many specific product categories established by the IECEE, for example, IEC 60065: Safety requirements for mains operated electronic and related apparatus for household and similar general use, or IEC 60335: Safety for household and similar electrical appliances.

If a product is already UL-Listed, Classified, or Recognized (see Section 12.2.1), or has a test mark, such as a GS or VDE Mark (see Section 12.3.3), the required testing for the CB Scheme is mostly already completed and accepted. In some cases, additional testing is necessary to comply with national deviations required in countries to which the products are to be marketed.

12.2 Fire Regulations in North America (UL, CSA)

Electrotechnical products are tested worldwide according to established safety standards. Whereas in many countries the testing regulations and performance requirements are based on standards developd by the IEC, in the United States and Canada, national standards that sometimes deviate from the IEC standards are commonly used and recognized. In the United States, safety testing of electrotechnical products is often based on Underwriters Laboratories Inc. (UL). In Canada, the standards issued by Canadian Standards Association (CSA) are applied.

The ongoing harmonization of UL and CSA testing standards and the corresponding alignment of existing standards and regulations led to a situation in 1993 in which the two organizations entered into a Memorandum of Understandig (MoU) granting each other acceptance of test data, with verification, for the purpose of listing or recognition.

12.2.1 Test and Approval Procedures of the Underwriters Laboratories

Underwriters Laboratories Inc. (UL) was founded in 1894 with the aim of carrying out technical safety tests for the protection of people and goods. As a not-for-profit organization without capital stock, UL must use the profits to establish, maintain, and operate laboratories for examination and testing of devices, systems, and materials to determine their relationship to hazards to life and property, and to elaborate standards, classifications, and specifications, for example, devices, products, and equipment [81].

UL started their work by carrying out technical fire protection tests. Later, their field of activity expanded and today it includes product safety testing in the field of, for example, fire protection, electrical engineering, fire and burglary protection, heating and air conditioning, chemical safety, marine safety, automotive safety, and certification.

UL has fire testing laboratories located in Northbrook, IL; Melville, NY; Santa Clara, CA; Research Triangle Park, NC and Camas, WA, and subsidiaries in Mexico, Denmark, Germany, England, Italy, India, Singapore, Taiwan, Hong Kong, and Japan.

Additionally UL has numerous international, affiliate and representative offices as well as field representatives located throughout the world. The UL websites [83] give detailed information about e.g. global resources, UL news, testing and certification services or information about UL-certified products.

UL Standards for Safety are developed under a procedure that provides for participation and comment from the affected public as well as industry.

Thus, manufacturers, consumers, or consumer-oriented organizations, governmental officials, inspection authorities, and insurance interests provide input to UL before UL draws up a final version of the UL Standard taking into account the comments made. The UL Standards were often taken over by other bodies and standards organizations. The UL Standards Department has developed a separate Web site that contains information on UL-Standards, for example the Catalogue of Standards and the scopes of the standards. Today, a total number of 86 UL Standards are harmonized with CSA, IEC and ISO. Three types of UL certifications exist: Listing, Classification, and Recognition.

Successful completion of a Listing Certification allows a finished product to carry the UL listing mark, which is similar to the German VDE test mark. In contrast, Classification covers specific characteristics of the finished product designated by the manufacturer. It results in an UL Mark incorporating wording for a specific hazard, or if evaluated to a non-UL standard, for example, ASTM, OSHA, and so forth, recognition is a preselection certification for materials and components and results backwards in a UL Mark. The UL categories of materials are very important in connection with the use of plastics in electrotechnical products, because specific ratings are often a prerequirement of the Listing of a product by UL. By itself, neither a part nor a material can gain a UL Mark, but the testing of a certified product is simplified if materials that are already UL-Recognized are used.

On behalf of UL, field representatives and authorized representatives visit periodically and unannounced the manufacturers' facilities for inspecting and monitoring the manufacturers' production and verifying if the product is still in compliance with the UL Report requirements. The field representative randomly selects production samples for Follow-Up testing at UL. The UL product directories help to find products that meet UL requirements. They contain the names of companies that manufacture products, components, devices, materials or systems in accordance with UL safety requirements.

These are products eligible to carry UL's Listing Mark, Recognized Component Mark, or Classification Marking. The Plastics Recognized Component Directory contains the category of polymeric materials components QMFZ2. Recognition cards, also known as "Yellow Cards," are an excerpt of the Recognized Component Directory. They give the manufacturers the documentary evidence that their products meet the relevant UL requirements.

12.2.1.1 UL 94. Flammability tests

UL 94 "Test for Flammability of Plastic Materials for Parts in Devices and Appliances" is one of the most important UL standards relating to fire safety test methods and requirements and contains several fire tests [42].

The various flaming tests according to UL 94 that have been adopted by national or international standardization bodies such as the International Electrotechnical Commission (IEC) can be classified as exclusively material tests. They involve standard-size specimens and are intended to serve as a preliminary indication of their acceptability with respect to flammability for a particular application. These methods are not intended to provide correlation with performance under actual end-use conditions. The final acceptance of the material is dependent on its use in complete equipment that conforms with the standards applicable to such equipment.

The requirements of UL 94 together with a few other tests described in UL 746 A, B, and C form the basis for the "Recognition" of plastics as summarized in the "Recognized Component Directory." UL 94 applies not only to electrical parts, appliances, and consumer and office equipment but also to all areas of application except the use of plastics in buildings. UL 94 is particularly significant for the use of plastics in electrotechnical products, as a UL Listing of the product frequently requires a specific flammability classification of the materials used.

Testing for Classification in Class HB

Depending on the fire requirements of the UL standards for the end-product, materials have to meet the horizontal burning test (class HB) or the more stringent vertical burning tests (class UL V-0, V-1, or V-2). These test methods and their criteria are identical to IEC 60695-11-10 (see Section 12.1.2.3). Successful completion of the horizontal test leads to classification in class HB. Horizontally oriented test specimens are exposed to a 20 ($\pm$ 1) mm (0.79 in.) flame source for 30 s. UL® is a registered trademark of Underwriters Laboratories, Inc.

The test layout is illustrated in Fig. 12.7 in Section 12.1.2.4. Classification in class HB requires that the burning rate of the test specimen should not exceed a maximum value dependent on its thickness or that the specimen extinguishes after removal of the flame. The test specifications are summarized in Table 12.27.

Table 12.27   UL 94 Testing for classification in class HB

Specimen	Three bar-shaped test specimens 125 ($\pm$ 5) mm × 13.0 ($\pm$ 0.5) mm with two reference marks 25 ($\pm$ 1) mm and 100 ($\pm$ 1) mm from the end that is to be ignited. Maximum thickness is 13 mm. Depending on test performance, it may be necessary to test a further set of three samples.
Specimen conditioning	At least 48 h at 23 ($\pm$ 2) °C and 50 ($\pm$ 5) % relative humidity
Specimen position	Longitudinal axis of specimen approximately horizontal, traverse axis inclined 45 ($\pm$ 2)° to horizontal
Ignition source	Laboratory burner according to IEC 60695-11-4 flames A, B or C 20 ($\pm$ 2) mm non-luminous flame [ISO 10093 describes the burner as ignition source P/PF 2 (50 W)]. The longitudinal axis of the burner is tilted at 45° toward the horizontal during application of flame.
Application of flame	30 s. If the flame front reaches the 25 mm mark on the specimen before 30 s, the burner is removed.
Conclusion	Classification in class HB 40 if: • There is no visible burning with a flame after removal of the flame. • The test specimen extinguishes before the 100 mm mark is reached. • The burning rate between the reference marks does not exceed 40 mm/min. Classification in class HB 75 if: • The burning rate between the reference marks does not exceed 75 mm/min.

Testing for Classification in Classes V-0, V-1, and V-2

This well-known test for classifying solid plastic specimens in classes V-0, V-1, and V-2 is a small-scale laboratory screening procedure for comparing the relative burning behavior of vertically oriented specimen, exposed to a small-flame ignition source. The test layout is illustrated in Fig. 12.8 in Section 12.1.2.4. The vertical test takes the same specimens as are used for the HB test. A flame is applied twice to the lower end of the test specimen for 10 s. Burning and/or glowing times, when dripping occurs, and whether or not the cotton beneath is ignited are noted. The best rating, V-0, is achieved if the mean afterflame time of five samples after 10 applications of the flame does not exceed 5 s. The material is placed in class V-1 if the mean afterflame time is < 25 s. If flaming particles ignite the surgical cotton, the material is classified in V-2. The test specifications are summarized in Table 12.20 of Section 12.1.2.4.

Testing for Classification in Classes VTM-0, VTM-1, and VTM-2

UL 94 contains a method for testing materials that, because of their thinness, distort, shrink, or are consumed up to the holding clamp when tested using the UL 94 Vertical Burning Test. The test method and criteria are identical to ISO 9773 [84].

Instead of solid test specimens, 200 mm (8 in.) long cylindrically wound rolls of films are used. Oriented specimens are tested vertically. The film roll is produced by winding a 200 mm x 50 mm (8 in. x 2 in.) strip of the test sample around a 13 mm (0.5 in.) diameter metal rod. A piece of adhesive is used to prevent unrolling. After the rod is withdrawn and the film roll clamped vertically, the flame is applied twice for 3 s to the lower end of the film roll. The test specifications are summarized in Table 12.28. The main assessment criteria are the same as in the vertical test for solid specimens, that is, afterflame time and igniting of the surgical cotton.

Testing for Classification in Classes 5 V

Testing for classification in classes 5 V is the most stringent of all UL small-scale burning tests. The test for vertically positioned solid plastic specimens differs from the UL 94 V test described in the preceding mainly by the five applications of flame to each specimen. The test method and criteria are identical to IEC 60695-11-20 (Table 12.21 in Section 12.1.2.4). The 5 VA classification is specified for fire enclosures on office machines (see UL Standard 1950 in Section 12.2.1.2).

Testing for Classification in Classes HBF, HF-1, or HF-2

UL 94 contains a method for testing foam materials that are oriented in a horizontal position. The test method and criteria are identical to ISO 9772 [85]. The test apparatus is illustrated in Fig. 12.11. A flame from a special burner is applied to one side of the foam sample lying horizontally on a wire grid. Class HBF is achieved if the burning rate does not exceed 40 mm/min (1.57 in./min). If the specimen extinguishes within 2 s of removal of the ignition source, it is classified in classes HF-2 or HF-1 depending on whether burning drippings occur. The test specifications are summarized in Table 12.29.

Table 12.28   UL 94 test specifications for classification in classes 94 VTM-0, 94 VTM-1and 94 VTM-2

Specimen	Two sets of five film rolls 200 mm long, 12.7 mm internal diameter, marking 125 mm from one end. The film rolls are to be prepared by wrapping a specimen 200 mm long by 50 mm around the axis of a 12.7 ($\pm$ 0.5) mm diameter mandrel. Depending on test performance further sets of samples may have to be tested.
Specimen conditioning	• Storage of first set of samples at least 48 h in 23/50 standard atmosphere, • Storage of second set of samples for 168 h in the air-circulating oven at 70 ° C followed by 4 h cooling to room temperature in desiccator.
Specimen position	Film roll clamped vertically with reference mark 125 mm above bottom of sample, film roll closed by spring clamp at top and open at bottom, horizontal layer of surgical cotton 50 mm × 50 mm × 6 mm located 300 ($\pm$ 10) mm beneath bottom of sample, mass of the cotton ≤ 0.08 g
Ignition source	Bunsen or Tirrill burner with 20 ($\pm$ 1) mm high nonluminous flame
Application of flame	Twice 3 ($\pm$ 0.5) s for each sample, the second application starts as soon as the sample extinguishes.
Conclusions	classification in class 94 VTM-0 if: • Afterflame time ≤ 10 s, • Sum of the afterflame times for 10 applications of flame does not exceed 50 s, • No burning drops, • No material consumed as far as the 125 mm mark, • Individual test specimen afterflame plus afterglow time not more than 30 s after second flame application. Classification in class 94 VTM-1 if: • Afterflame time ≤ 30 s, • Sum of the afterflame times for 10 applications of flame does not exceed 250 s, • Individual test specimen afterflame plus afterglow time not more than 60 s after second flame application, • Other criteria as for class 94 VTM-0. Classification in class 94 VTM-2 if: • Ignition of surgical cotton by flaming particles or burning drops. • Other criteria as for class 94 VTM-1.

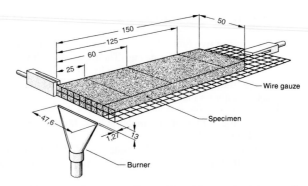

Figure 12.11   Test layout for classification in classes HBF, HF-1, and HF-2 according to UL 94

Table 12.29   UL 94 HBF, HF-1 and HF-2: Horizontal burning foamed material test specifications

Specimen	Two sets of five specimens 150 (± 1) mm × 50 (± 1) mm in the minimum and maximum thickness of intended application with three reference marks 25 mm, 60 mm, and 125 mm from the end to be ignited, should be tested. Maximum thickness 13 mm. Depending on performance, it may be necessary to test further sets of samples.
Specimen conditioning	• Storage of one set of specimens for at least 48 h at 23 (± 2) °C and 50 (± 5)% relative humidity, • Storage of one set of specimens for 168 (± 2) h in the air-circulating oven at 70 (± 2) °C followed by 4 h cooling to room temperature in a desiccator.
Specimen position	Specimen lies horizontally on a wire grid of specified mesh, horizontal layer of surgical cotton approx. 50 mm × 50 mm × 6 mm placed 175 (±25) mm below specimen
Ignition source	Burner with a special wing tip to give a 38 (± 1) mm nonluminous flame
Application of flame	60 s
Conclusions	Classification in class HBF if: • Burning rate between the 25 mm and 125 mm marks does not exceed 40 mm/min, or the specimen extinguishes before the 125 mm mark and the requirements for classes HF-1 and HF-2 are not fulfilled Classification in class HF-1 if: • Afterflame time ≤ 2 s for at least four of five specimens, • Afterflame time does not exceed 10 s for any specimen, • No destruction of sample past 60 mm mark, • No incandescence for more than 30 s for each individual specimen after ignition source is removed, • No ignition of the cotton indicator. Classification in class HF-2 if: • Ignition of the cotton indicator, • Other criteria as for class HF-1.

Radiant Panel Flame Spread Test

In the case of large housing parts, it is necessary to determine their possible contribution to flame spread. UL 94 requires that the spread of flame be evaluated in accordance with ASTM E 162 [86] (Table 12.30). The end-product criteria establish the level of flame spread accepted. The test setup is shown in Fig. 12.12. The test specimen with the dimension 6 in. x 10 in. (150 mm x 460 mm) is exposed to a radiant panel and ignited at its upper hot end, after which the extent of combustion toward the bottom of the specimen is evaluated.

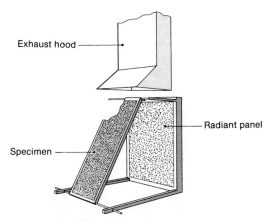

Figure 12.12   Radiant panel test apparatus

Table 12.30   Radiant panel flame spread, test specifications

Specimen	Four test specimens 460 (± 3) mm × 150 (± 3) mm in the minimum and maximum thickness, six specimens if the average flame spread is < 50	
Test layout	The flame spread index shall be determined to ASTM E 162	
Conclusions	Radiant panel flame spread classes	
	Flame spread	*Flame spread class*
	(average of four specimens)	
	15 maximum	RP 15
	25 maximum	RP 25
	50 maximum	RP 50
	75 maximum	RP 75
	100 maximum	RP 100
	150 maximum	RP 150
	200 maximum	RP 200

12.2.1.2 Fire Regulations in other UL Standards

UL Standard 746A

The safety of electrical equipment depends on the correct selection of materials, design, and processing of parts as well as the assembly, mounting, and relative positions of these parts. The properties required for individual parts are defined by the function or functions of the part.

UL 746A contains short-term test procedures to be used for the evaluation of materials used for parts intended for specific applications in electrical end-products [87].

The evaluations according to UL 746A provide data with respect to the physical, electrical, flammability, thermal, and other properties of the materials under consideration and are intended to provide guidance for the material manufacturer and other interested parties (Table 12.31).

Programs for the evaluation of material modifications, such as plating of plastics or the use of flame-retardant paints, are contained in the Standard UL 746C. Only when it is not practical to obtain test specimens from the finished article, the manufacturer shall furnish molded test specimens, or sample sheets as required in the individual test method. Because the flammability of plastics may vary with thickness and with orientation of the material, consideration is to be given to the testing specimens that they are representative of the orientation and both the thickest and thinnest sections of the finished article.

Comparative Tracking Index

A risk of fire may develop within electrical equipment as a result of electrical tracking of insulating material that is exposed to various contaminating environments and surface conditions. The comparative tracking index (CTI) provides a comparison of the performance of insulating material under wet and contaminated conditions.

The CTI is the numerical value of the highest voltage (max. 600 V) at which an electrical insulation material will withstand 50 drops of electrolytic test solution A (398 Ω cm) being dripped onto it without permanent electrically conductive tracking. The test is carried out under the conditions specified in ASTM D 3638-85 [88] or IEC 112 [89]. The Comparative Tracking Performance Level Category (PLC) is to be assigned on the basis of the CTI (Table 12.31).

Hot Wire Ignition Test

Under certain conditions of operation or malfunctioning of electrical equipment, wires, other conductors, resistors, or other parts become abnormally hot. When these overheated parts are in intimate contact with insulating materials, the insulating materials may ignite. The hot wire ignition test to UL 746A, Section 30, is intended to determine the relative resistance of insulation materials to ignition under such conditions. For a given material the hot wire performance level is to be assigned based on the determined mean time for ignition (Table 12.31). Hot wire ignition tests are carried out on five bar-shaped specimens of dimensions 125 ($\pm$ 5) mm (4.9 in.) long, 130 ($\pm$ 0.3) mm (5.1 in.) wide, and 3.0 ($\pm$ 0.1) mm (0.12 in.) thick. For more details, please see the description of the technically equipment hot wire coil ignitability test on materials in Section 12.1.2.4.

High-Current Arc Ignition

Under certain normal or abnormal operation of electric equipment, insulating material might be in proximity to arcing. If the intensity and duration of the arcing are severe, the insulating material can become ignited. The test according to UL 746A Section 31 is intended to simulate such a condition. In this test, the specimen is located in the direct vicinity of a high-current arc. The number of arc ignitions up to specimen ignition is determined.

The PLC is assigned on the basis on this number (Table 12.31).

Table 12.31 UL 746A Polymeric materials short-term property evaluations

High-voltage, low-current, dry arc resistance	Comparative tracking level [CTI]	High-voltage arc-tracking rate [HVTR]	Hot wire ignition [HWI]	High-current arc ignition [HAI]	High voltage arc resistance to ignition [HVAR]	Assigned PLC
ASTM D 495 range-mean time of arc resistance [TAR] (s)	Range-tracking index [TI] (v)	Range-tracking rate [TR] (mm/min)	Range-mean ignition time [IT] (s)	Range-mean number of arcs to cause ignition [NA]	Range-mean time to ignition [TI] (s)	
420 ≤ TAR	600 ≤ TI	0 < TR ≤ 10	120 ≤ IT	120 ≤ NA	300 ≤ TI	0
360 ≤ TAR < 420	400 ≤ TI < 600	10 < TR ≤ 25.4	60 ≤ IT < 120	60 ≤ NA < 120	120 ≤ TI < 300	1
300 ≤ TAR < 360	250 ≤ TI ≤ 400	25.4 < TR ≤ 80	30 ≤ IT < 60	30 ≤ NA < 60	30 ≤ TI < 120	2
240 ≤ TAR < 300	175 ≤ TI ≤ 250	80 < TR ≤ 150	15 ≤ IT < 30	15 ≤ NA < 30	0 ≤ TI < 30	3
180 ≤ TAR < 240	100 ≤ TI ≤ 175	150 < TR	7 ≤ IT < 15	0 ≤ NA < 15		4
120 ≤ TAR < 180	0 ≤ TI ≤ 100		0 ≤ IT < 7			5
60 ≤ TAR < 120						6
0 ≤ TAR < 60						7

Performance level categories (PLC)

High-Voltage Arc Tracking Rate

High-voltage arcs are ignited on the surface of the specimen for a period of 2 min. The electrodes are moved away from each other until such time as the arc has been extinguished. Once it has been extinguished the electrodes are moved together again until it ignites. After a period of 2 min, the length of the conductive tracking path established in this way is measured and divided by two. Table 12.31 indicates the conductive path generated in mm/min and the corresponding TR values and PLC classes.

High-Voltage, Low-Current, Dry Arc Resistance

The specimen is exposed to a high-voltage arc for 7 min (max.) while the load is increased.

The time is measured, in seconds, up to the point at which the arc is extinguished between the ends of the test unit through a low-resistance tracking path on the surface. The assessment is made on the basis of the eight PLC classes shown in Table 12.31. The results obtained with material approx. 3 mm (0.12 in.) thick are regarded as representative of the behavior of materials in any thickness.

High-Voltage Arc Resistance to Ignition

A high-voltage arc is generated on the surface of the specimen for a maximum of 5 min. The time is measured, in seconds, up to the point at which a hole forms or the specimen ignites. The result is summarized in the performance level classes 0–3 set out in Table 12.31 and included in the UL card as the Performance Level Class.

Glow-Wire Ignitability Test

The test method for the determination of ignitability of an insulating material from an electrically heated wire is described in the Glow-wire ignitability test method on materials IEC 60695-2-13 (see Section 12.1.2.4). The Glow-Wire Ignition Temperature (GWIT) is assigned as the temperature that is 25 °C higher than the temperature of the tip of the glow wire that does not cause ignition during three consecutive tests.

UL Standard 746B

UL 746B contains long-term test procedures to be used for the evaluation of materials used for parts intended for specific applications in end-products [90]. The relative thermal index, which is an indication of the material's ability to retain a particular property when exposed to elevated temperatures for an extended period of time, is assessed on an evaluation of long-term thermal-aging data. The tests include mechanical, electrical, and flammability (vertical burning UL 94) properties.

UL Standard 746C

UL 746C contains requirements with respect to the mechanical, electrical, thermal, and flammability properties of plastics depending on their application in electrical equipment, devices, and appliances, for example, as material for polymeric enclosures, internal barriers or as carriers for live parts [91]. The service conditions of the electrotechnical product, for example portable or fixed or stationary equipment as well as the interaction of the materials properties is considered and the individual requirements are chosen so that in combination they ensure an adequate safety level for the polymeric part in the finished electrotechnical product. The test procedures include reference to the data obtained from the UL 94 flammability tests, the standard property tests in UL 746A, as well as other means of evaluation. The standard contains also programs for the investigation of material part modifications, such as plating of plastics or the use of flame-retardant paints.

Because the methods described in UL 94 are intended to be used to measure and describe the flammability properties of materials in the form of standard-size specimens in response to heat and flame under controlled laboratory conditions, UL 746C includes tests of finished components to assess the influence of the structural makeup of the finished part on the actual response to heat and flame of the electrotechnical product.

For example, enclosures for electrotechnical appliances must pass a fire performance test with a flame ignition source as a finished product if the material does not fulfill the relevant UL 94 requirements 5 VA, 5 VB, V-0, V-1, or V-2 for testing flammability of plastic materials. The test conditions depend on the use of the appliance. Portable appliances, which are usually under supervision, are tested with 12 mm (0.47 in.) high needle flame or a 20 mm (0.79 in.) high flame applied twice for 30 s with a break of 1 min. The results are acceptable if all flames are extinguished within 1 min after the ignition source is removed. Electrical equipment that is fixed or stationary and not easily carried or conveyed by hand is tested with a 127 mm (5 in.) high flame applied five times for 5 s with intervals of 5 s.

After the fifth flame application, the afterburning time must not exceed 1 min. In addition, the housing must not be destroyed in the area of the test flame to such an extent that the integrity is affected with regard to the containment of fire.

Flaming drops or flaming or glowing particles that ignite the surgical cotton below the test specimen are not allowed. Alternatively, the material is to be considered equivalent if it complies with the requirements for the relevant UL 94 material classification. UL 746C also requires flame retardant coatings to pass the enclosure flammability tests with the flaming ignition sources described in the preceding if the coated test specimens do not fulfill the requirements of the relevant UL 94 material specifications. UL 746C contains in section 50 the enclosure flammability 746 5VS test for coated materials. Test plaques in the dimensions 150 mm x 150 mm (5.9 in. x 5.9 in.) and in the minimum and maximum thickness are tested with a 125 mm (4.9 in.) high flame with an inner blue cone applied for 60 s.

Several test methods have been employed by UL to judge the burning behavior of external plastic materials of appliances and devices depending on their type and size, the likelihood of ignition of the material due to its proximity to one ignition source within the product or by an external ignition source and the amount (surface area) of plastic involved.

For intermediate and larger size products, consideration should be given to the probability of ignition of the material by sources within the equipment or by external sources. UL uses primarily two larger laboratory-scale fire tests to assess the burning behavior of such materials. Generally, the plastic surfaces associated with intermediate size products are exposed to the radiant heat and flaming pilot ignition of the ASTM E 162 Radiant Panel Test (see Fig. 12.12). The test specimen is exposed to an external gas-fired heat radiator and ignited at its upper, hot end, after which the extent of combustion toward the bottom of the specimen is evaluated. The rate of flame spread and the temperature rise in the overhead exhaust stack of the apparatus are used in calculating a flame spread index. A maximum value of 200 is allowed.

The surface plastic materials used in larger size [projected surface $> 0.93 \, m^2$ or a single linear dimension $> 1.83 \, m$ (6 ft)] products are exposed to the large flame of the UL 723 (ASTM E 84) 25 ft Tunnel Test [92]. This provides for comparative measurement of the surface flame spread of materials on a classification scale, where red-oak lumber exhibits one approximate flame spread value of 100, and inorganic cement board exhibits a flame spread value of zero. Large specimens, normally 0.51 m (21 in.) wide x 7.3 m (24 ft) in length form the top of a horizontally positioned test furnace. Details of the test setup and specifications can be taken from Section 10.2 Building USA. Larger size products shall have a maximum in flame spread rating of 200.

UL 746 C also requires that polymeric materials used in electrical equipment pass two tests with electrical ignition sources, the hot wire ignition test and the high current arc ignition test. The requirements in both tests depend on the UL 94 flammability classification of the polymeric material. According to UL 746 C, polymeric materials used in enclosures for portable appliances should not ignite before 7 s when tested by the hot wire ignition test. Materials that do not comply with the minimum-hot-wire ignition levels may be evaluated by an abnormal overload test or the glow-wire end-product test.

The abnormal overload test passes abnormal high currents through current-carrying parts of the equipment or representative sections of it. Overcurrent values and times depend on the overcurrent protective device rating. The glow-wire temperature requirements are based on the functional end-use application of the product. As an example fixed or stationary equipment is tested at 750 °C.

In the high current arc test that is described in UL 746 A, Section 31, the number of arc ignitions required to set the specimen in fire is determined. According to UL 746 C, polymeric materials in contact with uninsulated live parts of a thickness of < 0.8 mm (0.03 in.) should survive at least 15 cycles without igniting.

UL Standard 746 D

The Standard for Polymeric Materials – Fabricated Parts, UL 746 D, contains requirements for traceability and performance of parts molded and fabricated from polymeric materials as well as requirements for materials that have been modified to match the requirements of a specific application, including the use of recycled and regrind materials, the use of additives, for example, flame retardants and colorants and the blending of two or more materials [93].

UL Standard 1694

Small components, that is, encapsulated integrated circuits, which contain materials that cannot be fabricated into standardized specimens in the minimum use thickness and subjected to applicable preselection tests such as UL 94, are tested with a needle flame to simulate the effect of small flames, which may result from fault conditions within equipment. The fire hazard is assessed by a simulation technique. The burner and the confirmatory test procedure are in accordance with IEC 60695-2-2 which is referenced in Section 12.1.2.3. The test layout corresponds to Fig. 12.5 in Section 12.1.2.3.

The 12 (± 1) mm (0.47 in.) long needle flame is applied to that part of the surface of the specimen that is likely to be affected by flames resulting from fault conditions or from any source of ignition accidentally applied.

Whether a small component ignites, depends, to a large extent, on its mass or volume and how long the flame is applied. Four different times from 5 to 30 s depending on the volumetric range of the component are given in UL 1694 [94].

Assessment criteria are the same as in the UL 94 vertical test for solid specimens, that is, after-flame time and burning drops. Additional material classifications exist for small components, which after ignition by the test flame are totally consumed but have a limited burning time. The test specifications are summarized in Table 12.32. Small components that are molded from materials that are classed 5 VA, 5 VB, V-0, V-1 or V-2 in the minimum use thickness need not be subjected to the needle-flame test.

Table 12.32   UL 1694: Flammability test for small components, test specifications

Specimens	Five samples of the small component
Specimen conditioning	Storage of the specimens for a minimum of 24h in a forced draft air circulating oven followed by 2h cooling to room temperature in a desiccator
Specimen position	Specimen arranged in its most unfavorable position of normal use. A horizontal layer of surgical cotton approx. 50 mm × 50 mm × 6 mm is placed 200 (± 5) mm below the bottom edge of the small component
Ignition source	Needle flame burner according to IEC 695-11-7 12 (± 1) mm high

Application of flame	*Volumetric range (mm^3)*	*Time of each (twice) flame application time (s)*
	$V \le$ 250	5
	$250 < V \le$ 500	10
	$500 < V \le 1750$	20
	$1750 < V \le 2500$	33

The second application starts as soon as the specimen, ignited by the first application, extinguishes.

Conclusions	Classification in class SC-0 if: • Afterflame time ≤ 10s, • Sum of the afterflame times for 10 flame applications does not exceed 50s, no burning drops, • Individual test specimen afterflame plus afterglow time not more than 30s after second flame application, • No complete consumption of component. Classification in class SC-1 if: • Afterflame time ≤ 10s, • Sum of the afterflame times for 10 flame applications does not exceed 250s, • Individual test specimen afterflame plus afterglow time not more than 60s after second flame application, • Other criteria as for class SC-0. Classification in class SC-2 if: • Ignition of surgical cotton by flaming particles or burning drops, • Other criteria as for class SC-1. Classification in SC-TC 0; SC-TC 1 and SC-TC 2 if: • Small component is totally consumed; • Other criteria as for classes SC-0, SC-1, SC-2.

UL Standard 1950

The common UL and CSA Standard for Safety of Information Technology Equipment, including Electrical Business Equipment UL 1950, CSA 22.2 No. 950 [95], is based on IEC 950 [96] (also under Section 12.6.2.1).

The standard contains basic requirements for products covered by UL under its Follow-Up Service and its application is intended to prevent injury or damage due to, for example, electric shock, mechanical and heat hazards, and fire. Temperatures that could cause a fire risk in information technology equipment such as visual display units and data processing equipment may result from overloads, component failure, insulation breakdown, high resistance, or loose connections. However, fires originating within the equipment should not spread beyond the intermediate vicinity of the source of fire or cause damage to the surrounding of the equipment.

To fulfill this protective goal, according to UL 1950, the following design objectives should be met:

- Taking all reasonable steps to avoid high temperature that might cause ignitions,
- Controlling the position of combustible materials in relation to possible ignition sources,
- Limiting the quantity of combustible material used,
- Ensuring that, if combustible materials are used, they have the lowest flammability practicable,
- Using enclosures or barriers, if necessary, to limit the spread of fire within the equipment,
- Using suitable materials for the outer enclosures of the equipment.

Materials and components inside a "fire enclosure" and air-filter assemblies should have a flammability class according to UL 94 of V-2, HF-2 or better, or they should pass the flammability test for fire enclosures of movable equipment having a mass $\leq$ 18 kg (39.16 lb), which is described in detail below.

The standard contains numerous exemptions, for which these requirements do not apply. Examples are materials and components that either meet the flammability requirements of the relevant IEC component standard or which are encapsulated within a metal enclosure with a volume of $0.06 \mathrm{m}^3$ or less without any ventilation openings or within a sealed unit filled with an inert gas.

Materials used for enclosures of equipment and decorative parts should be such that the risk of ignition and the spread of fire or flames are minimized. For movable equipment having a total mass not exceeding 18 kg, fire enclosures are considered to comply without a flammability test on the finished product if, in the smallest thickness used, the material fulfills the requirements of class V-1 or better. Alternatively, the complete part of the fire enclosure, a section of the part or a test plaque representing the thinnest thickness is tested with a 20 mm (0.79 in.) high Bunsen burner flame, or a 12 mm (0.47 in.) needle flame according to IEC 60605-2-2 (Section 12.1.2.3), which is applied to an inside surface of the sample at a point judged to be likely to become ignited because of its proximity to possible sources of ignition. The flame is applied twice for 30s with a break of 1 min after the second flame application; the afterburning time must not exceed 1 min.

For movable equipment having a total mass exceeding 18 kg and for stationary equipment, fire enclosures are considered to comply without test if, in the smallest thickness used, the material is of flammability class 5 V. Alternatively three samples, each consisting of either a complete fire enclosure or a section representing the least wall thickness and including any ventilation opening are tested with a 130 mm (5.1 in.) high flame with an inner cone of 40 mm (1.58 in.). The flame is applied five times for 5s with intervals of 5s to an inside surface at the most critical place considering internal ignition sources. After the fifth flame application, the afterburning time must not exceed 1 min. In addition, flaming drops or glowing particles that ignite the surgical cotton 300 mm (11.8 in.) below the test specimen are not allowed.

12.2.2 CSA Test and Approval Procedures

In describing product fire performance similarly to the UL Yellow Card, the Canadian Standards Association (CSA) rates polymeric materials or finished parts of an electrical equipment according to a number of standardized tests. Flammability standards are summarized in C22.2, No. 0.6-M 1982: Flammability testing of polymeric materials [97]. This standard consists of 10 different (tests A–J) test procedures. Most of them are identical with the various UL 94, IEC-60695-11, or UL 746A test methods.

The CSA C22.2 test standard contains two flame tests for housings for live parts. One of these tests is carried out by applying a 19 mm (0.75 in.) high luminous Bunsen burner flame twice for 30s with an interval of 1 min to the test component or to a corresponding thick sheet of insulating material. Depending on the type and use of the electrical equipment, the test specimen is positioned horizontally or vertically. As a rule, the test is passed if the after-flame time after both flame applications does not exceed 1 min, and if the test specimen does not exhibit any great damage. CSA C22.2 Test A is carried out in a similar manner to the UL 94 5V test. However, the test is more severe: each application of the 127 mm (5 in.) long Bunsen burner flame to the finished housing or to a part of a sheet of insulating material of the same thickness is of 15s duration. In addition, during the first four flame applications, the afterflame time has to be not more than 30s, and after the fifth application not more than 60s.

12.3 Fire Regulations and Standards in Europe

12.3.1 European Committee for Electrotechnical Standardization (CENELEC)

CENELEC is the European Committee for Electrotechnical Standardization. It was set up in 1973 as a nonprofit organization under Belgian Law, and has been officially recognized as the only European Standards Organization in its field by the European Commission. CENELEC supports the policies of the European Union and EFTA for free trade, worker and consumer safety, and public procurement. The members of CENELEC are the National Electrotechnical Committees of the 19 member countries of the European Union and EFTA. All interested parties are consulted during the CENELEC standards drafting, through involvement in technical meetings at national and European levels and through inquiries conducted by the members. The Technical Board (BT) coordinates the work of the technical bodies, for example, Technical Committees (TC) or Sub-Committees (SC). The TCs are responsible for the preparation of the European Standards (EN). TCs take into account any ISO/IEC work coming within their scope together with such data as may be supplied by members and by other relevant international organizations, and work on related subjects in any other Technical Committee. CEN, CENELEC, and ETSI, the three European Standards Organizations, have jointly developed a common Web site [98] for the European Committee for Electrotechnical Standardization.

Reporting Secretariats (SRs) exist to provide information to the Technical Board on any ISO or IEC work that could be of concern to CENELEC. Once CENELEC has started work by selecting an international standard or any other document to develop into a European Standard, all national work on the same subject is stopped. This suspension of national work is called "Standstill." New standardization initiatives originating in Europe are offered to IEC with the request that they will be undertaken at international level.

Only if IEC does not want to undertake the work, or if it cannot meet CENELEC target dates, does the work continue at European level. Afterwards, the result is offered to IEC again. All drafts of an IEC standard are the object of a parallel voting procedure in CENELEC and IEC.

Globalization of the markets affects the volume of CENELEC work due to progressively shifting emphasis toward international standardization in IEC. Today, in more than 80% of cases, the initial document comes from the IEC.

European Standards are established as a general rule because it is important that members' national standards become identical wherever possible.

Members are obliged to implement European Standards by giving them the status of a national standard. A list of the membership of CENELEC, their affiliates, cooperating partners, and other principal electrotechnical standardization organizations is given in this book in Section 6 of the Appendix.

12.3.2 Electrotechnical Products and EU Directives

The primary purpose of the several Directives laid down by the European Commission was to ensure the free movement of goods [99]. To support this objective each Directive established "essential requirements" (ER) and required these requirements be incorporated in European harmonized standards.

In the frame of the future European Technical Approval (ETA), every product standard shall have, as far as possible, a harmonized indication related to the level of performance or classes under the individual Directives' scope. Where necessary, existing standards may be amended for fully meeting the ER of a European Directive. The following EU Directives are important for electrotechnical products:

- Construction Products Directive,
- Low Voltage Equipment Directive,
- Safety of Machinery,
- Directive on the interoperability of the European high-speed train network.

Construction Products Directive and Low Voltage Directive

At present, the Member States of the EU are in the process of changing over to a harmonized reaction-to-fire classification system based on the Construction Products Directive (CPD) [100], relating to construction products. In general, products permanently installed in buildings and constructions are governed by the CPD. Table 10.27 in Section 10.4, European Union, shows the future Euroclasses of the uniform classification system according to Commission Decision 2000/147/EC together with the corresponding test methods and classification criteria for construction products, except floorings. "Linear" products, such as pipes, pipe insulation, and cables, were excluded from the current Euroclasses decision. More work is needed for these product families before a final solution can be proposed, because problems have arisen when using both the SBI and Room/Corner tests (Section 10.4). Electrotechnical products such as cables may appear under more than one application. This can lead to different levels of performance of a product in a given structure because of the different fire scenarios associated with the intended use. Therefore, products need to be assessed according to end-use conditions in relevant scenarios.

In general, low-voltage cables, optical fibers, and cables, permanently installed in buildings and constructions are governed by the Low-Voltage Directive (LVD) [101], which essentially ensures the electrical safety of products.

It applies to products with 50–1000 VAC or 75–1500 VDC input. Products may include computers, information technology equipment and household products. This directive is a requirement only for the EEA member nations and not required for products sold outside this community.

The nature and use of power, control, and communication cables are sufficiently different from other construction products to warrant a separate treatment. This treatment consists in a separate classification for such products, based on test methods specifically adapted to take account of these differences.

Consequently, a new classification entitled "Classes of reaction-to-fire performance for power, control and communication cables" needs to be added to Decision 2000/147/EC. The October 2000 draft for discussion with the EC Fire Regulators Group is shown in Table 12.33 [102]. Most of the limit values shown here still have to be fixed.

Table 12.33 Classes of reaction-to-fire performance for power, control and communication cables

Class	Test methods	Classification criteria[a]	Additional classification
A_c	EN ISO 1716	PCS $\leq$ 2 MJ·kg[a), b)]	–
B_c	EN 50266-2-x[c)], EN 50265-2-1	FS $\leq$ m, THR $\leq$ MJ, Peak RHR $\leq$ kW, FIGRA $\leq$ W·s^{-1}, TSP $\leq$ m^2, Peak SPR $\leq$ m^2/s, H $\leq$ 425 mm	Acidity/corrosivity[e)] and flaming droplets/particles
C_c	EN 50266-2-y[d)], EN 50265-2-1		
D_c			
E_c	EN 50265-2-1	H $\leq$ 425 mm	
F_c	No performance determined		

a) PCS = gross calorific potential, FS = flame spread, THR = total heat release, RHR = rate of heat release, FIGRA = fire growth rate, TSP = total smoke production, SPR = smoke production rate, H = ignitability
b) Mineral insulated cables without a polymeric sheath, as defined in HD 50386, are deemed to satisfy the class A_c requirement without the need for testing
c) EN 50266-2-4 modified on the basis of FIPEC scenario 2 and to include heat release and smoke measurements
d) EN 50266-2-4 modified to include heat release and smoke measurements
e) EN 50267-2-3: a1 = conductivity < 2.5 µS/mm and pH > 4.3; a2 = conductivity < 10 µS/mm and pH > 4.3; a3 = not a1 or a2

Machinery Directive

The Council Directive on the approximation of the laws of the EU member states relating to machinery (89/ 392/EEC) demands in its Annex I, Section 1.5.6 that machinery shall be so designed and constructed to avoid any risk of fire [103]. CEN and CENELEC are producing a set of standards to assist designers, manufacturers, and other interested bodies to interpret this essential safety requirement to achieve conformity with European Legislation. The Draft European Standard prEN 13478 drawn up by WG 16 of CEN/TC 114 "Safety of machinery" describes methods of identification of the fire hazard from machinery and the performance of a corresponding risk assessment [104].

It describes the basic concepts and methodology of technical measures for fire prevention and protection during design and construction of machinery. Risk reduction is primarily achieved by design/engineering measures, which eliminate or minimize the fire hazard, for example, use of flame-retardant materials in the construction of the machine, minimization of the risk of overheating or the limitation of fire effects by using proper shields or enclosures. The safety of machinery against fire involves fire prevention, protection, and fighting. In general, these include technical, structural, organizational efforts, and public fire-fighting measures. Effective fire safety of machinery may require the implementation of a single measure or a combination of measures.

Fire Hazard Assessment of Electrotechnical Equipment of Railway Vehicles

The railway industry is generally concerned with the movement of people as well as goods. It is therefore essential that safety of rolling stock is achieved when failures, including fires, occur. A mandate (M024) was given to CEN by the Commission of the European Communities and the European Free Trade Association to prepare a European standard to support essential requirements of the Directive on the interoperability of the European high-speed train network [105].

The European Standard prEN 45545-1 "Fire protection on railway vehicles," has been prepared by CEN/TC 256, "Railway applications," in cooperation with CENELEC/TC 9X, "Electrical and electronic in railway applications." It is a part of a series of seven interlinked standards [106] regarding: "Railway applications – Fire protection on railway vehicles." The subject draft standard specifies the measures on railway vehicles for fire protection, fire safety objectives and general requirements for fire protection measures. It is a general document that defines operation categories, design categories, and vehicles classification to satisfy the fire protection measures. It takes into account potential hazards and operation categories in the aim to ensure fire protection and limit fire spreading and effects of fire and smoke (thermal, toxic fumes, obscuration of escape routes).

Part 2 is in preparation and specifies the fire behavior requirements for selecting materials and products for railway vehicles.

FIRESTARR has been a 3-year EU-research project (completed in 2001) to assist the standardization work of CEN/TC 256 and CENELEC/TC 9X on the "Fire protection on railway vehicles" [107].

The major aim of the project is to develop the most representative and comprehensive test methods that provide relevant information on the fire behavior of materials and components and a classification proposal for Part 2 of the standard. Part 5 prescribes the fire safety requirements for electrical equipment. The standardization work of railway cables rolling stock is prescribed in Section 12.5.5.

12.3.3 Approval Procedures in Europe

CE Marking

CE marking is a declaration from the manufacturer that the product conforms to a specific Directive adopted in the EEA (European Economic Area). It further symbolizes the fact that the product has been subjected to an appropriate conformity assessment procedure contained in the directive.

For proof of the overall safety of a product by virtue of compliance with relevant safety standards, the voluntary testing and certification by independent institutions such as the VDE Institute is necessary. Only marks for safety and quality attestations such as the VDE Mark License Certificate or the GS Mark reassure the manufacturer about compliance with legal requirements of product safety and environmental issues, as well as the end user that the products comply with the acknowledged rules of technology.

VDE Regulations and Approval Procedures

Electrical safety requirements in Germany are laid down in the VDE (Association for Electrical, Electronic & Information Technologies), which is a nonprofit association for electrical science and technology founded in 1893. One of VDE's main tasks is the promotion of technical progress and the application of electrical engineering/electronics, information technology, and associated technologies [108]. VDE is one of the largest technical and scientific associations in Europe.

The German Electrotechnical Commission (DKE) [109] is responsible for the development of standards and safety requirements in the field of electrical engineering. DKE is an executive of DIN (Deutsches Institut für Normung e.V.) and VDE (Verband der Elektrotechnik, Elektronik, Informationstechnik e.V.). DKE represents Germany in the International and European Standardization Organizations IEC, CENELEC, and ETSI (the European Telecommunications

Standards Institute). The results of the DKE standardization work are extensively harmonized to European and international standards and are published as DIN EN standards and also marked as VDE regulations and VDE guidelines. Because the VDE regulations are based on civil law they do not have the status of binding legal provisions such as state laws or the safety regulations of the underwriters for statutory accident insurance. However, they are accepted by presumption of law as standard technical practice and are thus often referred to in legal provisions and safety regulations. In the case of failure or accident involving electrical equipment that does not comply to the VDE regulations, it has to be shown that the design is at least equal to accepted standard technical practice. The DKE Committee K 133, formerly the UK 131.3 subcommittee, is responsible for the development of fire safety standards.

K 133 is a horizontal committee with a pilot safety function within the DKE to give guidance on the fire hazard assessment and testing to other DKE Committees responsible for finished components and electrotechnical products. K 133 is the German mirror committee to IEC Technical Committee (TC) 89 Fire Hazard Testing.

As a neutral and independent institution, the VDE Testing and Certification Institute, founded in 1920 in Berlin, and, since 1968, located in Offenbach am Main, carries out testing of electrotechnical products, components, and systems and awards the VDE test mark, which is recognized worldwide [110]. The VDE Institute is accredited on a national and international level. The VDE Mark indicates conformity with the VDE standards or European or internationally harmonized standards, respectively, and confirms compliance with protective requirements of the applicable EC Directive. The VDE Mark is a symbol for testing based on the assessment of electrical, mechanical, thermal, toxic, fire, and other hazards.

For cables, insulated cords, installation conduits, and ducts, a special VDE Cable Mark is applicable. A lot of other VDE Marks may be applied, e.g., the VDE Component Mark for electronic components. The VDE Institute also certifies test results obtained from other certified bodies, for example, CB Test Certificates, CCA Notifications of Test Results, or EC Type-Examination Certificates, and is a Notified Body for the scope of the Low-Voltage Directive and the Medical Devices Directive. Through partner organizations worldwide and the presence of the VDE Authorized Office and VDE experts, the VDE Institute is an international partner for safety and quality.

The GS Mark

The GS Mark, the German safety approval mark, shows conformity with the German Equipment Safety Law. It is a recognized mark for electrotechnical products, such as office equipment, household appliances and industrial equipment and is widely accepted throughout Europe. The GS Mark signals to the buyer, customer, and consumer that the product, as well as the production process, have been tested by an authorized institution such as TÜV Rheinland [111] and is supported by regular surveillance audits.

12.4 Fire Regulations and Standards in Asia

Most of the consumer goods to be sold in the Asian markets (Japan, Hong Kong, Philippines, China, Malaysia, Singapore, Indonesia, Thailand, Taiwan, and Korea) need an approval. Because many Asian states are on the way to implementing new testing and classification systems, obtaining appropriate safety compliance licenses and approvals can be complex and confusing. This is particularly true for safety and security regulations for E&E Products. Although these regulations are somewhat in line with IEC Standards, the written rules and current practices are often at odds and require some interpretation. In many cases, the safety tests have to be carried out by "Notified Bodies" of the corresponding Asian state, such as national or seminational Institutes. Some of the Asian countries have joined the CB Scheme, for example, Japan, Korea, and Singapore, which simplifies the testing and certification process. Special companies, for example, FEMAC [112], offer their assistance for companies, which try to enter the Asian markets. In the following, some flame retardant regulations and test standards in Japan and China are exemplified.

Flame Retardant Regulations and Test Standards in Japan

The Japanese standards, which are relevant for the characterization of the flammability and fire safety assessment of plastics, electrotechnical parts, and equipment, are based on the relevant tests developed by IEC/TC 89.

Table 12.34   Fire hazard test methods for electric and electronic products in Japan

ITS C0060, Part 2 [113]	Fire hazard testing Test methods, glow-wire test and guidance
ITS C0061, Part 2 [114]	Fire hazard testing Needle-flame test

Flame retardant sheaths of telecommunication cables are tested according to ITS C3521.

Flame Retardant Regulations and Test Standards in China

Certification for China

Because China participates in the CB Scheme, the evaluation process is straightforward; testing is based on IEC standards. There are currently two certification schemes for the Chinese market, CCIB (China Commodity Inspection Bureau) and CCEE (China Commission for Conformity Certification of Electrical Equipment). The CCIB certification is compulsory for a wide range of product categories, including data processing or household products. CCEE is the CB Scheme participant for China. For some products (e.g., refrigerators and televisions), both CCIB and CCEE certifications are required.

Most of the fire safety regulations in China were issued by the related ministries of the central government, such as the Ministry of Public Security. At the time being, there are few mandatory regulations for electric and electronic products. The fire hazard test methods were set up on the basis of the corresponding IEC standards shown in Table 12.35.

Table 12.35   Fire hazard test methods for electric and electronic Products in China [115]

Test method	Equivalent IEC Standard
Glow-wire test, GB 5169.4-85 [116]	IEC 60695-2-1
Needle-flame test, GB 5169.5-85 [116]	IEC 60695-2-2
Bad-connection test, GB 5169.6-85 [116]	IEC 60695-2-3
Bunsen-flame test, GB 5169.7-85 [117]	IEC Bunsen-flame test according to IEC SC 50D (sec) 35, January 1983

12.5 Fire Hazard Assessment on Cables

A variety of standards for the fire hazard testing of cables and wires are in use around the world. In the past, most product specifications quoted national standards for fire testing. For example, in the United States the National Fire Protection Agency (NFPA) regulates electrical cables through NFPA 70 [118], which is the National Electrical Code [NEC]. A comprehensive survey of fire testing of electrical cables in North America has been given by Hirschler [119]. A comparison between US and European fire test standards for cables and a proposal regarding worldwide harmonization of the flame methods is given by Richter and Schmidt [120].

The "Muster-Leitungsanlagen-Richtlinie MLAR" [121] regulates the conducting wire and pipes systems in Germany. However, in view of the globalization of the markets, the standardization bodies increasingly refer to ISO/IEC work. Particularly in Europe, a comprehensive conversion to international European Standards (EN) takes place in the field of fire hazard testing of cables and wires.

IEC/TC 20 "Burning characteristics of electric cables" is responsible for the international standardization of fire hazard testing with a working group (WG) on the safety function for fire-related test methods. After modernization of its structure, standards for fire hazard testing will now be prepared in WG 18.

In fire testing of cables and wires, a section of the finished end-product is exposed to a small flame under defined conditions. The degree to which it burns during and after exposure and the rate of combustion are assessed. In most cases, testing is restricted to laboratory small-scale tests.

Increased attention has been devoted to real fires in which the fire behavior of cables and wires may be greatly influenced by the installation conditions and preheating caused by the combustion of long vertical runs of bundles of cables on cable trays. As a consequence, the classification of single cables or wires is not generally transferable to bundled cables. For this reason, many national and international standards for intermediate or large-scale tests have been developed, especially dealing with the fire behavior of cable bundles.

12.5.1 Small-Scale Testing

Vertical Burning Test on a Single Insulated Wire or Cable

A nonexhaustive list of the most important small-scale test methods for wire and cable including the test specifications and classifications is shown in Table 12.36. The principle of these tests is demonstrated by the flame propagation test according to IEC 60332-1 and IEC 60332-2 or their technically equivalent European standards EN 50265-2-1 and EN 50265-2-2.

In the flame propagation test to IEC 60332-1 [122] a vertically oriented test specimen, a piece of finished cable or wire 600 (± 25) mm (23.6 in.) long, is exposed to a Bunsen burner 175 mm (6.9 in.) flame with 40 mm (1.58 in.) inner blue cone. The test layout is illustrated in Fig. 12.13. Wires and cables of diameter greater than 50 mm (1.97 in.) have to be tested with two gas burners. The flame is continuously applied for a period of time, which is related to the overall diameter of the test specimen (Table 12.36). The damaged length is a measure of its resistance to flame propagation. After all burning has ceased, the charred or affected portion shall not have reached within 50 mm of the lower edge of the top clamp.

Table 12.36 Summary of test conditions for cable and wire test methods in laboratory scale

Test method	Ignition source				Test enclosure specified	No. of test runs
	Burner	Power (kW)	Flame length (mm)	Flame application time t (s)		
EN 50265-1	Part 1	Describes the test apparatus for parts 2–1 and 2–2			Metallic screen 1200 (± 25) mm high, 300 (± 25) mm wide, 450 (± 25) deep (if failure in recorded 2 more shall be carried out)	
EN 50265-2-1	According to IEC 60695-2-4/1	1		Depending on outer diameter D of the sample: $D \leq 25$ mm: $t = 60$ s $25 < D \leq 50$: $t = 120$ s $50 < D \leq 75$: $t = 240$ s $D > 75$: $t = 480$ s		1
IEC 60332-1						
VDE 0482, Part 265-1						
EN 50265-2-2	Propane burner		125 (± 25)	20 (± 1) (should the conductor prematurely melt at $t < (20\pm1)$ s, the test shall be repeated on sample No. 2, to a duration of $(t$-2) s		2
IEC 60332-2						
VDE 0482, Part 265-2						
ABD 0031 ≤ FAR Part 25, § 25.1359 (d), Amdt. 25-72				30		3

[a] For single core with paint or lacquer coating, the sample is stored 4 h at 60 (± 2) °C before conditioning.

Specimen			Performance requirements / Maximum allowable		
Condition-ing	Length [mm]	Position	Char length	Flame time [s]	Flame time of drips [s]
min. 16h 50 (± 20)% 23 (± 5)° C a)	600 (± 25)	Vertical	Distance between the lower edge of the top support and the onset of charring is > 50 mm, burning does not extend downwards to a point > 540 mm from the lower edge of the top support.		
a)			Distance between the lower edge of the top clamp and the charred portion is > 50 mm.	–	–
	500 (± 25)		76	30	3

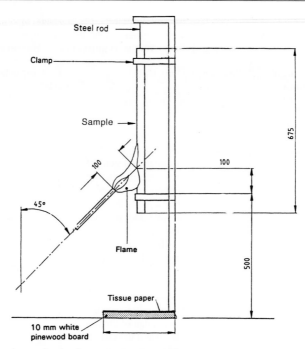

Figure 12.13   Test layout for the flame propagation test to IEC 60332-1

Vertical Burning Test on a Single Small Copper Wire and Cable

IEC 60332-2 [123], which has been prepared by IEC/TC 20 and IEC/TC 46, specifies a fire test method on small insulated copper cables or wires when the method specified in IEC 60332-1 is not applicable because the specimen melts during the application of the flame. The test specimen consists of a piece of finished copper cable or wire 600 (± 25) mm (23.6 in.) long with a range of application of 0.4–0.8 mm (0.016–0.032 in.) for solid copper conductor

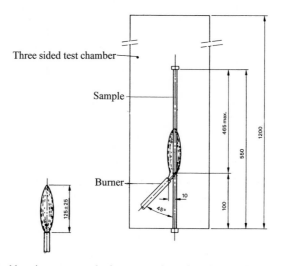

Figure 12.14   Vertical burning test on a single copper wire and cable according to IEC 60332-2

and 0.1 mm^2 to 0.5 mm^2 cross section for stranded conductors. The test specimen is held in a vertical position and a load of 5 N/mm^2 of conductor area is attached to the lower part of the specimen. A calibrated propane burner is used to ignite the specimen. The 125 ($\pm$ 25) mm (4.9 in.) flame is continuously applied for a maximum period of 20 s. After all burning has ceased, the charred or affected portion shall not have reached within 50 mm (2.0 in.) of the lower edge of the top clamp (Fig. 12.14).

12.5.2 Large-Scale Testing

The small-scale fire tests according to IEC 60332-1 or 60332-2 are not appropriate for scenarios with cables or wires of high or concentrated fire load and with ignition sources of very high energy. To meet the requirements of such scenarios, several national and international standards with more stringent large-scale flame spread tests better reflecting end-use conditions have been developed. They can be divided into vertical and horizontal ladder test methods.

Vertical Ladder Test Methods

A nonexhaustive list of the most popular vertical ladder test methods including a comparison of the test specifications and classifications is shown in Table 12.37. The principle of these large-scale tests is demonstrated by the bundled wires or cables tested to IEC 60332-3, UL 1666, and NF C32-070.

Cable Test According to IEC 60332-3

IEC 60332-3 (2nd edition), which was published as a Technical Report in 1992 [124], has been superseded at the end of 2000 by a new series of International Standards published in six separate parts [125]. These standards are technically equivalent to the European standards EN 50266-2-1 to EN 50266-2-5. All preexisting categories of test are retained and updated. IEC 60332-3-10 details the apparatus and its arrangement and calibration.

The test chamber is 1 m (3.3 ft) wide; 2 m (6.6 ft) deep and 4 m (13.2 ft) high. The test rack is placed vertically into the chamber with the cables and wires pointing to the burner. The distance of the test rack to the rear chamber wall is 150 mm (5.9 in.).

The ribbon gas burner, specified as the ignition source one or two S/PF 5 to ISO/FDIS 10093, is placed horizontally in front of the test rack with its front edge at 7.5 ($\pm$ 5) mm (0.3 in.) distance from the specimens surface and at least 500 ($\pm$ 5) mm (19.7 in.) distance from the lower end of the testing rack. The test bench with an installed cable ladder and ribbon burner is shown in Fig. 12.15. The energy supply of the ribbon gas burner working with a mixture of propane/air according to IEC 60332-3 is 73.7($\pm$ 1.68) MJ/h. Combustion air is admitted at the base of the test chamber at a rate of 5 ($\pm$ 0.5) m^3/min at 20 ($\pm$ 10)°C. The test sample consists of a number of cable or wire test pieces with same length, each having a minimum length of 3.5 m (11.5 ft).

The number of 3.5 m test pieces depends on the volume per meter of nonmetallic material of one test piece.

The categories A, B, C, and D correspond to a nominal volume of nonmetallic material of 7, 3.5, 1.5, and 0.5 l/m. Two methods of mounting are applicable to category A. A summary of the test conditions is given in Table 12.38. Depending on the category, the flame application time is 20 min or 40 min. After burning has ceased, the maximum charred length of the test sample should not exceed 2.5 m (8.2 ft) above the bottom edge of the burner, neither at the front or the rear of the ladder. If burning has not ceased 1 h after completion of flame application, the flames are extinguished and the extent of the damage is measured. IEC/TC 20C continues to develop the procedure for large-scale test standards.

Table 12.37 Summary of test conditions of vertical ladder cable test methods

Test method	Heat source			Application time of heat source (min)	Test enclosure	
	Burner	Power (kW)	Alternate source		Specified	Air flow rate (m³/s)
IEC 60332-3	Yes[a]	21	No	Category C: 20 min; category A + B: 40 min	Yes	0.083
CEI 20-22	No	30	Electrical furnace; two radiant plates 500 mm × 500 mm	60		0.7
IEC 383; Section 2.5; gas burner procedure according to Section 2.5.4.4.4	Yes[a]	21	Oily ray		No	Not applicable
UL 1581; Section 1160						
UL 1581; Section 1164	Yes[b]	20	No	20	Yes	≥ 0.17
FT 4 test according CSA C22.2. clause 4.1						
UL 1685; Section 4–11: UL Flame exposure	Yes[a]	21				0.6 to 0.7
UL 1685; Section 12–19: FT 4/ IE EE 1202: type of flame exposure	Yes[b]					
UL 1666	Yes[c]	154.5		30		–
NF C 32-070; class C 1; UIC 895 VE	No	–	Electrical ring[g], [h] furnace			0.5
DIN 4102. Part 16 "Brandschacht-test"	Yes	21	No	10		0.166

[a] Angle of the burner horizontal
[b] Angle of the burner 20° up
[c] Diffusion burner plate at the bottom of the cable tray
[d] Char length measured from horizontal line of burner

[e] Second set depends on the variance of the test results
[f] Second and third set depend on the variance of the test results

No. of test runs	Specimen				Conclusion		
	Length (m)	Width of test sample (m)	Mounting techniques	Thin size cables to be bundled	Maximum allowable char length from bottom (m)	Peak smoke release rate (m² s⁻¹)	Total smoke released (m²)
1	3.5	0.5/0.8	Front or front and back	Mounted flush with no spaces	3.1		
	4.5	0.2			4.1	No requirement	
3	2.4	0.15		No	2.4		
2	2.3	0.25	Front only	If D < 13 mm	$1.5^{d)}$		
–							
$2^{e)}$	2.44	0.15		No	2.44	0.25	95
						0.40	150
$2^{f)}$	5.33	0.31	Single layer		3.66	No requirement	
2	1.6	–	–	Yes	$0.25^{f)}$		
3	1.0	$0.19^{i)}$	Single layer $^{j)}$ double layer $^{k)}$	No	0.85		

g) Two 20 mm long tangential pilot flames are located above the top of the furnace (upper position)
h) Temperature of the furnace after stabilization 780 °C – 880 °C

i) One specimen consists of four test samples each 1 m long
j) Distance between the cables of diameter D: $D/2$ but minimum 5 mm
k) Mounted flush without spaces

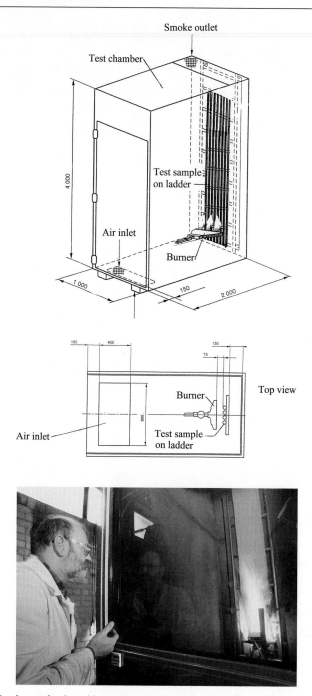

Figure 12.15   Test layout for the cable test according to IEC 60332-3

Table 12.38 Summary of test conditions[a]

Category according to the 2000 Series of the standard	A F/R (IEC 60332-3-21)	A (IEC 60332-3-22)		B (IEC 60332-3-23)		C (IEC 60332-3-24)		D (IEC 60332-3-25)
Category and designation according to IEC 60332-3-1992	A F/R	B F		C F		A F		
Range of conductor cross sections (mm²)	> 35[b]	≤ 35[c]	> 35[b]	≤ 35[c]	> 35[b]	≤ 35[c]	> 35[b]	–
Nonmetallic volume per meter of test samples (l)	7	7		3.5		1.5		0.5
Number of layers:								
For the standard ladder: max. width of test sample: 300 mm	2	≥ 1	1	≥ 1	1	≥ 1	1	≥ 1
For the wide ladder: max. width of test sample: 600 mm		–	1					
Position of test pieces	spaced	touching	spaced	touching	spaced	touching	spaced	touching
Flame application time (min)	40	40		20		20		20
Number of burners	1	2		1		1		1

[a] From IEC 60332-2 2nd Edition 1992-03; Categories of future series of standard included

[b] At least one conductor > 35 mm²

[c] No conductor cross-section > 35 mm²

Cable Tests According to UL 1666

The test rig a according to UL 1666 [126] consists of a concrete test cabinet, with two chambers in two floors, connected by a slot. A simplified scheme for the flame test is shown in Fig. 12.16. The total height is 5.7 m (18.7 ft) and the inner space 2.44 m x 1.22 m (8 ft x 4 ft). The cables are mounted on a metal frame running vertically through the two floors and a burner diffusion plate on the floor ignites the cables for 30 min. The test conditions are shown in Table 12.39.

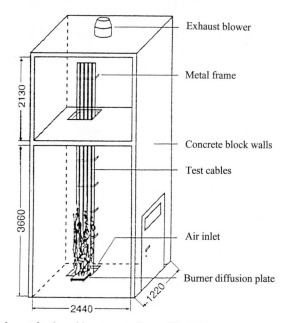

Figure 12.16   Test layout for the cable test according to UL 1666

Cable Test According to NF C 32-070

The fire test methods of NF C 32-070 [127] differ in some cases considerably from those in other countries or those prescribed by IEC/TC 20.

The standard contains test methods for the reaction to fire of the cable and its fire resistance. The cables are divided into three classes of combustibility: C1, C2, and C3 and two classes of fire resistance: CR1 and CR2. To differentiate between classes C2 and C3 a test is carried out with one, or in the case of thicker cables, two Bunsen burners according to IEC 60695-2-4/1. If the test is passed, the cable is classified as C2; otherwise it is classified in the least favorable class C3. Cables used in applications involving stringent fire safety requirements must be of class C1 and must pass a special test developed in France using the fire test rig illustrated in Fig. 12.17.

A 160 cm (63 in.) long sample of the finished cable (or bundle of cables in the case of thin cables) is suspended vertically from a frame in a test chamber with internal dimensions of 0.8 m x 0.7 m x 2.0 m (2.6 ft x 2.3 ft x 6.6 ft). An annular furnace which can be slid up and down is fixed to the lower part of the frame. In its top position, it surrounds the lower part of the cable, and in its lowest position it surrounds a probe for checking its heat output, the secondary calibration device. Two 20 mm (0.79 in.) long, tangential pilot flames are located above the top furnace position, 10–15 mm from the surface of the cable. Above the pilot flames, the cable is enclosed concentrically by an 80 mm (3.15 in.) long metallic chimney,

Table 12.39 Test specifications for determining the flame propagation height of cables in shafts to UL 1666

Specimen	Two sets of specimens of electrical or optical-fiber cables. Each set is to consist of multiple (5.33 m) specimen length of cable
Specimen conditioning	*Examples:* Storage 48 h in 23/50 standard atmosphere
Test layout	Fire test chamber 5.79 m height and 1.22 m width. One slot 0.305 m by 0.61 m in the first floor. An identical slot directly above in the first floor. The cable lengths are to be installed through both slots. The number (N) of the cable is computed by: $N = 12 \times 25.4 / D$, where D = outside diameter of the round cable in mm. For nonround cables D is the smallest dimension of the cable diameter.
Ignition source	Burner apparatus consisting of a piping and a steel diffuser plate. 76 mm above the bottom of the first test chamber, propane gas flame, power 154.5 kW
Test duration	• Until nonmetallic cable parts are completely consumed for the full length • Maximum 30 min
Conclusion	Passed if: • Flame propagation height (combustible material having been softened, partially or completely consumed) of each set has not equaled or exceeded 3.66 m • The difference between the propagation heights of cable specimens (two sets) does not exceed 15 % (if yes, a third set has to be tested)

which serves to conduct the flames upwards from the cable below. An air stream flowing upwards at approx. 2 m/s is generated with a fan in the flue of the chamber. Before the start of the test, the radiation output of the furnace is first adjusted at its upper position by placing a measuring probe instead of the cable. This is the primary calibration device. Once the output of the furnace as determined by the rise in temperature of this measuring probe has reached the required value, the furnace is slid to the lower position. The cable is then suspended into the test chamber. The actual test starts when the furnace is brought to its upper position again and lasts 30 min during which ventilation is interrupted for 1 min after 10 min. The cable achieves class C1 if in two tests the section of cable above the chimney remains undamaged.

Horizontal Ladder Test Methods

Business, organizations and institutions of different kind are more and more dependent on communications.

The plenum is an attractive area to place electrical and optical-fiber communication cables. These (horizontal) plenums include the area above a suspended ceiling and under a raised floor. In the United States, the rapid growth of computers on local area networks (LAN) and the resulting increased mass of communication cables in such plenum areas has led to a concern about the accumulation of combustibles in these spaces, when installed in return-air plenums [128]. These cables must be listed as having passed the flame and smoke requirements of UL 910 [129]. The test is performed in the Steiner Tunnel, in which cable bundles are placed on a horizontal tray and exposed for 20 min to two burners with 87.9 kW output. The test is essentially the same as the test described in NFPA 262. Any resulting fire in the cable must spread no more than 162 cm (63.78 in.) with strict limitations on the amount of smoke permitted. Figure 12.18 shows the test layout for cables in the Steiner Tunnel (see also Section 10.2 Building USA). The test specifications are summarized in Table 12.40.

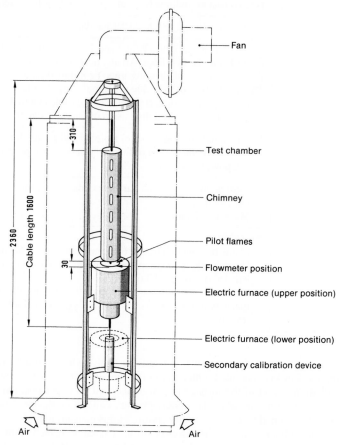

Figure 12.17   C1 test according to NF C 32-070

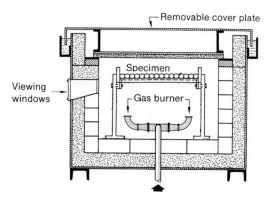

Figure 12.18   Test layout for the Steiner Tunnel test for cables according to UL 910

Table 12.40   Test specifications for determining flame propagation and smoke density of cables to UL 910

Specimen	Cable specimens length 7.32 m installed without any space between adjacent specimens in a single layer across the bottom of a cable tray
Specimen position	Horizontal
Ignition source	Two gas burners (methane gas) with 87.9 kW output located 47.6 mm below the cable tray
Test duration	20 min
Conclusion	Passed if: • Maximum flame-propagation distance is not to exceed 162 cm beyond the initial (137 cm) test flame. • Peak optical smoke density ≤ 0.5 (32 % light transmission) • Average optical density ≤ 0.15

The need for an International Fire Test Standard for communication cable installations is being studied by IEC/TC 20 (flammability of wire and cable).

12.5.3 Side Effects of Cable Fires

Smoke from burning cables is an identified risk in the early stages of a fire and addressed by IEC 61 034/EN 50 268. Both standards are published in two parts. Part 1 gives details of the test apparatus and verification procedure to be used for the measurement of the smoke density of the products of combustion of electric or optical cables burned under defined conditions in a static smoke test [130]. It includes details of a test cube of $27\,m^3$ volume; a photometric system for light measurement, a qualification procedure that also defines the standard fire source, and a smoke-mixing method. The details of the test rig are shown in Fig. 12.19. The test method was first developed in the United Kingdom by scientists of London Underground Ltd.

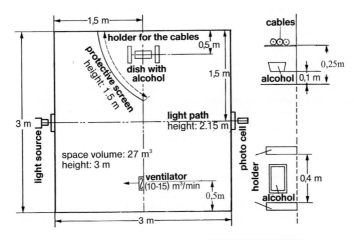

Figure 12.19   Test layout for smoke density of cables to IEC 61 034/EN 50 268

IEC 61034 Part 2 describes the test procedure and recommended requirements for use, when no specific requirements are given in the particular cable standard or specification [131]. The details for the preparation and assembly of the cables for test are shown in Table 12.41. The standard fire source consists of 1 l of alcohol placed below a 1 m long, horizontally oriented test piece consisting of cables or cable bundles. Observations of absorbance within the cubic room are made by measuring the smoke emission with time. Because of its design and performance criteria, the test is most appropriately used for cables with a low smoke-emitting potential.

Table 12.41   IEC 61034 Part 1 and Part 2 cable smoke density test specifications

Specimen	Electric or optical cables		
Number of test pieces	Cable test pieces shall consist of one or more samples of cable 1.00 ($\pm$ 0.05) m long		
	Overall diameter of the cable (D)		Number of test pieces
	mm	Cables	Bundles[a]
	$D > 40$	1	–
	$20 < D \leq 40$	2	–
	$10 < D \leq 20$	3	–
	$5 < D \leq 10$	$N_1 = 45/D$ cables[b]	–
	$2 \leq D \leq 5$	–	$N_2 = 45/3 D$[b]
Test piece position	Cables and bundles in horizontal position above the tray		
Conditioning	At least 16h at 25 ($\pm$ 5) °C		
Assembly of test pieces	Cables or bundles shall be bound together at the ends and at 0.3 m from each end		
Test layout	Cubic enclosure; inside dimensions 3,000 ($\pm$ 30) mm; photometric system		
Ignition source	Standard fire source 1.00 ($\pm$ 0.01) l alcohol with the composition: • Ethanol: 90 ($\pm$ 1) % • Methanol: 4 ($\pm$ 1) % • Water: 6 ($\pm$ 1) %		
Conclusion	Recommended minimum value of the light transmittance: 60 %		

[a] Each bundle shall consist of seven cables twisted together
[b] Values of N_1 and N_2 to be rounded downwards to the integer

12.5.4 Fire Resisting Characteristics of Cables

In the event of a fire, it may be vital to the safety of the building occupants that certain electrical systems remain functioning until the rescue has been terminated. Examples are

• Fire detectors and alarm systems,
• Emergency escape lighting,

- Smoke extraction systems,
- Power supply for extinguishing systems.

The Standard IEC 60331-11, a revision of IEC 331-1970, introduces a range of improvements based on practical experience gained [132]. The heat source is specified more precisely by fuel rates and located in such a way that no interference with the test flame will arise from debris falling from the sample. The standard has been extended by three parts in that the test apparatus may be used to test low voltage power cables and control cables with a rated voltage up and including 0.6/1.0 kV, electric data cables and optical fiber cables.

The apparatus to be used for testing cables required to maintain circuit integrity when subjected to flame with a controlled heat output corresponding to a temperature of at least 750 °C specified in Part 11 of the standard, is shown in Fig. 12.20. The 1200 mm (47.2 in.) long cable sample is held horizontally by supports. The heat source, a ribbon-type propane burner with a nominal burner face length of 500 mm (19.7 in.) with a Venturi mixer, is aligned with the test sample. Annex A of the standard contains a verification procedure to check the burner and control system and the determination of the exact burner location during the cable testing. Further sections for Part 11 are under consideration, for example, for fire at higher temperatures, fire with mechanical shock or fire with water sprays. During the test, a current for continuity checking is passed through all conductors of the cable.

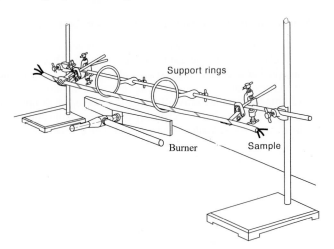

Figure 12.20   Test layout for the circuit integrity test for cables to IEC 60331-11

Figure 12.21 shows as an example the basic circuit diagram for 0.6/1.0 kV cables. The connection of the conductors to the transformer and the test voltage are prescribed in the standards. If not specified in the relevant cable standard, a 90 min flame application is recommended. A cable possesses the characteristics for providing circuit integrity as long as, during the course of the test, the voltage is maintained, that is, no fuse fails and no lamp is extinguished. In the case of optical fiber cables, the fibers are connected to an optical apparatus to determine the change in optical transmittance.

The acceptance criterion is the maximum increase in light attenuation, which is prescribed in the relevant specification. At present, the major work of TC 20 is an extension of IEC 60331 for introducing mechanical shock and an enhanced flame. This will be covered by new Parts 12 and 13.

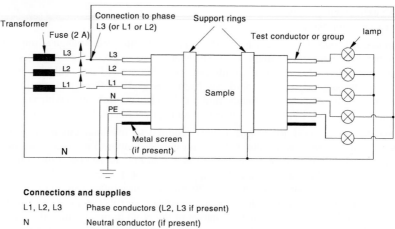

Connections and supplies

L1, L2, L3   Phase conductors (L2, L3 if present)

N   Neutral conductor (if present)

PE   Protective earth (if present)

Figure 12.21   Basic circuit diagram for 0.6 / 1.0 kV cables for the circuit integrity test to IEC 60331-11

Circuit Integrity Maintenance of Electric Cable Systems to DIN 4102-12

In contrast to the insulation and integrity tests of a single cable according to IEC 60331-11 or the related harmonized EN Standards, the test to German Standard DIN 4102 Part 12 [133] is designed to establish whether a complete electrical cable system is able to maintain circuit integrity over a defined period of time in the heating regime in the test chamber according to the ISO 834 standard time-temperature curve (Fig. 12.22). The cable samples are mounted to the walls and the ceiling inside the fire test chamber in end-use application using the recommended fixing devices, cable trays, ladders, and so forth. For the measurement of the circuit

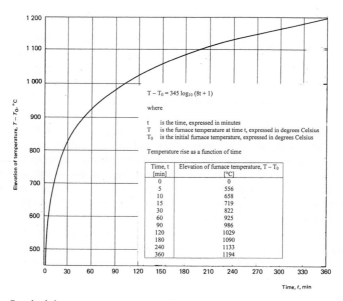

Figure 12.22   Standard time-temperature curve to ISO 834

integrity, the cables are protruded through the outer walls of the fire test chamber and connected to the relevant test voltage (110 V for telecommunication cables; 400 V for power cables). Dependent on the measured duration of the circuit integrity, a distinction is made between the following classes:

Circuit integrity maintenance of electric cable systems	Classes	Time [min]
	E 30	≥ 30
	E 60	≥ 60
	E 90	≥ 90

BS 6387 (1994) "Specification for performance requirements for cables required to maintain circuit integrity under fire conditions" only exposes single cables without any mounting system to a gas flame [134].

12.5.5 Fire Hazard Assessment on Railway Rolling Stock Cables

Railway rolling stock cables are intended for transmission and distribution of electricity in monitoring, control and power circuits. The Draft of the European Standard prEN YYY [135] gives guidance in the safe use of rolling stock cables specified in prEN 50306 [136], prEN 50264 [137], and prEN XXS [138]. It was prepared for Technical Committee CENELEC/TC 20, Electric cables, by Working Group 12, Railway cables, as part of the overall program of work in CENELEC/TC 9 X, Electrical and electronic applications for railways.

PrEN 50264 covers a range of cables, with standard wall thickness of insulation and sheath rated at 3.6/6 kV with conductor sizes 1.0 mm^2 up to 400 mm^2. The standard prEN 50306 covers cables with thin wall insulation, restricted to a rating of 300 V to earth and a maximum conductor size of 2.5 mm^2. Both standards cover cables based on halogen-free materials. prEN YYY gives recommendations on how to select, locate, and install cables so that they do not present a fire hazard to adjacent materials [139].

Based on the protective goal of the standard, insulating and sheathing materials referred to in prEN 50264 and prEN 50306 are specifically selected and tested to give the spread of flame, emission to smoke, and toxicity consistent with the required performance of the cable to suit the operation categories and hazard levels given in prEN 45545-1, "Railway applications – Fire protection of railway vehicles – Part 1: General rules" [106].

These standards will include specifications according to reaction-to-fire, flame propagation, smoke emission, corrosive and acid gas emission, toxic potency of the combustion gases, and resistance to fire. prEN 50305 [140] gives particular test methods applicable to the cables covered by prEN 50264 and prEN 50306. The standard describes detailed reaction-to-fire tests for determining flame propagation according to EN 50265-2-1 [141] for single cables and EN 50266-2-4 for cable bundles [142] depending on the cable diameter and the nominal total volume of nonmetallic material.

12.5.6 Future Developments in Cable Testing

A comprehensive fire research program named FIPEC (Fire Performance of Electric Cables) was initiated by the Commission of the European Communities DG XII, the Directorate for Research and Development [143].

One of the objectives was to identify improvements of existing test methods (e.g., IEC 60332-3) in order to use them for classification purposes. The FIPEC program comprised the development of fire test procedures and investigation of mathematical fire models to predict fire development in real-scale scenarios from smaller scale experiments.

The new measurement techniques for assessing the fire performance of electric cables, which includes heat release measurement, have been developed for use with existing test methods to provide an assessment system that is sensitive enough to differentiate between cables with reasonable fire safety properties and those with very good properties needed for high hazard installations or for high-density telecommunications installations.

The new techniques elaborated in the FIPEC-program can be used as a basis for new standardization work in CENELEC. The draft EN 50266 specifies methods of test for the assessment of vertical flame spread of vertically mounted bunched wires or cables, electrical or optical, under defined conditions. The CENELEC/TC 20 working group is currently in an early stage of the elaboration of a new standard that details a new apparatus for the measurement of the heat release during the fire test according to EN 50266. The system for collecting the combustion products consists of a hood installed above the top of the test cabin as defined in EN 50266-1 (Fig. 12.23). During the test, oxygen consumption, carbon dioxide concentration, and the volume flow in the exhaust duct are measured continuously and the values of the rate of heat release versus time calculated [144].

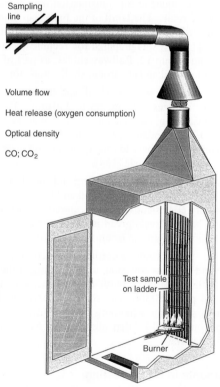

Figure 12.23   Test layout for a heat release and smoke measuring apparatus in EN 50266-1 cable fire test

12.6 Other Fire Safety Test Methods

12.6.1 Fire Hazard Testing of Cable Management Systems

Cable management products cover a family of products/systems providing various purposes such as:

- Accommodation and support of electrical cables and fiberoptic cables,
- Mechanical protection,
- Electrical protection and electrical separation between circuits,
- Protection against ingress of solid objects or fluids,
- Fire resistance.

Depending on product types, these systems include components to join profile lengths in various installation types including changes in direction and of level, separation in branches, termination, and closing of the enclosure. Some of the cable management accommodate, support, and protect electrical accessories such as socket outlets, switches, and protection devices. For most of the applications, the cable management products represent small mass and fire load. Where applications include significant quantities of cables, the cable management products represent much smaller mass and fire load than the cables alone.

These cable management products have been covered by the Low Voltage Directive (LVD) [101] for more than 25 years, the conformity of the products being certified by the manufacturer or the responsible vendor. The LVD is a total safety Directive and covers all the requirements of the Construction Products Directive [100] applicable to cable management products.

Technical standards continuously evolve in CENELEC-relevant Technical Product Committees such as CENELEC/TC 213 or installation rules Committee CENELEC/TC 64. At the present time, CENELEC/TC 213 is restructuring its standards to better show that fire hazard is duly covered. A nonexhaustive list of the important fire test methods in laboratory scale for cable management products is shown in Table 12.42.

12.6.2 Fire Hazard Testing of Office and Consumer Equipment

12.6.2.1 Office equipment

The explosion in global communication during the 1980s and 1990s has been made possible through the use of plastics. Plastics used in office equipment generally have to meet fire performance requirements. These are part of standards such as IEC 950 [96], EN 60950 [145], UL 1950 [95], and CSA C 22.2 [97]. The UL and CSA standards are identical and are described in Section 12.2.1.2.

The IEC, EN, and UL standards essentially have the same flammability requirements with minor differences. The IEC 60950 based standards are used worldwide, whereas UL 1950 is more commonly used in the United States and the CSA Standard in Canada.

To ensure further the safety of their office machines, some producers have additional requirements, which are more stringent than those appearing in the above mentioned standards. IBM, for example, requires fire enclosures if the available power in the machine exceeds 15 W [146]. Resins for fire enclosures must meet the flammability class V-0 if the total mass of the machine is < 18 kg (39.6 lb). The above mentioned standards require only V-1 class plastics for such machines.

Table 12.42   Summary of test conditions for fire test methods in laboratory scale for cable trunking systems, cable ducting and conduit systems for electrical installations

Test method	Ignition source		Test enclosure specified
	Burner	**Flame application time (s)**	
EN 50085-1	1 kW burner according to IEC 60695-2-4/1	60 (± 2) Place of flame impingement not specified in detail	Metallic screen 1200 (± 25) mm wide; 300 (± 25) mm high; 450 (± 25) mm deep
VDE 0604 Part 1 (1998)			
EN 50085-2-1			
VDE 0604, Part 2–1			
EN 50085-2-3	Needle flame according to IEC 60695-2-2	60	
VDE 604, Part 2–3			
DIN/VDE 0604 Part 1 (1986)	Propane gas burner Flame length: total: 175 mm inner blue core: 55 mm	Depending on the material thickness D of the sample: $D \leq 3$ mm: t = 60 $3 < D \leq 5$ mm: $t = 120$ $D > 5$ mm: $t = 180$	
EN 50086-1	1 kW burner according to IEC 60695-2-4/1	depending on the material thickness D of the sample: $0 < D \leq 0.5$: $t = 15^{+1/-0}$ $0.5 < D \leq 1.0$: $t = 20$ $1.0 < D \leq 1.5$: $t = 25$ $1.5 < D \leq 2.0$: $t = 35$ $2.0 < D \leq 2.5$: $t = 45$ $2.5 < D \leq 3.0$: $t = 55$ $3.0 < D \leq 3.5$: $t = 65$ $3.5 < D \leq 4.0$: $t = 75$ $4.0 < D \leq 4.5$: $t = 85$ $4.5 < D \leq 5.0$: $t = 130$ $5.0 < D \leq 5.5$: $t = 200$ $5.5 < D \leq 6.0$: $t = 300$ $6.0 < D \leq 6.5$: $t = 500^{+1/-0}$	
VDE 0605-1			

Specimen			Dripping indicator	Conclusion
No. of test runs	Type	Length (mm)		
3	Cable trunking systems; cable ducting systems	675 (± 10)	Tissue paper on white pinewood board	After burning time < 30 s No ignition of the tissue paper or any scorching of the pinewood board
	Slotted cable trunking			
	Trunking	300		
	Conduit systems	675 (± 10)		Non-flame-spreading conduit system: After-burning time < 30 s No evidence of burning or charring within 50 mm of the lower extremity of the upper clamp and also within 50 mm of the upper extremity of the lower clamps; no ignition of the tissue paper

12.6.2.2 Consumer Equipment

The fire performance requirements for plastic material used in the manufacture of consumer products, for example, TV sets are different in US, IEC, and European standards. In the United States, plastics used for enclosures are virtually always required to pass one of the vertical UL 94 flammability tests. The requirements for TV sets in the United States are set in UL 1410 [147]. IEC 60065 [148] and EN 60065, "Audio, video and similar electronic apparatus safety requirements" [149] allow major plastic parts in TV set backplates and housings to be made from materials that fulfill the requirements of the UL 94 horizontal test. No flammability requirements are required for plastic materials, which are well protected from internal ignition sources using fire enclosures or specified distances from potential ignition sources. Several studies were carried out to assess the role of flame retardants in plastics for preventing or delaying fires. In studies focusing on TV sets it was found that flame retarded plastics guarantee high resistance to external ignition sources such as matches or candles [150].

12.6.2.3 Household Appliances

An elegant appearance and a high performance value are a must when it comes to household appliances and commodities. Safety requirements, including those of fire safety, are set out in standards, recommendations, and guidelines.

IEC/TC 61 aims at producing and maintaining international standards relating to the safety of household and similar appliances in a manner that is timely, efficient, and that keeps pace with modern technology [151]. WG 4 of TC 61 deals with the revision of IEC 60335-1 Clause 30, which contains the requirements for the resistance to heat and fire. Consequential testing has been a part of IEC 60335-1 for a long time. Additional testing that is required in certain circumstances, following a special result in a previous test, for example, if during a glow-wire test flames are emitted from a test specimen or flames exceed a specified length. Then, additional testing, for example, using a needle flame, may be required to determine whether the primary flame could ignite the surrounding materials/components. The basic discussion in TC 61 and in other committees such as TC 74 is, whether that consequential testing or other means of fire protection can be the way to come to end-products with minimized fire hazard. Special discussion points in IEC/TC 61 and TC 89 are the questions how to simulate statistically relevant ignition sources with the bad-connection test or the glow-wire test and the needle-flame test for the surrounding parts of the test specimen emitting flames, or to use other fire-protective measures.

A possible solution could be found in using intelligent ways of designing the equipment, for instance the use of fire enclosures similar to those recommended in several standards, or the use of smoke detector electronic circuits or of an Arc Fault Circuit Interrupter (AFCI). The AFCI incorporates a circuit breaker and is designed to detect an undesirable arc fault and shutoff power to the circuit before a fire can initiate [152].

Studies have shown that in home fires, most casualties occur at night and that people could have been saved if they had woken up early enough [153]. Investigations on the influence of smoke detectors on escape times in a fire compartment have been carried out by Wittbecker [154]. The aim of this exercise was to calculate escape times in an assumed realistic domestic dwelling. The results of this study demonstrate the great improvement in survival rates that could be achieved through the use of a fire alarm system.

Television programs and public information [155] drew attention to the great benefit of inexpensive smoke detectors to life and property and give guidance on their purchase, function, and installation.

12.7 Future Developments

There is a rapid global increase in the use of electrotechnical products in all application areas. As a result of the globalization of trade, industry, and politics, in the future, national standards and classification systems for fire hazard assessment in the E&E sector will more and more be substituted by internationally valid harmonized systems. This will dramatically reduce the innumerable existing national tests and costs, guarantee the free movement of certified materials and electrotechnical products and parts, and provide more fire safety by uniform classification and testing systems worldwide.

IEC/TC 89 has added fire safety engineering to its scope, as far as it applies to electrotechnical products. Advances in the understanding and use of fire safety engineering will have a positive impact on the demand for TC 89 standards, especially for product test methods, which provide performance-based data in a format suitable for use in fire safety engineering. The output of the two ISO groups, TC 92/SC 4 "Fire safety engineering" and ISO/TC 92/SC 3 "Fire threat to people and the environment," is expected to influence a further revision of IEC 60695-1-1, "Fire hazard assessment." The introduction of fire safety engineering into the field of electrical engineering will lead to improved fire safety and save lives and money. IEC/TC 89 has prepared the tools and the guidance to use them. It is now at the product committees to follow this guidance and to overcome the historical dilemma of including end-use testing and preselection testing into their product standards. The other task is to solve the problems that arise from fire hazard assessment and product certification, if different concepts yield different classifications for electrical equipment and built-in plastics.

References

[1] G. Zelesny: Corporate Market Research. Bayer AG; Intranet: Markt & Umfeld, 2001
[2] Plastics and electrical & electronic products, 2001: Association of Plastics Manufacturers in Europe (APME) (http://www.apme)
[3] UL-Fire Test, 2000: Facility (http://www.ul.com/fire/fire2.html)
[4] F.W. Wittbecker, B. Müller: 25 Jahre Brandtechnologie bei der Bayer AG, Unfallfrei (brandtechnologie@ bayer-ag.de)
[5] Siemens Axiva GmbH & Co. KG, Brandhaus Höchst, Industriepark Höchst, Geb. C369, D-65926 Frankfurt am Main (mitzlaff@axiva.com), 2001
[6] S.E. Goodall: ETZ-A 94 (1973), p. 314
[7] R. Leber: Elektronorm 26 (1972), p. 513
[8] IEC-Website, 2000 (http://www.iec.ch)
[9] IEC/TC 89, Nov. 1999; Fire Hazard Testing. Strategic Policy Statement CA/1648/R
[10] IEC 60695-1-1, 3rd Ed., Nov. 1999: Fire hazard testing: Guidance for assessing the fire hazard of electrotechnical products. General guidelines
[11] ISO/TR 13387-1, Oct. 1999: Fire safety engineering. Application of fire performance concepts to design objectives
ISO/TR 13387-2, Oct. 1999: Fire safety engineering. Design fire scenarios and design fires
ISO/TR 13387-3, Oct. 1999: Fire safety engineering. Assessment and verification of mathematical fire models
ISO/TR 13387-4, Oct. 1999: Fire safety engineering. Initiation and development of fire and generation of fire effluent
ISO/TR 13387-5, Oct. 1999: Fire safety engineering. Movement of fire effluents
ISO/TR 13387-6, Oct. 1999: Fire safety engineering. Structural response and fire spread beyond the enclosure of origin
ISO/TR 13387-7, Oct. 1999: Fire safety engineering. Detection, activation and suppression
ISO/TR 13387-8, Oct. 1999: Fire safety engineering. Life safety. Occupant behaviour, location and condition

[12] G.I. Smelkov, V.A. Pekhotikow: On the Assessment of an Electrical Fire Risk VNIIPO MVD RF Moscow. IEC/TC 89 plenary meeting Milan, Sept. 1998 (89/330/INF)

[13] ISO/TR 11696-2, May 1997: Fire Tests, Reaction to Fire. Guide to the use of Test Results in Fire Hazard Analysis of Construction Products

[14] P.J. Briggs, A.J. Morgan: Reaction to Fire Behaviour of Thermoplastic Diffuser Panels In Lightning Assemblies. Flame Retardants '98, p. 257–267

[15] A.E. Cote, J.L. Linville: Fire Protection Handbook. National Fire Protection Association Quincy, Massachusetts; Origins of Electrical Fires in Buildings; p. 2–22, 2–23, 1992

[16] Home Office, Apr. 1997: Fire Statistics United Kingdom, London, 1995

[17] M. de Poortere, C. Schonbach, M. Simonson: Fire. Mater. 24, p. 53–60, 2000

[18] M. Simonson, H. Stripple: LCA Study of Flame Retardants in TV Enclosures, p. 159–169, 2000

[19] H. Arlt: Risiken und einige typische Schäden bei Erweiterungen und Erneuerungen von Schaltanlagen. Allianz Report 68 (1995) 2, p. 60–64

[20] H. Urbig, VFDB Apr. 1998: Brände durch elektrische Energie; ihre Ursache und statistische Verteilung, p. 167-168

[21] Fachtagung, 18./19. Okt. 1977: Brandschadenverhütung in elektrischen Anlagen. Düsseldorf, Verband der Sachversicherer e.v. Köln. Der Elektromeister, Deutsches Elektrohandwerk

[22] Zahlenspiegel der Versicherungswirtschaft, Nov. 1997, GVD

[23] Sterbefälle (Unfälle) durch Feuer und Flammen 1988–1996, 1998: Pos.-Nr. E 890 – E 899; Statistisches Bundesamt, Wiesbaden

[24] IEC 60695-10-2, 1st Ed., Aug. 1995: Fire hazard testing: Guidance and test methods for the minimization of the effects of abnormal heat on electrotechnical products involved in fires. Section 2: Method for testing products made from non-metallic materials for resistance to heat using the ball pressure test

[25] IEC 60695-2-10, Oct. 2000: Fire hazard testing: Glowing hot-wire based methods. Glow-wire apparatus and common procedure

[26] IEC 60695-2-11, 1st Ed., Oct. 2000: Fire hazard testing: Test methods, section 1/sheet 1: Glow-wire end-product test and guidance

[27] Vollautomatische Flammenanalyse für Isolierstoffe; IITB-Mitteilungen 1999. Fraunhofer-Institut, Fraunhofstrasse 1, 76131 Karlsruhe / Germany

[28] K.H. Schwarz: ETZ-B 14, p. 273, 1962

[29] R. Holm: Handbuch der elektrischen Kontakte. 3. Auflage, Springer, p. 130–146; specially Table 27.04 on page 135, tests 3–5, 1958

[30] IEC 60695-2-3, 1st Ed., Jan. 1984: Fire hazard testing: Test methods, Bad-connection test with heaters

[31] IEC 60695-2-2, 2nd Ed., May 1991: Fire hazard testing: Test methods, section 2: Needle-flame test. IEC 60695-2-2, Amd. 1, 2nd Ed., March 1994: Amendment No. 1

[32] IEC 60695-2-4/0, 1st Ed., Aug. 1991: Fire hazard testing. Part 2: Test methods, section 4/sheet 0: Diffusion type and premixed type flame test methods

[33] IEC 60695-2-4/1, 1st Ed., Apr. 1991: Fire hazard testing. Part 2: Test methods, section 4/sheet 1: 1 kW nominal pre-mixed test flame. IEC 60695-2-4/1, Amd. 1, 1st Ed., May 1994: Amendment No. 1

[34] IEC 60695-1-30 2nd Ed.: Fire hazard testing. Guidance for assessing fire hazard of electrotechnical products. Use of preselection test procedures (89/399/CD)

[35] IEC 60695-3-1, 1st Ed., Jan. 1982: Fire hazard testing. Examples of fire hazard assessment procedures and interpretation of results. Combustion characteristics and survey of test methods for their determination

[36] IEC 60695-2-12, 1st Ed., Oct. 2000: Fire hazard testing. Test methods, section 1/sheet 2: Glow-wire flammability test on materials.

[37] IEC 60695-2-13, 1st Ed., Oct. 2000: Fire hazard testing. Test methods, section 1/sheet 3: Glow-wire ignitability test on materials

[38] IEC 60695-2-20, 1st Ed., Aug. 1995: Fire hazard testing. Glowing/Hot wire based test methods, section 20: Hot-wire coil ignitability test on materials.

[39] IEC 60707, 1st Ed., 1981: Methods of test for the determination of the flammability of solid electrical insulating materials when exposed to an igniting source

[40] ISO 1210, Aug. 1992: Plastics. Determination of flammability characteristics of plastics in the form of small specimens in contact with a small flame

[41] ISO 10351, 1996: Plastics. Determination of the combustibility of specimens using a 125 mm flame source

[42] UL 94, 5th Ed., 1997: Test for Flammability of Plastic Materials for Parts in Devices and Appliances

[43] IEC 60707, 2nd Ed., Mar. 1999: Flammability of solid non-metallic materials when exposed to flame sources. List of test methods

[44] IEC 60695-11-10, 1st Ed. Mar. 1999: Fire hazard testing. Test flames, 50 W horizontal and vertical flame test methods

[45] IEC 60695-11-20, Mar. 1999: Fire hazard testing. Test flames, 500 W flame test methods

[46] IEC/TS 60695-11-3, Mar. 2000: Fire hazard testing. Test flames: 500 W flames. Apparatus and confirmational test methods. IEC/TS 60695-11-4, Feb. 2000: Fire hazard testing. Test flames: 50-W flames. Apparatus and confirmational test methods

[47] IEC 60695-9-1, 1st Ed., Dec. 1998: Fire hazard testing. Surface spread of flame. General guidance

[48] IEC 60695-9-2 TS, 1st Ed.: Fire hazard testing: Surface spread of flame. Summary and relevance of test methods

[49] ISO/TR 9122-1, 1989: Toxicity testing of fire effluent. General

[50] C. Hugget: Fire Mater. (1980) 4, p. 61–65

[51] IEC 60695-8-1, 1st Ed., Jan. 2001: Fire hazard testing. Heat release, General guidance

[52] IEC 60695-8-2 TS, 1st Ed., July 2000: Fire hazard testing. Heat release – Summary and relevance of test methods

[53] UL 2043: Fire Test for Heat and Visible Smoke Release for Discrete Products and Their Accessories Installed in Air-Handling Spaces; with supplements up to Feb. 1998

[54] V. Babrauskas, S.J. Grayson (Eds.): Heat Release in Fires. Elsevier Applied Science Publishers, London, 1992

[55] M.M. Hirschler: Electric Cable Fire Hazard Assessment with the Cone Calorimeter (p. 44–65) in Fire Hazard and Risk Assessment (ASTM STM 1150). American Society for Testing and Materials, Philadelphia, 1992

[56] V. Babrauskas: Ten Years of Heat Release Research with the Cone Calorimeter (http://www.doctor-fire.com/cone.html), 2000

[57] A. Ahonen, M. Kokkola, H. Weckmann: Burning Characteristics of Potential Ignition Sources of Room Fires. VTT Research Reports

[58] H.K. Hasegawa, N.J. Alvares, J.A. White: Fire Tests of Packaged and Pallefized Computer Products, Fire Technology, Vol. 35, No. 4, p. 291, 1999

[59] IEC 60695-6-1, 1st Ed., Febr. 2001: Fire hazard testing. Smoke opacity, General guidance

[60] IEC 60695-6-2, Febr. 2001: Fire hazard testing. Smoke obscuration. Summary and relevance of test methods

[61] IEC 60695-6-30 TR 2, 1st Ed., Oct. 1996: Fire hazard testing. Guidance and test methods on the assessment of obscuration hazard of vision caused by smoke opacity from electrotechnical products involved in fires. Small scale static method. Determination of smoke opacity. Description of the apparatus

[62] IEC 60695-6-31 TS, 1st Ed., Apr. 1999: Fire hazard testing. Smoke obscuration, small-scale static test, materials

[63] IEC 60695-7-1, 1st Ed., Nov. 1993: Fire hazard testing. Guidance on the minimization of toxic hazards due to fires involving electrotechnical products. General

[64] IEC 60695-7-2 TS, 1st Ed.: Toxicity of fire effluent. Summary and relevance of test methods (89/408/CD).

[65] IEC 60695-7-3 TR 2, 1st Ed., Dec. 1998: Fire hazard testing. Toxicity of fire effluent. Use and interpretation of test results

[66] ISO/TR 9122-1, 1st Ed. 1993: Toxicity testing of fire effluents. General.
ISO/TR 9122-2, 1st Ed. 1993: Toxicity testing of fire effluents. Guidelines for biological assays to determine the acute inhalation toxicity of fire effluents (basic principles, criteria and methodology)
ISO/TR 9122-3, 1st Ed. 1993: Toxicity testing of fire effluents. Methods for the analysis of gases and vapours in fire effluents
ISO/TR 9122-4, 1st Ed. 1993: Toxicity testing of fire effluents. The fire model. Furnaces and combustion apparatus used in small-scale testing
ISO/TR 9122-5, 1st Ed. 1993: Toxicity testing of fire effluents. Prediction of toxic effects of fire effluents

[67] IEC 60695-7-50, TS, 1st Ed.: Fire hazard testing: Toxicity of fire effluent. Estimation of toxic potency: Apparatus and test method (89/461/CD)

[68] DIN 53436: Erzeugung thermischer Zersetzungsprodukte von Werkstoffen unter Luftzufuhr und ihre toxikologische Prüfung. Teil 1, 1981: Zersetzungsgerät und Bestimmung der Versuchstemperatur. Teil 2, 1986: Verfahren der thermischen Zersetzung. Teil 3, 1989: Verfahren zur inhalationstoxikologischen Untersuchung

[69] F.H. Prager, H.J. Einbrodt, J. Hupfeld, B. Müller, H. Sand: *J. Fire Sci.*, (Sep./Oct. 1987 5), p. 308–325

[70] J. Pauluhn, G. Kimmerle, T. Märtins, F. Prager, W. Pump: *J. of Fire Sci.* (Jan./Feb. 1994) 12, p. 63

[71] IEC 60695-7-51, TS, 1st Ed.: Fire hazard testing: Toxicity of fire effluent. Estimation of toxic potency test: Calculation and interpretation of test results (89/462/CDV)

[72] J. Pauluhn: *J. of Fire Sci.*, (Mar./Apr. 1993) 11, p. 110/ 130

[73] F.W. Wittbecker, W. Klingsch, B. Bansemer: Ernst & Sohn, *Bauphysik* (2000) 22, Heft 1

[74] IEC 60695-5-1, 1st Ed., June 1993: Fire hazard testing. Assessment of potential corrosion damage by fire effluent. General guidance

[75] IEC 60695-5-2 TR 2, 1st Ed., Dec. 1994: Fire hazard testing. Assessment of potential corrosion damage by fire effluent. Guidance on the selection and use of test methods

[76] E. Barth, B. Müller, F.H. Prager, F.W. Wittbecker: *J. of Fire Sci.*, (Sept./Oct. 1992) p. 432

[77] I.T. Chapin, P. Gandhi, L.M. Caudill: Comparison of Communications LAN Cable Smoke Corrosivity by US and IEC Test Methods. Fire Risk and Hazard Research Assessment Research Application Symposium, San Francisco, CA, Jun. 25–27, 1997 (http://www.wireville.com/firesafe/corrosiv.htm) July 2000,

[78] T.J. Tanaka: *Fire Material* (1999) 23, p. 103–108

[79] ISO 5660-1, Jan. 1996: Fire tests: Reaction to fire. Rate of heat release from building products (cone calorimeter method)

[80] F.W. Wittbecker: Zur szenarioabhängigen Ermittlung und Beurteilung des brandtechnologischen Materialverhaltens. Heft 9, Schriftenreihe des Lehr- und Forschungsgebietes Baustofftechnologie und Brandschutz; Institut für konstruktiven Ingenieurbau, Berg. Universität, Gesamthochschule Wuppertal, 2000

[81] Testing for Public Safety, 1978. Underwriters Laboratories Inc.

[82] International Electrotechnical Committee for Conformity Testing to Standards for Electrical Equipment (IECEE), Geneva, Switzerland (www.iecee.org/cbscheme)

[83] Underwriters Laboratories Inc., 2000 (http://www.ul.com)

[84] ISO 9773, 1998: Plastics. Determination of burning behaviour of flexible vertical specimens in contact with a small-flame ignition source

[85] ISO 9772, 1998: Cellular plastics. Determination of horizontal burning characteristics of small specimens subjected to a small flame

[86] ASTM E 162-98: Standard Test Method for Surface Flammability of Materials Using a Radiant Heat Energy Source

[87] UL 746A, 4th Ed.: Polymeric Materials. Short Term Property Evaluations; with supplements up to March 1997

[88] ASTM D 3638-85: Test Method for Comparative Tracking Index of Electrical Insulating Materials

[89] IEC 112, Jun. 1984: Method for determining the comparative and the proof tracking indices of solid insulating materials under moist conditions

[90] UL 746B, Aug. 1996: Polymeric Materials. Long Term Property Evaluations; with supplements up to Feb. 2000

[91] UL 746C: Polymeric Materials. Use in Electrical Equipment. Evaluations; with supplements up to July 1997

[92] ASTM E 84-98: Standard Test Method for Surface Burning Characteristics of Building Materials

[93] UL 746D: Polymeric Materials. Fabricated Parts; with supplements up to Oct. 1995

[94] UL 1694, Jul. 1997: Tests for Flammability of Small Polymeric Component Materials

[95] UL 1950; CSA 22.2, No. 950

[96] IEC 950: Safety of Information Technology Equipment; with supplements up to Jul. 1997

[97] CSA Stand. C22.2 No. 0.6-M1982: Flammability Testing of Polymeric Materials

[98] New Approach Standardization in the European internal market (http://www.New Approach.org)

[99] The New Approach Directives (http://www.NewApproach.org/directiveList.asp)

[100] Official Journal of the European Communities Commission Decision of 8 Feb. 2000 implementing Council Directive 89/106/EEC as regards the classification of the reaction to fire performance of construction products [document number C (2000) 133]

[101] EU-Low Voltage Directive (LVD) 73/23 EEC

[102] RG N 208, 11. Oct. 2000: Draft Commission Decision xx/xx/2000 amending Commission Decision 2000/147/ EC implementing Council Directive 89/106/EEC as regards the classification of the
reaction to fire performance of construction products (http://www.intercomm.dial. pipex.com/com-
missiondoc.htm)

[103] EU-Safety of Machinery Directive 89/392/EEC

[104] prEN 13478, May 1999: Safety of machinery. Fire prevention and protection

[105] Directive on the interoperability of the European high speed train network, 96/48 EEC, 23 July
1996

[106] prEN 45545-1 Railway applications – Fire protection on railway vehicles: General.
prEN 45545-2: Requirements for fire behaviour of materials and components (in prep.).
prEN 45545-3: Fire resistance requirements for fire barriers and partitions.
prEN 45545-4: Fire safety requirements for railway rolling stock design.
prEN 45545-5: Fire safety requirements for electrical equipment including that of trolley buses,
track guided buses and magnetic levitation vehicles.
prEN 45545-6: Fire control and management systems.
prEN 45545-7: Fire safety requirements for flammable liquid and flammable gas installations.

[107] EU project FIRESTARR „Fire protection on Railway vehicles". Contract no. SMT4-CT97-2164,
Final Report April 2001

[108] Association for Electrical, Electronic & Information Technologies (VDE) (http://www.vde.com/
vde/html /e/online/online.htm)

[109] German Electrical Commission (DKE) (http://www.dke.de)

[110] VDE Testing and Certification Institute, Merianstrasse 28, D-63069 Offenbach (Main), Germany.

[111] TÜV Rheinland, Am Grauen Stein, D-51101 Köln, Germany (http://www.us.tuv.com)

[112] FEMAC (Far East Market Access Services) GmbH, Friedrichstrasse 10, D-70174 Stuttgart, Germany (http://www.femac.com)

[113] ITS C 0060, Part 2: Fire hazard testing. Test methods. Glow-wire-test and guidance

[114] ITS C 0061, Part 2: Fire hazard testing. Needle-flame test

[115] Ou Yuxiang: Flame Retardant Regulations and Test Standards in China. National Laboratory of
Flame-retarded Materials. Institute of Technology, Beijing 100081, China

[116] GB 5169.4-85: Glow-wire test
GB 5169.5-85: Needle-flame test
GB 5169.6-85: Bad-connection test

[117] GB 5169.7-85: Bunsen-flame test

[118] NFPA 70, 1990: National Electrical Code, NEC, National Fire Protection Association, Batterymarch Park, Quincy, MA (1989)

[119] M.M. Hirschler: Fire Mater. (1992) 16, p. 107–118

[120] S. Richter, R. Schmidt: Testing of Cables Designed for Fire Resistance. Proceedings of 46th International Wire and Cable Symposium, Philadelphia, Nov. 1997

[121] Muster-Richtlinie über brandschutztechnische Anforderungen an Leitungsanlagen (Muster-Lei-
tungs-anlagen-Richtlinie MLAR). Fachkommission Bauaufsicht, Mar. 2000

[122] IEC 60332-1: 1993, Apr. 1993: Tests on electric cables under fire conditions. Test on a single vertical insulated wire or cable

[123] IEC 60332-2: 1989, March 1989: Tests on electric cables under fire conditions. Test on a single
small vertical insulated copper wire or cable

[124] IEC 60332-3, 1992: Tests on electric cables under fire conditions. Tests on bunched wires or cables

[125] IEC 60332-3-10, 1st Ed., Oct. 2000: Tests on electric cables under fire conditions. Test for vertical
flame spread of vertically-mounted bunched wires or cables. Apparatus
IEC 60332-3-21, 1st Ed., Oct. 2000: Tests on electric cables under fire conditions. Test for vertical
flame spread of vertically-mounted bunched wires or cables – Category A F/R
IEC 60332-3-22, 1st Ed., Oct. 2000: Tests on electric cables under fire conditions. Test for vertical
flame spread of vertically-mounted bunched wires or cables – Category A
IEC 60332-3-23, 1st Ed., Oct. 2000: Tests on electric cables under fire conditions. Test for vertical
flame spread of vertically-mounted bunched wires or cables – Category B
IEC 60332-3-24, 1st Ed., Oct. 2000: Tests on electric cables under fire conditions. Test for vertical
flame spread of vertically-mounted bunched wires or cables – Category C
IEC 60332-3-25, 1st Ed., Oct. 2000: Tests on electric cables under fire conditions. Test for vertical
flame spread of vertically-mounted bunched wires or cables – Category D

[126] UL 1666, 3rd Ed. 1997: Tests for Flame Propagation. Height of Electrical and Optical-Fiber Cables Installed Vertically in Shafts

[127] NF C 32-070, 1. Jul. 1992: Insulated cables and flexible cords for installations. Classification tests non cables and cords with respect to their behaviour to fire (HD 405=1 S1 Amendment 1)

[128] L. Caudill: Fire Testing of Communications Cables in Plenums; International Fire Safety Conference in San Francisco 1996 (http://www.datacable.org/firesafe.htm)

[129] UL 910, 1998: Test for Flame-Propagation and Smoke-Density Values for Electrical and Optical-Fiber Cables Used in Spaces Transporting Environmental Air; with supplements up to May 1995

[130] IEC 61034-1, 2nd Ed., Aug. 1997: Measurement of smoke density of cables burning under defined conditions: Test apparatus

[131] IEC 61034-2, 2nd Ed., Sept. 1997: Measurement of smoke density of cables burning under defined conditions: Test procedure and requirements

[132] IEC 60331-11, 1st Ed., Apr. 1999: Tests for electric cables under fire conditions, Circuit integrity. Apparatus. Fire alone at a flame temperature of at least 750 °C
IEC 60331-21, 1st Ed., Apr. 1999: Tests for electric cables under fire conditions, Circuit integrity. Procedures and requirements. Cables of rated voltage up to and including 0.6/1.0 kV
IEC 60331-23, 1st Ed., Apr. 1999: Tests for electric cables under fire conditions. Circuit integrity Procedures and requirements. Electric data cables
IEC 60331-25, 1st Ed., Apr. 1999: Tests for electric cables under fire conditions. Circuit integrity Procedures and requirements. Optical fibre cables

[133] DIN 4102-12: Fire behaviour of building materials and components: Reliability of electric systems, requirements and testing

[134] BS 6387, 1983: Performance requirements for cables required to maintain circuit integrity under fire conditions

[135] Draft prEN YYY, Mar. 1999: Railway Applications. Railway Rolling Stock Cables having special fire performance. Thin Wall and Standard Wall. Guide to use

[136] EN 50306-1, Apr. 1999, Draft: Railway rolling stock cables having special fire performance and special requirements for increased safety. Thin wall: General requirements
EN 50306-2, Apr. 1999, Draft: Railway rolling stock cables having special fire performance and special requirements for increased safety. Thin wall: Single core cables
EN 50306-3, Apr. 1999, Draft: Railway rolling stock cables having special fire performance and special requirements for increased safety. Thin wall: Single core and multicore cables (Pairs, triples and quads) Screened and thin wall sheathed
EN 50306-4, Apr. 1999, Draft: Railway rolling stock cables having special fire performance and special requirements for increased safety. Thin wall: Multicore and multipair cables. Standard wall sheathed

[137] EN 50264-1, 1997, Draft: Railway applications. Railway rolling stock cables having special fire performance. Standard wall. Guide to use
EN 50264-2, 1997, Draft: Railway applications. Railway rolling stock cables having special fire performance. Standard wall: Single core cables
EN 50264-3, 1998, Draft: Railway applications. Railway rolling stock cables having special fire performance. Standard wall: Multicore cables

[138] prEN XXS: Railway Applications: Railway Rolling Stock Cables having special fire performance. Special Types

[139] prEN YYY, Mar. 1999: Railway Rolling Stock Cables having special fire performance. Thin Wall and Standard Wall. Guide to use [CENELEC (TC20) (SEC) 1202]

[140] prEN 50305, Apr. 1999: Railway applications. Railway rolling stock cables having special fire performance. Test methods

[141] DIN EN 50265-2-1 (VDE 0482, Part 265-2-1), 1998: Common test methods for cables under fire conditions. Test for resistance to vertical flame propagation for a single insulated conductor or cable: Procedures, 1 kW pre-mixed flame

[142] prEN 50266-2-4, Nov. 1999: Allgemeine Prüfverfahren für Kabel und isolierte Leitungen im Brandfall. Prüfung der senkrechten Flammenausbreitung von senkrecht angeordneten Bündeln von Kabeln und isolierten Leitungen. Prüfverfahren, Prüfart C

[143] Fire Performance of Electric Cables. The FIPEC Report. Discussion Seminar 15th Sept. 2000; Royal Institute of British Architects RIBA

[144] prEN 50XXX, Febr. 2000: Common test methods for cables under fire conditions; Heat release measurement on cables during flame spread test

[145] EN 60950, 1997: Safety of information technology equipment

[146] IBM: Fire safety concerns play key role in the material selection process. Modern Plastics Encyclo-pedia-Electronics (http://www.modplas.com/encycloped...es/arlicles_indu-stry_electron.htm)

[147] UL 1410, 15th Ed. 1997: Television Receivers and High-Voltage Video Products

[148] IEC 60065, 6th Ed. 1998: Audio, video and similar electronic apparatus. Safety requirements

[149] EN 60065, VDE 0860, Oct. 1999: Audio, Video- und ähnliche elektronische Geräte, Sicherheit-sanforderungen

[150] M. Simonson, H. Stripple: The incorporation on fire considerations in the life-cycle assessment of polymeric composite materials. A preparatory study INTERFLAM'99

[151] IEC/TC 61: Strategic Policy Statement CA/1903/R

[152] D.A. Lee, A.M. Trotta, W.H. King, Jr.: New Technology for Preventing Residential Electrical Fires. Arc-Fault Circuit Interrupters (AFCIs)

[153] National Commission on fire prevention and Control, 1973; America Burning, Washington

[154] F.W. Wittbecke: FIRE Europe V (Mar. 1997) p. 4–6

[155] Rauchmelder retten Leben. Ministerien für Bauen und Wohnen sowie für Inneres und Justiz des Landes Nordrhein-Westfalen (http://www.mbw.nrw.de; http://www.im.nrw.de)

[156] ABD 0031: Airbus Directives (ABD) and Procedures, Pressurized Section of Fuselage, Fire-wor-thiness Requirements

[157] NES 713, 1985: Naval Engineering Standard. Determination of the toxicity index of the products of combustion from small test specimens of materials

[158] NF C20-454, 1998: Test methods. Fire behaviour. Analysis and titration of gases evolved during pyrolysis or combustion of materials used in electrotechnical products. Exposure to abnormal heat and fire. Tube furnace method

[159] CEI 20-37-7, 1998: Tests on gas evolved during combustion of electric cables and their com-pounds. Determination of toxicity index of gases evolved during combustion of electric cables

[160] IMO, Resolution MSC.41 (64), 5 Dec. 1994: Interim Standard for Measuring Smoke and Toxic Products of Combustion

[161] B.C. Levin et al, 1982: Further Development of a Test Method for the Assessment of the Acute Inhalation Toxicity of Combustion Products. NBSIR 82-2532 Washington: US-National Bureau of Standards

[162] Y.C. Alaric, R.C. Anderson, 1979: Toxicol. Appl. Pharmacol., 51, p. 341–362

[163] Office of Fire Prevention and Control. Combustion Toxicity Testing. In: New York State Uniform Fire Prevention and Building Code. Albany, New York, State: Department of State, 1985, articles 15, part 1120

[164] ASTM E 1678-97: Standard Test Method for Measuring Smoke Toxicity for use in Fire Hazard Analysis

13 Furniture and Furnishings

K.T. Paul, K.A. Reimann, and B. Sundström

13.1 Introduction

Fires generally start in the contents of buildings, and the amount of ignitable materials introduced into buildings, for example, furniture and furnishings, has greatly increased over the past few decades. The furniture and furnishings in a building may constitute a major fire hazard because even if building regulations are strict in most countries, the technical fire requirements for the building contents may be modest. The result can be that the fire hazard from the furniture is much greater than the fire hazard from the building materials. Furniture and furnishings include all moveable items such as cupboards, wardrobes, chests, tables, desks, seating, bedding, and so forth. Other building contents considered as furnishings include household textiles (decorations, curtains) or floor coverings (the use of which is controlled by the building inspectorate in some countries), but these are not discussed in this chapter. Upholstered furniture used in transportation (mainly as seating) is also dealt with elsewhere. Upholstered furniture has been identified as the most fire hazardous furniture in terms of ease of ignition and rapid fire growth. Most research and test method development has therefore focused on these products.

Measures to limit the fire hazards of the contents of buildings, that is furniture, have been introduced relatively recently and even now are applied only in some countries. This may be considered strange, as in many instances the building structure will be required to withstand the fire generated by the building contents.

Furniture and furnishings pose differing fire hazards depending on the surroundings in which they are used. Decisive factors that affect fire hazard are the type (private household or auditorium), size (single family dwelling or multistory block), and purpose (barracks with young, highly trained soldiers or hospital with infirm patients) of a building as well as its contents (high or low fire load). The assessment of fire hazard is dealt with in greater detail in Chapter 8.

A survey of the furniture fire situation in Europe (EU countries) by de Boer [1] in 1990 reached the following major conclusions:

Fires in bedding and furniture make up only 15 % of all fires in private dwellings. They cause 49 % of all fatal and 30 % of all nonfatal injuries in these premises. In public buildings, bedding and furniture fires account for 4 % of all fires, but they cause 55 % of all fatal and 30 % of all nonfatal injuries.

The main ignition sources in bedding and furniture fires are smoking materials (32 %), matches, lighters (10 %), and, where they are commonly used, electric blankets (20 %).

A significant hazard in bedding and furniture fires in private dwellings arises from people being trapped by a rapidly growing bedding and furniture fire with the consequential smoke spread that gives less time to escape. This situation resulted in more victims occurring in the room of origin of the fire. Tables 13.1 and 13.2 show the UK statistics for 1989 [2], which is the last year before the UK 1988 Furniture and Furnishings, (Fire) (Safety) Regulations [3] became effective.

Table 13.1 Fires in Dwellings in the UK in 1989, Casualties by source of ignition [2]

Ignition source	Fatalities	Nonfatal injuries
Total	583	916
Smokers materials	234	1974
Matches	65	674
Cooking	40	3900
Heating	87	767
Electrical	52	913
Other	105	882

Table 13.2 Fires in dwellings and occupied buildings in the UK in 1989, casualties by item first ignited and by item responsible for fire development [2]

Furnishings	Item first ignited			Item responsible for growth		
	Total	Fata-lities	Non-fatal injuries	Total	Fata-lities	Non-fatal injuries
Bedding, beds, and mattresses	4387	113	1210	5011	120	1384
Upholstery covers and furniture	7317	328	2974	8964	402	3414
Curtains and blinds	545	2	106	701	1	141
Floor coverings	742	13	150	978	10	189
Structure and fittings	2895	3	104	5724	37	555

About 30 % of the ignition sources are smokers' materials and matches, which makes these the most common ignition source involved in fire deaths in the United Kingdom. UK statistics also show that textiles, upholstery and furnishings mainly account for almost 50 % of the fire growth.

In 1988, the UK introduced regulations [3] to improve the fire properties of upholstered furniture suitable for domestic use.

The UK Department of Trade and Industry (DTI) published a detailed assessment of the effectiveness of the Furniture and Furnishings (Fire) (Safety) Regulations 1988 [4]. Figures 13.1 and 13.2 have been reproduced from this publication and show the trends in upholstered furniture-related fire casualties. These data have been corrected to allow for the effect of smoke alarms, which in the UK has been relatively small, with 1–2 % of fires being detected by alarms in 1995 and 10 to 12 % in 1997. The DTI report shows that there were 247 fire deaths caused by upholstered furniture in 1988, compared to 95 in 1998. Before 1988, there was a reduction in fire deaths due to burns, but an increase due to smoke inhalation giving a generally constant value, but after 1988 deaths from both causes showed a decrease. An apparent increase in nonfatal injuries was seen in smoke injuries, which may be due to the precautionary policy of referring possible casualties for medical assessment.

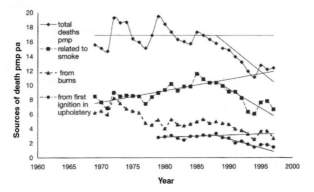

Figure 13.1   Pre- and post-1988 UK trends in fatal fire injuries [4]

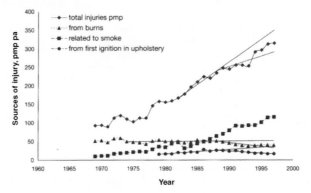

Figure 13.2   Pre- and post-1988 UK trends in nonfatal fire injuries [4]

The significant dates within this period are 1980, in which regulations requiring upholstered seating composites to resist a lighted cigarette were introduced, and 1988, which saw the introduction of the Furniture and Furnishings (Fire) (Safety) Regulations. The latter essentially require upholstered seating to be resistant to lighted cigarettes and small flames and for the fillings of all seating, mattresses, and other items of upholstered furniture to be ignition resistant and for polyurethane (PU) foams to be of the combustion-modified type. This regulation does not require mattresses to be ignition resistant as such but under the General Product Safety Regulations [5], it would be reasonable to expect mattresses to comply with the relevant British Standard, that is, to the low hazard category of BS 7177 [6]. The latter requires resistance to cigarette and small flame tests of EN 597-1 and EN 597-2 [7] and also to the covered cigarette test of Annex B of BS 7177.

Information concerning upholstered furniture fire deaths in the United States is given in Figs. 13.3–13.5 [8]. Many fire fatalities are caused by mattress fires, which are started mainly by the effects of glowing smoking materials. Data from Fig. 13.4 indicate that fire deaths for the years 1980–1984 compared to 1991–1995 were down by about 35 % for mattresses and 19 %

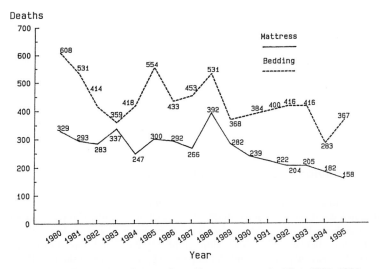

Figure 13.3 Trend in US fatal home fires starting with mattresses or bedding: 1980–1995

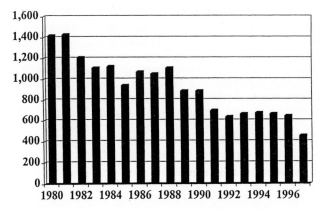

Figure 13.4 US upholstered furniture fire deaths, cigarette ignition: 1980–1997

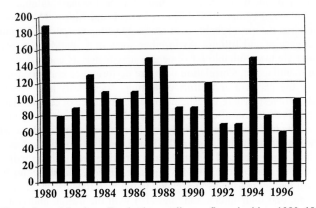

Figure 13.5 US upholstered furniture fire deaths, small open flame ignition: 1980–1997

for bedding when that was the item first ignited. Mattresses complying with the standard have made a contribution to reducing the hazards of a smoldering fire (see Fig. 13.3).

In the late 1970s, the voluntary UFAC cigarette smoldering standard became effective and the number of fire deaths has exhibited a downward trend (see Fig. 13.4). The total number of deaths from cigarette ignition is higher than those from open flame ignition and the latter have stayed relatively constant over the same time period (see Fig. 13.5).

In the United States, it is considered that about 50% of domestic fires are first detected by a smoke alarm. First introduced in 1976, it is estimated that 90% of US homes now contain alarms, although not all may be in working order.

13.2 Fire Behavior of Upholstered Furniture

During the past three decades, a large number of comprehensive studies concerning the fire hazard of upholstered furniture were undertaken in United States, United Kingdom, Scandinavia, and elsewhere. It is not proposed to review these in detail and readers are referred to the literature and in particular to the NIST Monograph [9], the CBUF report [10] and the recent book published by Krasny, Parker, and Babrauskas [11], which comprehensively deal with the historical research and the state of the art today. A detailed bibliography of the early research and test developments may also be found in the previous edition of this book. Because of this, only the more important aspects of upholstered flammability are outlined here.

Early investigations showed very clearly that for many upholstery composites used for seating and bedroom furniture, smoking materials could cause smoldering and flaming combustion. Depending on the materials used in the upholstery composite, a cigarette causes the covering to char or melt near the glowing tip. In the former case the cover material starts to char and incandescence occurs in the vicinity of the cigarette embers. The underlying cushioning material is heated and starts to smolder thus liberating further heat and causing the smoldering to spread. Depending on the situation, this could suddenly burst into flame after a period lasting for up to several hours. In the second case, the cover material melts away from the incandescent region exposing the underlay. The intensity of the ignition source (cigarette) will be reduced by means of heat deprivation effects in the smoldering zone.

The application of an open flame to an upholstery composite can result in the cover melting or glowing and charring. If the cover material catches fire, it can act as a high-intensity secondary ignition source and ignite the cushion filling, after which the entire combination will burn.

Laboratory tests on individual materials have proved to be the least suitable for assessing the fire hazard of upholstered composites, that is, of cover and filling. Numerous studies have confirmed this and that the most reliable tests are those using composites because of the interaction between coverings and fillings. As in other areas (e.g., building, transportation, etc.) laboratory fire tests are most suitable for screening new materials under development and for quality control in manufacture. Correlation with fire performance under end-use conditions is required to justify the expense and effort of testing. These tests also serve as a valuable extension of model fire tests where investigation of the influence of certain material characteristics is required. This does not, of course, eliminate the need for model fire tests on finished parts or full-scale tests in, for example, a fully furnished room, as these show whether a correlation exists between the individual test methods and real fire situations. Research programs have also shown that a major hazard of upholstered furniture fires is the rate of fire growth and in particular the rate of heat, smoke, and toxic gas generation. Many studies have confirmed this and established that these aspects are related to the materials used and in particular to the covering, the filling, and their interaction, and also with the design and construction of the furniture itself. Furniture is a complicated product comprising many different materials and assemblies. Pool fires may occur underneath the furniture. Fire may also develop in cavities beneath the fabric layer and beneath the seat platform while the supporting frame may collapse.

13.3 Regulations and Test Methods

Official requirements regarding the fire performance of furniture are currently found in relatively few countries. The trend toward the establishment of fire safety requirements in public buildings, with regard to both the ignition risk and the contribution to flame spread and heat release rate, is obvious. A number of general and specific requirements concerning the fire hazard classification for furniture already exist. The emphasis is to eliminate the ignition risk by primary sources of ignition. Few countries have regulations relating to the fire performance of furniture and furnishings because they form part of the mobile contents of buildings and are thus not generally subject to building codes. The authorities may demand performance requirements for furniture in the case of buildings of special types and occupation. Owing to the lack of suitable test methods, these requirements may sometimes be satisfied by passing the tests specified in the building regulations. For example, in France, upholstery components may need to comply with prescribed classifications of M3 or M2 depending on the assessment of fire hazard (for further details concerning the French test and classification procedures, see Section 10.8).

In the early 1970s, a number of significant investigations concerning the fire performance of furniture were carried out in the United Kingdom and the United States. At present the most stringent regulations and test procedures for determining the fire hazard of furniture exist in the United Kingdom and the United States. For this reason, only the tests and regulations of the United Kingdom and the United States are dealt with in detail.

Major national and international efforts have been directed toward producing ignitability tests using cigarette and the small flame ignition sources because these are the most common causes of upholstery fires. A large number of the standard tests are based on BS 5852 Part 1 [12], and these are described in the next section. Slight variations of the original tests have become international standards, ISO 8191-1, ISO 8191-2 [13], EN 1021-1, EN 1021-2 [14], and IMO resolution A.652 (16) [15]. Textile working groups have published cigarette and small flame ignition resistance tests for bedding [24, 24a].

13.3.1 EC Draft Directive and European Tests

In the early 1990s, the European Commission initiated work for a European Directive for Upholstered Furniture [16], partly because of concern over the fire safety of upholstered furniture. The aims were to improve the ignition resistance to small flame and cigarettes of domestic type furniture, and to larger flaming sources for furniture used in public areas. The second aim was to limit the burning behavior of upholstery such that it did not cause a hazard, in terms of heat, smoke, and fire gases within a given time, to persons within the room of the fire. CEN/TC 207, Furniture, WG6, Test Methods for Fire Behaviour, was mandated to develop suitable ignition tests that would use cigarette and small flame tests for low-risk furniture and sources equivalent to one and five to six double sheets of crumpled broadsheet newspaper for furniture for use in public areas. The European Commission also commissioned a research program entitled the Combustion Behaviour of Upholstered Furniture CBUF [10] to investigate the burning behavior of upholstered furniture and relevant test methods.

13.3.2 Tests and Test Development

The European Standards Organisation CEN has published four ignition tests for Upholstered Furniture. EN 1021-1 and EN 1021-2 are cigarette and small-flame tests for upholstered seating, and EN 597-1 and EN 597-2 are cigarette and small-flame ignition tests for mattresses. The seating tests for the ignitability of upholstered seating are based on BS 5852 Part 1 (1979) and use a glowing cigarette and a small flame intended to simulate a match. The test apparatus is shown in Fig. 13.6 and consists of two steel frames hinged together, capable of being locked at right angles to each other. The frames are equipped with expanded metal mesh. The horizontal frame measures 450 mm x 150 mm (17.7 in. x 5.9 in.) while the vertical frame is 450 mm x 300 mm (17.7 in. x 11.8 in.). The test apparatus is opened out so that both parts are horizontal for inserting the covering material, which is stretched over the entire frame and behind the retaining hinge rod. The covering material measures 800 mm x 650 mm (31.5 in. x 25.6 in.). The upholstery fillings are laid under the cover and measure 450 mm x 150 mm x 75 mm (17.7 in. x 5.9 in. x 3.0 in.) and 450 mm x 300 mm x 75 mm (17.7 in. x 11.8 in. x 3 in.) for the seat and back, respectively. The cover material is held on the steel frames with clips. Finally the frames are returned to their original perpendicular positions. An important aspect of the test rig is that it enables fabric tension and filling compression to be simulated. It also simulates the crevice that occurs between the seat and back or seat and arm of actual furniture, which is considered to be a critical feature for the ignition of seating. The test specimen comprises the outer cover, interliner if used, and the soft upholstery fillings. It does not include trims such as piped or ruched edges or features such as deep buttoning.

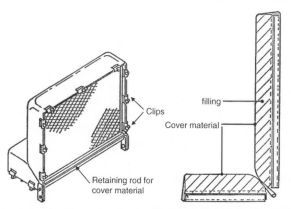

Figure 13.6   Apparatus for testing the ignitability of upholstered seating according to BS 5852 Part 1, ISO 8191-1, ISO 8191-2, EN 1021-1, EN 1021-2, and IMO resolution A.652

The glowing cigarette [68 mm (2.7 in.) long, 8 mm (0.3 in.) diameter, mass 1 g (0.035 oz.)] is placed at the junction between the vertical and horizontal components at least 50 mm (1.97 in.) from the nearest side edge. If progressive smoldering for more than 1 h or if flaming occurs, the material fails the test. If no such phenomena occur, the test is repeated with a new cigarette placed at least 50 mm away from the position of the first cigarette. If no progressive smoldering or flaming is observed in both tests, a pass result is recorded.

The small flame test uses a 65 mm internal diameter tube, 200 mm long, connected to flexible tubing as a burner that produces a butane flame approx. 35 mm high. The burner tube is laid along the joint between the vertical and horizontal components of the test rig and at least 50 mm from the nearest edge and the damage from any previous test. It is applied for 15 s (the BS 5852 Part 1 test uses 20 s). If flaming or glowing continues for more than 120 s after the ignition source is removed or if progressive smoldering occurs, the material fails the test. If this does not occur with two tests, a pass result is recorded. In the CEN standards, a water soak is applied to covers to prevent the use of easily removed flame retardant treatments. This is similar to that used in the UK Furniture and Furnishings (Fire) (Safety) Regulations, see next section.

The procedure described in the preceding, is essentially the same as for ISO 8191-1 and ISO 8191-2, and IMO Resolution A.652 for international shipping. The ignition flame time varies between tests, 15 s or 20 s, which may give different results in some cases.

EN 597-1 and EN 597-2 are cigarette and small-flame ignition tests for mattresses (Fig. 13.7). The latter uses a test specimen of minimum size 350 mm x 450 mm x nominal thickness. The test specimen reproduces the features of the full-size mattress and in the tests the ignition sources are applied to the mattress surface, to tufts, along quilt lines, and also to tape edges if present.

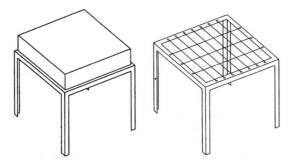

Figure 13.7   Test rig and specimen used for EN 597-1 and EN 597-2 mattress tests

CEN carried out an interlaboratory trial with seating and mattress composites using two large multijet gas flames simulating the 20 g (0.7 oz) and 100 g (3.5 oz) paper ball ignition sources. The latter were considered to be equivalent to the 1 and 5/6 double sheets of broadsheet newspaper required by the EC. The flames were generated by burning propane in a square, multihole burner, which is pivoted to allow it to follow any burning and melting of the test composites. The burner was 100 mm square and was based on the design of the burner used in the California TB 133 test [17]. The BS 5852 Part 2 [18] seating test rig and the EN 597 mattress rig were used to test the specimens. The trial showed that neither the 20 g or 100 g equivalent gas flames with either the seating or mattress test specimens achieved the repeatability or reproducibility required for a standard test. Only the 20 g equivalent flame with the seating test achieved the severity required. The protocol for the latter is to be published as a CEN report. Further development of this interesting approach will probably depend on the future availability of support.

13.3.3 UK Regulations and Tests for Furniture Suitable for Domestic Use

The Upholstered Furniture (Safety) Regulations introduced in 1980 in Great Britain [19] required that upholstered seating should pass the BS 5852 Part 1 cigarette test. Furniture that did not also pass the BS 5852 Part 1 flame test was labeled accordingly. More stringent requirements for furniture suitable for domestic use were introduced in 1988 with the Furniture and Furnishings (Fire) (Safety) Regulations defined in SI 1324 (1988), SI 2358 (1989) and SI 207 (1993) [4]. These Statutory Instruments are Acts of Parliament that define the application, enforcement, and test procedures and requirements for upholstered furniture suitable for domestic purposes within the United Kingdom. The tests used are essentially BS 5852 Part 1 (1979), BS 5852 Part 2 (1982), and BS 6807 [20]. Details of these tests are given in the preceding, (BS 5852 Part 1 is similar to EN 1021-1 and EN 1021-2), or in the next section, which deals with British tests.

A very important aspect of the United Kingdom Regulations is that the tests are used to classify materials used for soft upholstery. Upholstery components that pass the relevant test can be interchanged and used with other components that pass their relevant test. This applies to fabrics, interliners, and to all fillings. The importance of this is that manufacturers can classify their materials without excessive testing. The exception is the cigarette test, which is applied to the actual upholstery composite. In practice, the cigarette test may be carried out with the fabric over the standard PU foam used for the flame test as this represents a worst case PU foam filling when compared to the specified combustion-modified PU foams. This enables a further reduction in testing. The various test schedules are summarized in Table 13.3 and are discussed in the next section.

Table 13.3   Component tests within SI 1324, 1988, SI 2358, 1989 and SI 207, 1993

Test schedule	Test component
SI 1324 (1988)	
Schedule 1 Part I	PU block foam
Schedule 1 Part II	PU crumb foam
Schedule 1 Part III	Latex rubber foam
Schedule 2 Part I	Single nonfoam fillings
Schedule 2 Part II	Composite fillings other than mattresses, bed bases, and cushions and pillows
Schedule 2 Part III	Pillows and cushions with primary covers
Schedule 2 Part IV	Composites fillings of mattresses and bed bases
Schedule 3	Interliners
Schedule 4	Cigarette test for composites
Schedule 5 Part I	Match test for covers
Schedule 5 Part II	Match test for stretch covers
SI 2358 (1989)	
Schedule 4 Part II	Cigarette test for invisible parts (no water soak)
Schedule 5 Part III	Match test for invisible parts of covers
SI 207, 1993	Extends the scope, does not alter the tests

13.3.3.1 Fillings

Essentially all fillings are tested using a specified flame retarded polyester cover with the test method of BS 5852 Part 2 (1982) with ignition source 2. This is applied to the fillings listed with the exception of PU block foams, which are tested to crib 5 with modified pass/fail criteria of a weight loss of < 60 g without the flame spread and penetration criteria. Composite fillings of mattresses and bed bases are tested to the following:

- Schedule 2 Part IV uses ignition source 2 with the BS 6807 mattress-type test specimen.
- Schedule 1 Part II PU Crumb Foam also requires that the crumb is made from foam, which complies with Schedule 1 Part I.

13.3.3.2 Covers and Interliners

Schedule 3 tests interliners for use with designated types of non-flame-retarded and natural fabrics over a standard PU foam.

Schedule 5 Part I tests coverings and Part II stretch covers with a standard PU foam filling using the small flame (20 s duration) of BS 5852 Part 1 (1979). This essentially means that satisfactory fabrics are ignition resistant and protective, that is, nonmelting. Tests for covers and interliners require that these are first subjected to a water soak procedure to ensure that easily soluble treatments that could be removed by simple cleaning or spillage are not used in the manufacture of actual furniture.

SI 2358 (1989) requires that invisible fabrics, that is those used beneath cushions, on seat platforms, beneath chairs, and so forth, are tested to Schedule 4 Part II cigarette, and Schedule 5 Part III match flame, which are the equivalent of SI 1324 (1988), Schedule 4 and Schedule 5, respectively. Both tests do not require the water soak for the covering while the small-flame test is carried out using a combustion-modified PU foam complying with Schedule 1 Part I.

13.3.3.3 Composites

Schedule 4 is unique in that it tests the actual furniture composite with the cigarette of BS 5852 Part 1 (1979). Tests are frequently carried out using a standard PU filling to represent a worst case PU foam to reduce the testing load. This often gives a more reasonable test than using PU foam complying with Schedule 1 Part I, but is not used as an alternative to more easily ignitable fillings, for example, cotton wadding.

13.3.4 UK Regulations and Tests for Furniture Used in Public Areas

The fire safety of persons in public areas in the United Kingdom is essentially controlled by a number of measures directly relating to the assessment of the threat to life from fire that is assessed by local authorities or fire prevention officers. The recommendations of these officers are applied by a number of mechanisms but essentially follow the measures recommended in the Home Office Guides for Fire Safety Measures for Public Areas. These Guides apply to Places of Work, Certain Multiple Occupancies, Hotels and Boarding Houses, Places of Entertainment and Like Premises, and so forth [21]. The fire performance levels recommended in these publications essentially formed the basis for the medium and high hazard levels of BS 7176 [22] and BS 7177 [6]. In addition, it is usual to require that the fillings of upholstered furniture for public area use should comply with the requirements of SI 1324.

BS 5852 Section 4 (1990) [23] is essentially the same as BS 5852 Part 2 (1982) and is generally similar in principle and operation to BS 5852 Part 1 but is larger and is used with larger ignition sources. The test rig dimensions are 450 mm x 450 mm for the vertical and 450 mm x 350 mm for the horizontal component. The ignition sources comprise two gas flames and four wooden cribs. The No. 5 crib, together with the cigarette and the small flame of EN 1021-1 and EN 1021-2, respectively, are used for upholstered furniture for use in public buildings. The No. 7 wooden crib (126 g) is used with the cigarette and small-flame tests to define upholstery for high-risk areas. Section 5 of this test is applied to actual upholstered furniture and the sources are applied to the upper surface at the seat/back junction and below or against the furniture depending on the distance between the base of the furniture and the floor. The No. 5 wood crib (17.5 g) is used to define block PU fillings, while the No. 2 gas flame is used to define all other fillings within the UK Furniture and Furnishings (Fire) (Safety) Regulations (see previous section).

BS 6807 [20] is similar to the EN 597 standards but uses the two large gas flames and the four wood cribs described in the preceding for BS 5852 (1990). These sources are used to test mattresses for similar end-use hazard applications as those described for seating tests for similar applications. The BS 6807 test method has a number of sections and may be used to test mattresses on their own and also made up beds with different combinations of bedding and mattress. Tests may be carried out on the test mattress test specimen of EN 597 or on actual mattresses.

BS 7175 [24] is similar to BS 6807 but tests bedding combinations over a noncombustible fiber mat, which is used to simulate a mattress. This test is used with the cigarette and small flame and also with the two large gas flames and the four wooden cribs of BS 5852 (1990). Textile working groups have published cigarette and small flame ignition resistance tests for bedding [24, 24a].

BS 7176 and BS 7177 list ignition test performance requirements for upholstered seating and mattresses, respectively. Furniture is specified according to low, medium, high, and very high hazard end-use applications. The standards also contain examples of the hazard applications, for example, low hazard is typically domestic, office, some hotels, and requires cigarette and small flame resistance to EN 1021-1 and EN 1021-2, respectively. (Note: The regulations of SI 1324 and amendments also apply to furniture suitable for domestic use). Medium hazard typically relates to cigarette and No. 5 crib and is applied to public areas while the cigarette and No. 7 crib are applied to high hazard areas. Ignition sources for the very high hazard category are not specified in the British Standards. However, a test for mattresses for use in locked accommodation uses four No. 7 cribs to test a fully vandalized mattress and is published by the Home Office Prison Service.

Research has shown that the combination of ignition resisted, protective, That is, nonmelting fabrics with combustion-modified PU foams will give a significant increase in the period between ignition at the start of rapid growth, that is, an increase in the potential escape time.

13.3.5 ISO Ignition Tests

ISO/TC 136 Furniture, SC1 Test Methods, WG 4 Fire Tests for Furniture, published two ignition tests for upholstered seating, namely ISO 8191-1, and ISO 8191-2, which are cigarette and small-flame ignition tests for upholstered seating composites. These are essentially the same as the tests of BS 5852 Part 1 and use the 20 s flame ignition time. ISO was unable to agree on a standard type of large flaming ignition source but an interlaboratory trial using large-flame ignition sources with seating composites was carried out using the test rig of BS 5852 Part 2 with ignition flames 2 and 3, cribs 5 and 7, and 20 g and 100 g paper bags [25]. An important aspect of this study was the comparison of the different types of ignition source. Little difference in reproducibility was observed between the wood and paper sources, which were rather less reproducible than the simple gas flames. This implied that specimen/source interaction was a possible cause of variability. The ISO also considered similar tests for mattresses but these were not finalized. The ISO working group decided to await developments within CEN/TC 207/WG6 and agreed to use CEN standards as the basis for new ISO standards. Test development essentially ceased within ISO/TC 136/SC1/WG4 and the working group was later disbanded. Textile working groups have published cigarette and small-flame ignition resistance tests for bedding [24, 24a].

13.3.6 US Regulations and Tests

There is only one national regulation in the United States concerning the fire behavior of upholstered furniture and that concerns mattress flammability requirements for residential applications. There are no national requirements for upholstered seating although the voluntary UFAC cigarette smoldering standard for upholstered seating composites was introduced in the 1970s. Some individual states, notably California, have fire performance requirements for upholstered seating and mattresses.

13.3.6.1 Regulatory Test Methods for Mattresses

Federal Mattress Flammability Standard [26] is defined in 16 CFR, Part 1632 (FF 4-72, as amended). CFR refers to the Code of Federal Regulations. The standard, first introduced in 1972, is now enforced by the US Consumer Product Safety Commission. It was implemented because of the number of fire fatalities caused by mattress fires that were largely started by glowing smoking materials. The test method for determining the ignition resistance of a mattress to a glowing cigarette is used for products that are in development and not yet on the market as well as for quality control of commercially available mattresses.

The mattress meets specified performance requirements if glowing cigarettes, placed in certain spots on the horizontally positioned mattress illustrated in Fig. 13.8 do not cause charring of its surface extending more than 5.1 cm (2.0 in.) in any direction. A glowing cigarette is first laid on the bare mattress surface, then on the upper edge of the mattress where in the absence of a depression, it is held in place by three pins. Finally, various spots such as quilted surfaces are tested. These tests are repeated using two sheets placed on half of the mattress. The glowing cigarette is laid on a stretched sheet tucked under the mattress and a loose sheet laid immediately over it. At least 18 cigarettes are burned on each mattress, 9 on the bare mattress and 9 under the cotton sheeting material.

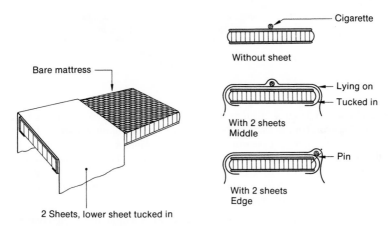

Figure 13.8   Various arrangements for the 16 CFR Part 1632 flammability test for mattresses

13.3.6.2 Voluntary UFAC Standard Test for Seating

The Upholstered Furniture Action Council (UFAC) set up by furniture manufacturers has developed several tests for determining the resistance of materials used in upholstered furniture and composites to glowing smokers materials [27]. There are now six UFAC furniture construction criteria:

- Polyurethane foam/filling materials (UFAC Filling/Padding Component Test Method – 1990, Part A or B)
- Decking materials (UFAC Decking Materials Test Method – 1990)
- Barrier/interliner materials (UFAC Barrier Test Method – 1990)
- Covering fabric (UFAC Fabric Classification Test Method – 1990)
- Welt cords (UFAC Welt Cord Test Method – 1990)

- Decorative trim or edging (UFAC Standard Test Methods for Decorative Trims, Edging, and Brush Fringes – 1993)

The tests on foams, barrier materials, fabric covers, and welts are carried out with the apparatus illustrated in Fig. 13.9. The test rig in this figure is made of plywood. Three tests on each are carried out with a glowing cigarette, 85 mm long weighing 1.1 g laid in the gap or on the welt between the back rest and seat cushion so that it is in contact with both. A piece of sheeting, 125 mm x 125 mm, is laid over the cigarette.

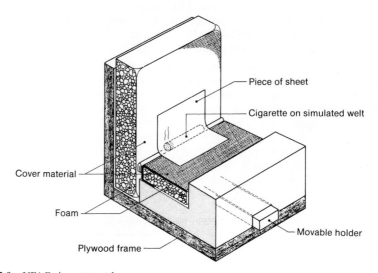

Figure 13.9   UFAC cigarette test layout

Polyurethane foam, 203 mm x 203 mm x 51 mm (8.0 in. x 8.0 in. x 2.0 in.) is used as cushioning for the backrest (in the case of the welt test a cotton filling is used). The seat cushion also consists of PUR foam and measures 203 mm x 127 mm x 51 mm (8.0 in. x 5.0 in. x 2.0 in.) [for the welt test 203 mm x 114 mm (8.0 in. x 4.5 in.)]. The covering fabric used in the tests is 203 mm x 381 mm (8.0 in. x 15 in.) for the back rest except with the PUR foam test and barrier test in which it is 305 mm x 305 mm and 203 mm x 305 mm, respectively. The seat cushion cover is 203 mm x 203 mm.

The PUR foam and barrier tests are passed if the charred length above the cigarette does not exceed 50 mm. Depending on the test performance, covering materials are classified in classes I and II. The former is achieved if the charred length above the cigarette is < 38 mm. If this length is exceeded, class II applies. In the welt test, the cotton filling acting as a backrest must not be charred for more than 37 mm above the cigarette.

Decking materials are tested using the apparatus illustrated in Fig. 13.10. The decking is laid on a plywood base and the cover fabric laid over it. The dimensions of all three components are 533 mm x 343 mm (21 in. x 13.5 in.). A plywood frame with the same external measurements and internal dimensions of 406 mm x 216 mm (16 in. x 8.5 in.) is laid over this assembly. Three glowing cigarettes are placed on the remaining exposed surface of the cover fabric at equal intervals (see Fig. 13.10) and covered with a piece of sheeting. The test is passed if the charred length on the base is < 38 mm measured from each cigarette.

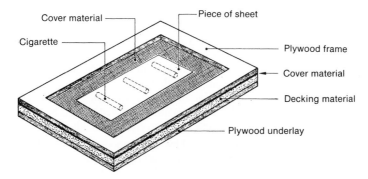

Figure 13.10   UFAC test rig for decking materials

13.3.6.3 California Tests

California T.B. 116 [28] and *T.B. 117* [29] specify testing of individual upholstery materials using an open gas flame as ignition source as well as a glowing cigarette. T.B. 116 gives requirements for cigarette smoldering resistance of full-scale (ready for sale to the consumer) furniture or mockups made with actual components of the furniture as designed for sale. The T.B. 117 standard gives requirements for both open flame and cigarette smoldering of furniture filling materials. The following are test procedures used in the T.B. 117 standard:

- *Resilient foams* are tested vertically with a Bunsen burner as in Federal Test Method Standard (FTMS) 191 Method 5903.2. The specimen dimensions are 305 mm x 76 mm x 12.7 mm (12 in. x 3.0 in. x 0.5 in.) and the number of specimens is at least 10, five of which are conditioned at 104 °C for 24 h. All five specimens of both test series must pass the test. If one specimen fails, five new specimens must be tested and pass. If one specimen still fails, the material fails the test. The test requirements are met if:
  - The mean char length of all specimens does not exceed 152 mm or 203 mm for any individual specimen,
  - The afterflame time of all specimens is not longer than 5 s on average (10 s for any individual specimen), and
  - The mean afterglow time does not exceed 15 s.
- *Shredded resilient foams* are tested by filling a cushion 330 mm x 330 mm (13 in. x 13 in.), made of the intended covering material using approximately the intended packing density. The cushion is positioned horizontally 19 mm (0.75 in.) above a 38 mm (1.5 in.) long Bunsen burner flame so that an area of at least 254 mm diameter is exposed to the flame. The test is passed if the weight of the specimen does not decrease by more than 5 % after having applied the flame for 12 s. The cushion must meet the test requirement before and after aging for 24 h at 104 °C. In addition, 20 specimens of the intended covering material are tested in FTMS 191 Method 5903.2; a flame is applied for 3 s and 12 s to five specimens each in both the warp and fill directions.
- *Expanded polystyrene beads* used as fillings are tested in a wire mesh test basket 203 mm x 203 mm x 76 mm (8.0 in. x 8.0 in. x 3.0 in.) high filled to the top, after aging for 48 h at 65 °C and then conditioned for 24 h at 21 °C. The ignition source is a methenamine Tablet lit with a match and placed in the middle of the material with crucible tongs. The test is passed if the weight loss in five successive tests does not exceed 5 %.
- *Nonsynthetic upholstery fillings* are tested vertically by the method described earlier for resilient foams.
- *Feathers and down* must be contained in a flame-retardant cover. The cover material is tested by a method based on FTMS No 191, Method 5903.2 under the conditions described

earlier for resilient foams. However, 20 specimens are tested; in this case five each are flamed for 3 s and 12 s in the warp and fill directions.

- *Man-made fiber filling materials* are tested according to a modified version of Commercial Standard (CS) 191-53. Again, the test materials must be conditioned for 24h at 21 °C and 55 % humidity. The fibers must be tested both with and without any attached woven or nonwoven materials such as scrims, and so forth. In this test a small flame (16 mm long) is applied for 5 s near the lower edge of five specimens (152 mm x 76 mm) inclined at 45° to the horizontal. The time required for the flame to proceed up the batting a distance of 127 mm is recorded and must be less than 10 s on average (individual specimens ≤ 7 s).

The effect of glowing cigarettes is investigated by two methods in the Cal. T.B. 117 standard. Cushioning materials are divided into resilient materials (excluding foams) and resilient foams. At least six specimens 305 mm x 305 mm (12.0 in. x 12.0 in.) are tested by burning cigarettes on their surface at the center. Three specimens are tested without any covering, and with the other three specimens, a piece of sheet is laid loosely over the cigarette. The test is passed if the maximum charred length does not exceed 51 mm (2.0 in.) in all directions measured from the cigarette.

The resistance of resilient foams to glowing cigarettes is assessed in a test rig that simulates an upholstered armchair. The test mockup, illustrated in Fig. 13.11 is made of plywood. The foam parts that make up the cushioning are laid in to form a backrest (184 mm x 203 mm x 51 mm) and seat cushion (203 mm x 101 mm x 51 mm). The foam is covered by a standard, defined upholstery fabric. An unfilter-tipped cigarette weighing 1.1 g, 85 mm long is used as an ignition source and placed at the crevice of the seat and back on the mockup after its ignition. A piece of cotton sheet (152 mm x 152 mm) is then laid over the cigarette and upholstered composite.

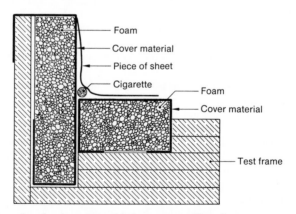

Figure 13.11   Test mockup for cigarette smoldering test on resilient foams

The test is continued until all evidence of combustion has ceased for at least 5 min. The remaining foam is weighed. The test is passed if in three successive runs at least 80 % of the foam is not destroyed (nonsmoldered). If significant destruction occurs in one test, three further tests may be carried out. If all three give satisfactory results the test is considered passed.

California TB 121 [30] was introduced in addition to the preceding regulation in the State of California by the Bureau of Home Furnishings. T.B. No. 121 describes the fire performance testing of mattresses used in buildings subject to high fire hazard. The regulation is mandated in the California Department of Corrections and in other areas of the country. The standard is viewed as outdated with its newspaper ignition source and lack of heat release measurements but it is still in the California law. Its use is not encouraged for general use. The bed frame and test piece are placed in a fire room [3.0m x 3.7m x 2.4m high (9.8 ft x 12.1 ft x 7.9 ft)] as shown in Fig. 13.12.

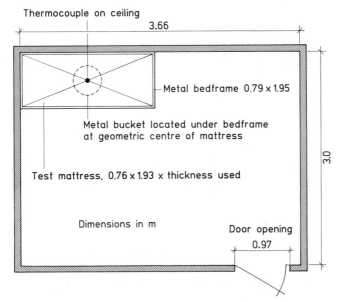

Figure 13.12   Burn room for testing mattresses in high risk occupancies, Cal. T.B. 121

A metal bucket containing 10 double sheets of crumpled newspaper is located under the bed. A thermocouple is fixed to the ceiling above the geometric center of the mattress surface. The material passes the test, if, after the ignition source has been consumed:

- The mattress has not suffered a weight loss of more than 10 % in the first 10 min from the start of the test,
- the temperature at the thermocouple has not reached 260 °C, and
- the carbon monoxide concentration in the test chamber has not exceeded 1000 ppm.

California TB 129, a more widely used and updated mattress standard, was developed in concert with and issued by California as Technical Bulletin 129 [31]. The standard was developed and initially written with the help of the mattress industry and NIST. At this point, Cal. T.B. 129 is a voluntary standard in the State of California but has been rewritten into the NFPA 101 Life Safety Code and exists essentially unchanged as the standards NFPA 267 and ASTM 1590. It is required in some jurisdictions throughout the United States, for example, it is necessary in some health care facilities and some state university systems requiring dormitory mattresses to comply with the standard.

The test uses a propane 'T' shaped ignition burner illustrated in Figs. 13.13 and 13.4 with a heat output of approximately 17–19 kW. Propane gas is used at a rate of 12 l/min for a total of 180 s. Unlike in the California T.B. 121 standard, the ignition burner is positioned at the side of the mattress (Fig. 13.14). The primary test criteria are heat release measurements using oxygen consumption calorimetry.

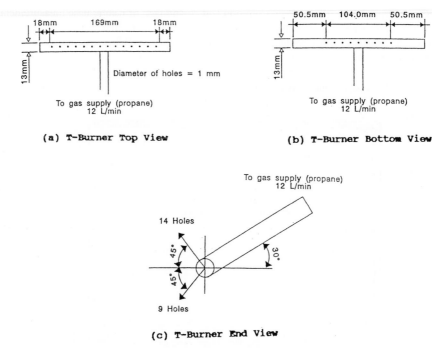

Figure 13.13   View of T-shaped Cal. T.B. 129 gas ignition burner

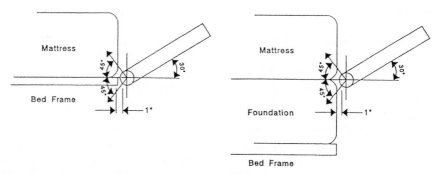

Figure 13.14   Orientation of T-burner to test mattress in Cal. T.B. 129

There are three room layouts permitted in the test. In option A the test room dimensions are 2.44m x 3.66m x 2.44m high (8 ft x 12 ft x 8 ft). For option B the room is 3.05m x 3.66m x 2.44m high (10 ft x 12 ft x 8 ft). In option C an open calorimeter layout is used. In all cases combustion gases are collected in an exhaust hood or collection system. Oxygen concentration is monitored throughout the test to calculate heat release. Also during the test, concentrations of CO and CO_2 are recorded as well as smoke optical density and mass loss measurements.

A single mattress fails to meet the requirements of this test procedure if any of the following criteria are exceeded:

- Weight loss due to combustion of 1.62 kg (3 lbs) or greater in the first 10 min of the test.
- A maximum rate of heat release of 100 kW or greater.
- A total heat release rate of 25 MJ or greater in the first 10 min of the test.

A further test procedure for the fire performance of mattresses used in high risk areas such as prisons is the so-called rolled-up Michigan test. The test was developed, in part, with the help of prisoners to simulate a worst-case fire scenario. The ignition source consisting of crumpled sheets of newspaper is located inside the rolled-up mattress. To pass this test weight loss should not exceed specified limits.

California T.B. 133 [17] describes the test procedure for seating furniture in high-risk and public occupancies. Such facilities might include prisons, health care facilities, public auditoriums, and hotels. Compliance with the standard in California is mandatory when these occupancies do not have full fire sprinkler protection. Even when NFPA compliant sprinkler protection is in place the use of T.B. 133 compliant furniture is highly recommended by the state. The standard has also been adopted by certain other state and local jurisdictions. Even if not mandated by law, some furniture manufacturers offer T.B. 133 compliant furniture for these types of public occupancies. T.B. 133 provides a severe ignition source for upholstered furniture. The full-scale room test was originally developed and issued in 1984 with an ignition source of crumpled newspaper and test requirements of the temperature level increase, the smoke opacity, CO concentration of the fire effluents, and the weight loss. The 1991 version of the standard includes a square propane ignition burner and oxygen consumption calorimetry test criteria. The burner, illustrated in Fig. 13.15, generates about 18 kW of energy for a period of about 80 s. The test can be run either on a finished piece of upholstered furniture or a smaller scale mockup.

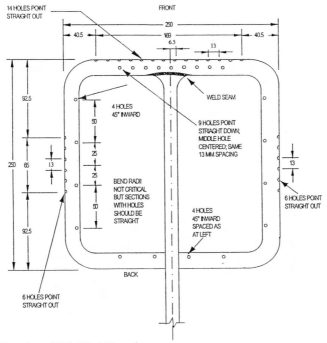

Figure 13.15 Top view of Cal. T.B. 133 gas burner

The currently written standard allows for either conventional test criteria or oxygen consumption calorimetry measurements to be met. In the case of the conventional criteria, seating furniture fails to meet the test requirements if any of the following are exceeded in the room test:

- A temperature increase of 93 °C or greater at the ceiling thermocouple,
- A temperature increase of 10 °C or greater at the 3.3 m (10.8 ft) thermocouple,
- Greater than 75 % opacity at the 3.3-m smoke opacity monitor,
- CO concentration in the room of 1000 ppm or greater for 5 min,
- Weight loss due to consumption of 1362 g or greater in the first 10 min of the test.

Alternately, using oxygen consumption calorimetry, seating furniture fails to meet the test requirements if any of the following are exceeded:

- A maximum rate of heat release of 80 kW or greater,
- A total heat release of 25 MJ or greater in the first 10 min of the test,
- Greater than 75 % opacity at the 3.3 m smoke opacity monitor,
- CO concentration in the room of 1000 ppm or greater for 5 min.

13.4 Fire Growth Rate, Heat Release Rate, Smoke Production Rate and Toxicity Tests

Ignition tests relate to the initiation of the fire but once burning, the burning rate and the production of heat, smoke, and toxic gas species are of prime interest and different test procedures are needed to determine these parameters. Probably the most well-known in Europe are the cone calorimeter for small-scale and the furniture calorimeter for large-scale testing. These are not normally used for regulatory purposes. In the United States, large-scale fire growth tests based on the California system may be used for regulatory purposes and are included in Section 13.3.6.

The cone calorimeter, defined in ISO 5660 [32] and shown in Fig. 13.16, is described in Section 10.23 ISO. The standards ISO 5660, ASTM E 1474, NFPA 264 A [33], and the CBUF test protocol [10, 34] provide details of the test and its use for the evaluation of upholstered furniture composites.

Specimen preparation is important for achieving good reproducible and representative test results. The specimen is made up from the cover fabric, interliner (if relevant), and filling. The specimen size is 100 mm x 100 mm x 50 mm. Fabric and interliner also fully cover the sides of the specimen. The specimen is wrapped in aluminum foil except for the exposed upper

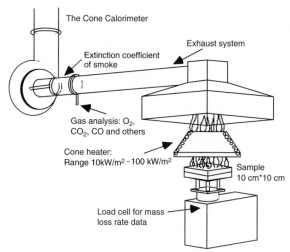

Figure 13.16   Cone calorimeter to ISO 5660

surface. The specimen is positioned on the standard specimen holder and exposed to an irradiance of 35 kW/m^2. The reported measurements are heat release rate versus time curve; time to ignition; effective heat of combustion; average heat release rate 180 s after ignition; peak heat release rate; total heat release; mass loss; smoke yield; and yields of carbon monoxide, carbon dioxide, and other gases if required. There are no pass/fail criteria and these data are used for fire engineering applications.

13.4.1 The Furniture Calorimeter

The furniture calorimeter NT FIRE 032 [35] (see Fig. 13.17), as defined by NORDTEST was first published in 1987. It was backed up with testing work on real furniture [36]. NT FIRE 032 was revised in 1991. Further improvements of the test method were made during the CBUF program, which gives a complete description of the test.

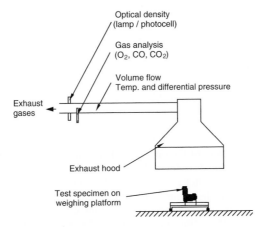

Figure 13.17   The furniture calorimeter

A full-size piece of furniture is placed on a weighing platform beneath an exhaust hood. The test specimen burns without any restriction of air supply and the combustion gases from the product are extracted through a 3 m x 3 m exhaust hood. Gas species concentrations, gas flow rates and smoke optical densities are measured in the duct, and rates of heat release, smoke production, gas species production, and mass loss are determined. Rate of heat release is determined using the oxygen consumption method. Smoke optical density is measured with a white light system having a detector with the same spectral responsivity as the human eye. As mass loss rate from the sample is measured simultaneously, yield data of smoke can also be determined. Fire gases may be measured by various techniques, but Fourier Transform Infrared (FTIR) is becoming more popular because it will determine gases on a concentration-time basis [38].

The furniture calorimeter is extremely well ventilated and tests have also been carried out in which furniture is burned within the ISO room to ISO 9705 [37] (see Section 10.23 ISO), where the ventilation is restricted by the doorway. This test is essentially a small room, the door of which is positioned beneath a hood of the type used for the furniture calorimeter. The test arrangement is shown in Fig. 13.18.

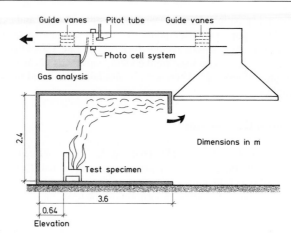

Figure 13.18   ISO 9705 Room test also used as a room calorimeter to test furniture

13.4.2 Furniture Fire Models

In recent years, considerable efforts have been made to model the fire growth of upholstered furniture. Early models correlated the measured peak rates of heat release with the mass of the upholstery. These were improved by the incorporation of factors for the cover, filling, and for the design and construction of the furniture. The next stage was to replace the cover and filling factors by a direct measurement of the rate of heat release for the composite. The cone calorimeter has been the main source for these data but some workers have used the flame spread parameters from the LIFT apparatus [39]. The development of furniture flammability models is difficult because the burning behavior of actual furniture is strongly influenced by its design and construction. Furniture is a complicated product comprising a number of different materials and assemblies. Pool fires may occur underneath the furniture. Fire may also develop in cavities beneath the fabric and beneath the seat platform, while the supporting frame may collapse. These effects are very difficult to model. An important aspect is the correlation of results from bench-scale tests to those of full-scale tests and then to the observed burning behavior of actual furniture. Figure 13.19 illustrates this using a series of standard tests and models. An important aspect is the use of the intermediate stages to test composites and furniture and also to confirm the validity of the models used to link the different stages.

Models using fire physics (and ignoring the phenomena described in the preceding), statistical correlation of parameters important for the fire growth, and combinations of the two methodologies have been used to predict fire behavior.

Correlations may be used to predict peak heat release rate for a furniture item based on cone calorimeter test data [40]. They generally may be described as follows.

Peak HRR of real furniture is given by the product of:

- *function*
- *HRR* and *ignition data* from the cone calorimeter
- *mass* factor of furniture item
- *frame* factor of furniture item
- *style* factor of furniture item.

The reliability of such correlations is directly dependent on the database used to develop the correlations and changes in materials, design, and so forth, may seriously alter the validity of predictions. Figure 13.20 shows a result from a correlation study on European furniture reported in [10].

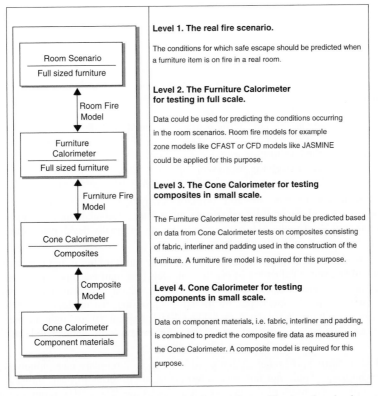

Figure 13.19   Scheme for estimating the fire hazards from an item of furniture burning in a room

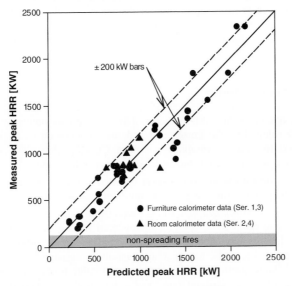

Figure 13.20   Prediction of peak HRR compared to measured HRR using a correlation formula based on data from the cone calorimeter [10]

The use of fire physics as the direct tool to predict the burning behavior is difficult because of the nature of upholstered furniture, but it has been used for simple geometries such as mattresses. Models based on fire physics are very complicated and all of the required input parameters may not be known or very difficult to measure accurately.

Combination models of fire physics and correlations may be very useful. One of the models in the CBUF project predicts the full heat release curve from a full-scale furniture item based on data from the cone calorimeter [40]. It is assumed that the heat release rate can be predicted as the convolution integral of the burning area rate and the heat release rate from the burning area. The heat release rate from the burning area is assumed to be the same as in cone calorimeter tests with an irradiance of 35 kW/m^2. The effects of falling burning pieces, collapsing, and pool burning in the furniture calorimeter are not individually represented but are indirectly and partly represented through the "deconvolved" area. The effective area functions used and the way they were determined are described further in the CBUF project report [10].

Combination models of fire physics and correlations may be very useful. One of the models in the CBUF project predicts the full heat release curve from a full-scale furniture item based on data from the cone calorimeter. It is based on a technique originally developed for representing flame spread over wall and ceiling surfaces [41].

The CBUF work relies on a large database of full-scale furniture burns and bench-scale tests. The empirical and statistical nature of the three models (the mattress model has a more scientific basis) reflects the complexity of the science involved and also that of upholstered furniture itself but, as acknowledged by the CBUF program, the models are limited by the database from which they were developed. The CBUF program has given the industry a series of very useful methods of predicting the burning behavior of upholstered furniture but acknowledges the need for further model development.

The draft EC directive [16] requires that in addition to ignition resistance, upholstered furniture, when burning, should not represent a hazard to life in terms of heat, smoke, and toxic gases to people in the room of the fire within a given time. The CBUF program provides a means of predicting this for many, but not all, items of upholstered furniture from heat release measurements of fabrics and fillings (currently work is limited to PU foams). However, the CBUF system in its present form is probably insufficiently well developed and robust to form the basis of European furniture fire safety regulations and it is a large step from a valuable comprehensive research program to legislative or other controls for the fire safety of people and products.

13.5 Future Developments

Test method development within ISO has ceased and future ISO standards will probably be similar to those developed by the European standards organization CEN. The ISO working group had decided to await developments within CEN/TC 207/WG6 and to use CEN standards as the basis for future ISO standards before it was disbanded.

In the early 1990s, the European Commission initiated work for a European Directive for upholstered furniture [16] but this has yet to be implemented and its future is uncertain.

The *European Standards Organisation* CEN is currently revising EN 1021-1 and EN 1021-2 which are cigarette and small-flame tests for upholstered seating: Apart from the use of three instead of two small-flame applications, no major changes are likely although modifications to the test rig are proposed to make test specimen assembly easier. EN 597-1 and EN 597-2, cigarette and small flame ignition tests for mattresses, are due for revision but major changes are not anticipated. A report describing a protocol using a multi-jet gas flame burner as a large flaming ignition source for seating tests will probably be published. The development of this and other larger flame tests for seating and mattress standard tests is unlikely without support from European Commission and industry.

Although the *European draft directive* requires that in addition to ignition resistance, burning upholstered furniture should not represent a hazard to life in terms of heat, smoke, and toxic gases to people in the room of the fire within a given time, it appears that regulations to limit the growth rate of fires are very unlikely. Although the furniture calorimeter and ISO room provide test methods for the determination of the fire growth rate, their use as regulatory tools is unlikely. The CBUF Program provides a means of predicting the growth of furniture fires, but is probably insufficiently well developed to form the basis of European furniture fire safety regulations.

The EC formulated the General Product Safety Directive in 1992, which has been published recently. In essence, the directive requires that products must be safe in normal and foreseeable conditions of use. Upholstered furniture could become incorporated into the scope of this directive. Standard fire tests could be published in the official journal of the EC, and if so, then compliance with these standards would be mandatory. Alternatively, upholstery may need to comply with European or National Standards.

In the summer of 2001, the EC issued an Invitation to Tender for Technical Assistance concerning the safety of a number of products including the fire performance of stuffed furniture. The outcome of this recent development is awaited.

In the *UK, British Standards* are currently revising BS 5852, BS 6807, BS 7176, and BS 7177. It is anticipated that the revisions will be relatively minor and will largely concern editorial changes. Major technical changes are unlikely.

The Furniture and Furnishings (Fire) (Safety) Regulations are being reexamined. It has been suggested that they could be simplified, but it is considered that major changes are unlikely in view of the recent publication of the DTI report showing their effectiveness in reducing fire casualties in the United Kingdom.

Although many of the established *US* flammability tests have not dramatically changed in the last 10 years, there is considerable current activity in test method development and possible regulation or legislation for both mattress and furniture areas. This has been initiated by the National Association of State Fire Marshals (NASFM), numerous state legislative bodies, the federal government, regulatory bodies, and by industry itself.

Mattresses for use in various public buildings must meet the cigarette resistance requirements of the Federal Mattress Flammability Standard but are often expected to comply with additional flammability requirements to satisfy state and local authorities. Difficulties arise because of the differences between the test methods, test requirements, and inconsistencies between the standards.

The industry, through the International Sleep Products Association (ISPA) [42], and a related group, the Sleep Products Safety Council (SPSC), proactively sponsored and developed the California T.B. 129 test. Although adopted by ASTM and NFPA, it is still essentially a voluntary standard but was intended to specify all mattresses for public occupancies nation wide.

Further analysis of mattresses fires showed that more deaths were associated with bedding-initiated fires than with the mattress itself. This prompted the SPSC to undertake a research project with NIST to establish the hazards of flaming bedding and mattress. The initial report, issued by NIST [43], investigated the fire contributions from various sets of bedclothes in conjunction with an inert mattress. The results, shown in Fig. 13.21, led to the development of a new ignition burner with an output of about 70 kW. This has two T-type perforated tubes, 25.4 cm (10.0 in.) and 30.48 cm (12.0 in.) long. One is positioned at the side of a mattress and the other just on top to simulate burning bedclothes. The burner is able to follow any retreating of mattress during the test. This work may form the basis of a new national mattress flammability standard for the regulation of residential use mattresses possibly by the Consumer Products Safety Commission (CPSC). The next phase of research work is to evaluate smaller scale mattress tests and their correlation with existing full-scale tests.

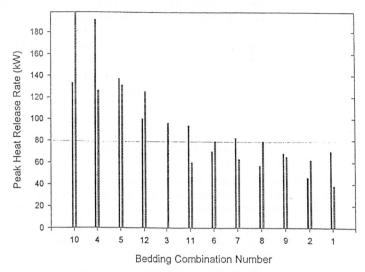

Figure 13.21   Peak heat release rates for various bedclothes combinations on an inert mattress

Attempts to pass a federal law and state laws for mattress flammability have yet to be successful. The National Association of State Fire Marshals and the Children's Coalition for Fire-Safe Mattresses [44] have both submitted petitions for better fire resistance of mattresses and/or bedclothes.

Considerable activity has also occurred for furniture. In 1993, the NASFM petitioned CPSC to address flammability hazards associated with upholstered furniture, taking into account small and large open flame ignition as well as cigarette ignition. CPSC granted the part dealing with small flame ignition with the possible intention of developing a test procedure. One reason for this was that the fire deaths due to small open flame ignition have stayed relatively constant. The large-flame petition was not granted partly because of the large cost of compliance and partly because the fire statistics did not support this approach. Action on the cigarette ignition component was deferred partly because of the apparent consequences of the voluntary UFAC standard in reducing cigarette-initiated fires

The composite test now recommended by CPSC uses the BS 5852 Part 1 (also ISO 8191) test rig. The same butane flame is applied, and in one test requirement a specimen would pass the test if charring did not reach the top of the vertical part of the test specimen. A provision for use of flame retardant interliners has also been made that may allow the use of more flammable fabrics. This proposal has not completed the regulatory process and may be changed or completely disregarded in the end.

The State of California is currently reviewing Cal. T.B. 117. Although the final version of the standard has yet to be decided, requirements for individual filling components will probably remain similar to the current standard but a composite test, based on the BS 5852 Part 1 test rig, may be included.

Other research work is focusing on a small flame furniture resistance test based on the BS 5852 Part 1 test rig that would allow the limited burning of furniture if the rate of heat release is low enough to permit likely escape from a residential building. This work could result in a voluntary standard similar to the UFAC cigarette smoldering standards or perhaps in a regulatory action with CPSC.

At least eight states have seen attempts to pass legislation for upholstered furniture flammability but only California has laws. In view of this activity, many believe a single national furniture flammability standard would be better.

The development of self-extinguishing or cigarettes of reduced ignition propensity or power has continued [45]. A large number of papers concerning cigarettes of low smoldering propensity have been published, particularly in the Journal of Fire Sciences. Interested readers are referred to that journal. In at least one state case, New York, an unsuccessful attempt to legislate the use of so-called "fire safe cigarettes" has been made. In addition to focusing on the ignition of upholstered furniture, this approach would affect other items. However, cigarettes of low smoldering propensity may be less effective in reducing fire than ignition resistant furniture [46].

Although many of these developments have yet to be finalized, it is clear that many US activities are likely to result in improved fire performance for furnishings resulting in fewer fire deaths.

References for Chapter 13

[1] J.A. de Boer: Fire and Furnishing in Buildings and Transport – statistical data on the existing situation in Europe, Conference proceedings on Fire and Furnishing in Building and Transport, Luxembourg, 6–8 November 1990. A statistical study by CFPA Europe

[2] Home Office Statistical Bulletin. Issue 8/97, Home Office Research, Development and Statistics Directorate, London

[3] The Furniture and Furnishings (Fire) (Safety) Regulations 1988 SI 1324,
The Furniture and Furnishings (Fire) (Safety) (Amendments) Regulations 1989 SI 2358,
The Furniture and Furnishings (Fire) (Safety) (Amendments) Regulations 1993 SI 207,
Department of Trade and Industry, London

[4] Effectiveness of the Furniture and Furnishings (Fire) (Safety) Regulations 1988, Consumer Affairs Directorate, Department of Trade and Industry, June 2000, London, UK.

[5] General Product Safety Regulations 1994, SI 2328, Department of Trade and Industry, London

[6] BS 7177:1996, British Standard Specification for Resistance to ignition of mattresses, divans and bed bases

[7] EN 597-1:1994, Furniture – Assessment of the ignitability of mattresses and upholstered bed bases – Part 1: Ignition source: Smouldering cigarette
EN 597-2:1994, Furniture – Assessment of the ignitability of mattresses and upholstered bed bases – Part 2: Ignition source: Match flame equivalent

[8] J.H. Hall Jr.: Home Fires Starting with Mattresses or Bedding – Patterns, Trends and Their Implications for Strategies, National Fire Protection Association, 1998

[9] V. Babrauskas: Fire behaviour of upholstered furniture. NBS Monograph 173, Nov. NIST, USA, 1985

[10] CBUF Fire Safety of Upholstered Furniture – the final report of the CBUF research programme. Edited by B. Sundström. European Commission, Measurements and Testing, Report EUR 16477 EN

[11] J. Krasny, W. Parker, V. Babrauskas: Fire Behaviour of Upholstered Furniture and Mattresses, Noyes Publications/William Andrew Publishing, USA, ISBN 0-8155-1457-3

[12] BS 5852: Part 1:1979, Fire tests for furniture, Part 1. Methods of test for the ignitability by smokers' materials of upholstered composites for seating

[13] ISO 8191, Furniture – Assessment of the Ignitability of Furniture,
Part 1:1987, Ignition Source: Smouldering cigarette,
Part 2:1988, Ignition Source: Match flame equivalent

[14] EN 1021-1:1994, Furniture – Assessment of ignitability of upholstered furniture – Part 1: Ignition source: Smouldering cigarette

EN 1021-2:1994, Furniture – Assessment of ignitability of upholstered furniture – Part 2: Ignition source: Match flame equivalent

[15] IMO resolution A.652 (16), FTP Code, International Code for Application of Fire Test Procedures, IMO International Maritime Organization, London, 1998

[16] L. Gravigny: "Draft European Directive on the Fire Behaviour of Upholstered Furniture", Flame Retardants '92, pp 238–242, Conference. Elsevier Science Publishers Ltd, ISBN 1-85166-758-x.

[17] Technical Bulletin 133, Flammability Test Procedure for Seating Furniture for Use in Public Occupancies, State of California, January 1991

[18] BS 5852 Part 2:1982, Fire tests for furniture. Methods of test for the ignitability of upholstered composites for seating by flaming sources

[19] The Upholstered Furniture (Safety) Regulations, 1980, and (Amendment) Regulations 1983 Department of Trade and Industry, London

[20] BS 6807:1996, Assessment of the Ignitability of Mattresses, Divans and Bed Bases with Primary and Secondary ignitions Sources

[21] Fire Precautions Act 1971, Guide to Fire Precautions in:
Existing Places of entertainment and like premises;
Premises used as hotels, and boarding houses which require a fire certificate;
Existing places of work that require a fire certificate Note other guides have also been published.

[22] BS 7176:1996, British Standard Specification for resistance to ignition of upholstered furniture for non-domestic seating by testing composites.

[23] BS 5852:1990, Methods of test for assessment of the ignitability of upholstered seating by smouldering and flaming ignition sources

[24] BS 7175:1989, Method of Test for the ignitability of bedcovers and pillows by smouldering and flaming ignition sources

[24a] BS EN ISO 12952: Parts 1 to 4: 1999. Textiles – Burning behaviour of bedding items

[25] K.T. Paul: Textile Flammability, Current and Future Issues, Textile Institute Conference, Manchester, UK, 1999

[26] Federal Mattress Flammability Standard, 16 CFR, Part 1632 (FF 4-72, as amended)

[27] The Upholstered Furniture Action Council (UFAC), PO Box 2436. High Point, NC 27261, USA. The following test methods were introduced by UFAC in 1979:
Polyurethane foam test method
Decking materials test method
Barrier test method
Fabric classification test method
Welt cord test method

[28] Technical Bulletin 116, Requirements, Test Procedure and Apparatus for Testing the Flame Retardance of Upholstered Furniture, State of California, January 1980

[29] Technical Bulletin 117, Requirements, Test Procedure and Apparatus for Testing the Flame Retardance of Resilient Filling Materials Used in Upholstered Furniture, State of California, March 2000

[30] Technical Bulletin 121, Flammability Test Procedure for Mattresses for use in High Risk Occupancies, State of California, April 1980

[31] Technical Bulletin 129, Flammability Test Procedure for Mattresses for use in Public Buildings, State of California, October 1992

[32] ISO 5660-1:1993, Fire Test, Reaction to fire – Part 1, Rate of heat release for from building products (Cone calorimeter method)

[33] ASTM E 1474-01, Test method for determining the Heat Release Rate of Upholstered and Mattress Components or Composites Using a bench Scale Oxygen calorimeter, ASTM, Philadelphia, USA
NFPA 264A, Standard Method of Test of Heat Release Rates of Upholstered Furniture Components or Composites and Mattresses Using an Oxygen consumption Calorimeter, NFPA, Quincy, MA, USA

[34] V. Babrauskas and I .Wetterlund: The CBUF Cone Calorimeter Test Protocol: Results from International Round Robin Testing, SP report 1996:12, Swedish National Testing and Research Institute, Borås, Sweden (1996)

[35] Upholstered Furniture: Burning Behaviour – Full Scale Test, Nordtest method NT FIRE 032, 1991, Helsinki

[36] B. Sundström: Full-Scale Fire Testing of Upholstered Furniture and the use of Test Data, Reprinted

from New Technology to Reduce Fire Losses and Costs, Elsevier, SP Technical Report SP-RAPP 1986:47

[37] ISO 9705:1993, Fire tests- Full scale room tests for surface products

[38] Combustion products, smoke gas concentration, continuous FTIR analysis, Nordtest method, NT FIRE 047, Helsinki 1993

[39] M. Dietenberger in: Heat Release in Fires (V. Babrauskas, S.J. Grayson (Eds.), Elsevier Science, London 1992

[40] V. Babrauskas, J.F. Krasny: Prediction of Upholstered Chair Heat Release Rates from Bench-Scale Measurement, Fire safety and Engineering, ASTM SP 882, pp. 268-284, ASTM Philadelphia, PA 1985

[41] U. Wickström, U. Göransson in: Heat Release in Fires, V. Babrauskas, S.J. Grayson, (Eds.), Elsevier Applied Science, London, pp. 461–477, 1992

[42] International Sleep Products Association (ISPA) and Sleep Products Safety Council (SPSC), 501 Wythe Street, Alexandria, VA 22314

[43] T.J. Ohlemiller, J.R. Shields, R. McLane, R.G. Gann: Flammability Assessment Methodology for Mattresses, National Institute of Standards and Technology, 2000

[44] Children's Coalition for Fire-Safe Mattresses, 1515 K Street, Suite 300, Sacramento, CA 95814

[45] D.J. Barillo, P.A. Brigham, D.A. Kayden, R.T. Heck, A.T. McManus: J. Burn Care Rehab., 21, (2000) pp. 164–170

[46] K.T. Paul: Cigarettes of reduced smouldering power and their role to reduce the fire risks of upholstered seating, mattresses and bed assemblies. J of Fire Sciences, Vol. 18, No 1, Jan/Feb 2000

Fire Effluents

14 Preliminary Remarks

H.U. WERTHER

This section of the book is devoted to fire effluents. In general, these are understood to be the phenomena accompanying a fire that are not a result of the immediate, direct effect of the flames. However, this definition is imprecise. The effects considered here concern the consequences of heat and particularly the impact of the combustion products, that is, fire gases, aerosols, acidic components, and smoke particles on persons and objects. They may result in short-term exposure and cause long-term damage, which can immediately be detected or initiate gradual processes initially not recognized or taken seriously. They concern health and survival chances of persons remaining alive after having been exposed to fire effluents. They also affect objects that were not destroyed by direct fire involvement but were contaminated by smoke gases.

The three most important topics of this complex will be treated here: Chapter 15 deals with the main smoke testing technologies and the evaluation of smoke density. Chapter 16 describes the basics and testing of the toxicity of fire effluents. The concluding Chapter 17 discusses causes and effects of the corrosivity of fire effluents.

The toxicity of fire effluents is dealt with at some length, as for many years this subject has received growing attention. At the time the first edition of this book was published, only a small number of methods for determining the fire gas toxicity of products was available, the most important being NF F 16–101 and DIN 53 436. However, such measurements often only recorded additional or accompanying data and were limited to special areas of application. This approach has dramatically changed following spectacular catastrophes and the growing use of plastics and their involvement in fires.

In the meantime, leading specialists of the working group "Toxic hazards in fire" in ISO/TC 92/SC1 have developed the foundations for a general assessment of the toxic effects from fire effluents. They present the basis for appropriate classifications in the context of European harmonization activities. As an example, in the framework of the "Firestarr" project, the European Commission is creating uniform criteria to evaluate products that, in the future, will be used in the construction of rail vehicles throughout Europe. One of the key criteria for their approval will be the testing and assessment of the toxicity of the combustion products generated by these materials.

Significant innovations in the area of corrosion have arisen since the first edition of this book. This does not primarily apply to new testing methods, but rather to the causes and treatment of corrosive effects. It has been shown that in case of fire, with the exception of poly(vinyl chloride) (PVC), all major commercial plastics show a contribution to corrosion approximately comparable to that of natural materials. Owing to its high chlorine content, PVC is inherently flame retarded, but has the disadvantage of initiating high corrosion damage following fires. However, procedures on how to deal correctly with these detrimental effects have been made available in the meantime.

The widespread use of electronics as controls for machinery and in data processing has given rise to a particularly corrosion-susceptible technology. In case of fire, it requires selective and rapid action by specialists. If, after a fire, sanitation measures are correctly carried out, today there are good opportunities to avoid total losses, even for very sensitive electronic systems used, for example, in space technology. Sanitation measures are therefore particularly dealt with in the following chapter on corrosion, as they may significantly contribute to lower financial losses.

15 Smoke Development of Fire Effluents

M. KOPPERS

15.1 Introduction

In hazard assessment, one of the key factors is the potential hazard due to obscuration of light caused by smoke production from burning material. A realistic assessment can be obtained only by a real-scale test. An isolated small-scale test, not representative for the final use of the product, can only indicate the response of a product to the fire model selected. However, small-scale smoke tests, as described in this chapter, will provide information to assist in the determination and subsequent control of smoke hazards.

In recent years, major advances have been made in the analysis of fire effluents. Fire effluents are defined as the totality of gases and/or aerosols including suspended particles created by combustion or pyrolysis.

In this chapter, methods to determine the obscuration caused by smoke, that is, the visible part of the fire effluents, are described.

It is recognized that the composition of the mixture of combustion products is dependent on the nature of the combusting materials, the prevailing temperatures, and the ventilation conditions, especially access of oxygen in the seat of the fire. To these must be added the various physical and chemical processes taking place on a microscopic scale during formation of smoke, which impart certain characteristics to it (see also Chapter 7). Smoke development in a fire is thus a nonreproducible individual event that can be quantified only with great difficulty, if at all, by measuring techniques.

A further problem is to develop techniques for measuring smoke density that correspond as closely as possible to the perception of the human eye or the vision obscured by smoke. Various investigations have shown that the product of visual range and smoke density is approximately constant. It is, however, affected by the luminosity of the object and local light conditions. In the case of lacrimatory smoke, these relationships apply only to low smoke densities.

Photocells that cover the same range of wavelengths as the human eye are used in the photometric equipment available at present. However, they cannot measure human psychological or physiological reactions. It is hoped that such imperfect measuring techniques will nevertheless provide knowledge, for example, on the use of certain materials that, when translated into practice, will increase the chances of escape in a fire and thus improve safety levels.

In the following, an introduction is given to the means of measuring smoke and subsequently the techniques for measuring smoke density are presented.

15.2 Fundamental Principles of Smoke Measurement

Smoke consists of an aerosol of particles. It can either be measured as a function of its gravimetric properties (the mass of smoke particles), its light-obscuring properties, or a mixture of the two. Obscuring properties are a function of the number, size, and nature of the particles in the light path. If the particles are considered opaque, the capacity of the smoke to obscure light is related to the sum of the cross-sectional areas of the particles in the light path. It is measured in units of area, for example, m^2.

The measurements may be made in small-, intermediate-, or full-scale tests. They may be performed in closed systems called cumulative or static methods. They may also be performed in flow-through systems, and these are called dynamic methods. Static methods are all small or intermediate in scale. Dynamic methods are used at all scales.

Smoke density is measured by various techniques including purely visual recognition of smoke measuring rods in a room during a model fire test. The most frequently used tests are those based on optical methods, that is, attenuation of a light beam. There are also mechanical (separation of liquid and solid aerosol particles from the smoke effluents) and electrical (generation of electrical charges in an ionization chamber) methods. The latter methods are considered briefly below while the more important optical tests are discussed in detail subsequently.

15.2.1 Optical Methods

Bouguer's Law

Optical smoke measurements are derived from Bouguer's Law, which describes the attenuation of monochromatic light by an absorbing medium:

$$\frac{I}{T} = e^{kL}$$

$$k = \left(\frac{1}{L}\right)\ln\left(\frac{I}{T}\right)$$

The units of k are reciprocal length (m^{-1})

where:

ln   = natural logarithm,
T = intensity of transmitted light,
I = intensity of incident light,
L = light path length through the smoke,
k = linear Napierian absorption coefficient (or extinction coefficient).

The attenuation of light principle through an absorbing medium is shown in Fig. 15.1.

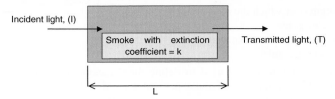

Figure 15.1   Attenuation of light by smoke

Extinction Area

A useful measurement of the amount of smoke is the total effective cross-sectional area of all the smoke particles. This is known as the extinction area of the smoke (S), as shown in Fig. 15.2.

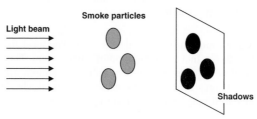

Figure 15.2   Extinction area

The extinction area can be thought of as the total area of the shadows of the smoke particles.

The extinction area is related both to the extinction coefficient of the smoke and to the volume that the smoke is contained within, by the equation:

$$S = kV$$

where V is the volume of the chamber in which the smoke is contained.

This equation only applies if the smoke is homogeneous.

Specific Extinction Area

In a test where the mass loss of the specimen is measured, the specific extinction area, σ_f can be calculated.

$$\sigma_f = \frac{S}{\Delta m}$$

where:

Δm = the mass loss of the specimen.

The units of σ_f are area/mass, for example $m^2 kg^{-1}$. σ_f has units of area per unit mass because it is a measure of the extinction area of smoke per unit mass of fuel burnt.

The specific extinction area σ_f is a basic measurement of smoke that can be made in both large and small-scale tests. It is independent of:

- The path length over which the measurement is taken,
- The flow rate of gases,
- The surface area of product exposed,
- The mass of the specimen.

The specific extinction area σ_f is used to define the "smokiness" of fuels. A fuel with a larger σ_f produces larger amounts of smoke per unit mass of fuel.

It is important to realize that σ_f does not give information on either the amount of smoke generated in a fire or the rate of smoke generation in a fire. To obtain this information, either the mass loss of the specimen (Δm) or the mass loss rate of the specimen ($\dot{m}$) must also be known. Then the extinction area of smoke generated is given by:

$$S = \sigma_f \Delta m$$

In dynamic systems, the specific extinction area may be obtained from:

$$\sigma_f = k\frac{\dot{V}}{\dot{m}}$$

where:
$\dot{V}$ = the volume flow rate,
$\dot{m}$ = the mass loss rate,
and the smoke production rate is given by:

$$\dot{S} = \sigma_f \dot{m}$$

Log10 Units

In some studies base 10 logarithms are used to calculate the optical density per unit light path length, D, which is properly named the linear decadic absorption coefficient, and like k (the linear Napierian absorption coefficient), also has units of reciprocal length (e.g., m^{-1}):

$$\frac{I}{T} = 10^{DL} \quad D = (\frac{1}{L})\log_{10}(\frac{I}{T})$$

$$k = D\ln 10 \quad \text{or} \quad k = 2.303\,D$$

The extinction area of smoke (S) can also be calculated from D using the equation

$$S = 2.303\,DV$$

Several variants of base 10 units can be found in the literature. A commonly used quantity is the dimensionless optical density $D' = \log_e (1 / T)$. For a given amount of smoke D' is proportional to the light path length, and is thus apparatus dependent. Results from one apparatus cannot be directly compared to results from other apparatusses.

Mass Optical Density

When working in $\log_{10}$ units, the equivalent variable to σ_f is called the mass optical density (D_{mass}) and it is related to σ_f as follows:

$$D_{mass} = \frac{\sigma_f}{\ln 10} = \frac{\sigma_f}{2.303}$$

The units of D_{mass} are area/mass, for example, $m^2 \cdot kg^{-1}$
In static systems the relationship between light obscuration and specimen mass loss is:

$$D_{mass} = \frac{D'V}{\Delta mL}$$

where:

D_{mass} = mass optical density
D' = optical density
V = volume of the chamber
Δm = mass loss of the specimen
L = light path length

In dynamic systems mass optical density can be obtained from:

$$D_{mass} = D\,\frac{\dot{V}}{\dot{m}}$$

Light Sources

Both white light and monochromatic laser light sources are used for smoke measurement.

Because light attenuation through smoke is dependent on the absorption and the scattering of light, and because the latter is dependent on wavelength, caution should be exercised when comparing data obtained from measuring systems using different light sources.

Visibility

Correlations have been established between visibility distance in smoke (ω) and the magnitude of the extinction coefficient (k) of the smoke (often expressed as the $\log_{10}$ analogue, i.e., optical density per unit light path length).

The general relationship is that the product of visibility and extinction coefficient is a constant, but the value of the constant depends on the contrast and illumination of the target being reviewed.

Figure 15.3 shows two relationships. Note that graphs comparing visibility against extinction coefficient usually have wide error bands because several types of light and targets are used in the studies.

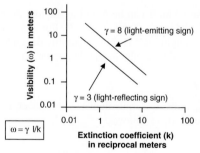

Figure 15.3   Visibility (ω) versus extinction coefficient (k)

15.2.1.1 Static Methods

In a static smoke test, the specimen burns in a closed chamber and the smoke produced builds up over time. In some tests, a fan stirs the smoke to prevent layering and to make it homogeneous.

The amount of smoke is measured by monitoring the attenuation of a light beam shining through the smoke. The extinction area of the smoke is a useful measure of the amount of smoke produced and is a function of the opacity of the smoke, the volume of the chamber, and the light path.

$$S = (\frac{V}{L}) \ln(\frac{I}{T})$$

This equation applies only if the smoke is homogeneous.

15.2.1.2 Dynamic Methods

In dynamic tests, the smoke from the specimen is drawn through an exhaust system at a measured flow rate and the opacity of the smoke stream is measured at regular intervals by monitoring the transmitted intensity (T) of a light beam (I) shining through the smoke. L is the path length. The principle is shown in Fig. 15.4.

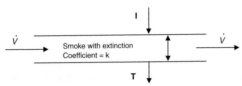

Figure 15.4   Dynamic smoke measurement

The smoke production rate ($\dot{S}$) at a given moment is calculated using the equation:

$$\dot{S} = k\dot{V} \text{ (area/time, m}^2\text{s}^{-1})$$

where
$\dot{V}$ = the volume flow rate of the exhaust gases.

15.2.2 Mechanical Methods

Smoke can be measured mechanically by depositing liquid and solid aerosol particles from smoke gases on a surface, usually a filter. The deposited smoke particles are evaluated gravimetrically or optically. The amount of smoke (in %) deposited for a specific material is calculated from the weight of smoke deposited and the weight of the original material.

Mechanical methods are not really suitable for evaluating smoke development in a fire incident and the associated hazard for humans because they give no prediction of sight obscuration. In the past, they were used as screening tests to eliminate certain materials and required for building materials in the United States (ASTM E 162 flammability test). Mechanical methods are virtually no more in use today.

15.2.3 Electrical Methods

Smoke production may also be determined with ionization chambers. This principle is widely used in smoke and fire detectors, particularly in industrial plants, public buildings, and also in private dwellings.

Ionization chambers work on the following principle: oxygen and nitrogen molecules in the air between the electrodes are partially ionized by a radiation source (normally α-radiators such as radium [^{226}Ra]). The charge carriers give rise to a low current. If smoke particles penetrate the ionization chamber, they combine with the ions whose mobility is then reduced, thus lowering the flow of current. The smallest smoke particles, which are invisible, are ionized preferentially. Such ionization chambers are thus used for detecting the very first stages of an initiating fire when no smoke is visible.

As with the mechanical methods, only a limited indication of sight obscuration is provided. Only those optical methods that show some correlation with the perception of the human eye are used in practice for the actual measurement of smoke density.

15.3 Test Apparatus and Test Procedures in Practical Applications

15.3.1 Validity and Differences of Individual Methods

Although all the optical techniques for measuring smoke density described below are based on the principle of extinction measurements, they differ, in some cases considerably, as a result of special conditions laid down for each method. The test results are thus usually only valid for the particular method.

The individual methods differ principally as follows:

- They can be carried out under static or dynamic conditions. In the static case, all smoke generated is contained in a closed system (chamber) where it is measured. In the dynamic situation, the system is open; the smoke is measured as it escapes from the apparatus. Both procedures reflect a certain situation in a fire incident: the static method simulates smoke production in a closed room, while the dynamic method corresponds to smoke production in an escape route for example.
- The length of the light path through the smoke layer varies in nearly all methods. The measuring device can be arranged horizontally or vertically, although the former is usual in dynamic tests. In static tests it can be either, although there is a danger of stratification (i.e., the formation of smoke layers of differing optical density) and hence of distorted results. Stratification can be avoided by arranging the light path vertically or circulating the air.
- The position of the test specimens with respect to the heat source may be horizontal or vertical. With thermoplastics, vertical positioning is disadvantageous because the specimen withdraws from the flame or radiant heat source due to melting and flowing away.
- Energy can be applied to the test specimen by a radiant heat source and/or open flame. Decomposition of the specimen takes place under smoldering or flaming conditions, depending on the type and level of energy supply. Very different results can thus be obtained for the same material. Smoke methods involve smoldering and flaming conditions to take both these forms of smoke production, which almost always occurs in fires, into account.
- Results are interpreted in different ways for each test method. For example, by calculating the maximum smoke density or percentage extinction. The methods of calculation are described for each test below.

It is not possible to describe smoke production in a fire incident quantitatively. To obtain defined information, the numerous variables must be eliminated by working under standardized conditions. This can be achieved most conveniently by working on a laboratory scale. The conditions selected and laid down for a particular test method enable reproducible results to be achieved. It is thus possible, at least under test conditions, to obtain specific information on smoke density of a material. It is impossible, however, to draw quantitative conclusions regarding the smoke generated from a material in a real fire. Nevertheless, limited predictions using laboratory results can be made on a purely empirical basis, that is, from long experience of smoke development in real fires and in laboratory tests. Besides the references given in the following, further information for standards related to smoke measurement can be found in the reference section to this chapter.

15.3.2 Tests Based on Static Methods

15.3.2.1 NBS Smoke Chamber Determination of Specific Optical Density

Few methods exist for testing uniquely the smoke production of burning materials. The principal one is the "NBS smoke chamber." It was developed in the United States by the National Bureau of Standards. The test specifications, general view and principle of the NBS smoke chamber are shown in Table 15.1 and Figs. 15.5 and 15.6.

Several test methods are based on the NBS smoke chamber such as IEC 60695-6-30 [1], ISO 5659-2 (in Japan JIS K 7242-2) [2], BS 6401 [3], NF C 20-902-1 [4], NF C 20-902-2 [5], ASTM E-662 [6], and NFPA 258 [7].

The attenuation of a light beam caused by smoke collection in the test chamber is measured. The smoke is generated by pyrolysis (smoldering conditions) or combustion (flaming conditions). The results are expressed as specific optical smoke density D_s derived from a geometric factor and the measured optical density, a measurement characteristic of the concentration of smoke. The specific optical density D_s is calculated from:

$$D_s = \frac{V}{A \cdot L} \ \log_{10}\left(\frac{100}{T}\right)$$

where:

V = volume of the closed test chamber,
A = exposed area of the test specimen,
L = length of the light path through the smoke,
T = light transmittance (in %) read from the photosensitive instrument.

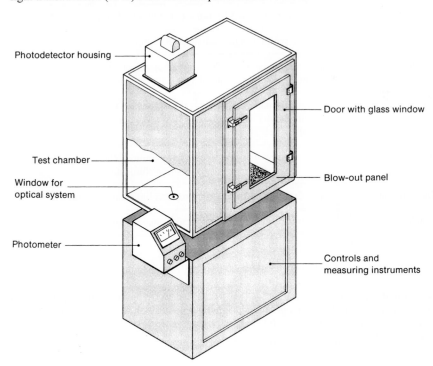

Figure 15.5   General view of the NBS smoke chamber

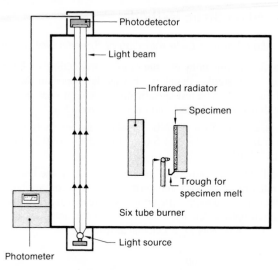

Figure 15.6   Principle of the NBS smoke chamber

Table 15.1   Test specifications to ASTM E 662

Specimens	At least six samples 76 mm × 76 mm × max. 25 mm, three of these samples are tested under flaming conditions and three under smoldering conditions
Test chamber	Internal dimensions 914 mm × 610 mm × 914 mm. Volume 0.51 m³
Specimen position	Vertical, parallel to the radiant heat source
Radiant heat source	Vertical furnace with a 76 mm diameter opening, radiation flux on specimen 25 kW/m² (smoldering conditions)
Ignition source	Six-tube propane microburner positioned 6.4 mm away from and 6.4 mm above the lower edge of the specimen (irradiation and application of flame = flaming conditions)
Photometric system	Light source with optics (color temperature 2200 ± 100 K) vertical light beam, length 914 mm, photosensitive element
Test duration	Max. 20 min or 3 min after minimum light transmission
Conclusion	Determination of specific optical smoke density

The expression $\dfrac{V}{A \cdot L} = 132$ is constant for the NBS smoke density chamber.

The maximum specific optical density D_m is calculated from the preceding equation using the minimum light transmittance T obtained during the test. Further details and examples of calculations are given in IEC 60695-6-30 [1].

The NBS smoke density chamber to ASTM E 662 is referenced in some building codes in the USA (see Chapter 10.2).

In transportation, the NBS chamber is used in the following areas.

Aircraft

- In prEN 2824 [8] and prEN 2825 [9] there is a smoke density requirement for materials used in passenger compartments. During the smoke density test, sampling of gases is carried out to determine further the toxicity of certain compounds (prEN 2826) [10] (see also Section 11.3.2.3 Aircraft).

Railways

- In the French NF F 16-101 [11], the NBS smoke density chamber is used to determine, besides the maximum smoke density, VOF_4, the smoke obscuration during the first 4 min of measurement (NF X 10-702-1) [12] (see also Section 11.2.2.3 Rail vehicles).

Ships

- IMO (International Maritime Organization) has adopted ISO 5659-2 [2] in their FTP-code (Fire Test Procedures) [28] (see also Section 11.4.3 Ships).

ISO 5659-2 uses the same NBS smoke chamber but the specimen is in a horizontal position and the heat source is a conical electrical radiator with an incident flux of 25 kW/m^2 or 50 kW/m^2 and with or without a pilot flame.

15.3.2.2 XP2 Smoke Density Chamber

In the United States, the so-called XP2 smoke density chamber to ASTM D 2843 [13] is specified for the determination of smoke production from plastics in various building codes. A diagram of the test equipment and the test specifications are presented in Fig. 15.7 and Table 15.2, respectively.

At least three tests must be carried out and the arithmetic mean, on which the classification depends, calculated. The specimen is laid on the holder (wire grid) and the burner is pivoted away and lit. The intensity of the light beam is adjusted so that the photometer reads 0 %. The burner pivots back automatically to the specimen and the light absorption is recorded as a function of time. The flame is applied for 4 min.

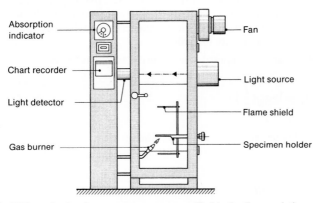

Figure 15.7   The XP2 smoke density test apparatus as specified in the fire regulations of Switzerland

Table 15.2   Test specifications

Specimens	Six samples • Solid materials: 30 mm × 30 mm × 4 mm or thickness used • Foams: 60 mm × 60 mm × 25 mm or thickness used • Coverings: 30 mm × 30 mm × original thickness
Test chamber	Internal dimensions 790 mm × 308 mm × 308 mm Air circulation 6.0–6.5 l/s
Specimen position	Horizontal on wire grid 60 mm × 60 mm
Ignition source	Propane gas pivotable burner, flame 150 mm long impinges on specimen 45° to horizontal, distance of burner orifice to centre of specimen 45 mm
Photometric system	Light source with optics (color temperature app. 5 500 K) horizontal light path, length 308 mm, photometer with narrow band photocell (max. sensitivity 550 nm)
Test duration	4 min
Conclusion	Typical code requirement is a Smoke Density Rating of 75 or less

The light source and fan are located on the right hand side of the test chamber while the photometer is on the left. A 90 mm x 150 mm (3.5 in. x 5.9 in.) "EXIT" sign is fixed to the back of the inside of the chamber 480 mm (18.9 in.) above the floor.

The mean of the three tests is obtained and the resultant mean light absorbance plotted as a function of time. The maximum smoke density is read off as the peak of the curve.

The smoke density is expressed in percent calculated from the area under the experimental curve divided by the total area (consisting of the interval 0–4 min multiplied by 0–100 % light absorbance). The result is multiplied by 100 and is designated as the Smoke Density Rating.

15.3.2.3 Three Meter Cube Smoke Chamber

IEC 61034-1 [14] and IEC 61034-2 [15] describe a test mainly for electric cables which is also used in transportation (BS 6853 [16] – railways in the UK). The test arrangement is shown in Fig. 12.19 of Chapter 12 Electrical engineering and the specifications for the determination of specific optical density in Table 15.3.

The measured absorbance A_m (optical density) is calculated as follows:

$$A_m = \log_{10}\left(\frac{I}{T}\right)$$

results are reported in terms of a parameter A_0 where:

$$A_0 = A_m \frac{V}{nL}$$

where:

V = volume of the closed test chamber ($27 \, m^3$),
L = length of the light path through the smoke ($3 \, m$),
n = the number of units of test specimen.

Table 15.3   Three meter cube smoke chamber test specifications

Specimens	Materials or products, amount depending on cable diameter
Test chamber	Internal dimensions $3\,m \times 3\,m \times 3\,m$. Volume $27\,m^3$. Smoke is mixed by a fan.
Specimen position	Horizontally oriented, positioned above ignition source
Ignition source	Pan of 1 l of alcohol
Photometric system	Photometric system using white light (color temperature 2800–3200 K), shining horizontally across the chamber at a height of 2.15 m
Test duration	End if there is no decrease in light transmittance for 5 min after the fire source has extinguished or when test duration reaches 40 min
Conclusion	Determination of specific optical smoke density

15.3.2.4 ISO Dual Chamber Box

In the Netherlands, the ISO dual chamber box according to ISO/TR 5924 [17] is in use as NEN 6066 Bepaling van rookproduktie bij brand van bouwmateriaal (combinaties) [18]. This method has been withdrawn in ISO, and with the initiation of the European Euroclasses for building products, it is expected that the requirement of smoke measurement with NEN 6066 will be withdrawn in due time. The general arrangement of the smoke chamber is shown in Fig. 15.8.

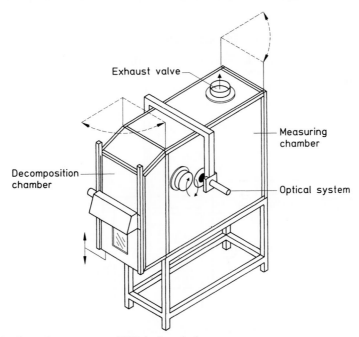

Figure 15.8   General arrangement of ISO dual smoke box

15.3.3 Tests Based on Dynamic Methods

The Methods listed below for determining smoke production of materials are supplementary tests to techniques determining fire performance or burning characteristics. The smoke density is measured in the exhaust duct by means of an optical system of which the principle is described in DIN 50055 [19] or a laser-based system.

Especially in the building codes, reference is made to smoke measurement as an optional test, mostly not part of the classification system but quoted under special observations in the test report.

15.3.3.1 DIN 4102-15 Brandschacht

A device for measuring smoke (see Chapter Building Section 10.9) is located 100 mm above the thermocouple for determining the smoke gas temperature in the "Brandschacht" [20]. The measuring device consists of a light source which emits a horizontal beam of light [light path 500 mm (19.7 in.)] through the constricted upper part of the "Brandschacht" leading to the exhaust flue. The beam of light, attenuated by the smoke, is measured by a light detector.

The smoke density determined this way is not a prescribed criterion for classifying building materials but is quoted under "special observations" in the test report and referred to in the test mark.

15.3.3.2 ISO 5660-2 Cone Calorimeter

In ISO 5660-2: Reaction-to-fire tests – Heat release, smoke production and mass loss rate from building products – Part 2: Smoke production rate (dynamic measurement) [21], smoke is determined by dynamic measurement in the exhaust of the cone calorimeter by means of a laser-based optical system. The test apparatus and specifications are described in Section 10.23 ISO.

15.3.3.3 DIN 5510-2 Using DB Brandschacht

In the DIN 5510 standards for German railways [22], the fire tests are performed in the "DB Brandschacht" according to DIN 54 837 [29] (see Section 11.2 Rail vehicles) and smoke development is measured in the exhaust, as shown in Fig. 15.9.

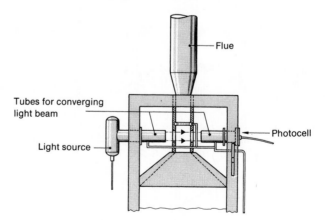

Figure 15.9   Smoke density measurement in the DB Brandschacht

There are two classes of smoke production as shown in Table 15.4.

Table 15.4   Smoke development classes according to DIN 5510-2

Smoke development class	Integral of light decrease (% · min)
SR1	< 100
SR2	< 50

15.3.3.4 ASTM E 84 Steiner Tunnel

In the United States, smoke production for building materials is usually tested according to ASTM E 84 [23]. The smoke development is measured at the "vent end" of the Steiner tunnel (see Section 10.2). Details are shown in Figs. 15.10 and 15.11.

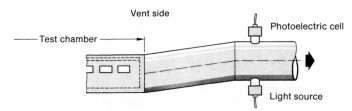

Figure 15.10   ASTM E84 Apparatus to measure smoke development

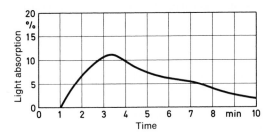

Figure 15.11   Time absorption curve for smoke density of red oak

The light beam is aimed vertically up the vent pipe and the distance between the light source and photoelectric cell is 914 mm ± 102 mm (36.0 in. ± 4.0 in.). The measured absorption is recorded continuously and displayed graphically as a function of time. A typical time-absorption curve for smoke density of red oak is shown in Fig. 15.11. The area under the test curve is divided by that of the curve for red oak (obtained by calibrations in each individual tunnel) and multiplied by 100. The resultant value for the material under test is compared with a scale on which cement board and red oak have been arbitrarily set at 0 and 100, respectively.

15.3.3.5 Smoke Generation According to ISO/TR 5659-3

This technical report was published by ISO/TC 61 Plastics [24] to have, besides the static test as described in ISO 5659-2 [25], a possibility to measure smoke in a dynamic way.

The heat source is the same conical radiator as in ISO 5659-2 (see Section 10.23 ISO), but the smoke evolved is conducted into an exhaust system consisting of a hood duct and fan. The duct contains both an orifice plate, the pressure across which is used to monitor the speed of air flow along the duct and photometric equipment for measuring the optical density of the smoke effluent stream throughout the test. The results are reported in terms of the measured optical density over the period of the test.

15.3.3.6 Single Burning Item Test

In the European Union, based on the Construction Products Directive 89/106/EEC [26], the Single Burning Item (SBI) test to EN 13823 [27] (see Section 10.4 European Union) was developed and recently introduced to allow the classification of the reaction-to-fire of building products into the Euroclasses. It is possible to measure and to classify the smoke density of building products in the SBI, but there are no mandatory smoke requirements at the moment.

15.4 Future Developments

In the European Construction Products Directive, smoke density may become one of the additional factors in the classification. In particular the RSP (Rate of Smoke Production) and SMOGRA (Smoke Growth Rate Index) will become important.

In ISO/TC 92/SC4 "Fire Safety Engineering," there is a strong interest in smoke obscuration as one of the means to assess the safety of occupants in a building or other compartments (transportation: bus, train, underground). As smoke accumulates in an enclosure, it becomes increasingly difficult for occupants to find their way out. This results in a significant effect on the time required for their escape. Moreover, at some degree of smoke intensity, occupants can no longer discern boundaries and will become unaware of their location relative to doors, walls, windows, and so forth, even if they are familiar with the occupancy. When this occurs, occupants may become so disoriented that they are unable to manage their own escape.

References

[1] IEC 60695-6-30 Fire hazard testing – Part 6–30: Smoke opacity – Small-scale static method
[2] ISO 5659-2 Plastics – Smoke generation – Part 2: Determination of optical density by a single chamber test
[3] BS 6401Method for measurement in the laboratory of the specific optical density of smoke generated by materials
[4] NF C 20-902-1 Essais relatives aux risques de feu – Méthodes d'essais – Détermination de l'opacité des fumées an atmosphère non renouvelée – Méthodologie et dispositif d'essai
[5] NF C 20-902-2 Essais relatifs aux risques de feu – Méthodes d'essais – Détermination de l'opacité des fumées an atmosphère non renouvelée – Méthodes d'essais pour matériaux utilisés dans les câbles électriques et dans les câbles à fibres optiques
[6] ASTM E 662 Test method for specific optical density of smoke generated by solid materials
[7] NFPA 258 Smoke generation of solid materials
[8] prEN 2824 Aerospace series – Burning behaviour of non-metallic materials under the influence of radiating heat and flames – Determination of smoke density and gas components in the smoke of materials – Test equipment, apparatus and media
[9] prEN 2825 Aerospace series – Burning behaviour of non-metallic materials under the influence of radiating heat and flames – Determination of smoke density
[10] prEN 2826 Aerospace series – Burning behaviour of non-metallic materials under the influence of radiating heat and flames – Determination of gas components in the smoke
[11] NF F 16-101 Matériel roulant ferroviaire – Comportement au feu – Choix des matériaux
[12] NF X 10-702-1Détermination de l'opacité des fumées en atmosphère non renouvelée – Partie 1- Description du dispositif d'essais et méthode de vérification et de réglage du dispositif d'essai

[13] ASTM D 2843 Test method for density of smoke from burning or decomposition of plastics
[14] IEC 61034-1 Measurement of smoke density of electric cables burning under defined conditions – test apparatus
[15] IEC 61034-2 Measurement of smoke density of electric cables burning under defined conditions – test procedures and requirements
[16] BS 6853 Code of practice for fire precautions in the design and construction of passenger carrying trains
[17] ISO/TR 5924 Fire tests – Reaction to fire – Smoke generated by building products (dual chamber test)
[18] NEN 6066 Bepaling van rookproductie bij brand van bouwmaterialen (combinaties)
[19] DIN 50055 Light measuring system for testing smoke development
[20] DIN 4102-15 Brandverhalten von Baustoffen und Bauteilen; Brandschacht
[21] ISO 5660-2 Reaction to fire tests – Heat release, smoke production and mass loss rate from building products – Part 2: Smoke production rate (dynamic measurement)
[22] DIN 5510-2 Preventive fire protection in railway vehicles; Fire behaviour and fire side effects of materials and parts; Classification, demands and test methods
[23] ASTM E 84 Test method for surface burning characteristics of building materials
[24] ISO/TC 61 Plastics – SC4 Burning behavior
[25] ISO/TR 5659-3 Plastics – Smoke generation – Part 3
[26] CPD Construction Products Directive 89/106/EEC
[27] EN 13823 Reaction to fire tests for building products: Exposure to the thermal attack by a single burning item (SBI)
[28] IMO FTP International Code for Application of Fire Test Procedures (Resolution MSC.61(67))
[29] DIN 54837 Testing of materials, small components and component sections for rail vehicles; Determination of burning behaviour using a gas burner

Further Reading

IEC 60695-6-1 Fire hazard testing – Part 6–1: Smoke opacity – General guidance
ISO 5659-1 Plastics – Smoke generation – Part 1: Guidance on optical density testing
BS 7904 Smoke measurement units – Their basis and use in smoke opacity test methods
NT Fire 004 Building products: Heat release and smoke generation
NT Fire 007 Floorings: Fire spread and smoke generation
DIN E 65600-1 Luft- und Raumfahrt; Brandverhalten; Bestimmung der Rauchdichte und der Rauchgaskomponenten von Werkstoffen unter Einwirkung von strahlender Wärme und Flammen; Prüfeinrichtung, Prüfgeräte und Prüfmittel
DIN E 65600-2 Luft- und Raumfahrt; Brandverhalten; Bestimmung der Rauchdichte und der Rauchgaskomponenten von Werkstoffen unter Einwirkung von strahlender Wärme und Flammen; Bestimmung der Rauchdichte
NF T 51-073 Matières plastiques – comportement au feu – Méthode d'essai pour la mesure de l'opacité des fumées

16 Toxicity of Fire Effluents

J. Pauluhn

16.1 Introduction

This chapter addresses principles of hazard identification of smoke released from products under specified laboratory- or bench-scale conditions. Such conditions are more often designed to meet optimally the objective of test, which is the comparison of the relative toxic potency of materials under highly standardized and reproducible conditions. This is to achieve stable "steady-state" exposure atmosphere amenable to a reproducible testing, including the analytical characterization of atmospheres, rather than an exact duplication of any fire scenario most likely to occur under real-life conditions. This source of difficulty lies in doubts concerning the atmospheres of fire effluents derived from the various systems used and their relevance to the yield and quality of atmospheres found in real fires. Part of this concern has centered around unrealistic expectations of what such a system can achieve.

As there are several characteristics of fires, it is unreasonable to expect any single piece of apparatus or testing approach to mimic all situations possible. The major requirement of any combustion apparatus is that the user can demonstrate a relationship to a specific state of a fire so that the performance of material in those conditions can be predicted. Through comprehensive work, attempts have been made to define or categorize a limited number of thermal decomposition conditions found in fires by the character of the atmospheres for their content of carbon monoxide (CO), carbon dioxide (CO_2), and oxygen (O_2) as well as the temperature of combustion. More aptly, the effects of fire ventilation on combustion products are expressed in terms of relationship between concentration of products and equivalence ratio, ϕ. The equivalence ratio, defined as the ratio of the amount of gasified material (fuel) to the amount of air, normalized by the stoichiometric fuel-to-air ratio, has recently been used successfully to correlate the generation of heat and products of complete and incomplete combustion and to predict the concentrations of products of fires (for details see Tewarson [1]). For well-ventilated fires, $\phi < 1.0$, where mostly heat and products of complete combustion (such as CO_2 and water) are generated. For ventilation-controlled fires, $\phi > 1.0$, where mostly products of incomplete combustion are generated. Such products are often toxic and corrosive and reduce visibility, and thus are dangerous to life and property. The currently valid ISO Standards addressing these issues are referenced under [2–17].

Another unreasonable expectation surrounding combustion toxicology in general, and the combustion apparatus in particular, is a requirement for the test to be a hazard evaluation over and above a toxicology assessment. This does not occur in other fields of toxicology, where qualitative and quantitative toxicology data are generated. The risk that the chemical or product may present is considered after the toxicological investigation, not as part of the investigation. The hazard is first defined, that is, can lethality or other serious toxicity occur? If so, by which agent is it caused and at what concentration? In this context, one major paradigm in toxicology is the establishment of both the concentration-effect relationship and the concentration x duration of exposure ($C \times t$) – effect relationship (Fig. 16.1), which provides the basis of hazard characterization and assessment. The risk, that is, the probability of the hazard occurring, can then be estimated taking into account the physical properties and end-use configuration of the product. For combustion of materials, this requires several essential prerequisites, for example, the ignition and propagation of fire and the likelihood of exposure. These aspects, however, are beyond the scope of the process of toxicological hazard identification.

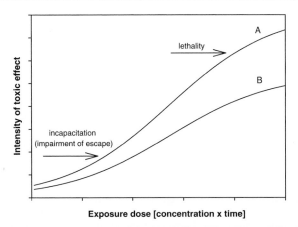

Figure 16.1   Analysis of the total exposure dose – effect relationship, the basic paradigm of toxicology. The total exposure dose is the product of concentration and duration of exposure (Haber's Rule). This rule may be of limited validity when the duration of exposure is very short and the concentration is very high and vice versa (see Fig. 16.2). The relationships of A and B are contingent on the mode of action and the relative toxic potency of agents.

This appears to be one major drawback of laboratory-scale tests because the generated fire effluents may not necessarily be predictive for product-specific use patterns and, accordingly, the most prevalent fire scenario. Moreover, the results of such tests have to be judged cautiously with respect of the kind of fire effluent released, its yield, and potential interactions which may be additive, synergistic, or antagonistic. Furthermore, dosimetric concepts used in conventional inhalation toxicology appear to be difficult to apply in combustion toxicology because the extent of exposure in "real-life" fires depends heavily on the release rate of individual smoke components as well as the propagation of fire which, in themselves, provide a constantly changing environment of exposure. Therefore, in combustion toxicology, the concept of "fractional effective dose" per "increment of duration of exposure" may be a more appropriate means for hazard assessment than assuming a "total dose" per "entire duration of exposure" which has traditionally been used for chemicals.

Thus, the hazard identification of fire effluents appears to be much more complex and prone to misjudgments as well as misunderstandings when compared to the conventional procedures used in inhalation toxicology testing, for instance, of chemicals. Basically, the objective of test is to determine the "acute lethal toxic potency" under standardized conditions so that relative comparisons can be made. When exposure is by inhalation, the dose metrics of choice is the actual airborne concentration to calculate the *median lethal concentration* (LC_{50}) which is the concentration of a toxic gas in ppm (ml/m^3 air) or fire effluent in $g \cdot m^{-3}$ statistically calculated from concentration-response data to produce lethality in 50 % of test animals within a specified exposure and post-exposure time (for more details see Sections 16.5.3 and 16.5.4 of this chapter). At present, for the evaluation of chemicals for acute toxicity it is necessary to meet current national classification and labeling requirements. Acute inhalation toxicity testing usually requires exposure of rats to an inhalation chamber equilibrium concentration for a period of at least 4 h [18–20]. Classification for other purposes, however, requires a LC_{50} value based on exposure duration ranging from 0.5 h (combustion toxicology), 1 h (transportation of dangerous goods), 4 h (chemicals) or up to 8 h (AEGL-3) [21] and may take into account endpoints ranging from "notable discomfort" (such as sensory irritation or odor perception), impaired ability to escape, the latter being contingent on a variety of disturbances ranging from alterations in neuromuscular functions, cognitive deficits to overwhelming apprehension due to the presence of irritant smoke and fire, and irreversible or other long-lasting effects up to lethality [21]. Current testing Guidelines do not encourage the use of animals solely for the calculation of an LC_{50} and favor approaches minimizing the numbers of animals and take full account of their welfare. However, especially for the investigation and assessment of new product entities, such experimental approaches are considered to be indispensable because analytical procedures are prone to artifacts due to the interference of agents present. Commonly, not all agents considered to be of potential health concern may be addressed by the analytical procedures used. This issue is even complicated further by the fact that agents may be present in their volatile or particulate form (aerosol) or absorb onto the surface of particulates (soot). This, in turn, might have greater impacts on the toxic hazard. All these aspects are assessed in an integrated manner in combustion toxicology.

16.2 Toxic Principles of Toxic Agents from Smoke

Toxicology is the study of the adverse effects of chemicals on living organisms. Because toxicologists and those who apply toxicological principles are often involved in the assessment of hazard and the projection of effects in a specific population, an alternate definition could be "the science that defines limits of safety of chemical agents." Toxicity, on the other hand, is defined as the inherent ability of a chemical to affect living systems adversely, and is expressed when the characteristics of exposure and the spectrum of effects come together in a correlative relationship customarily referred to as the dose-response relationship [22].

Important aspects in the analysis of this relationship are

- The spectrum of the toxic dose that, in terms of LD_{50} values (oral dosage causing death in 50 % of treated animals), ranges from 0.00001 mg/kg body weight for *Botulinum* toxin to 1 mg/kg body weight for nicotine or 10,000 mg/kg body weight for ethyl alcohol,
- Route and site of exposure,
- Duration and frequency of exposure,
- The spectrum of toxic effects (pathomechanisms), including the respective dose-response relationship.

To extrapolate data obtained in animal models, uncertainty factors (UFs) must be applied to account for recognized uncertainties in the extrapolations from the experimental data conditions to an estimate appropriate to the assumed human scenario.

16.2.1 Selection of Test Species for Bioassays

The first concern in extrapolating rodent data to humans is whether or not the rodent responds to the decomposition products of materials in a manner comparable to human, that is, whether there is a common pathomechanism. Available data indicate that the mechanisms of toxicity of both CO and hydrogen cyanide (HCN) are the same in rodents and humans. However, there is evidence that rodents and humans may not respond in a similar manner to the irritant gases, such as hydrogen chloride/bromide/fluoride (HCl/HBr/HF), sulfur dioxide (SO_2), nitrogen oxides (NO_x) and organic aldehydes (e.g., formaldehyde, acrolein) to mention but a few.

Testing guidelines recommend specific animal models for the evaluation of toxicological endpoints. The use of common laboratory animals for inhalation toxicity studies continually supplements the database and furthers the understanding of toxicity data in experimental animals and their relevance for man. However, specific modes of action of the substance under investigation, the evaluation of specific endpoints, or the clarification of equivocal findings in common rodent species may require the use of animal species not commonly used in inhalation toxicology. The choice of animal models for inhalation toxicity testing is usually based on guideline requirements and practical considerations rather than validity for use in human beings. An animal species must be small enough to allow handling and exposure in sufficient numbers in relatively small inhalation chambers. An animal species, however, must be large enough to allow measurement of all endpoints relevant to identifying the inherent toxicity of the substance under investigation.

The evolution of an animal model in inhalation toxicology is a slow process. Comparison of different species is sometimes difficult to perform because the mode of exposure used for small rodents and for larger laboratory animals such as dogs or monkeys is different. Exposure conditions may differ in cooperative and noncooperative animal subjects and may also not be interchangeable from smaller to larger animal species due to variability in exposure paradigms. The dose-response relationship is conventionally evaluated by exposure to variable steady-state concentrations for a specified duration of time. Confounding factors are more likely to occur using larger species because larger animals usually identify any physical weakness of the exposure system and will abuse it.

In general, the laboratory animal of choice is the rat. A great deal of information is available on rat respiratory physiology, biochemistry, morphology, and genetic data when compared with other species. A considerable sum of toxicological data obtained and a number of compounds has been investigated in rats and could be conveniently used for reference purposes. The size of the rat allows extensive laboratory investigations. Techniques are available that allow special surgical procedures and isolation of several purified cell populations from the rat lung.

Bioassays for the quantification of the irritant potency of inhaled toxicants give preference to mice. The higher respiratory rate of mice ($\approx$300 breaths/min) makes mice more susceptible to many inhaled irritants than rats ($\approx$120 breaths/min). This high respiratory rate of small laboratory rodents makes the uptake of inhaled agents faster than in slower breathing humans and provides an additional, implicit safety factor. On the other hand, elimination of toxic gases, such as CO, may be substantially faster which has to be accounted for when the determination of this toxic species in blood is considered (see later). Despite these obvious species differences, the results from acute inhalation toxic potency tests for chemical agents likely to occur in fires are highly predictive for humans, that is, assumed human susceptibility does not differ by a factor of 3 from the respective bioassays (Table 16.1).

Table 16.1   Lethal toxic potencies of potential fire effluents (duration of exposure: 30 min). Data reproduced from Hartzell (1989) [23]

Chemical agent	Mice	Rats	Primates/humans[a]
CO (ppm)	3 500	5 300–6 600	2500–4000
HCN (ppm)	166[b]	110–200	170–230
NO_2 (ppm)	289[c]	127[d]–200[e]	–
HCl (ppm)	2644	3800	5000
Low oxygen (%)	6.7	7.5	6–7

[a] These values are estimates based on available data,
[b] Matijak-Schaper and Alarie, 1982 [24],
[c] Value was calculated on the basis of an available 5 min value (denominator used: 6.5),
[d] Hartzell (1989) [23],
[e] Levin (1997) [25]

In summary, no animal species mimics human in all respects. Therefore, animal models are, at best, a necessary compromise and must be used because they offer the advantages of experimental control and reproducibility. In general, the uncertainties of extrapolation of toxicological results across different species are minimized if a maximum of mechanistic understanding is gained from a study. This is achieved by measurement of a sufficient number of procedures and endpoints. This primary objective is usually not achieved in poorly designed and performed animal studies with several animal species and a limited number of parameters investigated.

16.2.2 Dose-Response and Time-Response Relationships

A major issue of inhalation toxicity is that of dose. Inhaled dose is more difficult to determine than the dose from other routes of administration. With oral or parenteral routes, a discrete amount is given in a bolus. With inhalation, the delivered dose depends on the exposure concentration and time, physical characteristics such as particle size or water solubility and chemical reactivity, as well as breathing pattern of the test species, namely the respiratory minute volume (RMV). Deposition patterns within the various regions of the respiratory tract also are important. Over the past few decades, the concept of dose as applied to toxicological studies has changed considerably [26, 27]. Initially, "dose" simply meant the concentration in the atmosphere in inhalation studies multiplied by the duration of exposure (C x t), or the amount ingested or instilled into the gastrointestinal tract in oral dosing studies. Empirical correlations were then used to evaluate the relationship of dose and response with little appreciation of the detailed biological interactions of the test agent. The extrapolation from one route to another is subject to tremendous errors and caution is advised when doing so. Default values are therefore not recommended and conversion factors must be calculated for each individual situation, making appropriate assumptions about body weight, respiratory minute volume, percentage deposition, retention, and absorption, also taking into account pulmonary and extrapulmonary pathomechanisms.

Especially in combustion toxicology, two exposure scenarios of humans have to be envisaged: one is the accidental, very high level exposure of humans directly involved in fires in the range of minutes; the other is the more passive, low-level exposure of people in the vicinity of fires, which may result in exposure duration of hours. Experimental exposure-response data and data from human exposure incidents often involve exposure duration differing from those obtained from animal studies. Therefore, extrapolation from the reported exposure period and chemical concentration of a toxic endpoint to an equivalent concentration for another specified period is often required.

The dependency of LC_{50}s on exposure time is one of the most widely substantiated phenomena in combustion toxicology. Historically, the relationship according to Haber, commonly called Haber's Rule (i.e., C x $t = k$, where C = exposure concentration, t = time, and k = a constant) has been used to relate exposure concentration and duration to a toxic effect [28, 29]. Basically, this concept states that exposure concentration and exposure duration may be reciprocally adjusted to maintain a cumulative exposure constant (k) and that this cumulative exposure constant will always reflect a specific toxic response. This inverse relationship of concentration and time may be valid when the toxic response to a chemical is equally dependent on the concentration and the exposure duration. However, work by ten Berge et al. (1986) [30] with acutely toxic chemicals revealed chemical-specific relationships between exposure concentration and exposure time that were often exponential rather than linear. This relationship can be expressed by the equation C^n x $t = k$, where n represents a chemical specific, and even a toxic endpoint specific, exponent. The relationship described by this equation is basically the form of a linear regression analysis of the log-log transformation of a plot of C vs. t. This work examined the airborne concentration (C) and short-term exposure time (*t*) relationship relative to lethal responses for approx. 20 chemicals and found that the empirically derived value of *n* ranged from 0.8 to 3.5 among this group of chemicals [30]. Hence, these authors showed that the value of the exponent *(n)* in the equation $C^n \times t = k$ quantitatively defines the relationship between exposure concentration and exposure duration for a given chemical and for a specific toxic or health effect endpoint. Haber's Rule is the special case where $n = 1$.

The dependency of LC_{50}s on exposure time is illustrated in Fig. 16.2 for CO and HCN. The hyperbolic relationship was calculated from the linearized function. The significance of the $C \times t$ curve to combustion toxicity testing and research is that it reveals that every material has two distinct toxicological properties that are, in turn, dependent on test conditions and biological endpoints: (1) the minimum amount required to elicit the toxicological response, that is, the concentration threshold and (2) the maximum rate at which the effect is induced, that is, the time-to-effect material performance threshold. Changes in respiratory patterns as a result of upper respiratory tract sensory irritation appear to be contingent on concentration rather than total dose ($C \times t$) [31–33] while at the inception of exposure it may be more dose-rate dependent.

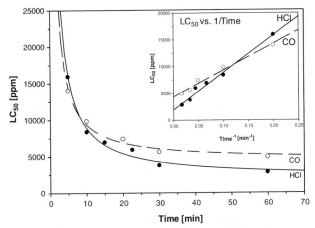

Figure 16.2 Duration of exposure – lethality relationship of rats exposed to either hydrogen chloride (HCl) or carbon monoxide (CO). Data from Hartzell et al. [34, 35]

The uptake of CO and the ensuing formation of carboxyhemoglobin (COHb), a measure of the body burden of CO, exhibit a concentration x time relationship while the occurrence of toxicity is dependent further on the time point at which the critical concentration of COHb in blood is attained. As delineated by the Stewart-Peterson equation (see Fig. 16.3; [36, 37]), the rate of the fractional loading of hemoglobin is highly contingent on the respiratory minute ventilation (RMV).

The RMV per mass body weight increases over proportionally with smaller body masses, that is, it is greater in smaller laboratory animals than in humans. It also is dependent on the presence of respiratory tract irritants that may evoke a concentration-dependent depression in respiration (Fig. 16.4). This, in turn, has major impacts on the extent of COHb formation even when exposed to equal concentrations (Fig. 16.3). Thus, in the presence of irritant chemicals, dosimetric adjustments are prone to erroneous assumptions. As illustrated in Fig. 16.4, from the lowest to the highest concentration a factor of ≈10 is covered, whereas the ensuing decrease in respiratory minute ventilation ranges from 10% to 75%. This demonstrates that dose calculations without appreciation of the actual breathing patterns of the test species may lead to false assumptions. Hence, stimulation of irritant receptors within the respiratory tract may have a marked impact on the respiratory minute volume of the test species so that time-related corrections of respiratory minute volume must be considered. This may also have some impact on the deposition pattern of the test agent within the respiratory tract. These considerations demonstrate that in combustion toxicology the prediction of the toxicity of complex smoke atmospheres is by far not a simple process, especially when made solely on the basis of analytical data.

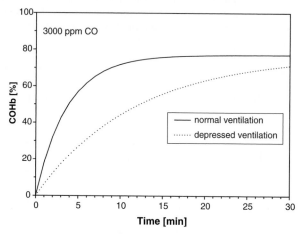

Figure 16.3 Computer-modeled time-dependence of loading curves of carboxyhemoglobin (COHb) of rats exposed to 3000 ppm CO using the Stewart-Peterson equation at either normal ventilation or a depressed ventilation due to the presence of irritant gases

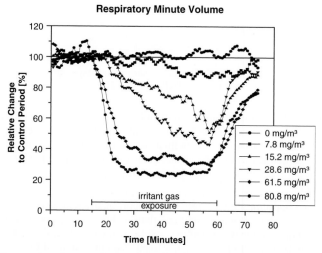

Figure 16.4 Change of respiratory minute volume in laboratory rodents (mice) as a result of exposure to a volatile *upper respiratory tract irritant*. Animals were exposed to air for 15 min (collection of baseline data) and then to an irritant gas for 45 min. Recovery of changes was observed during exposure to air for 15 min. All data are calculated relative to the baseline data. Measurements were made in nose-only volume-displacement plethysmographs as described elsewhere (Pauluhn [38])

One additional feature of dosimetry is that portal-of-entry (respiratory tract) effects may either be dose or concentration dependent. As shown for an irritant gas in Figure 16.4, effects related to *upper respiratory tract sensory irritation* evoke a decrease in RMV, the onset of which is almost immediate following exposure to high concentrations and often less immediate when exposed to lower concentrations. Nonetheless, the magnitude of response is apparently more concentration rather than dose dependent, because no appreciable exacerbation of effect occurs with increased duration of exposure. The time-effect relationship shown in Fig. 16.4 appears to suggest that at the onset of exposure, Haber's Rule may be valid whereas for a longer exposure duration it is time independent. As accidental exposure to fire effluents is commonly short, this dependency needs to be accounted for.

16.2.3 Species Differences and Uncertainty Factors

In cases in which there is little interspecies variability, conventionally, the most sensitive species, and/or the species most closely related to humans is selected, the interspecies uncertainty factor (UF) is typically reduced from 10 to 3. It should be noted that in all cases the mechanism of action, usually direct acting irritation, is not expected to vary significantly between species because the mode of action is local at the portal-of-entry. This means, for this mode of action neither toxicodynamic nor appreciable toxicokinetic species differences have to be accounted for [21].

In cases where the mechanism of action is unknown or there are likely to be differences in metabolic and physiological response among species, an interspecies UF of 10 is conventionally applied. This default UF of 10 is used when there is insufficient information about the chemical or its mechanism of action to justify a lower value, or if there are data suggesting a large degree of variability between species.

16.3 Endpoints in Acute Toxic Potency Tests of Combustion Products

As emphasized in the preceding, the determination of the acute toxic potency serves several objectives:

- Qualitative pathomechanism of toxic effluents apply according to the following classification scheme: irritation, asphyxiation, uncommon toxicity,
- Determination of relative acute toxic potency: relative to existing products or to reference substances such as wood or wool,
- Assessment of impairment of escape (incapacitation),
- To validate/refute theoretical assessment of the acute toxic potency predicted solely on the basis of analytical data.

Especially for new product types, this approach provides a scientifically sound means to verify the assumptions made, that is, as to whether the selection of analytical components is adequate, that analytical interference of toxicants present in complex combustion atmospheres does not occur to the extent considered to be of biological relevance, and that agents present as vapor or liquid (condensation aerosol) or adhering onto soot particulates exert their biological activity in the way predicted. For highly water-soluble agents normally deposited in the upper respiratory tract, soot particles may act as carrier to penetrate the lower respiratory tract. In the respiratory tract, both the vulnerability as well as the respective clearance mechanisms differs from one region of the tract to the other. Accordingly, a change in major deposition patterns may also change the toxic mechanism and potency of agents.

The effects of two agents given simultaneously produce a response that may simply be additive of their individual responses or may be greater (synergistic) or less (antagonistic) than that expected by addition of their individual responses. Due to the presence of multiple agents in combustion atmospheres, antagonistic or synergistic effects are difficult to assess without a state-of-the-art bioassay that suitably integrates all potential uncertainties. For example, antagonistic effects may ostensibly occur as a result of upper respiratory irritation due to stimulation of the trigeminal nerve (see above) and decreased uptake of agent (dispositional antagonist). Such antagonistic effects may also occur by chemical interactions, for example, inhalation of nitrogen dioxide (NO_2) may form methemoglobin (oxidized form of hemoglobin contained in the erythrocytes) and, accordingly, provides an endogenously formed antidote for blood cyanide. On the other hand, one major „synergist" in combustion laboratory-scale tests appears to be oxygen deficiency, especially in the presence of CO poisoning. Moreover, a slowing respiration resulting from the induced state of narcosis may exacerbate effects related to asphyxia. CO_2 is a known respiratory stimulant that increases the rate of uptake of other toxicants, thus producing a faster rate of formation of carboxyhemoglobin (COHb) from inhalation of CO (Fig. 16.3). Conversely, irritants, such as HCl, slow respiration in rodents due to upper respiratory tract sensory irritation, but increase the respiratory minute volume in primates from pulmonary irritation because of the absence of scrubbing of irritant gases by the nasal turbinates in obligate nasal breathing rodents.

Although the LC_{50} provides a means to rank the relative acute toxic potency of materials (e.g., see Table 16.1), it should be kept in mind, however, that the time-related onset of mortality is another important variable for the understanding of pathomechanism of concern (Fig. 16.2). Toxicants that cause asphyxia (such as HCN and CO) commonly evoke a mortality pattern occurring immediately at the time point reaching unbearable levels of asphyxiation. Irritants, in turn, may damage in a concentration- or dose-dependent manner the upper or lower respiratory tract, including airways. Highly water-soluble irritant gases (e.g., HCl, SO_2) may evoke airway inflammation with ensuing *obliterating bronchiolitis*. The inflammatory response requires time to develop and is accompanied by thickening of airway mucosa, bronchoproliferative response and airway plugging due to mucus production. As depicted in Fig. 16.5, this potential mode of delayed onset reaction is readily detectable in appropriately performed acute toxic lethal potency tests (see also Fig. 16.12). Eventually, increased airway obstruction leads to a mismatch of the ventilation: perfusion relationship, which means that the lung parenchyma (alveoli) is still intact and perfused by blood but not ventilated anymore due to airway plugging. Arterialized blood will then be diluted with venous blood (venous admixture)

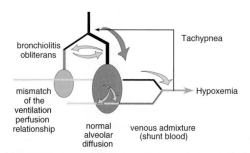

Figure 16.5   Schematic representation of the response of the lung to irritant gases causing a delayed onset of mortality. In brief, irritant gases at high enough concentrations elicit an inflammation with subsequent obstruction in the bronchial airways (*bronchiolitis obliterans*) without changing the diffusing capacity of the gas-exchange region (alveoli). As alveoli are still perfused by venous blood (low oxygen) but not ventilated (no access to oxygen due to obstructed airways) oxygenated (arterial) blood will eventually mix with venous blood. The resultant hypoxemia is compensated for by an increase in ventilation (tachypnea) which, in turn, decreases the efficiency of the alveolar gas exchange (uptake of oxygen and exhalation of CO_2). In commonly used laboratory rodents, this vicious circle needs 1–2 weeks to develop

resulting in hypoxemia and death. On the other hand, pulmonary irritants may damage the thin walls between blood capillaries and alveoli (blood/air barrier) resulting in flooding of alveoli (edema).

16.3.1 Selection of Endpoints

The endpoints selected in these types of studies are related to clinical findings (toxic signs, including their onset and duration, and lethality) and assessment of behavioral and physiological alterations, usually including determinations of toxic species in blood. Lethality has traditionally been used as the most robust biological endpoint in acute laboratory toxicity tests. Moreover, the incapacitating effect of smoke in preventing escape has become increasingly recognized. Such an effect may impair escape from the fire environment, and thereby result in lethality from thermal burns, oxygen deficiency, injury, or continued inhalation of toxic decomposition products. Consequently, the objective of some laboratory combustion toxicity tests involved the assessment of the potential of smoke to impair escape and to cause lethality to provide input data for an overall hazard assessment of materials in real fires. Admittedly, measurement of incapacitation appears to be a more appropriate endpoint as compared to lethality. However, it is more complex to determine as it depends on a number of variables difficult to quantify unequivocally in rodent bioassays [39]. These variables depend on the complexity of the test procedure and endpoints selected to quantify effect. They may range from measurements that detect deficiencies on neuromuscular function and behavioral changes up to altered cognitive functions. The impact on test results in relation to the development of fear, reduced visibility due to smoke preventing escape, and eye or respiratory tract irritation by irritant gases and aerosols remain issues difficult to address in animal bioassays.

In early studies to assess smoke incapacitation, methods included the shuttle box and tumble cage (traditional behavioral response methods), the motorized exercise wheel, and even an animal's ability to avoid drowning after exposure (a much less conventional method). Another method, that of shock avoidance by hind leg flexion [40, 41] became popular. This technique had numerous advantages. It was a simple method, requiring relatively little instrumentation or training of animals. It permitted exposure of animals in a "head-only" mode which reduced side effects (such as heat stress and condensation of smoke onto fur) of whole-body exposure. Also, the instrumentation was outside the exposure chamber which reduced corrosion. The method was reproducible, and the external location of the animals even allowed for an assessment of death or withdrawal of blood samples before a test was over. A major criticism of this method, and of nearly all incapacitation methods, is that the endpoint is too close to death. However, the added expense of training animals to perform a more complicated learned function (e.g., traversing a maze) was simply not practical [40].

Many behavioral and physiological tests have been used over the years. In addition to those techniques mentioned in the preceding, others include the rotorod, the pole-jump, the righting reflex, the electrocardiogram (ECG), and respiration rate [40]. Indices such as onset of changes in motoric activity and the occurrence of apneic periods have been determined for mixed gases in running wheels [42]. It is interesting to note that the behavioral methods continue to change, while lethality remains as a standard by which most other techniques are judged.

16.3.2 Assessment of Toxic Effects from Irritancy

Irritant fire products cause sensory stimulation of the eyes and upper respiratory tract, with pain and breathing difficulties, which may add to the asphyxiant effects of the narcotic gases. Moreover, in context with investigations of the combined effects of the asphyxiant CO and the irritant gas acrolein, the authors conclude that evidence does exist for an "antagonistic-like" effect when acrolein is present in concentrations less toxic (based on individual gas toxicity) than the CO in the gas mixture, a condition that seems likely to exist in most fire conditions [43]. These effects occur immediately on exposure and are for an extended duration of exposure concentration-related (see above), while lung inflammation and edema occur when a specific exposure dose is exceeded. Lung edema usually occurs a few hours after exposure. All fire atmospheres contain many irritant chemicals, including organic products (such as aldehydes) and inorganic acid gases, such as HCl, HF, HBr, SO_2, and NO_x partitioned between the gas phase, condensation aerosols and/or adsorbed onto carbonaceous smoke particles. Some 20 or so irritant fire products have been identified and there is evidence that unidentified irritant species are also present [44], so that the irritancy of combustion product atmospheres cannot be fully predicted from even a comprehensive chemical analysis.

Animal models are available to evaluate the most likely region of the respiratory tract in which the inhaled agent will be deposited. For upper respiratory tract irritants, the upper respiratory tract sensory irritancy can be assessed in terms of reflexively induced respiratory depression (RD_{50}, exposure concentration that decreases the respiratory rate by 50%). This concept of reflex reactions stems from stimulation of trigeminal nerve endings located in the respiratory epithelium of the nasal mucosa which is characterized by a pause between end of inspiration and expiration (Fig. 16.6). Lower respiratory tract irritants, in turn, evoke an apneic period between end of expiration and next breathing cycle. These changes could most suitably be detected and quantified in mice using the ASTM test method [45]. Further details have been described elsewhere [38, 46–51]. Overall, this methodology can be readily adapted to rats which, however, may experience a lesser sensitivity and stable response than mice [52]. It provides a means to assess the relative irritant potency of combustion gases, including the assessment of the lung region most prone to damage.

Exposure period

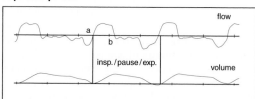

Control period

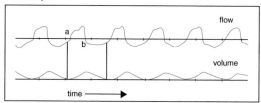

Figure 16.6   Analysis of respiratory pattern in rodents as a result of irritation. Inspiration: positive flow, expiration: negative flow, upper curve: flow, lower curve: volume. Volume is digitally derived from the flow signal. The flow algorithm identifies the start of each breath (a), including the bradypneic period (pause) beginning at the end of inspiration (b). This bradypneic period is a characteristic for „upper respiratory tract irritants" and is the cause of the depression in respiration shown in Fig. 16.4. X-axis ticks: 200 ms. Control data were collected during the air-preexposure period.

16.3.3 Toxic Species in Blood

CO analysis is among the first data to be obtained from fire victims as evidence for the cause of death. Fortunately, the body fluid, blood, to be sampled is not extremely difficult to obtain, and the stability of the carboxyhemoglobin (COHb) is reasonably good. For CO analysis, the use of the CO-oximeter is wide spread, and it has been noted to be the primary instrument employed by UK forensic laboratories [53]. The acceptance of this instrument stems from the convenience of the methodology.

CO is also endogenously generated through the degradation of porphyrines resulting in maximum background levels of COHb of $\approx 1\,\%$. The binding of CO to hemoglobin is reversible and competitive to oxygen with a 250 times higher binding affinity of CO than oxygen. Relationships between COHb concentrations and clinical symptoms have been established in health subjects [54]. In brief, concentrations up to 10 % did not coincide with specific symptoms; 10–20 % resulted in tension in forehead and dilation of skin vessels; 20–30 % in headache and pulsation in the sides of the head; 30–40 % in severe headache, ennui, dizziness, weakening of eyesight, nausea, vomiting, prostration; 40–50 % in the same as previous symptoms, and an increase in breathing rate and pulse, asphyxiation and prostration; 50–60 % in the same as previous symptoms, and coma, convulsions, and Cheyne-Stoke respiration; 60–70 % coma, convulsions, weak respiration and pulse, possible death; 70–80 % slowing and stopping of respiration, death within hours; 80–90 % death in less than an hour; > 90 % death within a few minutes. Individuals inflicted with cardiovascular diseases may be more susceptible to CO [55].

Also in controlled animal studies, especially the measurement of CO in blood provides a means for assessment of the "internal dose" of this toxicant. External exposure concentrations can readily be correlated with the respective concentration of COHb. Empirical data suggest that rats exposed to varying concentrations of CO for a fixed duration of exposure (30 min) elaborated an exponential relationship of inhalation chamber CO and the concentration of COHb (Fig. 16.7).

This relationship can be used to back-calculate the average concentration of CO in the exposure atmosphere to estimate whether appreciable effects on breathing patterns or interference within the analytical equipment used have occurred. In rodent species experiencing a

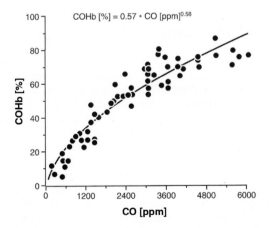

Figure 16.7 Association of exposure concentration of CO in the breathing zone and COHb formation (rats, exposure duration 30 min). Due to the rapid reconstitution of COHb in rats (see Fig. 16.8), blood was sampled immediately after cessation of exposure.

high respiratory rate, however, in poorly designed studies this endpoint is subject to artifacts often leading to a marked underestimation of carboxyhemoglobinemia. The COHb determinations illustrated in Fig. 16.8 were made from blood samples taken from surviving rats immediately after cessation of exposure. Thus, data shown in Fig. 16.8 suggest that CO dissociation from COHb and elimination via exhalation is fast in rats (elimination halftime approx. 12 min). This means that the time elapsed between the cessation of exposure and blood sampling is decisive for the outcome of this determination.

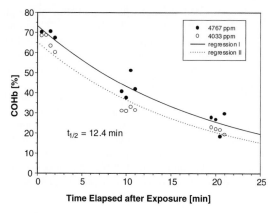

Figure 16.8   Kinetics of elimination of CO from COHb after cessation of inhalation exposure (rats, exposure duration 30 min)

Another toxic constituent amenable to determination in blood is cyanide. Blood cyanide analysis in fire victims present several challenges to prevent losses and artifacts from obscuring the correct values. The chemical nature of cyanide complicates these analyses. Whole blood has higher cyanide concentrations than plasma or serum, which is probably due to the sequestration of cyanide in the red blood cells [56–59]. Artificial concentrations of cyanide in blood may be observed readily due to many variables which have been discussed elsewhere [60–71].

16.4 Life-Threatening Responses to Toxic Agents from Fires

The aim of most toxicity evaluations is to provide data to predict the consequences of exposure in humans. Suitable available data concerning the effect of similar substances on humans must be considered in determining the relevance of animal studies. Data exist concerning the effect of fire effluent atmospheres, which have been derived from studies on many fire victims, especially postmortem examinations. Victims are generally found with high COHb levels, indicating exposure to CO, and in some cases HCN exposure has been implicated. CO and HCN are both known to cause progressive central nervous system depression leading to unconsciousness and death. This type of toxic effect has been termed "narcosis" or "asphyxia" and is considered to be very important in the response of humans to fire effluents. There are many reports of fire effluents being described as "irritant," causing coughing, choking, and an inability to see. Chemical pneumonitis has also been reported in both fire survivors and fatalities. Irritancy, both sensory, that is, occurring in the upper airways of the respiratory tract, and pulmonary, that is, occurring at the alveolar level, is considered to be a major factor in the response of humans to fire effluents. There have been few, if any, fire casualty reports due to other significant toxicological effects apart from narcosis and irritancy.

The effects seen in experimental animals have been very similar to those seen in humans. Death has been attributed to the presence of "narcotic" gases such as CO and HCN. Especially in small laboratory-scale animal models, factors such as oxygen vitiation have to be accounted for not to confound effects caused by the mere depletion of oxygen rather than specific agents of concern. Irritants have been shown to be present, detected by clinical observations of salivation, nasal discharge, lachrimation, and measurements of respiratory rate. Pulmonary damage has also been confirmed by protein extravasation in bronchoalveolar lavage fluid or histopathological examination of the lungs of animals exposed to high concentrations of corrosive irritants. In general, direct-acting chemicals appear to have the same spectrum of activity across the species, including humans, because specific toxicokinetic and toxicodynamic aspects do not play a major role. In contrast, these aspects may play a role for agents that cause toxic effects remote from the respiratory tract (systemic effects). Depending on the physicochemical properties of a volatile agent (e.g., water solubility, chemical reactivity, and blood-air partition) or the aerodynamic properties of particulates, within the respiratory tract the agent is deposited at different locations. In this context, the three regions of interest are the extrathoracic, the tracheobronchial, and the pulmonary. Respiratory toxicologists typically delineate these areas because of differences in size, structure, function, and cell types. The dose of agent absorbed is also contingent on the level of physical activity of the exposed individual. Active physical work or exercise (escape) increases the depth and frequency of breathing and therefore the total amount of toxic gas to which the lung is exposed. In addition, an increased heart rate increases blood flow which can lead to greater systemic absorption of inhaled gases. However, the factors governing systemic absorption and distribution in body tissues are beyond the scope of the chapter.

16.4.1 Acute Responses to Toxic Agents

Gases and vapors may be classified according to the nature of their toxic action. Such a classification is often more appropriate than one based on chemical properties per se. Gases as chemically diverse as nitrogen and methane have similar toxic actions where the toxic action of other, chemically similar gases such as NO and NO_2 differ considerably. Irritants produce inflammation in the mucus membranes with which they come in contact. They are sometimes subdivided into upper respiratory and pulmonary irritants.

Asphyxiants are materials that deprive the body of oxygen. Simple asphyxiants are physiologically inert gases that are present in the atmosphere in sufficient quantity to exclude an adequate supply of oxygen (*hypoxemic anoxia*). Chemical asphyxiants are materials that render the body incapable of utilizing an adequate supply of oxygen. They are thus toxic in concentrations far below the level needed for damage from simple asphyxiants. Two classic examples of chemical asphyxiants are CO and cyanide. CO interferes with the transport of oxygen to the tissues by its affinity for hemoglobin (*anemic anoxia*). Cyanide does not interfere with the transfer of oxygen to the tissues, but does alter cellular use of oxygen in energy production (*cytotoxic anoxia*). More examples are provided in the section addressing toxic endpoints. For general discussion of toxic principles of toxicant likely to occur in fires, see refs. [22, 54].

Particulates produced by combustion are commonly termed "smoke." More precisely, particulates consist of a variety of concentrated, visible aerosols formed in large part by condensation or supersaturated vapors. Smoke usually results from combustion of organic materials and may contain a variety of solids, liquids and gases. Owing to their high particle and gas concentrations, a dynamic equilibrium between the various phases may occur. An aerosol is an assembly of liquid or solid particles suspended in air long enough to enable observation and measurement. Especially in combustion toxicology the generation of particulates is likely to occur. For instance, due to the presence of soot, high concentrations of supersaturated vapors subjected to a temperature gradient may form condensation aerosols while cooling. Under such conditions, particle size is difficult to predict because of coagulation of aerosol and subsequent growth. Thus, the physical and chemical interactions between gases and particles are extremely complex and depend on a number of factors, including water vapor. Moreover, gases and vapors may be adsorbed onto the surface of aerosols or dissolved within droplets, rendering the prediction of the most critically exposed area within the respiratory system difficult. For example, polycyclic aromatic hydrocarbons (PAH) frequently occur on the surface of atmospheric particles. During coal combustion, a variety of trace metals (Zn, As, Sb) as well as sulfuric acid are concentrated on the surface of the ultrafine fraction of the aerosol. These particles are of particular toxicological importance because they are difficult to remove from the effluent and, when inhaled, penetrate to the sensitive pulmonary region of the lung. There, they may act locally, may accumulate in alveolar macrophages, or become dissolved in the lining fluids with rapid clearance from the lung (but into the systemic circulation).

16.4.2 Chronic Toxicity and Environmental Exposure of Persistent Agents from Smoke

In practically all fires involving natural products or plastics, a variety of products with high persistence in the environment may be formed. For instance, polycyclic aromatic hydrocarbons as well as polyhalogenated dibenzodioxins and -furans pertain to an ubiquitous class of chemicals produced during the combustion of fossil fuels. During a fire, temperatures are reached, which favor the generation of dioxins, if halogens, suitable carbon-containing molecules and/or traces of special heavy metals, are available. Natural products, including so-called halogen-free products, may also fulfill these conditions as they normally contain halogenated compounds as a result of impurities. Owing to the common toxic principle of the class of halogenated dioxins/furans/biphenyls, their concentrations are often expressed in terms of equivalents of 2,3,7,8-tetrachlorodibenzo-p-dioxin (TCDD). The ubiquitous release of such agents of high persistence in the environment has resulted in a widespread pollution with accumulation in adipose tissue in aquatic or terrestrial animal species and bioaccumulation/-magnification in the food chain. The health risk of chronic low-level exposure of human beings to these ubiquitous class of agents has been dealt with in detail elsewhere [22, 72].

16.5 Principles of Acute Combustion Toxicity Tests

16.5.1 Exposure Systems

Inhalation exposure systems are designed and operated to provide defined and analytically well characterized test atmospheres in the breathing zone of laboratory animals or human volunteers. In combustion toxicity testing, there are two basic types of exposure systems, based on how the test material is delivered: static, with no flow of test atmosphere into the chamber, and dynamic, with a single-pass flow-through system. Descriptions of the individual components of these systems are given in the following subsection.

Definition of Exposure Concentration. For an individual chemical agent, the acute toxic lethal potency is defined by the analytically determined concentration of the agent under consideration in the vicinity of the breathing zone of the test species. In contrast, in combustion toxicology, commonly "product-related" concentrations are used. These are either the "mass charge" or "mass loss" concentration, that is, the amount (mass), of a test specimen placed into a combustion device or consumed during combustion per unit of total volume of air of an inhalation chamber (static exposure) or passed through an inhalation chamber (dynamic exposure). Notwithstanding, "toxicity-related" concentrations are invariably expressed in terms of actual, analytically determined concentrations of individual components present in the breathing zone of the test animals.

Static Exposure Systems. In a static system, the test atmosphere is produced by introducing a finite amount of material into a closed exposure chamber. The test animals remain in this closed system for the duration of test without any replacement of air. The major advantage of a static system is that consumption of test material is minimal, and continuous generation of aerosols and gases is not required. The difficulties with this type of system are depletion of oxygen concentration, rising temperature and carbon dioxide, and decreasing concentration of test material. These factors limit the practical exposure duration to about 1 h. Additional short-comings of static systems are that the analytical characterization of chamber concentrations is usually limited because sampling affects the concentration of the material inside the chamber. The decay of toxicant concentration in the chamber is due to the deposition on surfaces by various aerosol-removing mechanisms, absorption of vapors on the surfaces, and deposition in the animals. Thus, especially in combustion toxicology studies, the particular disadvantages of static inhalation chambers are that the combustion atmospheres vary with time and that they cannot be diluted with secondary air to maintain viable conditions. This may, in certain circumstances, render the interpretation of the outcome of the test difficult because test results are predetermined by the vitiation of oxygen or high level of carbon dioxide – which also acts as a stimulus for an increase in RMV – rather than by toxic species considered to be „product-specific.“

Dynamic Exposure Systems. In dynamic systems, the test atmosphere is being continuously delivered to and exhausted from the animal exposure chamber in a flow-through manner; test material is not recirculated. After an initial rise, the chamber concentration will approach and maintain a stable equilibrium concentration if the generation rate is constant (Fig. 16.9). This stable concentration is usually reproducible after careful test runs have been made. Prediction of this equilibrium concentration requires accurate information on generation rate, losses of test material in various parts of the system, and flow rates. In the dynamic system, temperature and relative humidity are well regulated, oxygen is replenished/substituted in a controlled manner, and vapors exhaled from animals are removed continuously. The dose delivered to the

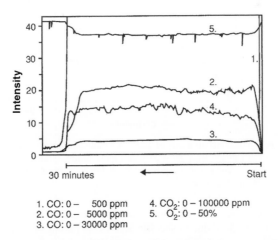

1. CO: 0 –   500 ppm   4. CO_2: 0 – 100000 ppm
2. CO: 0 – 5000 ppm   5. O_2: 0 – 50%
3. CO: 0 – 30000 ppm

Figure 16.9 Real-time monitoring of fire effluents and oxygen during exposure of experimental animals using the decomposition apparatus according to DIN 53436. Measurements were made in parallel using equipment of different sensitivities for CO, to quantify short-term fluctuations of CO most precisely. This illustration demonstrates that using the DIN 53436 apparatus temporally stable equilibrium concentrations are attained rapidly after the start of test.

animal is easier to determine from the stable airborne concentrations attained. Ideally, the exposure is to an inhalation chamber steady-state concentration that facilitates repeated measurements of gaseous agents, condensate, and soot.

Exposure Chambers. Experimental animals, depending on their number and size, are either exposed in an inhalation chamber (whole-body) or their head or nose protrudes into an inhalation chamber (nose-only). Either mode of exposure chambers can be operated using static or dynamic systems. Distinctive in context with combustion toxicity testing, whole-body chambers have several disadvantages because of the time to attain steady-state and the potential loss of selected agents from the exposure atmosphere with time (Fig. 16.10). In such chambers low- to semi-volatile agents (aerosols, condensate) may behave differently when compared to volatile or gaseous materials. The former may adsorb onto surfaces and disappear from the exposed atmosphere whereas volatile agents appear to dominate the outcome of test. Data depicted in Fig. 16.10 illustrate that in static whole-body inhalation chamber an increasing concentration of CO (generated by wood combustion) does not produce proportional increases of aerosol (TSP: total suspended particulate matter), demonstrating that an artificial fractionation of exposure atmosphere occurred. Thus, it appears that smoke and condensation aerosols exist in an equilibrium between production and condensation/adhesion onto the surface of the whole-body inhalation chamber. Conversely, when the mode of exposure is dynamic in a nose-only exposure chamber, an almost linear relationship of volatile airborne particulates and volatile agents or gases could be established (Fig. 16.10). Furthermore, when exposure is in larger whole-body chambers the time to attain the inhalation chamber equilibrium concentration has to be accounted for. Ideally, for all types of chambers, the equilibrium concentration should be attained shortly after commencement of exposure. By definition, „duration of exposure" is related to the time animals are exposed to the equilibrium concentration.

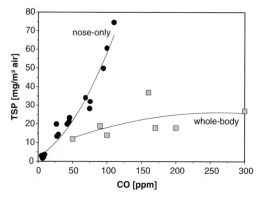

Figure 16.10   Relationship of concentrations of airborne volatile/gaseous (surrogate endpoint CO) and total particulate matter (TSP) in dynamic nose-only and quasi-static, whole-body inhalation chambers. The mismatch of volatile gases and particulates is considered to be a shortcoming in whole-body inhalation exposure chambers. (Reproduced from [73])

Further advantages of the nose-only, head-only, or head/nose-only modes of exposure are that a reduction of the exposure dose due to specific behavioral patterns (e.g., by burying the nose into the animals' fur to use it as a filter) or uncontrolled uptake of agents by non-inhalation routes via preening activities do not play any role. Furthermore, this mode of exposure allows physiological measurements, for example respiratory rate or hind-leg flexion, during exposure. An example of a dynamic, nose-only exposure system is depicted in Fig. 16.11. Using such system, controlled and highly reproducible steady-state concentrations of combustion atmospheres are produced and can aptly be used for both the nose-only exposure of rodents and the characterization by various analytical procedures. Owing to the requirements alluded to earlier, in combustion toxicity testing, a dynamic exposure system provides many advantages over the static system. Further details have been reviewed elsewhere [39, 74–76]. Accordingly, in combustion toxicology, the preferred mode of exposure is dynamic, nose-only. These considerations demonstrate that the primary objective of laboratory-scale test methods is to establish a robust and reproducible means to analyze and compare the relative acute toxic potencies of materials rather than establishing any true „simulation of selected human exposure scenarios".

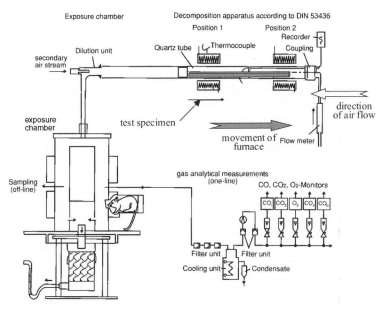

Figure 16.11 Decomposition apparatus according to DIN 53436 and nose-only inhalation exposure chamber. The test specimen fulfilling specified volumetric criteria is located in a boat within the quartz tube with counter current directions of the movement of oven and primary air flow. While moving the oven from position 1 to 2 (30 cm), the test specimen is combusted with 1 cm/min commonly using primary air flows in the range of 100–300 l/h. The reference temperature can be measured and recorded during the test. To prevent undue testing conditions occurring in the inhalation chamber with respect to temperature and oxygen vitiation, secondary air is used prior to entering the inhalation chamber. During the exposure of experimental animals, analytical characterization of the test atmosphere uses real-time monitoring of concentrations of the most salient fire effluents (see Fig. 16.9), including oxygen vitiation. Off-line sampling for specific agents, for example, airborne particulates (TSP), hydrogen cyanide, or halogenated agents.

16.5.2 DIN 53436 Tube Furnace and Nose-Only Exposure of Animals

Laboratory-scale tests using the DIN 53436 tube furnace [77] are amenable to a „rectangular" type of exposure pattern. As illustrated in Fig. 16.11, combustion atmospheres are generated under specified and reproducible conditions. To allow multiple analytical procedures to be used subsequently during the exposure of animals, attainment of a „steady-state" (Fig. 16.9) is an essential prerequisite. Prior to entering the inhalation chamber, atmospheres are diluted with air to allow testing of atmospheres in the absence of undue oxygen depletion, excessive levels of CO_2 or heat stress. Accordingly, this methodology is capable of measuring product-specific features without the overriding effects inherent with any smoke from fire. One additional advantage of this system is that analytical exposure indices can readily be used to predict the acute lethal toxic potency of the atmospheres generated.

16.5.3 Acute Inhalation Toxicity

Purpose of Test: The "intrinsic acute inhalation toxicity" is the adverse effect caused by substance following a single uninterrupted exposure by inhalation over a short period of time (24h or less) to a substance capable of being inhaled. At present, the evaluation of a chemical for its acute inhalation toxicity is focused on toxic lethal potency and is necessary to meet current classification and labeling requirements. For the testing of fire effluents, a fixed duration of exposure of 30 min is considered to be appropriate. Determination of acute toxicity is usually an initial step in the assessment and evaluation of the toxic characteristics of a substance that may be inhaled such as a gas, volatile substance, or aerosol/particle. An evaluation of acute toxicity data should include the relationship, if any, between the animals' exposure to the test substance and the incidence and severity of all abnormalities, including behavioral and clinical abnormalities, the reversibility of observed abnormalities, gross lesions, body weight changes, effects on mortality, and any other toxic effects. For chemical substances, internationally harmonized and standardized testing procedures are available (e.g., OPPTS 870.1300; OECD guideline 403 [18, 19]). Acute inhalation toxicity data from an acute study may serve as a basis for classification and labeling of individual chemical substances. In combustion toxicology, however, it has to be kept in mind that the product under consideration, for example, a piece of plastic or foam does in itself not contain any inherently acutely toxic substance. Depending on the conditions of thermal degradation, the quality and yield of a toxic species may change. Therefore, in combustion toxicology the assessment of the overall toxic hazard, this means the potential for harm resulting from exposure to toxic species of combustion in sufficient high concentration, is more important than the intrinsic toxic properties of a highly toxic substance in very low concentration.

In retrospect, the development of combustion toxicology testing procedures has probably suffered from being considered as part of "Fire Testing" rather than part of "Toxicology." The philosophies behind the development and, especially, the interpretation of test results in the two disciplines are different. In fire testing, a particular property is considered to be important in fire safety, for example, ability to ignite, surface spread of flame or smoke production. In mainstream toxicology, the nature of the toxicity, including analysis of the concentration-response relationship, is addressed. If it is considered that humans are likely to show similar adverse effects to those found in laboratory animals, then an assessment of the potential risk to health in any given situation can be made. This takes into account the likely human exposure and compares it with the effect levels seen in animals. Thus, a substance does not simply "pass" or "fail" its toxicology assessment, although some classification schemes do use quantitative toxicological data in such an unsophisticated way. From the regulatory standpoint, the question to be asked is not, "What is the LC_{50} of a material X, and is it below or above a value Y?" but rather, "Have the major toxicants been identified and does the projected use of the material represent a hazard to life?" This question must be answered by details of the likelihood of ignition, the speed of flame spread, the heat release, and the rate of buildup of smoke and toxic gases, including oxygen depletion. Thus, it can be seen – provided a material does not produce unknown or extraordinary toxicants and its decomposition process is understood – that the true "toxicity test" for toxic hazard may actually be an ignitability or surface spread of flame test.

Toxic Potency. Toxicity of the smoke from a specimen of material or product is based on a per-unit-specimen mass relation. When the testing is for a lethality end-condition, the toxic potency is determined as an LC_{50}. The unit is in concentration, that is, mg/l of air. It must be emphasized, however, that toxic potency is not an inherent property of a material. The LC_{50} is a variable that depends on the test animal, the test apparatus, the specimen combustion condition, and the length of the exposure time. The latter is particularly important, as for many substances the value of the LC_{50} tends to be inversely dependent on the exposure time. For fire toxicity, a 30 min exposure time (along with a 14-day post-exposure observation period) has often been specified.

Toxic Fire Hazard. This term is a subset of "fire hazard," occurring when the hazard being considered is due to inhalation of toxic combustion products alone. Toxic potency is one, but not the only, factor in toxic fire hazard. Mass loss rate and flame spread rate, among other factors, also play a dominant role.

16.5.4 Calculation of the Median Lethal Concentration (LC_{50})

For each substance, a dosage- or concentration-effect relationship exists, which is assumed to be characteristic for a specific effect and species. To quantify this relationship, the term "minimum lethal dose" was originally suggested as a measure of toxicity, but it was subject to differing interpretations. Trevan (1927) [78] reviewed the ambiguous nature of this term and suggested adoption of the "median lethal dose" or LD_{50} as the unit of acute toxicity. The median lethal dose was defined as the dose that kills half of a suitably large number of subjects. For an inhalation study, the "median lethal concentration" or LC_{50} for a specified duration of exposure is the equivalent.

Determination of the LC_{50} requires a mathematical description of the concentration-effect curve, but dealing with a sigmoid curve in this manner is cumbersome and complex. Therefore, in most instances, a probit analysis of lethality data is considered most appropriate. In brief, the concentration-effect curve assumed the character of an exponential function, easily converted to linearity by considering the concentration in terms of its logarithm. Bliss explicitly described the transformation of the dosage-effect curve into a linear function [79] and also explained its derivation in a situation where only a limited amount of experimental data is available [80]. Once the dosage-effect curve has been determined, the calculation of the LC_{50} and its error is possible [81–83]. Other mathematical transformations that have been employed to linearize the concentration-effect curve include the use of the logistic function [84], angular transformation [85], and moving averages and interpolation [86]. Algorithms for computerized solutions using only a few animals per group are published [83–87]. One example of an LC_{50} determination (SO_2, nose-only exposure of rats for 30 min) is depicted in Fig. 16.12. As alluded to in the preceding, the type of onset of mortality is as important to know as reducing the result of a complex test to a single numerical value, the LC_{50}.

Thus, with respect to endpoints, bioassays addressing all aspects of concern are not attainable at the present time because of two major limitations of test methods. The first limitation is the inability to generate, in the laboratory, a "highly standardized" smoke that is indeed representative of that produced in real fires. Because of the continuously varying factors of heat, ventilation, and fuel in actual fires, caution is advised to extrapolate directly laboratory-scale methodologies to real fire situations. However, results of recent verification tests with full-scale fires have indicated that laboratory methods may be capable of generating smoke with comparable toxicity to that produced in certain stages of actual fires. The second limitation in the use of laboratory tests is the relevancy and validity of presently used test animals for complex behavioral tests that attempt to quantify the impairment of escape. For the latter it has been proposed to use one third of the LC_{50} to estimate the respective effective concentration (EC_{50}) that incapacitates 50 % of animals (ISO/TR 9122/4 [5]). This estimation, however, is based mainly on impairment of neuromuscular function.

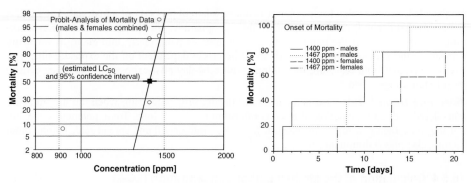

Figure 16.12   Left panel: Bliss-corrected mortality as a function of "breathing-zone" exposure concentration of sulfur dioxide (rats, nose-only exposure to SO_2, duration of exposure 30 min). Right panel: Time-related cumulative mortality occurring in a delayed fashion as a result of *bronchiolitis obliterans*

16.6 Prediction of the Actual Lethal Potency of Fire Effluents

Although smoke is a complex mixture of airborne solid, liquid particles, and gases evolved when a material undergoes pyrolysis or combustion, approaches have been made to predict effects from smoke from fires. It is desirable to model, mathematically, the effects caused by common toxicants from fire, and therefore obviate the use of large numbers of laboratory animals in smoke toxicology testing. Accordingly, today, the use of animal bioassays to evaluate the relative toxic potency of combustion products from new materials has decreased to a minimum. Prevailing experimental data collated from many experimental laboratory tests in the past [25, 88–95] provide ample evidence that despite variability of materials, the number of toxic species demonstrating an appreciable acute health hazard is relatively limited. Depending on the chemical elements contained in the product, the main toxic smoke components are CO, CO_2, SO_2, HCN, HCl, (HF, HBr), NO_X, or ammonia (NH_3) often in combination with atmospheres depleted of oxygen or enriched with CO_2. More than 100 chemical species have been identified. In addition, the smoke evolved is highly dependent on the combustion conditions; thus, the study of smoke toxicity is much more difficult than the study of conventional inhalation toxicity of one or more specific chemicals. Therefore, comparison of predicted data from the above mentioned main components by mathematical modeling with observed data from integrated bioassays provide a scientifically sound means to judge as to whether the great number of minority components contained in smoke bear indeed any acute threat to health.

Laboratory toxicity tests have the capability of serving an extremely important function, namely the identification of materials that produce unusually or, more commonly, unexpected toxic smoke, that is, toxic fire effluents not explicitly taken into account at the outset of study. Accordingly, neither the strategy devised for the analytical sampling of aerosols and volatile agents in exposure atmospheres nor the analytical procedure itself may specifically address such potentially toxic chemical species. In order to accomplish the objectives of a laboratory combustion toxicity test, a test method that addresses toxic endpoints should have five basic components. These are:

- A combustion device for generating variable steady-state concentrations of smoke for a duration long enough to allow both the application of reliable analytical methods and animal exposure,
- An exposure chamber for exposure of test animals to the smoke, if considered to be appropriate for validation/confirmatory purposes,

- A method for measurement and quantification of toxicity,
- State-of-the-art analytical characterization of exposure atmospheres that show the type of chemicals and concentrations and physical aggregates to which animals are exposed,
- An integrated assessment of animal exposure data, such as clinical findings, lethality, and toxic species in blood, and analytical data to model the toxicological effects of concern likely to occur.

The requirements for each component that will accomplish the objectives stated above are discussed in the various chapters of this book.

The first step in the modeling of toxicological effects is to determine what constitutes the exposure dose associated with a given response, for example, lethality caused by asphyxiants or irritants, incapacitation, and so forth. Furthermore, it is recommended to determine the dependence of that effective exposure dose on the concentration of a toxicant, since Haber's Rule may not be valid over the range of concentration (or duration of exposure) of interest. This is done from a concentration time-response data base for the toxicant being studied. It has been found that, in general, the exposure dose required to cause a particular response decreases with increasing concentration of a toxicant [23, 34, 36, 96, 97].

Once effective exposure doses are characterized, extension of the concept to that of a fractional effective dose (FED) and the fractional effective concentration (FEC), along with the summation or integration of FEDs or FECs, result in workable tools in combustion toxicology [17, 96]. From toxicant concentration as a function of time data, incremental exposure doses are calculated and related to the specific Ct exposure dose required to produce the given toxicological effect at each particular incremental concentration. Thus FED is calculated for each small time interval. Continuous summation of these FEDs is carried out and the time at which this sum becomes unity (100 %) represents the time at which 50 % of the exposed subjects would be expected to incur the toxicological effect.

For asphyxiants, the overall principle of the model in its simplest form is shown in Equation (16.1):

$$FED = \sum_{i=1}^{n} \sum_{t_1}^{t_2} \frac{C_i}{(Ct)_i} \Delta t \tag{16.1}$$

where C_i is the average concentration of an asphyxiant gas "i" over the chosen time increment, Δt, in ppm and $(Ct)_i$ is the specific exposure dose in ppm $\cdot$ min that would prevent occupants' safe escape. FEDs are determined for each asphyxiant at each discrete increment of time. The time at which their accumulated sum exceeds a specified threshold value represents the time available for escape relative to chosen safety criteria. An expanded form of Equation (16.1) is shown as Equation (16.2), where [CO] = average concentration of CO (ppm) over the time increment, t [HCN] = average concentration of HCN (ppm) over the time increment, t (time increment (min)).

$$FED = \sum_{t_1}^{t_2} \frac{[CO]}{35000 ppm \cdot min} \Delta t + \sum_{t_1}^{t_2} \frac{\exp([HCN]/43)}{220 min} \Delta t \tag{16.2}$$

The effects of sensory/upper respiratory and, to some extent, pulmonary irritation may be assessed using the FEC concept shown in Equation (16.3). As a first-order assumption, direct additivity of the effects of the different irritant gases is employed. It is also assumed that the concentration of each irritant gas reflects its presence totally in the vapor phase. FECs are determined for each irritant at each discrete increment of time. The time at which their sum exceeds a specified threshold value represents the time available for escape relative to chosen safety criteria.

$$FEC = \frac{[HCl]}{F_{HCl}} + \frac{[HBr]}{F_{HBr}} + \frac{[HF]}{F_{HF}} + \frac{[SO_2]}{F_{SO_2}} + \frac{[NO_2]}{F_{NO_2}} +$$
$$\frac{[acrolein]}{F_{acrolein}} + \frac{[formaldehyde]}{F_{formaldehyde}} + \sum \frac{[Irritant]_i}{F_{C_i}}$$

(16.3)

where:

[ ] = irritant gas concentrations in ppm,
F   = irritant gas concentrations in ppm that are expected to seriously compromise occupants' ability to take effective action to accomplish escape.

The following LC_{50} values are used (rats, duration of exposure 30 min): CO 5500 ppm, HCN: 170 ppm, HCI: 3800 ppm, NO_2: 120 ppm, SO_2: 1400 ppm [35, 97, 98]. For NO_x, the assumption was made that the LC_{50} of NO is equal NO_2 which is a more conservative approach.

Example calculations for synthetic materials releasing HCN or sulfur dioxide as combustion gases of concern are summarized in Table 16.2. For each pathomechanism likely to occur (asphyxiation, irritancy) a separate LC_{50} is calculated. Moreover, based on the allometric relationship of inhalation chamber concentrations of CO and blood COHb level (see Fig. 16.7), actually observed and predicted COHb can readily be compared to judge the "quality" of prediction.

Table 16.2   Prediction of acute lethal toxic potency of plastic materials, one releasing hydrogen cyanide, the other sulfur dioxide

Substance: Plastic A (HCN release)		
Primary flow-l/h	100.00	(through furnace)
Sec. flow-l/h	700.00	(dilution)
Nomin.-conc.- mg/l	24.50	(furnace load)
% combusted	97.50	
% O_2	18.00	
CO_2	20,000.00	ppm
CO	1400.00	ppm
HCN	130.00	ppm
NH_3	0.00	ppm
NOx	7.00	ppm
SO_2	0.00	ppm
HCl	0.00	ppm
% COHb	38.00	
% Mortality	20.00	
CO-calculated	1556.50	ppm (based on COHb)
COHb-calculated	35.74	% (based on CO)
CO_2/CO-ratio	14.29	

Table 16.2   (Continuation)

Substance: Plastic A (HCN release)		
Modeling of acute lethal toxic potency		
CO_2	808.08	ppm/mg comp./l air
CO	56.57	ppm/mg comp./l air
HCN	5.25	ppm/mg comp./l air
NH_3	0.00	ppm/mg comp./l air
NOx	0.28	ppm/mg comp./l air
SO_2	0.00	ppm/mg comp./l air
HCl	0.00	ppm/mg comp./l air
LC_{50}-asphyxiants	24.28	mg/l
LC_{50}-irritants	424.29	mg/l
LC_{50}-observed	26.00	mg/l

Substance: Plastic B (Sulfur dioxide release)		
Primary flow-l/h	100.00	(through furnace)
Sec. flow-l/h	1500.00	(dilution)
Nomin.-conc.-mg/l	18.50	(furnace load)
% combusted	30.80	
% O_2	19.50	
CO_2	7000.00	ppm
CO	450.00	ppm
HCN	0.00	ppm
NH_3	0.00	ppm
NOx	3.00	ppm
SO_2	1167.00	ppm
HCl	0.00	ppm
% COHb	19.00	
% Mortality	10.00	
CO-calculated	468.98	ppm (based on COHb)
COHb-calculated	18.56	% (based on CO)
CO_2/CO	15.56	
Modeling of acute lethal toxic potency		
CO_2	378.38	ppm/mg comp./l air
CO	24.32	ppm/mg comp./l air
HCN	0.00	ppm/mg comp./l air
NH_3	0.00	ppm/mg comp./l air
NOx	0.16	ppm/mg comp./l air
SO_2	63.08	ppm/mg comp./l air
HCl	0.00	ppm/mg comp./l air
LC_{50}-asphyxiants	226.11	mg/l
LC_{50}-irritants	21.55	mg/l
LC_{50}-observed	23.00	mg/l

Comp.: furnace mass load of product tested

Predictions obtained with smoke atmospheres containing CO and HCN using the Finney equation [99] were compared with the NIST N-Gas model [88] (Fig. 16.13). Opposite to the model devised, the NIST N-Gas model takes also into account the additional effects attributable to O_2 depletion and high CO_2 concentrations more likely to occur in the static NIST test chamber than under conditions present using the DIN 53436 method. Owing to the dynamic mode of exposure and the use of secondary dilution air, such corrections appear not necessary when using the DIN 53436 method. Using this method, the ratio of concentration of carbon dioxide and carbon monoxide (CO_2/CO) was close to 10 or smaller, that is, these conditions resemble more the characteristic of a ventilation-controlled condition. Low ratios of CO_2/CO, usually occurring under O_2 vitiated conditions, favor the formation of HCN whereas at higher ratios oxidized species of nitrogen, for example, NO and NO_2, are observed.

The model prediction is based on the following empirical mathematical relationship shown in Equation (16.4)

$$\text{FED} = \frac{m\,[\text{CO}]}{[\text{CO}_2] - b} + \frac{[\text{HCN}]}{d} + \frac{21 - [\text{O}_2]}{21 - \text{LC}_{50}\text{O}_2} \tag{16.4}$$

The terms m and b define empirically the synergistic interaction of CO_2 and CO and equal 18 and 122,000 if the CO_2 concentrations are 5 % or less. For studies in which the CO_2 concentrations are above 5 %, m and b equal 23 and 38,600, respectively. The term d is the LC_{50} value of HCN and is in the N-Gas model used for comparison 110 ppm for 30 min exposures plus 14-day post-exposure deaths. (Note: In more recent modifications of this model, a LC_{50} value of HCN of 150 ppm is proposed [94]). Furthermore, this model has been extended to modeling of seven gases, including HBr and nitrogen dioxide [25]. Ideally, when the N-Gas value equals one, 50 % of the animals should die. Examination of animal lethality data for the three and four gas combinations indicate that the mean N-Gas value where animal deaths occur is 1.1 with a standard deviation of ± 0.1. The authors found in the pure gas work that one half of the animals are likely to die when the N-Gas value is approx. 1.1, no animals usually die below 0.9 and all the animals usually die above 1.3. The N-Gas model contains empirically determined parameters related to the NIST Toxicity Test Method (static exposure of six male rats, duration of exposure 30 min).

From experimental data using the DIN 53436 method, three approaches were used to analyze the data (Fig. 16.13) [98]. The first considered the LC_{50} prediction based on CO and HCN or SO_2 and NO_x assuming no additive interaction between asphyxiants and irritants. In the second, the LC_{50} prediction was based on CO, HCN or SO_2 and NO_x assuming additivity, and in the third, LC_{50} values were predicted according to the N-Gas model which takes into account CO, HCN, O_2 depletion, and CO_2. Data summarized in Fig. 16.13 resulted in a fairly good correlation. The Finney equation revealed to be useful to predict the lethal toxic potency of materials releasing CO and HCN as well as SO_2. The best prediction was achieved when toxic potencies of asphyxiants and irritants were additively combined [slope = 0.96, $r = 0.82$, Fig. 16.13(b)]. The correlation was insignificantly better when each mode of action (asphyxia and irritation) was not considered to be additive [slope = 1.12, $r = 0.88$, Fig. 16.13(a)]. This approach, however, resulted in predicted LC_{50} values higher than the observed ones demonstrating that toxicity is dependent on the additive action from either class. In some experiments from the historical database, predicted LC_{50} values appeared to be considerably smaller than the observed ones. This result prompted further experiments to assess the selectivity of the analytical method used for HCN determinations. More recent experimental evidence suggests that indicator tube or electrochemical procedures for the determination of HCN do not provide the required selectivity for HCN and chamber HCN concentrations may be significantly overestimated using that method, particularly when HCl and/or SO_2 is present in the atmosphere.

This experimental error was eliminated by using a gas chromatographic method for HCN determinations.

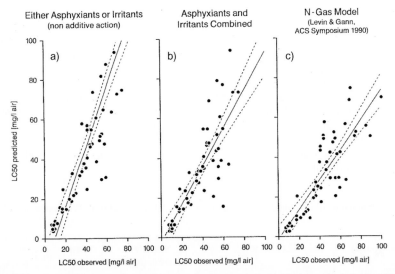

Figure 16.13   Predicted versus observed LC_{50} values using the DIN 53436 method. Dashed lines represent the 95 % interval; (a) LC_{50} prediction was based on CO and HCN or SO_2 and nitrogen oxides (NO_x). (b) LC_{50} prediction was based on CO, HCN or SO_2, and NO_x. Thus, in (a) the predicted LC_{50} assumed an independent mode of action (no additivity) between the asphyxiants and irritants while in (b) a dependent mode of action (all agents combined) was assumed. (c) LC_{50} prediction according to the N-Gas model which takes into account the concentrations of CO, HCN, CO_2, including O_2 depletion

The comparison of the LC_{50} values obtained by the N-Gas model also resulted in a reasonably good correlation [slope = 0.74, r = 0.85, Fig. 16.13(c)]. The shallower slope indicates that predicted LC_{50}s are smaller than observed ones. This conservative nature of the NIST N-Gas model most likely is attributable to the lower LC_{50} values for CO and in particular HCN used in this model (LC_{50} values used for HCN: FED/Finney model = 170 ppm, N-Gas model = 110 ppm). Also the use of male rats only in the NIST protocol might have contributed to the more conservative nature of the N-Gas model. In none of the studies conducted clinical observation of animals revealed unusual toxic effects (e.g., neurotoxicity) [95].

16.7 Toxic Potency of Materials and Toxic Hazard from Fire Situations

The toxic potency is dependent on the intrinsic toxic properties of the incriminated agent while the toxic hazard also takes into account the exposure intensity of a potentially toxic product released by thermal decomposition of a given material. In combustion toxicology, toxic potency data are derived from short-term inhalation tests with a duration of exposure of small laboratory animals of usually 30 min. Endpoints that express the acute toxic potency are discussed in the preceding. Because of its robustness, the endpoint most commonly used is the median acute lethal concentration (LC_{50}). Materials are decomposed using a temperature range of 350–600°C and after the dilution, the decomposition products are passed to the inhalation exposure chamber. Decomposition products may not necessarily generate the quantity (yield) and quality (toxic agent produced) of fire effluents generated under real-life conditions. Therefore, relative toxic potencies obtained in small-scale tests may differ significantly when compared to large-scale ones. Therefore, product-specific default values for extrapolation from small- to large-scale tests are usually not recommended and caution is advised when so doing.

Several laboratory-scale methods have been used for inhalation toxicity studies (Table 16.3). One of the methods most often used are the static radiant heat smoke toxicity NIST method [94] and the dynamic tubular furnace DIN 53436 method. Briefly, these methodologies can be described as follows: in the NIST method a sample approximately one forth the largest sample permissible or 10g (whichever is smaller) is exposed to the radiant heat flux for 15 min in a 200 l chamber. Concentration of the principal gaseous smoke components (CO, CO_2, HCN, HCI , HBr, reduced O_2) are measured for 30 min. Based on the concentration of the measured gases and the N-Gas model, the mass of material necessary to produce an LC_{50} is evaluated. This value can be calculated for death that occur during the 30-min exposure and for the 30 min plus 14 day post-exposure observation period for both the mass of the material loaded into the furnace or the mass consumed. For more details of this method, see Table 16.3 and Kaplan et al. (1983) [40].

Table 16.3   Internationally standardized decomposition models (adapted from Wittbecker et al. 1999 [96])

Test (standard) (S – static D – dynamic)	Thermal impact	Sample dimensions (mm)	Sample volume (cm^3)	Test duration (min)
ASTM E662 (S) NBS chamber	25 kW/m^2 with and without pilot flame	76 – 76 ≤ 25	144	6 20
ISO 5659 (S) NBS modified	25 kW/m^2 with and without pilot flame	76 – 76 ≤ 25	144	20
ASTM D 2843 (S) XP2 chamber ÖN B3800	Burner	30 – 30 4 60 – 60 25	3.6 90	≤ 15
GOST (S) 12.1044.89	Radiant panel 400–700 °C	40 – 40 ≤ 10	16	
IEC/TC 89 (S) cable test	Fuel ignition source	300 300 – 2 – 25 2 – 400 – 4 0 – 25 1000 – 50 0 – 2 – 4	225/1 80	2
ISO 5924 (S) ISO Smoke Box	10 kW/m^2 50 kW/m^2	165 – 165 – < 70 165 – 165 – < 70	< 190 6	20
DIN 53436 (D)	100–900 °C	270 – 5 – 2	2.7	30

Table 16.3 (Continuation)

Test (standard) (S – static D – dynamic)	Thermal impact	Sample dimensions (mm)	Sample volume (cm³)	Test duration (min)
NF F 16–101 (D)	100–900 °C	270 – 5 – 2	2.7	10 – 15
DIN 54837 (D)	Burner	500 – 190 – d	190	5
DIN 4102-15 Brandschacht	Burner	1000 – 190 – 80	60800	10
ORE 14	Radiant panel/burner	50 cm² – 2 mm	10	5
DIN 4102 Part 14	Radiant panel and flame	1050 – 230 – d		12 – 30
ISO 5660 Cone Calorimeter	Radiant cone	100 – 100 – ≤ 50	500	60

The method and apparatus used according to DIN 53436 are very suitable for bioassays because this method provides a steady-state concentration of 30 min (Fig. 16.9) and tenable concentrations of oxygen and carbon dioxide. All exposures summarized in Fig. 16.13 were conducted using the DIN 53436 combustion apparatus and a nose-only animal exposure system as depicted in Fig. 16.11. The animal exposure chamber volume was approx. 10 l. Depending on the density of the test specimen, the tube was charged with 0.3-0.45 ml specimen/cm, total length 30 cm (11.8 in.). Specimens undergoing flaming combustion can be tested in a segmented form to prevent an uncontrolled flame propagation within the tube. The rate of annular furnace travel is 1 cm/min (0.39 in./min), and the direction was countercurrent to that of the primary air flow. The lead time (annular furnace in operation but without exposure of the rats) is 10 min. A reproducible decomposition and/or combustion condition is ensured throughout the animal experimental study to permit material-specific decomposition. Conventionally, the temperature for decomposition is selected on the basis of maximum yield of most toxic decomposition products rather than attempting to simulate a specific fire type. Owing to the countercurrent principle of furnace and air movements a reproducible inhalation chamber equilibrium concentration can be achieved both under "flaming" and "non-flaming" decomposition. Conventionally, the primary air flow through furnace is 100 l/h. A different inhalation chamber concentration is attained by a secondary dilution (see Fig. 16.11).

16.8 Combustion Toxicology and Clinical Experience

Fire survivors show signs of narcosis including dissociation, faintness, and loss of consciousness, and have CO and cyanide in their blood [44]. They also suffer from eye irritation and painful difficult breathing, leading to breath holding, and faintness when a breath is taken. However, the irritant and narcotic effects of smoke vary between different types of fires and different individuals. In some cases, thick smoke is found to be mildly irritant but breathable, while in others it has been reported that a single breath resulted in faintness or chest pain. In some cases, it is possible to relate the onset of narcosis after several minutes exposure to the gradual loading of narcotic gases, while in other cases dead and unburned victims have been discovered seated at tables with food, apparently having been overcome so rapidly that they were unable to make a move, a phenomenon not easily explainable in terms of classical toxicology [68].

Fire survivors who have breathed smoke often go through a respiratory crisis some hours after exposure, with a low partial pressure of arterial oxygen, due to a transient pulmonary edema (a phenomenon which is also common in toxicology studies). There is also pathological evidence for "chemical burns" (lung inflammation), although the picture may be complicated by the presence of thermal respiratory tract or body surface burns.

A difficulty in understanding human fire deaths is that for nonburned victims the mean COHb concentration (approximately 50 % COHb) is in general similar to that found in nonfire deaths from CO poisoning (from poorly vented heaters and car exhaust suicides [68, 101], whereas toxicology studies clearly show that the various narcotic and irritant gases are apparently additive in causing death. Of course, these sources also contain other toxins besides CO, and it is possible to identify many cases where smoke deaths have occurred in the presence of sublethal COHb levels. One problem may be that COHb is the only toxicant simply and routinely measured and, as alluded to above, HCN is subject to marked changes in cadavers or upon the time of storage of blood.

16.9 Future Directions

There is a thrust for the advancement of internationally recognized standards intended to address the consequences of human exposure to the life threat components of fire *at the position* or *location* of a building's occupants, rather than within the environment generated in the immediate vicinity of a fire. Both people movement and smoke movement, along with resulting changes in thermal conditions and toxicant concentrations with time and distance, must all be considered to assess actual exposure to occupants. Such standards should provide procedures for evaluation of the life threat components of fire hazard analysis in terms of the status of exposed human subjects at discrete time intervals, allowing for an assessment of their ability to escape safely. The life threat components to be addressed should include fire effluent toxicity, heat, and visual obscuration due to smoke. The fire effluent toxicity component should consider the effects of the asphyxiant toxicants as well as the effects of both sensory/upper respiratory irritation and pulmonary irritation and their impact on escape.

In light of current knowledge, this appears to be hampered by the (to some extent limited) knowledge of the relationships between lethality or sublethal physiological/psychological (behavioral) effects and exposure to heat, thermal radiation, narcotic and irritant gases, aerosols, and their combinations. To achieve the aforementioned objective, for the various constituents of fire effluents, the kinetics of transport and decay must be known to arrive at data amenable for risk calculation.

This risk calculation makes it necessary to refer to internationally recognized acute short-term exposure indices for hazardous substances. In this context, exposure duration as short as 10 min should be considered. It appears that development of Acute Exposure Guideline Levels (AEGLs) [21, 102] provides adequate socially accepted short-term, high-level exposure data amenable for such risk calculation. AEGLs represent ceiling exposure values for the general public and are applicable to emergency exposure periods ranging from 10 min to 8 h. Three AEGLs are developed for each of five exposure periods (10 and 30 min, 1, 4, and 8 h) and will be distinguished by varying degrees of severity of toxic effects. The three AEGLs have been defined as follows [21]:

- *AEGL-3* is the airborne concentration (expressed as ppm and mg/m^3) of a substance at or above which it is predicted that the general population, including "susceptible" but excluding "hyper susceptible" individuals, could experience life-threatening effects or death.

- *AEGL-2* is the airborne concentration (expressed as ppm and mg/m³) of a substance at or above which it is predicted that the general population, including "susceptible" but excluding "hyper susceptible" individuals, could experience irreversible or other serious, long-lasting effects or impaired ability to escape.
- *AEGL-1* is the airborne concentration (expressed as ppm and mg/m³) of a substance at or above which it is predicted that the general population, including "susceptible" but excluding "hypersusceptible" individuals, could experience notable discomfort. Airborne concentrations below AEGL-1 represent exposure levels that could produce mild odor, taste, or other sensory irritation.

References

[1] A. Tewarson, *Toxicology*, 115 (1996) p.145–156
[2] ISO/TR 9122–1:1989. Toxicity testing of fire effluents – Part 1: General
[3] ISO/TR 9122–2:1993. Toxicity testing of fire effluents – Part 2: Guidelines for biological assays to determine the acute inhalation toxicity of fire effluents (basic principles, criteria and methodology)
[4] ISO/TR 9122–3:1993. Toxicity testing of fire effluents – Part 3: Methods for the analysis of gases and vapours in fire effluents
[5] ISO/TR 9122–4:1993. Toxicity testing of fire effluents – Part 4: The fire model (furnaces and combustion apparatus used in small-scale testing).
[6] ISO/TR 9122–5:1993. Toxicity testing of fire effluents – Part 5: Prediction of the toxic effect of fire effluents
[7] ISO/TR 13387–1:1999. Fire safety engineering – Part 1: Application of fire performance concepts to design objectives
[8] ISO/TR 13387–2:1999. Fire safety engineering – Part 2: Design fire scenarios and design fires
[9] ISO/TR 13387–3:1999. Fire safety engineering – Part 3: Assessment and verification of mathematical fire models
[10] ISO/TR 13387–4:1999. Fire safety engineering – Part 4: Initiation and development of fire and generation of fire effluents
[11] ISO/TR 13387–5:1999. Fire safety engineering – Part 5: Movement of fire effluents
[12] ISO/TR 13387–6:1999. Fire safety engineering – Part 6: Structural response and fire spread beyond the enclosure of origin
[13] ISO/TR 13387–7:1999. Fire safety engineering – Part 7: Detection, activation and suppression
[14] ISO/TR 13387–8:1999. Fire safety engineering – Part 8: Life safety – occupant behaviour, location and condition
[15] ISO 13344:1996. Determination of the lethal toxic potency of fire effluents
[16] ISO 13943:2000. Fire Safety – Vocabulary
[17] ISO/Proposed Draft TR 13571 (2001): Life threat from fires – guidance on the estimation of the time available for escape using fire data
[18] OPPTS 870. 1300 (1998). US-EPA Health Effects Test Guidelines 870. 1300 – Acute Inhalation Toxicity. United States Environmental Protection Agency, Office of Prevention, Pesticides and Toxic Substances, EPA 712C-98–193, August 1998
[19] Organization for Economic Cooperation and Development (OECD) (1981). Guideline for Testing of Chemicals No. 403. "Acute Inhalation Toxicity", adopted May 12, 1981
[20] EEC Directive 92/69/EWG. Journal of the European Community – Legal Specifications L 383A, <u>35</u>, December 29, 1992. B. 2. Acute Toxicity-Inhalation. p. 121
[21] AEGL (1999). Standing Operating Procedures for the Developing Acute Exposure Guideline Levels for Hazardous Chemicals. National Academy Press, Washington, DC (2001)
[22] C. Klaassen (Ed): Casarett & Doull's Toxicology – The Basic Science of Poisons, McGraw-Hill, New York 1995
[23] G.E. Hartzell, A.F. Grand, W.G. Switzer, In: G. L. Nelson (Ed.), Fire and Polymers: Hazards Identification and Prevention. ACS Symposium Series 425, American Chemical Society, Washington, DC, p. 12-20, 1990
[24] M. Matijak-Schaper, Y. Alarie, *Combust. Toxicol.* (1982) 9, p. 21–61
[25] B.C. Levin, *Drug Chem. Toxicol.* (1997) 20, p. 271–280

[26] M.E. Andersen, *Inhal. Toxicol.* (1995) 7, p. 909–915
[27] A.M., Jarabek, *Inhal. Toxicol.* (1995) 7, p. 927–946
[28] W.E. Rinehart, T. Hatch, *Ind. Hyg. J.* (1964) 25, p. 545–553
[29] S.C. Packham, Quantitating dose in combustion toxicology, ISO/TC92/SC3/WG-3, 1984
[30] W.F. Ten Berge, A. Zwart, L.M. Appelman, *J. Hazard Materials* (1986) 13, p. 301–309
[31] P.M.J. Bos, A. Zwart, P.G.J. Reuzel, P.C. Bragt, *Crit. Rev. Toxicol.* (1992) 21, p. 423-450
[32] G.D. Nielsen, Y. Alarie, *Toxicol. Appl. Pharmacol.* (1982) 65, p. 459-477
[33] M. Schaper, *Amer. Ind. Hyg. Assn. J.* (1993) 54, p. 488–544
[34] G.E. Hartzell, W.G. Switzer, D.N. Priest, *J. Fire Sci.* (1985) 3, p. 330-342
[35] G.E. Hartzell, S.C. Packham, A.F. Grand, W.G. Switzer, *J. Fire Sci.* (1985) 3, p. 196-207
[36] G.E. Hartzell, H.W. Stacy, W.G. Switzer, D.N. Priest, S.C. Packham, *J Fire Sciences,* (1985) 3, p. 263–279
[37] R.D. Stewart, J.E. Peterson, T.N. Fisher, M.J. Hosko, E.D. Baretta, H.C. Doff, A.A. Herrmann, *Archives of Environmental Health* (1973) 26, p. 1–7
[38] J. *Pauluhn,* In: Use of mechanistic information in risk assessment, H. M. Bolt, B. Hellman, L. Dencker (Eds.), *EUROTOX proceedings, Arch. Toxicol. Suppl.* (1994) 16, p. 77–86
[39] J.W.Klimisch, JDoe, G.E. Hartzell, S.C. Packham, J. Pauluhn, D.A. Purser, *Fire Sciences* (1987) 5, p. 73-104
[40] H.L. Kaplan, A.F. Grand, G.E. Hartzell, Combustion Toxicology: Principles and Test Methods. Technomic Publishing, Lancaster, PA, 1983
[41] S.C. Packham, Behavior and Physiology: Tools for the assessment of relative toxicity. Paper presented at 17th National Cellular Plastics Conference on Safety and Product Liability, Washington D C (1974)
[42] T. Sakurai, *J. Fire Sci.* (1989) 7, p. 22–77
[43] C.R. Crane, D.C. Sanders, B.R. Endecott, *J. Fire Sci.* (1992) 10, p. 133–159
[44] D.A. Purser, Toxicity assessment of combustion products. In: C. L. Beyler (Ed.), SFPE Handbook of Fire Protection Engineering. Part 1. National Fire Protection Association, Quincy, MA, p. 200-245, 1988
[45] ASTM (1984). Standard Test Method for Estimating Sensory Irritancy of Airborne Chemicals. ASTM Targetation: E 981–84. American Society for Testing and Materials, Philadelphia, PA
[46] Y. Alarie, *Arch. Environ. Health* (1966) 13, p. 433–449
[47] Y. Alarie, *Crit. Rev. Toxicol.* (1973) 2, p. 299
[48] Y. Alarie, *Environ. Health Perspect.* (1981) 42, p. 9–13
[49] Y. Alarie, *Adv. Environ. Toxicol.* (1984) 8, p. 153-164
[50] Y. Alarie, J.E. Luo, In: Toxicology of the Nasal Passages, C. E. Barrow (Ed.), Hemisphere Publishing Corp. , Washington, New York, London, p. 91–100, 1986
[51] J. Pauluhn, *Toxicol. Lett.* (1996) 86, p. 177–185
[52] J. Pauluhn, Respiratory Toxicology and Risk Assessment: Species Differences: Impact on Testing. IPCS Joint Series No. 18, Wissenschaftliche Verlagsgesellschaft mbH Stuttgart, p. 145–1173, 1994
[53] S.D. Christian, Thesis: Factors affecting the life threat to aged persons in domestic dwelling fires. South Bank University, 1993
[54] G. Kimmerle, JFF/Combust. Toxicol. (1974) p. 4–51
[55] IRK/AGLMB Richtwerte für die Innenraumluft: Kohlenmonoxid. Bundesgesundheitsblatt p. 425–428, 1997
[56] B. Ballantyne, J. Bright, P. Williams, *Med. Sci. Law.* (1970) 10, p. 225–229
[57] B. Ballantyne, *J. Forens. Sci. Soc.* (1976) 16, p. 305-310
[58] S. Barr, *Analyst* (1966) 9, p. 268-272
[59] B. Ballantyne, In: Forensic Toxicology, B. Ballantyne, (Ed.) Bristol, Wright, p. 99-113, 1974
[60] F.J. Baud, P. Barriot, V. Toffis, et al*., New Eng. J. Med.* (1991) 325, p. 1761-1766
[61] S. Goenechea, *Z. Rechtsmed.* (1982) 88, p. 97-101
[62] D.E. McMillan, A.C. Svoboda, *J. Pharmacol. Exp. Ther.* (1982) 221, p. 37-42
[63] B. Ballantyne, In: Developments in the Science and Practice of Toxicology, A. W. Hayes, R.C. Schnell, T. S. Miya, (Eds.) Elsevier, Amsterdam, p. 583-586, 1983
[64] B. Ballantyne, *Fund. Appl. Toxicol.* (1983) 3, p. 400-408
[65] A.C. Maehly, A. Swensson*, Int. Arch. Arbeitsmed.* (1970) 27, p. 195-209
[66] B. Ballantyne, *Clin. Toxicol.* (1977) 11, p. 173 -193
[67] I.S. Symington, R.A. Anderson, J.S. Oliver, I. Thomson, W.A. Harland, W.J. Kerr, *Lancet* (July 8, 1978) p. 91-92

[68] D.A. Purser, *Toxicol. Lett.* (1992) 64/65, p. 247–255

[69] A.S. Curry, D.E. Price, E.R. Rutter, *Acta Pharmacol. Toxicol.* (1967) 25, p. 339-344

[70] A.S. Curry, Poison Detection in Human Organs, 3rd ed. Springfield, Thomas, p. 9798, 1976

[71] A. Bernt, Ch. Kerde, O. Proko, *Dtsch. Z. Gericht. Med.* (1961) 51, p. 522-534

[72] K. Steenland, J. Deddens, L. Piacitelli,. *Am. J. Epidemiol.* (2001) 154, p. 451–458

[73] U.F. Achmadi, J. Pauluhn, *Exp. Toxicol. Pathol.* (1998) 50, p. 67–72

[74] J. Pauluhn, *Toxicol. Lett.* (1992) 64/65, p. 265–271

[75] F.H. Prager, G. Kimmerle, T. Maertins, M. Mann, J. Pauluhn, *Fire Mat.* (1994) 18, p. 107–109

[76] R.F. Phalen, Inhalation Studies: Foundations and Techniques. CRC Press, Boca Raton, FL, 1984

[77] DIN 53436 - Erzeugung thermischer Zersetzungsprodukte von Werkstoffen unter Luftzufuhr und ihre toxikologische Prüfung. Teil 1: Zersetzungsgerät und Bestimmung der Versuchstemperatur. Teil 2: Verfahren der thermischen Zersetzung. Teil 3: Verfahren zur inhalationstoxikologischen Untersuchung

[78] J.W. Trevan, *Proc. R. Soc. Lond. Ser. B* (1927) 101, p. 483-514

[79] C.I. Bliss, *Ann. Appl. Biol.* (1935) 22, p. 134-167

[80] C.I. Bliss, *Q. J. Pharm. Pharmacol.* (1938) 11, p. 192–216

[81] D.J. Finney, Statistical method in biological assay. Hafner, New York, pp. 471-472, 1964

[82] D.J. Finney, Probit analysis. Cambridge Univ. Press , London, 1971

[83] A.P. Rosiello, J.M. Essigmann, G.N. Wogan, *J. Tox. and Environ.Health* (1977) 3, p. 797–809

[84] J. Berkson, *J. Am. Stat. Assoc.* (1944) 39, p. 357-365

[85] L.F. Knudsen, J.M. Curtis, *J. Am. Stat. Assoc.* (1947) 42, p. 282–296

[86] W.R. Thompson, *Bacteriol. Rev.* (1947) 11, p. 115-145

[87] M. Schaper, R.D. Thompson, C.S. Weil, *Arch. Toxicol.* (1994) 68, p. 332–337

[88] B.C. Levin, R.G., Gann, in:. G.L. Nelson (Ed.), Fire Polymers: Hazard Identification and Prevention. ACS Symposium Series No. 425, American Chemical Society, 1990, p. 3-11

[89] B.C. Levin, J.L. Gurman, M. Paabo, L. Baier, T. Holt, Toxicological effects of different time exposures to the fire gases: Carbon monoxide or hydrogen cyanide or to carbon monoxide combined with hydrogen cyanide or carbon dioxide. In: Proceedings of the Ninth Joint Panel Meeting of the US-Japan (UJNR) Panel on Fire Research and Safety. NBSIR 88-3753, National Institute of Standards and Technology, Gaithersburg, MD, 1988

[90] B.C. Levin, M. Paabo, J.L Gurman, S.E. Harris, *Fundam. Appl. Toxicol.* (1981) 9, p. 236-250

[91] B.C. Levin, M. Paabo, C. Bailey, S.E. Harris, J.L. Gurman, *Toxicologist* (1985) 5, p. 127

[92] B.C. Levin, M. Paabo, J.L Gurman, S.E. Harris, *Fundam. Appl. Toxicol.* (1987) 9, p. 236-250

[93] B.C. Levin, M. Paabo, L. Highbarger, E N. Eller, *Toxicologist* (1990) 10, p. 84

[94] B.C. Levin, *Toxicol. Lett.* (1992) 64/65, p. 257–264

[95] J. Pauluhn, G. Kimmerle, T. Märtins, F.H. Prager, W. Pump, *J. Fire Sci.* (1994) 12, p. 63-104

[96] G.E Hartzell: *J. Fire Sci.* (1989) 7, p. 179-193

[97] G.E. Hartzell, A.F. Grand, W.G. Switzer, In: G. E. Hartzell (Ed.) Advances in Combustion Toxicology. Vol. 2. Technomic, Lancaster, PA. p. 285-308, 1989

[98] J. Pauluhn, *J. Fire Sci.* (1993) 11, p. 109–130

[99] D.J. Finney, Probit Analysis. Cambridge University Press, London, 1952

[100] W. Wittbecker, D. Daems, U. Werther, Performance of Polyurethane (PUR) Building Products in Fires. ISOPA, Brussels, 1999

[101] Consumer Product Safety Commission Fed. Register 45:182:61883–61927, 1980

[102] NRC (National Research Council). Guidelines for Developing Community Emergency Exposure Levels for Hazardous Substances. National Academy Press, Washington, DC (1993)

17   Corrosivity of Fire Effluents

T. KOPPERS, H.U. WERTHER, AND H.GRUPP

17.1 Introduction

According to DIN 50900-2, corrosion is defined as the reaction of metallic materials with its surroundings, resulting in a measurable alteration of the materials and their functions. A distinction is made between the most frequent atmospheric corrosion, soil corrosion and corrosion caused by aqueous electrolytes. In all types of corrosion, moisture is the main potential damage factor. Below a concentration of 30 % humidity and 1.5 °C, corrosion is not to be expected. Therefore, the amount of corrosion damage depends mainly on the atmospheric humidity and air temperature, followed by the type of agent causing the corrosion, its concentration and duration of action, the type of material or component attacked, and last but not least on the cleaning-up procedures applied.

Using the above definition, plastics are not subjected to corrosion. However, in case of fire, all burning organic materials – plastics included – will create species with a corrosive potential, as water and acid components are always part of fire effluents. Therefore, after fire accidents, one has to act to avoid atmospheric corrosion by chemical and/or electrochemical reactions on the surface of metallic materials.

The corrosion damage discussed here is exclusively related to fire accidents and is concerned with buildings and contents not subjected to direct fire effects, because corrosion effects are negligible compared with thermal damage in the direct region of fire influence. More precisely, the corrosion damage of interest can be roughly divided into damage to buildings caused by the long-term corrosive effects of fire effluents and damage to equipment caused by the corrosive attack to industrial machinery as well as electrical and electronic installations [1].

As mentioned in the preceding, all combustible materials – wood, plastics, chipboard and cotton included – will liberate corrosive fire effluents [1]. Generally, a growing extent of corrosion may be expected with a higher content of elements such as chlorine, bromine, sulfur and nitrogen in the burning materials involved. These elements exist in the fire gases as compounds able to react with water – an inherent component of all fire effluents – to acids with a different, but usually the highest possible corrosive potential. Therefore, in case of plastics, the primary corrosion effect results from the decomposition of poly(vinyl chloride) (PVC). This material may contain up to 57 weight-% of chlorine.

Corrosion arises from the formation of hydrogen halide (HCl or HBr) from chlorine- or bromine-containing materials. For plastics containing halogenated flame retardants, the corrosive potential is markedly less in comparison to PVC because of the lower content of halides, which is often < 10 % of weight and sometimes even < 1 %. Flame retardants usually contain bromine, which is much more effective than chlorine. Consequently, compared to chlorine, a lower bromine content is required for achieving a comparable fire safety level.

In addition one must be careful not to equate high chlorine content of a material with high corrosion risk. In a fire, generally, incomplete combustion occurs and only a certain amount of the fire load is transformed into corrosive products. For example, in practice, floor coverings are only partially destroyed because of their position. Densely packed stored materials are also less affected due to the formation of a charred layer, which protects the underlying material. In general, the effect of flame retarded materials hinders the fire propagation resulting in incomplete combustion.

When heated over 160 °C, PVC starts to decompose and to release HCl without formation of visible soot, the precursors of which, however, react as activated carbon. This means corrosion takes place at contamination levels that are lower than under the presence of soot particles, which carry a high amount of corrosive and toxic/hazardous decomposition products.

Therefore, a plastic such as PVC with a high content of chlorine plays a special role in corrosion. Investigations of corrosion damage are limited mainly to the effects of PVC fire effluents. It was first thought that there might be a danger to the strength of reinforced concrete structures from these gases but many investigations have shown that these fears are unfounded [2–5] or can be avoided by using suitable cleaning and restoration methods [16].

17.2 Principles of Corrosion

A short explanation of the principles of corrosive attack of fire gases on metals is now given using PVC as an example. An introduction to fundamentals and types of corrosion can be found in two comprehensive reviews [6, 7].

The HCl produced from PVC in the region of the fire combines with the ample supply of water vapor to give hydrochloric acid, which then condenses on the cooler metallic parts away from the immediate region of the fire. The actual corrosion occurs as a phase-boundary reaction between the solid metal phase and the liquid condensate phase. A "corrosion element" results, in which the anodic process of metal dissolution and the cathodic process of reduction of hydrogen ions to hydrogen and of oxygen to hydroxyl ions occur. The following reactions take place:

Anodic process (metal dissolution; oxidation):

$$M \rightarrow M^{2+} + 2e^-$$

Cathodic process (reduction)

$$2H^+ + 2e^- \rightarrow H_2 + (H^+ \text{ reduction})$$
$$O_2 + 2H_2O + 4e^- \rightarrow 4OH^- (O_2 \text{ reduction})$$

Depending on whether the dissolution of the metal occurs by hydrogen ion reduction or by oxygen reduction, the process is termed acid or oxygen corrosion. If the HCl produced from PVC fire gases impinges on a steel surface, acid corrosion takes place formally according to the following reactions:

$$Fe + 2HCl \rightarrow FeCl_2 + H_2$$
$$FeCl_2 + 2H_2O \rightarrow Fe(OH)_2 + 2HCl$$

HCl reacts like a promoter (catalyst): it starts the chemical reaction and remains free for further attack. This cycle may take place more than 20 times until FeOOH or similar chloride-free species are eventually formed.

However, the HCl condensate always contains dissolved atmospheric oxygen. Therefore, oxygen corrosion occurs simultaneously:

$$2Fe + O_2 + 2H_2O \rightarrow 2Fe(OH)_2$$

The actual rust formation process is complicated and can be formulated as follows:

$$Fe \rightarrow Fe^{2+} + 2e^-$$
$$2Fe^{2+} + O \rightarrow 2Fe^{3+} + O^{2-}$$

$$
\begin{aligned}
O + 2e^- &\rightarrow O^{2-} \\
O^{2-} + H_2O &\rightarrow 2\,OH^- \\
Fe^{2+} + 2\,OH^- &\rightarrow Fe(OH)_2 \\
2\,Fe(OH)_2 + O + H_2O &\rightarrow 2\,Fe(OH)_3 \\
Fe(OH)_3 &\rightarrow FeO(OH) + H_2O \text{ (rust)}
\end{aligned}
$$

The extent of corrosion depends in the main on the prevailing atmospheric humidity. As Fig. 17.1 shows, there is a sudden increase in corrosion rate in the region between 40 % and 70 % relative humidity. For HCl the critical relative humidity is 56 %, for HBr it is 39.5 %.

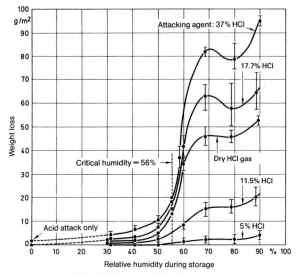

Figure 17.1 Corrosion rate of nonalloyed steel stored over different HCl concentrations as a function of atmospheric humidity (after Längle, [13], Part I, p. 35)

17.3 Methods for Assessing Corrosivity

Test methods regarding corrosivity of fire gases have been developed in several countries and implemented by international organizations such as IEC and ISO to become international standards. One can distinguish between indirect and direct methods.

The indirect methods are based on the principle that corrosivity is due to the content of either hydrogen ions or metallic ions formed in an aqueous medium, caused by effluents created from a fire. The IEC 60754 series reflect this principle, whereas the IEC 11907 series stand for a direct method.

The IEC 60754 series are related to cable tests and consist of two parts. In Part 1, the material under test is heated in a stream of dry air and the gases absorbed in 0.1 M sodium hydroxide solution. The amount of halogen acid is then determined by acidifying the solution with nitric acid adding a measured volume of 0.1 M silver nitrate solution and back titrating the excess with 0.1 M ammonium thiocyanate using ferric ammonium sulfate as an indicator, or by any other equivalent analytical method having at least the same accuracy. The test method is not recommended for use where the amount of halogen acid is < 5 mg/g of the sample taken. The method is not suitable for defining compounds or materials described as "zero halogen".

In IEC 60754 Part 2, the method is similar to DIN/VDE 0472 Part 813 [8]. It works with a combustion tube of 700–800 mm (27.5 – 31.5 in.) length and 32–45 mm (1.26–1.77 in.) internal diameter, placed in a tube furnace, 500–600 mm (19.7–23.6 in.) long. A sample of 1 g of the material to be tested in a porcelain crucible is placed in the middle of the furnace preheated to minimum 935 °C. The decomposition products are transported out of the tube by an air stream of 10 l/h. The gas mixtures passing through washing bottles, filled with distilled water, where pH-value and conductivity are measured. The recommended pH should not be less than 4.3 and the conductivity should not exceed 10 μS/mm. In the French method to NF C 20-453 [9] combustion products are generated and determined in a similar manner.

The direct methods measure metal loss caused by the corrosive potential of the fire effluents. They are summarized in the ISO 11907 series, which consist of four parts under the general title: Plastics – smoke generation – Determination of the corrosivity of fire effluents.

Part 1 is the guidance paper. Part 2 describes a static method originally developed by CNET (Centre National d'Etudes des Télécommunications). Test apparatus and test piece are shown in Fig. 17.2. In this test, a predetermined quantity of material is burnt in a leak-tight test chamber with a constant temperature and well-defined relative humidity [10]. The combustion products condense on a printed circuit board with a defined copper track, which is kept at constant temperature below the dew point of smokes. The change in electric resistance is a measure of corrosivity. Because of limitations of scale and ventilation, this test is suitable for the comparative testing of materials only, and not for a testing of finished products under various simulated fire conditions.

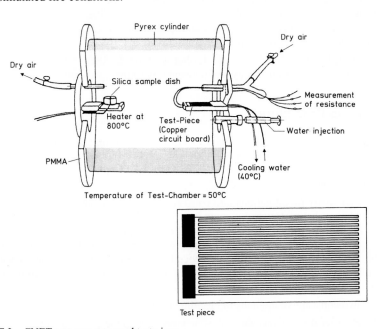

Figure 17.2   CNET test apparatus and test piece

Part 3 uses a DIN traveling furnace according to DIN 53436 [11]. The furnace is moved over the test specimen located in a quartz cuvette inside a quartz glass tube through which air is passed at a specific flow rate. The conditions can be chosen to simulate a smoldering or developing fire. The corrosion target, a copper printed wire board, is exposed to the fire effluents, and condensation is enhanced by using a cooling system. The change in resistance of the target is used to denote the corrosion hazard.

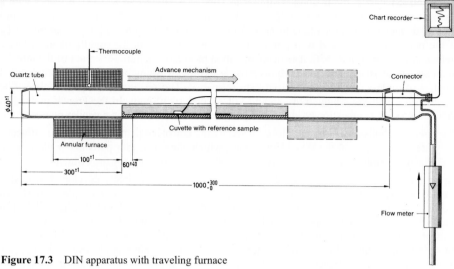

Figure 17.3   DIN apparatus with traveling furnace

Part 4 uses the cone calorimeter according to ISO 5660-1, as described in Chapter 10.23 ISO. In the exhaust duct, gas is sampled and led to the exposure chamber with the corrosion target. The target consists of two identical circuits on a printed board. One circuit is exposed to the effluent gases and the other one is protected by a coating and is used as a reference.

In ISO/TC 61/SC4, it was decided to stop further work on the ISO 11907 series. Hardly any data is available on the reproducibility and repeatability of these methods.

17.4 Clean-up Procedures

This seems an appropriate place to note the work carried out on this topic by Grupp whose publications [2, 12–15] contain more than 100 references. Another very important contribution was made by H.P.Wollner [16], R. Pentenrieder, and others [17–20], representing the current state of the art.

There have been many investigations into corrosion damage of materials and equipment. The results have shown that materials and equipment do not usually become a total writeoff and that there is no impairment of functional efficiency, provided an appropriate cleaning procedure is used as soon as possible. Relevant procedures are described by H.P. Wollner [16] and are described here briefly. There is no general rule as to which of the methods described here would be applied in a particular case. This depends on the requirements of the particular restoration.

After a fire, two consecutive methods of cleaning must be implemented:

- Immediately after the fire, measures must be taken to stop the corrosion attack.
- In a second step, the fire condensates and any corrosion products must be removed using special cleaning procedures for electronics, machines and buildings.

17.4.1 Methods to Stop Corrosion Attack After a Fire

To stop corrosion attack, which usually takes place above 1.5 °C and 30 % atmospheric humidity, an obvious immediate measure is the lowering of humidity. Essentially, this can be done either by raising the temperature of the surroundings or by the use of certain substances as drying agents. Raising the temperature, which involves heating the enclosed air, is often not feasible in large buildings. The warm and humidity-carrying air must be blown out and exchanged against dry air; otherwise, there is no reduction of the relative humidity. Powerful dehumidifiers are therefore preferred. These can lower the relative atmosphere humidity to about 30 %, which is below the critical point of 56 % for HCl condensates and 39.5 % for HBr condensates, in a few hours even in large buildings. Silica gel and calcium chloride are used as drying agents; silica gel is preferred because of its easy regenerative properties, whereas calcium chloride is less commonly used because of its restrictive regenerative ability and its tendency to fuse and cake.

A further immediate measure consists in the application of anticorrosion oils as rust inhibitors, which work by chemically inhibiting corrosion attack and by repelling water. The oil flows under the moisture film and so keeps it away from the metal surface. Another effect of the oil preservation is to prevent further penetration of humidity into the contaminated surface. This is known as the "dewatering effect". The disadvantages are the high cost of such oils and the fact that their removal is labor intensive. The method can be helpful in case of industry machines, but not recommended for buildings and electronic devices.

17.4.2 Special Clean-up Procedures

Various measures are used for cleaning electrical as well as electronic devices, machines and buildings to remove fire residues, corrosion attack stoppers, and fire extinguishers. For instance, one must carefully remove ABC type extinguishing powder, which may create corrosion on metal surfaces because of its hygroscopic nature and its phosphate and ammonium sulfate salts content. In general, all cleaning methods have to be selected depending on the damage extent and other factors, in particular, however, on the product involved and its function.

17.4.2.1 Electrical and Electronic Equipment

The most demanding cleaning procedures are necessary for electrical and electronic equipment. Electronic devices have small distances between parts of different functions and metals. Their three-dimensional structure is difficult to clean by hand and the quality requirements are high.

The function of such devices will be already impaired by low-level contamination from soot and acid particles. Leakage current and electromigration are the result. Nevertheless, today, suitable state-of-the-art cleaning methods not only revamp simple electronic parts of industry machines, but highly sophisticated electronic parts of nuclear power stations and satellites as well. Even after fire damages and spills of acids in semiconductor plants, restoration is considered as a measure for loss mitigation and is approved by manufacturers of semiconductor manufacturing equipment. The limit values as defined in relevant standards to guarantee the function of such devices (DIN E 1000 15,1; IPS-SIU-840B; IEC 61-340-5-1; J-STD-001B; ANSI/PC-A-610B) are lowered permanently. In the meantime, the high standards for military equipment are transferred to a public Joint Industry Standard (MIL 28809a to J-STD-001B). Using the correct procedure, it is possible, in case of halogen contamination, to lower the limit value of 1 µg of chloride equivalent/cm^2 (3 µg/cm² in case of simple industrial electronic parts) to a level of ca. 0.2 µg/cm^2, provided an exact analysis clarifies the type and amount of damage. Afterwards, the decision has to be made on the best available cleaning method in knowledge of the function and purpose of the material and systems involved.

Dry cleaning methods are not reliable enough and therefore only usable as a precleaning procedure. Sonography is not suitable because of the high risk to damage chips and transistors. Third, fine degreasing cleaners as well as corrosion attack stoppers are not recommended, either.

Fully satisfying results will be obtained only with a completely water-based wet cleaning procedure. The process conditions should be selected carefully (alkaline medium pH 9–12) and the procedure carried out by specialists. Today, these procedures should include the total removal of flux agents. Such flux agents are extremely water sensitive and after the water-based cleaning procedure, electromigration cannot be excluded.

Chemicals and procedures used for restoration work must be adapted to produce the predicted and necessary standard for the restoration job. Industrial companies have certified and audited the restoration procedures necessary after a fire as well as the restoration companies applying these procedures. H.P. Wollner ([16], p. 94) summarizes the most important restoration procedures for electrical and electronic equipment in Table 17.1.

17.4.2.2 Industry Machines

In the case of industry machines, the cleaning procedure is less complicated than the methods used for electronic and electrical devices. It is easy to detect damages and to exclude late effects. This leads to a better cost/efficiency ratio. Nevertheless, simple wiping-by-hand or blowing methods without demounting are not acceptable. The best results are obtained after demounting the machines to single pieces. Irreparably damaged parts can be exchanged; the other parts cleaned using an ultrasonic bath. Suitable chemicals (degreasing cleaner, corrosion removal cleaner, etc.) guarantee dry and excellent surfaces. Where corrosion happened, abrasive blasting techniques can help: Dry-, hot steam-, high pressure- or CO_2-blasting should preferably be combined with a following process of dragging, polishing, and/or surface blackening.

17.4.2.3 Buildings

In cleaning buildings, the method to be adopted mainly depends on whether the contaminated materials are porous or nonporous. Furthermore, it is necessary to determine the chloride distribution in the structure subjected to the fire gases. This is achieved by determining the water-soluble chloride content and the depth of chloride penetration in the structure.

Drycleaning methods without any abrasive effect such as blowing and sucking can be accepted only as precleaning measures for dust and loose soot fixed on the surface of materials. In the case of porous substrates like concrete and if high concentrations of chloride are to be removed from depths up to 5 mm (0.2 in.), the method of choice is the application of steam or high pressure hot water combined with chemical additives. In principle, steam and hot water will only extract the chloride from the surface up to a depth of 2–3 mm (0.08–0.12 in.). In the following drying period, the chloride ions in the deeper part of the structure will diffuse to the surface leading to a uniform concentration in the whole concrete. Then, a second washout step will further remove the chloride ions, lowering again the concentration in the upper part of the concrete, and so forth. Such washout and drying procedures must be repeated several times to obtain the desired effect. Care should be taken for a correct procedure in order to avoid that the chloride ions are transferred in deeper concrete areas by an incorrectly handled water steam application. Furthermore, one should add that in special cases, the method of electrodiffusion allows to reach concrete contaminated in a region deeper than 5 mm.

Table 17.1 Standard restoration procedures for electronics

	Dry Restoration	Organic One-Cycle Restoration	Selective Hydrous Restoration	Organic/Hydrous Restoration	Completely Hydrous Restoration
Contamination	Loose dust, etc	Adhesive dust, oil, tar, no corrosion	Dust, light fire, water	Light to medium strong fire, oil	Strong fire or water damage
Cleaning modules	Dry with fine cleaning	Dipping bath, high pressure if needed	Dipping bath, high pressure (hydrous)	Dipping bath, high pressure (hydrous)	Dipping bath, high pressure
Cleaning cabinet area	Dry with fine cleaning	Wiping procedure, high pressure cleaning as needed	Dry with fine cleaning	Wiping procedure, high pressure cleaning as needed(organic)	Wiping procedure and high pressure cleaning
Cleaning agents	Organic and hydrous fine cleaners	Organic fine cleaners	Hydrous electronics and organic fine cleaner	Hydrous electronics and organic fine cleaner	Hydrous electronics cleaner
Dismounting depth	Minor (coverings, housings)		All assemblies, individual inserts and components as needed		
Fluxing agent removal	No	Yes	Yes	Yes	Yes
Corrosion removal	No (possibly local)	No	Module: yes; cabinet: local	Total	Total
Rinsing	Not applicable	Not applicable	After hydrous cleaning	Total	Total
Residual contamination with chloride	$< 10\ \mu g/cm^2$	$< 5\ \mu g/cm^2$	Hydrous: $< 1\ \mu g/cm^2$; dry: $< 10\ \mu g/cm^2$	Hydrous: $< 1\ \mu g/cm^2$; organic: $< 5\mu g/cm^2$	$< 1\ \mu g/cm^2$
Residual contamination with anions/ NaCl equivalents	$< 16\ \mu g/cm^2$	$< 8\ \mu g/cm^2$	Hydrous: $< 1.56\ \mu g/cm^2$; dry: $< 16\ \mu g/cm^2$	Hydrous: $< 1.56\ \mu g/cm^2$; organic: $< 8\ \mu g/cm^2$	$< 1.56\ \mu g/cm^2$
Surface insulation resistance after restoration (moist climate)	Varying	$> 10^7 \Omega$	Hydrous: $> 10^{11} \Omega$ dry: varying	Hydrous: $> 10^{11} \Omega$ organic: $> 10^7 \Omega$	$> 10^{11} \Omega$
Quality target (electronics class)	Operability for simple objects (class 1)	Operability: simple to medium complex objects (class 1 + 2)	High dependability for sensitive sub-systems (class 2)	High dependability (class 2)	Highest dependability (class 2 + 3); approval according to J-STD-001B

Dry cleaning methods with an abrasive effect are necessary if the contamination is high and not to handle by wet cleaning systems as described. In Table 17.2, H.P. Wollner ([16], p. 94) summarizes the most suitable cleaning methods. Sand blasting allows abrasion only to a depth of 2–3 mm. When supported by high pressure up to 800 bar, thermal damaged concrete can be removed without influencing the nondamaged concrete areas.

The most expensive and therefore least used method is "torcreting" or highest pressure blasting (2000 bar), which involves removing the concrete contaminated up to the steel reinforcement followed by spraying fresh concrete of appropriate quality.

Cleaning methods using chalk and calcium hydroxide with hot water – often recommended in the 1980s – are outdated and have been substituted by methods that are more efficient.

Table 17.2 Comparison of standard and special restoration procedures for buildings

Procedure group	Procedures/methods	Area efficiency on walls (m^2/h) or efficiency parameter	Total effort (preparation, procedure, equipment)
Sucking	Sucking/blowing Spraying extraction Vacuum extraction	15–30 5–20 (floor) 5–15 (walls, ceilings)	Minor Minor Medium
High pressure	Hot water-hpc hp rotation procedure Hot steam procedure (98 °C) Wet blasting	5–20 5–16 10–20 2–8	Medium Medium Medium High
Super pressure	Up to 600 bar 1000–2000 bar	2–20 2–15	Medium Very high
Manual procedures	Hand wiping procedure Chem. graffiti removal	10–15 2–5	Minor Minor
Stripping	Grinding, milling, needling, Chipping, poking, chiseling Stripping of floor coverings	1–3 5–15	Medium Medium
Dry or vapor blasting	Dry blasting Vapor blasting JOS-procedure Suction head blasting CO$_2$-blasting	2–10 2–10 2–8 2–8 1–20	High High High High High
Drying	Air dehumidification (sorption, condensation) Warm air drying Condensation heat drying Freeze drying	50–150 m^3/machine 60–100 m^3/machine 20–60 m^3/machine 7.5–10 €/kg paper	Minor Minor Minor High
Others	Ozonization Electro(diffusion)restoration Sprayed concrete	max. 60 m^3/machine 75–100 €/m^2 1–2 m^3/h; 4–10 m^3/h	Medium Very high Very high

17.5 Conclusions

Any corrosive contamination caused by fire effluents can be removed using special cleaning methods without any loss of reliability of the materials or systems involved. Depending on the damage and the products involved, the cost is often high, in most cases, however, lower than the price for new equipment. It is important to note that the success of restoration is best and at lowest cost, the earlier adequate clean-up measures are started after a fire. In all cases, specialists should be contacted, because electronic devices, machinery and buildings are becoming increasingly complex, necessitating more and more differentiated combinations of cleaning methods adapted to the respective components, parts, and materials.

References

[1] K. Fischer: Schadenprisma 5 (1976) 4, pp. 61/80
[2] P.H. Efferts, H.Grupp, W.Jach: Der Maschinenschaden 43 (1970)3, p. 89/99
[3] K. Fischer: Zeitschrift VFDB 20 (1971) 4, pp.140/144; 21 (1972), p.1/11
[4] W.A. Morris, J.S. Hopkinson: Fire International 47 (1973), pp.71–81
[5] C. Reiter: Versicherungswirtschaft 28 (1973) 9, pp. 492–497
[6] H .Gräfen, E.M. Horn, U. Gramberg: Korrosion in Ullmanns Enzyklopädie der Technischen Chemie, 4th edit., Vol. 15, Verlag Chemie, Weinheim, p. 1–59, 1978
[7] R.T. Foley, B.F. Brown, in: Kirk-Othmer (Eds.): Encyclopedia of Chemical Technology, 3rd edit., Vol. 7, John Wiley & Sons, New York, London, p. 113/143, 1979
[8] DIN 57 472 Part 813, August 1983. Testing of cables, wires and flexible cords. Corrosivity of combustion gases (VDE specification)
[9] NF C 20-453, Jan. 1985. Conventional determination of corrosivity of smoke
[10] P.Rio, TJ. O'Neill: Flame retardants '87 Conference. London, 26–27 Nov. 1987
[11] DIN 53436, 1966. Prüfung von Kunststoffen. Gerät für die thermische Zersetzung von Kunststoffen unter Luftzufuhr
[12] P.H. Effertz, H. Grupp: Der Maschinenschaden 45 (1972) 1, p. 7/19
[13] H. Grupp: Der Maschinenschaden 53 (1980) 1, p.20–28; 53 (1980) 2, p. 56–63
[14] H. Grupp: Brandschutzausgerüstete Kunststoffe und ihre Brandfolgeschäden sowie Brandschutzmaßnahmen. Symposium Kriechstromfestigkeit und Brennbarkeit von Kunststoffen, Haus der Technik, Essen, 6.5.1981
[15] F.K. Behrens, H. Renz, H. Grupp: Der Maschinenschaden 54 (1981) 3, p. 77–86
[16] H.P. Wollner: Schadenminderung durch Sanierung, Band 2, RKH Verlag Tondok, München, 1999
[17] M. Tuchschmid, R. Werner: Vergleichende Untersuchungen korrosiver Brandgaskondensate, Allianz Report 5, p. 204–211, 2001
[18] H. Fuchs, R. Pentenrieder, W. Streul, H.-P. Wollner: Kostenmanagement heisst Kostenbewusstsein und Umweltschutz, Teil 1: Sanierung elektrischer und elektronischer Systeme, Allianz report 70, 1, p. 32–40, 1997
[19] H.D. Hager: Der Maschinenschaden, 63 (1990) 4, p. 162–164
[20] R. Pentenrieder: Zuverlässigkeit elektronischer Geräte und Anlagen nach einer Sanierung, QZ, Qualität und Zuverlässigkeit 34, 1989

IV

Appendix

1 Suppliers of Flame Retardants and Smoke Suppressants

Manufacturers and suppliers	Products
3 V Deutschland GmbH Siemensstraße 3/1 D-71263 Weil der Stadt, Germany www.3Vsigma.com	Melamine cyanurate
Adeka Palmarole Rue de Strasbourg F-68300 Saint-Louis, France	Phosphate esters, phosphorus compounds
Akzo Nobel Chemicals Inc. 5 Livingstone Avenue Dobbs Ferry, NY 10522-3407, USA http://phosphorus.akzonobelusa.com	Phosphate esters, halogenated phosphorus compounds
Albemarle Corporation Bromine Chemicals Division, 451 Florida Blvd., Baton Rouge, LA 70801, USA www.albemarle.com	Brominated compounds Phosphate esters
Alcoa Industrial Chemicals Alcoa Technical Center 7th Street Road, Route 780 Alcoa Center, PA 15069-0001, USA www.alumina.alcoa.com	Aluminum trihydrate
Ameribrom, Inc., a subsidiary of the Dead Sea Bromine Group www.dsbgfr.com	Brominated compounds Magnesium hydroxide Melamine cyanurate
Asahi Denka Kogyo KK, Furukawa Building 8, 2-chome, Nihonbashi-Murmachin, Chuo-ku, Tokyo, Japan	Halogenated compounds, Phosphate esters
Bayer AG SP-PCH-PHCH Building K10 D-58368 Leverkusen, Germany www.experts4additives.de	Phosphoric acid-alkyl/aryl-ester Phosphonic acid-ester
Campine N. V. Nijverheidsstraat 2, B-2340 Beerse, Belgium www.campine.be	Antimony trioxide, masterbatches

Chemische Fabrik Budenheim
Rheinstraße 2 A
D-55 257 Budenheim, Germany
www.budenheim-cfb.com

Phosphorus and nitrogen compounds
Ammonium polyphosphate
Melamine cyanurate

Chisso Co.,
7-3 Marunouchi 2-chome,
Chiyoda-ku,
Tokyo 100, Japan
www.chisso.co.jp

Phosphorus and nitrogen compounds

Ciba Specialty Chemicals
CH-4002 Basel, Switzerland
www.cibasc.com

Flamestab NOR 116, Tinuvin FR
Melamine cyanurate, melamine polyphosphate

Clariant GmbH
Pigments & Additives Division
D-65 840 Sulzbach, Germany
http://pa.clariant.com

Organic phosphorus compounds, red phosphorus, ammonium polyphosphate

Climax Molybdenum Co.,
1600 Huron Pkwy.,
Ann Arbor, MI 48 106, USA
www.climaxmolybdenum.com

Molybdenum compounds
Smoke suppressants

Daihachi Chemical Industry Co. Ltd.,
Hiranomachi Yachiyo Bldg. 8-13
Hiranomachi 1-chome, Chuo-ku
Osaka 541-0046, Japan

Halogenated and halogenfree phosphate esters
Phosphorus compounds

Dead Sea Bromine Group
P.O. Box 180,
Beer Sheva 84 101, Israel
www.dsbgfr.com

Brominated compounds,
Magnesium hydroxide, melamine cyanurate
Phosphate esters

Dover Chemical Corporation
3 676 Davis Road N.W.
Dover, OH 44 622-0040, USA
www.doverchem.com

Chlorinated paraffins

DSM Melamine BV
Poststraat 1, Sittard
P.O. Box 43, 6 130 AA Sittard
The Netherlands
www. dsm.com

Melamine

C.H. Erbslöh
Düsseldorfer Straße 103
D-47 809 Krefeld, Germany
www.cherbsloeh.de

Brominated compounds, zinc borate
Magnesium hydroxide,
Chloroparaffins, expandable graphite

Eurobrom B.V., a subsidiary of
the Dead Sea Bromine Group
Verrijn Stuart Laan 1 c
NL-2288 EK Rijswijk, The Netherlands
www.dsbgfr.com

Brominated compounds
Magnesium hydroxide, melamine cyanurate,
Phosphate esters

Grafitbergbau Kaisersberg GmbH Bergmannstraße 39 A-8713 St. Stefan ob Leoben, Austria www.grafit.at	Expandable graphite
Graftech Inc. PO Box 94637 Cleveland, OH 44101, USA www.graftech.com	Expandable graphite
Great Lakes Chemical Corporation One Great Lakes Boulevard P.O. Box 2200 West Lafayette IN 47906-0200, USA www.pa.greatlakes.com *Europe*: Great Lakes Sales (Germany) GmbH Sattlerweg 8 D-51429 Bergisch Gladbach, Germany *Asia*: Great Lakes Chemical Singapore Pte. Ltd 65 Chulia Street #37-05 OCBC Centre Singapore 049513	Brominated compounds, phosphate esters Antimony trioxide, zinc borate
Harwick Chemical Corp., 60 S. Seiberling St., Akron, OH 44305, USA	Phosphate esters, halogenated phosphorus compounds, chlorinated paraffins, aluminum hydroxide, antimony oxide, zinc borate
Incemin AG Schachen 82 CH-5113 Holderbank, Switzerland www.ankerpoort.com	Aluminum trihydrate Magnesium hydroxide
Italmatch Chemicals Spa v. P. Chiesa 7/13, Torri Piane, San Benigno, 16149 Genova, Italy www.italmatch.it	Red phosphorus
J.M. Huber Engineered Materials, 4401 Northside Parkway, Suite 600 Atlanta, GA 30327, USA www.huberemd.com	Aluminum trihydrate Magnesium hydroxide
Kisuma Chemicals BV, NL-9640 AK Veendam, The Netherlands www.kyowa-chem.co.jp	Magnesium hydroxide
Kyowa Chemical Ind.Co., Ltd., 4035, Hayashida-machi Sakaide-shi, Kagawa 762, Japan www.kyowa-chem.co.jp	Magnesium hydroxide
Lehmann & Voss & Co. Schimmelmannstr. 103 D-22043 Hamburg, Germany www.lehvoss.de	Brominated and halogenfree flame retardant masterbatches

Martin Marietta
P.O. Box 15470
Baltimore, MD USA 21220-0470, USA
www.magspecialties.com

Magnesium hydroxide

Martinswerk GmbH,
a subsidiary of Albemarle Corp.
Kölnerstraße 110,
D-50127 Bergheim, Germany
www.martinswerk.de

Aluminum trihydrate
Magnesium hydroxide

Nabaltec GmbH
Alustraße 50–52
D-92421 Schwandorf, Germany
www.nabaltec.de

Aluminum trihydrate
Aluminum monohydrate (Boehmit)
Magnesium hydroxide

NRC Nordmann, Rassmann GmbH
Kajen 2
D-20459 Hamburg, Germany
www.nrc.de

Brominated compounds,
Phosphorus compounds, zinc borate
Nitrogen compounds, expandable graphite

Nyacol Inc.,
Megunco Rd., P.3O.3Box 349,
Ashland, MA 01721, USA
www.nyacol.com

Antimony oxide

Oxychem, Occidental Chem. Corp.
P.O. Box 809050
Dallas, Texas 75380-9050, USA
www.oxychem.com

Chlorinated compounds

Rhodia Consumer Specialities Ltd.
F.A.D.P. Building, Oldbury,
West Midlands, PO Box 7870
B69 4HX, UK
www.rhodia.com

Phosphate esters, phosphorus and
nitrogen compounds

Rinkagaku Kogyo Co. Ltd.
34 Shin-Bori, Shin-Minato
Toyoma, Japan
www.rinka.co.jp

Red phosphorus

Sanko Chemical Co. Ltd.,
16 Tohri-cho 8-chome, Kurume City,
Fukuoka 830, Japan

Phosphorus compounds

Schill + Seilacher Aktiengesellschaft
Schoenaicher Str. 205
D-71032 Boeblingen, Germany
www.schillseilacher.de

Flacavon, flame retardants for textiles

Sherwin-Williams Company
101 Prospect Avenue
Cleveland, Ohio 44115, USA
www.kemgard.com

Molybdenum compounds,
Smoke suppressants

Süd-Chemie AG BU Plastic Additives Ostenrieder Str. 15 D-80368 Moosburg, Germany www.sud-chemie.com	Nanocomposites Nanofil 15 and 32
Sumitomo Chemical Co. Ltd., New Sumitomo Building, 15 5-chome, Kitahama, Higashi-ku, Osaka, Japan www.sumitomo-chem.co.jp	Antimony oxide, phosphorus compounds, Halogenated compounds
U.S. Borax Inc. 26877 Tourney Road Valencia, CA 91355, USA www.borax.com	Zinc borate
Teijin Chemicals Ltd., Hibiya Daibiru Bldg. 2-2, Uchisaiwaicho 1-chome Chiyoda-ku, Tokyo 100-0011, Japan www.teijinkasei.co.jp	Brominated compounds
Tosoh Corporation 1-7-7, Akasaka, Minato-ku, Tokyo 107, Japan www.tosoh.com	Brominated compounds
UCAR Carbon Co. Inc. PO Box 94637 Cleveland, Ohio 44101, USA www.ucar.com	Expandable graphite

2 Abbreviations for Plastics and Additives

2.1 Abbreviations for Plastics

Thermoplastics

ABS	acrylonitrile/butadiene/styrene terpolymer
ASA	acrylonitrile styrene acrylate
EMA	ethylene methylacrylate copolymer
EPS	expandable polystyrene
ETFE	ethylene tetrafluoroethylene copolymer
FEP	perfluoroethylene propylene
HIPS	high impact polystyrene
PA	polyamide
PAA	poly(acrylic acid)
PA-6	polyamide-6
PAN	polyacrylonitrile
PBD	polybutadiene
PBI	polybenzimidazoles
PBO	poly(p-phenylene-2,6-benzo-bisoxazole) (polyoxazoles)
PBT	polybutylene terephthalate
PBZT	polybenzothiazoles
PC	polycarbonate
PCTFE	poly(chlorotrifluoro-ethylene)
PE	polyethylene
PEDEK	poly(etherdiphenylether-ketone)
PEEK	poly(etheretherketone)
PEEKK	poly(etheretherketone-ketone)
PE-HD	high density polyethylene (HDPE)
PEK	poly(etherketone)
PE-LD	low density polyethylene (LDPE)
PE-LLD	linear low density poly-ethylene (LLDPE)
PES	polyether sulfones
PET	polyethylene terephthalate
PI	polyimides
PIP	polyisoprene
PIPD	poly(2,6-diimidazol[4,5-b:4',5'-e]pyridinylene-1,4 (2,5-dihydroxy)phenylene)
PMAA	poly(methacrylic acid)
PMA	polymethacrylonitrile
PMMA	poly(methylmethacrylate)
POM	polyoxymethylene
PP	polypropylene
PPE	polyphenylene ether
PPO	polyphenylene oxide
PPS	polyphenylene sulfide
PPTA	p-aramids
PSO	polysulfones
PTFE	polytetrafluoroethylene
PVC	poly(vinylchloride)
PVCC	chlorinated poly(vinyl chloride)
PVDC	poly(vinylidene chloride)
PVF	poly(vinyl fluoride)
PVDF	poly(vinylidene fluoride)
SAN	styrene/acrylonitrile copolymer

Thermosets

EP	Expoxy resin
MF	melamine formaldehyde resin
PUR	polyurethane
PIR	polyisocyanurate
PF	phenol formaldehyde resin
UF	urea formaldehyde resin
UP	unsaturated polyester resin

Inorganic Resins

PCS	polycarbosilane copolymers
PDMS	polydimethylsiloxane
POSS	polyhedral oligomeric silsesquioxane
PSS	polysilsesquioxane
VTS	vinyl-triethoxy-silane

Elastomer/Thermosets

ACM	acrylic rubbers
AFMU	carboxy-fluoro-nitroso rubbers
AU/EU	polyurethane
BIIR	bromobutyl rubbers
BR	1,4-polyutadiene
CIIR	chlorobutyl rubbers
CM	chlorinated polyethylene
CO / ECO	epichlorohydrin rubbers
CR	neoprene rubbers (polychloroprene)
CSM	chlorosulfonated polyethylene
EAM	ethylene acrylate rubbers
EPM	ethylene propylene rubbers (EPDM)
EVM	ethylene vinyl acetate rubbers (EVA)
FPM	fluorocarbon rubbers (FKM)
FFKM	perfluoro rubbers
FMQ	fluoromethyl(vinyl)silicone rubbers (FVMQ)
FZ	polyfluoroalkoxyphosphazene rubbers
GPO	propylene oxide rubbers
HNBR	hydrogenated acrylonitrile-butadiene rubbers
IIR	butyl rubbers
IR	1,4-polyisoprene
MQ	methyl(vinyl)silicone rubbers (VMQ)
NBR	acrylonitrile butadiene rubbers
NR	natural rubber
PNR	polynorbornene
PZ	polyaryloxyphosphazene rubbers
SBR	styrene butadiene rubbers
TM(T)	polysulfide rubbers
XNBR	carboxylated acrylonitrile butadiene rubbers

Elastomer/Thermoplastics

CPE	polyether ester copolymers
EVA/VC	EVA/chlorinated polyolefin blends
MPR	melt-processible rubbers
NBR/PVC	NBR/PVC blends
PEBA	polyether amide copolymers
PP/NR-VD	polypropylene/cross-linked NR blends

PP/EPDM-VD	polypropylene/cross-linked EPDM blends
PP/IIR-VD	polypropylene/cross-linked IIR blends
PP/NBR-VD	polypropylene/cross-linked NBR blends
PUR	polyurethane
SBS	styrene butadiene styrene block copolymers
SEBS	styrene ethylene butylene styrene block copolymers
SIS	styrene isoprene styrene block copolymers
TPO	thermoplastic polyolefin elastomers
TPU	thermoplastic polyurethane elastomers
TPV	thermoplastic vulcanizates

2.2 Abbreviations for Additives

APB	ammonium pentaborate
APP	ammonium polyphosphate
ATH	aluminum trihydrate
BCOH	β-cyclodextrine
CP	chloroparaffin
DBDPO	decabromodiphenyloxide
DMMP	dimethyl methane phosphonate
EBTPI	ethylene bis(tetrabromophthalimide)
FB (415, ZB)	zinc borates
HBCD	hexabromocyclododecane
HET	chlorendic acid
INDAN	octabromotrimethylphenylindan
MOH	mannitol
PBB-PA	poly(pentabromobenzyl-acrylate)
PER	pentaerythritol
PY	diammonium diphosphate
SOH	d-sorbitol
TBBA	tetrabromobisphenol A
TCEP	tris-(chloroethyl) phosphate
TCPP	tris-(chloropropyl) phosphate
TDCPP	tris-(dichloropropyl) phosphate
THP	tetrakis(hydroxymethyl)phosphonium salts
XOH	xylitol
ZHS	zinc hydroxystannate

3 Thermal Characteristics of Selected Thermoplastics

Table 3.1 Thermal characteristics of selected thermoplastics

Poymer	Bulk density [g/cm³]	Temperature resistance short term [°C]	Temperature resistance long term [°C]	Vicat-softening point B [°C]	Decomposition range [°C]	Flash-ignition temperature[3] [°C]	Self-ignition temperature[3] [°C]	Heat of combustion ΔH [kJ/kg]
Polyethylene LD[1]	0.91	95	75	–	340–440	340	350	48000
Polyethylene HD	0.96	110	80	75–80	340–440	340	350	46500
Polypropylene	0.91	140	110	80–95	330–410	350–370	390–410	44000
Polystyrene	1.05	90	80	85–95	300–400	345–360	490	40000
ABS	1.04	85	80	85–100	–	390	400	35000
SAN	1.08	95	85	100–110	–	370	455	36000
PVC rigid	1.40	70	60	75–80	200–300	390	455	18000
Polytetrafluoro-ethylene	2.20	260	200	110	510–540	560	580	4600
Polymethyl methacrylate	1.18	95	70–80	85–110	170–300	300–350	450	26000
Polyamide 6	1.13	170	80–100	200–220	300–350	420	450	32000
Polyethylene terephthalate	1.34	150	130	80	285–305	440	480	21500
Polycarbonate	1.20	140	100	150–155	350–400	520	2)	31000
Polyoxymeth-ylene	1.42	140	90–110	170	220	325–400	375	17000

[1] LD = Low density HD = High density
2) no ignition
[3] by ASTM D 1929

4 Terminology of Fire Protection

The definitions and terms listed in Sections 3.1 and 3.2 are based on the ISO/TAG 5 glossary of fire terms, AFNOR X 65-020, DIN 50 060 standards [1–4] , the bilingual ISO 3261 and ISO 4880 and the more recent trilingual ISO 13943 [5–7] as well as numerous discussions by the author with English, French and German colleagues. The fire protection standards described in this book and relevant specialized literature have also been taken into account.

The definitions of fire protection terms given in 4.1 have been taken from ISO/TAG 5, but have been altered where necessary.

The three language glossary in Section 4.2 is considerably more extensive than the above standards and contains expressions which are not specific to fire protection but are widely used in the field. The glossary is not claimed to be complete but is intended to assist the reader acquainting himself with the subject as well as those consulting literature in other languages. It is hoped that it will provide the impetus for a thesaurus of fire protection terminology in several languages.

References

[1] ISO/TAG 5/N 50 rev., Feb. 1988. Glossary of fire terms (draft ISO/IEC guide)
[2] Norme expérimentale X 65-020. Essais de comportement au feu. Vocabulaire. December 1978
[3] ISO 8421-1-1987. General terms and phenomena of fire
[4] DIN 50060. Aug. 1985. Prüfung des Brandverhaltens von Werkstoffen und Erzeugnissen. Begriffe
[5] ISO 4880-1997 Burning behaviour of textiles and textile products – Vocabulary
[6] ISO 3261-1975 Fire Tests – Vocabulary
[7] ISO 13943-2000 Fire Safety – Vocabulary

4.1 Definitions of terms connected with fire protection

Term	Definition
Afterflame	Persistence of flaming of a material after the ignition source has been removed
Afterglow	Persistence of glowing of a material after cessation of flaming or, if no flaming occurs, after the ignition source has been removed
Burn	To undergo combustion
Burning behavior	All the physical and/or chemical changes that take place when a material or product is exposed to a specified ignition source
Calorific potential	Calorific energy which could be released by the complete combustion of a unit mass of a material (ISO 8421-1)
Char (n.)	Carbonaceous residue resulting from pyrolysis or incomplete combustion
Char (v.)	To form carbonaceous residue during pyrolysis or incomplete combustion
Combustible	Capable of burning
Combustion	Exothermic reaction of a substance with an oxidizer generally accompanied by flames and/or glowing and/or emission of smoke
Damaged area	Total of the area of a material permanently affected by thermal phenomena under specified test conditions: loss of material, shrinking, softening, melting, charring, combustion, pyrolysis, etc.
Ease of ignition	The ease with which a material can be ignited under specified test conditions
Fire	a) A process of combustion characterized by the emission of heat accompanied by smoke and/or flame b) Rapid combustion spreading uncontrolled in time and space
Fire behavior	All the physical and/or chemical changes that take place when a material, product and/or structure is exposed to an uncontrolled fire
Fire effluent	The total gaseous, particulate or aerosol effluent from combustion or pyrolysis
Fire load	The sum of the calorific energies which could be released by the complete combustion of all the combustible materials in a space, including the facings of the walls, partitions, floors and ceilings
Fire performance	See fire behavior
Fire resistance	The ability of an element of building construction to fulfil for a stated period of time the required load bearing function, integrity and/or thermal insulation specified in the standard fire resistance test (see ISO 834)
Fire retardant	A substance added, or a treatment applied to a material in order to suppress, significantly reduce or delay the combustion of the material
Flame (n.)	Zone of combustion in the gaseous phase from which light is emitted
Flame (v.)	To undergo combustion in the gaseous phase with emission of light
Flame retardance	The property of a material either inherent or by virtue of a substance added or a treatment applied to suppress, significantly reduce or delay the propagation of flame
Flame retardant	See fire retardant
Flame spread	Propagation of a flame front
Flame spread rate	Distance travelled by a flame front during its propagation per unit time under specified test conditions

Flame spread time	Time taken by a flame on a burning material to travel over a specified distance or surface area under specified test conditions
Flammability	Ability of a material or product to burn with a flame under specified test conditions
Flammable	Capable of burning with a flame under specified test conditions
Flash-over	Transition to a state of total surface involvement in a fire of combustible materials within an enclosure
Flash point	The minimum temperature at which, under specified test conditions, a substance emits sufficient flammable gas to ignite momentarily on application of an ignition source
Full fire development	The transition to a state of full involvement of combustible materials in a fire
Fully developed fire	The state of total involvement of combustible materials in a fire
Glowing combustion	Combustion of a material in the solid phase without flame but with emission of light from the combustion zone
Heat of combustion	See calorific potential
Heat release rate	The thermal energy released per unit time by a material during combustion under specified test conditions
Ignite (vt)	To initiate combustion
Ignite (vi)	To catch fire with or without the application of an external heat source
Ignition temperature	Minimum temperature of a material at which sustained combustion can be initiated under specified test conditions Emission of light produced by a material when intensely heated. It can be produced with or without combustion (ISO/TAG 5)
Incandescence	Glowing produced without combustion or other chemical reaction, e.g. by electrical heating of a tungsten filament (ISO/TC 21)
Light (v.)	To initiate flaming combustion
Lighted (a.)	The state of a material after ignition and during persistence of flame
Lighting (n.)	First appearance of flame
Mass burning rate	Mass of a material burnt per unit time under specified test conditions
Melting behaviour	Phenomena accompanying the softening of a material under the influence of heat (including shrinking, dripping, burning of molten material, etc.)
Opacity of smoke	The ratio (I/T) of incident luminous flux (I) to transmitted luminous flux (T) through smoke under specified test conditions (ISO/TAG 5)
Optical density	Measure of degree of opacity usually expressed as the common smoke logarithm of the ratio of the incident light intensity to the transmitted light intensity (ISO/TC 38) The common logarithm of the opacity of smoke (ISO/TAG 5)
Pyrolysis	Irreversible chemical decomposition of a material due to an increase in temperature without oxidation
Pyrophoric material	A material capable of spontaneous ignition when brought into contact with air
Radiation	The thermal energy transfer by electromagnetic waves
Reaction to fire	The response of a material under specified test conditions in contributing by its own decomposition to a fire to which it is exposed
Scorch (v.)	To modify the surface of a material by limited carbonisation due to heat
Self heating	An exothermic reaction within a material resulting in a rise in temperature of the material
Self ignition	Ignition resulting from self heating

Self propagation of flame	The propagation of a flame front after removal of applied heat source
Smoke	A visible suspension of solid and/or liquid particles in gases resulting from combustion or pyrolysis
Smoke obscuration	The reduction in luminous intensity due to passage through smoke
Smoldering	The slow combustion of a material without light being visible and generally evidenced by an increase in temperature and/or by smoke
Soot	Finely divided particles, mainly carbon, produced and/or deposited during the incomplete combustion of organic materials
Spontaneous ignition	The minimum temperature at which ignition is obtained by temperature heating, under specified test conditions in the absence of any additional ignition source
Temperature-Time curve	The time-related variation of temperature measured in a specified way during the standard fire resistance test

4.2 English, German and French terms frequently used in connetion with fire protection

English	German	French
actual calorific value	effektive Verbrennungswärme (f)	potentiel (m) calorifique réel
ad-hoc fire experiment	Ad-hoc-Brandversuch (m)	essai (m) au feu ad hoc
afterflame	Nachbrennen (n) mit Flamme	flamme (f) résiduelle
afterflame time	Nachbrenndauer (f)	durée (f) de la flamme résiduelle
afterglow	Nachglimmen (n)	incandescence (f) résiduelle
alcohol flame	Alkoholflamme (f)	flamme (f) d'alcool
to apply a flame	beflammen	appliquer une flamme
approval	Anerkennung (f)	homologation (f)
arc	Lichtbogen (m)	arc (m) électrique
area burning rate	Flächenabbrennrate (f)	vitesse (f) de combustion en surface
arson	Brandstiftung (f)	incendie (m) criminel
arsonist	Brandstifter (m)	incendiaire (m)
ash	Asche (f)	cendre (f) ou cendres (f pl.)
building	Bauwesen	bâtiment (m) ou construction (f)
building component	Bauteil (n)	élément (m) de construction
building material	Baustoff (m)	matériau (m) de construction
building regulations	Bauvorschriften (f pl.)	réglementation (f) (bâtiment)
building structure	Bauwerksteil (n)	structure (f)
to burn	brennen	brûler
burned area	verbrannte Flächenanteile (m pl.)	surface (f) ou zone (f) brûlée
burner	Brenner (m)	brûleur (m)
burning behavior	Brennverhalten (n)	comportement (m) au feu
to burn out	ausbrennen, ausglühen	calciner, réduire en cendres
Bunsen burner	Bunsenbrenner (m)	brûleur (m) ou bec (m) Bunsen
calorific potential	Heizwert, spezifische Verbrennungswärme	pouvoir (m) calorifique

Char	Verkohlungsrückstand (m)	résidu (m) charbonneux
to char	verkohlen	carboniser
chimney	Rauchabzug (m)	conduit (m) de fumée
combustibility	Brennbarkeit (f)	combustibilité (f)
combustible	brennbar	combustible
combustion	Verbrennung (f)	combustion (f)
combustion chamber	Brennkasten (m)	chambre (f) de combustion
combustion products	Verbrennungsprodukte (n pl.)	produits (m pl.) de combustion
composite building materials	Verbund-Baustoffe (m pl.)	matériaux (m pl.) de construction composites ou matériaux multi-couches
conveyor belt	Fördergurt (m)	courroie (f) transporteuse
to cool	abkühlen, kühlen	refroidir
corrosion	Korrosion (f)	corrosion (f)
corrosivity	Korrosivität (f)	corrosivité (f)
course of fire	Brandverlauf (m)	déroulement (m) de l'incendie
damaged area	beschädigte Fläche (f)	surface (f) ou zone (f) endommagée
damaged length	Länge der Beschädigung (f)	longueur (f) endommagée
decomposition	Zersetzung (f)	décomposition (f)
decreasing fire	abklingender Brand (m)	incendie (m) décroissant
degradation	Abbau (m)	dégradation (f)
thermal –	thermischer –	– thermique
oxidative –	oxidativer –	– oxidative
diffusion flame	Diffusionsflamme (f)	flamme (f) de diffusion
ease of ignition	Entzündbarkeit (f)	facilité (f) d'allumage
edge application of flame	Kantenbeflammung (f)	application (f) de la flamme sur la tranche (de l'éprouvette)
exposure time (of an ignition source)	Einwirkungsdauer (f) (einer Zündquelle)	temps (m) d'exposition (d'une source d'allumage)
extinction	Verlöschen (n)	extinction (f)
to extinguish	löschen	éteindre
fire	Brand (m), Schadenfeuer (n), Feuer (n)	feu (m), incendie (m)
fire behavior	Brandverhalten (n)	comportement (m) au feu ou tenue (f) au feu
fire brigade	Feuerwehr (f)	sapeurs-pompiers (m pl.)
fire development	Brandentwicklung (f)	développement (m) de l'incendie
fire effluents	flüchtige Brandprodukte (n pl.), Brandgase (n pl.)	effluents (m) du feu
fire experiment	Brandversuch (m)	essai (m) au feu
fire exposure	Feuerbeanspruchung (f)	exposition (f) au feu
fire gases	Brandgase (n pl.)	gaz (m pl.) de combustion
fire hazard	Brandrisiko (n)	danger (m) du feu
fire insurance	Feuerversicherung (f)	assurance (f) incendie
fire load	Brandlast (f)	charge (f) calorifique
fire-load density	flächenbezogene Brandlast (f) (Brandbelastung)	charge (f) calorifique par unité de surface
fire loss	Brandschaden (m)	dommage (m) causé par l'incendie

fire performance	Brandverhalten (n)	comportement (m) au feu ou tenue (f) au feu
place of fire origin	Brandherd (m)	foyer (m) de l'incendie
fire precaution	vorbeugender (baulicher) Brandschutz (m)	prévention (f) incendie (bâtiment)
fire propagation	Brandausbreitung (f)	propagation (f) de l'incendie
fire protection	Brandschutz (m)	sécurité (f) ou protection (f) contre l'incendie
fire resistance	Feuerwiderstandsfähigkeit (f)	résistance (f) au feu
fire retardant	abbrandverzögernd	ignifuge
fire-retardant treatment	Brandschutzausrüstung (f)	ignifugation (f)
fire room (test)	Brandraum (m) (Versuch)	local (m) pour essais au feu
fire safety	Brandschutz (m)	sécurité (f) ou protection (f) contre l'incendie
fire scenario	Brand-Fallstudie (f) (Szenario [n])	scénario (m) d'incendie
fire test	Brandprüfung (f)	essai (m) au feu
flame	Flamme (f)	flamme (f)
to flame	brennen (mit Flamme)	flamber
flame application	Beflammung (f)	application (f) de la flamme
flame front	Flammenfront (f)	front (m) de flamme
flame height	Flammenhöhe (f)	hauteur (f) de flamme
flameless combustion	Verbrennung (f) ohne Flammenerscheinung	combustion (f) sans flamme
flame retardant	Flammschutzmittel (n)	ignifugeant (m) ou retardateur (m) de flamme
flame spread	Flammenausbreitung (f)	propagation (f) de la flamme
flame-spread rate	Flammenausbreitungs-geschwindigkeit (f)	vitesse (f) de propagation de la flamme
flame-spread time	Flammenausbreitungsdauer (f)	durée (f) de propagation de la flamme
flame tip	Flammenspitze (f)	pointe (f) de la flamme
flammability	Entflammbarkeit (f)	inflammabilité (f)
flammable	entflammbar	inflammable
flash-over	schlagartige Flammen-ausbreitung (f)	embrasement (m) généralisé
flash point	Flammpunkt (m)	point (m) d'éclair
flooring	Fußbodenbelag (m)	revêtement (m) de sol
fuel	Brennstoff (m), Kraftstoff (m)	combustible (m), carburant (m)
fully developed fire	vollentwickelter Brand (m) Vollbrand (m)	feu (m) généralisé
furnace	Ofen (m)	four (m)
gas	Gas (n)	gaz (m)
gas burner	Gasbrenner (m)	brûleur (m) … gaz
to glow	glimmen	brûler avec incandescence
glowing combustion	Glimmen (n)	combustion (f) incandescente
heat	Wärme (f)	chaleur (f)
heating	Erwärmung (f)	échauffement (m)
heat of combustion	spezifische Verbrennungs-wärme (f)	pouvoir (m) calorifique (calorific potential)

heat release rate	Wärmeabgaberate (f)	débit (m) thermique
to ignite	entzünden, sich entzünden	allumer, prendre feu
ignition	Zündung (f)	allumage (m)
ignition device	Zündvorrichtung (f)	dispositif (m) d'allumage
ignition point	Entzündungspunkt (m)	point (m) d'allumage
ignition source	Zündquelle (f)	source (f) d'allumage
ignition temperature	Entzündungstemperatur (f)	température (f) d'allumage
ignition time	Zünddauer (f)	temps (m) d'allumage
incandescence	Glühen (n)	incandescence (f)
to be incandescent	glühen	être incandescent
initiating fire	Entstehungsbrand (m)	incendie (m) naissant
intumescence	Intumeszenz (f)	intumescence (f)
intumescent	intumeszierend, aufschäumend	intumescent
irradiation	Bestrahlung (f)	irradiation (f)
laboratory test	Laborprüfverfahren (n)	essai (m) en laboratoire
to light	anzünden	enflammer
light absorption	Lichtabsorption (f)	absorption (f) de lumière
lighting	Aufflammen (n)	inflammation (f)
light source	Lichtquelle (f)	source (f) lumineuse
luminous flame	leuchtende Flamme (f)	flamme (f) lumineuse
mass burning rate	Abbrandrate (f)	vitesse (f) massique de combustion
to melt	schmelzen	fondre
melting behaviour	Schmelzverhalten (n)	comportement (m) thermofusible
method of test	Prüfmethode (f)	méthode (f) d'essai
model fire test	Modellbrandversuch (m)	essai (m) d'incendie en maquette
multiple flame burner	Reihenbrenner (m)	brûleur (m) multiflamme
non-combustibility	Nichtbrennbarkeit (f)	incombustibilité (f)
non-combustible	nichtbrennbar	incombustible
optical density of smoke	optische Rauchdichte (f)	densité (f) optique de la fumée
photocell	Photozelle (f)	cellule (f) photo-électrique
pilot flame	Zündflamme (f)	flamme-pilote (f)
premixed flame	vorgemischte Flamme (f)	flamme (f) prémélangée
post-combustion	Nachverbrennung (f)	post-combustion (f)
to put out	löschen	éteindre
pyrolysis	Pyrolyse (f)	pyrolyse (f)
phyrophoric material	pyrophorer Stoff (m)	pyrophore (m)
radiant flux	Strahlungsfluss (m)	flux (m) de rayonnement
radiant panel	Strahlerplatte (f)	panneau (m) radiant
radiation	Wärmestrahlung (f)	rayonnement (m) thermique
radiation intensity	Strahlungsintensität (f)	intensité (f) de rayonnement
radiator	Strahler (m)	radiateur (m), épiradiateur (m)
radiometer	Wärmestrahlungsmessgerät (m)	radiomètre (m)
rate of burning	Brenngeschwindigkeit (f)	vitesse (f) de combustion
reaction to fire	Brandverhalten	réaction (f) au feu
recognition	Anerkennung (f)	reconnaissance (f)
regulations	Vorschriften (f pl.)	réglementation (f)
sample	Probekörper (m)	éprouvette (f)
to scorch	versengen	roussir

self-heating	Selbsterwärmung (f)	auto-chauffage (m)
self-ignition	Selbstentzündung (f)	inflammation (f) spontanée
self-propagation of flame	selbständigeFlammen-ausbreitung (f)	autopropagation (f) de flamme
short-circuit	Kurzschluss (m)	court-circuit (m)
smoke	Rauch (m)	fumée (f)
smoke chamber	Rauchkammer (f)	chambre (f) ... fumée
smoke density	Rauchdichte (f)	densité(f) de fumée
smoke development	Rauchentwicklung (f)	développement (m) ou dégagement (m) de fumée
smoke generation	Rauchentwicklung (f)	développement (m) ou dégagement (m) de fumée
smoke layer	Rauchschicht (f)	couche (f) de fumée
smoke obscuration	Verdunkelung durch Rauch	obscurcissement (m) par la fumée
smoke particles	Rauchteilchen (-partikel) (n pl.)	particules (f pl.) de fumée
smoke production	Rauchentwicklung (f)	développement (m) ou dégagement (m) de fumée
smoke suppressant	Rauchverminderer (m)	agent (m) réducteur de fumée
smouldering	Schwelen (n)	feu (m) couvant
to soften	erweichen	ramollir
soot	Ruß (m)	suie (f)
spark	Funke (m)	étincelle (f)
specific optical density	materialbezogene optische Rauchdichte (f)	densité (f) optique spécifique
specimen	Probe (f)	échantillon (m)
specimen holder	Probenhalterung (f)	porte-éprouvette (m)
spontaneous ignition	Selbstentzündung (f)	inflammation (f) spontanée
spontaneous ignition	Selbstentzündungstemperatur (f)	température (f) d'inflammation spontanée
spread of flame	Flammenausbreitung (f)	propagation (f) de la flamme
structural fire precautions	vorbeugender, baulicher Brandschutz	prévention (f) incendie (bâtiment)
surface burn	oberflächiges Abbrennen (n)*	combustion (f) en surface
surface flash	oberflächiges Abflammen (n)*	effet (m) éclair de surface
temperature resistant	temperaturbeständig	thermostable
testing laboratory	Prüflaboratorium (n)	laboratoire (m) d'essais
test report	Prüfbericht (m)	procès-verbal (m) ou compte-rendu (m) d'essais
test result	Prüfergebnis (n)	résultat (m) d'essai
thermal	thermisch	thermique
thermocouple	Thermoelement (n)	thermocouple (m)
time-temperature curve	Temperatur-Zeit-Kurve (f)	courbe (f) température temps
toxicity	Toxizität (f)	toxicité (f)
to vaporize	verdampfen	volatiliser
wicking	Dochtwirkung (f)	effet (m) mèche
wood crib	Holzkrippe (f)	foyer (m) de bois

* = no identical German term

5 International and National Standards Organizations

5.1 International Organizations

International Electrotechnical Commission
(IEC)
3, rue de Varembé
P.O. Box 131
CH-1211 Geneva 20

Tel.   + 41 22 919 02 11
Fax   + 41 22 919 03 00
Mail   info@iec.ch
Internet: www.iec.ch

International Organisation for
Standardisation (ISO)
1, rue de Varembé
P.O. Box 56
CH-1211 Geneva

Tel.   + 41 22 749 01 11
Fax   + 41 22 733 34 30
Mail   central@iso.ch
Internet: www.iso.ch

Comité Européen de Normalisation/
Europäisches Komitee für Normung/
European Committee for Standardisation
(CEN)
36, rue de Stassart
B-1050 Brussels

Tel.   +32 2 550 08 19
Fax   +32 2 550 08 11
Mail   infodesk@cenorm.be
Internet: www.cenorm.be

5.2 National Organizations

Australia

Standards Australia International Ltd. (SAI)
286, Sussex Street
GPO Box 5420
AU-Sidney, NSW 2001

Tel.   +61 2 82 06 60 00
Fax   +61 2 82 06 60 01
Mail   intsect@standards.com.au
Internet: www.standards.com.au

Austria

Österreichisches Normungsinstitut (ON)
Heinestr. 38
Postfach 130
AT-1021 Wien

Tel.   +43 1 213 00
Fax   +43 1 213 00 650
Mail   sales@on-norm.at
Internet: www.on-norm.at

Belarus

State Committe for Standardisation,
Metrology and Certification of Belarus
(BELST)
Starovilensky Trakt 93
BY-Minsk 220053

Tel.   +375 172 37 52 13
Fax   +375 172 37 25 88
Mail   belgiss@mail.belpak.by

Belgium

Institut Belge de Normalisation (IBN)
Av. de la Brabançonne 29
BE-1000 Bruxelles

Tel.   +32 2 550 08 19
Fax   +32 2 550 08 11
Mail   voohof@ibn.be
Internet: www.ibn.be

Brazil

Associacao Brasileira de Normas Tecnicas
(ABNT)
Av. 13 de Maio, n° 13, 28° andar
BR-20003-900- Rio de Janeiro-RJ

Tel.   +55 21 210 31 22
Fax   +55 21 220 17 62
Mail   abnt@abnt.or.br
Internet: www.abnt.or.br

Bulgaria

State Agency for Standardisation and
Metrology (BDS)
21, 6th September Str.
BG-1000 Sofia

Tel.   +359 2 989 84 88
Fax   +359 2 986 17 07
Mail   standards@sasm.orbitel.bg

Canada

Standards Council of Canada (SCC)
270 Albert Street, Suite 200
CA-Ottawa, Ontario K1P 6N7

Tel.   +1 613 238 32 22
Fax   +1 613 569 78 08
Mail   info@scc.ca
Internet: www.scc.ca

Chile

Instituto Nacional de Normalización (INN)
Matías Cousino 64 - 6° piso
Casilla 995 – Correo Central
CL-Santiago

Tel.   +56 2 441 03 30
Fax   +56 2 441 04 27
Mail   inn@entelchile.net
Internet: www.inn.cl

China

China State Bureau of Quality and
Technical Supervision (CSBTS)
4, Zichun Road, Haidian District
P.O. Box 8010
CN-Beijing 100088

Tel.   +86 10 6 203 24 24
Fax   +86 10 6 203 37 37
Mail   csbts@mail.csbts.cn.net
Internet: www.csbts.cn.ne

Croatia

State Office for Standardisation and
Metrology (DZNM)
Ulica grada Vukovara 78
HR-10000 Zagreb

Tel.   +385 1 610 63 20
Fax   +385 1 610 93 20
Mail   ured.ravnatelja@dznm.hr
Internet: www.dznm.hr

Czech Republik

Czech Standards Institute (CSNI)
Biskupsky dvur 5
CZ-110 02 Praha 1

Tel.   +420 2 21 80 21 11
Fax   +420 2 21 80 23 11
Mail   internat.dept@csni.cz
Internet: www.csni.cz

Denmark

Dansk Standard (DS)
Kollegievej 6
DK-2920 Charlottenlund

Tel.   +45 39 96 61 01
Fax   +45 39 61 02
Mail   dansk.standard@ds.dk
Internet: www.ds.dk

Estonia

Eesti Standardikeskus (ESK)
10, Aru Street
EE-10317 Tallinn

Tel.   +372 651 92 00
Fax   +372 651 92 20
Mail   info@evs.ee
Internet: www.evs.ee

Finland

Finnish Standards Association (SFS)
P.O. Box 116
FI-00241 Helsinki

Tel.   +358 9 149 93 31
Fax   +358 9 146 49 25
Mail   sfs@sfs.fi
Internet: www.sfs.fi

France

Association Française de Normalisation
(AFNOR)
11, av. Francis de Pressensé
F-93571 Saint-Denis La Plaine Cedex

Tel.   +33 1 41 62 80 00
Fax   +33 1 49 17 90 00
Mail   uari@afnor.fr
Internet: www.afnor.fr

Germany

Deutsches Institut für Normung (DIN)
Burggrafenstr. 6
D-10772 Berlin

Tel.   +49 30 26 01 0
Fax   +49 30 26 01 12 31
Mail   directorate.international@din.de
Internet: www.din.de

Greece

Hellenic Organisation for Standardisation
(ELOT)
313, Acharnon Street
GR-111 45 Athens

Tel.   +30 1 21 20 100
Fax   +30 1 21 20 131
Mail   elotinfo@elot.gr
Internet: www.elot.gr

Hungary

Magyar Szabványügyi Testület (MSZT)
Pf. 24
HU-1450 Budapest 9

Tel.   +36 1 456 68 00
Fax   +36 1 456 68 23
Mail   isoline@mszt.hu
Internet: www.mszt.hu

India

Bureau of Indian Standards (BIS)
Manak Bhavan
9 Behadur Shah Zafar Marg
IN-New Dehli 110002

Tel.   +91 11 323 79 91
Fax   +91 11 323 93 99
Mail   bis@vsnl.com
Internet: www.bis.org.in

Indonesia

Badan Standardisasi Nasional (BSN)
Manggala Wanabakti Blok IV Lt. 4
Jl. Gatot Subroto, Senayan
ID-Jakarta 10270

Tel.   +62 21 574 70 43
Fax   +62 21 574 70 45
Mail   bsn@bsn.or.id
Internet: www.bsn.or.id

Ireland

National Standards Authority of Ireland
(NSAI)
Glasnevin
IE-Dublin-9

Tel.   +353 1 807 38 00
Fax   +353 1 807 38 38
Mail   nsai@nsai.ie
Internet: www.nsai.ie

Israel

Standards Institution of Israel (SII)
42 chaim Levanon Street
IL-Tel Aviv 69977

Tel.   +972 3 646 51 54
Fax   +972 3 641 96 83
Mail   iso/iec@sii.org.il
Internet: www.sii.org.il

Italy

Ente Nazionale Italiano di Unificazione
(UNI)
Via Battistotti Sassi 11/b
I-20133 Milano

Tel.   +39 02 70 02 41
Fax   +39 02 70 10 61 49
Mail   uni@uni.com
Internet: www.uni.com

Japan

Japanese Industrial Standards Committee
(JISC)
Technical Regulation, Standards and
Conformity Assessment Policy Unit
Ministry of Economy, Trade and Industry
JP-Tokyo 100-8901

Tel.   +81 3 35 01 94 71
Fax   +81 3 35 80 86 37
Mail   isojisc@meti.go.jp
Internet: www.jisc.org

Republic of Korea

Korean Agency for Technology and Standards
(KATS)
Ministry of Commerce, Industry and Energy
2, Joongang-dong, Kwachon-city
KR-Kyuungi-do 427-010

Tel.   +82 2 509 73 99
Fax   +82 2 509 79 77
Mail   standard@ats.go.kr
Internet: www.ats.go.kr

Latvia

Latvian Standard (LVS)
157, Kr. Valdemara Street
LV-Riga 1013

Tel.   +371 7 37 13 08
Fax   +371 7 37 13 24
Mail   lvs@lvs.lv
Internet: www.lvs.lv

Lithuania

Lithuanian Standards Board (LSD)	Tel.	+370 2 70 93 60
T. Kosciuskos g. 30	Fax	+370 2 22 62 52
LT-2600 Vilnius	Mail	istboard@lsd.lt
	Internet: www.lsd.lt	

Luxembourg

Service de l'Energie de l'Etat (SEE)		
Organisme Luxembourgeois de		
Normalisation	Tel.	+352 46 97 46 1
34, av. De la Porte-Neuve	Fax	+352 46 97 46 39
B.P. 10	Mail	see.nomalisation@eg.etat.lu
L-2010 Luxembourg	Internet: www.etat.lu/see	

Mexico

Direción General de Normas (DGN)		
Calle Puente de Tecamachalco N° 6	Tel.	+52 5 729 94 80
Lomas de Tecamachalco	Fax	+52 5 729 94 84
Sección Fuentes	Mail	cidgn@economia.gob.mx
MX-53 950 Naucalpan de Juárez	Internet: www.economia.gob.mx/normas	

Marocco

Service de Normalisation Industrielle		
Marocaine (SNIMA)		
Ministère de l'industrie, du commerce,		
de l'énergie et des mines	Tel.	+212 37 71 62 14
Angle Av. Kamal Zebdi et Rue Dadi	Fax	+212 37 71 17 98
Secteur 21 Hay Riad	Mail	snima@mcinet.gov.ma
MA-10100 Rabat	Internet: www.mcinet.gov.ma	

Netherlands

Nederlands Normalisatie-Instituut (NEN)	Tel.	+31 15 2 69 03 90
P.O. Box 5059	Fax	+31 15 2 69 01 90
NL-2600 GB Delft	Mail	info@nen.nl
	Internet: www.nen.nl	

New Zealand

Standards New Zealand (SNZ)	Tel.	+64 4 498 59 90
Radio New Zealand House	Fax	+64 4 498 59 94
155 The Terrace	Mail	snz@standards.co.nz
NZ-Wellington	Internet: www.standards.co.nz	

Norway

Norges Standardiseringsforbund (NSF)	Tel.	+47 22 04 92 11
Drammensveien 145A	Fax	+47 22 04 92 11
Postboks 353 Skoyen	Mail	firmapost@standard.no
N-0213 Oslo	Internet: www.standard.no	

Poland

Polish Committee for Standardisation (PKN)	Tel.	+48 22 620 54 34
Ul. Elektoralna 2	Fax	+48 22 620 54 34
P.O. Box 411	Mail	intdoc@pkn.pl
PL-00-950 Warszawa	Internet:	www.pkn.pl

Portugal

Instituto Portuguès da Qualidade (IPQ)	Tel.	+351 21 294 8100
Rua António Giao, 2	Fax	+351 21 294 81 01
P-2869-513 Caparica	Mail	ipq@mail.ipq.pt
	Internet:	www.ipq.pt

Romania

Asociatiade Standardizare din Romania	Tel.	+40 1 211 32 96
(ASRO)	Fax	+40 1 210 08 33
Str. Mendeleev 21–25	Mail	irs@kappa.ro
RO-70168 Bucuresti 1		

Russian Federation

State Committee of the Russian Federation		
for Standardisation and Metrology	Tel.	+7 095 236 40 44
(GOST R)	Fax	+7 095 237 60 32
Leninsky Prospekt 9	Mail	info@gost.ru
RU-Moskva 117049	Internet:	www.gost.ru

Slovakia

Slovak Institute for Standardisation	Tel.	+421 7 60 29 44 74
(SUTN)	Fax	+421 7 65 41 18 88
P.O. Box 246	Mail	ms post@sutn.gov.sk
SK-840 00 Brastislava 4	Internet:	www.sutn.gov.sk

Slovenia

Slovenian Institute for Standardization (SIST)	Tel.	+386 1 478 30 00
Smartinska 140	Fax	+386 1 478 30 94
SI-1000 Ljubljana	Mail	sist@sist.si
	Internet:	www.sist.si

South Africa

South African Bureau of Standards		
(SABS)	Tel.	+27 12 428 79 11
1 Dr Lateggan Rd, Groenkloof	Fax	+27 12 344 15 68
Private Bag X191	Mail	info@sabs.co.za
ZA-Pretoria 0001	Internet:	www.sabs.co.za

Spain

Asociación Espanola de Normalización	Tel.	+34 91 432 60 00
y Certicación (AENOR)	Fax	+34 91 31049 76
Génova, 6	Mail	aenor@aenor.es
E-28004 Madrid	Internet:	www.aenor.es

Sweden

Swedish Standards Institute (SIS)	Tel.	+46 8 555 520 00
Sankt Paulgatan	Fax	+46 8 555 520 01
S-11880 Stockholm	Mail	info@sis.se
	Internet:	www.sis.se

Switzerland

Schweizerische Normen-Vereinigung	Tel.	+41 52 224 54 54
(SNV)	Fax	+41 52 224 54 74
Bürglistr. 29	Mail	info@snv.ch
CH-8400 Winterthur	Internet:	www.snv.ch

Turkey

Türk Standardlari Enstitüsü (TSE)	Tel.	+90 312 417 83 30
Necatibey Cad. 112	Fax	+90 312 425 43 99
Bakanlikar	Mail	usm@tse.org.tr
TR-06100 Ankara	Internet:	www.tse.org.tr

USA

American National Standards Institute	Tel.	+1 212 642 49 00
(ANSI)	Fax	+1 212 398 00 23
1819L Street, NW	Mail	info@ansi.org
US-Washington, DC 20036	Internet:	www.ansi.org

Ukraine

State Committee of Standardisation,		
Metrology and Certification of Ukraine	Tel.	+380 44 226 29 71
(DSTU)	Fax	+380 44 226 29 70
174 Gorkiy Street, 03680	Mail	dstu@issi.kiev.ua
UA-Kyiv-680	Internet:	www.dstu.gov.ua

United Kingdom

British Standard Institution (BSI)	Tel.	+44 208 996 90 00
389 Chiswick High Road	Fax	+44 208 996 74 00
GB-London W44AL	Mail	standards.international
		@bsi-global.com
	Internet:	www.bsi-global.com

Yugoslavia

Savezni zavod za standardizaciju (SZS)	Tel.	+381 11 361 31 50
Kneza Milosa 20	Fax	+381 11 361 73 41
Post fah 609	Mail	jus@szs.sv.gov.yu
YU-11000 Beograd		

6 Electrotechnical Safety and Standard Organizations

Member	Status	Member	Status
Australia:	1, 4	Korea:	1, 5
Austria:	1, 5, 6	Latvia:	1, 2
Belarus:	1	Lithuania:	1, 2, 7
Belgium:	1, 4, 6	Luxemburg:	1, 6
Bosnia & Herzegovina:	1, 2	Malaysia:	1
Brazil:	1, 8	Mexico:	1
Bulgaria:	1, 5, 7	Netherlands:	1, 4, 6
Canada:	1, 4	New Zealand:	1
China:	1, 4	Norway:	1, 5, 6
Colombia:	1, 3	Pakistan:	1
Croatia:	1, 7	Philipines, Rep. of the:	1
Cuba:	1, 3	Poland:	1, 4
Cyprus:	1, 2, 7	Portugal:	1, 6
Czech Republic:	1, 4, 6	Romania:	1, 5, 7
Denmark:	1, 4, 6	Russian Federation:	1, 4
Egypt:	1	Saudi Arabia:	1
Eritrea:	1, 3	Singapore:	1, 5
Estonia:	1, 2, 7	Slovakia:	1, 5, 7
Finland:	1, 5, 6	Slovenia:	1, 7
France:	1, 4, 6	South Africa:	1, 5
Germany:	1, 4, 6	Spain:	1, 4, 6
Greece:	1, 5, 6	Sweden:	1, 4
Hungary:	1, 5, 7	Switzerland:	1, 6
Iceland:	1, 2, 6	Thailand:	1, 4
India:	1, 5	Turkey:	1, 7
Indonesia:	1	Ukraine:	1, 5
Ireland:	1, 6	United Kingdom:	1, 4, 6
Israel:	1	United States of America:	1, 4
Italy:	1, 4, 6	Uruguay:	1, 3
Japan:	1, 4	Yugoslavia:	1, 5
Kenya:	1, 3		

Member status:
1) IEC member
2) IEC associate member
3) IEC pre-associate member
4) IEC/TC 89 participating country
5) IEC/TC 89 observer country
6) CENELEC member state
7) CENELEC affiliates
8) Suspended October 1st, 1999

Source: IEC-Webside

Australia

Australian National Committee of IEC
Standards Australia
P.O. Box 1055
AU – Strathfield NSW 2135

Telephone: +61 2 9746 4700
Telefax: +61 2 9746 8450
E-mail: warren.miller@standards.com.au
WWW: http://www.standards.com.au

Austria

Austrian Electrotechnical Committee
c/o Oesterreichischer Verband für
Elektrotechnik
Eschenbachgasse 9
AT – 1010 Wien

Telephone: +43 (1) 587 63 73
Telefax: +43 (1) 586 74 08
E-mail: p.rausch@ove.at
WWW: http://ove.e2i.at

Belarus

Belarus National Committee of the IEC
Belstandart
Starovilensky Trakt, 93
BY – 220053 Minsk

Telephone: +375 172 37 52 13
Telefax: +375 172 37 25 88
Telex: 2521170 shkala
E-mail: belst@mcsm.belpak.minsk.by

Belgium

Comité Electrotechnique Belge
9 A Av. Frans Van Kalken
Boîte 2
BE – 1070 Bruxelles

Telephone: +32 (2) 556 01 10
Telefax: +32 (2) 556 01 20
E-mail: centraloffice@bec-ceb.be
WWW: http://www.bec-ceb.be

Bosnia Herzegovina

IEC Nat. Com. of Bosnia & Herzegovina
Institute for Standardization, Metrology &
Patents of B & H,
Hamdije Cemerlica 2 (ENERGOINVEST
building)
BA – 71000 Sarajevo

Telephone: +387 71 652 765
Telefax: +387 71 652 757
E-mail: zsmp@bih.net.ba

Brazil

Brazilian National Committee of the IEC
COBEI – ABNT / CB – 03
Rua Libero Badaro, 496 – 10° andar
BR – 01008.000 – Sao Paulo – SP

Telephone: +55 11 239 11 55
Telefax: +55 11 3104 0192
WWW: http://www.abnt.org.br

Bulgaria

Bulgarian National Committee of the IEC
Committee for Standardization and
Metrology
21, 6th September Street
BG – 1000 SOFIA

Telephone: +359 (2) 875 950
Telefax: +359 (2) 986 1707
Telex: 22570 dks bg
E-mail: csm@techno-link.com

Canada

Canadian National Committee of the IEC   Telephone: +1 (613) 238 32 22
Standards Council of Canada   Telefax: +1 (613) 995 45 64
International Standardization Division   Telex: 053–4403 stancan ott
45, O'Connor Street, Suite 1200   Telegrams: Stancan, Ottawa
CA – Ottawa, ONT. K1P 6N7   WWW: http://www.scc.ca

China

Chinese National Committee of the IEC   Telephone: +86 (10) 6202 2288
CSBTS   Telefax: +86 (10) 6203 3737
4 Zhichun Road   Telegrams: 1918 bejing
Haidian District, P.O. Box 8010   E-mail: ieccn@mail.csbts.cn.net
CN – Beijing 100088   WWW: http://www.csbts.cn.net/

Colombia

Instituto Colombiano de Normas
Técnicas y Certificacion (ICONTEC)   Telephone: +57 1 315 03 77
Carrera 37 No. 52–95   Telefax: +57 1 222 14 35
Edificio ICONTEC, P.O. Box 14237   Telex: 4 25 00 icont co
CO – Santafe de Bogota, D.C.   E-mail: sicontec@coll.telecom.com.co

Croatia

State Office for Standardization and   Telephone: +385 1 610 11 11
Metrology   Telefax: +385 1 610 93 24
Ulica Grada Vukovara 78   ou/or Tel.: +385 1 610 60 35
HR – 10000 Zagreb   E-mail: pisarnica@dznm.hr

Cuba

Cuban National Bureau of Standards (NC)   Telephone: +53 7 30 0835
Oficina Nacionale de Normalizacion   Telefax: +53 7 33 8048
Calle E No. 261 entre 11 y 13   et/and Tel.: +53 7 30 0022
CU – Vedado, Ciiudad de la Habana 10400   E-mail: ncnorma@ceniai.inf.cu

Cyprus

IEC National Committe of Cyprus
Cyprus Organization for Standards &
Control of Quality   Telephone: +357 (2) 30 01 92
Ministry of Commerce, Industry &   Telefax: +357 (2) 37 51 20
Tourism   Telex: 2283 mincomin
CY – 1421 Nicosia   Tel.:   +357 (2) 37 50 53

Czech Republic

Czech National Committee of the IEC
Czech Standards Institute (CSNI)   Telephone: +420 2 21 802 100
Biskupsky dvùr 5   Telefax: +420 2 21 802 311
CZ – 110 02 Praha 1   WWW: http://www.csni.cz/

Denmark

Dansk Standard
Danish Standards Association
Kollegievej 6
DK – 2920 Charlottenlund

Telephone: +45 (39) 96 61 01
Telefax: +45 (39) 96 61 02
E-mail: dansk.standard@ds.dk
WWW: http://www.ds.dk

Egypt

The Egyptian National Committee
Ministry of Electricity & Energy
Abbassia Post Office
EG – Cairo

Telephone: +20 (2) 83 06 41
Telefax: +20 (2) 261 65 12
Telex: 92 097 power un
E-mail: iecegypt@link.com.eg

Eritrea

Eritrean Standards Institution (ESI)
P.O. Box 245
ER – Asmara

Telephone: +291 1 11 56 24
Telefax: +291 1 12 05 86
ou/or Tel.: +291 1 12 02 45
E-mail: akberom@ecl.doe.gov.er

Estonia

Estonian National Committee of the IEC
Estonian Electrotechnical Committee
10, Aru Street
EE – 10317 Tallinn

Telephone: +372 651 9219
Telefax: +372 651 9220
E-mail: eek@evs.ee

Finland

Finnish Electrotechnical Standards
Association (SESKO)
P.O. BOX 134
FI – 00211 Helsinki

Telephone: +358 9 696 391
Telefax: +358 9 677 059
E-mail: finc@sesko.fi
WWW: http://www.sesko.fi

France

Union Technique de l'Electricité et de la
Communication (UTE)
Comité Electrotechnique Français
33, av. du Général Leclerc, BP 23
FR – 92262 Fontenay-Aux-Roses Cedex

Telephone: +33 1 40 93 62 00
Telefax: +33 1 40 93 44 08
E-mail: frenchnc@ute.asso.fr
WWW: http://www.ute-fr.com

Germany

Deutsches Komitee der IEC
Deutsche Elektrotechnische Kommission
im DIN und VDE
Stresemannallee 15
DE – 60596 Frankfurt am Main

Telephone: +49 (69) 630 80
Telefax: +49 (69) 96 31 52 18
E-mail: dke.zbi@t-online.de
WWW: http://www.vde.de

Greece

Hellenic Organization for Standardization
(ELOT)
313, Acharnon St.
GR – 111 45 Athens

Telephone: +30 (1) 21 20 100
Telefax: +30 (1) 21 20 430
E-mail: elotinfo@elot.gr
WWW: http://www.elot.gr

Hungary

Magyar Szabvanyügyi Testület
Hungarian Standards Institution
Ülloi ut 25
HU – 1091 Budapest

Telephone: +36 (1) 218 30 11
Telefax: +36 (1) 218 51 25
WWW: http://www.mszt.hu

Iceland

IEC National Committee of Iceland
Icelandic Council for Standardization
(STRI)
Holtagardar
IS – 104 Reykjavik

Telephone: +354 520 7150
Telefax: +354 520 7171
E-mail: stri@stri.is
WWW: http://www.stri.is

India

Bureau of Indian Standards
Manak Bhavan
9, Bahadur Shah Zafar Marg
IN – New Delhi 110 002

Telephone: +91 (11) 323 01 31
Telefax: +91 (11) 323 40 62
Telex: 031–65870 Answerback 'BIS/IN'
E-mail: bisind@del2.vsnl.net.in
WWW: http://wwwdel.vsnl.net.in/bis.org/

Indonesia

Badan Standardisasi Nasional (BSN)
Manggala Wanabakti Blok 4, 4th Floor
Jl. Jenderal Gatot Subroto, Senayan
ID – Jakarta 10270

Telephone: +62 21 574 70 43
Telefax: +62 21 574 70 45
ou/or Tel.: +62 21 574 70 44
E-mail: bsn-std@rad.net.id

Ireland

Electro-Technical Council of Ireland
Ballymun Road
Glasnevin
IE – Dublin 9

Telephone: +353 1 807 3800
Telefax: +353 1 807 3838
E-mail: nsai@nsai.ie
WWW: http://www.nsai.ie

Israel

The Standards Institution of Israel
42, Chaim Levanon Street
IL – Tel-Aviv 69977

Telephone: +972 3 64 65 154
Telefax: +972 3 64 19 683
WWW: http://www.sii.org.il

Italy

Comitato Elettrotechnico Italiano
Viale Monza, 259
IT – 20126 Milano

Telephone: +39 02 25 77 31
Telefax: +39 02 25 77 3210
WWW: http://www.ceiuni.it

Japan

Japanese Industrial Standards Committee
c/o International Standards Division
AIST/MITI
3–1, Kasumigaseki 1-chome, Chiyoda-ku
JP – Tokyo 100

Telephone: +81 3 3501 2096
Telefax: +81 3 3580 8637
E-mail: jisc_iec@jsa.or.jp
WWW: http://www.jisc.org/

Kenya

Kenya Bureau of Standards (KEBS)
Off Mombasa Road
Behind Belle Vue Cinema
P.O. Box 54974
KE – Nairobi

Telephone: +254 2 50 22 10/19
Telefax: +254 2 50 32 93
Telex: 25252 VIWANGO
E-mail: kebs@africaonline.co.ke

Korea

Korean National Committee of IEC
Korean Agency for Technology and
Standards (KATS), MOCIE
2, Joongang-dong, Kwachon
KR – Kyunggi-do, 427–010

Telephone: +82 2 507 4369
Telefax: +82 2 503 7977
ou/or Tel.: +82 2 509 7396 / 97 / 98
E-mail: standard@ats.go.kr
WWW: http://www.ats.go.kr

Latvia

Latvian National Committee of the IEC
Latvian Electrotechnical Commission
(LEC)
Pulkveza Brieza Street 12
LV – 1230 Riga

Telephone: +371 732 82 19
Telefax: +371 733 13 30
E-mail: gerke@guru.energo.lv

Lithuania

Lithuanian National Committee of the IEC
Lithuanian Standards Board (LST)
T. Kosciuskos g. 30
LT – 2600 Vilnius

Telephone: +370 2 70 93 60
Telefax: +370 2 22 62 52
E-mail: LSTBOARD@LSD.LT

Luxemburg

Comité National CEI du Luxembourg
Service de l'Energie de l'Etat (SEE)
B.P. No. 10
LU – 2010 Luxembourg

Telephone: +352 46 97 46 – (1)
Telefax: +352 22 25 24
E-mail: see.normalisation@eg.etat.lu
WWW: http://www.etat.lu/SEE/

Malaysia

Malaysian National Committee of the IEC
Dept. of Standards Malaysia (DSM)
21st Floor, Wisma MPSA
Persiaran Perbandaran
MY – 40675 Shah Alam, Selangor

Telephone: +60 3 559 80 33
Telefax: +60 3 559 24 97
E-mail: central@dsm.gov.my
WWW: http://www.dsm.gov.my

Mexico

Direccion General de Normas
Direcc. de Asuntos Internacionales (CEM)
Av. Puente de Tecamachalco No. 6
Col. Lomas de Tecamachalco Secc.
Fuentes
MX – 53950 Naucalpan DE Juarez,
Edo. de Mexico

Telephone: +52 (5) 729 94 80
Telefax: +53 (5) 729 94 84
WWW: http://www.secofi.gob.mx/
           dgn1.html

Netherlands

Netherlands National Committee of the	Telephone:	+31 (15) 2 690 390
IEC	Telefax:	+31 (15) 2 690 190
Kalfjeslaan 2	Telex:	38 144 nni nl
Post Box 5059	Telegrams:	Normalisatie Delft
NL – 2600 GB Delft	E-mail:	Nec@nni.nl
	WWW:	http://www.nni.nl

New Zealand

New Zealand Electrotechnical Committee	Telephone:	+64 (4) 498 5990
Standards New Zealand	Telefax:	+64 (4) 498 5994
Private Bag 2439	E-mail:	snz@standards.co.nz
NZ – Wellington 6020	WWW:	http://www.standards.co.nz/

Norway

Norsk Elektroteknistk Komite (NEK)	Telephone:	+47 22 52 69 50
Harbitzalléen 2A	Telefax:	+47 22 52 69 61
Postboks 280 Skoyen	E-mail:	nek@nek.no
NO – 0212 Oslo	WWW:	http://www.nek.no

Pakistan

Pakistan National Committee		
of the IEC EDC	Telephone:	+92 (21) 772 65 01
Pakistan Standards Institution	Telefax:	+92 (21) 772 81 24
39, Garden Road, Saddar	Telegrams:	Peyasai
PK – Karachi 3	E-mail:	pakqltyk@super.net.pk

Philippines

IEC Nat. Committee of the Philippines		
Bureau of Product Standards (BPS)	Telephone:	+63 2 890 - 4965
Trade and Industry Building	Telefax:	+63 2 890 - 4926
361 Sen. Gil J. Puyat Avenue	ou/or Fax:	+63 2 890 - 5130
PH – Makati City 1200, Metro Manila	E-mail:	dtibpsrp@mnl.sequel.net

Poland

Polish National Committee of the IEC		
Polish Committee for Standardization	Telephone:	+48 (22) 620 54 34
Ul. Elektoralna 2	Telefax:	+48 (22) 620 54 34
P.O. Box 411	ou/or Fax:	+48 (22) 620 07 41
PL – 00-950 Warszawa	E-mail:	intdoc@pkn.pl

Portugal

Portuguese National Committee of the IEC	Telephone:	+351 (1) 294 81 00/02
Instituto Portugues da Qualidade	Telefax:	+351 (1) 294 81 01
Rua C à Avenida dos Très Vales	ou/or Tel.:	+351 (1) 294 81 02
PT – 2825 Monte de Caparica	E-mail:	Nore@mail.kpq.pt
	WWW:	http://www.ipq.pt

Romania

Romanian National Committee of the IEC
I.C.P.E.
313, Splaiul Unirii   Telephone: +40 1 321 72 63
P.O. Box 104   Telefax: +40 1 321 37 69
RO – 74204 Bucharest 3   E-mail: cer@icpe.ro

Russia

Russian Federation Committee for the IEC   Telephone: +7 (095) 236 40 44
Gosstandart of Russia   Telefax: +7 (095) 237 60 32
Leninsky pr. 9   Telex: 411378 gost ru
RU – 117049 Moscow M-49   Telegrams: Moscva Gosstansart
                                           WWW: http://www.gost.ru

Saudi Arabia

Saudi Arabian National Committee of the
IEC Saudi Arabian Standards Org. (SASO)   Telephone: +966 1 452 00 00
P.O. Box 3437   Telefax: +966 1 452 00 86
SA – Riyadh 11471   Telex: 40 16 10 saso sj

Singapore

Singapore National Committee of the IEC
c/o Singapore Productivity and Standards
Board   Telephone: +65 778 7777
1 Science Park Drive   Telefax: +65 776 1280
SG – Singapore 118 221   WWW: http://www.psb.gov.sg/

Slovakia

Slovensky Elektrotechnicky Vybor (SEV)
Slovak Office of Standards, Metrology and
Testing (UNMS)
Stefanovicova 3, P.O. Box 76   Telephone: +421 7 52494 728
SK – 810 05 Bratislava 15   Telefax: +421 7 52491 050

Slovenia

Slovenian IEC National Committee
Ministrstvo za Znanost in Tehnologijo
Standards & Metrology Institute   Telephone: +386 61 178 30 00
Kotnikova 6   Telefax: +386 61 178 31 96
SI – 1000 Ljubljana   WWW: http://www.usm.mzt.si

South Africa

South African National Committee of the   Telephone: +27 (12) 428 79 11
IEC South African Bureau of Standards   Telefax: +27 (12) 344 15 68
Private Bag X 191   Telex: 3–21308 sa bs sa
ZA – Pretoria 0001   Telegrams: Comparator Pretoria

Spain

Comite Nacional Espanol de la	Telephone:	+34 91 432 60 00
CEI AENOR	Telefax:	+34 91 310 45 96
Génova, 6	E-mail:	norm.clciec@aenor.es
ES – 28004 Madrid	WWW:	http://www.aenor.es

Sweden

Svenska Elektriska Kommissionen	Telephone:	+46 8 444 14 00
Box 1284	Telefax:	+46 8 444 14 30
SE – 164 29 Kista	E-mail:	snc@sekom.se
	WWW:	http://www.sekom.se

Switzerland

Swiss Electrotechnical Committee (CES)	Telephone:	+41 (1) 956 11 80
Swiss Electrotechnical Association (SEV)	Telefax:	+41 (1) 956 11 90
Luppmenstraße 1	E-mail:	ces@sev.ch
CH – 8320 Fehraltorf	WWW:	http://www.sev.ch

Thailand

Thai National Committee of the IEC	Telephone:	+66 2 202 35 01
Thai Industrial Standard Institute (TISI)	Telefax:	+66 2 247 87 41
Ministry of Industry	ou/or Fax:	+66 2 202 35 11
Rama VI Street	E-mail:	thaistan@tisi.go.th
TH – 10400 Bangkok	WWW:	http://www.tisi.go.th

Turkey

Turkisch National Committee of the IEC	Telephone:	+90 (312) 417 83 30
Türk Standardlari Enstitüsü	Telefax:	+90 (312) 425 43 99
Necatibey Caddesi, 112	Telex:	tse tr 42047
TR – Bakanliklar/Ankara	Telegrams:	Standard Ankara
	E-mail:	didb@tse.org.tr
	WWW:	http://www.tse.org.tr

Ukraine

Ukrainian National Committee of the IEC		
State Committee of Ukraine		
for Standardization,	Telephone:	+380 (44) 226 29 71
Metrology & Certification	Telefax:	+380 (44) 226 29 70
Gorkiy St. 174	Telegrams:	131033 'Megom'
UA – 252650, GSP, Kiev-6	E-mail:	iec@dstul.kiev.ua

United Kingdom

British Electrotechnical Committee		
British Standards Institution	Telephone:	+44 181 996 9000
389 Chiswick High Road	Telefax:	+44 181 996 7799
GB – LondonW4 4AL	WWW:	http://www.bsi.org.uk

United States of America

U.S. National Committee of the IEC ANSI	Telephone:	+1 (212) 642-4900
11, West 42nd Street, 13th Floor	Telefax:	+1 (212) 398-0023
US – New York, NY 10036	WWW:	http://www.ansi.org

Uruguay

Instituto Uruguayo de Normas Tecnicas
(UNIT)

Plaza Independencia 812	Telephone:	+598 2 901 20 48
2o Piso	Telefax:	+598 2 902 16 81
UY – Montevideo	ou/or Tel.:	+598 2 901 16 80

Yugoslavia

Federal Institution for Standardization

Department of IEC Standards	Telephone:	+381 11 361 73 14
National Committee of IEC	Telefax:	+381 11 361 74 28
Kneza Milosa 20, P.O. Box 933	et/and Fax:	+381 11 361 73 41
YU – 11000 Belgrad	E-mail:	Jus@szs.sv.gov.yu

Other Organizations

CEN

European Committee for Standardization	Telephone:	+32 2 550 08 11
Rue de Stassart, 36	Telefax:	+32 2 550 08 19
B – 1050 Brussels	WWW:	http://www.cenorm.be

ETSI

European Telecommunications Standards	Telephone:	+33 4 92 94 42 00
Institute	Telefax:	+33 4 93 65 47 16
F – 06921 Sophia Antipolis Cedex	WWW:	http://www.etsi.fr

IEC

International Electrotechnical Commission

Rue de Varembé, 3	Telephone:	+41 22 919 02 11
P.O. Box 131	Telefax:	+41 22 919 03 00
CH – 1211 Geneva 20	WWW:	http://www.iec.ch

ISO

International Organisation for	Telephone:	+41 22 749 01 11
Standardization	Telefax:	+41 22 733 72 56
Rue de Varembé, 1	WWW:	http://www.iso.ch
CH – 1211 Geneva 20		

ITU

International Telecommunication Union	Telephone:	+41 22 730 51 51
Place des Nations	Telefax:	+41 22 733 72 56
CH – 1211 Geneva 20	WWW:	http://www.itu.ch

VDE

VDE Testing and Certification Institute
Merianstraße 28
D – 63 069 Offenbach
Germany

Telephone: +49 69 8 306 0
Telefax: +49 69 8 306 555
WWW: http://www.vde.de

UL

Underwriters Laboratories Inc.
333 Pfingsten Road
Northbrook, IL 60062
USA

Telephone: +1 847 272 8 800
Telefax: +1 847 272 2020
WWW: http://www.ul.com

7 Abbreviations

The following list the principal abbreviations used for plastics, fire safety and associated fields. It is arranged alphabetically without regard to individual topics or a logical relationship. The field and application where the abbreviation appears are given.

Abbreviation	Meaning	Field	Application
ABNT	Associacao Brasileira de Normas Tecnicas	Standards organization	Brazil
AEA	Asociación Electrotécnica Espanola	Electrotechnical standards organization	Spain
AFNOR	Association Française de Normalisation	Standards organization	France
AIA	American Insurance Association	Building	USA
ANSI	American National Standards Institute	Standards organization	USA
APME	Association of Plastic Manufactures in Europe	Plastics	EU
AS	Australian Standard	Standard	Australia
ASTM	American Society for Testing and Materials	Standards organisation	USA
BAM	Bundesanstalt für Material-prüfung und Forschung	Federal establishment for materials testing and research	Germany
BPF	The British Plastic Federation	Plastics Association	UK
BRE/FRS	Building Research Establishment Fire Research Station	Building	United Kingdom/ UK
BRI	Building Research Institute, Ministry of Construction	Building	Japan
BS	British Standard	Standard	UK
BSEF	Bromine Science and Environment Forum	Flame retardants	Brussels
BSI	British Standards Institution	Standards	UK
BVD	Brand-Verhütungsdienst für Industrie und Gewerbe	Fire protection, Buildings etc.	Switzerland
CAA	Civil Aviation Authority	Aviation	UK

Abbreviation	Meaning	Field	Application
CABO	Council of American Building Officials	Building	USA
CAS	China Association for Standardization	Standards organization	China
CCE	Commission des Communautés Européennes	European organization	Economics incl. fire protection
CEE	Commission for Conformity Certification of Electrical Equipment	Electrical engineering	Recognition of test results
CEFIC	European Chemical Industry Council	Association	European Union
CEN	Comité Européen de Normalisation	Standardization	European Union
CENELEC	Comité Européen de Normalisation Electrotechnique	Electrotechnical standardization	European Union
CEP	Comissao Electrotécnica Portuguesa	Standards organization	Electrical engineering, Portugal
CEPMC	Council of European Producers of Materials for Construction	Association	European Union
CFFA	Chemical Fabrics and Film Association	Association	USA
CFR	Code of Federal Regulations	Legal	USA
CHF	Critical Heat Flux	Test method criterion	Flooring Radiant Panel
CIB	Conseil International du Bâtiment pour la Recherche et la Documentation	Building	International Council for Research and Documentation
CIRFS	Comité International de la Rayonne et des Fibres Synthethiques	International man made fibres association	Brussels
CNR	Consiglio Nazionale delle Ricerche	Research institute	Italy, fire protection
CPSC	Consumer Product Safety Commission	Consumer protection	USA
CS	Commercial Standards	Standards	USA
CSA	Canadian Standard Organisation	Standards organization	Canada
CSE	Centro Studi ed Espirienze die Vigili del Fuoco	Research institute	Fire brigade, fire protection, Italy

Abbreviation	Meaning	Field	Application
CSIRO	Commonwealth Scientific and Industrial Research Organisation, Division of Building Research	Research institute	Building, fire protection, Australia
CSTB	Centre Scientifique et Technique du Bâtiment	Research and test institute	Building, fire protection, France
DACH	Deutsche Akkreditierstelle Chemie	Accreditation, inter alia plastics, fire testing	Germany
DAP	Deutsche Akkreditierstelle Prüfwesen	Accreditation, inter alia fire testing	Germany
DAR	Deutscher Akkreditierungs Rat	Accreditation	German accreditation council
DB	Deutsche Bahn AG	Railways	Transportation, German railway
DEFRA	Department of the Environment, Food & Rural Affairs (previously DOE)	Governmental	UK
DEPA	Danish Environmental Protection Agency	Governmental	Denmark
DGN	Direccion General de Normas	Standards organization	Mexico
DGQ	Deutsche Gesellschaft für Qualität	Quality assurance	Germany
DIBt	Deutsches Institut für Bautechnik	Building, governmental institute implementing technical rules	German institute for construction technology, Berlin
DIN	Deutsches Institut für Normung e.V.	Standards organization	Germany
DIS	Draft International Standard	Standardization	ISO
DKE	Deutsche Elektrische Kommission	Electrical engineering, Standardization	Germany
DOC	Department of Commerce	Governmental	Ministry of trade, USA
DOE	Department of the Environment (now: DEFRA)	Governmental	UK
DP	Draft Proposal	Standardization	ISO
DS	Dansk Standardiseringsrad	Standard organization	Denmark
DSC	Direction de la Sécurité Civile	Governmental	Inter alia fire protection, France
EBA	Eisenbahn-Bundesamt	Governmental railway agency	Railway, Germany
EBFRIP	European Brominated Flame Retardants Industry Panel	Flame retardants association	Brussels

Abbreviation	Meaning	Field	Application
EBS	Experimental Building Section	Building	Material testing, inter alia fire protection, Australia
EC	European Commission	Governmental	European Union
ECE	Economic Commission for Europe	International organization	Economics, United Nations (UN)
EDF	Electricité de France	Electrical engineering	France
EEC	European Economic Community	International organization	Inter alia trade, fire protection
EFRA	European Flame Retardants Association	Flame retardants	European producers of flame retardants
EFTA	European Free Trade Organisation	Trade	Europe
EG	Europäische Gemeinschaft	European community	European Union
EGOLF	European Group of Official Laboratories for Fire testing	European organization	Fire testing
ELOT	Hellenenic Organization for Standardization	Standards organization	Greece
EMPA	Eidgenössische Materialprüfungs- und Versuchsanstalt	Material testing	Federal Testing institute, inter alia fire protection, Switzerland
EN	Norme Européenne	European standard	Europe
EOTA	European Organization for Conformity Assessment	Organization	Europe
EOTC	European Organization for Technical Approvals	Organization	Europe
ETA	European Technical Approval	Approval	Europe
ETAG	European Technical Approval Guidelines	Approval	Europe
EUFAC	European Upholstered Furniture Action Council	Furniture organization	Inter alia fire protection
EURATEX	European Association of the Textile Industry	Textile association	Brussels
EuPC	European Plastics Converters	Plastics association	Plastics converters inter alia building products
FAA	Federal Aviation Administration	Governmental	USA
FAR	Federal Aviation Regulations	Governmental	USA

Abbreviation	Meaning	Field	Application
FIGRA	Fire Growth Rate Index	Test method criterion	SBI (EN 13823), ISO 9705
FIT	Flash-Ignition Temperature	Test method criterion	Plastics, USA
FKT	Fachausschuss Kraft-fahrzeugtechnik	Transportation (vehicles)	Germany
FM	Factory Mutual	Insurance, testing	Inter alia fire protection, USA
FMVSS	Federal Motor Vehicle Safety Standards	Transportation	USA
FNK	Fachnormenausschuss Kunststoffe im DIN	Standardization (plastics)	Germany
FNM	Fachnormenausschuss Materialprüfung im DIN	Standardization (materials testing)	Germany
FPA	Fire Protection Association	Fire prevention, fire protection	UK
FR	Federal Register	Legal	Official gazette, USA
FR	Flame Retardant	Plastics additive	
FRCA	Fire Retardant Chemicals Association	Flame retardants association	USA
FRCJ	Flame Retardant Chemicals Association Japan	Flame retardants association	Japan
FRS	Fire Research Station	Fire research and testing	UK, part of BRE
FSC	Flame Spread Classification	Test method criterion	Building ASTM E 84, USA
FTMS	Federal Test Method Standard	Standardization	USA
GOST	USSRV State Committee for Standards	Standards organization	USSR
HEN	Harmonised European Standard	Standard	Europe
HHS	Department of Health and Human Services	Governmental	USA
HUD	Department of Housing and Urban Development	Governmental	USA
IATA	International Air Transpor-tation Association	Transportation	Aviation
IBN	Institut Belge de Normalisation	Standards organization	Belgium
ICAO	International Civil Aviation Organization	Transportation	Aviation
ICBO	International Conference of Building Officials	Building codes	USA

Abbreviation	Meaning	Field	Application
IEC	International Electrotechnical Commission	Electrical engineering	Standardization
IEEE	Institute of Electrical and Electronic Engineers	Electrical engineering	USA
IIRS	Institute for Industrial Research and Standards	Standard organization	Eire
IL	Association of Fire Testing Laboratories of European Industries	Fire protection association	Europe
IMO	International Maritime Organization	Transportation	Sea transport
INN	Instituto National de Normalización	Standards organization	Chile
IRANOR	Instituto National de Racionalizatión y Normalización	Standards organization	Spain
ISI	Indian Standards Institute	Standards organization	India
ISO	International Standardization Organisation	Standards organization	
JAA	Joint Airworthiness Authorities	Governmental	Europe
JAR	Joint Airworthiness Requirements	Governmental	Europe
JIS	Japanese Industrial Standards	Standardisation	Japan
JISC	Japanese Industrial Standards Committee	Standards organisation	Japan
JO	Journal Officiel	Legal	Official gazette, France
JSA	Japanese Standards Organization	Standards	Japan
KBA	Kraftfahrt-Bundesamt	Governmental	Transport (vehicles), inter alia fire protection
KOMO	Stichting voor Onderzoek, Beoordeling en Keuring van Materialen en Constructies	Building	Netherlands
KVS	Kunststoffverband Schweiz	Plastics association	Swiss Plastics Association
LBA	Luftfahrt-Bundesamt	Governmental	Transport (aircrafts), inter alia fire protection
LBO	Landesbauordnung	Building	Federal building code, inter alia fire protection, Germany

Abbreviation	Meaning	Field	Application
LCPP	Laboratoire Central de la Préfecture de Police	Test institute	Building, fire tests, etc. France
LNE	Laboratoire National d'Essais	Test institute	Building, fire tests, etc. France
LOI	Limiting Oxygen Index	Test method	ASTM D 2863, USA
MBO	Musterbauordnung	Building	Model building code, inter alia fire protection, Germany
MCSR	Motor Carrier Safety Regulations	Transportation	USA
MITI	Ministry of International Trade and Industry	Governmental	Japan
MPA	Materialprüfanstalt	Test institute	Germany
MSZH	Magyar Szabványugyi Hivatal	Standards organization	Hungary
NAUBau	Normenausschuss Bauwesen im DIN	Standardization (building)	Germany
NAFTA	North American Free Trade Agreement	Trade	North America
NATA	National Association of Testing Authorities	Test institutes	Fire testing, etc., Australia
NBC	National Building Code of Canada	Building	Canada
NBN	Norme Belge	Standardization	Belgium
NBS	National Bureau of Standards (now NIST)	Standards, research	USA
NEN	Nederlandse Norm	Standardization	Netherlands
NFPA	National Fire Protection Association	Fire protection	USA
NHTSA	National Highway Transport Safety Association	Governmental	Transport, USA
NIST	National Institute of Standards and Technologies (former NBS)	Standards, research	Test methods fire protection, etc., USA
NKB	Nordiska Kommitten för Byggbestämmelser	Building	Europe; Nordic Countries
NMP	Normenausschuss Materialprüfung im DIN	Standardization (materials testing)	Germany
NNI	Nederlands Normalisatie-Institut	Standardization organization	Netherlands
NPRM	Notice of Proposed Rule Making	Legal	USA

Abbreviation	Meaning	Field	Application
NS	Norsk Standard	Standardization	Norway
NSF	Norges Standardiserings-forbund	Standardization organization	Norway
NT	Nordtest	Test method	Europe; Nordic countries
OLG	Official Laboratories Group for fire testing	Building	EU
ON	Österreichisches Normungsinstitut	Standardization organization	Austria
ÖNORM	Österreichische Norm	Standardization	Austria
OTSZ	Országos Tüvédelmi Szabályzat	Fire protection building regulations	Hungary
PA	Prüfzeichen mit Auflagen	Test mark (building)	Germany
pR	Prenormative	Standard	European Union
PSA	Property Services Agency	Governmental	UK
RG	Regulator Group	Building	European Union
RHR	Rate of Heat Release	Test method criterion	SBI (EN 13823) or ISO 5660 (Cone calorimeter)
RILEM	Réunion Internationale des Laboratoires d'Essais et de Recherche sur les Matériaux et les Constructions	Building	Association of test laboratories
SAA	Standards Association of Australia	Standards organization	Australia
SABS	South African Bureau of Standards	Standards organization	South Africa
SANZ	Standards Association of New Zealand	Standards organization	New Zealand
SBB	Schweizerische Bundesbahn	Transport (railway)	Swiss railway
SBG	Seeberufsgenossenschaft	Transport (shipping)	Germany
SBI	Single Burning Item	Test method	European Union, CEN
SBN	Svensk Byggnorm	Building	Building regulations, Sweden
SC	Sub-committee	Standardization	ISO and IEC
SCC	Standards Council of Canada	Standards organization	Canada
SCC	Standing Committee on Construction	Building	EU
SFS	Suomen Standardisoimis-litto r.y.	Standards organization	Finland

Abbreviation	Meaning	Field	Application
SIA	Schweizerischer Ingenieur- und Architektenverein	Private organization	Fire protection, Switzerland
SII	Standards Institution of Israel	Standards organization	Israel
SIS	Standardiseringskommissionen i Sverige	Standards organization	Sweden
SIT	Self-Ignition Temperature	Test method criterion	Plastics, ASTM D 1929, USA
SMOGRA	Smoke Growth Rate Index	Test method criterion	SBI (EN 13823)
SNCF	Société Nationale des Chemins de Fer	Transport (railway)	France
SNPE	Société Nationale des Poudres et Explosifs	Test institute	Building, fire testing, etc., France
SNV	Schweizerische Normenvereinigung	Standards organization	Switzerland
SOLAS	International Convention for the safety of Life and	Shipping	Regulations
SP	Statens Provningsanstalt	Test institute	Sweden
SPMC	Syndicat des Producteurs de Matières Plastiques	Plastics association	Plastic Producers, France
SPR	Smoke Production Rate	Test method criterion	SBI (pr EN 13823)
SPV	Statens Planverk	Governmental	Building, Sweden
SSV	Schiffssicherheitsverordnung	Shipping	Germany
STG	Schiffsbautechnische Gesellschaft	Shipping	Germany
TC	Technical Committee	Standardisation	CEN, ISO and IEC
TGA	Trägergemeinschaft für Akkreditierung	Accreditation	Germany
THR	Total Heat release	Test method criterion	SBI (EN 13823) or ISO 5669
TNO	Toegepast Natuurwetenschappelijk Onderzoek	Research and test institute	Building, fire protection, etc., Netherlands
TOSCA or TSCA	Toxic Substances Control Act	Legal	USA
TR	Technical Report	Standardization	ISO
TSP	Total Smoke Production	Test method criterion	SBI (EN 13823)
TTI	Time to Ignition	Test method criterion	SBI (EN 13823)

Abbreviation	Meaning	Field	Application
UBA	Umweltbundesamt	Governmental	Environmental protection agency, Germany
UBC	Uniform Building Code	Building	Model building code, USA
UEA	Union Européenne d'Ameublement	European federation of furniture manufacturers	Brussels
UFAC	Upholstered Furniture Action Council	Furniture industry	Fire safety specifications, USA
UIC	Union Internationale des Chemins de Fer	Transportation railways	Specifications including fire safety
UKAS	United Kingdom Accreditation Service	Accreditation	Building, fire behavior UK
UL	Underwriters' Laboratories Inc.	Test methods, test institute, test marks, follow up service	Fire testing, electrical engineering, building, etc. USA
ULC	Underwriters' Laboratories of Canada	Test methods, test institute, test marks, follow up service	Fire test, electrical engineering, building, etc. Canada
UMTA	Urban Mass Transportation Administration	Governmental, transportation	USA
UNI	Ente Nazionale Italiano di Unificazione	Standards organisation	Italy
UTAC	Union Technique de l'Automobile du Motocycle et du Cycle	Transportation road vehicles	Specifications, inter alia fire testing, France
UTE	Union Technique de l'Electricité	Electrical engineering association	France
VDE	Verband Deutscher Elektrotechniker e.V.	Electrical engineering Association	Germany
VKE	Verband Kunststofferzeugende Industrie e.V.	Plastics association	Germany
VKF	Vereinigung kantonaler Feuerversicherungen	Building	Fire insurers' association, Switzerland
VTT	Valtion Teknillinen Tutkimuskeskus	Test institute	Fire test, building, etc., Finland
WG	Working Group	Standardization	CEN, ISO and IEC

8 Journals and Books

The following list contains only those journals and monographs devoted exclusively to fire protection. No claim is made regarding completeness and sources concerned with fire fighting and the fire services have been disregarded. For relevant papers in plastics journals and pertinent sections in monographs on plastics the reader should consult the references at the end of each section in this book.

8.1 Journals

CA (Chemical Abstracts) Service Selects: Flammability. Fortnightly. American Chemical Society, 2540 Olentangy River Rd., P.O. Box 3012, Columbus Ohio 43210, USA
www.cas.org

Combustion and Flame. Three volumes per annum each consisting of three issues. Elsevier North-Holland, Inc., 52 Vanderbilt Avenue, New York, N.Y. 10017, USA
www.web-editions.com/issues_flame_1.htm

Fire and Materials. 6 issues. Wiley InterScience, Inc., Hoboken, NJ 07030, USA
www.wiley.com

Fire International. Quarterly. Harrow HA1 2EW, London, UK
www.dmgworldmedia.com

Fire Safety Journal. Quarterly. Elsevier Science, Amsterdam, NL
www.elsevier.nl

Fire Technology. Quarterly. National Fire Protection Association (NFPA), 1 Batterymarch Park, P.O. Box 9101, Quincy, Massachusetts, USA
www.nfpa.org

Fire and Flammability Bulletin. Monthly. Interscience Communications Ltd., Greenwich, London SE10 8JT, UK
http://dspace.dial.pipex.com

VFDB-Zeitschrift, Forschung und Technik im Brandschutz (Fire protection research and engineering). Quarterly. W. Kohlhammer GmbH, 70549 Stuttgart, Germany
www.vfdb.de

8.2 Books

R.M. Aseeva, G.E. Zaikov: Combustion of polymeric materials. Hanser Publishers, Munich, New York, 1985.
www.hanser.de

C.F. Cullis, M.M. Hirschler: The combustion of organic polymers. Clarendon Press, Oxford, 1981

A.F. Grand, C.A. Wilkie (Eds.): Fire retardancy of polymeric materials. Marcel Dekker, Inc., 270 Madison Avenue, New York, NY 10016, 2000
www.dekker.com

C.J. Hilado: Flammability Handbook for Plastics. 5th ed., CRC Press, 2000 N.W. Corporate Blvd., Boca Raton, FL 33431-9868, 1998
www.crcpress.com

A.R. Horrocks, D. Price (Eds.): Fire retardant materials. CRC Press, 2000 N.W. Corporate Blvd., Boca Raton, FL 33431-9868, 2001
www.crcpress.com

W.C. Kuryla, A.J. Papa (Eds.): Flame Retardancy of Polymeric Materials, Five volumes (1973–1979). Marcel Dekker, Inc., New York

M. Le Bras, G. Camino, S. Bourbigot, R. Delobel: Fire retardancy of polymers. The use of intumescence. The Royal Society of Chemistry, Cambridge, UK, 1998
www.rsc.org

M. Lewin, S.M. Atlas, E.M. Pearce (Eds.): Flame Retardant Polymeric Materials. Three volumes (1975–1982). Plenum Press, New York

G. Nelson, C. Wilkie (Eds.): Fire and polymers. Oxford University Press, UK, 2001
www.oup-usa.org/acs

Author Index

Index of Standards

Key Word Index